CHAPTER 2   Fractions with Variables and Algebraic Expressions

Evaluating an expression:
1. Combine like terms, if possible.
2. Substitute the values given for any variables.
3. Follow the rules for order of operations and simplify.

Fractions: See Chapter R

Placement of Negative Signs: $-\dfrac{a}{b} = \dfrac{-a}{b} = \dfrac{a}{-b}$

The Fundamental Principle of Fractions:
$\dfrac{a}{b} = \dfrac{a \cdot k}{b \cdot k}$ where $b, k \neq 0$

Decimals: See Chapter R

Order of Operations: See Chapter R

CHAPTER 3   Solving Equations and Inequalities

Linear Equation in *x* (first-degree equation in *x*):
$ax + b = c$, where $a \neq 0$

Addition Principle of Equality:
$A = B$ and $A + C = B + C$ have the same solutions (where A, B, and C are algebraic expressions).

Multiplication (or Division) Principle of Equality:
$A = B$ and $AC = BC$ and $\dfrac{A}{C} = \dfrac{B}{C}$

Intervals of Real Numbers:

Name	Symbolic Representation	Graph
Open Interval	$a < x < b$	
Closed Interval	$a \leq x \leq b$	
Half-Open Interval	$a < x \leq b$ $a \leq x < b$	
Open Interval	$x > a$ $x < a$	
Half-Open Interval	$x \geq a$ $x \leq a$	

Consecutive Integers:
$n, n + 1, n + 2, \ldots$

Consecutive Odd Integers:
$n, n + 2, n + 4, \ldots$ where n is an odd integer

Consecutive Even Integers:
$n, n + 2, n + 4, \ldots$ where n is an even integer

Percent Formula:
$R \cdot B = A$

Geometric Formulas:

SQUARE
Perimeter: $P = 4s$
Area: $A = s^2$

RECTANGLE
Perimeter: $P = 2l + 2w$
Area: $A = lw$

PARALLELOGRAM

Perimeter: $P = 2a + 2b$
Area: $A = bh$

TRIANGLE
Perimeter: $P = a + b + c$
Area: $A = \dfrac{1}{2}bh$

CIRCLE
Circumference: $C = 2\pi r = \pi d$
Area: $A = \pi r^2$

TRAPEZOID
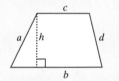
Perimeter: $P = a + b + c + d$
Area: $A = \dfrac{1}{2}h(b + c)$

Volume Formulas:
Rectangular Solid: $V = lwh$

Rectangular Pyramid: $V = \dfrac{1}{3}lwh$

Right Circular Cylinder: $V = \pi r^2 h$

Right Circular Cone: $V = \dfrac{1}{3}\pi r^2 h$

Sphere: $V = \dfrac{4}{3}\pi r^3$

CHAPTER 4   Graphing Linear Equations and Inequalities in Two Variables

Cartesian Coordinate System:

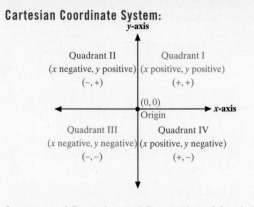

The **domain**, **D**, of a relation is the set of all first coordinates in the relation.

The **range**, **R**, of a relation is the set of all second coordinates in the relation.

Function:

A **function** is a relation in which each domain element has a unique range element.

Vertical Line Test:

If **any** vertical line intersects the graph of a relation at more than one point, then the relation graphed is **not** a function.

Summary of Formulas and Properties of Straight Lines:

1. $Ax + By = C$, where A and B do not equal 0.   Standard form
2. $m = \dfrac{y_2 - y_1}{x_2 - x_1}$, where $x_1 \neq x_2$.   Slope of a line
3. $y = mx + b$ Slope-intercept form
4. $y - y_1 = m(x - x_1)$ Point-slope form
5. $y = b$ Horizontal line, $m = 0$
6. $x = a$ Vertical line, m is undefined
7. Parallel lines have the same slope ($m_1 = m_2$).
8. Perpendicular lines have slopes that are negative reciprocals of each other $\left(m_2 = \dfrac{-1}{m_1} \text{ or } m_1 m_2 = -1 \right)$.

Relation, Domain, and Range:

A **relation** is a set of ordered pairs of real numbers.

Linear Inequality Terminology:

Half-plane: A straight line separates a plane into two **half-planes**.
Boundary line: The line itself is called the **boundary line**.
Closed half-plane: If the boundary line is included, then the half-plane is said to be **closed**.
Open half-plane: If the boundary line is not included, then the half-plane is said to be **open**.

Graphing Linear Inequalities:

1. Graph the boundary line, dashed if the inequality is $<$ or $>$, solid if the inequality is $\leq$ or $\geq$.
2. **a. Method 1**: Test any point obviously on one side of the line. If the test point satisfies the inequality, shade the half-plane on the same side of the line. Otherwise, shade the other half-plane.
   b. Method 2: Solve the inequality for y. If the solution shows $y < $ or $y \leq$, then shade the half-plane below the line. If the solution shows $y > $ or $y \geq$, then shade the half-plane above the line.

CHAPTER 5   Systems of Linear Equations

Solutions of Systems of Linear Equations:

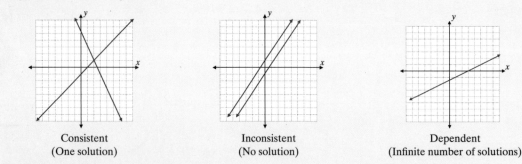

Consistent
(One solution)

Inconsistent
(No solution)

Dependent
(Infinite number of solutions)

Solving a System of Two Linear Inequalities:

1. Graph both half planes.
2. Shade the region that is common to both of these half-planes, called the **intersection**. (If there is no intersection, then the system is inconsistent and has no solution.)
3. To check, pick one test-point in the intersection and verify that it satisfies *both* inequalities.

INTRODUCTORY
ALGEBRA

D. FRANKLIN WRIGHT
CERRITOS COLLEGE

HAWKES
PUBLISHING

Editor: Nina Miller
Project Editor: Larry Wadsworth Jr.
Developmental Editor: Marcel Prevuznak
Production Editors: Phillip Bushkar, Kimberly Cumbie
Answer Key Editors: Larien Acosta, Eric Wilder
Editorial Assistants: Bethany Bates, Mandy Glover, D. Kanthi, B. Syam Prasad
Layout: QSI (Pvt.) Ltd.: U. Nagesh, E. Jeevan Kumar
Art: Ayvin Samonte
Cover Art and Design: Johnson Design

HAWKES
PUBLISHING

A division of Quant Systems, Inc.

Library of Congress Control Number: 2007934775

Printed in the United States of America

ISBN:
Student: 978-1-932628-32-6
Student Bundle: 978-1-932628-33-3

CONTENTS

Preface   vii

Hawkes Learning Systems: Introductory Algebra   xxi

CHAPTER R Review of Basic Topics   1

R.1   Exponents and Order of Operations   2
R.2   Prime Numbers, Factoring, and LCM   14
R.3   Fractions (Multiplication and Division)   24
R.4   Fractions (Addition and Subtraction)   37
R.5   Decimals and Percents   47
Chapter R Index of Key Ideas and Terms   64
Chapter R Chapter Review   69
Chapter R Chapter Test   72

CHAPTER 1 Integers and Real Numbers   75

1.1   The Real Number Line and Absolute Value   76
1.2   Addition with Integers   90
1.3   Subtraction with Integers   96
1.4   Multiplication and Division with Real Numbers   106
1.5   Properties of Real Numbers   118
Chapter 1 Index of Key Ideas and Terms   125
Chapter 1 Chapter Review   129
Chapter 1 Chapter Test   133

CHAPTER 2 Fractions with Variables and Algebraic Expressions   135

2.1   Simplifying and Evaluating Algebraic Expressions   136
2.2   Multiplication and Division with Fractions   143
2.3   Addition and Subtraction with Fractions   153
2.4   Decimal Numbers and Fractions   162
2.5   Order of Operations with Negative Numbers and Fractions   171
2.6   Translating English Phrases and Algebraic Expressions   179
Chapter 2 Index of Key Ideas and Terms   186
Chapter 2 Chapter Review   190
Chapter 2 Chapter Test   194
Chapter 2 Cumulative Review   196

Contents

CHAPTER 3 Solving Equations and Inequalities — 199

3.1 Solving Linear Equations: $x + b = c$ and $ax = c$ 200
3.2 Solving Linear Equations: $ax + b = c$ 213
3.3 More Linear Equations: $ax + b = dx + c$ 220
3.4 Solving Linear Inequalities and Applications   228
3.5 Working with Formulas   240
3.6 Applications: Number Problems and Consecutive Integers   251
3.7 Applications: Percent Problems
(Discount, Taxes, Commission, Profit, and Others)   261
3.8 Formulas in Geometry   271
Chapter 3 Index of Key Ideas and Terms   282
Chapter 3 Chapter Review   288
Chapter 3 Chapter Test   293
Chapter 3 Cumulative Review   296

CHAPTER 4 Graphing Linear Equations and Inequalities in Two Variables 301

4.1 The Cartesian Coordinate System and Reading Graphs   302
4.2 Graphing Linear Equations in Two Variables: $Ax + By = C$ 320
4.3 The Slope-Intercept Form: $y = mx + b$ 333
4.4 The Point-Slope Form: $y - y_1 = m(x - x_1)$ 350
4.5 Introduction to Functions and Function Notation   361
4.6 Graphing Linear Inequalities: $y < mx + b$ 373
Chapter 4 Index of Key Ideas and Terms   383
Chapter 4 Chapter Review   387
Chapter 4 Chapter Test   393
Chapter 4 Cumulative Review   396

CHAPTER 5 Systems of Linear Equations — 401

5.1 Systems of Equations: Solutions by Graphing   402
5.2 Systems of Equations: Solutions by Substitution   414
5.3 Systems of Equations: Solutions by Addition   420
5.4 Applications: Distance-Rate-Time, Number Problems, Amounts and Costs   429
5.5 Applications: Interest and Mixture   440
5.6 Graphing Systems of Linear Inequalities   449
Chapter 5 Index of Key Ideas and Terms   454
Chapter 5 Chapter Review   457
Chapter 5 Chapter Test   460
Chapter 5 Cumulative Review   462

CHAPTER 6 Exponents and Polynomials 467

6.1 Exponents 468
6.2 Exponents and Scientific Notation 481
6.3 Introduction to Polynomials 495
6.4 Addition and Subtraction with Polynomials 502
6.5 Multiplication with Polynomials 508
6.6 Special Products of Binomials 515
6.7 Division with Polynomials 525
Chapter 6 Index of Key Ideas and Terms 534
Chapter 6 Chapter Review 537
Chapter 6 Chapter Test 541
Chapter 6 Cumulative Review 543

CHAPTER 7 Factoring Polynomials and Solving Quadratic Equations 547

7.1 Greatest Common Factor and Factoring by Grouping 548
7.2 Factoring Trinomials: $x^2 + bx + c$ 559
7.3 More on Factoring Trinomials: $ax^2 + bx + c$ 566
7.4 Factoring Special Products: Difference of Two Squares and Perfect Square Trinomials 579
7.5 Solving Quadratic Equations by Factoring 586
7.6 Applications of Quadratic Equations 594
7.7 Additional Applications of Quadratic Equations 602
Chapter 7 Index of Key Ideas and Terms 606
Chapter 7 Chapter Review 611
Chapter 7 Chapter Test 615
Chapter 7 Cumulative Review 617

CHAPTER 8 Rational Expressions 621

8.1 Reducing Rational Expressions 622
8.2 Multiplication and Division with Rational Expressions 633
8.3 Addition and Subtraction with Rational Expressions 639
8.4 Complex Algebraic Fractions 647
8.5 Solving Proportions and Other Equations with Rational Expressions 654
8.6 Applications 665
8.7 Additional Applications: Variation 675
Chapter 8 Index of Key Ideas and Terms 682
Chapter 8 Chapter Review 686
Chapter 8 Chapter Test 691
Chapter 8 Cumulative Review 693

Contents

CHAPTER 9   Real Numbers and Radicals   697

9.1   Real Numbers and Evaluating Radicals   698
9.2   Simplifying Radicals   711
9.3   Addition, Subtraction, and Multiplication with Radicals   720
9.4   Rationalizing Denominators   726
9.5   Solving Equations with Radicals   732
9.6   Rational Exponents   738
9.7   The Pythagorean Theorem   744
Chapter 9 Index of Key Ideas and Terms   758
Chapter 9 Chapter Review   762
Chapter 9 Chapter Test   766
Chapter 9 Cumulative Review   768

CHAPTER 10   Quadratic Equations   771

10.1   Quadratic Equations: The Square Root Method   772
10.2   Quadratic Equations: Completing the Square   781
10.3   Quadratic Equations: The Quadratic Formula   788
10.4   Applications   796
10.5   Quadratic Functions: $y = ax^2 + bx + c$ 808
Chapter 10 Index of Key Ideas and Terms   821
Chapter 10 Chapter Review   823
Chapter 10 Chapter Test   827
Chapter 10 Cumulative Review   829

APPENDIX   833

A.1   Difference of Two Cubes and Sum of Two Cubes   833
A.2   Pi   838

ANSWERS   841

INDEX   884

PREFACE

Purpose and Style

Introductory Algebra (sixth edition) provides a smooth transition from arithmetic (or prealgebra) to the more abstract skills and reasoning abilities developed in a beginning algebra course. With feedback from users, insightful comments from reviewers, and skillful editing and design by the editorial staff at Hawkes Publishing, I have confidence that students and instructors alike will find that this text is indeed a superior teaching and learning tool. In particular, new in this edition, Chapter R (*Review of Basic Topics*) is designed to help students review basic arithmetic knowledge in exponents, order of operations, prime numbers, factoring, least common multiple (LCM), fractions, decimals, and percents. This chapter can be treated as part of the course or simply review for the students at their convenience. The text may be used independently or in conjunction with the software package ***Hawkes Learning Systems: Introductory Algebra*** developed by Quant Systems/Hawkes Learning Systems.

The writing style gives carefully worded, thorough explanations that are direct, easy to understand and mathematically accurate. The use of color, boldface, subheading, and shaded boxes helps students understand and reference important topics. Each topic is developed in a straightforward step-by-step manner. Each section contains many detailed examples to lead students successfully through the exercises and help them develop an understanding of the related algebraic concepts. Practice problems with answers are provided in almost every section to allow students to "warm up" and to provide instructors with immediate classroom feedback.

Reading graphs and topics from geometry are integrated within the discussions and problems. For example circle graphs and bar graphs are used in discussions as early as Chapter R and in Chapter 1.

Students are encouraged to use calculators when appropriate and explicit directions and diagrams are provided as they relate to a TI-84 Plus calculator. The emphasis is on the use of calculators as aids with the understanding that calculators do not take the place of critical thinking and analysis.

The NCTM and AMATYC curriculum standards have been taken into consideration in the development of the topics throughout the text. In particular:

- there is emphasis on reading and writing skills as they relate to mathematics,

- techniques for using a graphing calculator are discussed early and detailed instructions are included to help in graphing linear inequalities (Section 4.6), in finding the solutions to systems of two linear equations (Section 5.1), and in graphing systems of two linear inequalities (Section 5.6),

- a special effort has been made to make the exercises motivating and interesting, and

- special exercises titled "Writing and Thinking About Mathematics" are designed to help students develop writing skills related to mathematical thinking.

Features

Integers and Real Numbers

CHAPTER

1

Did You Know?

Arithmetic operations defined on the set of positive integers, negative integers, and zero are studied in this chapter. The integer zero will be shown to have interesting properties under the operations of addition, subtraction, multiplication, and division.

Curiously, zero was not recognized as a number by early Greek mathematicians. When Hindu scientists developed the place-value numeration system we currently use, the zero symbol was initially a place holder but not a number. The spread of Islam transmitted the Hindu number system to Europe where it became known as the Hindu-Arabic system and replaced Roman numerals. The word zero comes from the Hindu word meaning "void," which was translated into Arabic as "sifr" and later into Latin as "zephirum," hence the derivation of our English words "zero" and "cipher."

Almost all of the operational properties of zero were known to the Hindus. However, the Hindu mathematician Bhaskara the Learned (1114 – 1185?) asserted that a number divided by zero was zero, or possibly infinite. Bhaskara did not seem to understand the role of zero as a divisor since division by zero is undefined and hence, an impossible operation in mathematics.

Albert Einstein, in his development of a proof that the universe was stable and unchangeable in time, divided both sides of one of his intermediate equations by a complicated expression that under certain circumstances could become zero. When the expression became zero, Einstein's proof did not hold and the possibility of a pulsating, expanding, or contracting universe had to be considered. This error was pointed out to Einstein and he was forced to withdraw his proof that the universe was

1.1 The Real Number Line and Absolute Value

1.2 Addition with Integers

1.3

1.4

1.5

Introduction:

Presented before the first section of every chapter, this feature provides an introduction to the subject of the chapter and its purpose.

Did You Know?:

A feature at the beginning of every chapter presents some interesting math history related to the chapter at hand.

Objectives:

The objectives provide the students with a clear and concise list of skills presented in each section.

CHAPTER 1   Integers and Real Numbers

Numbers and number concepts form the foundation for the study of algebra. In this chapter, you will learn about positive and negative numbers and how to operate with these numbers. Believe it or not, the key number is 0. Pay particularly close attention to the idea of the magnitude of a number, called its absolute value, and the terminology used to represent different types of numbers.

1.1   The Real Number Line and Absolute Value

Objectives

After completing this section, you will be able to:

1. Identify types of numbers.
2. Determine if given numbers are greater than, less than, or equal to other given numbers.
3. Determine absolute values.

The study of algebra requires that we know a variety of types of numbers and their names. The set of numbers

$$\mathbb{N} = \{1, 2, 3, 4, 5, 6, 7, 8, 9, 10, 11, ...\}$$

is called the **counting numbers** or **natural numbers**. The three dots (an ellipsis) indicate that the pattern is to continue without end. Putting 0 with the set of natural numbers gives the set of **whole numbers**.

$$\mathbb{W} = \{0, 1, 2, 3, 4, 5, 6, 7, 8, 9, 10, 11, ...\}$$

Thus mathematicians make the distinction that 0 is a whole number but not a natural number.

To help in understanding different types of numbers and their relationships to each other, we begin with a "picture" called a **number line**. For example, choose some point on a horizontal line and label it with the number 0 (Figure 1.1).

Figure 1.1

Now choose another point on the line to the right of 0 and label it with the number 1 (Figure 1.2). This point is arbitrary and once it is chosen a units scale is determined for the remainder of the line.

    0   1

Figure 1.2

76

Section 2.4 **Decimal Numbers and Fractions**

Dividing Two Decimal Numbers

To find the quotient of two decimal numbers:

1. *Move the decimal point in the divisor to the right to get a whole number,*
2. *Move the decimal point in the dividend the same number of places,*
3. *Place the decimal point in the quotient above the new place in the dividend, then*
4. *Divide as with whole numbers.*

Example 2: Division with Decimals

Find the quotient $3.2\overline{)51.52}$

Solution: $3.2\overline{)51.52}$

$32.\overline{)515.2}$ Decimal point in quotient.

$32.\overline{)515.2}$ We are actually multiplying both the divisor and

dividend by 10: $\dfrac{51.52}{3.2} \cdot \dfrac{10}{10} = \dfrac{515.2}{32}$.

$$\begin{array}{r} 16.1 \\ 32\overline{)515.2} \\ \underline{32} \\ 195 \\ \underline{192} \\ 32 \\ \underline{32} \\ 0 \end{array}$$ Divide.

Changing Fractions to Decimal F...

both
...erator

Definition Boxes:

Definitions are presented in highly visible boxes for easy reference.

Examples:

Examples are denoted with titled headers indicating the problem solving skill being presented. Each section contains many carefully explained examples with lots of tables, diagrams, and graphs. Examples are presented in an easy to understand step-by-step fashion and annotated for additional clarification.

Notes:

Notes highlight common mistakes and give additional clarification to more subtle details.

CHAPTER 2   **Fractions with Variables and Algebraic Expressions**

These rules are very explicit and should be studied carefully. Note that in Rule 3, neither multiplication nor division has priority over the other. Whichever of these operations occurs first, **moving left to right**, is done first. In Rule 4, addition and subtraction are handled in the same way. Unless they occur within grouping symbols, **addition and subtraction are the last operations to be performed.**

A well-known mnemonic device for remembering the rules for order of operations is the following:

Please	Excuse	My	Dear	Aunt	Sally
↓	↓	↓	↓	↓	↓
Parentheses	Exponents	Multiplication	Division	Addition	Subtraction

NOTES   Even though the mnemonic **PEMDAS** is helpful, remember that multiplication and division are performed as they appear, left to right. Also, addition and subtraction are performed as they appear, left to right.

For example

$$12 \div 3 \cdot 4 = 4 \cdot 4 = 16,$$

but

$$12 \cdot 3 \div 4 = 36 \div 4 = 9.$$

As discussed in Section 2.1, a negative sign in front of a variable indicates that the coefficient is –1. For example,

$$-x^2 = -1 \cdot x^2.$$

This is consistent with the rules for order of operations (indicating that exponents come before multiplication) and is particularly useful in determining the values of expressions involving negative numbers and exponents. For example, each of the expressions

$$-7^2 \quad \text{and} \quad (-7)^2$$

has a different value. By the order of operations,

$$-7^2 = -1 \cdot 7^2 = -1 \cdot 49 = -49,$$

but

$$(-7)^2 = (-7)(-7) = 49.$$

In the second expression the base is a negative number. Remember that if the base is a negative number, then the negative number must be placed in parentheses.

Section 4.2 **Graphing Linear Equations in Two Variables:** $Ax + B = C$

Using a TI-84 Plus Calculator to Graph Straight Lines

By following the steps outlined here you will be able to see the graphs of all straight lines that are not vertical. The calculator will not graph vertical lines using the methods described here.

To have the calculator graph a nonvertical straight line, you must first solve the equation for y. For example,

given the equation: $2x + y = 3$

solving for y gives: $y = -2x + 3$ (or $y = 3 - 2x$)

Step 1: Press the **MODE** key and set all the highlighted keys as shown in the diagram. If the mode is not as shown, press the down arrow until you reach the desired line and press **ENTER** .

It is particularly important that **Func** is highlighted. This stands for function. Function concepts and notation will be introduced later in this chapter.

Step 2: Press the ⬛ key and then press 6 to get graphs. This will give scales from –10 to 10 f If you later decide that you would like some ⬛ and set the window values to the app

---

Calculator Instruction:

Step-by-step instructions are presented to introduce students to basic graphing skills with a TI-84 Plus calculator along with actual screen shots of a TI-84 Plus for visual reference.

Practice Problems:

Practice Problems are presented at the end of almost every section with answers giving the students an opportunity to practice their newly acquired skills.

Section 4.3 **The Slope-Intercept Form:** $y = mx + b$

Practice Problems

1. Find the slope of the line determined by the point $(1, 5)$ and $(2, -2)$.
2. Find the equation of the line through the point $(0, 3)$ with slope $-\frac{1}{2}$.
3. Find the slope, m, and y-intercept, b, for the equation $2x + 3y = 6$.
4. Write an equation for the horizontal line through the point $(-2, 3)$.
5. Write an equation for the vertical line through the point $(-1, -2)$.

4.3 Exercises

Find the slope of the line determined by each pair of points in Exercises 1 – 20.

1. $(1, 4), (2, -1)$ 2. $(3, 1), (5, 0)$ 3. $(4, 7), (-3, -1)$

4. $(-6, 2), (1, 3)$ 5. $(-3, 8), (5, 9)$ 6. $(0, 0), (-6, -4)$

7. $(1, 5), (4, 8)$ 8. $(2, 6), (6, 2)$ 9. $(0, 0), (-5, 3)$

10. $(0, 0), (9, -3)$ 11. $(-2, 0), (5, 1)$ 12. $(-3, -2), (0, 4)$

13. $(-4, 0), (4, 4)$ 14. $(-1, 5), (4, -5)$ 15. $(-3, 4), (6, -14)$

16. $(-2, -2), (6, -18)$ 17. $(4, 2), \left(-1, \frac{1}{2}\right)$ 18. $\left(\frac{3}{4}, 2\right), \left(1, \frac{3}{2}\right)$

19. $\left(\frac{7}{2}, 3\right), \left(\frac{1}{2}, -\frac{3}{4}\right)$ 20. $\left(-2, \frac{4}{5}\right), \left(\frac{3}{2}, \frac{1}{10}\right)$

Write the equation and draw the graph of the line passing through the given y-intercept with the given slope in Exercises 21 – 40.

21. $(0, 0), m = \frac{2}{3}$ 22. $(0, 1), m = \frac{1}{5}$ 23. $(0, -3), m = -\frac{3}{4}$

24. $(0, 6), m = 0$ 25. $(0, 3), m = -\frac{5}{3}$ 26. $(0, 2), m = 4$

27. $(0, -1), m = 2$ 28. $(0, 4), m = -\frac{3}{5}$ 29. $(0, -7), m = 0$

Answers to Practice Problems: 1. $m = -7$ 2. $y = -\frac{1}{2}x + 3$ 3. $m = -\frac{2}{3}$ and $b = 2$ 4. $y = 3$ 5. $x = -1$

347

Exercises:

Each section includes a variety of paired and graded exercises to give the students much needed practice applying and reinforcing the skills learned in the section. More than 6000 carefully selected and graded exercises are provided in the sections. The exercises proceed from relatively easy to more difficult ones.

Chapter 1 Index of Key Ideas and Terms

Section 1.1 The Real number Line and Absolute Value

Types of Numbers

Counting numbers (or natural numbers)

$N = \{1, 2, 3, 4, 5, 6, 7, 8, 9, 10, 11, ...\}$ page 76

Whole numbers   page 76

$W = \{0, 1, 2, 3, 4, 5, 6, ...\}$

Integers   page 78

Integers: $\{..., -3, -2, -1, 0, 1, 2, 3, ...\}$

Positive integers: $\{1, 2, 3, 4, 5, ...\}$

Negative integers: $\{..., -4, -3, -2, -1\}$

The integer 0 is neither positive nor negative.

Rational numbers   page 79

A **rational number** is a number that can be written in the form $\frac{a}{b}$, where a and b are integers and $b \neq 0$.

OR,

A **rational number** is a number that can be written in decimal form as a terminating decimal or as an infinite repeating decimal.

Irrational numbers   page 80

Irrational numbers are numbers that can be written as infinite nonrepeating decimals.

Real numbers   page 80

All rational and irrational numbers are classified as **real numbers**.

Diagram of Types of Numbers   page 81

Inequality Symbols   pages 81 - 83

Read from left to right:

$<$ "is less than"

$>$ "is greater than"

$\leq$ "is less than or equal to"

$\geq$ "is greater than or equal to"

Note: Inequality symbols may be read from left to right right to left.

Absolute Value

The **absolute value** of a real number is its distance from 0.

Symbolically,

$|a| = a$ if a is a positive number or 0.

$|a| = -a$ if a is a negative number.

Index of Key Ideas and Terms:

Each chapter contains an index highlighting the main concepts and skills presented in the chapter along with full definitions and page numbers for easy reference.

Writing and Thinking:

These exercises provide the student an opportunity to independently explore and expand on concepts presented in the chapter.

Calculator Problems

A TI-84 Plus calculator has the absolute value command built in. To access the absolute command press **MATH**, go to NUM at the top of the screen and press **1** or **ENTER**. The command **abs(** will appear on the display screen. Then enter any arithmetic expression you wish and the calculator will print the absolute value of that expression. (**Note:** The negative sign is on the key marked **(−)** next to the **ENTER** key.)

Follow the directions above and use your calculator to find the value of each of the following expressions.

93. $|34 - 80|$ **94.** $|17.5 + 16.3 - 95.2|$ **95.** $-|10 - 16|$

96. $-|-10 - 11|$ **97.** $\dfrac{|-6| - |-3|}{|-6 - 3|}$ **98.** $\dfrac{|4| + |-4|}{|-8|}$

Writing and Thinking About Mathematics

99. Explain, in your own words, how a variable expression such as $-y$ might represent a positive number.

100. Explain, in your own words, the meaning of absolute value.

Hawkes Learning Systems: Introductory Algebra

Name That Real Number
Introduction to Absolute Values

Hawkes Learning Systems:

Each section's exercises are followed by a feature which highlights corresponding lessons in *HLS: Introductory Algebra* software for ease of professor-assigned or student-motivated assignments, review, and instruction.

Additional Features

Chapter Review: Review exercises from the chapter organized by section.

Chapter Test: Provides an opportunity for the students to practice the skills presented in the chapter in a test format.

Cumulative Review: As new concepts build on previous concepts, the cumulative review provides the student with an opportunity to continually reinforce existing skills while practicing newer skills.

Answers: Answers are provided for odd numbered section exercises and for all even and odd numbered exercises in the Chapter Tests and Cumulative Reviews.

Teachers' Edition:

Answers: Answers to all the exercises are conveniently located in the margins next to the problems.

Teaching Notes: Suggestions for more in-depth classroom discussions and alternate methods and techniques are located in the margins.

Also included in this edition:

- Review chapter covering basic arithmetic skills
- Chapter Reviews provide review exercises from the chapter organized by section.
- Arrangement of chapters and sections for better flow, continuity, and progression

Content

There is sufficient material for a three- or four-unit course. The topics in Chapters 1 – 9 form the core of the course. A new chapter, Chapter R (*Review of Basic Topics*) has been added to provide a review of topics from arithmetic and prealgebra. Depending on time and students' background, these topics can be treated as part of the course or to be studied as needed at the students' convenience. Chapter 10 (*Quadratic Equations*) and the Appendix (*Sums and Differences of Cubes*) provide additional flexibility in the course.

Chapter R, ***Review of Basic Topics***, provides reviews of fundamental topics from arithmetic and prealgebra important for a good start in algebra. In particular, students will find this chapter helpful if there has been a break since their last course in mathematics. Topics included are (R.1) Exponents and Order of Operations, (R.2) Prime Numbers, Factoring and LCM, (R.3) Fractions (Multiplication and Division), (R.4) Fractions (Addition and Subtraction), (R.5) Decimals and Percents. While the use of a TI-84 Plus graphing calculator is introduced to help in calculating with large numbers and with decimals, students must understand that a calculator is an aid to understanding and does not replace analytical thinking needed for success in mathematics.

Chapter 1, ***Integers and Real Numbers***, develops the algebraic concepts of integers and the basic skills of operating (adding, subtracting, multiplying, and dividing) with integers. Variables and absolute value are defined and number lines are used to aid in understanding addition and subtraction with integers. The chapter closes with a discussion of types of numbers and the properties of addition and multiplication with real numbers.

Chapter 2, ***Fractions with Variables and Algebraic Expressions***, introduces like terms and simplifying and evaluating algebraic expressions. Fractions involving variables in the numerator and denominator are now used to develop an in-depth understanding of fractions. The Rules for Order of Operations are used to evaluate expressions with positive and negative numbers and fractions. The last section involves translating English phrases and algebraic expressions as a lead in to interpreting and understanding word problems.

Chapter 3, ***Solving Equations and Inequalities***, takes three sections to show how to solve different forms of linear equations in one variable. This approach lays the ground work for solving linear inequalities and working with a variety of formulas. Formulas involve at least two variables. For example, $I = Prt$ relates to simple interest and $C = \dfrac{5}{9}(F - 32)$ relates temperatures measured in Celsius and Fahrenheit. Two sections are devoted to applications in topics such as consecutive integers and percent problems. The chapter closes with a discussion of formulas specifically related to geometric concepts: perimeter, area, and volume.

Chapter 4, *Graphing Linear Equations and Inequalities in Two Variables*, allows for the early introduction of a graphing calculator and the ideas and notation related to functions. The chapter begins with an introduction to the Cartesian coordinate system and graphing ordered pairs of real numbers. The sections on graphing linear equations include the standard form, $Ax + By = C$, the slope-intercept form, $y = mx + b$, and the point-slope form, $y - y_1 = m(x - x_1)$. Included are vertical and horizontal lines and parallel and perpendicular lines. Section 4.6 deals with half-planes and graphing linear inequalities in two variables. (Note that this topic was formerly in the appendix and many reviewers suggested that it be part of the main text.) Students should understand that the concept of functions is one of the most important and useful in all of mathematics and they will see it again and again.

Chapter 5, *Systems of Linear Equations*, shows three ways to solve systems of two linear equations: by graphing (including the use of a graphing calculator), by substitution, and by addition. Applications in this chapter are related to distance-rate-time, number problems, amounts and costs, interest, and mixture. Graphing systems of linear inequalities (including the use of a graphing calculator) is another topic included now that was formerly in the appendix.

Chapter 6, *Exponents and Polynomials*, begins with the basic properties of exponents and then develops into more rules of exponents and a discussion of scientific notation. This is followed by the definition of polynomials and operations with polynomials (addition, subtraction, multiplication, and division). The final topics of the chapter include the FOIL method of multiplication with two binomials, special products of binomials (differences of two squares and perfect square trinomials), and the division algorithm.

Chapter 7, *Factoring Polynomials and Solving Quadratic Equations*, introduces methods of factoring polynomials, including finding common monomial factors, factoring by grouping, factoring trinomials by trial-and-error, and factoring special products (difference of two squares and perfect square trinomials). Students are then taught to use these methods to solve quadratic equations by factoring. (Quadratic equations are discussed further in Chapter 10 and in Intermediate Algebra.) Applications of quadratic equations are developed in the last two sections.

Chapter 8, ***Rational Expressions***, provides still more practice with factoring and shows how to use factoring to operate with rational expressions (algebraic fractions with polynomials in the numerator and denominator). This includes simplifying complex algebraic fractions. The topics of solving equations and the use of rational expressions to analyze and solve word problems (proportions, distance-rate-time, work, and variation) complete the chapter.

Chapter 9, ***Real Numbers and Radicals***, discusses the real numbers in detail with emphasis on simplifying radicals along with operating with radical expressions and evaluating radical expressions with a graphing calculator. Methods for solving equations involving radicals are discussed along with fractional exponents. The Pythagorean Theorem is presented and related to finding the distance between two points.

Chapter 10, ***Quadratic Equations***, develops the quadratic formula over three sections by first discussing solving quadratic equations by the square root method and then by completing the square. Applications are related to the Pythagorean Theorem, work, and distance-rate-time. The last section presents quadratic functions and the graphs of parabolas. The graphing calculator provides a valuable aid to understanding the nature of these graphs.

Appendix A.1, ***Difference of Two Cubes and Sum of Two Cubes***, deals with multiplying to get the sums and differences of two cubes and then reversing the process by factoring the sums and differences of two cubes.

Appendix A.2, ***Pi***, contains a brief discussion of the irrational number π.

Acknowledgements

Special thanks to Nina Miller, Larry Wadsworth, Marcel Prevuznak, and the rest of the Hawkes Learning Systems staff for their hard work and invaluable assistance in the development and production of this text.

Many thanks go to the following manuscript reviewers who offered their constructive and critical comments for this Sixth Edition of *Introductory Algebra*:

Russ Baker	Howard Community College
Sue Beck	Morehead State University
Stephanie Burton	Holmes Community College - Goodman
Angie Carlsen	Danville Community College
Donna Densmore	Bossier Parish Community College
Virginia Eaves	Bossier Parish Community College
Mike Hall	Arkansas State University
Karol Kendrick	Bossier Parish Community College
Linda Messia	Ohlone College
Staci Phillips	Bossier Parish Community College
David Rule	Holmes Community College - Goodman
Consuelo Stewart	Howard Community College
Melinda Treadway	Carl Sandburg College - The Extension Center

Also, sincere thanks go to those reviewers of the previous editions.

Finally, again, thank you James Hawkes and Greg Hill for their faith in this sixth edition and their willingness to commit so many resources to guarantee a top-quality product for students and teachers.

D. Franklin Wright

TO THE STUDENT

The goal of this text and of your instructor is for you to succeed in introductory algebra. Certainly, you should make this your goal as well. What follows is a brief discussion about developing good work habits and using the features of this text to your best advantage. For you to achieve the greatest return on your investment of time and energy you should practice the following three rules of learning.

1. Reserve a block of time to study every day.
2. Study what you don't know.
3. Don't be afraid to make mistakes.

How to use this book

The following seven-step guide will not only make using this book a more worthwhile and efficient task, but it will also help you benefit more from classroom lectures or the assistance that you receive in a math lab.

1. Try to look over the assigned section(s) before attending class or lab. In this way, new ideas may not sound so foreign when you hear them mentioned again. This will also help you see where you need to ask questions about material that seems difficult to you.
2. Read examples carefully. They have been chosen and written to show you all of the problem-solving steps that you need to be familiar with. You might even try to solve example problems on your own before studying the solutions that are given.
3. Work the section exercises faithfully as they are assigned. Problem-solving practice is the single most important element in achieving success in any math class, and there is no good substitute for actually doing this work yourself. Demonstrating that you can think independently through each step of each type of problem will also give you confidence in your ability to answer questions on quizzes and exams. Check the Answer Key periodically while working section exercises to be sure that you have the right ideas and are proceeding in the right manner.
4. Use the "Writing and Thinking About Mathematics" questions as an opportunity to explore the way that you think about math. A big part of learning and understanding mathematics is being able to talk about mathematical ideas and communicate the thinking that you do when you approach new concepts and problems. These questions can help

you analyze your own approach to mathematics and, in class or group discussions, learn from ideas expressed by your fellow students.

5. Use the Chapter Index of Key Ideas and Terms as a recap when you begin to prepare for a Chapter Test. It will reference all the major ideas that you should be familiar with from that chapter and indicate where you can turn if review is needed. You can also use the Chapter Index as a final checklist once you feel you have completed your review and are prepared for the Chapter Test.

6. Chapter Reviews provide essential problems from each section of the chapter to help you review all pertinent material before practicing with the Chapter Test.

7. Chapter Tests are provided so that you can practice for the tests that are actually given in class or lab. To simulate a test situation, block out a one-hour, uninterrupted period in a quiet place where your only focus is on accurately completing the Chapter Test. Use the Answer Key at the back of the book as a self-check only after you have completed all of the questions on the test.

8. Cumulative Reviews will help you retain the skills that you acquired in studying earlier chapters. They appear after every chapter beginning with Chapter 2. Approach them in much the same manner as you would the Chapter Tests in order to keep all of your skills sharp throughout the entire course.

How to Prepare for an Exam

Gaining Skill and Confidence

The stress that many students feel while trying to succeed in mathematics is what you have probably heard called "math anxiety." It is a real-life phenomenon, and many students experience such a high level of anxiety during mathematics exams in particular that they simply cannot perform to the best of their abilities. It is possible to overcome this stress simply by building your confidence in your ability to do mathematics and by minimizing your fears of making mistakes.

No matter how much it may seem that in mathematics you must either be right or wrong, with no middle ground, you should realize that you can be learning just as much from the times that you make mistakes as you can from the times that your work is correct. Success will come. Don't think that making mistakes at first means that you'll never be any good at mathematics. Learning mathematics requires lots of practice. Most importantly, it requires a true confidence in yourself and in the fact that with practice and persistence the mistakes will become fewer, the successes will become greater, and you will be able to say, "I can do this."

Showing What You Know

If you have attended class or lab regularly, taken good notes, read your textbook, kept up with homework exercises, and asked for help when it was needed, then you have already made significant progress in preparing for an exam and conquering any anxiety. Here are a few other suggestions to maximize your preparedness and minimize your stress.

1. Give yourself enough time to review. You will generally have several days advance notice before an exam. Set aside a block of time each day with the goal of reviewing a manageable portion of the material that the test will cover. Don't cram!
2. Work through a lot of problems to refresh your memory and sharpen your skills. Go back to redo selected exercises from all of your homework assignments.
3. Reread your text and your notes, and use the Chapter Index of Key Ideas and Terms and the Chapter Test to recap major ideas and do a self-evaluated test simulation.
4. Be sure that you are well-rested so that you can be alert and focused during the exam.
5. Don't study up to the last minute. Give yourself some time to wind down before the exam. This will help you to organize your thoughts and feel more calm as the test begins.
6. As you take the test, realize that its purpose is not to trick you, but to give you and your instructor an accurate idea of what you have learned. Good study habits, a positive attitude, and confidence in your own ability will be reflected in your performance on any exam.
7. Finally, you should realize that your responsibility does not end with taking the exam. When your instructor returns your corrected exam, you should review your instructor's comments and any mistakes that you might have made. Take the opportunity to learn from this important feedback about what you have accomplished, where you could work harder, and how you can best prepare for future exams.

HAWKES LEARNING SYSTEMS: INTRODUCTORY ALGEBRA

Overview

This multimedia courseware allows students to become better problem-solvers by creating a mastery level of learning in the classroom. The software includes a(n) "Instruct", "Practice", "Tutor", and "Certify" mode in each lesson, allowing students to learn through step-by-step interactions with the software. These automated homework system's tutorial and assessment modes extend instructional influence beyond the classroom. Intelligence is what makes the tutorials so unique. By offering intelligent tutoring and mastery level testing to measure what has been learned, the software extends the instructor's ability to influence students to solve problems. This courseware can be ordered either seperately or bundled together with this text.

Minimum Requirements

In order to run **HLS: Introductory Algebra**, you will need:

1 GHz or faster processor
Windows® 2000 or later
128 MB RAM (256 MB recommended)
200 MB hard drive space
800x600 resolution (1024x768 recommended)
Internet Explorer 6.0 or later
CD-ROM drive

Getting Started

Before you can run **HLS: Introductory Algebra**, you will need an access code. This 30 character code is <u>your</u> personal access code. To obtain an access code, go to **http://www.hawkeslearning.com** and follow the links to the access code request page (unless directed otherwise by your instructor.)

Installation

Insert the ***HLS: Introductory Algebra*** Installation CD-ROM into the CD-ROM drive. Select the Start/Run command, type in the CD-ROM drive letter followed by \setup.exe. (For example, d:\setup.exe where d is the CD-ROM drive letter.)

The complete installation may use over 110 MB of hard drive space and will install the entire product, except the multimedia files, on your hard drive.

After selecting the desired installation option, follow the on-screen instructions to complete your installation of ***HLS: Introductory Algebra.***

Starting the Courseware

After you installed ***HLS: Introductory Algebra*** on your computer, to run the courseware select Start/Programs/Hawkes Learning Systems/Introductory Algebra.

You will be prompted to enter your access code with a message box similar to the following:

Type your entire access code in the box. When you are finished, press OK.

If you typed in your access code correctly, you will be prompted to save the code to disk. If you choose to save your code to disk, typing in the access code each time you run ***HLS: Introductory Algebra*** will not be necessary. Instead, select the [Load from File] button when prompted to enter your access code and choose the path to your saved access code.

Now that you have entered your access code and saved it to diskette, you are ready to run a lesson. From the table of contents screen, choose the appropriate chapter and then choose the lesson you wish to run.

Features

Each lesson in *HLS: Introductory Algebra* has four modes: Instruct, Practice, Tutor, and Certify.

Instruct:
Instruct provides an exposition on the material covered in the lesson in a multimedia environment. This same instruct mode can be accessed via the tutor mode.

Practice:
Practice allows you to hone your problem-solving skills. It provides an unlimited number of randomly generated problems. Practice also provides access to the Tutor mode by selecting the Tutor button located by the Submit button.

Tutor:
Tutor mode is broken up into several parts: Instruct, Explain Error, Step by Step, and Solution.

1. **Instruct**, which can also be selected directly from Practice mode, contains a multimedia lecture of the material covered in a lesson.

2. **Explain Error** is active whenever a problem is incorrectly answered. It will attempt to explain the error that caused you to incorrectly answer the problem.

3. **Step by Step** is an interactive "step through" of the problem. It breaks each problem into several steps, explains to you each step in solving the problem, and asks you a question about the step. After you answer the last step correctly, you have solved the problem.

4. **Solution** will provide you with a detailed "worked-out" solution to the problem.

Throughout the Tutor, you will see words or phrases colored green with a dashed underline. These are called Hot

Words. Clicking on a Hot Word will provide you with more information on these word(s) or phrases.

Certify: Certify is the testing mode. You are given a finite number of problems and a certain number of strikes (problems you can get wrong). If you answer the required number of questions, you will receive a certification code and a certificate. Write down your certification code and/or print out your certificate. The certification code will be used by your instructor to update your records. Note that the Tutor is not available in Certify.

Integration of Courseware and Textbook

Throughout this text, you will see this icon that helps to integrate the Introductory Algebra textbook and **HLS: Introductory Algebra** courseware.

This icon indicates which **HLS: Introductory Algebra** lessons you should run in order to test yourself on the subject material and to review the contents of a chapter.

Support

If you have questions about **HLS: Introductory Algebra** or are having technical difficulties, we can be contacted as follows:

Phone: (843) 571-2825
Email: support@hawkeslearning.com
Web: www.hawkeslearning.com

Our support hours are 8:30 a.m. to 5:30 p.m., Eastern Time, Monday through Friday.

Review of Basic Topics

Did You Know?

The **decimal system** we use is based on the use of ten digits (0, 1, 2, 3, 4, 5, 6, 7, 8, 9) and a place value system based on powers of 10 (1, 10, 100, 1000, and so on). This system is attributed to the Hindu-Arabic peoples circa 800 A. D. The decimal system makes arithmetic operations relatively easy compared to ancient numeration systems. Consider how you might try to operate (add, subtract, multiply, and divide) using the symbols of the following systems.

Egyptian Numerals (Hieroglyphics) (3500 B. C.)

Symbol	Name	Value
|	Staff (vertical stroke)	1
∩	Heel bone (arch)	10
૭	Coil of rope (scroll)	100
ⳡ	Lotus flower	1000
⌒	Pointing finger	10,000
⌒	Bourbot (tadpole)	100,000
⳨	Astonished man	1,000,000

Example:

represents the number one thousand six hundred twenty-seven, or
$1000 + 600 + 20 + 7 = 1627$

Roman Numeral System

Symbol	I	V	X	L	C	D	M
Value	1	5	10	50	100	500	1000

We still use the Roman system as hours on clocks and dates on buildings. The symbols were written largest to smallest, from left to right. The system used both addition and subtraction. The values of symbols were added except for the use of subtraction for 4, 9, 40, 90, 400, and 900. (For example, IV = 5 − 1 = 4.)

To understand the difficulties in operating in the Roman system, try simple addition:

$$\text{CCXXXII} \rightarrow 232$$
$$+ \quad \text{CDXIII} \rightarrow 413$$

R.1 **Exponents and Order of Operations**

R.2 **Prime Numbers, Factoring, and LCM**

R.3 **Fractions (Multiplication and Division)**

R.4 **Fractions (Addition and Subtraction)**

R.5 **Decimals and Percents**

"Mathematics is the door and key to the sciences."

Roger Bacon (1214 - 1294)

Exponents and Order of Operations

Objectives

After completing this section, you will be able to:

1. Understand the terms **exponent**, **base**, and **power**.
2. Evaluate expressions with exponents.
3. Follow the **rules for order of operations** to evaluate numerical expressions.

Exponents

The **whole numbers** consist of the number 0 and the *natural numbers* (also called the *counting numbers*). We use $\mathbb{N}$ to represent the set of natural numbers and $\mathbb{W}$ to represent the set of whole numbers.

Whole Numbers

The **whole numbers** are the natural (or counting) numbers and the number 0.

Natural numbers = $\mathbb{N}$ = { *1, 2, 3, 4, 5, 6, 7, 8, 9, 10, 11, … }*

Whole numbers = $\mathbb{W}$ = { *0, 1, 2, 3, 4, 5, 6, 7, 8, 9, 10, 11, …}*

Note that 0 is a whole number but not a natural number.

The three dots … are called an *ellipsis* and are used to indicate that the pattern continues without end.

The operation of addition with whole numbers can be indicated by writing the numbers horizontally separated by (+) signs. The numbers being added are called **addends** and the result is called the **sum**. For example,

$$7 + 4 + 8 = 19.$$

addends   sum

The operation of multiplication with whole numbers can be indicated by writing the numbers horizontally separated by a raised dot ($\cdot$), separated by a times sign ($\times$), or by writing the numbers next to parentheses.

The numbers being multiplied are called **factors** and the result is called the **product**. For example,

$$12 \cdot 3 = 36 \qquad\qquad 6 \times 8 = 48 \qquad\qquad 4(15) = 60$$

factors product   factors product   factors product

We know that repeated addition of the same number is shortened by using multiplication. For example,

$$3 + 3 + 3 + 3 = 4 \cdot 3 = 12 \quad \text{and} \quad 10 + 10 + 10 + 10 + 10 = 5 \cdot 10 = 50$$

factors product   factors product

In a similar manner, repeated multiplication by the same number can be shortened by using **exponents**. For example, if 3 is used as a factor 4 times, we can write

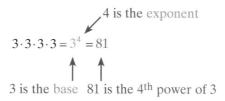

4 is the exponent

$$3 \cdot 3 \cdot 3 \cdot 3 = 3^4 = 81$$

3 is the base   81 is the 4th power of 3

In the equation $3^4 = 81$, 3 is the **base**, and 4 is the **exponent**. (Exponents are written slightly to the right and above the base.) The expression 3^4 is an **exponential expression** and is read "3 to the fourth power." We can also say that "81 is the fourth power of 3."

Example 1: Writing Exponents

With Repeated Multiplication	**With Exponents**
a. $5 \cdot 5 = 25$	$5^2 = 25$
b. $2 \cdot 2 \cdot 2 = 8$	$2^3 = 8$
c. $5 \cdot 5 \cdot 5 = 125$	$5^3 = 125$
d. $10 \cdot 10 \cdot 10 \cdot 10 = 10,000$	$10^4 = 10,000$

NOTES **Variables** are introduced here so that rules and definitions can be stated in a general form. Variables will be used in the examples and exercises beginning with Chapter 1.

Variable

*A **variable** is a symbol (generally a letter of the alphabet) that is used to represent an unknown number or any one of several numbers.*

The letters **n** and **a** are used as variables in the following discussion of exponents.

Exponent and Base

*A whole number **n** is an **exponent** if it is used to tell how many times another whole number **a** is used as a factor. The repeated factor **a** is called the **base** of the exponent. Symbolically,*

$$\underbrace{a \cdot a \cdot a \cdot \ldots \cdot a \cdot a}_{n \text{ factors}} = a^n.$$

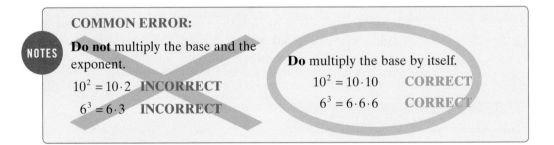

NOTES

COMMON ERROR:

Do not multiply the base and the exponent.

$10^2 = 10 \cdot 2$ **INCORRECT**

$6^3 = 6 \cdot 3$ **INCORRECT**

Do multiply the base by itself.

$10^2 = 10 \cdot 10$ **CORRECT**

$6^3 = 6 \cdot 6 \cdot 6$ **CORRECT**

In expressions with exponent 2, the base is said to be **squared**. In expressions with exponent 3, the base is said to be **cubed**.

Example 2: Translating Exponents

a. $8^2 = 64$ is read "eight squared is equal to sixty-four."

b. $5^3 = 125$ is read "five cubed is equal to one hundred twenty-five."

Expressions with exponents other than 2 or 3 are read as the base "to the _____ power." For example,

$2^5 = 32$ is read "two to the fifth power is equal to thirty-two."

The square of a whole number is called a **perfect square**. To help in factoring and working with fractions, you should memorize the squares of the whole numbers from 1 to 20. The following table lists these perfect squares.

Number (n)	1	2	3	4	5	6	7	8	9	10
Square (n^2)	1	4	9	16	25	36	49	64	81	100

Number (n)	11	12	13	14	15	16	17	18	19	20
Square (n^2)	121	144	169	196	225	256	289	324	361	400

Using a TI-84 Plus Graphing Calculator

The TI-84 Plus graphing calculator has keys marked as 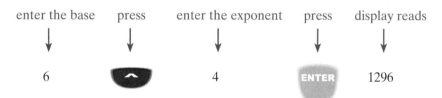 x^2 and a caret key. These keys can be used to find powers. The x^2 key will find squares and the caret key can be used to find squares and the values of other exponential expressions.

To find 6^4,

enter the base	press	enter the exponent	press	display reads
↓	↓	↓	↓	↓
6	^	4	ENTER	1296

The display will appear as follows:

To find 27^2,

enter the base	press	press	display reads
↓	↓	↓	↓
27			729

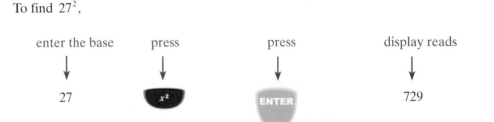

The display will appear as follows:

Alternatively, we could have entered:

enter the base	press	enter the exponent	press	display reads
↓	↓	↓	↓	↓
27		2		729

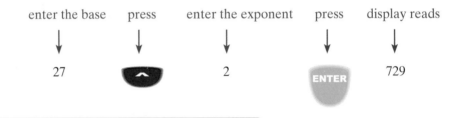

Example 3: Evaluating Exponents with a Calculator

Use a TI-84 Plus graphing calculator to find the value of each of the following exponential expressions.

   a. 9^5 **b.** 41^3 **c.** 152^0

Solutions for a TI-84 Plus Calculator

  a. **Step 1**:   Enter 9. (This is the base.)
      Step 2:   Press the caret key   .
      Step 3:   Enter 5. (This is the exponent.)
      Step 4:   Press the key ENTER .

The display should read **59049**.

b. Step 1: Enter 41.

   Step 2: Press the caret key 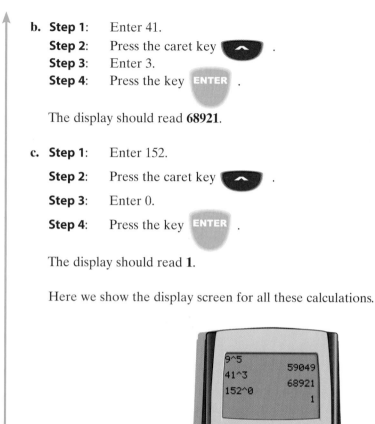 .

   Step 3: Enter 3.

   Step 4: Press the key ENTER .

The display should read **68921**.

c. Step 1: Enter 152.

   Step 2: Press the caret key .

   Step 3: Enter 0.

   Step 4: Press the key ENTER .

The display should read **1**.

Here we show the display screen for all these calculations.

```
9^5
            59049
41^3
            68921
152^0
                1
```

Rules for Order of Operations

Numerical expressions may have more than one operation indicated. There may be exponents, addition, subtraction [indicated with a minus sign (−)], division [indicated with a division sign (÷)] as well as symbols of inclusion, such as parentheses ( ), brackets [ ], and braces { }.

To evaluate an expression with more than one indicated operation, rules are needed so that everyone will arrive at the same answer. For example, consider how you would evaluate the following expression:

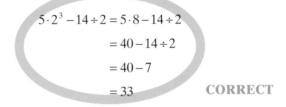

$$5 \cdot 2^3 - 14 \div 2 = 5 \cdot 8 - 14 \div 2$$
$$= 40 - 14 \div 2$$
$$= 40 - 7$$
$$= 33 \qquad \textbf{CORRECT}$$

However, if we were simply to proceed from left to right as in the following steps, we would get an entirely different answer:

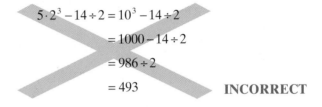

$$5 \cdot 2^3 - 14 \div 2 = 10^3 - 14 \div 2$$
$$= 1000 - 14 \div 2$$
$$= 986 \div 2$$
$$= 493 \qquad \textbf{INCORRECT}$$

By using the following rules for order of operations, we can conclude that the first answer (33) is the accepted correct answer and that the second answer (493) is incorrect.

Rules for Order of Operations

1. *First, simplify within grouping symbols, such as parentheses ( ), brackets [ ], or braces { }. Start with the innermost grouping.*

2. *Second, evaluate any numbers or expressions with exponents.*

3. *Third, moving from **left to right**, perform any multiplications or divisions in the order in which they appear.*

4. *Fourth, moving from **left to right**, perform any additions or subtractions in the order in which they appear.*

These rules are very explicit and should be studied carefully. **Note that in Rule 3, neither multiplication nor division has priority over the other.** Whichever of these operations occurs first, moving **left to right**, is done first. In Rule 4, addition and subtraction are handled in the same way. Unless they occur within grouping symbols, **addition and subtraction are the last operations to be performed.**

For example, consider the relatively simple expression $2 + 5 \cdot 6$. If we make the mistake of adding before multiplying, we get

$$2 + 5 \cdot 6 = 7 \cdot 6 = 42.$$ **INCORRECT**

The correct procedure yields

$$2 + 5 \cdot 6 = 2 + 30 = 32.$$ CORRECT

A well-known mnemonic device for remembering the Rules for Order of Operations is the following:

Please	**E**xcuse	**M**y	**D**ear	**A**unt	**S**ally
↓	↓	↓	↓	↓	↓
Parentheses	**Exponents**	**Multiplication**	**Division**	**Addition**	**Subtraction**

NOTES Even though the mnemonic **PEMDAS** is helpful, remember that multiplication and division are performed as they appear, left to right. Also, addition and subtraction are performed as they appear, left to right.

The following examples show how to apply these rules. In some cases, more than one step can be performed at the same time. This is possible when parts are separated by + or − signs or are within separate symbols of inclusion. **Work through each example step by step, and rewrite the examples on a separate sheet of paper.**

Example 4: Order of Operations

Evaluate the expression $24 \div 8 + 4 \cdot 2 - 6$.

Solution:

$$24 \div 8 + 4 \cdot 2 - 6$$ Divide before multiplying in this case.
Remember to move left to right.

$$= 3 + 4 \cdot 2 - 6$$ Multiply before adding or subtracting.

$$= 3 + 8 - 6$$ Add before subtracting in this case.
Remember to move left to right.

$$= 11 - 6$$ Subtract.

$$= 5$$

Example 5: Evaluating Expressions

Evaluate the expression $5 - 18 \div 9 \cdot 2 + 4(7)$.

Solution:

$$5 - \underbrace{18 \div 9} \cdot 2 + 4(7)$$ Divide.

$$= 5 - \underbrace{2 \cdot 2} + 4(7)$$ Multiply.

$$= 5 - \ 4 + \underbrace{4(7)}$$ Multiply.

$$= \underbrace{5 - \ 4} + \ 28$$ Subtract.

$$= \quad \underbrace{1 + \quad 28}$$ Add.

$$= \qquad 29$$

Example 6: Evaluating Expressions

Evaluate the expression $30 \div 10 \cdot 2^3 + 3(6 - 2)$.

Solution:

$$30 \div 10 \cdot 2^3 \quad + 3\underbrace{(6 - 2)}$$ Operate within parentheses.

$$= 30 \div 10 \cdot \underbrace{2^3} + \ 3(4)$$ Find the power.

$$= \underbrace{30 \div 10} \cdot 8 + \ 3(4)$$ Divide.

$$= \underbrace{3 \cdot 8} \qquad + \underbrace{3(4)}$$ Multiply in each part separated by +.

$$= \underbrace{24 \qquad + \quad 12}$$ Add.

$$= \qquad 36$$

Example 7: Evaluating Expressions with a Calculator

Use your TI-84 Plus calculator to evaluate the following expression.

$$1+2(5^2-1)-4+3\cdot2^3$$

Solution:

Now use pencil and paper to evaluate the expression. You should get the same answer because the calculator is programmed to use the rules for order of operations.

R.1 Exercises

For each of the following exponential expressions (a) name the base and (b) name the exponent, (c). Find the value of the exponential expression.

1. 2^3 **2.** 5^3 **3.** 4^2 **4.** 6^2

5. 9^2 **6.** 7^3 **7.** 11^2 **8.** 2^6

Find a base and exponent form for each of the following numbers without using the exponent 1.

9. 16 **10.** 4 **11.** 25 **12.** 32

13. 49 **14.** 121 **15.** 36 **16.** 100

17. 1000 **18.** 81 **19.** 8 **20.** 125

Rewrite the following products by using exponents.

21. $6\cdot6\cdot6\cdot6\cdot6$ **22.** $7\cdot7\cdot7\cdot7$ **23.** $11\cdot11\cdot11$

24. $13\cdot13\cdot13$ **25.** $2\cdot2\cdot2\cdot3\cdot3$ **26.** $2\cdot2\cdot5\cdot5\cdot5$

27. $2 \cdot 3 \cdot 3 \cdot 11 \cdot 11$ **28.** $5 \cdot 5 \cdot 5 \cdot 7 \cdot 7$ **29.** $3 \cdot 3 \cdot 3 \cdot 7 \cdot 7 \cdot 7$

30. $2 \cdot 2 \cdot 2 \cdot 2 \cdot 11 \cdot 11 \cdot 13 \cdot 13$

Find the following perfect squares. Write as many of them as you can from memory.

31. 8^2 **32.** 3^2 **33.** 11^2 **34.** 14^2 **35.** 20^2

36. 15^2 **37.** 13^2 **38.** 18^2 **39.** 30^2 **40.** 50^2

Use your calculator to find the value of each of the following exponential expressions. See Example 3.

41. 52^2 **42.** 35^2 **43.** 25^4 **44.** 32^4

45. 125^3 **46.** 47^5 **47.** 5^7 **48.** 2^{23}

Use the rules for order of operations to find the value of each of the following expressions. See Examples 4 through 7.

49. $6 + 5 \cdot 3$ **50.** $18 + 2 \cdot 5$

51. $20 - 4 \div 4$ **52.** $6 - 15 \div 3$

53. $32 - 14 + 10$ **54.** $25 - 10 + 11$

55. $18 \div 2 - 1 - 3 \cdot 2$ **56.** $6 \cdot 3 \div 2 - 5 + 13$

57. $2 + 3 \cdot 7 - 10 \div 2$ **58.** $14 \cdot 2 \div 7 \div 2 + 10$

59. $(2 + 3 \cdot 4) \div 7 - 2$ **60.** $(2 + 3) \cdot 4 \div 5 - 4$

61. $14(2 + 3) - 65 - 5$ **62.** $13(10 - 7) - 20 - 19$

63. $2 \cdot 5^2 - 8 \div 2$ **64.** $16 \div 2^4 + 9 \div 3^2$

65. $(2^3 + 2) \div 5 + 7^2 \div 7$ **66.** $(4^2 + 4) \div 2 \cdot 5 + 6^2 \div 6$

67. $(4 + 3)^2 + (2 + 3)^2$ **68.** $(2 + 1)^2 + (4 + 1)^2$

69. $8 \div 2 \cdot 4 - 16 \div 4 \cdot 2 + 3 \cdot 2^2$ **70.** $50 \div 2 \cdot 5 - 5^3 \div 5 + 5$

71. $(10 + 1)[(5 - 2)^2 + 3(4 - 3)]$ **72.** $(12 - 2)[4(6 - 3) + (4 - 3)^2]$

73. $100 + 2[\, 3(\, 4^2 - 6\,) + 2^3\,]$

74. $75 + 3[\, 2(\, 3 + 6\,)^2 - 10^2\,]$

Use your graphing calculator to evaluate each of the following expressions.

75. $16 + 3(\, 17 + 2^3 \div 2^2 - 4\,)$

76. $10^2 - 2(\, 16 \div 2^4 + 18 \div 3^2\,)$

77. $12^3 - 1\left[\left(3^2 + 5\right)^2\right]$

78. $10^4 - 1\left[\left(10 + 5\right)^2 - 5^2\right]$

79. $30 \div 2 - 11 + 2\left(5 - 1\right)^3$

80. $2\left(15 - 6 + 4\right) \div 13 \cdot 2 + 1$

Writing and Thinking About Mathematics

81. Use your calculator to evaluate the expression 0^0. What was the result? State, in your own words, the meaning of the result.

Hawkes Learning Systems: Introductory Algebra

Review of Exponents and Order of Operations

Prime Numbers, Factoring, and LCM

After completing this section, you will be able to:

1. *Understand the terms **prime number** and **composite number**.*
2. *Recognize the prime numbers less than 50.*
3. *Determine whether or not a number is prime.*
4. *Find the **prime factorization** of a composite number.*
5. *Find the **LCM** (least common multiple) of a set of counting numbers.*

Prime Numbers and Composite Numbers

Because of the relationship between multiplication and division, **factors** are also called **divisors** of the product. Division by a factor of a number gives a remainder of 0. For example,

$$\text{factors}$$
$$\downarrow \quad \downarrow$$

$$\begin{array}{r} 6 \\ 7{\overline{\smash{\big)}\,42}} \\ \underline{42} \\ 0 \leftarrow \text{remainder} \end{array}$$

$$7 \cdot 6 = 42$$

and

Thus 7 and 6 are divisors as well as factors of 42.

Every counting number, except the number 1, has **at least two** factors, as illustrated in the following list. Note that, in this list, every number has at least two factors, but 23, 19 and 5 have **exactly two** factors.

Examples of Counting Numbers	Factors
33	1, 3, 11, 33
23	1, 23
6	1, 2, 3, 6
19	1, 19
42	1, 2, 3, 6, 7, 14, 21, 42
5	1, 5
96	1, 2, 3, 4, 6, 8, 12, 16, 24, 32, 48, 96

In the list above 23, 19, and 5 have exactly two different factors. Such numbers are called **prime numbers**. The other numbers in the list (33, 6, 42, and 96) are called **composite numbers**.

Prime Number

*A **prime number** is a counting number greater than 1 that has exactly two different factors (or divisors), namely 1 and itself.*

Composite Number

*A **composite number** is a counting number with more than two different factors (or divisors).*

Example 1: Prime Numbers

Some prime numbers

2	2 has exactly two different factors, 1 and 2.
3	3 has exactly two different factors, 1 and 3.
11	11 has exactly two different factors, 1 and 11.
29	29 has exactly two different factors, 1 and 29.

Example 2: Composite Numbers

Some composite numbers

15	1, 3, 5, and 15 are all factors of 15.
39	1, 3, 13, and 39 are all factors of 39.
49	1, 7, and 49 are all factors of 49.
51	1, 3, 17, and 51 are all factors of 51.

Because we will be looking for factors in algebra, particularly in dealing with fractions, we need to be aware of the concept of a number being **exactly divisible** by another number. A number is exactly divisible by (or just **divisible by**) another number if the remainder in the division process is 0. For example,

$$
\begin{array}{r}
57 \\
5\overline{)285} \\
25 \\
\hline
35 \\
35 \\
\hline
\mathbf{0} \quad \text{remainder}
\end{array}
\qquad
\begin{array}{r}
142 \\
2\overline{)285} \\
2 \\
\hline
8 \\
8 \\
\hline
5 \\
4 \\
\hline
\mathbf{1} \quad \text{remainder}
\end{array}
$$

Thus 5 and 57 are both **factors** of 285 and both **divide exactly** into 285. The number 2 does not divide exactly into 285 and is not a factor of 285.

The related concepts of even and odd whole numbers are also useful.

Even and Odd Whole Numbers

*If a whole number is divisible by 2, it is **even**.*
*If a whole number is not divisible by 2, it is **odd**.*

The **even** whole numbers are

0, 2, 4, 6, 8, 10, 12, …

The **odd** whole numbers are

1, 3, 5, 7, 9, 11, 13, …

The prime numbers less than 50 are

2, 3, 5, 7, 11, 13, 17, 19, 23, 29, 31, 37, 41, 43, 47

For convenience and ease in working with fractions, you should memorize, or at least recognize, these primes. Also, you should note the following two facts about prime numbers:

1. 2 is the only even prime number; and
2. All other prime numbers are odd, but not all odd numbers are prime.
   (For example, 9, 15, and 33 are odd, but not prime.)

Prime Factorization of Composite Numbers

Using basic knowledge of multiplication and division, we can find factors of relatively small counting numbers. For example, basic multiplication facts give

$$63 = 9 \cdot 7.$$

However, for use in dealing with fractions, we need to find a factorization of 63 in which all of the factors are prime numbers. In this example, 9 is not a prime number. By factoring 9, we can write

$$63 = 3 \cdot 3 \cdot 7.$$

This last product $(3 \cdot 3 \cdot 7)$ contains all prime factors and is called the **prime factorization** of 63. Note that, because multiplication is a commutative operation (See Section 1.5 for a detailed discussion of the properties of real numbers.), the order of the factors is not important. Thus we could write $63 = 3 \cdot 7 \cdot 3$ as the prime factorization. However, for consistency, we will generally write the factors in order, from smallest to largest. Also, using exponents, we can write $63 = 3^2 \cdot 7$.

Regardless of the method used to find the prime factorization, **there is only one prime factorization for any composite number.** The following procedure can be used to find prime factorizations.

To Find the Prime Factorization of a Composite Number

1. Factor the composite number into any two factors.
2. Factor each factor that is not prime.
3. Continue this process until all factors are prime.
*The **prime factorization** is the product of all the prime factors.*

NOTES

Special Note:
You may have studied quick tests for divisibility by 2, 3, 5, 6, 9, and 10 in a previous course in mathematics. For example, a number is divisible by 2, and therefore even, if the units digit is 0, 2, 4, 6, or 8. We will make reference to some of these tests for divisibility in the examples.

Example 3: Prime Factorization

Find the prime factorization of 90.

Solution:

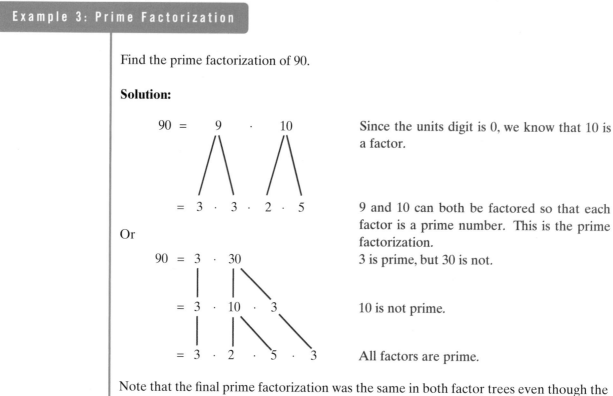

90 = 9 · 10   Since the units digit is 0, we know that 10 is a factor.

= 3 · 3 · 2 · 5   9 and 10 can both be factored so that each factor is a prime number. This is the prime factorization.

Or

90 = 3 · 30   3 is prime, but 30 is not.

= 3 · 10 · 3   10 is not prime.

= 3 · 2 · 5 · 3   All factors are prime.

Note that the final prime factorization was the same in both factor trees even though the first pair of factors was different.

Since multiplication is commutative, the order of the factors is not important. What is important is that **all the factors are prime**. Writing the factors in order, we can write

$$90 = 2 \cdot 3 \cdot 3 \cdot 5$$

or, with exponents,

$$90 = 2 \cdot 3^2 \cdot 5.$$

Example 4: Prime Factorization

Find the prime factorizations of each number:

a. 65   **b.** 72   **c.** 294

Solutions:

a. 65 = 5 · 13   5 is a factor because the units digit is 5. Since both 5 and 13 are prime, 5·13 is the prime factorization.

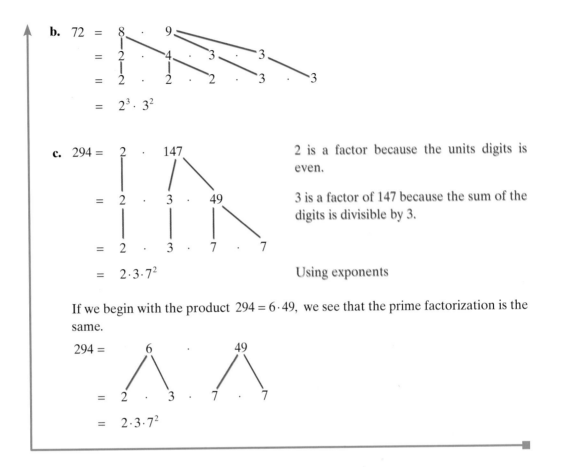

b. $72 = 8 \cdot 9$

$= 2 \cdot 4 \cdot 3 \cdot 3$

$= 2 \cdot 2 \cdot 2 \cdot 3 \cdot 3$

$= 2^3 \cdot 3^2$

c. $294 = 2 \cdot 147$ 2 is a factor because the units digits is even.

$= 2 \cdot 3 \cdot 49$ 3 is a factor of 147 because the sum of the digits is divisible by 3.

$= 2 \cdot 3 \cdot 7 \cdot 7$

$= 2 \cdot 3 \cdot 7^2$ Using exponents

If we begin with the product $294 = 6 \cdot 49$, we see that the prime factorization is the same.

$294 = 6 \cdot 49$

$= 2 \cdot 3 \cdot 7 \cdot 7$

$= 2 \cdot 3 \cdot 7^2$

Least Common Multiple (LCM)

The **multiples** of a number are the products of that number with the counting numbers. Thus the first multiple of any number is the number itself. All other multiples are larger than that number. We are interested in finding common multiples and more particularly the **least common multiple (LCM)** for a set of counting numbers. For example, consider the lists of multiples of 8 and 12 shown here.

Counting Numbers: 1,   2,   3,   4,   5,   6,   7,   8,   9,   10,   11,   12...

Multiples of 8:   8, 16, ㉔ 32, 40, ㊽ 56, 64, ㉒ 80, 88, ㊻..

Multiples of 12:   12, ㉔ 36, ㊽ 60, ㉒ 84, ㊻ 108, ⑫⓪ 132, ⑭⑭..

The common multiples of 8 and 12 are 24, 48, 72, 96, 120, ... The **least common multiple (LCM)** is 24.

Listing all the multiples, as we just did for 8 and 12, and then choosing the least common multiple (LCM) is not very efficient. The following technique involving prime factorizations is generally much easier to use.

To Find the LCM of a Set of Counting Numbers

1. *Find the prime factorization of each number.*
2. *List the prime factors that appear in any one of the prime factorizations.*
3. *Find the product of these primes using each prime the greatest number of times it appears in any one of the prime factorizations.*

Example 5: Least Common Multiple (LCM)

Find the least common multiple (LCM) of 8, 10, and 30.

Solution:

a. Prime factorizations:

$$8 \ = \ 2 \cdot 2 \cdot 2 \qquad \text{three 2's}$$
$$10 \ = \ 2 \cdot 5 \qquad \text{one 2, one 5}$$
$$30 \ = \ 2 \cdot 3 \cdot 5 \qquad \text{one 2, one 3, one 5}$$

b. Prime factors that are present are 2, 3, and 5.

c. The **most** number of times each prime factor is used in **any one** factorization:

Three 2's   (in 8)
One 3   (in 30)
One 5   (in 10 and in 30)

d. Find the product of these primes.
$$\text{LCM} = 2 \cdot 2 \cdot 2 \cdot 3 \cdot 5$$
$$= 2^3 \cdot 3 \cdot 5 = 120$$

120 is the LCM and therefore the smallest number divisible by 8, 10, and 30.

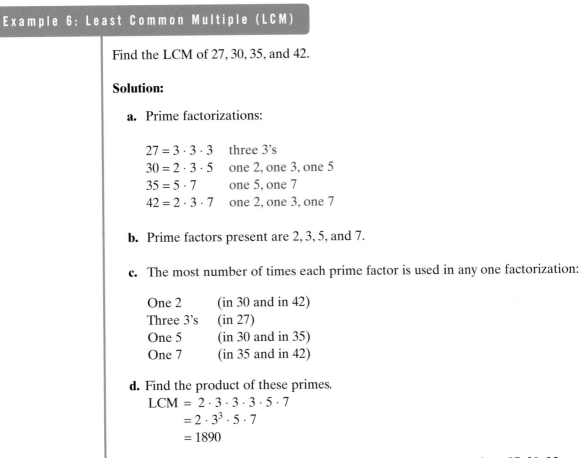

Example 6: Least Common Multiple (LCM)

Find the LCM of 27, 30, 35, and 42.

Solution:

a. Prime factorizations:

$27 = 3 \cdot 3 \cdot 3$ three 3's
$30 = 2 \cdot 3 \cdot 5$ one 2, one 3, one 5
$35 = 5 \cdot 7$ one 5, one 7
$42 = 2 \cdot 3 \cdot 7$ one 2, one 3, one 7

b. Prime factors present are 2, 3, 5, and 7.

c. The most number of times each prime factor is used in any one factorization:

One 2   (in 30 and in 42)
Three 3's   (in 27)
One 5   (in 30 and in 35)
One 7   (in 35 and in 42)

d. Find the product of these primes.
$$\text{LCM} = 2 \cdot 3 \cdot 3 \cdot 3 \cdot 5 \cdot 7$$
$$= 2 \cdot 3^3 \cdot 5 \cdot 7$$
$$= 1890$$

1890 is the smallest number divisible by all four of the numbers 27, 30, 35, and 42.

Tests for Divisibility

As mentioned in the special note on p. 17, here are the quick tests for divisibility.

An integer is divisible
By 2: *if the units digit is 0, 2, 4, 6, or 8.*
By 3: *if the sum of the digits is divisible by 3.*
By 5: *if the units digit is 0 or 5.*
By 6: *if the number is divisible by both 2 and 3.*
By 9: *if the sum of the digits is divisible by 9.*
By 10: *if the units digit is zero.*

R.2 Exercises

1. Define *prime number*.

2. List the prime numbers less than 50.

3. Define *composite number*.

4. True or False: (If a statement is false, explain why.)
   a. All prime numbers are even.
   b. All prime numbers are odd.
   c. All odd numbers are prime.
   d. 2 is the only even prime number.

5. List five prime numbers larger than 50.

6. Describe, in your own words, how to find the LCM of a set of counting numbers.

In Exercises 7 – 10, determine which numbers, if any, in each set of counting numbers are prime.

7. $\{13, 15, 17, 21\}$ **8.** $\{11, 19, 23, 51\}$

9. $\{2, 4, 6, 8, 10, 12, 14\}$ **10.** $\{7, 16, 25, 36, 47, 49\}$

In Exercises 11 – 20, find two factors of each number (other than 1 and the number itself) to determine that the number is composite. (Answers may vary.)

11. 72	**12.** 63	**13.** 68	**14.** 39
15. 502	**16.** 417	**17.** 170	**18.** 99
19. 444	**20.** 230		

Find the prime factorization of each of the numbers in exercises 21 – 40. If a number is prime, say "prime".

21. 52	**22.** 60	**23.** 616	**24.** 460
25. 308	**26.** 155	**27.** 79	**28.** 43
29. 289	**30.** 361	**31.** 125	**32.** 343
33. 400	**34.** 500	**35.** 120	**36.** 196
37. 231	**38.** 675	**39.** 1692	**40.** 2717

41. List the first twelve **multiples** of each number.

    a. 5   **b.** 6   **c.** 10   **d.** 15

42. From the lists you made in Exercise 41, find the least common multiple for each of the following sets of numbers.

    a. {5, 6}   **b.** {6, 10}   **c.** {5, 10, 15}   **d.** {6, 10, 15}

Find the LCM of each of the following sets of counting numbers.

43. {3, 5, 7}   **44.** {2, 7, 11}   **45.** {6, 10}

46. {9, 12}   **47.** {2, 3, 11}   **48.** {3, 5, 13}

49. {4, 14, 35}   **50.** {10, 12, 20}   **51.** {50, 75}

52. {30, 70}   **53.** {20, 90}   **54.** {50, 80}

55. {28, 98}   **56.** {45, 75}   **57.** {10, 15, 35}

58. {6, 24, 30}   **59.** {15, 45, 90}   **60.** {14, 28, 56}

61. {20, 50, 100}   **62.** {30, 60, 120}   **63.** {10, 15, 25}

64. {22, 44, 121}   **65.** {26, 28, 91}   **66.** {34, 51, 54}

67. {35, 40, 72}   **68.** {30, 35, 63}   **69.** {12, 21, 44}

70. {20, 28, 45}   **71.** {99, 121, 231}   **72.** {81, 225, 324}

73. {48, 120, 144, 192}   **74.** {125, 135, 225, 250}   **75.** {40, 56, 160, 196}

76. {35, 49, 63, 126}

Hawkes Learning Systems: Introductory Algebra

Review of Prime Numbers, Factoring, and LCM

R.3   Fractions (Multiplication and Division)

After completing this section, you will be able to:

1. *Understand the terms **numerator**, **denominator**, and **reciprocal**.*
2. *Raise fractions to higher terms.*
3. *Reduce fractions to lowest terms.*
4. *Multiply and divide with fractions.*

Introduction to Fractions

Fractions are numbers written in the form $\frac{a}{b}$ where the numerator (top number) and the denominator (bottom number) are whole numbers, with the denominator not 0. It is important to note that **the denominator cannot be 0**. We say that the denominator b is nonzero or that $b \neq 0$ (b is not equal to 0).

Fraction

*A **fraction** is a number that can be written in the form $\frac{a}{b}$ where a and b are whole numbers and $b \neq 0$.*

$$\frac{a}{b} \begin{array}{l} \leftarrow \text{ numerator} \\ \leftarrow \text{ denominator} \end{array}$$

Fractions can be used in two ways:

1. to indicate equal parts of a whole,
2. to indicate division, with the numerator divided by the denominator.

NOTES

Special Note
If 0 does occur in a denominator, we say that the fraction is **undefined** because division by 0 is undefined. For example, $\frac{13}{0}$ is undefined.

Example 1: Fractions

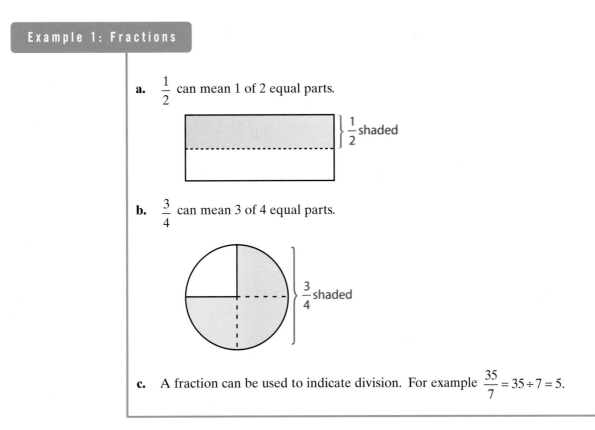

a. $\frac{1}{2}$ can mean 1 of 2 equal parts.

$\frac{1}{2}$ shaded

b. $\frac{3}{4}$ can mean 3 of 4 equal parts.

$\frac{3}{4}$ shaded

c. A fraction can be used to indicate division. For example $\frac{35}{7} = 35 \div 7 = 5$.

A very useful fact when dealing with fractions is that every whole number can be written in fraction form with denominator 1. Thus,

$$0 = \frac{0}{1}, \quad 1 = \frac{1}{1}, \quad 2 = \frac{2}{1}, \quad 3 = \frac{3}{1}, \quad 4 = \frac{4}{1}, \text{ and so on.}$$

Proper Fraction, Improper Fraction

*A **proper fraction** is a fraction in which the numerator is less than the denominator.*

*An **improper fraction** is a fraction in which the numerator is greater than or equal to the denominator.*

Examples of proper fractions: $\frac{7}{8}, \frac{10}{21}, \frac{3}{11}$, and $\frac{1}{2}$.

Examples of improper fractions: $\frac{4}{3}, \frac{28}{13}, \frac{79}{24}$, and $\frac{350}{250}$.

The term "improper fraction" is unfortunate and somewhat misleading because there is nothing "improper" about such fractions. Improper fractions are used to great advantage in much of mathematics.

Multiplying with Fractions

Now we state the rule for multiplying fractions and discuss the use of the word "of" to indicate multiplication by fractions. As shown in Example 3, to find a fraction "of" a number means to multiply.

To Multiply Fractions

1. *Multiply the numerators.*

$$\frac{a}{b} \cdot \frac{c}{d} = \frac{a \cdot c}{b \cdot d} \quad where \ b, d \neq 0.$$

2. *Multiply the denominators.*

Example 2: Multiplying Fractions

Find the product of $\frac{1}{3}$ and $\frac{2}{7}$.

Solution: $\frac{1}{3} \cdot \frac{2}{7} = \frac{1 \cdot 2}{3 \cdot 7} = \frac{2}{21}$

Example 3: Multiplying Fractions

Find $\frac{2}{5}$ of $\frac{7}{3}$.

Solution: $\frac{2}{5} \cdot \frac{7}{3} = \frac{14}{15}$ Note that the word "of" indicates multiplication.

Example 4: Multiplying Fractions

a. $\frac{4}{13} \cdot 3 = \frac{4}{13} \cdot \frac{3}{1} = \frac{4 \cdot 3}{13 \cdot 1} = \frac{12}{13}$

b. $\frac{9}{8} \cdot 0 = \frac{9}{8} \cdot \frac{0}{1} = \frac{9 \cdot 0}{8 \cdot 1} = \frac{0}{8} = 0$

c. $4 \cdot \frac{5}{3} \cdot \frac{2}{7} = \frac{4}{1} \cdot \frac{5}{3} \cdot \frac{2}{7} = \frac{40}{21}$

Building Fractions to Higher Terms and Reducing Fractions

The number 1 is called the **multiplicative identity** for whole numbers; that is, $a \cdot 1 = a$ for any whole number a. The number 1 is also the multiplicative identity for fractions since

$$\frac{a}{b} \cdot 1 = \frac{a}{b} \cdot \frac{1}{1} = \frac{a \cdot 1}{b \cdot 1} = \frac{a}{b}.$$

Now, to build a fraction to higher terms without changing its value, we need to know that 1 can be represented in fraction form with any counting number divided by itself:

$$1 = \frac{1}{1} = \frac{2}{2} = \frac{3}{3} = \frac{4}{4} = \frac{5}{5} = \frac{6}{6} = \frac{7}{7} \text{ and so on.}$$

This means that the numerator and denominator are to be multiplied by the same number.

Example 5: Raising Fractions to Higher Terms

Build $\frac{3}{4}$ to higher terms as indicated: $\frac{3}{4} = \frac{?}{36}$.

(That is, find a fraction that is equal to $\frac{3}{4}$ with denominator 36.)

Solution:

We know that $4 \cdot 9 = 36$, so we choose $1 = \frac{9}{9}$ and proceed as follows:

$$\frac{3}{4} = \frac{3}{4} \cdot \frac{9}{9} = \frac{27}{36}. \quad \text{(Thus } \frac{3}{4} \text{ and } \frac{27}{36} \text{ are two forms of the same number.)}$$

Example 6: Raising Fractions to Higher Terms

Build $\frac{11}{10}$ to higher terms as indicated: $\frac{11}{10} = \frac{?}{50}$.

Solution:

We know that $10 \cdot 5 = 50$, so we choose $1 = \frac{5}{5}$ and proceed as follows:

$$\frac{11}{10} = \frac{11}{10} \cdot \frac{5}{5} = \frac{55}{50}.$$

Changing a fraction to lower terms is the same as **reducing** the fraction. **A fraction is reduced to lowest terms if the numerator and the denominator have no common factor other than 1.** To reduce a fraction, we again use the fact that 1 is equal to any counting number divided by itself and "divide out" all common factors found in both the numerator and denominator. (Note that reduced fractions might be improper fractions.)

Example 7: Reducing Fractions to Lower Terms

Reduce $\dfrac{16}{28}$ to lowest terms.

Solution:

Find the prime factorization of the numerator and the denominator, then "divide out" the common factors.

$$\frac{16}{28} = \frac{2 \cdot 2 \cdot 2 \cdot 2}{2 \cdot 2 \cdot 7} = \frac{2}{2} \cdot \frac{2}{2} \cdot \frac{4}{7} = 1 \cdot 1 \cdot \frac{4}{7} = \frac{4}{7}$$

As a shortcut, we do not generally write the number 1 when reducing as we did in Example 7. We do use cancel marks to indicate dividing out common factors in numerators and denominators. **But remember that these numbers do not simply disappear. Their quotient is understood to be 1 even if the 1 is not written.**

Example 8: Reducing Fractions to Lower Terms

Reduce $\dfrac{45}{36}$ to lowest terms.

Solution:

a. Using prime factors, we have

$$\frac{45}{36} = \frac{\cancel{3} \cdot \cancel{3} \cdot 5}{2 \cdot 2 \cdot \cancel{3} \cdot \cancel{3}} = \frac{5}{4}.$$ Note that the answer is a reduced improper fraction.

b. Larger common factors can be divided out, but we must be sure that we have the largest common factor.

$$\frac{45}{36} = \frac{5 \cdot \cancel{9}}{4 \cdot \cancel{9}} = \frac{5}{4}$$

Multiplying and Reducing Fractions at the Same Time

Now we can multiply fractions and reduce all in one step by using prime factors (or other common factors). The advantage of using prime factors is that you can be sure that you have not missed a common factor and that your answer is reduced to lowest terms. Of course, if you "see" a common factor that is not prime, that factor may certainly be used in reducing.

Examples 9, 10, and 11 illustrate how to multiply and reduce at the same time by factoring the numerators and the denominators. **Note that if all the factors in the numerator or denominator divide out, then 1 must be used as a factor.**

Example 9: Multiplying and Reducing Fractions

Multiply and reduce to lowest terms.

$$\frac{18}{35} \cdot \frac{21}{12} = \frac{\cancel{2} \cdot \cancel{3} \cdot 3 \cdot 3 \cdot \cancel{7}}{5 \cdot \cancel{7} \cdot \cancel{2} \cdot 2 \cdot \cancel{3}} = \frac{9}{10}$$

Example 10: Multiplying and Reducing Fractions

Multiply and reduce to lowest terms.

$$\frac{17}{50} \cdot \frac{25}{34} \cdot 8 = \frac{17 \cdot 25 \cdot 8}{50 \cdot 34 \cdot 1}$$

$$= \frac{\cancel{17} \cdot \cancel{25} \cdot \cancel{2} \cdot \cancel{2} \cdot 2}{\cancel{2} \cdot \cancel{25} \cdot \cancel{2} \cdot \cancel{17} \cdot 1}$$

$$= \frac{2}{1} = 2$$

In this example, 25 is a common factor that is not a prime number.

Example 11: Multiplying and Reducing Fractions

Multiply and reduce to lowest terms.

$$\frac{3}{35} \cdot \frac{15}{4} \cdot \frac{7}{9} = \frac{\cancel{3} \cdot \cancel{3} \cdot \cancel{5} \cdot \cancel{7}}{\cancel{5} \cdot \cancel{7} \cdot 2 \cdot 2 \cdot \cancel{3} \cdot \cancel{3}} = \frac{1}{4}$$

Dividing with Fractions

If the product of two fractions is 1, then the fractions are called **reciprocals** of each other. For example,

$\dfrac{5}{8}$ and $\dfrac{8}{5}$ are reciprocals because $\dfrac{5}{8} \cdot \dfrac{8}{5} = \dfrac{40}{40} = 1.$

Reciprocal

*The **reciprocal** of $\dfrac{a}{b}$ is $\dfrac{b}{a}$ (where $a \neq 0$ and $b \neq 0$) and $\dfrac{a}{b} \cdot \dfrac{b}{a} = 1.$*

 NOTES   **The number 0 has no reciprocal.** That is $\dfrac{0}{1}$ has no reciprocal because $\dfrac{1}{0}$ is undefined.

Division with fractions is accomplished by multiplying by the reciprocal of the divisor. Thus a division problem can be changed into a multiplication problem as follows:

$$\frac{2}{3} \div \frac{5}{7} = \frac{2}{3} \cdot \frac{7}{5} = \frac{14}{15}.$$

Division

To divide by any nonzero number, multiply by its reciprocal.

In general,

$$\frac{a}{b} \div \frac{c}{d} = \frac{a}{b} \cdot \frac{d}{c} \quad \text{where } b, c, d \neq 0.$$

Example 12: Dividing Fractions

Divide $\dfrac{3}{4} \div 4$

Solution:

The reciprocal of 4 is $\dfrac{1}{4}$, so we multiply by $\dfrac{1}{4}$,

$$\frac{3}{4} \div 4 = \frac{3}{4} \div \frac{4}{1} = \frac{3}{4} \cdot \frac{1}{4} = \frac{3}{16}.$$

As with any multiplication problem, we reduce whenever possible by factoring numerators and denominators.

Example 13: Dividing and Reducing Fractions

Divide and reduce to lowest terms: $\dfrac{16}{27} \div \dfrac{8}{9}$.

Solution:

The reciprocal of $\dfrac{8}{9}$ is $\dfrac{9}{8}$. Reduce by factoring.

$$\frac{16}{27} \div \frac{8}{9} = \frac{16}{27} \cdot \frac{9}{8} = \frac{\cancel{8} \cdot 2 \cdot \cancel{9}}{3 \cdot \cancel{9} \cdot \cancel{8}} = \frac{2}{3}$$

Example 14: Applications: Dividing and Reducing Fractions

If you had $30 and you spent $26 to buy a computer keyboard, what fraction of your money did you spend for the keyboard? What fraction do you still have?

Solution:

a. The fraction you spent is $\dfrac{26}{30} = \dfrac{\cancel{2} \cdot 13}{\cancel{2} \cdot 3 \cdot 5} = \dfrac{13}{15}$

b. Since you still have $4, the fraction you still have is $\dfrac{4}{30} = \dfrac{\cancel{2} \cdot 2}{\cancel{2} \cdot 3 \cdot 5} = \dfrac{2}{15}$.

Example 15: Applications: Dividing Fractions

A truck is towing 3000 pounds. This is $\frac{2}{3}$ of the truck's towing capacity.

a. Can the truck tow more than 3000 pounds?

b. If you were to multiply $\frac{2}{3}$ times 3000, would the product be more or less than 3000?

c. What is the towing capacity of the truck?

Solution:

a. Yes, the truck can tow more than 3000 pounds because $\frac{2}{3}$ is less than 1.

b. The product would be less than 3000.

c. To find the towing capacity divide: $3000 \div \frac{2}{3} = \frac{3000}{1} \cdot \frac{3}{2} = 4500$.

The towing capacity of the truck is 4500 pounds.

R.3 Exercises

1. a. What is a proper fraction?   **b.** What is an improper fraction?

2. Define the term *reciprocal*.

3. List two uses of fractions.

4. Describe how to raise a fraction to higher terms.

5. Describe how to reduce a fraction to lowest terms.

In Exercises 6 – 9, answer the following questions:

a. What fraction of each figure is shaded?

b. What fraction of each figure is not shaded?

6.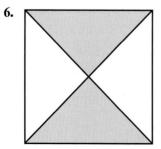

7.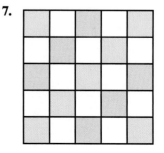

8.   **9.**

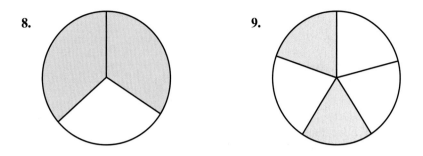

10. What is the value, if any, of each of the following expressions?

a. $\dfrac{0}{6} \cdot \dfrac{3}{4}$ **b.** $\dfrac{1}{7} \cdot \dfrac{0}{5}$ **c.** $\dfrac{3}{0} \cdot \dfrac{7}{8}$ **d.** $\dfrac{1}{0} \cdot \dfrac{0}{1}$

In Exercises 11 – 20, build each fraction to higher terms as indicated. Find the values of the missing numbers.

11. $\dfrac{3}{4} = \dfrac{3}{4} \cdot \dfrac{?}{?} = \dfrac{?}{12}$ **12.** $\dfrac{2}{3} = \dfrac{2}{3} \cdot \dfrac{?}{?} = \dfrac{?}{12}$

13. $\dfrac{6}{7} = \dfrac{6}{7} \cdot \dfrac{?}{?} = \dfrac{?}{14}$ **14.** $\dfrac{5}{8} = \dfrac{5}{8} \cdot \dfrac{?}{?} = \dfrac{?}{40}$

15. $\dfrac{3}{16} = \dfrac{3}{16} \cdot \dfrac{?}{?} = \dfrac{?}{80}$ **16.** $\dfrac{1}{17} = \dfrac{1}{17} \cdot \dfrac{?}{?} = \dfrac{?}{51}$

17. $\dfrac{7}{26} = \dfrac{7}{26} \cdot \dfrac{?}{?} = \dfrac{?}{52}$ **18.** $\dfrac{9}{10} = \dfrac{9}{10} \cdot \dfrac{?}{?} = \dfrac{?}{100}$

19. $\dfrac{18}{1} = \dfrac{18}{1} \cdot \dfrac{?}{?} = \dfrac{?}{3}$ **20.** $\dfrac{1}{5} = \dfrac{1}{5} \cdot \dfrac{?}{?} = \dfrac{?}{75}$

Find the products in Exercises 21 – 30. Reduce to lowest terms whenever possible.

21. $\dfrac{7}{8} \cdot \dfrac{4}{21}$ **22.** $\dfrac{23}{36} \cdot \dfrac{20}{46}$

23. $9 \cdot \dfrac{7}{24}$ **24.** $8 \cdot \dfrac{5}{12}$

25. $\dfrac{9}{10} \cdot \dfrac{35}{40} \cdot \dfrac{65}{15}$ **26.** $\dfrac{42}{52} \cdot \dfrac{27}{22} \cdot \dfrac{33}{9}$

27. $\dfrac{75}{8} \cdot \dfrac{16}{36} \cdot 9 \cdot \dfrac{7}{25}$

28. $\dfrac{17}{8} \cdot \dfrac{5}{42} \cdot \dfrac{18}{51} \cdot 4$

29. $\dfrac{3}{4} \cdot \dfrac{7}{2} \cdot \dfrac{22}{54} \cdot 18$

30. $\dfrac{69}{15} \cdot \dfrac{30}{8} \cdot \dfrac{14}{46} \cdot 9$

Find the quotients in Exercises 31 – 50. Reduce to lowest terms whenever possible.

31. $\dfrac{5}{8} \div \dfrac{3}{5}$

32. $\dfrac{2}{7} \div \dfrac{1}{2}$

33. $\dfrac{2}{3} \div \dfrac{1}{5}$

34. $\dfrac{2}{11} \div \dfrac{1}{7}$

35. $\dfrac{3}{14} \div \dfrac{3}{14}$

36. $\dfrac{5}{8} \div \dfrac{5}{8}$

37. $\dfrac{3}{4} \div \dfrac{4}{3}$

38. $\dfrac{9}{10} \div \dfrac{10}{9}$

39. $\dfrac{15}{20} \div 3$

40. $\dfrac{14}{20} \div 7$

41. $\dfrac{25}{40} \div 10$

42. $\dfrac{36}{80} \div 9$

43. $\dfrac{7}{8} \div 0$

44. $\dfrac{15}{64} \div 0$

45. $0 \div \dfrac{5}{6}$

46. $0 \div \dfrac{1}{2}$

47. $\dfrac{16}{35} \div \dfrac{2}{7}$

48. $\dfrac{15}{27} \div \dfrac{5}{9}$

49. $\dfrac{15}{24} \div \dfrac{25}{18}$

50. $\dfrac{36}{25} \div \dfrac{24}{20}$

51. If you have $20 and you spend $9 for a sandwich and drink, what fraction of your money have you spent? What fraction do you still have?

52. In a class of 40 students, 6 received a grade of A. What fraction of the class received an A? What fraction of the class did not receive an A?

53. You are planning a bicycle trip of 60 miles in the mountains and are told that $\dfrac{1}{5}$ of the trip is downhill. How many miles will be downhill?

How many miles will not be downhill? What fraction of the trip will not be downhill?

54. A study showed that $\dfrac{3}{14}$ of the students in an elementary school were left-handed. If the school had an enrollment of 700 students, how many were left-handed?

55. The product of $\dfrac{5}{6}$ with another number is $\dfrac{2}{5}$.
  a. Which number is the product?
  b. What is the other number?

56. The product of two numbers is 210.
  a. If one of the numbers is the fraction $\dfrac{2}{3}$, do you expect the other number to be larger or smaller than 210?
  b. What is the other number?

57. An airplane is carrying 90 passengers. This is $\dfrac{9}{10}$ of the capacity of the airplane.

  a. Is the capacity of the airplane more or less than 90?
  b. If you were to multiply 90 times $\dfrac{9}{10}$ would the product be more or less than 90?
  c. What is the capacity of the airplane?

58. The student senate has 75 members, and $\dfrac{7}{15}$ of these are women. A change in the senate constitution is being considered, and at the present time (before debating has begun), a survey shows that $\dfrac{3}{5}$ of the women and $\dfrac{4}{5}$ of the men are in favor of this change.
  a. How many women are on the student senate?
  b. How many women on the senate are in favor of the change?
  c. If the change requires a $\dfrac{2}{3}$ majority vote in favor to pass, would the constitutional change pass if the vote were taken today?
  d. By how many votes would the change pass or fail?

59. The tennis club has 250 members, and they are considering putting in a new tennis court. The cost of the new court is going to involve an assessment of $200 for each member. Seven-tenths of the members live quite near the club and $\frac{3}{5}$ of them are in favor of the assessment. However, $\frac{2}{3}$ of the members who do not live nearby are not in favor of the assessment.
   a. If a vote were taken today, would more than one-half of the members vote for or against the new court?
   b. By how many votes would the question pass or fail if more than one-half of the members must vote in favor for the question to pass?

60. There are 3000 students at Mountain High School and $\frac{1}{4}$ of these students are seniors. If $\frac{3}{5}$ of the seniors are in favor of the school forming a debating team and $\frac{7}{10}$ of the remaining students (not seniors) are also in favor of forming a debating team, how many students do not favor this idea?

Hawkes Learning Systems: Introductory Algebra

Review of Fractions (Multiplication and Division)

R.4

Fractions (Addition and Subtraction)

After completing this section, you will be able to:

1. Find the **LCD** (least common denominator) of a set of fractions.
2. Add fractions.
3. Subtract fractions.

Addition with Fractions

To add two (or more) fractions with the same denominator, we can think of the common denominator as the "name" of each fraction. The sum has this common name. Just as 3 oranges plus 2 oranges give a total of 5 oranges, 3 eighths plus 2 eighths give a total of 5 eighths. Figure R.1 illustrates how the sum of the two fractions $\frac{3}{8}$ and $\frac{2}{8}$ might be diagrammed.

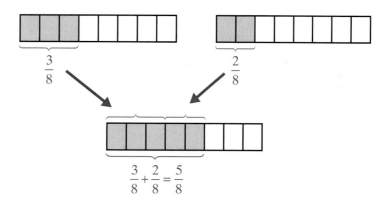

Figure R.1

To Add Two (or More) Fractions with the Same Denominator

1. Add the numerators.
2. Keep the common denominator. $\quad \dfrac{a}{b} + \dfrac{c}{b} = \dfrac{a+c}{b}\ $ where $b \neq 0$.
3. Reduce, if possible.

Example 1: Additions of Fractions

$$\frac{1}{5} + \frac{2}{5} = \frac{1+2}{5} = \frac{3}{5}$$

Example 2: Additions of Fractions

$$\frac{1}{11} + \frac{2}{11} + \frac{3}{11} = \frac{1+2+3}{11} = \frac{6}{11}$$

To add fractions with different denominators, we find the least common denominator of all the fractions, change each fraction to an equivalent fraction with this common denominator, and then add. For example, if the denominators are 3 and 5, the least common denominator is the least common multiple (LCM) which is $3 \cdot 5 = 15$. Thus, we proceed as follows to find the sum $\frac{1}{3} + \frac{2}{5}$:

$$\frac{1}{3} + \frac{2}{5} = \frac{1}{3} \cdot \frac{5}{5} + \frac{2}{5} \cdot \frac{3}{3} = \frac{5}{15} + \frac{6}{15} = \frac{11}{15}$$

In general, **the least common denominator (LCD) is the least common multiple (LCM) of the denominators.**

To Add Fractions with Different Denominators:

1. *Find the least common denominator (LCD).*
2. *Change each fraction into an equivalent fraction with that denominator.*
3. *Add the new fractions.*
4. *Reduce, if possible.*

Example 3: Adding Fractions with Different Denominators

Find the sum: $\frac{3}{8} + \frac{13}{12}$.

Solution:

a. Find the LCD. Remember that the least common denominator (LCD) is the least common multiple (LCM) of the denominators.

$$8 = 2 \cdot 2 \cdot 2 \atop 12 = 2 \cdot 2 \cdot 3 \Big\} \; LCD = 2 \cdot 2 \cdot 2 \cdot 3 = 24$$

Note: You might not need to use prime factorizations to find the LCD. If the denominators are numbers that are familiar to you, then you might be able to find the LCD simply by inspection.

b. Find fractions equal to $\dfrac{3}{8}$ and $\dfrac{13}{12}$ with denominator 24.

$$\dfrac{3}{8} = \dfrac{3}{8} \cdot \dfrac{3}{3} = \dfrac{9}{24} \qquad \text{Multiply by } \dfrac{3}{3} \text{ because } 8 \cdot 3 = 24.$$

$$\dfrac{13}{12} = \dfrac{13}{12} \cdot \dfrac{2}{2} = \dfrac{26}{24} \qquad \text{Multiply by } \dfrac{2}{2} \text{ because } 12 \cdot 2 = 24.$$

c. Add

$$\dfrac{3}{8} + \dfrac{13}{12} = \dfrac{9}{24} + \dfrac{26}{24} = \dfrac{9 + 26}{24} = \dfrac{35}{24}$$

d. The fraction $\dfrac{35}{24}$ is in lowest terms because 35 and 24 have only 1 as a common factor.

Example 4: Adding Fractions with Different Denominators

Find the sum: $\dfrac{7}{45} + \dfrac{7}{36}$.

Solution:

a. Find the LCD.

$$\begin{aligned} 45 &= 3 \cdot 3 \cdot 5 \\ 36 &= 2 \cdot 2 \cdot 3 \cdot 3 \end{aligned} \Big\} \; \begin{aligned} LCD &= 2 \cdot 2 \cdot 3 \cdot 3 \cdot 5 = 180 \\ &= (3 \cdot 3 \cdot 5)(2 \cdot 2) = 45 \cdot 4 \\ &= (2 \cdot 2 \cdot 3 \cdot 3)(5) = 36 \cdot 5 \end{aligned}$$

b. Steps b, c, and d from Example 3 can be written together in one step.

$$\begin{aligned} \dfrac{7}{45} + \dfrac{7}{36} &= \dfrac{7}{45} \cdot \dfrac{4}{4} + \dfrac{7}{36} \cdot \dfrac{5}{5} \\ &= \dfrac{28}{180} + \dfrac{35}{180} = \dfrac{63}{180} \\ &= \dfrac{\cancel{3} \cdot \cancel{3} \cdot 7}{2 \cdot 2 \cdot \cancel{3} \cdot \cancel{3} \cdot 5} = \dfrac{7}{20} \end{aligned}$$

Continued on next page...

Note that, in adding fractions, we also may choose to write them vertically. The process is the same.

$$\frac{7}{45} = \frac{7}{45} \cdot \frac{4}{4} = \frac{28}{180}$$

$$+\frac{7}{36} = \frac{7}{36} \cdot \frac{5}{5} = \frac{35}{180}$$

$$\frac{63}{180} = \frac{\cancel{3} \cdot \cancel{3} \cdot 7}{2 \cdot 2 \cdot \cancel{3} \cdot \cancel{3} \cdot 5} = \frac{7}{20}$$

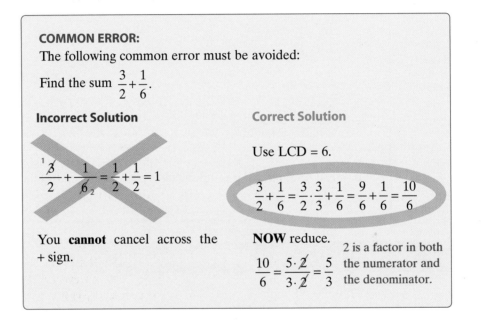

COMMON ERROR:

The following common error must be avoided:

Find the sum $\dfrac{3}{2} + \dfrac{1}{6}$.

Incorrect Solution

$$\frac{^1\cancel{3}}{2} + \frac{1}{\cancel{6}_2} = \frac{1}{2} + \frac{1}{2} = 1$$

You **cannot** cancel across the + sign.

Correct Solution

Use LCD = 6.

$$\frac{3}{2} + \frac{1}{6} = \frac{3}{2} \cdot \frac{3}{3} + \frac{1}{6} = \frac{9}{6} + \frac{1}{6} = \frac{10}{6}$$

NOW reduce.

$$\frac{10}{6} = \frac{5 \cdot \cancel{2}}{3 \cdot \cancel{2}} = \frac{5}{3}$$

2 is a factor in both the numerator and the denominator.

Subtraction with Fractions

Finding the difference between two fractions with a common denominator is similar to finding the sum. The numerators are simply subtracted instead of added. Just as with addition, the common denominator "names" each fraction. Figure R.2 shows how the difference between $\dfrac{4}{5}$ and $\dfrac{1}{5}$ might be diagrammed.

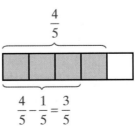

Figure R.2

To Subtract Two Fractions with the Same Denominator

1. *Subtract the numerators.*
2. *Keep the common denominators.* $\dfrac{a}{b} - \dfrac{c}{b} = \dfrac{a-c}{b}$ *where* $b \neq 0$.
3. *Reduce if possible.*

Example 5: Subtracting Fractions

Find the difference: $\dfrac{9}{10} - \dfrac{7}{10}$.

Solution:

$$\frac{9}{10} - \frac{7}{10} = \frac{9-7}{10} = \frac{2}{10} = \frac{\cancel{2} \cdot 1}{\cancel{2} \cdot 5} = \frac{1}{5}$$

The difference is reduced just as any fraction is reduced.

To Subtract Fractions with Different Denominators

1. *Find the least common denominator (LCD).*
2. *Change each fraction to an equal fraction with that denominator.*
3. *Subtract the new fractions.*
4. *Reduce, if possible.*

Example 6: Subtracting Fractions with Different Denominators

Find the difference: $\dfrac{24}{55} - \dfrac{4}{33}$.

Solution:

a. Find the LCD.

$$\left.\begin{array}{l} 55 = 5 \cdot 11 \\ 33 = 3 \cdot 11 \end{array}\right\} \text{LCD} = 3 \cdot 5 \cdot 11 = 165$$

Continued on next page...

b. Find equal fractions with denominator 165, subtract, and reduce.

$$\frac{24}{55} - \frac{4}{33} = \frac{24}{55} \cdot \frac{3}{3} - \frac{4}{33} \cdot \frac{5}{5} = \frac{72 - 20}{165}$$

$$= \frac{52}{165} = \frac{2 \cdot 2 \cdot 13}{3 \cdot 5 \cdot 11} = \frac{52}{165}$$

The answer $\dfrac{52}{165}$ does not reduce because there are no common prime factors in the numerator and denominator.

Example 7: Applications: Budget Proportions

If Keith's total income for the year was $48,000 and he spent $\dfrac{1}{5}$ of his income on rent and $\dfrac{1}{15}$ of his income on his car, what total amount did he spend on these two items?

Solution:

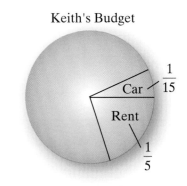

Keith's Budget

We can add the two fractions and then multiply the sum by $48,000. (Or, we can multiply each fraction by $48,000, and then add the results. We will get the same answer either way.) The LCD is 15.

$$\frac{1}{5} + \frac{1}{15} = \frac{1}{5} \cdot \frac{3}{3} + \frac{1}{15} = \frac{3}{15} + \frac{1}{15} = \frac{4}{15}$$

Now multiply $\dfrac{4}{15}$ times $48,000.

$$\frac{4}{\cancel{15}_1} \cdot \overset{3200}{\cancel{48,000}} = 4 \cdot 3200 = 12,800$$

Keith spent a total of $12,800 on rent and his car.

R.4 Exercises

Find the indicated products and sums. Reduce if possible.

1. a. $\dfrac{1}{3} \cdot \dfrac{1}{3}$ b. $\dfrac{1}{3} + \dfrac{1}{3}$ 2. a. $\dfrac{1}{5} \cdot \dfrac{1}{5}$ b. $\dfrac{1}{5} + \dfrac{1}{5}$

3. a. $\dfrac{2}{7} \cdot \dfrac{2}{7}$ b. $\dfrac{2}{7} + \dfrac{2}{7}$ 4. a. $\dfrac{3}{10} \cdot \dfrac{3}{10}$ b. $\dfrac{3}{10} + \dfrac{3}{10}$

5. a. $\dfrac{5}{4} \cdot \dfrac{5}{4}$ b. $\dfrac{5}{4} + \dfrac{5}{4}$ 6. a. $\dfrac{7}{100} \cdot \dfrac{7}{100}$ b. $\dfrac{7}{100} + \dfrac{7}{100}$

7. a. $\dfrac{8}{9} \cdot \dfrac{8}{9}$ b. $\dfrac{8}{9} + \dfrac{8}{9}$ 8. a. $\dfrac{5}{8} \cdot \dfrac{5}{8}$ b. $\dfrac{5}{8} + \dfrac{5}{8}$

9. a. $\dfrac{9}{12} \cdot \dfrac{9}{12}$ b. $\dfrac{9}{12} + \dfrac{9}{12}$ 10. a. $\dfrac{6}{10} \cdot \dfrac{6}{10}$ b. $\dfrac{6}{10} + \dfrac{6}{10}$

Find the indicated sums and differences, and reduce all answers to lowest terms.

11. $\dfrac{3}{14} + \dfrac{3}{14}$ 12. $\dfrac{4}{6} + \dfrac{4}{6}$ 13. $\dfrac{1}{10} + \dfrac{3}{10}$

14. $\dfrac{7}{15} + \dfrac{2}{15}$ 15. $\dfrac{14}{25} - \dfrac{6}{25}$ 16. $\dfrac{7}{16} - \dfrac{5}{16}$

17. $\dfrac{4}{15} - \dfrac{1}{15}$ 18. $\dfrac{7}{12} - \dfrac{1}{12}$ 19. $\dfrac{3}{8} - \dfrac{5}{16}$

20. $\dfrac{2}{5} + \dfrac{3}{10}$ 21. $\dfrac{2}{7} + \dfrac{4}{21} + \dfrac{1}{3}$ 22. $\dfrac{2}{39} + \dfrac{1}{3} + \dfrac{4}{13}$

23. $\dfrac{5}{6} - \dfrac{1}{2}$ 24. $\dfrac{2}{3} - \dfrac{1}{4}$ 25. $\dfrac{5}{7} - \dfrac{5}{14}$

26. $\dfrac{1}{3} - \dfrac{2}{15}$

27. $\dfrac{20}{35} - \dfrac{24}{42}$

28. $\dfrac{10}{25} - \dfrac{6}{15}$

29. $\dfrac{78}{100} - \dfrac{7}{10}$

30. $\dfrac{3}{10} - \dfrac{29}{100}$

31. $\dfrac{2}{3} + \dfrac{3}{4} - \dfrac{5}{6}$

32. $\dfrac{1}{5} + \dfrac{1}{10} - \dfrac{1}{4}$

33. $\dfrac{2}{27} - \dfrac{1}{54} + \dfrac{1}{45}$

34. $\dfrac{31}{90} - \dfrac{2}{45} + \dfrac{5}{42}$

35. $\dfrac{5}{6} - \dfrac{50}{60} + \dfrac{1}{3}$

36. $\dfrac{15}{45} - \dfrac{9}{27} + \dfrac{1}{5}$

37. $\dfrac{1}{2} + \dfrac{3}{4} + \dfrac{1}{100}$

38. $\dfrac{1}{4} + \dfrac{1}{8} + \dfrac{7}{100}$

39. $\dfrac{1}{5} + \dfrac{3}{10} - \dfrac{4}{15}$

40. $\dfrac{3}{8} + \dfrac{5}{12} - \dfrac{1}{4}$

41. $1 - \dfrac{13}{16}$

42. $1 - \dfrac{3}{7}$

43. $1 - \dfrac{5}{8}$

44. $1 - \dfrac{2}{3}$

45.
$$\begin{array}{r} \dfrac{3}{5} \\[6pt] \dfrac{7}{15} \\[6pt] +\dfrac{5}{6} \\ \hline \end{array}$$

46.
$$\begin{array}{r} \dfrac{4}{27} \\[6pt] \dfrac{5}{18} \\[6pt] +\dfrac{1}{9} \\ \hline \end{array}$$

47.
$$\begin{array}{r} \dfrac{7}{24} \\[6pt] \dfrac{7}{16} \\[6pt] +\dfrac{7}{12} \\ \hline \end{array}$$

48.
$$\begin{array}{r} \dfrac{5}{9} \\[6pt] \dfrac{2}{3} \\[6pt] +\dfrac{4}{15} \\ \hline \end{array}$$

49. Sam's income is $3300 a month and he plans to budget $\frac{1}{3}$ of his income for rent and $\frac{1}{10}$ of his income for food.

  a. What fraction of his income does he plan to spend on these two items?

  b. What amount of money does he plan to spend each month on these two items?

Sam's Budget

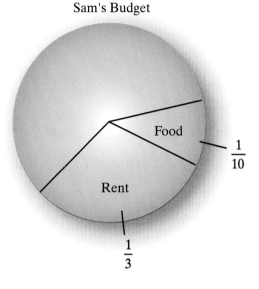

50. Of the personal computers (PCs) in use worldwide, the U. S. has $\frac{272}{1000}$, Japan $\frac{84}{1000}$, China $\frac{65}{1000}$, and Germany $\frac{56}{1000}$. What total fraction of the world's PCs are used in these four countries? What fraction is used in the rest of the world? (Source: *2006 World Almanac*)

PCs Worldwide

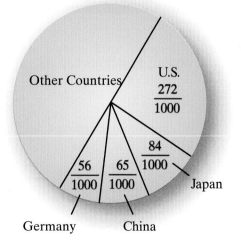

51. A watch has a rectangular-shaped display screen that is $\frac{3}{4}$ inch by $\frac{1}{2}$ inch. The display screen has a border of silver that is $\frac{1}{10}$ inch thick. What are the dimensions (length and width) of the face of the watch (including the silver border)?

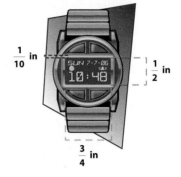

52. Four postal letters weigh $\frac{1}{2}$ ounce, $\frac{1}{5}$ ounce, $\frac{3}{10}$ ounce, and $\frac{9}{10}$ ounce. What is the total weight of the letters?

Hawkes Learning Systems: Introductory Algebra

Review of Fractions (Addition and Subtraction)

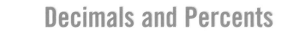

R.5 Decimals and Percents

After completing this section, you will be able to:

1. *Read and write decimal numbers.*
2. *Operate (add, subtract, multiply, and divide) with decimal numbers.*
3. *Round decimal numbers to designated places.*
4. *Understand percents.*
5. *Change decimals and fractions to percent form.*

Reading and Writing Decimal Numbers

The common **decimal notation** uses a **place value system** and a **decimal point** with whole numbers written to the left of the decimal point and fractions written to the right of the decimal point. We will say that numbers represented by decimal notation are **decimal numbers.** The values of several places in this decimal system are shown in Figure R.3 shown here.

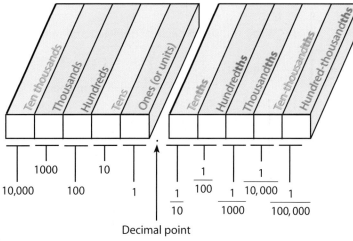

Figure R.3

In reading a fraction such as $\dfrac{247}{1000}$, we read the numerator as a whole number ("two hundred forty-seven") and then attach the name of the denominator ("thousand**ths**"). Note that **ths** (or **th**) is used to indicate the fraction. This same procedure is followed with numbers written in decimal notation:

$$\frac{247}{1000} = 0.247 \qquad \text{Read "two hundred forty-seven thousandths."}$$

A **mixed number** consists of a whole number and a fraction written next to each other with the understanding that the whole number and the fraction part are to be added together. For example, $2\dfrac{9}{100}$ is shorthand notation for $2 + \dfrac{9}{100}$. Now, in decimal form, the whole number is written to the left of the decimal point and the fraction part to the right. So, we can write

$$2\frac{9}{100} = 2.09 \qquad \text{Read "two and nine hundredths."}$$

In general, decimal numbers are read (and written) according to the following convention.

To Read or Write a Decimal Number:

1. *Read (or write) the whole number.*
2. *Read (or write) "**and**" in place of the decimal point.*
3. *Read (or write) the fraction part as a whole number with the name of the place of the last digit on the right.*

Example 1: Writing in Decimal Notation

Write $9\dfrac{63}{1000}$ in decimal notation and in words.

Solution:

$$\underset{\text{nine}}{9} \; . \; \underset{\textbf{and}\;\;\text{sixty-three}\;\;\text{thousandths}}{063} \quad \longleftarrow \quad \text{in decimal notation}$$

$\longleftarrow$ in words

"**and**" indicates the decimal point; the digit 3 is in the thousandths position.

Example 2: Writing in Decimal Notation

Write four hundred **and** two thousandths in decimal notation.

Solution:

Two 0's are inserted as placeholders.

The digit 2 is in the thousandths position.

400.002

Example 3: Writing in Decimal Notation

Write four hundred two thousandths in decimal notation.

Solution:

0.402

Note carefully how the use of **and** in the phrase in Example 2 gives it a completely different meaning from the phrase in this example.

Addition with Decimal Numbers

Addition with decimal numbers can be accomplished by writing the decimal numbers one under the other and keeping the decimal points aligned vertically. In this way, the whole numbers will be added to whole numbers, tenths added to tenths, hundredths to hundredths, and so on. The decimal point in the sum is in line with the decimal points in the addends. Thus,

Decimal points are aligned vertically.

$$
\begin{array}{r}
2.357 \\
+\ 6.140 \\
\hline
8.497
\end{array}
$$

← 0 may be written here to help align digits.

As in the number 6.140, 0's may be written to the right of the last digit in the fraction part to help keep the digits in the correct line. This will not change the value of any number or the sum.

To Add Decimal Numbers:

1. Write the addends in a vertical column.

2. Keep the decimal points aligned vertically.

3. Keep digits with the same position value aligned. (Zeros may be filled in as aids.)

4. Add the numbers, just as with whole numbers, keeping the decimal point in the sum aligned with the other decimal points.

Example 4: Adding Decimals

Find the sum: $17 + 4.88 + 50.033 + 0.6$.

Solution:

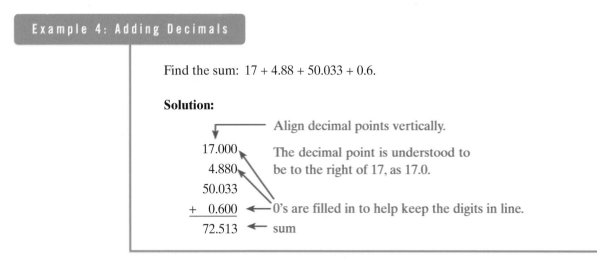

Align decimal points vertically.

$$
\begin{array}{r}
17.000 \\
4.880 \\
50.033 \\
+\ \ 0.600 \\
\hline
72.513
\end{array}
$$

The decimal point is understood to be to the right of 17, as 17.0.

0's are filled in to help keep the digits in line.

← sum

Subtraction with Decimal Numbers

To Subtract Decimal Numbers

1. *Write the numbers in a vertical column.*

2. *Keep the decimal points aligned vertically.*

3. *Keep digits with the same position value aligned. (Zeros may be filled in as aids.)*

4. *Subtract, just as with whole numbers, keeping the decimal point in the difference aligned with the other decimal points.*

Example 5: Subtracting Decimals

Find the difference: $21.715 - 14.823$.

Solution:

$$
\begin{array}{r}
21.715 \\
-\ 14.823 \\
\hline
6.892
\end{array}
$$

Example 6: Applications: Monetary Arithmetic

At the bookstore, Mrs. Gonzalez bought a text for $55, art supplies for $32.50, and computer supplies for $29.25. If tax was $9.34, how much change did she receive from a gift certificate worth $150?

Solution:

a. Find the total of her expenses including tax.

$55.00
 32.50
 29.25
+ 9.34
$126.09   total

b. Subtract the answer in part **a.** from $150.

$150.00
−   126.09
$23.91

She received $23.91 in change.

Example 7: Adding and Subtracting Decimals with a Calculator

Use a TI-84 Plus graphing calculator to perform the following operations:

a. $4852.621 + 5443.1$ **b.** $84.3613 − 46.98$

Solution:

After entering the values and operations into the calculator, the display will be as follows:

a.   **b.**

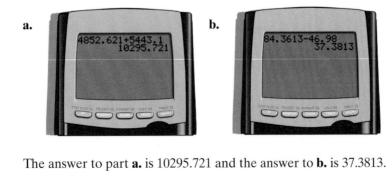

The answer to part **a.** is 10295.721 and the answer to **b.** is 37.3813.

Multiplication with Decimal Numbers

Decimal numbers are multiplied in the same manner as whole numbers are multiplied with the added concern of the correct placement of the decimal point in the product. Two examples are shown here, in both fraction form and decimal form, to illustrate how the decimal point is to be placed in the product.

Products in Fraction Form

$$\frac{4}{10} \cdot \frac{6}{100} = \frac{24}{1000}$$

$$\frac{5}{1000} \cdot \frac{7}{100} = \frac{35}{100,000}$$

Products in Decimal Form

$$
\begin{array}{r}
0.4 \quad \longleftarrow \text{1 place} \\
\times \quad 0.06 \quad \longleftarrow \text{2 places} \\
\hline
0.024
\end{array}
$$
total of 3 places (thousandths)

$$
\begin{array}{r}
0.005 \quad \longleftarrow \text{3 places} \\
\times \quad 0.07 \quad \longleftarrow \text{2 places} \\
\hline
0.00035
\end{array}
$$
total of 5 places (hundred-thousandths)

The following rule states how to multiply decimal numbers and place the decimal point in the product.

To Multiply Positive Decimal Numbers

1. *Multiply the two numbers as if they were whole numbers.*

2. *Count the total number of places to the right of the decimal points in both numbers being multiplied.*

3. *Place the decimal point in the product so that the number of places to the right is the same as that found in step 2.*

Example 8: Multiplying Decimals

Multiply: 2.435×4.1

Solution:

$$
\begin{array}{r}
2.435 \quad \longleftarrow \text{3 places} \\
\times \quad 4.1 \quad \longleftarrow \text{1 place} \\
\hline
2435 \\
9740 \\
\hline
9.9835 \quad \longleftarrow \text{4 places in the product}
\end{array}
$$
total of 4 places (ten-thousandths)

Division with Decimal Numbers

The process of division with decimal numbers is, in effect, the same as division with whole numbers with the added concern of where to place the decimal point in the quotient. This is reasonable because whole numbers are decimal numbers, and we would not expect a great change in a process as important as division.

To Divide Decimal Numbers

1. *Move the decimal point in the divisor to the right so that the divisor is a whole number.*
2. *Move the decimal point in the dividend the same number of places to the right.*
3. *Place the decimal point in the quotient directly above the new decimal point in the dividend.*
4. *Divide just as with whole numbers.*

Example 9: Dividing Decimals

Find the quotient: $63.86 \div 6.2$

Solution:

a. Write down the numbers.

$$6.2\overline{)63.86}$$

b. Move both decimal points one place to the right so that the divisor becomes a whole number. Then place the decimal point in the quotient.

$$6.2.\overline{)63.8.6} \longleftarrow \text{decimal point in quotient}$$

c. Proceed to divide as with whole numbers.

$$
\begin{array}{r}
10.3 \longleftarrow \text{quotient} \\
\text{divisor} \longrightarrow 62.\overline{)638.6} \longleftarrow \text{dividend} \\
\underline{62} \\
18 \\
\underline{0} \\
186 \\
\underline{186} \\
0 \longleftarrow \text{remainder}
\end{array}
$$

If the remainder is eventually 0, as it is in Example 9, then the quotient is a **terminating decimal.** We will see in Chapter 2 that if the remainder is never 0, then the quotient is an **infinite repeating decimal.** That is, the quotient will be a repeating pattern of digits. In division with decimal numbers, we generally agree to some place of accuracy for the quotient before the division is performed. If the remainder is not 0 by the time this place of accuracy is reached in the quotient, then we divide one more place and round the quotient.

Rounding to the Right of the Decimal Point

Rounding to the right of the decimal point in a decimal number is similar to rounding with whole numbers:

  1. *Look one digit to the right of the desired place of accuracy.*

  2. *If this digit is 5 or more, raise the digit in the desired place by 1. Otherwise, leave the digit as it is.*

  3. *Drop the remaining digits.*

(Note that when rounding with whole numbers these digits must be replaced by 0's)

Example 10: Dividing Decimals and Rounding

Find the quotient $82.3 \div 2.9$ to the nearest tenth.

Solution:

Divide until the quotient is in hundredths (one place more than tenths); then round to tenths.

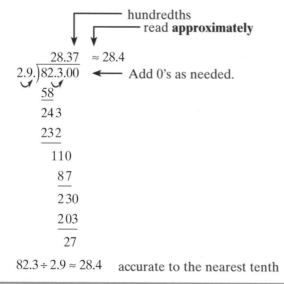

$$82.3 \div 2.9 \approx 28.4 \quad \text{accurate to the nearest tenth}$$

Example 11: Applications: Calculating Price

Suppose that the price of a gallon of gas is stated as $3.15 at the pump and the station owner tells you that the taxes you are paying are figured at 0.45 times the price he is actually charging for a gallon of gas. What price (to the nearest penny) is he actually charging for a gallon of gas? (Note: As we will see on page 57, 145% = 1.45.)

Solution:

To find the actual price of a gallon of gas, before taxes, divide the total price by 1.45.

$$
\begin{array}{r}
2.172 \\
1.45.)\overline{3.15.000} \\
2\ 90 \\ \hline
25\ 0 \\
14\ 5 \\ \hline
10\ 50 \\
10\ 15 \\ \hline
350 \\
290 \\ \hline
60
\end{array}
$$

So, the actual cost of the gas is about $2.17 per gallon before taxes.

Example 12: Multiplying and Dividing Decimals with a Calculator

Use a TI-84 Plus graphing calculator to perform the following operations and write the answer accurate to 4 decimal places:

a. 4.631(0.235)(2.3)   **b.** 19.36 ÷ 5.4

Solution:

After entering the values and operations into the calculator, the display will be as shown on the right:

The answers accurate to 4 decimal places:

a. 2.5031   **b.** 3.5852

Understanding Percent

The word **percent** comes from the Latin *per centum,* meaning "per hundred." So, **percent means hundredths.** The symbol % is called the **percent symbol** (or **percent sign**). This symbol can be treated as equivalent to the fraction $\dfrac{1}{100}$. For example,

$$\frac{25}{100} = 25\left(\frac{1}{100}\right) = 25\% \quad \text{and} \quad \frac{70}{100} = 70\left(\frac{1}{100}\right) = 70\%.$$

In Figure R.4 shown here, the large square is partitioned into 100 small squares, and each small square represents 1%, or $1 \cdot \dfrac{1}{100}$ of the large square. Thus the shaded portion of the large square is

$$\frac{40}{100} = 40 \cdot \frac{1}{100} = 40\%.$$

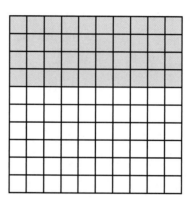

Figure R.4

Decimals and Percents

Now the relationship between decimals and percents can be seen by noting the relationship between decimals and fractions with denominator 100. For example,

Decimal Form		Fraction Form		Percent Form
0.33	=	$\dfrac{33}{100} = 33 \cdot \dfrac{1}{100}$	=	33%
0.74	=	$\dfrac{74}{100} = 74 \cdot \dfrac{1}{100}$	=	74%
0.2	=	$\dfrac{2}{10} = \dfrac{20}{100} = 20 \cdot \dfrac{1}{100}$	=	20%
0.625	=	$\dfrac{62.5}{100} = 62.5 \cdot \dfrac{1}{100}$	=	62.5%

By noting the way that the decimal point is moved in these four examples, we can make the change directly (and more efficiently) by using the following rule regardless of how many digits there are in the decimal number.

Example 13: Converting Decimals to Percents

Change each decimal to an equivalent percent.

    a. 0.254   **b.** 0.005   **c.** 1.5   **d.** 0.2

Solutions:

a. $0.254 = 25.4\%$

Decimal point % symbol added on
moved two places to the right

b. $0.005 = 0.5\%$ Note that this is less than 1%.

Decimal point % symbol added on
moved two places to the right

c. $1.5 = 150\%$ Note that this is more than 100%.

Decimal point % symbol added on
moved two places to the right

d. $0.2 = 20\%$

Decimal point % symbol added on
moved two places to the right

To Change a Decimal to a Percent:

Step 1. *Move the decimal point two places to the right.*

Step 2. *Add the % symbol.*

These two steps have the effect of multiplying by 100 and then dividing by 100. Thus the number is not changed. Just the form is changed.

To change percents to decimals, we reverse the procedure for changing decimals to percents. For example,

$$38\% = 38\left(\frac{1}{100}\right) = \frac{38}{100} = 0.38.$$

As indicated in the following statement, the same result can be found by noting the placement of the decimal point.

To Change a Percent to a Decimal:

Step 1. *Move the decimal point two places to the left.*

Step 2. *Delete the % symbol.*

Example 14: Converting Percents to Decimals

Change each percent to an equivalent decimal.

a. 38%
b. 16.2%
c. 100%
d. 0.25%

Solutions:

a. 38% = 0.38 ← % symbol deleted

 ↑ ↑
 Understood   Decimal point
 decimal point   moved two places left

b. 16.2% = 0.162

c. 100% = 1.00 = 1

d. 0.25% = 0.0025   The percent is less than 1%, and the decimal is less than 0.01.

Fractions and Percents

As we know, if a fraction has a denominator of 100, the fraction can be changed to a percent by writing the numerator and then writing the % symbol. However, if the denominator is not 100, a more general approach (easily applied with calculators) is to proceed as follows:

To Change a Fraction to a Percent

Step 1. *Change the fraction to a decimal*

(Divide the numerator by the denominator.)

Step 2. *Change the decimal to a percent.*

Example 15: Converting Fractions to Percents

Change $\dfrac{5}{8}$ to a percent.

Solution:

First divide 5 by 8 to get the decimal form. (This can be done with a calculator.)

$$
\begin{array}{r}
0.625 \\
8\overline{)5.000} \\
48 \\
\hline
20 \\
16 \\
\hline
40 \\
40 \\
\hline
0
\end{array}
$$

Now change 0.625 to a percent:

$$\frac{5}{8} = 0.625 = 62.5\%$$

Example 16: Calculating Percents

In the U.S. in 1900, there were about 27,000 college graduates (who received bachelor's degrees), of which about 22,000 were men. In 2003, there were about 1,300,000 college graduates, of which about 550,000 were men. Find the percentage of college graduates that were men in each of those years.

Solution:

For each year, divide (with a calculator) the number of male graduates by the total number of graduates:

Continued on next page...

a. For 1900, $\dfrac{22,000}{27,000} \approx 0.8148 = 81.48\%$ of college graduates were men.

b. For 2003, $\dfrac{550,000}{1,300,000} \approx 0.4231 = 42.31\%$ of college graduates were men.

R.5 Exercises

Write the following mixed numbers in decimal notation.

1. $7\dfrac{5}{10}$ **2.** $39\dfrac{8}{100}$ **3.** $14\dfrac{27}{1000}$ **4.** $100\dfrac{65}{100}$

Write the following decimal numbers in mixed number form. Do not reduce the fractional part.

5. 23.02   **6.** 200.8   **7.** 52.001   **8.** 6.034

Write the following words in decimal notation.

9. six tenths

10. fifteen thousandths

11. two and twenty-four hundredths

12. five and thirty-five thousandths

Write the following decimal numbers in words.

13. 0.2   **14.** 0.75   **15.** 260.003   **16.** 68.809

Round each of the following decimal numbers as indicated.

17. 72.06 (nearest tenth)

18. 0.0582 (nearest thousandth)

19. 0.295 (nearest hundredth)

20. 37.954 (nearest tenth)

Perform the indicated operations. Round any quotient to the nearest hundredth.

21. $0.6 + 0.4 + 0.4$

22. $7 + 5.1 + 0.8$

23. $0.79 + 4.92 + 0.05$

24. $4.005 + 0.056 + 0.9$

25. $5.4 - 3.76$

26. $17.83 - 9.9$

27. $39.6 - 13.71$

28. $55.002 - 53.008$

29.
$$57.3$$
$$52.08$$
$$+38.005$$

30.
$$1.007$$
$$30.442$$
$$+ 4.992$$

31.
$$21.007$$
$$- 1.543$$

32.
$$30.$$
$$- 6.45$$

33. $(0.2)(0.2)$

34. $8(0.125)$

35.
$$0.137$$
$$\times 0.08$$

36.
$$6.09$$
$$\times 0.11$$

37. $28 \div 5.6$

38. $35 \div 1.64$

39. $2.7\overline{)5.483}$

40. $2.54\overline{)45}$

41. Marshall wants to buy a new car for $28,000. He has talked to the loan officer at his credit union and knows that they will loan him $21,000. He must also pay $450 for a license fee and $2240 for taxes. What amount of cash will he need to buy the car?

42. Kathy wants to have a haircut and a manicure. She knows that the haircut will cost $62 and the manicure will cost $15.50. If she plans to tip the stylist $15, how much change will she receive from $100?

43. The eccentricity of a planet's orbit is the measure of how much the orbit varies from a perfectly circular pattern. Earth's orbit has an eccentricity of 0.017, and Pluto's orbit has an eccentricity of 0.254. How much greater is Pluto's eccentricity of orbit than Earth's?

44. Albany, New York, receives an average rainfall of 35.74 inches and 65.5 inches of snow. Charleston, South Carolina, receives an average rainfall of 51.59 inches and 0.6 inches of snow. On average, how much more rain is there in Charleston than in Albany? On average, how much more snow is there in Albany than in Charleston?

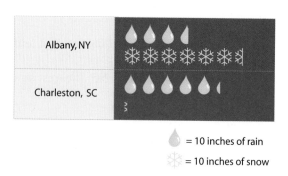

= 10 inches of rain

= 10 inches of snow

45. If the sale price of a new flat screen TV is $1200 and sales tax is figured at 0.08 times the price, what amount is paid for the television set?

46. To buy a used car, you can put $300 down and make 18 monthly payments of $114.20. How much will you pay for the car?

47. If a bicyclist rode 150.6 miles in 11.3 hours, what was her average speed in miles per hour (to the nearest tenth)?

48. Walter Payton played football for the Chicago Bears for 13 years. In those years he carried the ball 3838 times for a total of 16,726 yards. What was his average yardage per carry (to the nearest tenth)?

49. The dimensions of a rectangle are shown.

    a. Find the perimeter (distance around) of the rectangle.

    b. Find the area (length times width) of the rectangle (in square units).

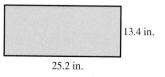

13.4 in.

25.2 in.

50. The dimensions of a triangle are shown.

    a. Find the perimeter (distance around) of the triangle.

    b. Find the area (base times height divided by 2) of the triangle (in square units).

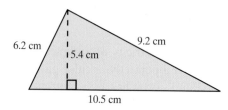

6.2 cm

5.4 cm

9.2 cm

10.5 cm

Change the following decimals to percents.

51. 0.03

52. 0.052

53. 3.0

54. 2.5

55. 1.08

56. 0.5

Change the following percents to decimals.

57. 6%

58. 11%

59. 3.2%

60. 12.5%

61. 120%

62. 80%

Change the following fractions and mixed numbers to percents.

63. $\dfrac{3}{20}$

64. $\dfrac{3}{10}$

65. $\dfrac{24}{25}$

66. $\dfrac{1}{4}$

67. $1\dfrac{5}{8}$

68. $2\dfrac{3}{4}$

69. Suppose that the state motor vehicle licensing fee is figured by multiplying the cost of your car by 0.009. Change 0.009 to a percent.

70. In the 1990 census, California ranked as the most populated state with about 30,000,000 people. In 2000 California again ranked first with about 33,872,000 people. By about what percent did the population of California grow in the ten years from one census to the other?

Use a TI-84 Plus graphing calculator to find the value of each of the following expressions (accurate to three decimal places).

71. $8.35 + 9.76 + 13.002$

72. $18.65 + 29.7 + 35.82 + 105.4$

73. $897.54 - 789.5$

74. $1894.76 - 1776.38$

75. $3.52(0.12)(2.5)$

76. $14.5 \cdot 72.16$

77. $5.7 \div 0.214$

78. $4.457 \div 0.12$

79. $\dfrac{67.15}{3.45}$

80. $\dfrac{4680}{19.64}$

Hawkes Learning Systems: Introductory Algebra

Review of Decimals and Percents

Review Chapter Index of Key Ideas and Terms

Section R.1 Exponents and Order of Operations

Whole Numbers   page 2

The **whole numbers** are the natural (or counting) numbers and the number 0.

Natural Numbers: $\mathbb{N} = \{\, 1, 2, 3, 4... \,\}$

Whole Numbers: $\mathbb{W} = \{\, 0, 1, 2, 3... \,\}$

Addends   page 2

The numbers being added.

Sum   page 2

The result of the numbers being added.

Factors   page 3

The numbers being multiplied.

Product   page 3

The result of the number being multiplied.

Exponents   page 2 - 7

   Base   page 3

   Perfect Square   page 5

Variable   page 4

A **variable** is a symbol that is used to represent an unknown number or any one of several numbers.

Rules for Order of Operations   page 7 - 11

1. Simplify within grouping symbols, such as parentheses ( ), brackets [ ], and braces { }, working from the inner most grouping outward.
2. Find any powers indicated by exponents.
3. Moving from **left to right**, perform any multiplications or divisions in the order they appear.
4. Moving from **left to right**, perform any additions or subtractions in the order they appear.

PEMDAS   page 9

Section R.2 Prime numbers, Factoring, and LCM

Prime Numbers and Composite Numbers   pages 14 - 16
> A **prime number** is a counting number greater than 1 that has
> exactly two different factors (or divisors), namely 1 and itself.
> A **composite number** is a counting number with more than two
> different factors (or divisors).
> **Even and Odd Whole Numbers**   page 16

Prime Factorization of Composite Numbers   pages 17 - 19
> **To Find the Prime Factorization of a Composite Number**   page 17
>> **1.** Factor the composite number into any two factors.
>> **2.** Factor each factor that is not prime.
>> **3.** Continue this process until all factors are prime.
> The **prime factorization** is the product of all the prime factors.

Least Common Multiple (LCM)   pages 19 – 21
> **To Find the LCM of a Set of Counting Numbers**   page 20
>> **1.** Find the prime factorization of each number.
>> **2.** List the prime factors that appear in any one of
>>   of the prime factorizations.
>> **3.** Find the product of these primes using each prime the
>>   the greatest number of times it appears in any one of
>>   the prime factorizations.

Section R.3 Fractions (Multiplication and Division)

Introduction to Fractions   pages 24 - 25
> **Fraction**   page 24
>> A **fraction** is a number that can be written in the form $\frac{a}{b}$
>> where a and b are whole numbers and $b \neq 0$.
> **Proper Fraction**   page 25
>> A **proper fraction** is a fraction in which the numerator is
>> less than the denominator.
> **Improper Fraction**   page 25
>> An **improper fraction** is a fraction in which the numerator is
>> greater than the denominator.

Multiplying with Fractions   page 26

Continued on next page...

Section R.3 Fractions (Multiplication and Division) (continued)

Raising Fractions to Higher Terms and Reducing Fractions   pages 27 – 28
    Multiplicative Identity   page 27
        The number 1 is called the **multiplicative identity**.

Multiplying and Reducing Fractions at the Same Time   pages 29

Dividing with Fractions   pages 30 – 32
    Reciprocal   page 30
        The **reciprocal** of $\dfrac{a}{b}$ is $\dfrac{b}{a}$ (where $a \neq 0$ and $b \neq 0$) and $\dfrac{a}{b} \cdot \dfrac{b}{a} = 1$.

    Division   page 30
        To **divide** by any nonzero number, multiply by its reciprocal.

Section R.4 Fractions (Addition and Subtraction)

Addition with Fractions   pages 37 - 40
    **To Add Two (or more) Fractions with the Same
Denominator**   pages 37 - 38
      1.   Add the like numerators.
      2.   Keep the common denominator.
      3.   Reduce, if possible.

    To Add Fractions with Different Denominators   pages 38 – 40
      1.   Find the least common denominator (LCD).
      2.   Change each fraction into an equivalent fraction with
          that denominator.
      3.   Add the new fractions.
      4.   Reduce, if possible.

Subtraction with Fractions   pages 40 - 42
    To Subtract Two Fraction with the Same Denominator   page 41
      1.   Subtract the numerators.
      2.   Keep the common denominator.
      3.   Reduce, if possible.
    To Subtract Fractions with Different Denominators   pages 41 - 42
      1.   Find the least common denominator (LCD).
      2.   Change each fraction to an equal fraction with the denominator.
      3.   Subtract the new fractions.
      4.   Reduce, if possible.

Section R.5 Decimals and Percents

Reading and Writing Decimal Numbers   pages 47 - 49
    To Read or Write a Decimal Number   pages 48 - 49

       1. Read (or write) the whole number.

       2. Read (or write) "**and**" in place of the decimal point.

       3. Read (or write) the fraction parts as a whole number with the name of the place of the last digit on the right.

Addition with Decimal Numbers   pages 49 - 50
    To Add Decimal Numbers   page 49

       1. Write the addends in a vertical column.

       2. Keep the decimal points aligned vertically.

       3. Keep digits with the same position value aligned. (Zeros may be filled in as aids.)

       4. Add the numbers, just as with whole numbers, keeping the decimal point in the sum aligned with the other decimal points.

Subtraction with Decimal Numbers   pages 50 - 51
    To Subtract Decimal Numbers   page 50

       1. Write the numbers in a vertical column.

       2. Keep the decimal points aligned vertically.

       3. Keep digits with the same position value aligned. (Zeros may be filled in as needed.)

       4. Subtract, just as with whole numbers, keeping the decimal point in the difference aligned with the other decimal points.

Multiplication with Decimal Numbers   page 52
    To Multiply Positive Decimal Numbers   page 52

       1. Multiply the two numbers as if they were whole numbers.

       2. Count the total number of places to the right of the decimal points in both numbers being multiplied.

       3. Place the decimal point in the product so that the number of places to the right is the same as that found in step 2.

Continued on next page...

Section R.5   Decimals and Percents   (continued)

Division with Decimal Numbers   pages 53 - 55

To Divide Decimal Numbers   page 53

1. Move the decimal point in the divisor to the right so that the divisor is a whole number.
2. Move the decimal point in the dividend the same number of places to the right.
3. Place the decimal point in the quotient directly above the new decimal point in the dividend.
4. Divide just as with whole numbers.

Rounding to the Right of the Decimal Point   page 54

1. Look one digit to the right of the desired place of accuracy.
2. If this digit is 5 or more, raise the digit in the desired place by 1. Otherwise, leave the digit as it is.
3. Drop the remaining digits.

Understanding Percent   pages 56 - 60

To Change a Decimal to a Percent   page 57

Step 1. Move the decimal point two places to the right.
Step 2. Add the % symbol.

To Change a Percent to a Decimal   page 58

Step 1. Move the decimal point two places to the left.
Step 2. Delete the % symbol.

To Change a Fraction to a Percent   page 59

Step 1. Change the fraction to a decimal.
Step 2. Change the decimal to a percent.

Hawkes Learning Systems: Introductory Algebra

For a review of the topics and problems from the Review Chapter, look at the following lessons from *Hawkes Learning Systems: Introductory Algebra*

Review of Exponents and Order of Operations
Review of Prime numbers, Factoring, and LCM
Review of Fractions (Multiplication and Division)
Review of Fractions (Addition and Subtraction)
Review of Decimals and Percents

Chapter R Review

Section R.1 Exponents and Order of Operations
Rewrite each product by using exponents.

1. $3 \cdot 3 \cdot 3 \cdot 5 \cdot 5$ **2.** $13 \cdot 13 \cdot 13 \cdot 17$ **3.** $2 \cdot 2 \cdot 11 \cdot 11$ **4.** $7 \cdot 11 \cdot 11 \cdot 19 \cdot 19 \cdot 19$

Find the value of each exponential expression. Use your calculator if necessary.

5. 12^2 **6.** 8^2 **7.** 6^7 **8.** 2^{15}

Find the value of each of the following expressions by using the rules for order of operations.

9. $19 + 2 \cdot 7$ **10.** $25 + 3 \cdot 4$ **11.** $40 - 4 \div 4$ **12.** $20 - 5 \div 5$

13. $4(2+3) \div 5 \cdot 2$ **14.** $(34+10) \div 11 \cdot 2$ **15.** $32 \div 2^4 + 27 \div 3^2$ **16.** $2 \cdot 5^3 - 8 \div 2^2$

17. $(4+5)^2 + (4+3)^2$ **18.** $40 - 9(3^2 - 2^3)$

19. $50 + 2\left[3(4^2 - 6) - 5^2\right]$ **20.** $(12-2)\left[5(7-3) + (5-3)^2\right]$

Section R.2 Prime Numbers, Factoring, and LCM

21. Show that the number 375 is divisible by 5.

22. Show that the number 912 is divisible by 3.

23. List the prime numbers less than 30.

24. What is a composite number?

Determine whether each of the following numbers is prime or composite.

25. 89 **26.** 189 **27.** 299 **28.** 301

Find the prime factorization of each number.

29. 175 **30.** 154 **31.** 195 **32.** 2475

Find the least common multiple (LCM) of each of the following sets of numbers.

33. $\{27, 30, 35\}$ **34.** $\{10, 27, 30\}$ **35.** $\{15, 24, 40\}$ **36.** $\{18, 36, 78\}$

37. $\{8, 10, 20, 60\}$ **38.** $\{11, 22, 66, 77\}$ **39.** $\{121, 169, 225\}$ **40.** $\{25, 49, 169\}$

Section R.3   Fractions (Multiplication and Division)

41. Find $\dfrac{1}{3}$ of 36.   **42.** Find $\dfrac{2}{3}$ of 36.   **43.** Find $\dfrac{1}{10}$ of $\dfrac{3}{4}$

Raise each fraction to higher terms as indicated. Find the values of the missing numbers.

44. $\dfrac{7}{8} = \dfrac{7}{8} \cdot \dfrac{?}{?} = \dfrac{?}{24}$ **45.** $\dfrac{15}{17} = \dfrac{15}{17} \cdot \dfrac{?}{?} = \dfrac{?}{51}$ **46.** $\dfrac{1}{16} = \dfrac{1}{16} \cdot \dfrac{?}{?} = \dfrac{?}{64}$

Reduce each fraction to lowest terms.

47. $\dfrac{15}{51}$ **48.** $\dfrac{72}{84}$ **49.** $\dfrac{20}{46}$ **50.** $\dfrac{6}{57}$

Perform the indicated operations. Reduce all answers.

51. $\dfrac{55}{44} \cdot \dfrac{8}{26}$ **52.** $\dfrac{52}{90} \cdot \dfrac{65}{26}$ **53.** $\dfrac{16}{25} \div \dfrac{4}{5}$ **54.** $\dfrac{19}{27} \div \dfrac{38}{54}$

55. $\dfrac{18}{35} \cdot \dfrac{1}{3} \cdot \dfrac{21}{12}$ **56.** $\dfrac{4}{9} \cdot \dfrac{2}{5} \cdot \dfrac{2}{3}$ **57.** $\dfrac{35}{33} \div \dfrac{7}{3}$ **58.** $\dfrac{3}{8} \div 4$

59. $\dfrac{2}{5} \cdot \dfrac{3}{4} \cdot \dfrac{11}{12} \cdot \dfrac{20}{26}$ **60.** $\dfrac{7}{16} \cdot \dfrac{9}{14} \cdot \dfrac{30}{45} \cdot \dfrac{4}{13}$

Section R.4   Fractions (Addition and Subtraction)

Perform the indicated operations. Reduce all answers.

61. $\dfrac{2}{3} + \dfrac{2}{3}$ **62.** $\dfrac{3}{5} + \dfrac{3}{5}$ **63.** $\dfrac{9}{16} - \dfrac{5}{16}$ **64.** $\dfrac{7}{10} - \dfrac{3}{10}$

65. $\dfrac{9}{100} + \dfrac{3}{10}$ **66.** $\dfrac{7}{50} + \dfrac{2}{25}$ **67.** $\dfrac{1}{2} + \dfrac{3}{4} + \dfrac{3}{100}$ **68.** $\dfrac{2}{3} + \dfrac{1}{9} + \dfrac{5}{18}$

69. $\dfrac{9}{10} - \dfrac{1}{20}$ **70.** $\dfrac{11}{12} - \dfrac{1}{2}$ **71.** $\dfrac{5}{8} - \dfrac{5}{16}$ **72.** $\dfrac{9}{16} - \dfrac{9}{32}$

73. $\dfrac{9}{45} + \dfrac{1}{36}$ **74.** $\dfrac{1}{8} + \dfrac{5}{12}$ **75.** $1 - \dfrac{1}{9}$ **76.** $1 - \dfrac{3}{10}$

77. $\begin{array}{r} \dfrac{3}{24} \\[6pt] \dfrac{1}{16} \\[6pt] + \dfrac{5}{12} \\ \hline \end{array}$ **78.** $\begin{array}{r} \dfrac{4}{5} \\[6pt] \dfrac{7}{15} \\[6pt] + \dfrac{1}{2} \\ \hline \end{array}$ **79.** $\begin{array}{r} \dfrac{4}{9} \\[6pt] \dfrac{2}{3} \\[6pt] + \dfrac{7}{15} \\ \hline \end{array}$ **80.** $\begin{array}{r} \dfrac{2}{5} \\[6pt] \dfrac{7}{15} \\[6pt] + \dfrac{5}{6} \\ \hline \end{array}$

Section R.5   Decimals and Percents

Write the following words in decimal notation.

81. fourteen hundredths

82. seven and six tenths

83. three hundred and three thousandths

84. three hundred three thousandths

Write each decimal number in words.

85. 75.09

86. 6.0023

Perform the indicated operations. Round any quotient to the nearest hundredth.

87. $86.34 + 70.58 + 16.7$

88. $100.54 - 37.6$

89. $18.3(7.55)$

90. $122.8 \div 3.1$

91. Fred bought a new car for \$45,000 and traded in his old car for \$15,250. How much cash did he need if taxes were \$2700 and license fees were \$560?

92. An architect's scale drawing shows a rectangle that measures 3.45 inches on one side and 4.3 inches on another side. What is the perimeter (distance around) of the rectangle in the drawing?

Change each decimal to percent form.

93. 0.035

94. 1.23

Change each percent to decimal form.

95. 83%

96. 0.5%

Change each fraction to percent form.

97. $\dfrac{1}{20}$

98. $\dfrac{3}{8}$

99. $\dfrac{45}{100}$

100. $\dfrac{7}{10}$

Chapter R Test

1. Write the following expression with exponents: $3 \cdot 3 \cdot 5 \cdot 5 \cdot 5 \cdot 5$

2. What is the value of the exponential expression 13^2? Is this number called "the square" or "the cube" of 13?

Use the rules for order of operations to find the value of each expression.

(Show your work in a step by step fashion.)

3. $17 + 25 \div 5 \cdot 2 - 3^2$

4. $8^2 - 2(16 \div 4 \cdot 2 + 4)$

5. $26 + 3(7^2 - 3 \cdot 5) - 11^2$

6. $24 \div 3 \cdot 2 - 18 \div 3^2 + 2 \cdot 5$

7. List the prime numbers less than 40.

8. Find the prime factorization of 940.

Find the least common multiple (LCM) of each set of numbers.

9. $\{9, 15, 24\}$

10. $\{20, 30, 40, 50\}$

Perform the indicated operations. Reduce each answer to lowest terms.

11. $\dfrac{1}{10} + \dfrac{3}{10}$

12. $\dfrac{5}{8} - \dfrac{3}{8}$

13. $\dfrac{25}{49} \cdot \dfrac{14}{30}$

14. $\dfrac{7}{18} \cdot \dfrac{3}{5} \cdot \dfrac{4}{9} \cdot \dfrac{1}{14}$

15. $\dfrac{24}{75} \div \dfrac{14}{15}$

16. $\dfrac{16}{19} \div \dfrac{8}{38}$

Perform the indicated operations.

17. $130.7 + 62.75 + 83.02$

18. $89.05 - 39.76$

19. $6.1(2.03)$

20. $\begin{array}{r} 107.5 \\ \times\ 3.02 \\ \hline \end{array}$

21. $16\overline{)49.6}$

22. $5.2\overline{)31.876}$

Write the following words in decimal notation.

23. seventeen and thirteen hundredths

24. six and one hundred seven thousandths

Write each decimal number in words.

25. 5.107

26. 100.001

27. The interest rate at the local bank is currently 4.32%. Change the percent to decimal notation.

Change each number to percent form.

28. 0.057

29. 0.03

30. 1.02

31. $\dfrac{1}{4}$

32. $\dfrac{1}{50}$

33. A bus is carrying 40 passengers. This is $\frac{5}{8}$ of its capacity. What is the capacity of the bus?

34. What is the quotient if $\frac{3}{8}$ of 64 is divided by $\frac{2}{5}$ of $\frac{15}{16}$?

35. The area of a rectangle is found by multiplying its width times its length. Find the area (in square feet) of a rectangle with width 17.2 feet and length 23.1 feet.

36. What is the average miles per gallon (to the nearest tenth) if a car is driven 250 miles on 12 gallons of gas?

Use your TI-84 Plus graphing calculator to find the value of each expression.

37. 15^4

38. $893.6 + 175.82 + 645.33$

39. $8476 + 7542 - 8319.7$

40. $393.635 \div 21.05$

Integers and Real Numbers

Did You Know?

Arithmetic operations defined on the set of positive integers, negative integers, and zero are studied in this chapter. The integer zero will be shown to have interesting properties under the operations of addition, subtraction, multiplication, and division.

Curiously, zero was not recognized as a number by early Greek mathematicians. When Hindu scientists developed the place-value numeration system we currently use, the zero symbol was initially a place holder but not a number. The spread of Islam transmitted the Hindu number system to Europe where it became known as the Hindu-Arabic system and replaced Roman numerals. The word zero comes from the Hindu word meaning "void," which was translated into Arabic as "sifr" and later into Latin as "zephirum," hence the derivation of our English words "zero" and "cipher."

Almost all of the operational properties of zero were known to the Hindus. However, the Hindu mathematician Bhaskara the Learned (1114 – 1185?) asserted that a number divided by zero was zero, or possibly infinite. Bhaskara did not seem to understand the role of zero as a divisor since division by zero is undefined and hence, an impossible operation in mathematics.

Albert Einstein, in his development of a proof that the universe was stable and unchangeable in time, divided both sides of one of his intermediate equations by a complicated expression that under certain circumstances could become zero. When the expression became zero, Einstein's proof did not hold and the possibility of a pulsating, expanding, or contracting universe had to be considered. This error was pointed out to Einstein and he was forced to withdraw his proof that the universe was stable. The moral of this story is that although zero seems like a "harmless" number, its operational properties are different from those of the positive and negative integers.

1.1	The Real Number Line and Absolute Value
1.2	Addition with Integers
1.3	Subtraction with Integers
1.4	Multiplication and Division with Real Numbers
1.5	Properties of Real Numbers

"How can it be that mathematics, being after all a product of human thought independent of experience, is so admirably adapted to the objects of reality?"

Albert Einstein (1879 – 1955)

Numbers and number concepts form the foundation for the study of algebra. In this chapter, you will learn about positive and negative numbers and how to operate with these numbers. Believe it or not, the key number is 0. Pay particularly close attention to the idea of the magnitude of a number, called its absolute value, and the terminology used to represent different types of numbers.

The Real Number Line and Absolute Value

Objectives

After completing this section, you will be able to:

1. *Identify types of numbers.*

2. *Determine if given numbers are greater than, less than, or equal to other given numbers.*

3. *Determine absolute values.*

The study of algebra requires that we know a variety of types of numbers and their names. The set of numbers

$$\mathbb{N} = \{\, 1, 2, 3, 4, 5, 6, 7, 8, 9, 10, 11, ... \,\}$$

is called the **counting numbers** or **natural numbers**. The three dots (an ellipsis) indicate that the pattern is to continue without end. Putting 0 with the set of natural numbers gives the set of **whole numbers**.

$$\mathbb{W} = \{\, 0, 1, 2, 3, 4, 5, 6, 7, 8, 9, 10, 11, ... \,\}$$

Thus mathematicians make the distinction that 0 is a whole number but not a natural number.

To help in understanding different types of numbers and their relationships to each other, we begin with a "picture" called a **number line**. For example, choose some point on a horizontal line and label it with the number 0 (Figure 1.1).

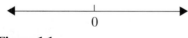

Figure 1.1

Now choose another point on the line to the right of 0 and label it with the number 1 (Figure 1.2). This point is arbitrary and once it is chosen a units scale is determined for the remainder of the line.

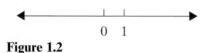

Figure 1.2

We now have a number line. Points corresponding to all the whole numbers are determined. The point corresponding to 2 is the same distance from 1 as 1 is from 0, 3 from 2, 4 from 3, and so on (Figure 1.3).

0 1 2 3 4

Figure 1.3

The **graph** of a number is the point that corresponds to the number and the number is called the **coordinate** of the point. We will follow the convention of using the terms "number" and "point" interchangeably. For example, a point can be called "seven" or "two". The graph of 7 is indicated by marking the point corresponding to 7 with a large dot (Figure 1.4).

6 7 8

Figure 1.4

The graph of the set $A = \{\, 2, 4, 6 \,\}$ is shown in Figure 1.5.

0 1 2 3 4 5 6

Figure 1.5

On a horizontal number line, the point one unit to the left of 0 is the **opposite** of 1. It is called **negative** 1 and is symbolized −1. Similarly, the point two units to the left of 0 is the opposite of 2, called negative 2, and symbolized −2, and so on (Figure 1.6).

The opposite of 1 is −1;	The opposite of −1 is −(−1) = +1;
The opposite of 2 is −2;	The opposite of −2 is −(−2) = +2;
The opposite of 3 is −3;	The opposite of −3 is −(−3) = +3;
and so on.	and so on.

NOTES   The negative sign (−) indicates the opposite of a number as well as a negative number. It is also used, as we will see in Section 1.3, to indicate subtraction. To avoid confusion, you must learn (by practice) just how the − sign is used in each particular situation.

Numbers and their Opposites

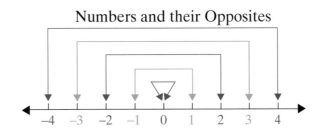

−4 −3 −2 −1 0 1 2 3 4

Figure 1.6

The set of numbers consisting of the whole numbers and their opposites is called the set of **integers**: $Z = \{ \, ... \, , -3, -2, -1, 0, 1, 2, 3, \, ... \, \}$.

The natural numbers are also called **positive integers**. Their opposites are called **negative integers**. **Zero is its own opposite and is neither positive nor negative** (Figure 1.7). Note that the opposite of a positive integer is a negative integer, and the opposite of a negative integer is a positive integer.

Integers:	$\{ \, ... \, , -3, -2, -1, 0, 1, 2, 3, \, ... \, \}$
Positive integers:	$\{ \, 1, 2, 3, 4, 5, \, ... \, \}$
Negative integers:	$\{ \, ... \, , -4, -3, -2, -1 \, \}$

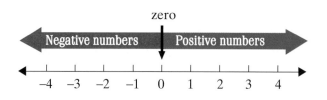

Figure 1.7

a. Find the opposite of 7.

Solution: -7

b. Find the opposite of -3.

Solution: $-(-3)$ or $+3$

In words, the opposite of -3 is $+3$.

Example 2: Number Line

a. Graph the set of integers $\{ -3, -1, 1, 3 \, \}$.

Solution:

$$\begin{array}{c}\text{—3 —2 —1   0   1   2   3}\end{array}$$

b. Graph the set of integers $\{ \, ... \, , -5, -4, -3 \, \}$.

Solution:

$$\begin{array}{c}\text{—6 —5 —4 —3 —2 —1   0}\end{array}$$

The three dots above the number line indicate that the pattern in the graph continues without end.

The integers are not the only numbers that can be represented on a number line. Fractions and decimal numbers such as $\frac{1}{2}$, $-\frac{4}{3}$, $\frac{3}{4}$, and -2.3, as well as numbers such as π and $\sqrt{2}$, can also be represented (Figure 1.8).

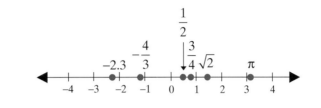

Figure 1.8

Numbers that can be written as fractions and whose numerators and denominators are integers have the technical name **rational numbers**. Positive and negative decimal numbers and the integers themselves can also be classified as rational numbers. For example, the following numbers are all rational numbers:

$$1.3 = \frac{13}{10}, \quad 5 = \frac{5}{1}, \quad -4 = \frac{-4}{1}, \quad \frac{3}{8}, \quad \text{and} \quad \frac{17}{6}.$$

Variables are needed to be able to state rules and definitions in general forms. The definition is restated here for easy reference and for emphasis.

Variable

*A **variable** is a symbol (generally a letter of the alphabet) that is used to represent an unknown number or any one of several numbers.*

The variables *a* and *b* are used to represent integers in the following definition of a **rational number**.

Rational Numbers

*A **rational number** is a number that can be written in the form of $\frac{a}{b}$ where a and b are integers and b ≠ 0.*

OR

*A **rational number** is a number that can be written in decimal form as a terminating decimal or as an infinite repeating decimal.*

Other numbers on a number line, such as $\sqrt{2}, \sqrt{3}, \pi$, and $\sqrt[3]{5}$, are called **irrational numbers**. These numbers can be written as **infinite, nonrepeating decimal numbers**. All rational numbers and irrational numbers are classified as **real numbers** and can be written in some decimal form. The number line is called the **real number line**.

We will discuss rational numbers (in both decimal form and fractional form) in detail in Chapter 2 and irrational numbers in Chapter 9. For now, we are only interested in recognizing various types of numbers and locating their positions on the real number line.

With a calculator, you can find the following decimal values and approximations:

Examples of Rational Numbers:

$\dfrac{3}{4} = 0.75$ This decimal number is terminating.

$\dfrac{1}{3} = 0.33333333...$ There is an infinite number of 3's in this repeating pattern.

$\dfrac{3}{11} = 0.27272727...$ This repeating pattern shows that there may be more than one digit in the pattern.

Examples of Irrational Numbers:

$\sqrt{2} = 1.414213562...$ There is an infinite number of digits with no repeating pattern.

$\pi = 3.141592653...$ There is an infinite number of digits with no repeating pattern.
(The TI-84 Plus calculator will show a 4 in place of the ninth place digit 3 because it rounds off decimal numbers.)

$1.41441444144441...$ There is an infinite number of digits and a pattern of sorts. However, the pattern is nonrepeating.

The following diagram (Figure 1.9) illustrates the relationships among the various categories of real numbers.

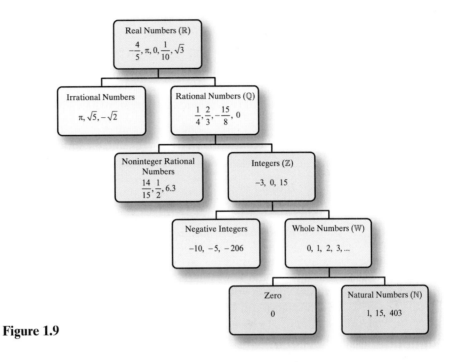

Figure 1.9

Inequality Symbols

On a horizontal number line, **smaller numbers are always to the left of larger numbers**. Each number is smaller than any number to its right and larger than any number to its left. Two symbols used to indicate order are

$$<, \quad \text{read "is less than"}$$
$$\text{and} \quad >, \quad \text{read "is greater than."}$$

Using the real number line in Figure 1.10, you can see the following relationships:

Using $<$		or	**Using** $>$	
$0 < 3$	0 is less than 3		$3 > 0$	3 is greater than 0
$-2 < 1$	-2 is less than 1		$1 > -2$	1 is greater than -2
$-7 < -4$	-7 is less than -4		$-4 > -7$	-4 is greater than -7
$\dfrac{1}{4} < \dfrac{9}{8}$	$\dfrac{1}{4}$ is less than $\dfrac{9}{8}$		$\dfrac{9}{8} > \dfrac{1}{4}$	$\dfrac{9}{8}$ is greater than $\dfrac{1}{4}$

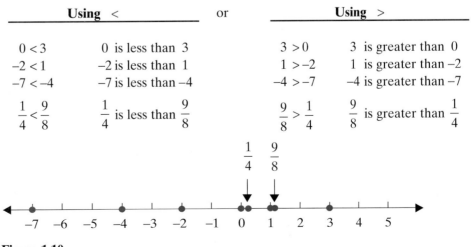

Figure 1.10

Two other symbols commonly used are

$$\leq, \qquad \text{read "is less than or equal to"}$$
$$\text{and} \quad \geq, \qquad \text{read "is greater than or equal to."}$$

For example, $5 \geq -10$ is true since 5 is greater than -10. Also, $5 \geq 5$ is true since 5 does equal 5.

Table of Symbols

=	*is equal to*		$\neq$	*is not equal to*
<	*is less than*		>	*is greater than*
$\leq$	*is less than or equal to*		$\geq$	*is greater than or equal to*

NOTES

Special Note About the Inequality Symbols.
Each symbol can be read from left to right as was just indicated in the Table of Symbols. However, each symbol can also be read from right to left. Thus any inequality can be read in two ways. For example, $6 < 10$ can be read from left to right as "6 is less than 10", but also from right to left as "10 is greater than 6." We will see that this flexibility is particularly useful when reading expressions with variables in Section 3.4.

Example 3: Inequalities

a. Determine whether each of the following statements is true or false.

$7 < 15$ True, since 7 is less than 15.

$3 > -1$ True, since 3 is greater than -1.

$4 \geq -4$ True, since 4 is greater than -4.

$2.7 \geq 2.7$ True, since 2.7 is equal to 2.7.

$-5 < -6$ False, since -5 is greater than -6.

(**Note**: $7 < 15$ can be read as "7 is less than 15" or as "15 is greater than 7.")

(**Note**: $3 > -1$ can be read as "3 is greater than -1" or as "-1 is less than 3.")

(**Note**: $4 \geq -4$ can be read as "4 is greater than or equal to -4" or as "-4 is less than or equal to 4.")

b. Graph the set of **real numbers** $\{ -\frac{3}{4}, 0, 1, 1.5, 3 \}$

Solution:

c. Graph all **natural numbers** less than or equal to 3.
Solution:

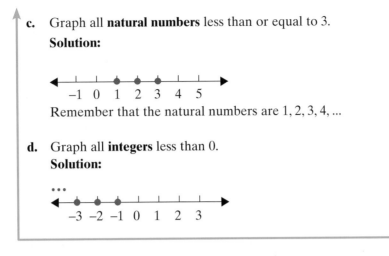

Remember that the natural numbers are 1, 2, 3, 4, ...

d. Graph all **integers** less than 0.
Solution:

Remember that the integers are ... –3, –2, –1, 0, 1, 2, 3, ...

Absolute Value

In working with the real number line, you may have noticed that any integer and its opposite lie the same number of units from 0 on the number line. For example, both +7 and –7 are seven units from 0 (Figure 1.11). The + and – signs indicate direction and the 7 indicates distance.

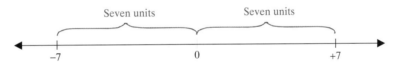

Figure 1.11

The **distance a number is from 0 on a number line** is called its **absolute value** and is symbolized by two vertical bars, $|\quad|$. Thus, $|+7| = 7$ and $|-7| = 7$. Similarly,

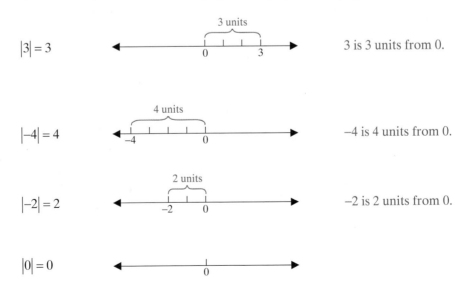

$|3| = 3$ 3 is 3 units from 0.

$|-4| = 4$ –4 is 4 units from 0.

$|-2| = 2$ –2 is 2 units from 0.

$|0| = 0$

$$\left| -\frac{4}{3} \right| = \frac{4}{3}$$

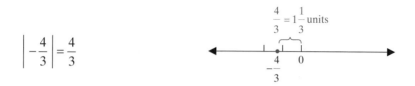

Since distance (similar to length) is never negative, the absolute value of a number is never negative. Or, the absolute value of a nonzero number is always positive.

Absolute Value

*The **absolute value** of a real number is its distance from 0. Note that the absolute value of a real number is never negative.*

$$|a| = a \qquad \text{if } a \text{ is a positive number or 0.}$$
$$|a| = -a \qquad \text{if } a \text{ is a negative number.}$$

NOTES

The symbol $-a$ should be thought of as the "opposite of a." Since a is a variable, a might represent a positive number, a negative number, or 0. This use of symbols can make the definition of absolute value difficult to understand at first. As an aid to understanding the use of the negative sign, consider the following examples.

If $a = -6$, then $-a = -(-6) = 6.$

Similarly,

If $x = -1$, then $-x = -(-1) = 1.$
If $y = -10$, then $-y = -(-10) = 10.$

Remember that $-a$ (the opposite of a) represents a positive number whenever a represents a negative number.

Example 4: Absolute Value

a. $|6.3| = 6.3$

The number 6.3 is 6.3 units from 0. Also, 6.3 is positive so its absolute value is the same as the number itself.

b. $|-5.1| = -(-5.1) = 5.1$

The number -5.1 is 5.1 units from 0. Also, -5.1 is negative so its absolute value is its opposite.

c. If $|x| = 7$, what are the possible values for x?

Solution: $x = 7$ or $x = -7$ since $|7| = 7$ and $|-7| = 7$.

d. If $|x| = 1.35$, what are the possible values for x?

Solution: $x = 1.35$ or $x = -1.35$ since $|1.35| = 1.35$ and $|-1.35| = 1.35$.

e. True or False: $|-4| \geq 4$

Solution: True, since $|-4| = 4$ and $4 \geq 4$.

f. True or False: $\left| -5\dfrac{1}{2} \right| < 5\dfrac{1}{2}$

Solution: False, since $\left| -5\dfrac{1}{2} \right| = 5\dfrac{1}{2}$ and $5\dfrac{1}{2} \not< 5\dfrac{1}{2}$.

($\not<$ is read "is not less than")

g. If $|x| = -3$, what are the possible values for x?

Solution: There are no values of x for which $|x| = -3$. The absolute value can never be negative. There is no solution.

h. If $|x| < 3$, what are the possible integer values for x? Graph these numbers on a number line.

Solution: The integers are within 3 units of 0: $-2, -1, 0, 1, 2$.

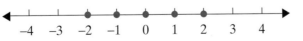

Continued on next page...

i. If $|x| \geq 4$, what are the possible integer values for x? Graph these numbers on a number line.

Solution: The integers must be 4 or more units from 0: $\ldots, -7, -6, -5, -4, 4, 5, 6, 7, \ldots$

Practice Problems

Fill in the blank with the appropriate symbol: <, >, or =.

1. -2 ____ 1

2. $1\dfrac{6}{10}$ ____ 1.6

3. $-(-4.1)$ ____ -7.2

4. *Graph the set of all negative integers on a number line.*

5. *True or False:* $3.6 \leq |-3.6|$

6. *List the numbers that satisfy the equation* $|x| = 8$.

7. *List the numbers that satisfy the equation* $|x| = -6$.

1.1 Exercises

In Exercises 1 – 10 graph each set of real numbers on a real number line. For decimal representations of fractions, use a calculator.

1. $\{1, 2, 5, 6\}$ **2.** $\{-3, -2, 0, 1\}$ **3.** $\{2, -3, 0, -1\}$

4. $\{-2, -1, 4, -3\}$ **5.** $\left\{0, -1, \dfrac{7}{4}, 3, 1\right\}$ **6.** $\left\{-2, -1, -\dfrac{1}{3}, 2\right\}$

7. $\left\{-\dfrac{3}{4}, 0, 2, 3.6\right\}$ **8.** $\left\{-3.4, -2, 0.5, 1, \dfrac{5}{2}\right\}$ **9.** $\left\{-\dfrac{7}{2}, -1.5, 1, \dfrac{4}{3}, 2\right\}$

10. $\left\{-4, -\dfrac{7}{3}, -1, 0.2, \dfrac{5}{2}\right\}$

Answers to Practice Problems: 1. < **2.** = **3.** > **4.**

5. True **6.** 8, −8 **7.** No solution

In Exercises 11 – 24, graph each set of integers on a real number line.

11. All positive integers less than 4

12. All whole numbers less than 7

13. All positive integers less than or equal to 3

14. All negative integers greater than or equal to –3

15. All integers less than –6

16. All integers less than 8

17. All integers greater than or equal to –1

18. All whole numbers less than or equal to 4

19. All negative integers greater than or equal to 2

20. All integers greater than or equal to –6

21. All integers more than 6 units from 0

22. All integers more than 3 units from 0

23. All integers less than 3 units from 0

24. All integers less than 8 units from 0

Find the value of each of the following absolute value expressions.

25. $|-10|$

26. $|-5.6|$

27. $-|-4|$

28. $-|-3.4|$

29. $-|11.3|$

30. $-|15|$

Fill in the blank in Exercises 31 – 46 with the appropriate symbol: <, >, or =.

31. $4 \underline{\hphantom{xx}} 6$

32. $-3 \underline{\hphantom{xx}} 1$

33. $-2 \underline{\hphantom{xx}} -4$

34. $-8 \underline{\hphantom{xx}} 0$

35. $5 \underline{\hphantom{xx}} -(-5)$

36. $-(-4.3) \underline{\hphantom{xx}} 4.3$

37. $5.6 \underline{\hphantom{xx}} -(-8.7)$

38. $2.3 \underline{\hphantom{xx}} 1.6$

39. $-\dfrac{3}{4} \underline{\hphantom{xx}} -1$

40. $-2.3 \underline{\hphantom{xx}} -2\dfrac{3}{10}$

41. $\dfrac{1}{3} \underline{\hphantom{xx}} \dfrac{1}{2}$

42. $-\dfrac{2}{3} \underline{\hphantom{xx}} \dfrac{1}{8}$

43. $-\dfrac{1}{2} \underline{\hphantom{xx}} -\dfrac{1}{3}$

44. $-\dfrac{4}{3} \underline{\hphantom{xx}} -\left(-\dfrac{1}{3}\right)$

45. $-\dfrac{2}{8} \underline{\hphantom{xx}} -\dfrac{1}{4}$

46. $\dfrac{9}{16} \underline{\hphantom{xx}} \dfrac{3}{4}$

Determine whether each statement in Exercises 47 – 72 is true or false. If a statement is false, rewrite it in a form that is a true statement. (There may be more than one way to correct a statement.)

47. $0 = -0$

48. $11 = -(-11)$

49. $-22 < -16$

50. $-6 < -8$

51. $-9 > -8.5$

52. $-2.3 < 1$

53. $-17 \le 17$

54. $-\dfrac{1}{3} \le 0$

55. $4.7 \ge 3.5$

56. $\dfrac{3}{5} > \dfrac{1}{4}$

57. $|-5| = 5$

58. $-|-6.2| = -6.2$

59. $-|-7| \ge -|7|$

60. $|-6| \ge 6$

61. $|-8| \ge 4$

62. $|-1.9| < 2$

63. $-|-3| < -|4|$

64. $-|5| > -|3.1|$

65. $\left|-\dfrac{5}{2}\right| < 2$

66. $|-3.4| < 0$

67. $\dfrac{2}{3} < |-1|$

68. $3 > \left|-\dfrac{4}{3}\right|$

69. $|-1.6| < |-2.1|$

70. $-|73| < |-73|$

71. $|2.5| = \left|-\dfrac{5}{2}\right|$

72. $|-1.75| = \left|\dfrac{7}{4}\right|$

List and then graph the numbers on a real number line that satisfy the equations in Exercises 73 – 82.

73. $|x| = 4$

74. $|y| = 6$

75. $|x| = 9$

76. $13 = |x|$

77. $0 = |y|$

78. $|x| = 0$

79. $-2 = |x|$

80. $|x| = -3$

81. $|x| = 3.5$

82. $|x| = 4.7$

On a number line, graph the integers that satisfy the conditions stated in Exercises 83 – 92.

83. $|x| \le 4$

84. $|x| \le 2$

85. $|x| > 5$

86. $|x| > 2$

87. $|x| < 7$

88. $|x| > 6$

89. $|x| \le x$

90. $|x| > x$

91. $|x| = x$

92. $|x| = -x$

Calculator Problems

A TI-84 Plus calculator has the absolute value command built in. To access the absolute command press **MATH**, go to **NUM** at the top of the screen and press **1** or **ENTER** . The command **abs(** will appear on the display screen. Then enter any arithmetic expression you wish and the calculator will print the absolute value of that expression. (**Note**: The negative sign is on the key marked **(–)** next to the **ENTER** key.)

Follow the directions above and use your calculator to find the value of each of the following expressions.

93. $|34 - 80|$

94. $|17.5 + 16.3 - 95.2|$

95. $-|10 - 16|$

96. $-|-10 - 11|$

97. $\dfrac{|-6| - |-3|}{|-6 - 3|}$

98. $\dfrac{|4| + |-4|}{|-8|}$

Writing and Thinking About Mathematics

99. Explain, in your own words, how a variable expression such as $-y$ might represent a positive number.

100. Explain, in your own words, the meaning of absolute value.

Hawkes Learning Systems: Introductory Algebra

Name That Real Number
Introduction to Absolute Values

1.2   Addition with Integers

Objectives

After completing this section, you will be able to:

1. Add integers.

2. Determine if given integers are solutions for specified equations.

Picture a straight line in an open field and numbers marked on a number line. An archer stands at 0 and shoots an arrow to +3, then stands at 3 and shoots the arrow 5 more units in the positive direction (to the right). Where will the arrow land? (Figure 1.12.)

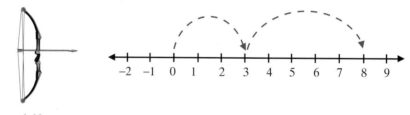

Figure 1.12

Naturally, you have figured out that the answer is +8. What you have done is add the two positive integers, +3 and +5.

$$(+3) + (+5) = +8 \quad \text{or} \quad 3 + 5 = 8$$

Suppose another archer shoots an arrow in the same manner as the first but in the opposite direction. Where would his arrow land? The arrow lands at −8. You have just added −3 and −5 (Figure 1.13).

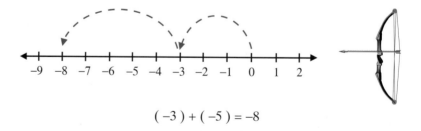

$$(-3) + (-5) = -8$$

Figure 1.13

If the archer stands at 0 and shoots an arrow to +3 and then the archer goes to +3 and turns around and shoots an arrow 5 units in the opposite direction, where will the arrow stick? Would you believe at –2? (Figure 1.14.)

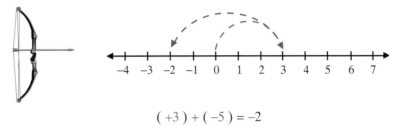

$$(+3) + (-5) = -2$$

Figure 1.14

For our final archer, the first shot is to –3. Then, after going to –3, he turns around and shoots 5 units in the opposite direction. Where is the arrow? It is at +2 (Figure 1.15).

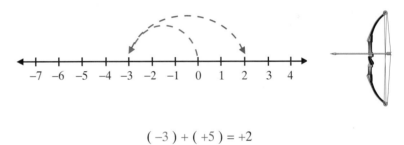

$$(-3) + (+5) = +2$$

Figure 1.15

In summary:

1. The sum of two positive integers is positive.

2. The sum of two negative integers is negative.

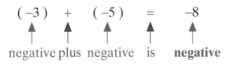

3. The sum of a positive integer and a negative integer may be negative or positive (or zero) depending on which number is further from 0.

Practice Problems

Find each sum. Add from left to right if there are more than two numbers.

1. $(-14) + (-6)$ **2.** $(+16) + (-10)$

3. $(-12) + (8)$ **4.** $11 + 7$

5. $(-13) + (+13)$ **6.** $(-11) + (-8)$

7. $(+6) + (-7) + (-1)$ **8.** $(+100) + (-100) + (+10)$

You probably did quite well and understand how to add integers. The rules can be written out in the following rather formal manner.

Rules for Addition with Integers

1. To add two integers with like signs, add their absolute values and use the common sign:

$$(+7) + (+3) = +(|+7| + |+3|) = +(7+3) = +10$$
$$(-7) + (-3) = -(|-7| + |-3|) = -(7+3) = -10$$

2. To add two integers with unlike signs, subtract their absolute values (the smaller from the larger) and use the sign of the number with the larger absolute value:

$$(-12) + (+10) = -(|-12| - |+10|) = -(12-10) = -2$$
$$(+12) + (-10) = +(|+12| - |-10|) = +(12-10) = +2$$
$$(-15) + (+15) = (|-15| - |+15|) = (15-15) = 0$$

An **equation** is a statement that two expressions are equal. Since equations in algebra are almost always written horizontally, you should become used to working with sums written horizontally. However, there are situations (as in long division) where sums (and differences) are written vertically with one number directly under another. We illustrate this technique in Example 1.

Answers to Practice Problems: 1. −20 **2.** +6 **3.** −4 **4.** 18 **5.** 0 **6.** −19 **7.** −2 **8.** 10

Example 1: Vertical Addition

Find each sum.

a.	-10	**b.**	-4	**c.**	-5	**d.**	-10
	$\underline{7}$		6		-8		3
	-3		$\underline{-15}$		$\underline{-9}$		$\underline{7}$
			-13		-22		0

Now that we know how to add positive and negative integers, we can determine whether or not a particular integer satisfies an equation that contains a variable. Recall, a **variable** is a letter (or symbol) that can represent one or more numbers. A number is said **to be a solution or to satisfy an equation** if it gives a true statement when substituted for the variable.

Example 2: Finding Solutions

Determine whether or not the given integer is a solution to the given equation by substituting for the variable and adding.

a. $x + 5 = -2$ given that $x = -7$
   Solution: $(-7) + 5 = -2$ is true, so -7 is a solution.

b. $y + (-4) = -6$ given that $y = -2$
   Solution: $(-2) + (-4) = -6$ is true, so -2 is a solution.

c. $14 + z = -3$ given that $z = -11$
   Solution: $14 + (-11) = -3$ is false since $14 + (-11) = +3$.
      So, -11 is **not** a solution.

1.2 Exercises

Find the sum in Exercises 1 – 50.

1. $4 + 9$

2. $8 + (-3)$

3. $(-9) + 5$

4. $(-7) + (-3)$

5. $(-9) + 9$

6. $2 + (-8)$

7. $11 + (-6)$

8. $(-12) + 3$

9. $-18 + 5$

10. $26 + (-26)$

11. $-5 + (-3)$

12. $11 + (-2)$

13. $(-2) + (-8)$

14. $10 + (-3)$

15. $17 + (-17)$

16. $(-7) + 20$

17. $21 + (-4)$

18. $(-15) + (-3)$

19. $(-12) + (-17)$ **20.** $24 + (-16)$ **21.** $-4 + (-5)$

22. $(-6) + (-8)$ **23.** $9 + (-12)$ **24.** $-12 + 9$

25. $38 + (-16)$ **26.** $(-20) + (-11)$ **27.** $(-33) + (-21)$

28. $(-21) + 18$ **29.** $-3 + 4 + (-8)$ **30.** $(-9) + (-6) + 5$

31. $(-9) + (-2) + (-5)$ **32.** $(-21) + 6 + 15$ **33.** $-13 + (-1) + (-12)$

34. $-19 + (-2) + (-4)$ **35.** $27 + (-14) + (-13)$ **36.** $-33 + 29 + 2$

37. $-43 + (-16) + 27$ **38.** $-68 + (-3) + 42$ **39.** $-38 + 49 + (-6)$

40. $102 + (-93) + (-6)$

41.
$$\begin{array}{r} -21 \\ \underline{-62} \end{array}$$

42.
$$\begin{array}{r} -12 \\ \underline{17} \end{array}$$

43.
$$\begin{array}{r} -15 \\ 8 \\ \underline{19} \end{array}$$

44.
$$\begin{array}{r} -7 \\ 23 \\ \underline{-9} \end{array}$$

45.
$$\begin{array}{r} -163 \\ 204 \\ \underline{-73} \end{array}$$

46.
$$\begin{array}{r} -93 \\ -87 \\ \underline{147} \end{array}$$

47.
$$\begin{array}{r} -16 \\ 34 \\ 2 \\ \underline{-31} \end{array}$$

48.
$$\begin{array}{r} 31 \\ -15 \\ -12 \\ \underline{5} \end{array}$$

49.
$$\begin{array}{r} 52 \\ -33 \\ 29 \\ \underline{-9} \end{array}$$

50.
$$\begin{array}{r} -29 \\ 35 \\ -17 \\ \underline{6} \end{array}$$

In Exercises 51 – 64, determine whether or not the given number is a solution to the given equation by substituting and then evaluating.

51. $x + 4 = 2$ given that $x = -2$ **52.** $x + (-7) = 10$ given that $x = -3$

53. $-10 + x = -14$ given that $x = -4$ **54.** $y + 9 = 7$ given that $y = -2$

55. $17 + y = 11$ given that $y = -6$ **56.** $z + (-1) = 9$ given that $z = 8$

57. $z + (-12) = 6$ given that $z = 18$

58. $x + 3 = -10$ given that $x = -7$

59. $|y| + (-10) = -12$ given that $y = -2$

60. $|x| + 15 = 17$ given that $x = -2$

61. $x + (-5) = -15$ given that $x = -10$

62. $-26 + |x| = -8$ given that $x = -18$

63. $42 + |z| = -30$ given that $z = -72$

64. $|x| + (-5) = -1$ given that $x = -4$

Calculator Problems

Use your TI-84 Plus calculator to find the value of each of the following expressions.

(Remember that the key marked **(−)** next to the **ENTER** key is used to indicate negative numbers.)

65. $47 + (-29) + 66$

66. $56 + (-41) + (-28)$

67. $2932 + 4751 + (-3876)$

68. $(-8154) + 2147 + (-136)$

69. $(-16,945) + (-27,302) + (-53,467)$

70. $(-12,299) + 15,631 + (-47,558)$

Writing and Thinking About Mathematics

71. Describe, in your own words, how the sum of the absolute values of two numbers might be 0. (Is this even possible?)

Hawkes Learning Systems: Introductory Algebra

Addition with Integers

1.3   Subtraction with Integers

Objectives

After completing this section, you will be able to:

1. Find the **additive inverse (opposite)** of an integer.

2. Subtract integers.

3. Determine if given integers are solutions for specified equations.

In basic arithmetic, subtraction is defined in terms of addition. For example, we know that the difference $32 - 25$ is equal to 7 because $25 + 7 = 32$. A beginning student in arithmetic does not know how to find a difference such as $15 - 20$, where a larger number is subtracted from a smaller number, because negative numbers are not yet defined and there is no way to add a positive number to 20 and get 15. Now, with our knowledge of negative numbers, we will define subtraction in such a way that larger numbers may be subtracted from smaller numbers. We will still define subtraction in terms of addition, but we will apply our new rules of addition with integers.

Before we proceed to develop the techniques for subtraction with integers, we will state and illustrate an important relationship between any integer and its opposite.

Additive Inverse

The **opposite** of an integer is called its **additive inverse**. The sum of a number and its additive inverse is zero. Symbolically, for any integer a,

$$a + (-a) = 0.$$

Example 1: Additive Inverse

a. Find the additive inverse (opposite) of 3.
Solution: The additive inverse of 3 is $-3, 3 + (-3) = 0$.

b. Find the opposite (additive inverse) of –7.
Solution: The additive inverse of -7 is $-(-7) = +7, (-7) + (+7) = 0$.

c. Find the additive inverse (opposite) of 0.
Solution: The additive inverse of 0 is $-0 = 0$.
That is, 0 is its own opposite, $(0) + (-0) = 0 + 0 = 0$.

Subtraction (Change in Value)

Intuitively, we can think of addition of numbers (positive or negative or both) as "*piling on*" or "*accumulating*" numbers. For example, when we add positive numbers (or negative) numbers, the sum is a number more positive (or more negative). Figure 1.16 illustrates these ideas.

(a.) Adding positive numbers "piles on" or "accumulates" the numbers in a positive direction.

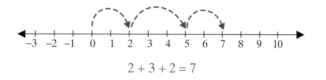

$$2 + 3 + 2 = 7$$

(b.) Adding negative numbers "piles on" or "accumulates" the numbers in a negative direction.

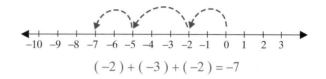

$$(-2) + (-3) + (-2) = -7$$

Figure 1.16

In **subtraction** we want to find the "*difference between*" two numbers. On a number line this translates as the "*distance between*" the two numbers **with direction considered**. As illustrated in Figure 1.17, the *distance between* 6 and 1 is five units. To find this *distance*, we subtract: $6 - 1 = 6 + (-1) = 5$. Note that to subtract 1, we add the opposite of 1 (which is –1).

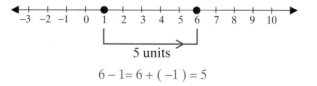

$$6 - 1 = 6 + (-1) = 5$$

Figure 1.17

In Figure 1.18 we see that the distance between 6 and –4 is ten units. Again, to find this distance, we subtract $6 - (-4)$. But we know that we must have 10 as an answer. To get 10, we add the opposite of –4 (which is +4) as follows: $6 - (-4) = 6 + (+4) = 10$.

To understand the "direction" in subtraction, think of subtraction on the number line as:

$$\textbf{(end value)} - \textbf{(start value)}.$$

This means that for $6 - (-4) = 6 + (+4) = 10$,

we have

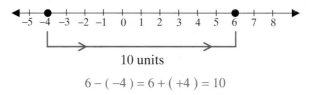

and, as illustrated in Figure 1.18, we would start at –4 and move 10 units in the **positive direction** to end at 6.

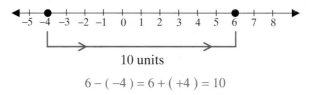

Figure 1.18

As illustrated in Figures 1.17 and 1.18, **subtraction is defined in terms of addition**.

To subtract, add the opposite of the number being subtracted.

If we reverse the order of subtraction, then the answer must indicate the opposite direction.

This means that for $-4 - 6 = -4 + (-6) = -10$,

we have

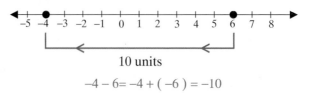

and, as illustrated in Figure 1.19, we would start at 6 and move 10 units in the **negative direction** to end at –4. We see that –4 and 6 are still ten units apart but subtraction now indicates a negative direction.

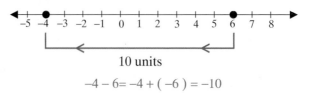

Figure 1.19

The formal definition of subtraction is as follows.

Subtraction

> *For any integers a and b,*
>
> $$a - b = a + (-b).$$
>
> *In words,*
>
> to subtract b from a, **add** the **opposite** of b to a.

$$(+1)-(-4) = (+1) + [-(-4)] = (+1)+(+4) = +5$$

add opposite

$$(-10)-(-3) = (-10) + [-(-3)] = (-10)+(+3) = -7$$

add opposite

Example 2: Subtracting Integers

  a. $(-1) - (-4) = (-1) + (+4) = +3$
  b. $(-1) - (-5) = (-1) + (+5) = +4$
  c. $(-1) - (-8) = (-1) + (+8) = 7$
  d. $(10) - (-2) = (10) + (+2) = 12$
  e. $(-10) - (-5) = (-10) + (+5) = -5$

In practice, the notation $a - b$ is thought of as addition of signed numbers. That is, because $a - b = a + (-b)$, we think of the plus sign, +, as being present in $a - b$. In fact, an expression such as $4 - 19$ can be thought of as "four plus negative nineteen." We have

$$4 - 19 = 4 + (-19) = -15$$
$$-25 - 30 = -25 + (-30) = -55$$
$$-3 - (-17) = -3 + (+17) = 14$$
$$24 - 11 - 6 = 24 + (-11) + (-6) = 7.$$

Generally, the second step is omitted and we go directly to the answer by computing the sum mentally.

$$4 - 19 = -15$$
$$-25 - 30 = -55$$
$$-3 - (-17) = 14$$
$$24 - 11 - 6 = 7$$

The numbers may also be written vertically, that is, one underneath the other. In this case, the sign of the number being subtracted (the bottom number) is changed and addition is performed.

Example 3: Addition and Subtraction

a.
Subtract	Add
43	43
$-(-25)$ $\xrightarrow[\text{change}]{\text{sign}}$	$+25$
	$\overline{68}$

b.
Subtract	Add
-38	-38
$-(+11)$ $\xrightarrow[\text{change}]{\text{sign}}$	-11
	$\overline{-49}$

c.
Subtract	Add
-73	-73
$-(-32)$ $\xrightarrow[\text{change}]{\text{sign}}$	$+32$
	$\overline{-41}$

d.
Subtract	Add
17	17
$-(+69)$ $\xrightarrow[\text{change}]{\text{sign}}$	-69
	$\overline{-52}$

Now, using subtraction as well as addition, we can determine whether or not a number is a solution to an equation of a slightly more complex nature.

To find the **change in value** between two numbers, take the final value and subtract the beginning value. Symbolically,

Change in Value = (Final Value) – (Beginning Value).

Example 4: Change in Value

a. On a winter day, the temperature dropped from 35° F at noon to 6° below zero (−6° F) at 7 p.m. What was the change in temperature?

35°F at noon
− 6°F at 7 pm

Solution: end temperature − beginning temperature = change in temperature

$$-6° \quad - \quad 35° \quad =$$
$$-6° \quad + \quad (-35°) \quad = \quad -41°$$

The change in temperature was −41°. (This means that the temperature *dropped* by 41°.)

b. A jet pilot flew her plane from an altitude of 30,000 ft to an altitude of 12,000 ft What was the change in altitude?

Solution: end altitude − beginning altitude = change in altitude

$$12{,}000 \quad - \quad 30{,}000 \quad = \quad -18{,}000 \text{ ft}$$

(This means that the plane *descended* 18,000 feet.)

The **net change** in a measure is the algebraic sum of several signed numbers. Example 5 illustrates how positive and negative numbers can be used to find the net change of weight (gain or loss) over a period of time.

Example 5: Net Change

Sue weighed 130 lbs. when she started to diet. The first week she lost 7 lbs., the second week she gained 2 lbs., and the third week she lost 5 lbs. What was her weight after 3 weeks of dieting?

Solution:
$$130 + (-7) + (+2) + (-5) = 123 + (+2) + (-5)$$
$$= 125 + (-5)$$
$$= 120\,\text{lbs}$$

Equations

Now, we can determine whether or not a number is a solution to an equation by using subtraction as well as addition.

Example 6: Evaluating Possible Solutions

Determine whether or not the given number is a solution to the given equation by substituting and then evaluating.

a. $x - (-5) = 6$ given that $x = 1$
Solution: $1 - (-5) = 1 + (+5) = 6$ is true, so 1 is a solution.

b. $5 - y = 7$ given that $y = -2$
Solution: $5 - (-2) = 5 + (+2) = 7$ is true, so -2 is a solution.
Note that parentheses were used around –2. This should be done whenever substituting negative numbers for the variable.

c. $z - 14 = -3$ given that $z = 10$
Solution: $10 - 14 = -4$ and $-4 = -3$ is false, so 10 is not a solution.

Practice Problems

1. *What is the additive inverse of 85?*
2. *Find the difference:* $-6 - (-5)$
3. *Simplify:* $-6 - 4 - (-2)$
4. *True or false:* $-5 + (-3) < -5 - (-3)$
5. *Is $x = 15$ a solution to the equation $x - 1 = -16$?*

1.3 Exercises

Find the additive inverse for each integer given in Exercises 1 – 10.

1. 11   **2.** 17   **3.** –6   **4.** –23   **5.** 47

6. –34   **7.** 0   **8.** 100   **9.** –52   **10.** –257

Simplify the expressions in Exercises 11 – 22.

11. $8 - 3$ **12.** $5 - 7$ **13.** $-4 - 6$ **14.** $-18 - 17$

15. $3 - (-4)$ **16.** $5 - (-7)$ **17.** $-8 - (-11)$ **18.** $0 - (-12)$

19. $-14 - 2$ **20.** $-8 - 7$ **21.** $16 - (-8)$ **22.** $15 - 23$

Perform the indicated subtraction in Exercises 23 – 30.

23. $\begin{array}{r} 27 \\ -(+42) \\ \hline \end{array}$ **24.** $\begin{array}{r} 19 \\ -(+26) \\ \hline \end{array}$ **25.** $\begin{array}{r} -23 \\ -(-7) \\ \hline \end{array}$ **26.** $\begin{array}{r} -41 \\ -(-8) \\ \hline \end{array}$

27. $\begin{array}{r} -21 \\ -(+36) \\ \hline \end{array}$ **28.** $\begin{array}{r} -47 \\ -(+13) \\ \hline \end{array}$ **29.** $\begin{array}{r} -27 \\ -(+27) \\ \hline \end{array}$ **30.** $\begin{array}{r} -19 \\ -(-26) \\ \hline \end{array}$

31. Find the difference between –5 and –6. (**Hint**: Subtract the numbers in the order given.)

32. Find the difference between 30 and –12. (**Hint**: Subtract the numbers in the order given.)

Answers to Practice Problems: 1. –85   **2.** –1   **3.** –8   **4.** True   **5.** Not a solution

33. Subtract −3 from −10.

34. Subtract −2 from 6.

35. Subtract 13 from −13.

36. Subtract 20 from −20.

37. From the sum of −4 and −17, subtract the sum of −12 and 6.

38. From the sum of 11 and −13, subtract the sum of 19 and −8.

Find the net change in value of each expression by performing the indicated operations.

39. $-6 + (-4) - 5$

40. $-2 - 2 + 11$

41. $6 + (-3) + (-4)$

42. $-3 + (-7) + 2$

43. $-5 - 2 - (-4)$

44. $-8 - 5 - (-3)$

45. $-7 - (-2) + 6$

46. $-3 - (-3) + (-6)$

47. $97 - 16 - (81)$

48. $-113 + 53 - 79$

Perform the operations on each side of the blank and then fill in the blank in Exercises 49 – 58 with the proper symbol: <, >, or =.

49. $-6 + (-2)$ _____ $3 + (-8)$

50. $-4 - (-3)$ _____ $-4 + (-3)$

51. $5 - 8$ _____ $8 - 5$

52. $7 - (-3)$ _____ $-3 - 7$

53. $11 + (-3)$ _____ $11 - 3$

54. $0 - 6$ _____ $0 - (-6)$

55. $-8 - (-8)$ _____ $-14 - 13$

56. $-7 - (-3)$ _____ $4 - 9$

57. $-151 - 86$ _____ $-(107 + 141)$

58. $25 - 62$ _____ $-11 - 23$

59. At 2 p.m. the temperature was 76°F. At 8 p.m. the temperature was 58°F. What was the change in temperature?

60. The temperature at 5 a.m. was 8°C below zero; at noon, the temperature was 27°C. What was the change in temperature?

61. Lotsa-Flavor Chewing Gum stock opened on Monday at $47 per share and closed Friday at $39 per share. Find the change in price of the stock.

62. The famous French mathematician René Descartes lived from 1596 to 1650. How long did he live?

63. If you travel from the top of Mt. Whitney, elevation 14,495 ft, to the floor of Death Valley, elevation 282 ft below sea level, what is the change in elevation?

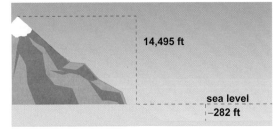

64. A submarine submerged 280 ft below the surface of the sea fired a rocket that reached an altitude of 30,000 ft. What was the change in altitude of the rocket?

65. The great English mathematician and scientist Isaac Newton was born in 1642 and died in 1727. How old was he when he died?

66. Mr. Meade is having a hard time selling his old car. He just slashed the price from $3500 to $2750. By how much did he change the price?

In Exercises 67 – 82, determine whether or not the given number is a solution to the given equation by substituting and then evaluating.

67. $x + 5 = -3$ given that $x = -8$

68. $x - 6 = -9$ given that $x = -3$

69. $15 - y = 17$ given that $y = -2$

70. $11 - x = 8$ given that $x = 3$

71. $x - 3 = -7$ given that $x = 4$

72. $y - 2 = 6$ given that $y = 4$

73. $-9 - x = -14$ given that $x = 5$

74. $x + 13 = 3$ given that $x = -10$

75. $x - 2 = -3$ given that $x = 1$

76. $-18 - y = -30$ given that $y = 12$

77. $|x| - (-10) = 14$ given that $x = -4$

78. $|y| - 12 = -3$ given that $y = -9$

79. $|z| - 5 = 0$ given that $z = -5$

80. $-16 - |x| = 0$ given that $x = 16$

81. $|x| - |-3| = 25$ given that $x = -28$

82. $|-2| + |x| = 13$ given that $x = -11$

Calculator Problems

In Exercises 83 – 86, use your TI-84 Plus calculator to find the value indicated on each side of the blank and then fill in the blank with the proper symbol: <, >, or =.

83. $648 - (-396)$ _____ $124 - 163$

84. $-19,824 - 23,417$ _____ $12,793 - (-14,387)$

85. $-43,931 - (-28,677)$ _____ $-(13,665 + 21,425)$

86. $-(24,295 + 13,107)$ _____ $-48,261 - (-16,276)$

Temperatures above 0 and below 0 as well as gains and losses can be thought of in terms of positive and negative numbers. Use positive and negative numbers to answer Exercises 87 – 90.

87. Harry and his wife went on a diet for 5 weeks. During those 5 weeks, Harry lost 5 pounds, gained 3 pounds, lost 2 pounds, lost 4 pounds, and gained 1 pound. What was his total loss (or gain) for the 5 weeks? If he weighed 210 pounds when he started the diet plan, what did he weigh at the end of the 5-week period? During the same time, his wife lost 10 pounds.

88. In a 5-day week the NASDAQ stock posted a gain of 38 points, a loss of 65 points, a loss of 32 points, a gain of 10 points, and a gain of 15 points. If the NASDAQ started the week at 2350 points, what was the market at the end of the week?

89. In ten running plays in a football game the fullback gained 5 yards, lost 3 yards, gained 15 yards, gained 7 yards, gained 12 yards, lost 4 yards, lost 2 yards, gained 20 yards, lost 5 yards, and gained 6 yards. What was his net yardage for the game?

90. Beginning at a temperature of $10°C$, the temperature in a scientific experiment was measured hourly for four hours. It dropped $5°$, dropped $8°$, dropped $6°$, then rose $3°$. What was the final temperature recorded?

Hawkes Learning Systems: Introductory Algebra

Subtraction with Integers

Multiplication and Division with Real Numbers

After completing this section, you will be able to:

1. Multiply integers.

2. Divide integers.

Multiplication with Real Numbers

Multiplication is shorthand for repeated addition. That is,

$$7 + 7 + 7 + 7 + 7 = 5 \cdot 7 = 35$$

and

$$(-6) + (-6) + (-6) = 3(-6) = -18.$$

Similarly,

$$(-2) + (-2) + (-2) + (-2) + (-2) = 5(-2) = -10.$$

Repeated addition with a negative integer results in a product of a positive integer and a negative integer. Since the sum of negative integers is negative, we have the following general rule.

The product of a positive integer and a negative integer is negative.

NOTES Multiplication can be represented by a raised dot, as in $5 \cdot 7$, or by a number next to a parenthesis as in $3(-6)$ or $(3)(-6)$.

Example 1: Products of Positive and Negative Integers

a. $5(-3) = (-3) + (-3) + (-3) + (-3) + (-3) = -15$

b. $7(-10) = -70$

c. $42(-1) = -42$

d. $3(-5) = -15$

The product of two negative integers can be explained in terms of opposites. We also need the fact that for any integer a, we can think of the **opposite of a**, $(-a)$, as the product of -1 and a. That is, we have

$$-a = -1 \cdot a.$$

Thus,

$$-4 = -1 \cdot 4 = -1(4).$$

and in the product −4(−7) we have,

$$-4(-7) = -1(4)(-7) = -1[(4)(-7)] = -1(-28) = -(-28) = 28.$$

Although one example does not prove a rule, this process can be used in general to arrive at the following correct conclusion:

The product of two negative integers is positive.

Example 2: Products of Negative Integers

a. $(-4)(-9) = +36$
b. $-7(-5) = +35$
c. $-2(-6) = +12$
d. $(-1)(-5)(-3)(-2) = 5(-3)(-2) = -15(-2) = +30$

What happens if a number is multiplied by 0? For example, $3(0) = 0 + 0 + 0 = 0$. In fact,

multiplication by 0 always gives a product of 0.

Example 3: Multiplication by 0

a. $6 \cdot 0 = 0$
b. $-13 \cdot 0 = 0$

Integers, decimals, and fractions are all part of a system of numbers called the **real number system**. (We will discuss real numbers and the properties of real numbers in more detail in Section 1.5.) The rules involving + and − signs for operating with decimal numbers and fractions are the same as those for operating with integers. Examples with addition, subtraction, and multiplication are shown in Example 4.

Example 4: Operating with Real Numbers

a. $(-46) + (-52) = -98$
b. $17.1 - (-4.2) = 17.1 + (+4.2) = 21.3$
c. $-5.2(-4) = +20.8$
d. $3(-3.5) = -10.5$
e. $\left(\dfrac{1}{2}\right)\left(-\dfrac{1}{5}\right) = -\dfrac{1}{10}$

The rules for multiplication with real numbers can be summarized as follows. Remember that these rules apply to integers, decimals, and fractions.

Rules for Multiplication with Real Numbers

If a and b are positive real numbers, then

1. The product of two positive numbers is positive: $a \cdot b = ab$.

2. The product of two negative numbers is positive: $(-a)(-b) = ab$.

3. The product of a positive number and a negative number is negative: $a(-b) = -ab$.

4. The product of 0 and any number is 0: $a \cdot 0 = 0$ and $(-a) \cdot 0 = 0$.

Practice Problems

Find the following products.

1. $5(-3)$

2. $-6(-4)$

3. $-8(4)$

4. $-12(0)$

5. $-9(-2)(-1)$

6. $3(-20)(5)$

7. $5(-3.7)$

8. $-4.1(-4.5)$

Division with Real Numbers

The rules for multiplication lead directly to the rules for division because division is defined in terms of multiplication. For convenience, division is indicated in fraction form.

Division with Real Numbers

For real numbers a, b, and x (where $b \neq 0$),

$$\frac{a}{b} = x \ \text{means that} \ a = b \cdot x.$$

For real numbers a and b (where $b \neq 0$),

$$\frac{a}{0} \ \textbf{is undefined}, \text{but} \ \frac{0}{b} = 0.$$

The following discussion explains why division by 0 is undefined.

Answers to Practice Problems: 1. −15 **2.** 24 **3.** −32 **4.** 0 **5.** −18 **6.** −300 **7.** −18.5 **8.** 18.45

Division by 0 is Undefined

1. *Suppose that $a \neq 0$ and $\dfrac{a}{0} = x$. Then, since division is related to multiplication, we must have $a = 0 \cdot x$. But this is not possible because $0 \cdot x = 0$ for any value of x and we stated that $a \neq 0$.*

2. *Suppose that $\dfrac{0}{0} = x$. Then, $0 = 0 \cdot x$ which is true for all values of x. But we must have a unique answer for x.*

*Therefore, in any case, we conclude that **division by 0 is undefined**.*

Example 5: Division with Integers

a. $\dfrac{36}{9} = 4$ because $36 = 9 \cdot 4$.

b. $\dfrac{-36}{9} = -4$ because $-36 = 9(-4)$.

c. $\dfrac{36}{-9} = -4$ because $36 = -9(-4)$.

d. $\dfrac{-36}{-9} = 4$ because $-36 = -9(4)$.

The rules for division can be stated as follows.

Rules for Division with Real Numbers

If a and b are positive real numbers,

1. *The quotient of two positive numbers is positive: $\dfrac{a}{b} = +\dfrac{a}{b}$.*

2. *The quotient of two negative numbers is positive: $\dfrac{-a}{-b} = +\dfrac{a}{b}$.*

3. *The quotient of a positive number and a negative number is negative:*

$$\dfrac{-a}{b} = -\dfrac{a}{b} \text{ and } \dfrac{a}{-b} = -\dfrac{a}{b}.$$

NOTES

The following common rules about multiplication and division with two non-zero real numbers are helpful in remembering the signs of answers.

1. If the numbers have the same sign, both the product and quotient will be positive.
2. If the numbers have different signs, both the product and quotient will be negative.

Practice Problems

Find the quotients.

1. $\dfrac{-30}{10}$ **2.** $\dfrac{40}{-10}$ **3.** $\dfrac{-20}{-10}$ **4.** $\dfrac{-7}{0}$

5. $\dfrac{0}{13}$ **6.** $\dfrac{7.5}{-3}$ **7.** $\dfrac{-4.32}{-4}$ **8.** $\dfrac{-5.2}{-2.6}$

Average (or Mean)

You may already be familiar with the concept of the **average** of a set of numbers. The average is also called the **arithmetic average** or **mean**. Your grade in this course may be based on the average of your exam scores. Newspapers and magazines report average income, average sales, average attendance at sporting events, and so on. The mean of a set of numbers is particularly important in the study of statistics. For example, scientists might be interested in the mean IQ of the students attending a certain university or the mean height of students in the fourth grade.

Average

*The **average** (or **mean**) of a set of numbers is the value found by adding the numbers in the set and then dividing the sum by the number of numbers in the set.*

Answers to Practice Problems: 1. –3 **2.** –4 **3.** 2 **4.** undefined **5.** 0 **6.** –2.5 **7.** 1.08 **8.** 2

Example 6: Average

a. At noon on five consecutive days in Aspen, Colorado the temperatures were $-5°$, $7°$, $6°$, $-7°$, and $14°$ (in degrees Fahrenheit). (Negative numbers represent temperatures below zero.) Find the average of these noon-day temperatures.

Solution: First, add the five temperatures.

$$(-5) + 7 + 6 + (-7) + 14 = 15$$

Now divide the sum, 15, by the number of temperatures, 5.

$$\frac{15}{5} = 3$$

The average noon temperature was 3° F.

b. In a placement exam for mathematics, a group of ten students had the following scores: 3 students scored 75, 2 students scored 80, 1 student scored 82, 3 students scored 85, and 1 student scored 88. What was the mean score for this group of students?

Solution: To find the total of all the scores, we multiply and then add. This is more efficient than adding all ten scores.

$$75 \cdot 3 = 225 \qquad\qquad\qquad \text{Multiply.}$$
$$80 \cdot 2 = 160$$
$$82 \cdot 1 = 82$$
$$85 \cdot 3 = 255$$
$$88 \cdot 1 = 88$$

$$225 + 160 + 82 + 255 + 88 = 810 \qquad \text{Add.}$$

$$810 \div 10 = 81 \qquad\qquad\qquad \text{Divide by the number of scores.}$$

The mean score on the placement test for this group of students was 81.

Continued on next page...

c. The following speeds (in miles per hour) of fifteen cars were recorded at a certain point on a freeway.

$$70 \quad 75 \quad 65 \quad 60 \quad 61$$
$$64 \quad 68 \quad 72 \quad 59 \quad 68$$
$$82 \quad 76 \quad 70 \quad 68 \quad 50$$

Find the average speed of these cars. (One car received a speeding ticket, while another had a broken muffler.)

Solution: Using a calculator, the sum of the speeds is 1008 mph.

Dividing by 15 gives the average speed:

$$1008 \div 15 = 67.2 \text{ mph}$$

1.4 Exercises

Find the product in Exercises 1 – 24.

1. $4 \cdot (-3)$ **2.** $(-5) \cdot 6$ **3.** $12 \cdot 4$ **4.** $19 \cdot 3$

5. $(-8)(-7)$ **6.** $(-11)(-2)$ **7.** $(-3)(7)$ **8.** $(5)(-6)$

9. $(-14)(-4)$ **10.** $(-11)(-6)$ **11.** $(-13)(-2)$ **12.** $(-8)(-9)$

13. $10(-7)$ **14.** $(-5)(12)$ **15.** $(-2)(-3)(-4)$ **16.** $(-6)(-3)(-9)$

17. $(-8) \cdot 4 \cdot 9$ **18.** $(-3) \cdot 2 \cdot (-3)$ **19.** $(-7)(-16) \cdot 0$ **20.** $(-9) \cdot 0 \cdot 4$

21. $(-2) \cdot 4.5$ **22.** $(-5) \cdot (-3.7)$ **23.** $4.3 \cdot (-1.7)$ **24.** $(-2.6)(-0.2)$

Find the quotient in Exercises 25 – 46.

25. $\dfrac{-8}{-2}$ **26.** $\dfrac{-20}{-10}$ **27.** $\dfrac{-30}{5}$ **28.** $\dfrac{-51}{3}$

29. $\dfrac{-26}{-13}$ **30.** $\dfrac{-91}{-7}$ **31.** $\dfrac{0}{6}$ **32.** $\dfrac{0}{-7}$

33. $\dfrac{-3}{0}$ **34.** $\dfrac{16}{0}$ **35.** $\dfrac{39}{-13}$ **36.** $\dfrac{44}{-4}$

37. $\dfrac{-34}{2}$ **38.** $\dfrac{-36}{9}$ **39.** $\dfrac{-60}{-12}$ **40.** $\dfrac{-48}{-16}$

41. $\dfrac{-4.8}{8}$ **42.** $\dfrac{-5.6}{7}$ **43.** $\dfrac{-4}{-0.2}$ **44.** $\dfrac{2.99}{-1.3}$

45. $\dfrac{-3}{-8}$ **46.** $\dfrac{2.8}{-1.4}$

Determine whether each statement in Exercises 47 – 56 is true or false. If a statement is false, rewrite it in a form that is true. (There may be more than one correct new form.)

47. $(-4) \cdot (6) < 3 \cdot 8$

48. $(-7) \cdot (-9) = 3 \cdot 21$

49. $(-12) \cdot (6) = 9(-8)$

50. $(-6)(9) = (18)(-3)$

51. $6(-3) \geq (-14) + (-4)$

52. $7 + 8 > (-10) + (-5)$

53. $(-7) + 0 \leq (-7) \cdot (0)$

54. $17 + (-3) < (-14) + (-4)$

55. $(-4)(9) = (-24) + (-12)$

56. $14 + 6 \leq (-2)(-10)$

57. Find the mean of the set of integers $-10, 15, 16, -17, -34,$ and -42.

58. Find the mean of the set of integers $-72, -100, -54, 82,$ and -96.

59. The temperature readings for 20 days at 3 p.m. at a local ski resort were recorded as follows:

$$24° \quad 11° \quad -5° \quad 14° \quad 15° \quad 5° \quad -6° \quad 13° \quad -2° \quad -8°$$
$$-10° \quad 32° \quad 31° \quad -7° \quad -9° \quad 4° \quad -18° \quad -9° \quad 5° \quad 20°$$

What was the average of the recorded temperatures for these 20 days?

60. On an exam in history, a class of twenty-one students had the following test scores: 4 scored 65, 3 scored 70, 6 scored 78, 2 scored 82, 1 scored 85, 3 scored 91, and 2 scored 95. What was the mean score (to the nearest hundredth) on this test for the class?

61. Fifteen students scored the following scores on an exam in accounting: 1 scored 67, 4 scored 73, 3 scored 77, 2 scored 80, 3 scored 88, and 2 scored 93. What was the average score for these students?

62. Twenty business executives made the following numbers of telephone calls during one week. Find the mean number of calls (to the nearest tenth) made by these executives.

| 20 | 16 | 14 | 11 | 51 | 18 | 16 | 42 | 49 | 12 |
| 40 | 36 | 28 | 52 | 25 | 18 | 22 | 33 | 9 | 19 |

63. The blood calcium level (in milligrams per deciliter) for 20 patients was reported as follows:

| 8.2 | 10.2 | 9.3 | 8.5 | 7.3 | 9.4 | 11.1 | 10.0 | 8.5 | 9.9 |
| 9.7 | 9.6 | 8.3 | 9.8 | 9.1 | 8.6 | 10.2 | 9.4 | 9.1 | 9.2 |

Find the mean blood calcium level (to the nearest tenth) for these patients.

64. Thirty students made the following scores on the final exam in an algebra course. Find the mean score (to the nearest tenth) on the final exam.

82	93	80	65	52	48	78	69	75	75
98	95	63	76	78	80	95	70	73	81
82	89	71	66	67	44	74	85	85	92

The **frequency** of a number is simply **a count of how many times that number appears**. In statistics, data is commonly given in the table form of a frequency distribution as illustrated in Exercises 65 and 66. To find the mean, multiply each number by its frequency, add these products, and divide the sum by the sum of the frequencies.

65. The heights of twenty-two men were recorded in the following frequency distribution. Find the mean height (to the nearest tenth of an inch) for these men.

Height (in inches)	Frequency
68	2
69	3
72	8
73	5
74	2
75	1
78	1

66. The students in a psychology class were asked the number of books that they had read in the last month. The following frequency distribution indicates the results. Find the mean number of books read (to the nearest tenth) by these students.

Number of Books	Frequency
0	3
1	2
2	6
3	4
4	2
5	1

67. The bar graph shows the approximate amounts of time per week spent watching TV for six groups (by age and sex) of people 18 years of age and older. What is the average amount of time per week people over the age of 18 spend watching TV? (Assume each group has the same number of people.)

Weekly TV Viewing by Age and Sex

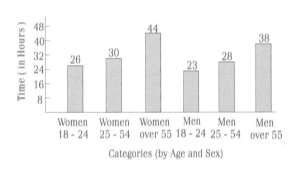

68. The following pictograph shows the number of phone calls received by a 1-hour radio talk show in Seattle during one week. (Not all calls actually get on the air.) What was the mean number of calls per show received that week?

Phone Calls Received by Seattle Talk Show

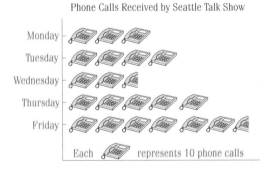

69. The following bar graph shows the area of each of the five Great Lakes. Lake Superior with an area of about 32,000 square miles is the world's largest fresh water lake. What is the mean size of these lakes?

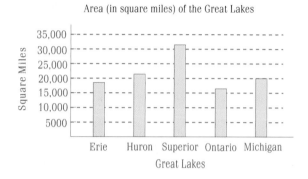

Area (in square miles) of the Great Lakes

Calculator Problems

Use a TI-84 Plus calculator to find the value of each expression in Exercises 70 – 81. Round quotients to the nearest hundredth. Remember the negative sign **(–)** is next to **ENTER** .

For example, $-14.8 \div (-5)$ would appear as follows:

70. $(273)(-24)(-180)$ **71.** $(-4613)(-45)(-166)$ **72.** $(54)(-17)(24)$

73. $(-77{,}459) \div 29$ **74.** $(-62{,}234) \div (-37)$ **75.** $(-35 - 45 - 56) \div 3$

76. $(-52 - 30 - 40 - 60) \div 4$ **77.** $72 \div (15 - 22)$ **78.** $95 \div (-3 - 7)$

79. $-15.3 \div (-5.4)$ **80.** $(-13.4)(-2.5)(-1.63)$ **81.** $(-2.5)(-3.41)(-10.6)$

Writing and Thinking About Mathematics

82. Explain the conditions under which the quotient of two numbers is 0.

83. Explain, in your own words, why division by 0 is not a valid arithmetic operation.

Hawkes Learning Systems: Introductory Algebra

Multiplication and Division with Real Numbers

1.5   Properties of Real Numbers

Objectives

After completing this section, you will be able to:

1. Apply the properties of real numbers to complete statements.

2. Name the real number properties that justify given statements.

Recognizing Types of Numbers

In mathematics, it is important to be able to recognize and name various types of numbers. The conditions stated or implied in certain word problems or applications allow for only positive solutions or only integer solutions. For example, the height of a person or the length of a piece of string cannot be a negative number although either measure could be a decimal number or mixed number. However, the count of the number of students in a class must be a positive number and an integer. On a multiple choice test where points are deducted for wrong answers, scores can possibly be negative integers. With these ideas in mind, we review some of the various types of numbers.

Integers

0, 4, −6, 14, and −35 are all integers.

Rational Numbers

$0, 4, -7, 100.72, \dfrac{3}{5}, 9\dfrac{3}{10}$, and 6.45454545... are all rational numbers.

In decimal form, every rational number is either

(1) a terminating decimal (such as 100.72 and $9\dfrac{3}{10} = 9.3$) or

(2) an infinite repeating decimal (such as 6.45454545...).

Remember that all integers are also rational numbers.

Irrational Numbers

$\sqrt{3}, \sqrt{5}, -\sqrt{10}, \pi$, and $\dfrac{2}{\sqrt{7}}$ are all irrational numbers.

In decimal form, every irrational number is an infinite nonrepeating decimal. For example, see the appendix with π to 3742 decimal places. There will never be a repeating pattern to the digits in π.

All of the numbers discussed here are **real numbers** and have corresponding points on a **real number line** as shown in Figure 1.20.

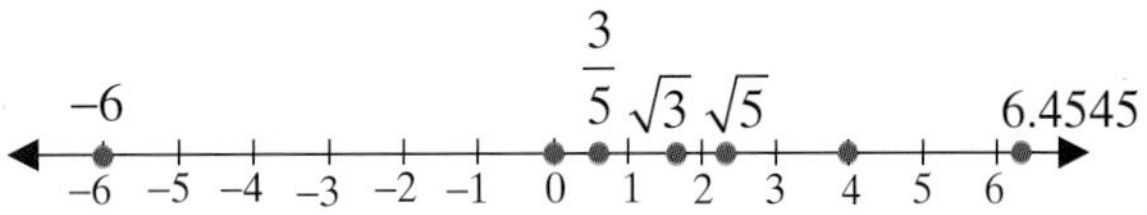

Figure 1.20

Example 1: Types of Numbers

Given the set of numbers $\{-43.2, -\sqrt{19}, -2, -\frac{3}{7}, 0, 0.58, \pi, \sqrt[3]{62}, 6, 13.5555...\}$, tell which numbers are

   a. Integers, **b.** Rational numbers, **c.** Irrational numbers, **d.** Real numbers.

Solutions:

a. The integers are $-2, 0$, and 6.

b. The rational numbers are $-43.2, -2, -\frac{3}{7}, 0, 0.58, 6$, and $13.5555...$

c. The irrational numbers are $-\sqrt{19}, \pi$, and $\sqrt[3]{62}$.

d. All of the numbers are real numbers.

In this chapter we have discussed real numbers and the order of real numbers and their placement on a real number line. However, the discussions of operations were limited to integers and a few decimals. In fact, **all the rules for operating with integers apply to all real numbers,** including decimal, fractions, and irrational numbers. Operations with positive decimals and fractions were covered in the Review Chapter and will be expanded in Chapter 2 to include negative decimals and fractions. As we continue to work with all types of real numbers and algebraic expressions, we will need to understand that while certain properties are true for addition and multiplication these same properties are not true for subtraction and division. For example,

the order of the numbers in addition ***does not*** change the result:

$$17 + 5 = 5 + 17 = 22$$

but, the order of the numbers in subtraction ***does*** change the result:

$$17 - 5 = 12 \quad \text{but} \quad 5 - 17 = -12.$$

We say that addition is **commutative** while subtraction is **not commutative**. The various properties of real numbers under the operations of addition and multiplication are summarized here. These properties are used throughout algebra and mathematics in developing formulas and general concepts.

Properties of Addition and Multiplication

In this table a, b, and c are real numbers.

Name of Property

For Addition		For Multiplication
$a + b = b + a$	Commutative Property	$ab = ba$
$3 + 6 = 6 + 3$		$4 \cdot 9 = 9 \cdot 4$
$(a + b) + c = a + (b + c)$	Associative Property	$a(bc) = (ab)c$
$(2 + 5) + 4 = 2 + (5 + 4)$		$(6 \cdot 2) \cdot 7 = 6 \cdot (2 \cdot 7)$
$a + 0 = 0 + a = a$	Identity	$a \cdot 1 = 1 \cdot a = a$
$20 + 0 = 0 + 20 = 20$		$-2 \cdot 1 = 1 \cdot (-2) = -2$
$a + (-a) = 0$	Inverse	$a \cdot \dfrac{1}{a} = 1 \left(for\ a \neq 0 \right)$
$10 + (-10) = 0$		$3 \cdot \dfrac{1}{3} = 1$

Zero Factor Law

$$a \cdot 0 = 0 \cdot a = 0 \qquad\qquad -5 \cdot 0 = 0 \cdot (-5) = 0$$

Distributive Property of Multiplication over Addition:

$$a(b + c) = ab + ac \qquad\qquad 3(x + 5) = 3 \cdot x + 3 \cdot 5$$

NOTES

The number **0** is called the **additive identity** because when 0 is added to a number the result is the same number. Likewise, the number **1** is called the **multiplicative identity** because when 1 is multiplied by a number the result is the same number. Also, the **additive inverse** of a number is its **opposite** and the **multiplicative inverse** of a number is its **reciprocal**.

The raised dot is optional when indicating multiplication between two variables or a number and a variable. For example,

$$x \cdot y = xy \quad \text{and} \quad 5 \cdot x = 5x.$$

In the case of a number and a variable, the number is called the **coefficient** of the variable. So in

$$5x \text{ the number 5 is the } \textbf{coefficient} \text{ of } x.$$

Thus, in an illustration of the distributive property involving a variable, we can write

$$5(x - 6) = 5x - 30.$$

Example 2: Properties of Addition and Multiplication

State the name of each property being illustrated.

a. $(-7) + 13 = 13 + (-7)$

Solution: Commutative Property of Addition

b. $8 + (9 + 1) = (8 + 9) + 1$

Solution: Associative Property of Addition

c. $(-25) \cdot 1 = -25$

Solution: Multiplicative Identity

d. $3(x + y) = 3x + 3y$

Solution: Distributive Property

e. $4(3 \cdot 2) = (4 \cdot 3) \cdot 2$

Solution: Associative Property of Multiplication

In each of the following equations, state the property illustrated and show that the statement is true for the value given for the variable by substituting the value in the equation and evaluating.

f. $x + 14 = 14 + x$ given that $x = -4$

Solution: The commutative property of addition is illustrated.
$$(-4) + 14 = 14 + (-4) = 10$$

Continued on next page...

121

g. $2x + 5 = 5 + 2x$ given that $x = 10$

Solution: The commutative property of addition is illustrated.
$$2 \cdot 10 + 5 = 5 + 2 \cdot 10 = 25$$

h. $12(y + 3) = 12y + 36$ given that $y = -2$

Solution: The distributive property is illustrated.
$$12(-2 + 3) = 12(1) = 12 \text{ and } 12(-2) + 36 = -24 + 36 = 12$$

Practice Problems

Determine the property being illustrated.

1. $(-2 \cdot 5) \cdot 2 = -2 \cdot (5 \cdot 2)$

2. $15 \cdot 0 = 0 \cdot 15 = 0$

3. $2 + 7 = 7 + 2$

4. $2(y + 5) = 2y + 10$

1.5 Exercises

1. Given the set of numbers $\{ -3.56, -\sqrt{8}, -\dfrac{5}{8}, -1, 0, 0.7, \pi, 4, \dfrac{13}{2}, \sqrt[3]{25} \}$, tell which numbers are
a. Integers   **b.** Rational numbers   **c.** Irrational numbers   **d.** Real numbers

2. Given the set of numbers $\{ -123.21, -\sqrt{118}, -\dfrac{32}{11}, -2\dfrac{1}{3}, 0.9, \dfrac{\pi}{2}, \dfrac{17}{10}, 5, \dfrac{\sqrt{107}}{5},$
$9.343434... \}$, tell which numbers are

a. Integers   **b.** Rational numbers   **c.** Irrational numbers   **d.** Real numbers

Complete the expressions in Exercises 3 – 20 using the given property.

3. $7 + 3 = $ _____   Commutative property of addition

4. $(6 \cdot 9) \cdot 3 = $ _____   Associative property of multiplication

5. $19 \cdot 4 = $ _____   Commutative property of multiplication

6. $18 + 5 = $ _____   Commutative property of addition

Answers to Practice Problems: 1. Associative Property of Multiplication **2.** Zero Factor Law
3. Commutative Property of Addition **4.** Distributive Property

7. $6(5 + 8) =$ ____   Distributive property

8. $16 + (9 + 11) =$ ____   Associative property of addition

9. $2 \cdot (3x) =$ ____   Associative property of multiplication

10. $3(x + 5) =$ ____   Distributive property

11. $3 + (x + 7) =$ ____   Associative property of addition

12. $9(x + 5) =$ ____   Distributive property

13. $6 \cdot 0 =$ ____   Zero factor law

14. $6 \cdot 1 =$ ____   Multiplicative identity

15. $0 + (x + 7) =$ ____   Additive identity

16. $0 \cdot (-13) =$ ____   Zero factor law

17. $2(x - 12) =$ ____   Distributive property

18. $(-5) + 5 =$ ____   Additive inverse

19. $6.3 + (-6.3) =$ ____   Additive inverse

20. $3 \cdot \dfrac{1}{3} =$ _____   Multiplicative inverse

Name the property of real numbers illustrated.

21. $5 + 16 = 16 + 5$ **22.** $5 \cdot 16 = 16 \cdot 5$

23. $32 \cdot 1 = 32$ **24.** $32 + 0 = 32$

25. $5 + (3 + 1) = (5 + 3) + 1$ **26.** $5 + (3 + 1) = (3 + 1) + 5$

27. $13(y + 2) = (y + 2) \cdot 13$ **28.** $13(y + 2) = 13y + 26$

29. $6(2 \cdot 9) = (2 \cdot 9) \cdot 6$ **30.** $6(2 \cdot 9) = (6 \cdot 2) \cdot 9$

31. $5 \cdot \dfrac{1}{5} = 1$ **32.** $14 \cdot \dfrac{1}{14} = 1$

33. $7.1 + (-7.1) = 0$

34. $(-9) + 9 = 0$

35. $1 \cdot 14.2 = 14.2$

36. $(5 \cdot 3) \cdot -7 = 5(3 \cdot -7)$

37. $5.68 \cdot 0 = 0 \cdot 5.68 = 0$

38. $0 + 5.68 = 5.68$

39. $2 + (x + 6) = (2 + x) + 6$

40. $2(x + 6) = 2x + 12$

In each of the following equations, state the property illustrated and show that the statement is true for the value of $x = 4, y = -2,$ or $z = 3$ by substituting the corresponding value in the equation and evaluating.

41. $6 \cdot x = x \cdot 6$

42. $19 + z = z + 19$

43. $8 + (5 + y) = (8 + 5) + y$

44. $(2 \cdot 7) \cdot x = 2 \cdot (7x)$

45. $5(x + 18) = 5x + 90$

46. $(2z + 14) + 3 = 2z + (14 + 3)$

47. $(6 \cdot y) \cdot 9 = 6 \cdot (y \cdot 9)$

48. $11 \cdot x = x \cdot 11$

49. $z + (-34) = -34 + z$

50. $3(y + 15) = 3y + 45$

51. $2(3 + x) = 2(x + 3)$

52. $(y + 2)(y - 4) = (y - 4)(y + 2)$

53. $5 + (x - 15) = (x - 15) + 5$

54. $z + (4 + x) = (4 + x) + z$

55. $(3x) \cdot 5 = 3 \cdot (x \cdot 5)$

56. $(x + y) + z = x + (y + z)$

Hawkes Learning Systems: Introductory Algebra

Properties of Real Numbers

Chapter 1 Index of Key Ideas and Terms

Section 1.1 The Real number Line and Absolute Value

Types of Numbers
    Counting numbers (or natural numbers)   page 76
        $\mathbb{N} = \{\,1, 2, 3, 4, 5, 6, 7, 8, 9, 10, 11, \ldots\,\}$
    Whole numbers   page 76
        $\mathbb{W} = \{\,0, 1, 2, 3, 4, 5, 6, \ldots\,\}$
    Integers   page 78

        Integers: $\{\,\ldots, -3, -2, -1, 0, 1, 2, 3, \ldots\,\}$
        Positive integers: $\{\,1, 2, 3, 4, 5, \ldots\,\}$
        Negative integers: $\{\,\ldots, -4, -3, -2, -1\,\}$
        The integer 0 is neither positive nor negative.
    Rational numbers   page 79

        A **rational number** is a number that can be written
        in the form $\dfrac{a}{b}$, where a and b are integers and $b \neq 0$.
        OR,
        A **rational number** is a number that can be written in
        decimal form as a terminating decimal or as an infinite
        repeating decimal.
    Irrational numbers   page 80

        Irrational numbers are numbers that can be written as
        infinite nonrepeating decimals.
    Real numbers   page 80
        All rational and irrational numbers are classified as **real
        numbers**.
        Diagram of Types of Numbers   page 81

Inequality Symbols   pages 81 - 83
    Read from left to right:
        $<$ "is less than"
        $>$ "is greater than"
        $\leq$ "is less than or equal to"
        $\geq$ "is greater than or equal to"
        Note: Inequality symbols may be read from left to right or from
        right to left.

Absolute Value   pages 83 - 86
    The **absolute value** of a real number is its distance from 0.
    Symbolically,

        $|a| = a$ if a is a positive number or 0.

        $|a| = -a$ if a is a negative number.

Section 1.2 Addition with Integers

Addition with Integers   pages 90 - 93

    1. To add two integers with like signs, add their absolute
        values and use the common sign.

    2. To add two integers with unlike signs, subtract their
        absolute values (the smaller from the larger) and use the
        sign of the number with the larger absolute value.

Section 1.3 Subtraction with Integers

Additive Inverse   page 96

    The opposite of an integer is called its **additive inverse**.
    Symbolically, for any integer a, $a + (-a) = 0$.

Subtraction with Integers   pages 97 - 99

    To subtract an integer, add its opposite.
    Symbolically, for integers a and b, $a - b = a + (-b)$.

Section 1.4 Multiplication and Division with Real Numbers

Rules for Multiplication with Real Numbers   page 108

    If a and b are positve real numbers, then

    1. The product of two positive numbers is positive: $a \cdot b = ab$.

    2. The product of two negative numbers is positive: $(-a)(-b) = ab$.

    3. The product of a positive number and a negative number is
        negative: $a(-b) = -ab$.

    4. The product of 0 and any number is 0: $a \cdot 0 = 0$ and $(-a) \cdot 0 = 0$.

Division with Real Numbers   page 108

    For real numbers a, b, and x, (where $b \neq 0$),

$$\frac{a}{b} = x \text{ means that } a = b \cdot x.$$

    For real numbers a and b, (where $b \neq 0$),

$$\frac{a}{0} \text{ is undefined, but } \frac{0}{b} = 0.$$

Division by 0 is undefined   page 109

Continued on next page...

Section 1.4 Multiplication and Division with Real Numbers (continued)

Rules for Division with Real Numbers page 109

If a and b are positive numbers

1. The quotient of two positive numbers is positive: $\dfrac{a}{b} = +\dfrac{a}{b}$.

2. The quotient of two negative numbers is positive: $\dfrac{-a}{-b} = +\dfrac{a}{b}$.

3. The quotient of a positive number and a negative number

   is negative: $\dfrac{-a}{b} = -\dfrac{a}{b}$ and $\dfrac{a}{-b} = -\dfrac{a}{b}$.

Average pages 110 - 112

The average (or **mean**) of a set of numbers is the value found by adding the numbers in the set and then dividing the sum by the number of numbers in the set.

Section 1.5 Properties of Real Numbers

Properties of Addition and Multiplication Table page 120

For Addition	Name of Property	For Multiplication
$a + b = b + a$	Commutative Property	$ab = ba$
$(a + b) + c = a + (b + c)$	Associative Property	$a(bc) = (ab)c$
$a + 0 = 0 + a = a$	Identity	$a \cdot 1 = 1 \cdot a = a$
$a + (-a) = 0$	Inverse	$a \cdot \dfrac{1}{a} = 1 \; (for \; a \neq 0)$

Zero Factor Law

$$a \cdot 0 = 0 \cdot a = 0$$

Distributive Property of Multiplication over Addition:

$$a\,(b + c) = ab + ac$$

Hawkes Learning Systems: Introductory Algebra

For a review of the topics and problems from Chapter 1, look at the following lessons from *Hawkes Learning Systems: Introductory Algebra*

Name That Real Number
Introduction to Absolute Values
Addition with Integers
Subtraction with Integers
Multiplication and Division with Real Numbers
Properties of Real Numbers

Chapter 1 Review

Section 1.1 The Real Number Line and Absolute Value

In Exercises 1 – 4, graph each set of real numbers on a real number line.

1. $\{-3, 0, 3, 4\}$ **2.** $\left\{-4, -1, -\frac{1}{4}, 2.5\right\}$

3. All positive integers less than or equal to 5.

4. All integers greater than –2.

Find the value of each absolute value expression.

5. $|-6|$ **6.** $-|7.3|$ **7.** $-|-4.1|$ **8.** $|-10|$

Determine whether each statement in Exercises 9 – 12 is true or false. If a statement is false, rewrite it in a form that is a true statement. (There may be more than one way to correct a statement.)

9. $-14 \le 14$ **10.** $|-1.5| \ge 1.5$ **11.** $|-2.3| < 0$ **12.** $7 = |-7|$

List the numbers, then graph the numbers on a real number line that satisfy the equations in Exercises 13 – 16.

13. $|x| = 5$ **14.** $|x| = -5$ **15.** $12 = |x|$ **16.** $3.5 = |y|$

On a real number line, graph the **integers** that satisfy the conditions stated in Exercises 17 – 20.

17. $|x| \le 3$ **18.** $|x| > 3$ **19.** $y = |y|$ **20.** $|y| < 2$

Section 1.2 Addition with Integers

Find the sum in Exercises 21 – 36.

21. $6 + 9$ **22.** $15 + 7$ **23.** $8 + (-3)$ **24.** $(-9) + 6$

25. $18 + (-18)$ **26.** $75 + (-75)$ **27.** $-23 + (-10)$ **28.** $-7 + (-14)$

29. $-12 + (-1) + (-5)$ **30.** $-20 + (-3) + (-2)$ **31.** $-42 + 20 + 3$

32. $-65 + (-4) + 40$

33.
$$\begin{array}{r} -8 \\ 23 \\ \underline{-7} \end{array}$$

34.
$$\begin{array}{r} -25 \\ 13 \\ \underline{-18} \end{array}$$

35. $\quad$ −22
$\quad$ −10
$\quad$ $\underline{-13}$

36. $\quad$ −6
$\quad$ 20
$\quad$ $\underline{32}$

In Exercises 37 – 40, determine whether or not the given number is a solution to the given equation by substituting and then evaluating.

37. $x + 5 = 2$ given that $x = -3$ $\qquad$ **38.** $x + (-8) = -10$ given that $x = 2$

39. $-10 + |y| = 10$ given that $y = -20$ $\quad$ **40.** $|y| + (-4) = -1$ given that $y = 3$

Section 1.3 Subtraction with Integers

Find the value of each expression in Exercises 41 – 50.

41. $-4 - 3$ $\qquad$ **42.** $6 - 8$ $\qquad$ **43.** $-15 - 4$ $\qquad$ **44.** $10 - 18$

45. $-7 + (-2) - (-5)$ $\qquad$ **46.** $-10 + (-7) - 14$ $\qquad$ **47.** $25 - (17) - (-13)$

48. $98 - 16 - (32)$ $\qquad$ **49.** $-100 + 126 - 20$ $\qquad$ **50.** $(-5) - (-5) + 22$

Subtract as indicated in Exercises 51 – 54.

51. $\quad$ 32
$\quad$ $\underline{-(40)}$

52. $\quad$ −24
$\quad$ $\underline{-(-7)}$

53. $\quad$ −16
$\quad$ $\underline{-(-25)}$

54. $\quad$ −29
$\quad$ $\underline{-(-36)}$

55. From the sum of −13 and −10 subtract the sum of 16 and −30.

56. Find the difference between 17 and −20.

57. Find the change in temperature if the temperature drops from 30°F at noon to −3°F at midnight.

58. An airplane drops in altitude from 30,000 feet to 24,000 feet. What is the change in altitude in terms of signed numbers?

In Exercises 59 and 60, determine whether or not the given number is a solution to the given equation by substituting and then evaluating.

59. $x - 3 = -4$ given that $x = -1$ $\qquad$ **60.** $12 - x = -5$ given that $x = -17$

Section 1.4 Multiplication and Division with Real Numbers

Find the value of each of the following expressions.

61. $(-5) \cdot 7$ **62.** $8(-3.2)$ **63.** $(-9)(-10)$ **64.** $(-7)(-13)$

65. $-3 \cdot 17 \cdot 5$ **66.** $-4 \cdot 3 \cdot (-2)$ **67.** $8(-9) \cdot 0$ **68.** $-6(-7)(-13)$

69. $\dfrac{-10}{-2}$ **70.** $\dfrac{-51}{-3}$ **71.** $\dfrac{-16}{0}$ **72.** $\dfrac{0}{-16}$

73. $\dfrac{3.2}{-8}$ **74.** $\dfrac{-4.5}{9}$ **75.** $\dfrac{-70}{3.5}$ **76.** $\dfrac{100}{-2.5}$

77. Find the mean of the following set of integers: $30, 25, 70, 85,$ and 100.

78. In a statistics class, two students scored 82 and three students scored 88. What was the average score for these five students?

79. The heights of twenty women were recorded as shown in the following frequency distribution. Find the mean height (to the nearest tenth of an inch) for these women.

Height (in inches)	Frequency
60	2
61	1
63	6
66	4
70	4
71	3

80. The bar graph shows the number of hurricanes during the Atlantic Hurricane Seasons from 2001 – 2006. What was the average number of hurricanes per season in these years? (Round your answer to the nearest integer.)

Number of hurricanes during the Atlantic Hurricane Seasons from 2001-2006
http://www.aoml.noaa.gov

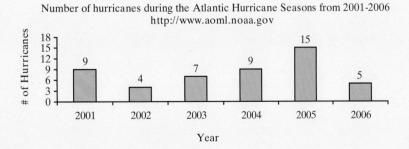

Section 1.5 Properties of Real Numbers

Name the property of real numbers illustrated.

81. $9 + 15 = 15 + 9$ **82.** $8 + (3 + 2) = (8 + 3) + 2$ **83.** $7 + 0 = 7$

84. $6 + (-6) = 0$ **85.** $2(3 \cdot 4) = (2 \cdot 3) \cdot 4$ **86.** $8(-7) = -7(8)$

87. $9.2 \cdot 1 = 9.2$ **88.** $19 \cdot \dfrac{1}{19} = 1$

89. $3(2 + x) = 6 + 3x$ **90.** $5(x - 7) = 5x - 35$

Calculator Problems

Use a TI-84 Plus calculator to find the value of each of the following expressions (accurate to hundredths).

91. $100.9 - (-32.8) + 104.73$ **92.** $-76.35 - (-14.7) + (-13)$

93. $7.2(-8.3)$ **94.** $93.75(17.2)(-25.1)$

95. $83.54 \div 16.3$ **96.** $-17.15 \div 0.15$

97. $\dfrac{98.5 - (-10)}{108.5}$ **98.** $\dfrac{-17.3 - (-20)}{-2.7}$

99. $\dfrac{|16.3| + |-16.3|}{-2}$ **100.** $\dfrac{-|-23.5| + |-20|}{-7}$

Chapter 1 Test

1. Given the set of numbers $\left\{-5, -\pi, -1, -\frac{1}{3}, 0, \frac{1}{2}, 3\frac{1}{4}, 7.121212...\right\}$, tell which numbers are
   a. Integers b. Rational numbers c. Irrational numbers d. Real numbers

2. a. True or False: All integers are rational numbers. (Explain your answer.)
   b. True or False: All rational numbers are integers. (Explain your answer.)

3. Fill in the blanks with the proper symbol: $<$, $>$, or $=$.
   a. -4 _____ -2 b. $-(-2)$ _____ 0 c. $|-9|$ _____ $|9|$

4. Graph the following set of numbers on a real number line:

   $$\left\{-2, -0.4, |-1|, 2\frac{1}{3}, 4.1\right\}$$

5. If x is a variable that represents any real number, state the definition of $|x|$.

6. What integers satisfy the equation $|y| = 7$?

7. List the **integers** that satisfy the following inequalities:
   a. $|x| < 3$ b. $|x| \geq 9$

Perform the indicated operations.

8. $15 - 45 + 17$

9. $13 - (-16) + (-6)$

10. $-29 - (-47)$

11. $(-7)(-16)$

12. $3(-42)$

13. $41(-5)(-3)(0)$

14. $\dfrac{-57}{19}$

15. $\dfrac{-140}{-20}$

16. $\dfrac{4.5}{-15}$

Name each property of real numbers illustrated.

17. $-3 + 0 = -3$

18. $6(x + 3) = 6x + 18$

19. $8 \cdot 2 = 2 \cdot 8$

20. $13 + (9 + 1) = (13 + 9) + 1$

21. $6 \cdot \dfrac{1}{6} = 1$

22. $-5 \cdot 0 = 0$

23. From the sum of -14 and -16, subtract the product of -6 and -5.

24. Find the product of the sum of -10 and 3 and the difference between -10 and 3.

25. In a 5-day week, the Dow Jones stock market average quote showed a gain of 32 points, a gain of 140 points, a loss of 30 points, a loss of 53 points, and a gain of 63 points. What was the net change in the stock market for the week?

26. The temperature at midnight in the mountain resort of Vail, Colorado was –10°F (10° below 0). At 6 a.m. the next morning the temperature was 5°F. What was the change in temperature?

27. Find the average of the following set of numbers: $\{-12, -15, -32, -31\}$.

28. Find the mean (to the nearest tenth) of the data shown in the following frequency distribution.

Bicycle Speeds (mph)	Frequency
19	4
20	5
22	7
25	3
28	1

Fractions with Variables and Algebraic Expressions

Did You Know?

Almost all of mathematics uses the language of sets to simplify notation and to help in understanding concepts. The theory of sets is one of the few branches of mathematics that was initially developed almost completely by one person, Georg Cantor.

Georg Cantor was born in St. Petersburg, Russia, but spent his adult life in Germany, first as a student at the University of Berlin and later as a professor at the University of Halle. Cantor's research and development of ideas concerning sets was met by ridicule and public attacks on his character and work. In 1885, Cantor suffered the first of a series of mental breakdowns, probably caused by the attacks on his work by other mathematicians, most notably a former teacher, Leopold Kronecker. It is suggested that Kronecker kept Cantor from becoming a professor at the prestigious University of Berlin.

We are not used to seeing intolerance toward new ideas among the scientific community, and as you study mathematics, you will probably have difficulty imagining how anyone could have felt hostile toward or threatened by Cantor's ideas. However, his idea of a set with an **infinite** number of elements was thought of as revolutionary by the mathematical establishment.

The problem is that infinite sets have the following curious property: a part may be numerically equal to a whole. For example,

$$N = \{\, 1, 2, 3, 4, \ 5, \ldots, n, \ldots \}$$
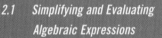
$$E = \{\, 2, 4, 6, 8, 10, \ldots, 2n, \ldots \}$$

Sets N and E are numerically equal because set E can be put in one-to-one correspondence (matched or counted) with set N. So there are as many even numbers as there are natural numbers! This is contrary to common sense and, in Cantor's time, contrary to usual mathematical assumptions.

2.1 Simplifying and Evaluating Algebraic Expressions

2.2 Multiplication and Division with Fractions

2.3 Addition and Subtraction with Fractions

2.4 Decimal Numbers and Fractions

2.5 Order of Operations with Negative Numbers and Fractions

2.6 Translating English Phrases and Algebraic Expressions

"The essence of mathematics lies in its freedom."

Georg Cantor (1845 - 1918)

$$\mathbf{A}$$ny Olympic athlete will tell you that practice is the key to success. Chapter 2 allows you to practice the basic arithmetic skills you already have with fractions, decimals, and percents. At the same time, you will find an integrated introduction to elementary algebra concepts using variables, formulas, and positive and negative numbers. Your success in this course may depend on how much you "practice" the basic skills represented in this chapter. Good luck and try your best. Keep a positive attitude.

2.1 Simplifying and Evaluating Algebraic Expressions

Objectives

After completing this section, you will be able to:

1. Simplify algebraic expressions by combining like terms.

2. Evaluate expressions for given values of the variables.

Simplifying Algebraic Expressions

A single number is called a **constant**. Any constant or variable or the indicated product and/or quotient of constants and variables is called a **term**. (**Note**: As discussed in Section R.1 of the Review Chapter, an expression with 2 as the exponent is read "squared" and an expression with 3 as the exponent is read "cubed." Thus $7x^2$ is read "seven x squared" and $-4y^3$ is read "negative four y cubed.") Examples of terms are

$$16, \ 3x, \ -5.2, \ 1.3xy, \ -5x^2, \ 14b^3, \ \text{and} \ -\frac{x}{y}.$$

The dot representing multiplication may be dropped when a variable is multiplied by a number or two or more variables are multiplied by each other. For example,

$$5 \cdot x = 5x \ \text{and} \ x \cdot y = xy.$$

The number written next to a variable is called the **numerical coefficient** (or the **coefficient**) of the variable. For example, in

$$5x^2, \ 5 \text{ is the coefficient of } x^2.$$

Similarly, in

$$1.3xy, \ 1.3 \text{ is the coefficient of } xy.$$

NOTES
If no number is written next to a variable, the coefficient is understood to be 1. If a negative sign $(-)$ is next to a variable, the coefficient is understood to be -1. For example,

$$x = 1 \cdot x, \ a^3 = 1 \cdot a^3, \ -x = -1 \cdot x, \text{and} \ -y^5 = -1 \cdot y^5.$$

Like Terms

Like terms (or *similar terms*) are terms that are constants or terms that contain the same variables raised to the same powers. (Note: The exponent on the variable is said to be the **degree** of the term.)

Like Terms

-6, 1.84, 145, $\dfrac{3}{4}$ are like terms because each term is a constant.

$-3a$, $15a$, $2.6a$, $\dfrac{2}{3}a$ are like terms because each term contains the same variable a, raised to the same power, 1. (Remember that $a = a^1$.) These terms are first-degree in a.

$5xy^2$ and $-3.2xy^2$ are like terms because each term contains the same two variables, x and y, with x first-degree in both terms and y second-degree in both terms.

Unlike Terms

$8x$ and $-9x^2$ are unlike terms (**not** like terms) because the variable x is not of the same power in both terms. $8x$ is first-degree in x and $-9x^2$ is second-degree in x.

Example 1: Like Terms

From the following list of terms, pick out the like terms:
$$-7,\ 2x,\ 4.1,\ -x,\ 3x^2y,\ 5x,\ -6x^2y,\ \text{and}\ 0$$

Solution: $-7, 4.1,$ and 0 are like terms. All are constants.
$2x, -x,$ and $5x$ are like terms.
$3x^2y$ and $-6x^2y$ are like terms.

To simplify expressions that contain like terms we want to **combine like terms**.

Combining Like Terms

To **combine like terms**, add (or subtract) the coefficients and keep the common variable expression.

The procedure for combining like terms uses the distributive property in the form

$$ba + ca = (b + c)a.$$

In particular, with numerical coefficients, we can combine like terms as follows:

$9x + 6x = (9 + 6)x$	By the distributive property
$= 15x$	Add the coefficients algebraically.
$3x^2 - 5x^2 = (3 - 5)x^2$	By the distributive property
$= -2x^2$	Add the coefficients algebraically.
$xy - 1.6xy = (1.0 - 1.6)xy$	By the distributive property (Note: $xy = 1xy = 1.0xy$)
$= -0.6xy$	Add the coefficients algebraically.

Example 2: Combine Like Terms

Combine like terms whenever possible.

a. $8x + 10x$

Solution: $8x + 10x = (8 + 10)x = 18x$ By the distributive property

b. $6.5y - 2.3y$

Solution: $6.5y - 2.3y = (6.5 - 2.3)y = 4.2y$

c. $4(n - 7) + 5(n + 1)$

Solution: $4(n - 7) + 5(n + 1) = 4n - 28 + 5n + 5$ Use the distributive property twice.

$= 4n + 5n - 28 + 5$ Use the commutative property of addition.

$= 9n - 23$ Combine like terms.

d. $2x^2 + 3a + x^2 - a$

Solution: $2x^2 + 3a + x^2 - a = 2x^2 + x^2 + 3a - a$ Use the commutative property of addition.

$= (2 + 1)x^2 + (3 - 1)a$ **Note**: $+x^2 = +1x^2$ and $-a = -1a$

$= 3x^2 + 2a$

e. $\dfrac{x+3x}{2}+5x$

  Solution: A fraction bar is a symbol of inclusion, like parentheses. So combine like terms in the numerator first.

$$\frac{x+3x}{2}+5x = \frac{4x}{2}+5x$$

$$= \frac{4}{2}\cdot x+5x$$

$$= 2x+5x$$

$$= 7x$$

Evaluating Algebraic Expressions

In most cases, if an expression is to be evaluated, like terms should be combined first and then the resulting expression evaluated by following the rules for order of operations.

Parentheses must be used around negative numbers when substituting.
Without parentheses, an evaluation can be dramatically changed and lead to wrong answers, particularly when even exponents are involved. We analyze with the exponent 2 as follows.

In general, except for $x = 0$,

  1. $-x^2$ **is negative** $[\,-6^2 = -1 \cdot 6^2 = -36\,]$

  2. $(-x)^2$ **is positive** $[\,(-6)^2 = (-6)(-6) = 36\,]$

  3. $-x^2 \neq (-x)^2$ $[\,-36 \neq 36\,]$

To Evaluate an Algebraic Expression

  1. Combine like terms, if possible.

  2. Substitute the values given for any variables.

  3. Follow the rules for order of operations. (See Section R.1)

Example 3: Evaluate Algebraic Expressions

a. Evaluate x^2 for $x = 3$ and for $x = -4$.

Solution: For $x = 3$, $x^2 = 3^2 = 9$.
For $x = -4$, $x^2 = (-4)^2 = 16$.

b. Evaluate $-x^2$ for $x = 3$ and for $x = -4$.

Solution: For $x = 3$, $-x^2 = -3^2 = -1 \cdot 9 = -9$.
For $x = -4$, $-x^2 = -(-4)^2 = -1(16) = -16$.

Simplify each expression below by combining like terms; then, evaluate the resulting expression using the given values for the variables.

c. Simplify and evaluate $2x + 5 + 7x$ for $x = -3$.

Solution: Simplify first:
$$2x + 5 + 7x = 2x + 7x + 5$$
$$= 9x + 5$$
Now evaluate:
$$9x + 5 = 9(-3) + 5$$
$$= -27 + 5$$
$$= -22$$

d. Simplify and evaluate $3ab - 4ab + 6a - a$ for $a = 2, b = -1$.

Solution: Simplify first:
$$3ab - 4ab + 6a - a = -ab + 5a$$
Now evaluate:
$$-ab + 5a = -1(2)(-1) + 5(2) \qquad \textbf{Note}: -ab = -1ab$$
$$= 2 + 10$$
$$= 12$$

e. Simplify and evaluate $\dfrac{5x + 3x}{4} + 2(x + 1)$ for $x = 5$.

Solution: Simplify first:

$$\frac{5x + 3x}{4} + 2(x + 1) = \frac{8x}{4} + 2x + 2$$
$$= 2x + 2x + 2$$
$$= 4x + 2$$

Now evaluate:

$$4x + 2 = 4 \cdot 5 + 2$$
$$= 20 + 2$$
$$= 22$$

Practice Problems

Simplify the following expressions by combining like terms.

1. $-2x - 5x$ **2.** $12y + 6 - y + 10$

3. $5(x - 1) + 4x$ **4.** $2b^2 - a + b^2 + a$

Simplify the expression, then evaluate the resulting expression if $x = 3$ *and* $y = -2$.

5. $2(x + 3y) + 4(x - y)$

2.1 Exercises

In Exercises 1 – 6, pick out the like terms in each list of terms.

1. $-5, \; \dfrac{1}{6}, \; 7x, \; 8, \; 9x, \; 3y$ **2.** $-2x^2, \; -13x^3, \; 5x^2, \; 14x^2, \; 10x^3$

3. $5xy, \; -x^2, \; -6xy, \; 3x^2y, \; 5x^2y, \; 2x^2$

4. $3ab^2, \; -ab^2, \; 8ab, \; 9a^2b, \; -10a^2b, \; ab, \; 12a^2$

5. $24, \; 8.3, \; 1.5xyz, \; -1.4xyz, \; -6, \; xyz, \; 5xy^2z, \; 2xyz^2$

6. $-35y, \; 1.62, \; -y^2, \; -y, \; 3y^2, \; \dfrac{1}{2}, \; 75y, \; 2.5y^2$

Find the value of each numerical expression in Exercises 7 – 10.

7. $(-8)^2$ **8.** -8^2 **9.** -11^2 **10.** $(-6)^2$

Simplify by combining like terms in each of the expressions in Exercises 11 – 42.

11. $8x + 7x$ **12.** $3y + 8y$ **13.** $5x - 2x$ **14.** $7x + (-3x)$

15. $-n - n$ **16.** $-x - x$ **17.** $6y^2 - y^2$ **18.** $16z^2 - 5z^2$

19. $23x^2 - 11x^2$ **20.** $18x^3 + 7x^3$ **21.** $4x + 2 + 3x$ **22.** $3x - 1 + x$

Answers to Practice Problems: **1.** $-7x$ **2.** $11y + 16$ **3.** $9x - 5$ **4.** $3b^2$ **5.** $6x + 2y; \; 14$

23. $2x - 3y - x$

24. $x + y + x - 2y$

25. $2x^2 + 5y + 6x^2 - 2y$

26. $4a + 2a - 3b - a$

27. $3(n + 1) - n$

28. $2(n - 4) + n + 1$

29. $5(a - b) + 2a - 3b$

30. $4a - 3b + 2(a + 2b)$

31. $3(2x + y) + 2(x - y)$

32. $4(x + 5y) + 3(2x - 7y)$

33. $2x + 3x^2 - 3x - x^2$

34. $2y^2 + 4y - y^2 - 3y$

35. $2(n^2 - 3n) + 4(-n^2 + 2n)$ **36.** $3n^2 + 2n - 5 - n^2 + n - 4$

37. $3x^2 + 4xy - 5xy + y^2$ **38.** $2x^2 - 5xy + 11xy + 3$ **39.** $\dfrac{x + 5x}{6} + x$

40. $2y - \dfrac{2y + 3y}{5}$

41. $y - \dfrac{2y + 4y}{3}$

42. $z - \dfrac{3z + 5z}{4}$

In Exercises 43 – 62, first (a) simplify the expressions and then (b) evaluate the simplified expression by using $x = 4$, $y = 3$, $a = -2$, and $b = -1$.

43. $5x + 4 - 2x$

44. $7x - 17 - x$

45. $x - 10 - 3x + 2$

46. $6a + 5a - a + 13$

47. $3(y - 1) + 2(y + 2)$

48. $4(y + 3) + 5(y - 2)$

49. $-5(x + y) + 2(x - y)$

50. $-2(a + b) + 3(b - a)$

51. $8.3x^2 - 5.7x^2 + x^2$

52. $3a^2 - a^2 + 4a - 5$

53. $5ab + b^2 - 2ab + b^3$

54. $5a + ab^2 - 2ab^2 + 3a$

55. $2.4(x + 1) + 1.3(x - 1)$

56. $1.3(y + 2) - 2.6(8 - y)$

57. $\dfrac{3a + 5a}{-2} + 12a$

58. $8a + \dfrac{5a + 4a}{9}$

59. $\dfrac{-4b - 2b}{-3} + \dfrac{2b + 5b}{7}$

60. $\dfrac{5b + 3b}{4} + \dfrac{-4b - b}{-5}$

61. $2x + 3\left[x - 2(9 + x)\right]$

62. $5x - 2\left[x + 5(x - 3)\right]$

Hawkes Learning Systems: Introductory Algebra

Variables and Algebraic Expressions
Simplifying Expressions
Simplifying Expressions with Parentheses
Evaluating Algebraic Expressions

2.2   Multiplication and Division with Fractions

Objectives

After completing this section, you will be able to:

1. *Reduce fractions to lowest terms.*

2. *Write fractions as equivalent fractions with specified denominators.*

3. *Multiply and divide fractions.*

Rational Numbers Defined

Sections R.3 and R.4 in the Review Chapter reviewed fractions and operating with fractions. (You may want to briefly review these sections.) In Chapter 2 we will continue to develop your skills and knowledge of fractions by including negative numbers and variables. As you will see, the methods and rules are the same but the ideas become a little more abstract and involve a deeper understanding. For example, in the Review Chapter we discussed fractions of the form $\dfrac{a}{b}$ where a and b were restricted to **whole numbers** with $b \neq 0$. Now we allow a and b to be **integers** but still with $b \neq 0$.

Rational Number

> A **rational number** *is a number that can be written in the form* $\dfrac{a}{b}$ *where a and b are* **integers** *and* $b \neq 0$.

NOTES   There are fractions that cannot be written with the numerator and denominator as integers. For example, $\dfrac{\pi}{3}$ is a fraction (very useful in trigonometry) but also an irrational number.

For now, when we deal with the fractional form $\dfrac{a}{b}$ we will assume the fraction represents a rational number and **use the terms fraction and rational number interchangeably**. The decimal forms of rational numbers will be discussed in Section 2.4.

Examples of rational numbers (fractions):

$$\frac{1}{2}, \ \frac{3}{10}, \ \frac{-11}{7}, \ -\frac{13}{2}, \ -10, \ \text{and} \ 15.$$

Note that every integer is also a rational number because integers can be written in fraction form with a denominator of 1. Thus,

$$0 = \frac{0}{1}, \; 1 = \frac{1}{1}, \; 2 = \frac{2}{1}, \; 3 = \frac{3}{1}, \text{ and so on.}$$

Also,

$$-1 = \frac{-1}{1}, \; -2 = \frac{-2}{1}, \; -3 = \frac{-3}{1}, \; -4 = \frac{-4}{1}, \text{ and so on.}$$

In general, fractions can be used to indicate:
1. Equal parts of a whole, or
2. Division.

Example 1: Fractions

a. $\frac{1}{2}$ can mean 1 of 2 equal parts.

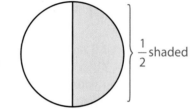

$\frac{1}{2}$ shaded

If you read $\frac{1}{2}$ of a book, then you can view this as having read one of two equal parts of the book. If the book has 50 pages, then you read 25 pages, since, as we will see, $\frac{25}{50} = \frac{1}{2}$.

b. $\frac{-45}{9}$ can mean to divide: $(-45) \div 9$.

Before actually operating with fractions, we need to understand the use and placement of negative signs in fractions. Consider the following three results involving fractions, negative signs, and the meaning of division:

$$-\frac{16}{8} = -2, \; \frac{-16}{8} = -2, \text{ and } \frac{16}{-8} = -2.$$

Thus, the three different placements of a negative sign give the same results, as noted in the general statement or rule on the next page.

Rules for the Placement of Negative Signs in Fractions

If a and b are real numbers, and $b \neq 0$, then

$$-\frac{a}{b} = \frac{-a}{b} = \frac{a}{-b}.$$

For example:

$$-\frac{3}{4b} = \frac{-3}{4b} = \frac{3}{-4b}.$$

Example 2: Negative Signs in Fractions

a. $\quad -\dfrac{12}{4} = \dfrac{-12}{4} = \dfrac{12}{-4} = -3$

b. $\quad -\dfrac{1}{5} = \dfrac{-1}{5} = \dfrac{1}{-5}$

Multiplication

We know from Section R.3 that multiplication with fractions is accomplished by multiplying the numerators and the denominators.

Multiplication with Fractions

To **multiply** two fractions, multiply the numerators and multiply the denominators.

$$\frac{a}{b} \cdot \frac{c}{d} = \frac{a \cdot c}{b \cdot d} \quad \text{where } b, d \neq 0.$$

For example:

$$\frac{2}{3} \cdot \frac{7}{5} = \frac{2 \cdot 7}{3 \cdot 5} = \frac{14}{15}.$$

Remember that the number 1 is called the **multiplicative identity** since the product of 1 with any number is that number. That is,

$$\frac{a}{b} \cdot 1 = \frac{a}{b}.$$

Thus if $k \neq 0$, we have

$$\frac{a}{b} = \frac{a}{b} \cdot 1 = \frac{a}{b} \cdot \frac{k}{k} = \frac{a \cdot k}{b \cdot k}$$

This relationship is called the **Fundamental Principle of Fractions**.

The Fundamental Principle of Fractions

$$\frac{a}{b} = \frac{a \cdot k}{b \cdot k}, \ \text{where} \ b, k \neq 0.$$

We can use the Fundamental Principle to find equal fractions with larger denominators (**build to higher terms**) or find equal fractions with smaller denominators (**reduce to lower terms**). Equal fractions are said to be **equivalent**.

To Build a Fraction to Higher Terms

To build a fraction to higher terms, use the Fundamental Principle and multiply both the numerator and denominator by the same nonzero number.

To Reduce a Fraction

To reduce a fraction, factor both the numerator and denominator and use the Fundamental Principle to "divide out" any common factors. If the numerator and the denominator have no common prime factors, the fraction has been reduced to *lowest terms*.

Example 3: Fundamental Principal of Fractions

a. Build $-\dfrac{5}{8}$ to higher terms with denominator of $16a$.

Solution:

Because we know that $8 \cdot 2a = 16a,$ use $k = 2a$ and

write $-\dfrac{5}{8} = -\dfrac{5}{8} \cdot \dfrac{2a}{2a} = -\dfrac{10a}{16a}.$

b. Write $-\dfrac{3}{7}$ in the form of an equivalent fraction with denominator 28.

Solution:

Because we know $7 \cdot 4 = 28,$ use $k = 4$ and

write $\dfrac{-3}{7} = \dfrac{-3}{7} \cdot \dfrac{4}{4} = \dfrac{-12}{28}.$

c. Reduce $\dfrac{-12}{20}$ to lowest terms by using prime factorization.

(**Note:** Section R.1 of the Review Chapter provides a thorough discussion of prime numbers.)

Solution: $\dfrac{-12}{20} = \dfrac{-1 \cdot \cancel{2} \cdot \cancel{2} \cdot 3}{\cancel{2} \cdot \cancel{2} \cdot 5} = \dfrac{-3}{5} = -\dfrac{3}{5}$ Note that the negative sign can be with the same numerator or in front of the fraction. We seldom use the negative sign in the denominator.

d. Find the product $\dfrac{15ac}{28b} \cdot \dfrac{4bc}{9a}$ in lowest terms. (Do not find the product directly. Factor and reduce as you multiply. Treat the variables as you would numbers.)

Solution:

$$\frac{15ac}{28b} \cdot \frac{4bc}{9a} = \frac{\cancel{3} \cdot 5 \cdot \cancel{a} \cdot c \cdot \cancel{2} \cdot \cancel{2} \cdot \cancel{b} \cdot c}{\cancel{2} \cdot \cancel{2} \cdot 7 \cdot \cancel{b} \cdot \cancel{3} \cdot 3 \cdot \cancel{a}} = \frac{5c^2}{21}$$

e. Find the product $\dfrac{4a}{12b} \cdot \dfrac{3b}{7a}$ in lowest terms. (Note that the number 1 is implied to be a factor even if it is not written.)

Solution:

$$\frac{4a}{12b} \cdot \frac{3b}{7a} = \frac{\cancel{4} \cdot \cancel{a} \cdot \cancel{3} \cdot \cancel{b} \cdot 1}{\cancel{4} \cdot \cancel{3} \cdot \cancel{b} \cdot 7 \cdot \cancel{a}} = \frac{1}{7}$$

Remember that if all factors are divided out (in the numerator or denominator), you must use 1 even if it is not written.

Division

We know from Section R.3 in the Review Chapter that division with fractions is accomplished by multiplication. The related concepts are very important and are reviewed again.

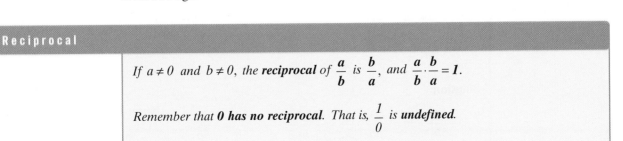

Reciprocal

If $a \neq 0$ and $b \neq 0$, the **reciprocal** of $\dfrac{a}{b}$ is $\dfrac{b}{a}$, and $\dfrac{a}{b} \cdot \dfrac{b}{a} = 1$.

Remember that **0 has no reciprocal**. That is, $\dfrac{1}{0}$ is **undefined**.

To divide by a number, we multiply by its reciprocal. For example,

$$\frac{2}{3} \div \frac{5}{6} = \frac{2}{3} \cdot \frac{6}{5}.$$

This gives the result (reduced) as

$$\frac{2}{3} \div \frac{5}{6} = \frac{2}{3} \cdot \frac{6}{5} = \frac{2 \cdot \cancel{3} \cdot 2}{\cancel{3} \cdot 5} = \frac{4}{5}.$$

Division

> *To divide by a nonzero fraction, multiply by its reciprocal:*
>
> $$\frac{a}{b} \div \frac{c}{d} = \frac{a}{b} \cdot \frac{d}{c}, \text{ where } b, c, d \neq 0$$

Example 4: Dividing Fractions with Negative Numbers

$$\left(-\frac{3}{4}\right) \div \left(-\frac{2}{5}\right)$$

Solution:

$$\left(-\frac{3}{4}\right) \div \left(-\frac{2}{5}\right) = \left(-\frac{3}{4}\right) \cdot \left(-\frac{5}{2}\right) = \frac{15}{8}$$

Note that, just as with integers, the product of two negative fractions is positive. In algebra, $\frac{15}{8}$, an **improper fraction** (a fraction with the numerator greater than the denominator), is preferred to the mixed number $1\frac{7}{8}$.

Improper fractions are perfectly acceptable as long as they are reduced, meaning the numerator and denominator have no common prime factors. We will discuss mixed numbers in more detail in the next section.

Example 5: Dividing Fractions with Variables

$$\frac{-21a}{5b} \div 3a$$

Solution:

$$\frac{-21a}{5b} \div 3a = \frac{-21a}{5b} \cdot \frac{1}{3a} = \frac{-1 \cdot \cancel{3} \cdot 7 \cdot \cancel{a}}{5 \cdot b \cdot \cancel{3} \cdot \cancel{a}} = \frac{-7}{5b} = -\frac{7}{5b}$$

Note: The reciprocal of $3a$ is $\frac{1}{3a}$, because $3a = \frac{3a}{1}$.

Example 6: Multiplication and Division of Fractions

Multiply the quotient of $\frac{3}{4}$ and $\frac{15}{8}$ by $\frac{7}{2}$.

Solution:

First find the quotient: $\frac{3}{4} \div \frac{15}{8} = \frac{\overset{1}{\cancel{3}}}{\underset{1}{\cancel{4}}} \cdot \frac{\overset{2}{\cancel{8}}}{\underset{5}{\cancel{15}}} = \frac{2}{5}$.

Now multiply: $\frac{\cancel{2}}{5} \cdot \frac{7}{\cancel{2}} = \frac{7}{5}$.

The answer is $\frac{7}{5}$.

Example 7: Multiplication and Division of Fractions

If $\frac{3}{8}$ of a number is 99, what is the number?

Solution:

Because we know the product, 99, the missing number can be found by dividing the product by $\frac{3}{8}$.

$$99 \div \frac{3}{8} = \frac{\overset{33}{\cancel{99}}}{1} \cdot \frac{8}{\cancel{3}} = 264$$

The other number is 264. That is $\frac{3}{8}$ **of** 264 is 99. $\left[\text{or } \frac{3}{8} \cdot 264 = 99 \right]$

(**Note**: As we will see in Chapter 3, this problem can be approached algebraically with a variable.)

NOTES

Important note about expressions involving fractions in algebra.

An expression with a fraction as the coefficient can be written in two different forms. For example, $\frac{9}{8}x$ and $\frac{9x}{8}$ have the same meaning. That is, for all values of x, we have $\frac{9}{8}x = \frac{9}{8} \cdot \frac{x}{1} = \frac{9x}{8}$.

2.2 Exercises

In Exercises 1 – 6, supply the missing numbers (or terms) so that each fraction will be raised to higher terms as indicated.

1. $\dfrac{5}{6} = \dfrac{5}{6} \cdot \dfrac{?}{?} = \dfrac{?}{48}$

2. $\dfrac{3}{13} = \dfrac{3}{13} \cdot \dfrac{?}{?} = \dfrac{?}{52}$

3. $\dfrac{0}{9} = \dfrac{0}{9} \cdot \dfrac{?}{?} = \dfrac{?}{63b}$

4. $\dfrac{-7}{24} = \dfrac{-7}{24} \cdot \dfrac{?}{?} = \dfrac{?}{72x}$

5. $\dfrac{-3}{5} = \dfrac{-3}{5} \cdot \dfrac{?}{?} = \dfrac{?}{35y}$

6. $\dfrac{8}{3} = \dfrac{8}{3} \cdot \dfrac{?}{?} = \dfrac{?}{36x}$

Reduce each fraction to lowest terms in Exercises 7 – 18.

7. $\dfrac{18}{45}$

8. $\dfrac{35}{63}$

9. $\dfrac{150xy}{350y}$

10. $\dfrac{60ab}{75a}$

11. $\dfrac{-12a}{-100ab}$

12. $\dfrac{-6x}{-51xy}$

13. $\dfrac{-30y}{45y}$

14. $\dfrac{-66ab}{88ab}$

15. $\dfrac{-28}{56x}$

16. $\dfrac{34}{-51y}$

17. $\dfrac{12xyz}{-35xy}$

18. $\dfrac{-8abc}{15ab}$

In Exercises 19 – 22, find the fraction **of** a number by multiplying.

19. Find $\dfrac{1}{2}$ of $\dfrac{3}{4}$. **20.** Find $\dfrac{2}{7}$ of $\dfrac{5}{7}$. **21.** Find $\dfrac{7}{8}$ of 40. **22.** Find $\dfrac{1}{3}$ of $\dfrac{1}{3}$.

In Exercises 23 – 52, multiply or divide as indicated and reduce each answer to lowest terms.

23. $\dfrac{-3}{8} \cdot \dfrac{4}{9}$

24. $\dfrac{4}{5} \cdot \dfrac{-3}{14}$

25. $\dfrac{9}{10x} \div \dfrac{10}{9x}$

26. $\dfrac{4}{5a} \div \dfrac{1}{5a}$

27. $\dfrac{-16y}{7} \div \dfrac{16}{y}$

28. $\dfrac{-20y}{9} \div \dfrac{10}{y}$

29. $\dfrac{-15x}{2} \cdot \dfrac{6x}{25}$

30. $\dfrac{-30a}{10} \cdot \dfrac{4a}{8}$

31. $\dfrac{26b}{51a} \cdot \dfrac{17b}{13a}$

32. $\dfrac{35x}{15y} \cdot \dfrac{14x}{8y}$

33. $\dfrac{-15}{4} \cdot \dfrac{5}{6} \cdot \dfrac{16}{9}$

34. $\dfrac{-20}{3} \cdot \dfrac{7}{6} \cdot \dfrac{9}{10}$

35. $\dfrac{9a}{15} \cdot \dfrac{10}{3b} \cdot \dfrac{1}{5a}$

36. $\dfrac{20}{3x} \cdot \dfrac{18}{5x} \cdot \dfrac{14x}{2}$

37. $\dfrac{-35y}{12} \cdot \dfrac{-6}{14y} \cdot \dfrac{4y}{3}$

38. $\dfrac{-4ab}{18a} \cdot \dfrac{-24a}{13b} \cdot \dfrac{39}{a}$

39. $0 \div \dfrac{37y}{21x}$

40. $0 \div \dfrac{5a}{3b}$

41. $\dfrac{6xy}{15x} \div 0$

42. $\dfrac{28x}{7xy} \div 0$

43. $\dfrac{92}{7a} \div \dfrac{46a}{77}$

44. $\dfrac{57x}{32} \div \dfrac{19}{16x}$

45. $\dfrac{21}{30} \div (-7)$

46. $\dfrac{18}{20} \div (-9)$

47. $\dfrac{45a}{30b} \cdot \dfrac{12ab}{18b} \div \left(\dfrac{10ac}{9c} \right)$

48. $\dfrac{15m}{3m} \cdot \dfrac{2mn}{28n} \div \left(\dfrac{14mn}{10} \right)$

49. $\dfrac{-35m}{21n} \cdot \dfrac{27n}{40mn} \cdot \dfrac{-9m}{40mn}$

50. $\dfrac{-72a}{4b} \cdot \dfrac{-16b}{36ab} \cdot \dfrac{8c}{20ac}$

51. $\dfrac{-9x}{40y} \cdot \dfrac{35x}{15y} \cdot \dfrac{5}{10} \cdot \dfrac{8xy}{30xyz}$

52. $\dfrac{-3x}{2xy} \cdot \dfrac{7}{44y} \cdot \dfrac{17}{y} \cdot \dfrac{22xy}{34x}$

53. Multiply the quotient of $\dfrac{19}{2}$ and $\dfrac{3}{4}$ by $\dfrac{13}{2}$.

54. Divide the product of $\dfrac{5}{8}$ and $\dfrac{9}{10}$ by the product of $\dfrac{9}{10}$ and $\dfrac{4}{5}$.

55. Find the product of $\dfrac{11}{4}$ with the quotient of $\dfrac{8}{5}$ and $\dfrac{11}{6}$.

56. Find the quotient if the product of $\dfrac{5}{8}$ and $\dfrac{36}{15}$ is divided by $\dfrac{4}{3}$.

57. The product of $\dfrac{3}{5}$ with another number is $-\dfrac{5}{8}$. What is the other number?

58. The product of $-\dfrac{1}{2}$ with another number is $-\dfrac{3}{7}$. What is the other number?

59. A glass of milk is 6 inches tall. If the glass is $\dfrac{1}{3}$ full of milk, what is the height of the milk in the glass?

60. In one elementary school $\dfrac{9}{10}$ of the students are over 4 feet tall. If the school had an enrollment of 500 students, how many are over 4 feet tall? How many are less than or equal to 4 feet tall?

61. A study showed that $\dfrac{7}{10}$ of the human body weight is water. If a person weighs 140 pounds, how many pounds is water?

62. A bus is carrying 60 passengers. This is $\dfrac{5}{6}$ of the capacity of the bus.

    a. Is the capacity of the bus more or less than 60?

    b. If you were to multiply 60 by $\dfrac{5}{6}$, would the product be more or less than 60?

    c. What is the capacity of the bus?

63. The continent of Africa covers approximately 11,707,000 square miles. This is $\frac{1}{5}$ of the land area in the world. What is the approximate total land area in the world?

64. If you have $20 and you spend $7 on a glass of milk and a piece of pie, what fraction of your money did you spend? What fraction of your money do you still have?

65. In a class of 40 students, 5 received a grade of A. What fraction of the class did not receive an A?

66. Suppose that a ball is dropped from a height of 30 ft. and that each bounce reaches to $\frac{3}{8}$ of the previous height. How high will the ball bounce on the second bounce?

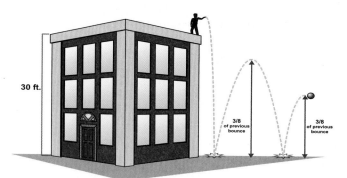

67. An airplane is carrying 180 passengers. This is $\frac{9}{10}$ of its capacity.

   a. Is the capacity more or less than 180?
   b. What is the capacity of the airplane?

68. The product of $\frac{5}{6}$ with another number is $\frac{1}{4}$. What is the other number?

Writing and Thinking About Mathematics

   69. Explain, in your own words, why 0 does not have a reciprocal.

   70. In your own words, explain why $\frac{3}{4}x$ and $\frac{3x}{4}$ have the same meaning.

Hawkes Learning Systems: Introductory Algebra

Multiplication and Division with Fractions

2.3	# Addition and Subtraction with Fractions

After completing this section, you will be able to:

1. Add and subtract fractions with like denominators.

2. Find the least common multiple (LCM) of two or more numbers.

3. Add and subtract fractions with unlike denominators.

From your review of Section R.4 in the Review Chapter you know that finding the sum of two or more fractions **with the same denominator** is similar to adding whole numbers of some particular item. For example, the sum of 5 *apples* and 6 *apples* is 11 *apples*. Similarly, the sum of 5 *seventeenths* and 6 *seventeenths* is 11 *seventeenths*:

$$\frac{5}{17} + \frac{6}{17} = \frac{11}{17}.$$

Algebraically, we can write

$$\frac{a}{b} + \frac{c}{b} = a \cdot \frac{1}{b} + c \cdot \frac{1}{b}$$

$$= (a + c) \cdot \frac{1}{b} \qquad \text{Use the distributive property.}$$

$$= \frac{a + c}{b}.$$

Adding Fractions with Like Denominators

*To **add** two (or more) fractions with like denominators, add the numerators and use the common denominator.*

$$\frac{a}{b} + \frac{c}{b} = \frac{a + c}{b} \qquad \textit{(where b is a nonzero number or algebraic expression)}$$

Example 1: Adding Fractions with Like Denominators

a. $\dfrac{3}{8} + \dfrac{4}{8} = \dfrac{3+4}{8} = \dfrac{7}{8}$ Here the common denominator is the number 8.

b. $\dfrac{3}{8x} + \dfrac{4}{8x} = \dfrac{3+4}{8x} = \dfrac{7}{8x}$ Here the common denominator is the algebraic term $8x$.

c. $\dfrac{2}{15} + \dfrac{3}{15} + \dfrac{1}{15} + \dfrac{6}{15} = \dfrac{2+3+1+6}{15} = \dfrac{12}{15} = \dfrac{\cancel{3} \cdot 4}{\cancel{3} \cdot 5} = \dfrac{4}{5}$ Reduce when possible.

d. $\dfrac{4x}{5} + \dfrac{3}{5} = \dfrac{4x+3}{5}$

To find the sum of fractions with different denominators, we need the concepts of **multiples** and **least common multiple (LCM)**. **Multiples** of a number are the products of that number with the counting numbers. For example, the multiples of 6 and 8 are listed:

Counting numbers: **1,   2,   3,   4,   5,   6,   7,   8,   9,   10,   11,   12,** ...
Multiples of 6:   6,   12,   18,   ⟨24⟩   30,   36,   42,   ⟨48⟩   54,   60,   66,   ⟨72⟩ ...
Multiples of 8:   8,   16,   ⟨24⟩   32,   40,   ⟨48⟩   56,   64,   ⟨72⟩   80,   88,   96, ...

We see that the common multiples of 6 and 8 are 24, 48, 72, 96, 120, and so on.

The smallest of these is 24, so 24 is the least common multiple (LCM) of 6 and 8. This LCM can be found by using the following procedure, restated here from Section R.2.

To Find the LCM

1. *Find the prime factorization of each number.*

2. *List the prime factors that appear in any one of the prime factorizations.*

3. *Find the product of these primes using each prime the greatest number of times it appears in any one of the prime factorizations.*

Note: In terms of exponents, the LCM is the product of the highest power of each of the prime factors.

Example 2: LCM with Numbers

Find the LCM of the numbers 27, 15, and 60.

Solution:

$$\left.\begin{array}{l} 27 = 9 \cdot 3 = 3 \cdot 3 \cdot 3 = 3^3 \\ 15 = 3 \cdot 5 \\ 60 = 10 \cdot 6 = 2 \cdot 5 \cdot 2 \cdot 3 = 2^2 \cdot 3 \cdot 5 \end{array}\right\} \quad \text{LCM} = 2^2 \cdot 3^3 \cdot 5 = 540$$

Example 3: LCM with Variables

Find the LCM for $4x$, $x^2 y$, and $6x^2$.

Solution:

We treat each variable in the same manner as a prime number;

$$\left.\begin{array}{l} 4x = 2^2 \cdot x \\ x^2 y = x^2 \cdot y \\ 6x^2 = 2 \cdot 3 \cdot x^2 \end{array}\right\} \text{LCM} = 2^2 \cdot 3 \cdot x^2 \cdot y = 12x^2 y$$

(See Section R.2 of the Review Chapter for a complete discussion of prime numbers and LCM.)

Adding Fractions with Unlike Denominators

1. *Find the LCM of the denominators.*

   *(This is called the **least common denominator** (**LCD**) and may involve variables.)*

2. *Change each fraction to an equivalent fraction with the LCM as the denominator.*

3. *Add the new fractions. Reduce if possible.*

Example 4: Adding Fractions with Unlike Denominators

a. $\dfrac{1}{4}+\dfrac{3}{8}+\dfrac{3}{10}$

Solution: $\left.\begin{array}{l}4=2^2 \\ 8=2^3 \\ 10=2\cdot5\end{array}\right\}$ $\text{LCM}=\text{LCD}=2^3\cdot5=40$

To get the common denominator in each fraction, we see
$$40=4\cdot10=8\cdot5=10\cdot4.$$

$$\dfrac{1}{4}+\dfrac{3}{8}+\dfrac{3}{10}=\left(\dfrac{1}{4}\cdot\dfrac{10}{10}\right)+\left(\dfrac{3}{8}\cdot\dfrac{5}{5}\right)+\left(\dfrac{3}{10}\cdot\dfrac{4}{4}\right)$$

Multiply each fraction by 1 in the form $\dfrac{k}{k}$.

$$=\dfrac{10}{40}+\dfrac{15}{40}+\dfrac{12}{40}$$

Each fraction has the same denominator.

$$=\dfrac{37}{40}$$

Add the fractions.

b. $\dfrac{5}{21a}+\dfrac{5}{28a}$

Solution: $\left.\begin{array}{l}21a=3\cdot7\cdot a \\ 28a=2^2\cdot7\cdot a\end{array}\right\}$ $\text{LCM}=\text{LCD}=2^2\cdot3\cdot7\cdot a=84a=21a\cdot4=28a\cdot3$

In each fraction, the numerator and denominator are multiplied by the same number to get 84a as the denominator.

$$\dfrac{5}{21a}+\dfrac{5}{28a}=\left(\dfrac{5}{21a}\cdot\dfrac{4}{4}\right)+\left(\dfrac{5}{28a}\cdot\dfrac{3}{3}\right)$$

$$=\dfrac{20}{84a}+\dfrac{15}{84a}=\dfrac{35}{84a}$$

$$=\dfrac{\cancel{7}\cdot5}{\cancel{7}\cdot12a}=\dfrac{5}{12a}$$

c. $\dfrac{4a}{6} + \dfrac{5}{4}$

Solution: $\left.\begin{array}{l} 6 = 2 \cdot 3 \\[4pt] 4 = 2 \cdot 2 \end{array}\right\}$ $\text{LCM} = 2 \cdot 2 \cdot 3 = 12 = 6 \cdot 2 = 4 \cdot 3$

Thus,

$$\dfrac{4a}{6} + \dfrac{5}{4} = \left(\dfrac{4a}{6} \cdot \dfrac{2}{2}\right) + \left(\dfrac{5}{4} \cdot \dfrac{3}{3}\right)$$

$$= \dfrac{8a}{12} + \dfrac{15}{12}$$

$$= \dfrac{8a + 15}{12}$$

Subtracting Fractions with Like Denominators

To subtract two fractions with like denominators, subtract the numerators and use the common denominator:

$$\dfrac{a}{b} - \dfrac{c}{b} = \dfrac{a - c}{b}$$ *(where b is a nonzero number or algebraic expression)*

Just as with addition, if the two fractions do not have the same denominator, find equivalent fractions with the LCD and subtract the new fractions.

Example 5: Subtracting Fractions

a. $\dfrac{1}{8a} - \dfrac{5}{8a}$

Solution: $\dfrac{1}{8a} - \dfrac{5}{8a} = \dfrac{1 - 5}{8a} = \dfrac{-4}{8a} = \dfrac{-1 \cdot \cancel{4}}{\cancel{4} \cdot 2 \cdot a} = \dfrac{-1}{2a}$ $\qquad \left(\text{or } -\dfrac{1}{2a}\right)$

b. $\dfrac{1}{45} - \dfrac{1}{72}$

Solution: $\left.\begin{array}{l} 45 = 3^2 \cdot 5 \\[4pt] 72 = 2^3 \cdot 3^2 \end{array}\right\}$ $\text{LCM} = \text{LCD} = 2^3 \cdot 3^2 \cdot 5 = 360 = 45 \cdot 8 = 72 \cdot 5$

$$\dfrac{1}{45} - \dfrac{1}{72} = \left(\dfrac{1}{45} \cdot \dfrac{8}{8}\right) - \left(\dfrac{1}{72} \cdot \dfrac{5}{5}\right) = \dfrac{8}{360} - \dfrac{5}{360} = \dfrac{3}{360} = \dfrac{\cancel{3} \cdot 1}{\cancel{3} \cdot 120} = \dfrac{1}{120}$$

Continued on next page...

c. $\dfrac{3}{x} - \dfrac{2}{5}$

Solution: In this case the LCD $= x \cdot 5 = 5x$

$$\frac{3}{x} + \frac{2}{5} = \left(\frac{3}{x} \cdot \frac{5}{5}\right) + \left(\frac{2}{5} \cdot \frac{x}{x}\right)$$

$$= \frac{15}{5x} + \frac{2x}{5x}$$

$$= \frac{15 + 2x}{5x} \qquad \text{This fraction cannot be reduced.}$$

Practice Problems

Perform the indicated operations and reduce all answers to lowest terms:

1. $-5 \cdot \dfrac{3}{10}$

2. $\dfrac{3}{4} \div \dfrac{4}{3}$

3. $\dfrac{4}{5} \cdot \dfrac{3}{20} - \dfrac{1}{2}$

4. $\dfrac{1}{2a} + \dfrac{1}{2a}$

5. $\dfrac{4}{x} - \dfrac{3}{5}$

6. $\dfrac{5}{24} + \dfrac{7}{36}$

2.3 Exercises

Find (a) all of the factors (divisors) of each number and (b) the first six multiples of each number.

1. 6   **2.** 12   **3.** 15   **4.** 28

In Exercises 5 – 10, find the LCM for each set of expressions. Treat each variable as you would a prime number.

5. $\{24, 15, 10\}$ **6.** $\{25, 35, 49\}$ **7.** $\{8x, 10y, 20xy\}$

8. $\{20xz, 24xy, 32yz\}$ **9.** $\{14x^2, 21xy, 35xy^2\}$ **10.** $\{60a, 105a^2b, 120ab\}$

Answers to Practice Problems: **1.** $-\dfrac{3}{2}$ **2.** $\dfrac{9}{16}$ **3.** $-\dfrac{19}{50}$ **4.** $\dfrac{1}{a}$ **5.** $\dfrac{20 - 3x}{5x}$ **6.** $\dfrac{29}{72}$

In Exercises 11 – 44, perform the indicated operations and reduce each answer to lowest terms.

11. $\dfrac{2}{9} + \dfrac{5}{9}$ **12.** $\dfrac{2}{7} + \dfrac{8}{7}$ **13.** $\dfrac{24}{23} - \dfrac{1}{23}$ **14.** $\dfrac{7}{15} - \dfrac{2}{15}$

15. $\dfrac{11}{12} - \dfrac{13}{12}$ **16.** $\dfrac{5}{16} - \dfrac{9}{16}$ **17.** $\dfrac{5}{6} + \dfrac{7}{10}$ **18.** $\dfrac{9}{20} + \dfrac{3}{8}$

19. $\dfrac{11}{15a} + \dfrac{5}{15a}$ **20.** $\dfrac{2}{3x} + \dfrac{5}{3x}$ **21.** $\dfrac{-19}{25x} + \dfrac{9}{25x}$ **22.** $\dfrac{-5}{13x} - \dfrac{8}{13x}$

23. $\dfrac{7}{10y} - \dfrac{1}{5y}$ **24.** $\dfrac{5}{6c} - \dfrac{1}{3c}$ **25.** $\dfrac{8}{9} + \left(-\dfrac{5}{12}\right)$ **26.** $\dfrac{3}{14} + \left(-\dfrac{5}{6}\right)$

27. $\dfrac{11}{24x} + \dfrac{5}{36x}$ **28.** $\dfrac{11}{12a} - \dfrac{7}{18a}$ **29.** $\dfrac{5}{y} - \dfrac{1}{3}$ **30.** $\dfrac{6}{y} + \dfrac{1}{2}$

31. $\dfrac{4}{x} + \dfrac{1}{4}$ **32.** $\dfrac{3}{5} - \dfrac{2}{b}$ **33.** $\dfrac{a}{5} - \dfrac{3}{4}$ **34.** $\dfrac{c}{10} + \dfrac{5}{8}$

35. $\dfrac{3}{4} + \left(-\dfrac{x}{8}\right)$ **36.** $\dfrac{5}{17} + \left(-\dfrac{x}{34}\right)$ **37.** $\dfrac{2}{3a} + \dfrac{1}{6a}$ **38.** $\dfrac{7}{16b} - \dfrac{9}{10b}$

39. $\dfrac{1}{3x} + \left(-\dfrac{5}{4x}\right)$ **40.** $\dfrac{7}{8y} + \left(-\dfrac{3}{6y}\right)$ **41.** $\dfrac{3}{4a} - \dfrac{6}{8a}$ **42.** $\dfrac{5}{6x} - \dfrac{10}{12x}$

43. $\dfrac{5x}{7} - \dfrac{10x}{14}$ **44.** $\dfrac{48y}{10} - \dfrac{24y}{5}$

Simplify each expression in Exercises 45 – 50.

45. a. $3a + 3a$ **b.** $3a \cdot 3a$ **c.** $\dfrac{1}{3a} + \dfrac{1}{3a}$ **d.** $\dfrac{1}{3a} \cdot \dfrac{1}{3a}$ **e.** $\dfrac{1}{3a} \div \dfrac{1}{3a}$

46. a. $5x + 5x$ **b.** $5x \cdot 5x$ **c.** $\dfrac{1}{5x} + \dfrac{1}{5x}$ **d.** $\dfrac{1}{5x} \cdot \dfrac{1}{5x}$ **e.** $\dfrac{1}{5x} \div \dfrac{1}{5x}$

47. a. $-7ab - 7ab$ **b.** $(-7ab) \cdot (-7ab)$ **c.** $\dfrac{1}{7ab} + \dfrac{1}{7ab}$ **d.** $\dfrac{1}{7ab} \cdot \dfrac{1}{7ab}$

     e. $\dfrac{1}{7ab} \div \dfrac{1}{7ab}$

48. a. $8y - 8y$ **b.** $8y \cdot (-8y)$ **c.** $\dfrac{1}{8y} - \dfrac{1}{8y}$ **d.** $\dfrac{1}{8y} \cdot \left(-\dfrac{1}{8y}\right)$

   e. $-\dfrac{1}{8y} \div \dfrac{1}{8y}$

49. a. $\dfrac{4x}{3} + \dfrac{4x}{3}$ **b.** $\dfrac{4x}{3} \cdot \dfrac{4x}{3}$ **c.** $\dfrac{4x}{3} - \dfrac{4x}{3}$ **d.** $\dfrac{4x}{3} \cdot \dfrac{3}{4x}$

   e. $\dfrac{4x}{3} \div \dfrac{3}{4x}$

50. a. $\dfrac{1}{6a} + \dfrac{1}{6a}$ **b.** $\dfrac{1}{6a} - \dfrac{1}{6a}$ **c.** $\dfrac{1}{6a} \cdot \dfrac{1}{6a}$ **d.** $\dfrac{1}{6a} \div \dfrac{1}{6a}$

51. The California lottery is based on choosing any six of the integers from 1 to 51. The probability of winning the lottery can be found by multiplying the fractions

$$\frac{6}{51} \cdot \frac{5}{50} \cdot \frac{4}{49} \cdot \frac{3}{48} \cdot \frac{2}{47} \cdot \frac{1}{46}.$$

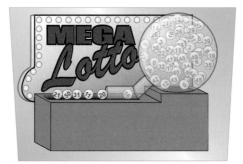

Multiply and reduce the product of these fractions to find the probability of winning the lottery in the form of a fraction with numerator 1.

52. The product of $\dfrac{9}{16}$ and $\dfrac{13}{7}$ is added to the quotient of $\dfrac{10}{3}$ and $\dfrac{7}{4}$. What is the sum?

53. Find the quotient if the sum of $\dfrac{4}{5}$ and $\dfrac{11}{15}$ is divided by the difference between $\dfrac{7}{12}$ and $\dfrac{4}{15}$.

54. The seventeenth hole at the local golf course is a par 4 hole. Ralph drove his ball 258 yards. If this distance was $\dfrac{3}{4}$ the length of the hole, how long is the seventeenth hole?

55. Delia's income is \$2700 a month and she plans to budget $\frac{1}{3}$ of her income for rent and $\frac{1}{10}$ of her income for food.

    a. What fraction of her income does she plan to spend each month on these two items?

    b. What amount of money does she plan to spend each month on these two items?

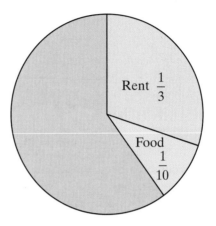

56. The tennis club has 400 members, and they are considering putting in a new tennis court. The cost of the new court is going to involve an assessment of \$250 for each member. Of the six-tenths of the members who live near the club, $\frac{4}{5}$ are in favor of the assesment. However, of the other members who live further away, only $\frac{3}{10}$ are in favor of the assessment.

    a. If a vote were taken today, would more than one-half of the members vote for or against the new court?

    b. By how many votes would the question pass or fail if more than one-half of the members must vote in favor for the question to pass?

Hawkes Learning Systems: Introductory Algebra

Addition and Subtraction with Fractions

Decimal Numbers and Fractions

Objectives

After completing this section, you will be able to:

1. *Operate with decimal numbers.*

2. *Change fractions to decimal form.*

3. *Change decimals to fraction form.*

Multiplication and Division with Decimals (Review)

While in Section R.5 in the Review Chapter we discussed the relationship between decimal numbers and percent, in this section we discuss the relationship between decimal numbers and fractions. Also, we show how the TI-84 Plus graphing calculator can be used to relate decimals and fractions. We begin with examples of multiplication and division with decimal numbers.

Multiplying Two Decimal Numbers

1. *Multiply the two numbers as if they were whole numbers.*

2. *Count the total number of places to the right of the decimal points in both numbers being multiplied.*

3. *Place the decimal point in the product so that the number of places to the right is the same as that found in step 2.*

Example 1: Multiplication with Decimals

Find the products.

a.   4.78
      $\times$ 0.3

b.   −16.4
      $\times$ 0.517

Solution: 4.78   2 digits to the right
          $\times$ 0.3   1 digit to the right
          1.434   3 digits to the right

Solution: −16.4
          $\times$ 0.517
          1148
          164
          820
          −8.4788

Dividing Two Decimal Numbers

To find the quotient of two decimal numbers:

1. Move the decimal point in the divisor to the right to get a whole number,

2. Move the decimal point in the dividend the same number of places,

3. Place the decimal point in the quotient above the new place in the dividend, then

4. Divide as with whole numbers.

Example 2: Division with Decimals

Find the quotient $3.2\overline{)51.52}$

Solution: $3.2\overline{)51.52}$

$32.\overline{)515.2}$ Decimal point in quotient.

We are actually multiplying both the divisor and dividend by 10: $\dfrac{51.52}{3.2} \cdot \dfrac{10}{10} = \dfrac{515.2}{32}$.

$$\begin{array}{r} 16.1 \\ 32\overline{)515.2} \\ \underline{32} \\ 195 \\ \underline{192} \\ 32 \\ \underline{32} \\ 0 \end{array}$$ Divide.

Changing Fractions to Decimal Form

A **rational number** (a fraction with integers for both numerator and denominator) can be written in decimal form by dividing the numerator by the denominator. The result will be either:

    1. a terminating decimal, or
    2. an infinite repeating decimal.

For example,

$$\frac{3}{4} = 0.75 \qquad \text{a terminating decimal}$$

$$\frac{-1}{3} = -0.333333... = -0.\overline{3} \qquad \text{an infinite repeating decimal}$$

(Note that a bar can be placed over the repeating digit(s).)

Irrational numbers written in decimal form are infinite nonrepeating decimals. For example,

$$\pi = 3.1415926535... \quad \text{an infinite nonrepeating decimal}$$

Example 3: Fraction to Decimal Form, Terminating Decimal

Change $3\frac{7}{8}$ to decimal form.

Solution: $3\frac{7}{8} = \frac{31}{8}$ so we divide. With a calculator $\frac{31}{8} = 3.875$.

Or, we can write

$$3\frac{7}{8} = 3 + \frac{7}{8} = 3 + 0.875 = 3.875$$

Long division will give the same result.

Example 4: Fraction to Decimal Form, Infinite Repeating Decimal

Write the fraction $\frac{1}{7}$ in decimal form as an infinite repeating decimal. Round the decimal to the nearest ten-thousandth (four decimal places).

Solution: By dividing with long division (or using a calculator) we will find that

$$\frac{1}{7} = 1 \div 7 = 0.142857142857... = 0.\overline{142857}.$$

This notation with the bar over the six digits 142857 indicates that they repeat indefinitely and in that order.

Thus to four decimal places we have $\dfrac{1}{7} \approx 0.1429$.

With a TI-84 Plus calculator you will find the following:

Changing Decimals to Fraction Form

To write a decimal in fraction form:

1. Write the fraction part of the decimal in the numerator and the position (place value) in the denominator.

2. Reduce the fraction to lowest terms.

Example 5: Decimal to Fraction Form

Write the decimal number 0.56 in fraction form, reduced.

Solution: $0.56 = \dfrac{56}{100} = \dfrac{4}{4} \cdot \dfrac{14}{25} = \dfrac{14}{25}$

Example 6: Calculating Average Speed

During three school days in one week, Matilde drove her mother's car to school and noted that her top speed each day was 25 mph, 35 mph, and 55 mph. When her mother asked what her average top speed was, what was her answer (to the nearest tenth)?

Solution: Matilde had taken this class and knew to add the numbers and divide by 3:

$$25 + 35 + 55 = 115$$

$$115 \div 3 = 38.3 \text{ mph (to the nearest tenth)}$$

Her average speed was 38.3 mph.

Using Calculators

Graphing Calculators and Fractions

The TI-84 Plus Graphing Calculator has commands that will:

1. *Change decimal numbers into fraction form.*

2. *Change fractions into decimal form (done by division).*

3. *Add two fractions with different denominators and find the result with a common denominator.*

To change decimal numbers into fraction form using the TI-84 Plus, locate the **>Frac** command by pressing **MATH** and noting that the first choice is **>Frac**. The use of this command is illustrated in Example 7.

Example 7: Evaluating Fractions with a Calculator

Use a TI-84 Plus calculator (or other graphing calculator) to perform the following:

a. Change the decimal number 0.82 into fraction form.

b. Change the fraction $\dfrac{5}{8}$ into decimal form.

c. Find the sum $\dfrac{3}{5}+\dfrac{3}{20}$ in fraction form.

Solutions: **a.** Proceed as follows:

Enter the numbers 0.82 (or just .82).

Press **MATH** .

Select **>Frac** by pressing **ENTER** .

Press **ENTER** again.

The screen will display:

Note that the fraction is reduced.

b. Proceed as follows:

Enter the fraction $\dfrac{5}{8}$ by entering $5 \div 8$.

Press ENTER .

The calculator will automatically give the answer in decimal form and the screen will appear as follows:

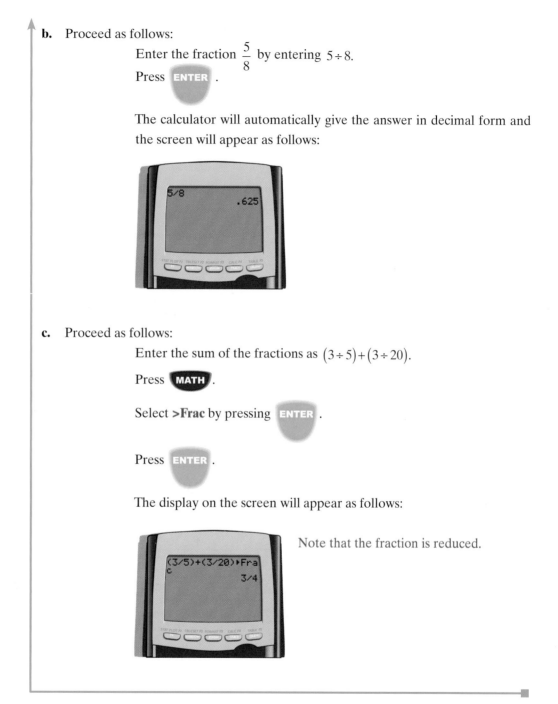

c. Proceed as follows:

Enter the sum of the fractions as $(3 \div 5) + (3 \div 20)$.

Press MATH .

Select **>Frac** by pressing ENTER .

Press ENTER .

The display on the screen will appear as follows:

Note that the fraction is reduced.

Example 8: Converting Fractions to Infinite Repeating Decimals

Write the fraction $\frac{3}{7}$ in decimal form as an infinite repeating decimal. Then round the decimal to the nearest ten thousandth (four decimal places).

Solution:

We can divide by using long division as in $7\overline{)3.000}$ or use a calculator as in $3 \div 7 = 0.428571428571$ (accurate to ten decimal places on a TI-84 Plus).

The calculator has automatically rounded the last digit. In fact the pattern of digits is an infinite pattern:

$$\frac{3}{7} = 0.428571428571...\quad \text{six digits in the pattern}$$

Or, we can put a bar over the pattern (indicating that the pattern repeats indefinitely):

$$\frac{3}{7} = 0.\overline{428571}.$$

Remember: The decimal form of every rational number is either a terminating decimal (as with $\frac{5}{8}$ in Example 7b) or an infinite repeating decimal (as with $\frac{3}{7}$ in Example 8.)

2.4 Exercises

Without using a calculator, find the indicated products in Exercises 1 – 8.

1. $(60.4)(1.8)$ **2.** $(21.6)(0.83)$ **3.** $(1.23)(-6.2)$

4. $(5.83)(-0.24)$ **5.** $(-5.12)(-30.7)$ **6.** $(-3.56)(-7.34)$

7. $(6.7)(3.2)(7.9)$ **8.** $(-1.5)(-1.5)(-2.5)$

Without using a calculator, find the indicated quotients in Exercises 9 – 16.

9. $5.1\overline{)77.01}$ **10.** $0.023\overline{)0.4922}$ **11.** $4.6\overline{)16.192}$

12. $0.66\overline{)9.7878}$ **13.** $0.256\overline{)0.09728}$ **14.** $0.175\overline{)0.40425}$

15. $15.3\overline{)0.3825}$ **16.** $21.4\overline{)342.614}$

In Exercises 17 – 24 use long division to change each fraction to decimal form. If the decimal is non-terminating, write it by using the bar notation over the repeating pattern of digits.

17. $\dfrac{3}{8}$ **18.** $\dfrac{4}{5}$ **19.** $\dfrac{1}{20}$ **20.** $\dfrac{3}{50}$

21. $\dfrac{2}{3}$ **22.** $\dfrac{5}{9}$ **23.** $\dfrac{7}{30}$ **24.** $\dfrac{3}{11}$

In Exercises 25 – 32, use a calculator to find the decimal form of each number. If the decimal is non-terminating, write it by using the bar notation over the repeating pattern of digits.

25. $\dfrac{7}{16}$ **26.** $\dfrac{5}{18}$ **27.** $-\dfrac{6}{11}$ **28.** $-\dfrac{1}{22}$

29. $1\dfrac{5}{27}$ **30.** $46\dfrac{2}{3}$ **31.** $6\dfrac{7}{12}$ **32.** $2\dfrac{1}{11}$

33. Find the average of the following set of decimal numbers (to the nearest hundredth): $\{86.46, 95.35, 99.07, 100.34\}$ [Do not use a calculator.]

34. Find the average of the following set of decimal numbers (to the nearest tenth): $\{17.6, 20.3, 24.5, 34.7, 38.9, 40.2\}$ [Do not use a calculator.]

35. Kathleen is remodeling her home and decided to buy some new furniture. She bought a new sofa for $1456.70, a new light fixture for the dining room for $815.00, and a used flat screen TV from her neighbor for $975.00. What was the average amount per item (to the nearest cent) that she spent?

36. Javier decided to calculate the average price per stock when he bought the following number of stocks: 50 shares of Yahoo for $1558.50, 30 shares of eBay for $1008.61 and 25 shares of Exxon for $1919.25. How many shares of stock did he buy? What was the average price per share (to the nearest cent) that he bought?

Use a TI-84 Plus graphing calculator to find the answers in Exercises 37 – 42 and then to change each of the answers into fraction form.

37. $0.57 + 0.73$

38. $1.58 - 2.36$

39. $-2.78 - 1.93 + 10.002$

40. $13.175 - 16.32 - 20.784$

41. $11.67 - 25.9 - 50$

42. $-14.3 + 20.01 + 6$

Use a TI-84 Plus graphing calculator to find the indicated sum or difference of fractions and write the answer in fraction form, reduced.

43. $\dfrac{3}{4} + \dfrac{1}{10}$

44. $\dfrac{7}{8} + \dfrac{7}{10}$

45. $\dfrac{3}{7} + \dfrac{1}{3} + \dfrac{2}{21}$

46. $\dfrac{2}{3} + \dfrac{1}{10} + \dfrac{17}{15}$

47. $\dfrac{3}{10} - \dfrac{27}{100} - \dfrac{133}{1000}$

48. $\dfrac{7}{100} - \dfrac{17}{100} - \dfrac{9}{10}$

Hawkes Learning Systems: Introductory Algebra

Decimals and Fractions

<table><tr><td>**2.5**</td></tr></table>

Order of Operations with Negative Numbers and Fractions

Objectives

After completing this section, you will be able to:

Evaluate integer, fractional, and decimal expressions using the rules for order of operations.

As we discussed in Section R.1 of the Review Chapter, mathematicians have agreed on a set of rules for the order of performing operations when evaluating numerical expressions that contain grouping symbols, exponents, and the operations of addition, subtraction, multiplication, and division. These rules (called the **rules for order of operations**) are used throughout all levels of mathematics and science (as well as calculators) so that there is only one correct answer for the value of an expression. For example, evaluate the following expression:

$$36 \div 4 + 6 \cdot 2^2.$$

Did you get 33 or 60? The correct answer is 33. In this section we will see how to apply the **rules for order of operations**, not only with whole numbers as in Section R.1, but with integers, decimals, and fractions as well.

Rules for Order of Operations

1. *Simplify within grouping symbols, such as parentheses ( ), brackets [ ], and braces { }, working from the inner most grouping outward.*
2. *Find any powers indicated by exponents.*
3. *Moving from **left to right**, perform any multiplications **or** divisions **in the order they appear**.*
4. *Moving from **left to right**, perform any additions **or** subtractions **in the order they appear**.*

NOTES

Other grouping symbols are the absolute value bars (such as $|3+5|$), the fraction bar (as in $\frac{4+7}{10+1}$), and the square root symbol (such as $\sqrt{5+11}$).

These rules are very explicit and should be studied carefully. Note that in Rule 3, neither multiplication nor division has priority over the other. Whichever of these operations occurs first, **moving left to right**, is done first. In Rule 4, addition and subtraction are handled in the same way. Unless they occur within grouping symbols, **addition and subtraction are the last operations to be performed**.

A well-known mnemonic device for remembering the rules for order of operations is the following:

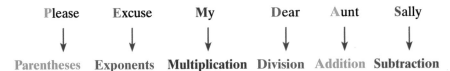

Please	**E**xcuse	**M**y	**D**ear	**A**unt	**S**ally
↓	↓	↓	↓	↓	↓
Parentheses	Exponents	Multiplication	Division	Addition	Subtraction

NOTES   Even though the mnemonic **PEMDAS** is helpful, remember that multiplication and division are performed as they appear, left to right. Also, addition and subtraction are performed as they appear, left to right.
For example

$$12 \div 3 \cdot 4 = 4 \cdot 4 = 16,$$

but

$$12 \cdot 3 \div 4 = 36 \div 4 = 9.$$

As discussed in Section 2.1, a negative sign in front of a variable indicates that the coefficient is –1. For example,

$$-x^2 = -1 \cdot x^2.$$

This is consistent with the rules for order of operations (indicating that exponents come before multiplication) and is particularly useful in determining the values of expressions involving negative numbers and exponents. For example, each of the expressions

$$-7^2 \quad \text{and} \quad (-7)^2$$

has a different value. By the order of operations,

$$-7^2 = -1 \cdot 7^2 = -1 \cdot 49 = -49,$$

but

$$(-7)^2 = (-7)(-7) = 49.$$

In the second expression the base is a negative number. Remember that if the base is a negative number, then the negative number must be placed in parentheses.

The following examples show how to apply the rules. In some cases, more than one step can be performed at the same time. This is possible when parts are separated by + or − signs or are within separate symbols of inclusion. **Work through each of the following examples step by step, and rewrite the examples on a separate sheet of paper.**

Example 1: Order of Operations with Integers

Use the rules for order of operations to evaluate each of the following expressions.

a. $36 \div 4 - 6 \cdot 2^2$ **b.** $2(3^2 - 1) - 3 \cdot 8$ **c.** $9 - 2[(3 \cdot 5 - 7^2) \div 2 + 2^2]$

Solutions:

a. $36 \div 4 - 6 \cdot 2^2 = 36 \div 4 - 6 \cdot 4$ Exponents

$\qquad\qquad\qquad = 9 - 24$ Divide and multiply, left to right.

$\qquad\qquad\qquad = -15$ Subtract.

b. $2(3^2 - 1) - 3 \cdot 2^3 = 2(9 - 1) - 3 \cdot 8$ Exponents

$\qquad\qquad\qquad = 2(8) - 3 \cdot 8$ Subtract inside the parentheses.

$\qquad\qquad\qquad = 16 - 24$ Multiply.

$\qquad\qquad\qquad = -8$ Subtract (or add algebraically).

c. $9 - 2[(3 \cdot 5 - 7^2) \div 2 + 2^2] = 9 - 2[(3 \cdot 5 - 49) \div 2 + 4]$ Exponents

$\qquad\qquad\qquad = 9 - 2[(15 - 49) \div 2 + 4]$ Multiply inside the parentheses.

$\qquad\qquad\qquad = 9 - 2[(-34) \div 2 + 4]$ Subtract inside the parentheses.

$\qquad\qquad\qquad = 9 - 2[-17 + 4]$ Divide inside the brackets.

$\qquad\qquad\qquad = 9 - 2[-13]$ Add inside the brackets.

$\qquad\qquad\qquad = 9 + 26$ Multiply.

$\qquad\qquad\qquad = 35$ Add.

Note: Because of the rules for order of operations at no time did we even subtract $9 - 2$.

Example 2: Order of Operations with Fractions

Use the rules for order of operations to evaluate each of the following expressions.

a. $2\dfrac{1}{3} \div \left(\dfrac{1}{4} + \dfrac{1}{3}\right)$ **b.** $\dfrac{3}{5} \cdot \dfrac{5}{6} + \dfrac{1}{4} \div \left(\dfrac{5}{2}\right)^2$ **c.** $\dfrac{1}{10} \div \dfrac{3}{4} \cdot \dfrac{1}{2} - \dfrac{3}{5} \cdot \dfrac{7}{30}$

Solutions:

a. $2\dfrac{1}{3} \div \left(\dfrac{1}{4} + \dfrac{1}{3}\right) = \dfrac{7}{3} \div \left(\dfrac{3}{12} + \dfrac{4}{12}\right)$ Change the mixed number to an improper fraction.

$= \dfrac{7}{3} \div \left(\dfrac{7}{12}\right)$ Add within parentheses.

$= \dfrac{\cancel{7}}{\cancel{3}} \cdot \dfrac{\overset{4}{\cancel{12}}}{\cancel{7}}$ Divide and reduce.

$= 4$

b. $\dfrac{3}{5} \cdot \dfrac{5}{6} + \dfrac{1}{4} \div \left(\dfrac{5}{2}\right)^2 = \dfrac{3}{5} \cdot \dfrac{5}{6} + \dfrac{1}{4} \div \left(\dfrac{25}{4}\right)$ Exponents

$= \dfrac{\cancel{3}}{\cancel{5}} \cdot \dfrac{\cancel{5}}{\underset{2}{\cancel{6}}} + \dfrac{1}{\cancel{4}} \cdot \dfrac{\cancel{4}}{25}$ Divide, then multiply and reduce.

$= \dfrac{1}{2} + \dfrac{1}{25}$

$= \dfrac{25}{50} + \dfrac{2}{50}$ Find the LCD.

$= \dfrac{27}{50}$ Add.

c. $\dfrac{1}{10} \div \dfrac{3}{4} \cdot \dfrac{1}{2} - \dfrac{3}{5} \cdot \dfrac{7}{30} = \dfrac{1}{10} \cdot \dfrac{4}{3} \cdot \dfrac{1}{2} - \dfrac{3}{5} \cdot \dfrac{7}{30}$ Divide.

$= \dfrac{1}{\underset{5}{\cancel{10}}} \cdot \dfrac{\cancel{4}}{3} \cdot \dfrac{1}{\cancel{2}} - \dfrac{\cancel{3}}{5} \cdot \dfrac{7}{\underset{10}{\cancel{30}}}$ Multiply and reduce.

$= \dfrac{1}{15} - \dfrac{7}{50}$ Simplify.

$= \dfrac{10}{150} - \dfrac{21}{150}$ Find the LCD.

$= -\dfrac{11}{150}$ Subtract.

Example 3: Order of Operations with Decimals (using Calculators)

Calculators are programmed to use the rules for order of operations. Use your TI-84 Plus graphing calculator to find the value of each of the following expressions.

a. $1.5 \div 6 - 4 \cdot 3.2^2$ **b.** $2.1(4.5^2 - 5^2) - 3(-1.6)^2$ **c.** $15.8 - 3.1(7^2 - 6.3^2) \div 5 \cdot 2.5$

Solutions:

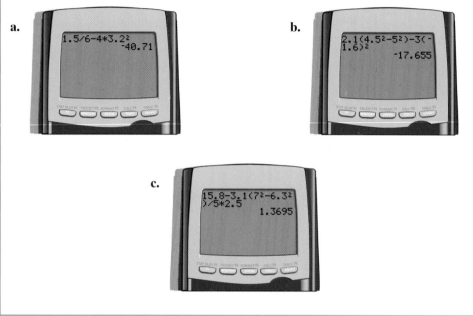

a.

```
1.5/6-4*3.2²
            -40.71
```

b.

```
2.1(4.5²-5²)-3(-
1.6)²
          -17.655
```

c.

```
15.8-3.1(7²-6.3²
)/5*2.5
           1.3695
```

Practice Problems

Use the rules for order of operations to evaluate each expression.

1. $36 \div 2 \cdot 6$ **2.** $8 \cdot 2 + 4(-2) - 4^2$ **3.** $9 \cdot 3 \div (2^2 - 5)$

4. $\left(\dfrac{1}{6}\right)^2 \div \dfrac{7}{12} - \dfrac{5}{8}$ **5.** $\left(-\dfrac{3}{4}\right) + \dfrac{3}{8} \cdot \dfrac{4}{5} \div \dfrac{2}{5} + \dfrac{2}{3}$ **6.** $-15 \div \left(\dfrac{1}{3} + \dfrac{1}{10}\right)$

Simplify.

7. $\dfrac{2}{5a} \div \dfrac{16}{3a} - \dfrac{2b}{7} \cdot \dfrac{7}{5b}$

Answers to Practice Problems: 1. 108 **2.** -8 **3.** -27 **4.** $-\dfrac{97}{168}$ **5.** $\dfrac{2}{3}$ **6.** $-\dfrac{450}{13}$ **7.** $-\dfrac{13}{40}$

2.5 Exercises

In Exercises 1 – 40, use the rules for order of operations to evaluate the expressions.

1. a. $24 \div 4 \cdot 6$ **b.** $24 \cdot 4 \div 6$ **2. a.** $20 \div 5 \cdot 2$ **b.** $20 \cdot 5 \div 2$

3. $15 \div (-3) \cdot 3 - 10$ **4.** $20 \cdot 2 \div 2^2 + 5(-2)$

5. $3^2 \div (-9) \cdot (4 - 2^2) + 5(-2)$ **6.** $4^2 \div (-8)(-2) + 3(2^2 - 5^2)$

7. $14 \cdot 3 \div (-2) - 6(4)$ **8.** $6(13 - 15)^2 \cdot 8 \div 2^2 + 3(-1)$

9. $-10 + 15 \div (-5) \cdot 3^2 - 10^2$ **10.** $16 \cdot 3 \div (2^2 - 5)$

11. $2 - 5[(-20) \div (-4) \cdot 2 - 40]$ **12.** $9 - 6[(-21) \div 7 \cdot 2 - (-8)]$

13. $(7-10)\left[49 \div (-7) + 20 \cdot 3 - (-10)\right]$

14. $(9-11)\left[(-10)^2 \cdot 2 + 6(-5)^2 - 10^2 + 3 \cdot 5\right]$

15. $8 - 9\left[(-39) \div (-13) + 7(-2) - (-2)^2\right]$

16. $6 - 20\left[(-15) \div 3 \cdot 5 + 6 \cdot 2 \div 3\right]$

17. $\left| 16 - 20 \right|\left[32 \div \left| 3 - 5 \right| - 5^2 \right]$

18. $\left| 10 - 30 \right|\left[4^2 \cdot \left| 5 - 8 \right| \div (-2)^2 + \left| 17 - 18 \right| \right]$

19. $\dfrac{(-10)+(-2)}{6^2 - 30} \div \left| 2 - 4 \right|$ **20.** $\left| 16 - 20 \right| + \dfrac{-10^2 + 5^2}{3 \cdot 4 - 7}$ **21.** $\dfrac{3}{8} \cdot \dfrac{4}{5} + \dfrac{1}{15}$

22. $\dfrac{1}{4} \cdot \dfrac{12}{15} + \dfrac{2}{7}$ **23.** $\dfrac{1}{3} \div \dfrac{1}{2} - \dfrac{5}{6} \cdot \dfrac{3}{4}$ **24.** $\dfrac{2}{9} \div \dfrac{14}{3} - \dfrac{1}{6} \cdot \dfrac{4}{7}$

25. $\left(\dfrac{5}{6}\right)^2 \div \dfrac{5}{12} - \dfrac{3}{8}$ **26.** $\left(\dfrac{2}{5}\right)^2 \cdot \dfrac{3}{8} + \dfrac{1}{5} \div \dfrac{3}{4}$ **27.** $\dfrac{7}{6} \cdot 2^2 - \dfrac{2}{3} \div 3\dfrac{1}{5}$

28. $\dfrac{3}{4} \div 3^2 - 4\left(1\dfrac{1}{2}\right)^2$ **29.** $\left(-\dfrac{3}{4}\right) \div \left(-\dfrac{3}{5}\right) \cdot \dfrac{7}{8} + \dfrac{3}{16}$

30. $\left(-\dfrac{2}{3}\right) \div \dfrac{7}{12} - \dfrac{2}{7} + \left(-\dfrac{1}{2}\right)^2$ **31.** $\left(-\dfrac{9}{10}\right) + \dfrac{5}{8} \cdot \dfrac{4}{5} \div \dfrac{6}{10} + \dfrac{2}{3}$

32. $-\dfrac{5}{9} - \dfrac{1}{3} \cdot \dfrac{2}{3} - 3\dfrac{1}{3}$ **33.** $-2\dfrac{1}{2} \cdot 3\dfrac{1}{5} \div \dfrac{3}{4} - \dfrac{7}{10}$ **34.** $\dfrac{5}{8} \div \dfrac{1}{10} + \left(-\dfrac{1}{3}\right)^2 \cdot \dfrac{3}{5}$

35. $\left(\dfrac{1}{2}-1\dfrac{3}{4}\right)\div\left(\dfrac{2}{3}+\dfrac{3}{4}\right)$ **36.** $\left(\dfrac{5}{8}+2\dfrac{1}{4}\right)\div\left(1-\dfrac{1}{4}\right)$ **37.** $-15\div\left(\dfrac{1}{4}-\dfrac{7}{8}\right)$

38. $-12\div\left(\dfrac{1}{2}+\dfrac{1}{10}\right)$ **39.** $\left(1\dfrac{1}{10}-3\dfrac{1}{5}\right)\div3\dfrac{1}{2}+1\dfrac{3}{10}$

40. $4\left(\dfrac{1}{2}\right)^{2}+\dfrac{7}{10}\div\dfrac{3}{5}\cdot\dfrac{5}{18}-\dfrac{9}{10}$

41. Find the average of the three numbers: $3\dfrac{2}{5}$, $5\dfrac{3}{4}$, and $6\dfrac{1}{10}$.

42. Find the average of the four numbers: $1\dfrac{1}{8}$, $2\dfrac{1}{2}$, $-5\dfrac{1}{2}$, and $-\dfrac{5}{8}$.

43. If the square of $\dfrac{7}{8}$ is subtracted from the square of $\dfrac{3}{10}$, what is the difference?

44. Find the quotient if the sum of $\dfrac{4}{5}$ and $\dfrac{2}{15}$ is divided by the difference between $\dfrac{7}{8}$ and $2\dfrac{1}{4}$.

Use your TI-84 Plus graphing calculator to evaluate each of the expressions in Exercises 45 – 50.

45. $3.4\div4+5\cdot8.32$ **46.** $8.1\div5+16.3\cdot7$

47. $0.75\div1.5+7\cdot3.1^{2}$ **48.** $1.05\div(-3)\cdot3.7-1.1^{2}$

49. $6.32\cdot8.4\div16.8+3.5^{2}$ **50.** $(82.7+16.2)\div(14.83-19.83)^{2}$

Use the rules for order of operations to simplify each of the following expressions.

51. $\dfrac{3}{4a}\div\dfrac{3}{16a}-\dfrac{2b}{3}\cdot\dfrac{3}{4b}$ **52.** $\dfrac{3}{4x}-\dfrac{5}{6}\div\dfrac{5x}{8}-\dfrac{1}{3x}$

53. $\dfrac{-5}{7y}-\dfrac{1}{2y}\cdot\dfrac{2}{3}-\dfrac{1}{21y}$ **54.** $\left(\dfrac{1}{5x}-\dfrac{2}{3x}\right)\div\left(\dfrac{5}{7y}-\dfrac{1}{3y}\right)$

55. $\left(\dfrac{7}{10x}+\dfrac{3}{5x}\right)\div\left(\dfrac{1}{2x}+\dfrac{3}{7x}\right)$ **56.** $\left(\dfrac{1}{5a}-\dfrac{2}{3a}\right)\div\left(\dfrac{7}{7b}-\dfrac{1}{3b}\right)$

57. $\left(\dfrac{3x}{2}-5x\right)\div\left(\dfrac{x}{3}+\dfrac{x}{15}\right)$ **58.** $\left(\dfrac{5y}{6}+\dfrac{2y}{3}\right)+\dfrac{y}{12}\div\dfrac{5}{8}$

59. $\dfrac{a}{8}\cdot\left(\dfrac{1}{3}\right)^{2}+\dfrac{2a}{16}\cdot\dfrac{1}{9}$ **60.** $\dfrac{y}{24}\cdot\left(\dfrac{4}{5}\right)^{2}+\dfrac{2y}{3}\div\left(\dfrac{2}{3}\right)^{2}$

Writing and Thinking About Mathematics

61. Explain, in your own words, why the following expression cannot be evaluated.

$$(24 - 2^4) + 6(3 - 5) \div (3^2 - 9)$$

62. Consider any number between 0 and 1. If you square this number, will the result be larger or smaller than the original number? Is this always the case? Explain.

63. Consider any number between −1 and 0. If you square this number, will the result be larger or smaller than the original number? Is this always the case? Explain.

Hawkes Learning Systems: Introductory Algebra

Order of Operations with Negative Numbers and Fractions

2.6

Translating English Phrases and Algebraic Expressions

After completing this section, you will be able to:

1. Write the meaning of algebraic expressions in words.

2. Write algebraic expressions for word phrases.

Translating English Phrases into Algebraic Expressions

Algebra is a language of mathematicians, and to understand mathematics, you must understand the language. We want to be able to change English phrases into their "algebraic" equivalents and vice versa. So if a problem is stated in English, we can translate the phrases into algebraic symbols and proceed to solve the problem according to the rules and methods developed for algebra.

Certain words are the keys to the basic operations. Some of these words are listed here and highlighted in boldface in Example 1.

Key Words to Look For in Translating Phrases

Addition	*Subtraction*	*Multiplication*	*Division*	*Exponent(Powers)*
add	subtract (from)	multiply	divide	square of
sum	difference	product	quotient	cube of
plus	minus	times		
more than	less than	twice		
increased by	decreased by	of (with fractions		
	less	and percent)		

The following examples illustrate how these key words used in English phrases can be translated into algebraic expressions. **Note that in each case "a number" or "the number" implies the use of a variable (an unknown quantity).**

Example 1: Use of Key Words

English Phrase	Algebraic Expression
a. 3 **multiplied by** the number represented by x the **product** of 3 and x 3 **times** x	$3x$
b. 3 **added to** a number the **sum of** z **and** 3 z **plus** 3 3 **more than** z z **increased by** 3	$z+3$
c. 2 **times** the quantity found by **adding** a number to 1 **twice** the **sum** of x and 1 the **product** of 2 with the **sum** of x and 1	$2(x+1)$
d. **twice** x **plus** 1 the **sum of twice** x and 1 2 **times** x **increased** by 1 1 **more than** the **product** of 2 and a number	$2x+1$
e. the **difference** between 5 **times** a number and 3 3 **less than** the **product** of a number and 5 5 **times** a number **minus** 3 3 **subtracted from** $5n$ 5 **multiplied by** a number **less** 3	$5n-3$
f. the **square** of a number	x^2
g. the **cube** of a number	n^3

NOTES

In Example 1b, the phrase "the sum of z and 3" was translated as $z + 3$. If the expression had been translated as $3 + z$, there would have been no mathematical error because addition is commutative. That is, $z + 3 = 3 + z$. However, in **e.**, the phrase "3 less than the product of a number and 5" must be translated as it was because subtraction is **not** commutative. Thus,

"3 less than 5 times a number" means $5n - 3$
while "5 times a number less than 3" means $3 - 5n$
and "3 less 5 times a number" means $3 - 5n$.

Therefore, be very careful when writing and/or interpreting expressions indicating subtraction. Be sure that the subtraction is in the order indicated by the wording in the problem. The same is true with expressions involving division.

The words **quotient** and **difference** deserve special mention because their use implies that the numbers given are to be operated on in the order given. That is, division and subtraction are done with the values in the same order that they are given in the problem. For example:

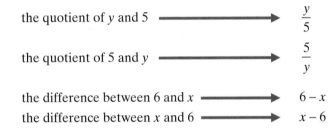

the quotient of y and 5 $\longrightarrow$ $\dfrac{y}{5}$

the quotient of 5 and y $\longrightarrow$ $\dfrac{5}{y}$

the difference between 6 and x $\longrightarrow$ $6 - x$
the difference between x and 6 $\longrightarrow$ $x - 6$

If we did not have these agreements concerning subtraction and division, then the phrases just illustrated might have more than one interpretation and be considered **ambiguous**.

An **ambiguous phrase** is one whose meaning is not clear or for which there may be two or more interpretations. This is a common occurrence in ordinary everyday language, and misunderstandings occur frequently. Imagine the difficulties diplomats have in communicating ideas from one language to another trying to avoid ambiguities. Even the order of subjects, verbs, and adjectives may not be the same from one language to another. Translating grammatical phrases in any language into mathematical expressions is quite similar. To avoid ambiguous phrases in mathematics, we try to be precise in the use of terminology, to be careful with grammatical construction, and to follow the rules for order of operations.

Translating Algebraic Expressions into English Phrases

Consider the three expressions to be translated into English:

$$7(n + 1), \quad 6(n - 3), \quad \text{and} \quad 7n + 1.$$

In the first two expressions, we indicate the parentheses with a phrase such as "the quantity" or "the sum of" or "the difference between." **Without the parentheses, we agree that the operations used in the expression are to be indicated in the order given.** Thus,

$7(n + 1)$ can be translated as   "seven times the sum of a number and 1,"

$6(n - 3)$ can be translated as   "six times the difference between a number and 3,"

while $7n + 1$ can be translated as   "seven times a number plus 1."

Example 2: Algebraic Expression to Phrase

Write an English phrase that indicates the meaning of each algebraic expression.

a. $5x$ **b.** $2n + 8$ **c.** $3(a - 2)$

Solutions:

Algebraic Expression	Possible English Phrase
a. $5x$	The product of a number and 5
b. $2n + 8$	Twice a number increased by 8
c. $3(a - 2)$	Three times the difference between a number and 2

Example 3: Phrase to Algebraic Expression

Change each phrase into an equivalent algebraic expression.

a. The quotient of a number and –4

b. 6 less than 5 times a number

c. Twice the sum of 3 and a number

d. The cost of renting a truck for one day and driving x miles if the rate is \$30 per day plus \$0.25 per mile

Solutions:

	Phrase	Algebraic Expression
a.	The quotient of a number and –4	$\dfrac{x}{-4}$
b.	6 less than 5 times a number	$5y - 6$

c. Twice the sum of 3 and a number $\qquad$ $2(3+n)$

d. The cost of renting a truck for one day
and driving x miles if the rate is \$30 per day $\qquad$ $30 + 0.25x$
plus \$0.25 per mile

Practice Problems

Change the following phrases to algebraic expressions.

1. 7 less than a number

2. The quotient of y and 5

3. 14 more than 3 times a number

Change the following algebraic expressions into English phrases.

4. $10 - x$ $\qquad$ *5. $2(y-3)$* $\qquad$ *6. $5n + 3n$*

2.6 Exercises

Translate each of the expressions in Exercises 1 – 12 into an equivalent English phrase.
(There may be more than one correct translation.)

1. $4x$ $\qquad$ **2.** $x + 6$ $\qquad$ **3.** $2x + 1$ $\qquad$ **4.** $4x - 7$

5. $7x - 5.3$ $\qquad$ **6.** $3.2(x + 2.5)$ $\qquad$ **7.** $-2(x - 8)$ $\qquad$ **8.** $10(x + 4)$

9. $5(2x + 3)$ $\qquad$ **10.** $3(4x - 5)$ $\qquad$ **11.** $6(x - 1)$ $\qquad$ **12.** $9(x + 3)$

Write each pair of expressions in words in Exercises 13 – 16. Notice the differences
between the expressions and the corresponding English phrases.

13. $3x + 7$; $3(x + 7)$ $\qquad$ **14.** $4x - 1$; $4(x - 1)$ $\qquad$ **15.** $7x - 3$; $7(x - 3)$

16. $5(x + 6)$; $5x + 6$

Answers to Practice Problems: **1.** $x - 7$ **2.** $\dfrac{y}{5}$ **3.** $3y + 14$ **4.** 10 decreased by a number **5.** twice the difference
between a number and 3 **6.** 5 times a number plus 3 times the same number

Write the algebraic expression described by each of the word phrases in Exercises 17 – 40. Choose your own variable.

17. 6 added to a number

18. 7 more than a number

19. 4 less than a number

20. A number decreased by 13

21. 5 less than 3 times a number

22. The difference between twice a number and 10

23. The difference between x and 3, all divided by 7

24. 9 times the sum of a number and 2

25. 3 times the difference between a number and 8

26. 13 less than the product of 4 with the sum of a number and 1

27. 5 subtracted from three times a number

28. The sum of twice a number and four times the number

29. 8 minus twice a number

30. The sum of a number and 9 times the number

31. 4 more than the product of 8 with the difference between a number and 6

32. Twenty decreased by 4.8 times a number

33. The difference between three times a number and five times the same number

34. Eight more than the product of 3 times the sum of a number and 6

35. Six less than twice the difference between a number and 7

36. Four less than 3 times the difference between 7 and a number

37. Nine more than twice the sum of 17 and a number

38. a. 5 less than 3 times a number
b. 5 less 3 times a number

39. a. 6 less than a number
b. 6 less a number

40. a. 20 less than a number
b. A number less than 20

Write the algebraic expression described by each of the word phrases in Exercises 41 – 56.

41. the number of hours in d days

42. the cost of x graphing calculators if one calculator costs $75

43. the number of seconds in *m* minutes

44. the cost of *x* gallons of gasoline if the cost of 1 gallon is $3.15

45. the number of days in *y* years (Assume 365 days in a year.)

46. The cost of *x* pounds of candy at $4.95 a pound

47. The annual interest on *x* dollars if the rate is 11% per year

48. The number of days in *t* weeks and 3 days

49. The number of minutes in *h* hours and 20 minutes

50. The points scored by a football team on *t* touchdowns (6 points each) and 1 field goal (3 points)

51. The cost of renting a car for one day and driving *m* miles if the rate is $20 per day plus 15 cents per mile

52. The cost of purchasing a fishing rod and reel if the rod costs *x* dollars and the reel costs $8 more than twice the cost of the rod

53. A sales person's weekly salary if he receives $250 as his base plus 9% of the weekly sales of *x* dollars

54. The selling price of an item that costs *c* dollars if the markup is 20% of the cost

55. The perimeter of a rectangle if the width is *w* centimeters and the length is 3 cm less than twice the width

3 cm less than twice the width

56. The area of a square with side *c* centimeters

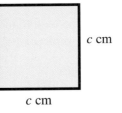

Hawkes Learning Systems: Introductory Algebra

Translating Phrases into Algebraic Expressions

Chapter 2 Index of Key Ideas and Terms

Section 2.1   Simplifying and Evaluating Algebraic Expressions

Algebraic Terms   page 136

> Any constant or variable or the indicated product and/or
> quotient of constants and powers of variables is called a **term**.

Like Terms   page 137

> **Like terms** (or similar terms) are terms that are constants or
> terms that contain the same variables that are of the same power.

> Combining Like Terms   pages 137 - 139

Evaluating Algebraic Expressions   pages 139 - 141

Section 2.2   Multiplication and Division with Fractions

Rational Number   page 143

> A **rational number** is a number that can be written in the form $\dfrac{a}{b}$
> where a and b are integers and $b \neq 0$

> OR

> A **rational number** is a number that can be written in decimal form as a
> terminating decimal or as an infinite repeating decimal.

Fraction   pages 143 - 144

> In general, fractions can be used to indicate:
>   **1.** Equal parts of a whole, or
>   **2.** Division

Placement of Negative Signs in Fractions   page 145

> If a and b are real numbers, and $b \neq 0$, then

$$-\frac{a}{b} = \frac{-a}{b} = \frac{a}{-b}$$

Multiplication with Fractions   page 145

$$\frac{a}{b} \cdot \frac{c}{d} = \frac{a \cdot c}{b \cdot d} \text{ where } b \neq 0 \text{ and } d \neq 0$$

Continued on next page...

Section 2.2 Multiplication and Division with Fractions (continued)

The Fundamental Principal of Fractions   pages 145 - 146

$$\frac{a}{b} = \frac{a \cdot k}{b \cdot k} \text{ where } k \neq 0 \text{ and } b \neq 0$$

Reciprocal   pages 147 - 148

If $a \neq 0$ and $b \neq 0$, the **reciprocal** of $\frac{a}{b}$ is $\frac{b}{a}$, and $\frac{a}{b} \cdot \frac{b}{a} = 1$.

Division with Fractions   page 148

To divide by a nonzero fraction, multiply by its reciprocal:

$$\frac{a}{b} \div \frac{c}{d} = \frac{a}{b} \cdot \frac{d}{c} \text{ where } b \neq 0 \quad c \neq 0, \text{ and } d \neq 0$$

Section 2.3 Addition and Subtraction with Fractions

Addition of Fractions with Like Denominators   page 153

To add two fractions with a common denominator add the numerators and use the common denominator:

$$\frac{a}{b} + \frac{c}{b} = \frac{a+c}{b}$$

Least Common Multiple   page 154

To find the LCM of a set of integers:
1. List the prime factorization of each number.
2. List the prime factors that appear in any one of the prime factorizations.
3. Find the product of these primes using each prime the greatest number of times that it appears in any one prime factorization.

Addition of Fractions with Different Denominators   page 155

To add two fractions with different denominators:
1. Find the LCM of the denominators (called the LCD).
2. Change each fraction to an equal fraction with the LCD.
3. Add the new fractions. Reduce if possible.
   (**Note:** Treat variables as prime factors.)

Continued on next page...

Section 2.3   Addition and Subtraction with Fractions   (continued)

Subtract with Fractions   page 157

To subtract two fractions with a common denominator, subtract
the numerators and use the common denominator.

$$\frac{a}{b} - \frac{c}{b} = \frac{a-c}{b} \text{ where } b \text{ is a nonzero number or}$$
an algebraic expression.

To subtract two fractions with different denominators, find the LCD,
change each fraction to an equal fraction with the LCD, and subtract.

Section 2.4   Decimal Numbers and Fractions

Multiplying Two Decimal Numbers   page 162

Dividing Two Decimal Numbers   page 163

Rational Numbers   pages 163 - 164

A **rational number** is a fraction with integers for both numerator
and denominator.

Irrational Numbers   page 164

Changing Decimals to Fraction Form   pages 165 - 168

Graphing Calculators and Fractions   pages 166 - 167

Section 2.5   Order of Operations with Negative Numbers and Fractions

Rules for Order of Operations   pages 171 - 175

1. Simplify within grouping symbols, such as parentheses ( ),
   brackets [ ], and braces { }, working from the inner most
   grouping outward.
2. Find any powers indicated by exponents.
3. Moving from **left to right**, perform any multiplications or
   divisions in the order they appear.
4. Moving from **left to right**, perform any additions or
   divisions in the order they appear.

Section 2.6 **Translating English Phrases and Algebraic Expressions**

Translating English Phrases into Algebraic Expressions   pages 179 - 181

Translating Algebraic Expressions into English Phrases   pages 182 - 183

Hawkes Learning Systems: Introductory Algebra

For a review of the topics and problems from Chapter 2, look at the following lessons from *Hawkes Learning Systems: Introductory Algebra*

Variables and Algebraic Expressions
Simplifying Expressions
Simplifying Expressions with Parentheses
Evaluating Algebraic Expressions
Multiplication and Division with Fractions
Addition and Subtractions with Fractions
Decimals and Fractions
Order of Operations with Negative Numbers and Fractions
Translating Phrases into Algebraic Expressions

Chapter 2 Review

2.1 Simplifying and Evaluating Algebraic Expressions

In Exercises 1 – 10, simplify each expression by combining like terms.

1. $9x + 3x + x$

2. $3y + 4y - y$

3. $18x^2 + 7x^2$

4. $-a^2 - 5a^2 + 3a^2$

5. $4x + 2 - 7x$

6. $-5y - 6 + 10y$

7. $4(n+1) - n$

8. $5(x-6) - 4$

9. $2y^2 + 5y - y^2 + 2y$

10. $12a + 3a^2 - 8a + 4a^2$

In Exercises 11 – 20, first (a) simplify the expression and then (b) evaluate the simplified expression by using $x = -1$, $y = 4$, and $a = -2$.

11. $3(x+5) + 2(x-5)$

12. $-4(y+6) + 2(y-1)$

13. $5a^2 + 3a - 6a^2 + a + 7$

14. $x^2 + 6x + 3x - 10$

15. $3a^2 - 4a + 2$

16. $5y^2 - 8y + 26$

17. $5x + x^2 - 2x + x^2 - 13$

18. $20 - 17 + 13x - 20 + 4x$

19. $12y + 6y^2 + y^2 - 136 - 3y$

20. $100 - 10x^2 + 90x^2 + 20x$

2.2 Multiplication and Division with Fractions

Perform the indicated operations and reduce all answers in Exercises 21 – 28.

21. $\dfrac{-7}{8} \cdot \dfrac{3}{14}$

22. $\dfrac{-5}{16} \cdot \dfrac{4}{15}$

23. $\dfrac{6}{31} \div \dfrac{9}{62}$

24. $\dfrac{15}{27} \div \dfrac{5}{9}$

25. $\dfrac{9a}{15} \div \dfrac{35a}{12}$

26. $\dfrac{6x}{32} \div \dfrac{5x}{8}$

27. Find $\dfrac{1}{2}$ of $\dfrac{8}{11}$.

28. Find $\dfrac{2}{3}$ of $\dfrac{21}{24}$.

29. In a class of 35 students, 6 received a grade of A. What fraction of the class did not receive an A?

30. If you have $20 and spend $12 on a soft drink and hamburger, what fraction of your money did you spend? What fraction of your money do you still have?

2.3 Addition and Subtraction with Fractions

In Exercises 31 – 34, find the LCM for each set of numbers or algebraic expressions.

31. $\{36, 15, 10\}$

32. $\{25, 35, 30\}$

33. $\{10xy, 21xz, 35yz\}$

34. $\{24a^2, 30ab, 40ab^2\}$

Perform the indicated operations and reduce all answers.

35. $\dfrac{3}{7} + \dfrac{6}{7}$

36. $\dfrac{9}{10} + \dfrac{6}{10}$

37. $\dfrac{7}{16} - \dfrac{9}{16}$

38. $\dfrac{9}{32} - \dfrac{15}{32}$

39. $\dfrac{5}{12} + \dfrac{7}{15}$

40. $\dfrac{1}{6} + \dfrac{1}{8}$

41. $1 - \dfrac{3}{5}$

42. $1 - \dfrac{11}{16}$

43. $\dfrac{1}{5} + \dfrac{3}{8} + \dfrac{3}{10}$

44. $\dfrac{5}{24} - \dfrac{5}{18} + \dfrac{3}{8}$

45. $\dfrac{7}{6} - \dfrac{9}{10} + \dfrac{4}{15}$

46. $\dfrac{1}{28} + \dfrac{8}{21} - \dfrac{5}{12}$

47. $\dfrac{x}{7} - \dfrac{1}{14}$

48. $\dfrac{y}{4} - \dfrac{1}{3}$

49. $\dfrac{5}{a} + \dfrac{2}{3}$

50. $\dfrac{1}{5} - \dfrac{10}{x}$

2.4 Decimal Numbers and Fractions

Perform the indicated operations.

51. $15.2 \cdot 3.5$

52. $8.6 \cdot 9.4$

53. $(1.22)(-5.1)$

54. $(-7.5)(-31)$

55. $21.2\overline{)76.32}$

56. $0.43\overline{)0.0215}$

57. $6.12\overline{)495.72}$

58. $15.1\overline{)228.01}$

Use a calculator to find the decimal form of each fraction. If the decimal is non-terminating, write it using the bar notation over the repeating pattern of digits.

59. $\dfrac{7}{16}$

60. $\dfrac{2}{9}$

61. $\dfrac{5}{11}$

62. $\dfrac{1}{6}$

63. $\dfrac{4}{13}$

64. $\dfrac{33}{8}$

Use a TI-84 Plus graphing calculator to find the answers in Exercises 65 – 70 and to change each answer into fraction form.

65. $0.83 + 0.19$

66. $15.58 - 26.31$

67. $-27.8 - 10.002$

68. $12.175 - 26.32$

69. $\dfrac{5}{8} + \dfrac{1}{4} + \dfrac{3}{10}$

70. $\dfrac{3}{4} - \dfrac{9}{10} - \dfrac{6}{16}$

2.5 Order of Operations with Negative Numbers and Fractions

Use the rules for order of operations to evaluate (or simplify) each expression in Exercises 71 – 84.

71. $16 \div 2 \cdot 8 + 24$

72. $32 - 8 \cdot 14 \div 7$

73. $-36 \div (-2)^2 + 25 - 2(16 - 17)$

74. $9(-1 + 2^2) - 7 - 6 \cdot 2^2$

75. $2(-1 + 5^2) - 4 - 2 \cdot 3^2$

76. $10 - 2[(19 - 14) \div 5 + 3(-4)]$

77. $\dfrac{1}{9} + \dfrac{1}{2} \div \dfrac{1}{6} \cdot \dfrac{5}{12}$

78. $\left(\dfrac{1}{2} - 2\right) \div \left(1 - \dfrac{2}{3}\right)$

79. $\left(\dfrac{1}{15} - \dfrac{1}{12}\right) \div 6$

80. $5\dfrac{1}{2} \div \left(\dfrac{1}{5} - \dfrac{1}{15}\right)$

81. $\dfrac{x}{2} - \dfrac{1}{4}$

82. $\dfrac{4}{7y} - \dfrac{3}{7} \div \dfrac{1}{2}$

83. $-5\dfrac{1}{7} \div (2 + 1)^2$

84. $\dfrac{2}{3} \div \dfrac{7}{12} - \dfrac{2}{7} + \left(\dfrac{1}{2}\right)^2$

85. If the square of $\dfrac{1}{10}$ is added to the square of $\dfrac{3}{5}$, what is the sum?

86. If the square of $\dfrac{2}{3}$ is subtracted from the square of $-\dfrac{5}{9}$, what is the difference?

87. If the product of $\dfrac{3}{5}$ and $\dfrac{3}{4}$ is divided by the product of $\dfrac{4}{5}$ and $-\dfrac{1}{2}$, what is the quotient?

88. Find the result if the sum of $\dfrac{7}{8}$ and $\dfrac{3}{10}$ is subtracted from 2.

89. If the difference between $\dfrac{5}{16}$ and $\dfrac{1}{6}$ is multiplied by $-\dfrac{9}{16}$, what is the product?

90. Find the quotient if the difference between 1 and $\dfrac{1}{15}$ is divided by the sum of 1 and $\dfrac{1}{15}$.

2.6 Translating English Phrases and Algebraic Expressions

Translate each expression in Exercises 91 – 100 into an equivalent English phrase. (There may be more than one correct translation.)

91. $5n + 3$ **92.** $6x - 5$ **93.** $8y + 3y$ **94.** $7n + 4n$

95. $-3(x + 2)$ **96.** $-2(n + 6)$ **97.** $5(x - 3)$ **98.** $4(x - 10)$

99. $\dfrac{7n}{33}$ **100.** $\dfrac{50}{6y}$

Write the algebraic expression described by each word phrase in Exercises 101 – 110.

101. 5 times a number increased by 3 times the same number

102. 4 times the sum of a number and 10

103. 17 more than twice a number

104. a number decreased by 32

105. 28 decreased by 6 times a number

106. 72 plus 8 times the sum of a number and 2

107. twice a number plus 3

108. the quotient of a number and 14

109. 22 divided by the sum of a number and 9

110. 32 less than the product of a number and 10

Chapter 2 Test

In Exercises 1 – 3, (a) simplify each expression by combining like terms, and then (b) evaluate each simplified expression by using $x = -2$ and $y = 3$.

1. $7x + 8x^2 - 3x^2$ **2.** $5y - y - 6 + 2y$ **3.** $2(x - 5) + 3(x + 4)$

4. Find the LCM of $10xy^2, 18x^2y$, and $15xy$.

In Exercises 5 – 10, perform the indicated operations. Simplify any fractional answers.

5. $(5.61)(3.2)$ **6.** $20.2\overline{)27.27}$ **7.** $\dfrac{-15x}{7} \cdot \dfrac{42}{20x}$

8. $\dfrac{6}{7a} \div \dfrac{3a}{28}$ **9.** $\dfrac{4}{15} + \dfrac{1}{3} + \dfrac{3}{10}$ **10.** $\dfrac{7}{2y} - \dfrac{8}{2y} - \dfrac{3}{2y}$

Use the rules for order of operations to evaluate (or simplify) each expression in Exercises 11 – 15.

11. $36 \div (-9) \cdot 2 - 16 + 4(-5)$ **12.** $8 - 10\left[\left(2 - 3^2\right) \div 7 + 5\right]$

13. $\dfrac{7}{8} - \dfrac{1}{3} \div \dfrac{5}{6} + \dfrac{1}{4}$ **14.** $\dfrac{7}{12} \cdot \dfrac{9}{56} \div \dfrac{5}{16} - \left(\dfrac{1}{3}\right)^2$

15. $\dfrac{7}{x} + \dfrac{1}{5}$

16. Use a TI-84 Plus graphing calculator to find the value of the expression $3.6(7.1) + (6.4)^2 - 82.5$.

17. Use a TI-84 Plus graphing calculator to find the decimal form of the fraction $\dfrac{3}{7}$. If the decimal is non-terminating, write it using the bar notation over the repeating pattern of digits.

18. Translate each expression into an equivalent English phrase. (There may be more than one correct translation.)

     a. $5n + 18$ **b.** $3(x + 6)$ **c.** $42 - 7y$

19. Write an algebraic expression described by each English phrase.

     a. the product of a number and 6 decreased by 3
     b. twice the sum of the number and 5
     c. 4 less than twice the sum of a number and 15

20. Multiply the sum of $\dfrac{3}{4}$ and $\dfrac{9}{10}$ by the quotient of $\dfrac{2}{3}$ and $-\dfrac{8}{15}$. What is the product?

21. The sale price of a jacket is $60. This is $\dfrac{3}{4}$ of the original price. What was the original price?

22. You know that the gas tank in your car holds 20 gallons of gas and the gauge reads $\dfrac{1}{4}$ full.

    a. How many gallon will be needed to fill the tank?

    b. If the price of gas is $3.10 per gallon and you have $40, do you have enough cash to fill the tank? How much extra do you have or how much are you short?

Cumulative Review: Chapters 1 – 2

Perform the indicated operations in Exercises 1 – 8. Reduce each answer to lowest terms.

1. $18 + 7 - 8 + 3$ **2.** $9 - 16 - 11 + 4$ **3.** $26 - 17$ **4.** $18 - 33$

5. $(14)(-9)$ **6.** $(-27)(-17)$ **7.** $\dfrac{273}{-7}$ **8.** $\dfrac{-744}{-6}$

9. Find the average of each set of numbers:

  a. $\{17, 15, 22\}$ **b.** $\{-18, -22, -27, -37\}$

  c. $\{23.7, 38.9, 47.6\}$ (to the nearest tenth)

  d. $\{9.89, 7.48, 6.52, 5.33\}$ (to the nearest hundredth)

Find the LCM for each set of numbers or expressions in Exercises 10 and 11.

10. $\{18, 27, 36\}$ **11.** $\{12x, 9xy, 24xy^2\}$

Name the property of real numbers illustrated in Exercises 12 – 20.

12. $(3x)y = 3(xy)$ **13.** $x \cdot 5 = 5 \cdot x$ **14.** $4(3x + 1) = 12x + 4$

15. $x + 17 = 17 + x$ **16.** $(y + 11) + 4 = y + (11 + 4)$ **17.** $6 \cdot 1 = 6$

18. $17 \cdot \dfrac{1}{17} = 1$ **19.** $16 + 0 = 16$ **20.** $8 + (-8) = 0$

21. Find the values of x that satisfy each equation:

  a. $|x| = 4$ **b.** $|x| = -5$

22. List, then graph the **integers** that satisfy the inequality $|y| \le 6$.

Simplify Exercises 23 – 26 by using the rules for order of operations.

23. $3 \cdot 2^2 - 5$ **24.** $(18 + 2 \cdot 3) \div 4 \cdot 3$

25. $(13 \cdot 5 - 5) \div 2 \cdot 3$ **26.** $4[6 - (2 \cdot 5 - 2 \cdot 3^2) - 11]$

Perform the indicated operations in Exercises 27 – 34. Reduce each answer to lowest terms.

27. $\dfrac{11}{9} \div \dfrac{22}{15}$

28. $\dfrac{7b}{15} \div \dfrac{2b}{6}$

29. $\dfrac{8x}{21} \cdot \dfrac{7}{12x}$

30. $\dfrac{3}{8} + \dfrac{7}{10}$

31. $\dfrac{4}{5y} + \dfrac{2}{5y}$

32. $\dfrac{3}{x} + \dfrac{1}{6}$

33. $\dfrac{5}{12} - \dfrac{4}{9}$

34. $\dfrac{7}{4a} - \dfrac{9}{4a}$

Simplify the expressions in Exercises 35 – 40 by using the rules for order of operations. Reduce all answers to lowest terms.

35. $\dfrac{1}{2} + \dfrac{3}{8} \div \dfrac{3}{4} - \dfrac{5}{6}$

36. $\left(\dfrac{3}{4} - \dfrac{5}{6}\right)\left(\dfrac{6}{5} + \dfrac{2}{3}\right)$

37. $\dfrac{7}{18} + \dfrac{5}{24} - \left(\dfrac{3}{8}\right)^2$

38. $\left(\dfrac{3}{5}\right)^2 \div \dfrac{8}{5} \cdot \dfrac{8}{3}$

39. $\dfrac{1}{3} + \dfrac{2}{5} \cdot \dfrac{5}{8} \div \dfrac{3}{2} - \dfrac{1}{2}$

40. $\left(\dfrac{1}{3} - \dfrac{3}{4}\right) \div \left(\dfrac{2}{3} + \dfrac{2}{7}\right)$

Perform the indicated operations in Exercises 41 – 46.

41. $5.2 \cdot 13.5$

42. $9.6 \cdot 5.4$

43. $(3.23)(-6.1)$

44. $31.2\overline{)199.68}$

45. $0.47\overline{)2.3594}$

46. $6.34\overline{)143.284}$

In Exercises 47 – 50, use a calculator to find the decimal form of each fraction. If the decimal is non-terminating, write it using the bar notation over the repeating pattern of digits.

47. $\dfrac{9}{16}$

48. $\dfrac{7}{9}$

49. $\dfrac{5}{22}$

50. $\dfrac{2}{7}$

Use a TI-84 Plus graphing calculator to find the answers in Exercises 51 – 54 and to change each answer into fraction form.

51. $1.83 + 0.28$

52. $5.38 - 6.35$

53. $-8.8 - 11.003$

54. $0.15 - 6.45$

In Exercises 55 – 58, first (a) simplify each expression, and then (b) evaluate the expression for $x = 5$ and $y = -2$.

55. $|x+y| + |x-y|$

56. $3|x| + 5|y| - 2(x-y)$

57. $3x + 2 - x$

58. $3y^2 - 5y + 7 - y^2 + 2y - 10 + y^3$

59. Translate each English phrase into an algebraic expression:
  a. Twice a number decreased by 3
  b. The quotient of a number and 6 increased by 3 times the number
  c. Thirteen less than twice the sum of a number and 10

60. Translate each algebraic expression into an English phrase:
  a. $6 + 4n$ **b.** $18(n - 5)$

61. If the difference between $\dfrac{1}{2}$ and $\dfrac{11}{18}$ is divided by the sum of $\dfrac{2}{3}$ and $\dfrac{7}{15}$, what is the quotient?

62. From the sum of –2 and –11, subtract the product of –4 and 2.

63. Lucia is making punch for a party. She is making 5 gallons. The recipe calls for $\dfrac{2}{3}$ cup of sugar per gallon of punch. How many cups of sugar does she need?

64. Find the mean (to the nearest tenth) of the data in the following frequency distribution.

Clocked Speeds of Bicycles (to nearest mph)	Frequency
19	4
20	5
22	7
25	3
28	1

65. There are 4000 registered voters in Greenville, and $\dfrac{3}{8}$ of these voters are registered Republicans. A survey indicates that $\dfrac{2}{3}$ of the registered Republicans are in favor of Bond Measure *XX* and $\dfrac{3}{5}$ of the other registered voters are in favor of this measure.

  a. How many of the registered Republicans are in favor of Bond Measure *XX*?
  b. How many of the registered voters are in favor of Bond Measure *XX*?

Simplify each expression in Exercises 66 – 68.

66. $6x + 3x - 5 + 4x$ **67.** $\dfrac{a}{7} + \dfrac{a}{14} - \dfrac{a}{21}$ **68.** $\dfrac{3}{y} + \dfrac{1}{4}$

Simplify each expression in Exercises 69 and 70.

69. **a.** $5x + 5x$ **b.** $5x \cdot 5x$ **c.** $\dfrac{1}{5x} + \dfrac{1}{5x}$ **d.** $\dfrac{1}{5x} \cdot \dfrac{1}{5x}$ **e.** $\dfrac{1}{5x} \div \dfrac{1}{5x}$

70. **a.** $6y - 6y$ **b.** $6y(-6y)$ **c.** $\dfrac{1}{6y} + \dfrac{1}{6y}$ **d.** $\dfrac{1}{6y}\left(-\dfrac{1}{6y}\right)$ **e.** $\dfrac{1}{6y} \div \left(-\dfrac{1}{6y}\right)$

Solving Equations and Inequalities

Did You Know?

Traditionally, algebra has been defined to mean generalized arithmetic where letters represent numbers. For example, $3 + 3 + 3 + 3 = 4 \cdot 3$ is a special case of the more general algebraic statement that $x + x + x + x = 4 \cdot x$. The name algebra comes from an Arabic word, al-jabr, which means "to restore."

A notorious Moslem ruler, Harun al-Rashid, the caliph made famous in *Tales of the Arabian Nights*, and his son Al-Mamun brought to their Baghdad court many renowned Moslem scholars. One of these scholars was the mathematician Mohammed ibn-Musa Al-Khowarizmi, who wrote a text (c. A.D. 800) entitled Ihm *al-jabar w' al-muqabal*. The text included instructions for solving equations by adding terms to both sides of the equation, thus "restoring" equality. The abbreviated title of Al-Khowarizmi's text, *Al-jabar*, became our word for equation solving and operations on letters standing for numbers, **algebra**.

Al-Khowarizmi's algebra was brought to western Europe through Moorish Spain in a Latin translation done by Robert of Chester (c. A.D. 1140). Al-Khowarizmi's name may have sounded familiar to you. It eventually was translated as "algorithm," which came to mean any series of steps used to solve a problem. Thus we speak of the division algorithm used to divide one number by another. Al-Khowarizmi is known as the father of algebra just as Euclid is known as the father of geometry.

One of the hallmarks of algebra is its use of specialized notation that enables complicated statements to be expressed using compact notation. Al-Khowarizmi's algebra used only a few symbols with most statements written out in words. He did not even use symbols for numbers! The development of algebraic symbols and notation occurred over the next 1000 years.

3.1   Solving Linear Equations:
      $x + b = c$ and $ax = c$

3.2   Solving Linear Equations:
      $ax + b = c$

3.3   More Linear Equations:
      $ax + b = dx + c$

3.4   Solving Linear Inequalities and
      Applications

3.5   Working with Formulas

3.6   Applications: Number Problems and
      Consecutive Integers

3.7   Applications: Percent Problems
      (Discount, Taxes, Commission, Profit,
      and Others)

3.8   Formulas in Geometry

"Algebra is generous. She often gives more than is asked of her."

Jean le Rond d' Alembert (1717? – 1783)

3.1   Solving Linear Equations: $x + b = c$ and $ax = c$

Objectives

After completing this section, you will be able to:

*1. Define the term **linear equation**.*

2. Solve equations of the form $x + b = c$.

3. Solve equations of the form $ax = c$.

4. Understand the four basic steps in solving applications.

One of the most important basic skills needed in mathematics is solving equations. That is, find the value of an unknown quantity that satisfies the conditions of a problem. Such skills are used in determining percents of profit, gas mileage, taxes, perimeters and areas of geometric figures, and in dealing with distance, rate, and time, and so on. We begin by learning to solve the simplest forms of equations and expand the related skills to solving equations more complex and difficult to solve. In any case, the basic skills and techniques we develop now are fundamental to the ideas related to solving a variety of types of equations. So, have patience and learn to follow the recommended steps in solving even the basic forms of equations. In this section, we will discuss solving linear equations in the following two forms:

$$x + b = c \quad \text{and} \quad ax = c.$$

In these equations we treat a, b, and c as constants and x as the unknown quantity.

Then, in the following section, we will combine the techniques developed here and discuss solving linear equations in the general form:

$$ax + b = c.$$

An **equation** is a statement that two algebraic expressions are equal. That is, both expressions represent the same number. If an equation contains a variable, any number that gives a true statement when substituted for the variable is called a **solution** to the equation. In a case where an equation has more than one solution, the solutions are said to form a **solution set**. The process of finding the solution is called **solving the equation**.

Linear Equation in x

If a, b, and c are **constants** and $a \neq 0$ then a **linear equation in x** is an equation that can be written in the form

$$ax + b = c.$$

(**Note:** A linear equation in x is also called a **first-degree equation in x** because the variable x can be written with the exponent 1. That is, $x = x^1$.)

Solving Equations of the Form $x + b = c$

To begin, we need the **Addition Principle of Equality**.

Addition Principle of Equality

If the same algebraic expression is added to both sides of an equation, the new equation has the same solutions as the original equation. Symbolically, if A, B, and C are algebraic expressions, then the equations

$$A = B$$

and

$$A + C = B + C$$

have the same solutions.

Equations with the same solutions are said to be **equivalent equations**.

The objective of solving linear (or first-degree) equations is to get the variable by itself (with a coefficient of $+1$) on one side of the equation and any constants on the other side. The following procedure will help in solving linear equations such as

$$x - 3 = 7, \quad -11 = y + 5, \quad 3z - 2z + 2.5 = 3.2 + 0.8, \quad \text{and} \quad x - \frac{2}{5} = \frac{3}{10}.$$

Procedure for Solving Linear Equations that Simplify to the Form $x + b = c$

1. Combine any like terms on each side of the equation.

2. Use the Addition Principle of Equality and add the opposite of the constant term to both sides. The objective is to isolate the variable on one side of the equation (either the left side or the right side) with a coefficient of $+1$.

3. Check your answer by substituting it for the variable in the original equation.

Every linear equation has exactly one solution. This means that once a solution has been found, there is no need to search for another solution.

Example 1: Solving $x + b = c$

Solve each of the following linear equations:

a. $x - 3 = 7$

b. $-11 = y + 5$

c. $3z - 2z + 2.5 = 3.2 + 0.8$

d. $x - \dfrac{2}{5} = \dfrac{3}{10}$

Solutions:

a. $x - 3 = 7$

$x - 3 + 3 = 7 + 3$ Add 3 (the opposite of -3) to both sides.

$x = 10$ Simplify.

Check: $x - 3 = 7$

$10 - 3 \overset{?}{=} 7$ Substitute $x = 10$.

$7 = 7$ True statement

b. $-11 = y + 5$ Note that the variable can be on the right side.

$-11 - 5 = y + 5 - 5$ Add -5 (the opposite of $+5$) to both sides.

$-16 = y$ Simplify.

Check: $-11 = y + 5$

$-11 \overset{?}{=} -16 + 5$ Substitute $y = -16$.

$-11 = -11$ True statement

c. $3z - 2z + 2.5 = 3.2 + 0.8$

$z + 2.5 = 4.0$ Combine like terms.

$z + 2.5 - 2.5 = 4.0 - 2.5$ Add -2.5 (the opposite of $+2.5$) to both sides.

$z = 1.5$ Simplify.

Check: $3z - 2z + 2.5 = 3.2 + 0.8$ Substitute $z = 1.5$.

$3(1.5) - 2(1.5) + 2.5 \overset{?}{=} 3.2 + 0.8$ Simplify.

$4.5 - 3.0 + 2.5 \overset{?}{=} 3.2 + 0.8$ True statement

$4.0 = 4.0$

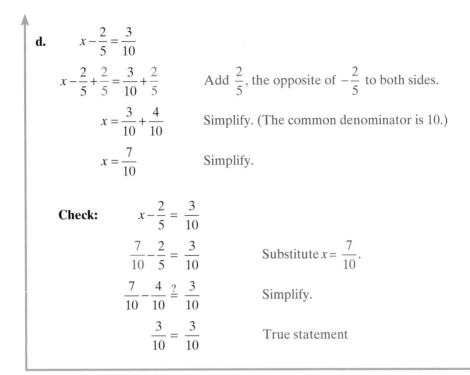

d.
$$x - \frac{2}{5} = \frac{3}{10}$$

$$x - \frac{2}{5} + \frac{2}{5} = \frac{3}{10} + \frac{2}{5} \qquad \text{Add } \frac{2}{5}, \text{ the opposite of } -\frac{2}{5} \text{ to both sides.}$$

$$x = \frac{3}{10} + \frac{4}{10} \qquad \text{Simplify. (The common denominator is 10.)}$$

$$x = \frac{7}{10} \qquad \text{Simplify.}$$

Check:
$$x - \frac{2}{5} = \frac{3}{10}$$

$$\frac{7}{10} - \frac{2}{5} = \frac{3}{10} \qquad \text{Substitute } x = \frac{7}{10}.$$

$$\frac{7}{10} - \frac{4}{10} \stackrel{?}{=} \frac{3}{10} \qquad \text{Simplify.}$$

$$\frac{3}{10} = \frac{3}{10} \qquad \text{True statement}$$

Note that, as illustrated in Example 1, variables other than x may be used.

Solving Equations of the Form $ax = c$

To solve equations of the form $ax = c$, where $a \neq 0$, we can use the idea of the reciprocal of the coefficient a. For example, as we studied in Section 2.1, the reciprocal of $\frac{3}{4}$ is $\frac{4}{3}$ and $\frac{3}{4} \cdot \frac{4}{3} = 1$. Also, we need the **Multiplication (or Division) Principle of Equality** as stated on the next page.

Multiplication (or Division) Principle of Equality

If both sides of an equation are multiplied by (or divided by) the same nonzero constant, the new equation has the same solutions as the original equation. Symbolically, if A and B are algebraic expressions and C is any nonzero constant, then the equations

$$A = B$$

and $AC = BC$ *where* $C \neq 0$

and $\dfrac{A}{C} = \dfrac{B}{C}$ *where* $C \neq 0$

have the same solutions. We say that the equations are equivalent.

Remember that the objective is to get the variable by itself on one side of the equation. That is, we want the variable to have +1 as its coefficient. The following procedure will accomplish this.

Procedure for Solving Linear Equations that Simplify to the Form $ax = c$

1. *Combine any like terms on each side of the equation.*

2. *Use the Multiplication (or Division) Principle of Equality and multiply both sides of the equation by the reciprocal of the coefficient of the variable. (**Note**: This is the same as dividing both sides of the equation by the coefficient.) Thus the coefficient of the variable will become +1.*

3. *Check your answer by substituting it into the original equation.*

Example 2: Solving $ax = c$

Solve the following equation.

a. $5x = 20$

Solution: $5x = 20$

$\dfrac{1}{5} \cdot (5x) = \dfrac{1}{5} \cdot 20$ Multiply by $\dfrac{1}{5}$, the reciprocal of 5.

$\left(\dfrac{1}{5} \cdot 5\right)x = \dfrac{1}{5} \cdot \dfrac{20}{1}$ Use the associative property of multiplication.

$1 \cdot x = 4$ Simplify.

$x = 4$

Check:
$$5x = 20$$
$$5 \cdot 4 \stackrel{?}{=} 20 \qquad \text{Substitute } x = 4.$$
$$20 = 20 \qquad \text{True statement}$$

Multiplying by the reciprocal of the coefficient is the same as **dividing** by the coefficient itself. So, we can multiply both sides by $\frac{1}{5}$, as we did, or we can divide both sides by 5. In either case, the coefficient of *x* becomes +1.

$$5x = 20$$
$$\frac{5x}{5} = \frac{20}{5} \qquad \text{Divide both sides by 5.}$$
$$\qquad\qquad \text{Simplify.}$$
$$x = 4$$

b. $1.1x + 0.2x = 12.2 - 3.1$

Solution: When decimal coefficients or constants are involved, you might want to use a calculator to perform some of the arithmetic.

$$1.1x + 0.2x = 12.2 - 3.1$$
$$1.3x = 9.1 \qquad\qquad \text{Combine like terms.}$$
$$\frac{1.3x}{1.3} = \frac{9.1}{1.3} \qquad\qquad \text{Use a calculator or pencil and paper to divide.}$$
$$x = 7.0$$

Check:
$$1.1x + 0.2x = 12.2 - 3.1$$
$$1.1(7) + 0.2(7) \stackrel{?}{=} 12.2 - 3.1 \qquad \text{Substitute } x = 7.$$
$$7.7 + 1.4 \stackrel{?}{=} 9.1 \qquad\qquad \text{Simplify.}$$
$$9.1 = 9.1 \qquad\qquad\qquad \text{True statement}$$

c. $\dfrac{4x}{5} = \dfrac{3}{10}$ (This could be written $\dfrac{4}{5}x = \dfrac{3}{10}$ because $\dfrac{4}{5}x$ is the same as $\dfrac{4x}{5}$.)

Solution:
$$\frac{4x}{5} = \frac{3}{10}$$
$$\frac{5}{4} \cdot \frac{4}{5}x = \frac{5}{4} \cdot \frac{3}{10} \qquad \text{Multiply both sides by } \frac{5}{4}.$$
$$1x = \frac{1 \cdot \cancel{5}}{4} \cdot \frac{3}{2 \cdot \cancel{5}} \qquad \text{Simplify.}$$
$$x = \frac{3}{8}$$

Continued on next page...

Check: $\dfrac{4x}{5} = \dfrac{3}{10}$

$\dfrac{4}{5} \cdot \dfrac{3}{8} \overset{?}{=} \dfrac{3}{10}$ Substitute $x = \dfrac{3}{8}$.

$\dfrac{3}{10} = \dfrac{3}{10}$ True statement

d. $-x = 4$

Solution: $-x = 4$

$-1x = 4$ -1 is the coefficient of x.

$\dfrac{-1x}{-1} = \dfrac{4}{-1}$ Divide by -1 so that the coefficient is $+1$.

$x = -4$

Check: $-x = 4$

$-(-4) \overset{?}{=} 4$ Substitute $x = -4$.

$4 = 4$ True statement

Problem Solving

George Pólya (1877 – 1985), a famous professor at Stanford University, studied the process of discovery learning. Among his many accomplishments, he developed the following four-step process as an approach to problem solving:

1. Understand the problem.
2. Devise a plan.
3. Carry out the plan.
4. Look back over the results.

For a complete discussion of these ideas, see *How To Solve It by Pólya* (Princeton University Press, 2nd edition, 1957). The following quote by Pólya illustrates his sense of humor and his understanding of students' dilemmas: "The traditional mathematics professor of the popular legend is absent-minded. He usually appears in public with a lost umbrella in each hand. He prefers to face a blackboard and to turn his back on the class. He writes *a*, he says *b*, he means *c*, but it should be *d*. Some of his sayings are handed down from generation to generation."

There are a variety of types of applications discussed throughout this text and subsequent courses in mathematics, and you will find these four steps helpful as guidelines for understanding and solving all of them. Applying the necessary skills to solve exercises, such as combining like terms or solving equations, is not the same as accumulating the knowledge to solve problems. **Problem solving can involve careful reading, reflection, and some original or independent thought**.

<hr>

Basic Steps for Solving Applications

1. *Understand the problem. For example,*

   a. *Read the problem carefully, maybe several times.*

   b. *Understand all the words.*

   c. *If it helps, restate the problem in your own words.*

   d. *Be sure that there is enough information.*

2. *Devise a plan. For example,*

   a. *Guess, estimate, or make a list of possibilities.*

   b. *Draw a picture or diagram.*

   c. *Represent the unknown quantity with a variable and form an equation.*

3. *Carry out the plan. For example,*

   a. *Try all the possibilities you have listed.*

   b. *Study your picture or diagram for insight into the solution.*

   c. *Solve any equation that you may have set up.*

4. *Look back over the results. For example,*

   a. *Can you see an easier way to solve the problem?*

   b. *Does your solution actually work? Does it make sense in terms of the wording of the problem? Is it reasonable?*

   c. *If there is an equation, check your answer in the equation.*

NOTES

You may find that many of the applications in this section can be solved by "reasoning," and there is nothing wrong with that approach. Reasoning is a fundamental part of all of mathematics. However, keep in mind that the algebraic techniques you are learning are important. They also involve reasoning and will prove very useful in solving more complicated problems in later sections and in later courses.

Example 3: Applications

a. When a $2.13 tax was added to the price of a skirt, the total bill was $37.63. What was the price of the skirt?

Solution: Let x = price of the skirt

Now, use the relationship:

$$\text{price of skirt} + \text{tax} = \text{total bill}$$

Set up the equation:

$$x + 2.13 = 37.63$$

Solve the equation:

$$x + 2.13 - 2.13 = 37.63 - 2.13$$ Use the Addition Principle by adding −2.13 to both sides.

$$x = 35.50$$ Simplify both sides.

The price of the skirt was $35.50.

b. The original price of a DVD player was reduced by $45.50. The sale price was $215.90. What was the original price?

Solution: Let y = original price of the DVD player.

Now, use the relationship:

$$\text{original price of DVD player} - \text{reduction} = \text{sale price}$$

Set up the equation:

$$y - 45.50 = 215.90$$

Solve the equation:

$$y - 45.50 + 45.50 = 215.90 + 45.50$$ Use the Addition Principle by adding 45.50 to both sides.

$$y = 261.40$$ Simplify both sides.

The original price of the DVD player was $261.40.

c. An exam is given with 15 problems and the students are allowed $1\frac{1}{2}$ hours to take it. How many **minutes** are allotted for each problem?

Solution: Here we let x = number of minutes allotted per problem. Then the product $15x$ will equal the total time for the exam. Since the time allowed is given in hours, we make the change

$$1\frac{1}{2} \text{ hours } = 90 \text{ minutes.}$$

Now use the relationship: number of problems times the time for each = total time:

$$15x = 90$$

$$\frac{15x}{15} = \frac{90}{15}$$

$$x = 6.$$

Each problem is allotted 6 minutes.

Practice Problems

Solve the following equations.

1. $-16 = x + 5$ **2.** $6y - 1.5 = 7y$ **3.** $4x = -20$

4. $\frac{3}{5}y = 33$ **5.** $1.7z + 2.4z = 8.2$ **6.** $-x = -8$

7. $3x = 10$ **8.** $5x - 4x + 1.6 = -2.7$ **9.** $\frac{4}{5} = 3x$

3.1 Exercises

Solve each of the equations in Exercises 1 – 50. (Remember that the objective is to isolate the variable on one side of the equation with a coefficient of +1.)

1. $x - 6 = 1$ **2.** $x - 10 = 9$ **3.** $y + 7 = 3$

4. $y + 12 = 5$ **5.** $x + 15 = -4$ **6.** $x + 17 = -10$

7. $22 = n - 15$ **8.** $36 = n - 20$ **9.** $6 = z + 12$

Answers to Practice Problems: **1.** $-21 = x$ **2.** $-1.5 = y$ **3.** $x = -5$ **4.** $y = 55$ **5.** $z = 2$ **6.** $x = 8$ **7.** $x = \dfrac{10}{3}$ **8.** $x = -4.3$ **9.** $\dfrac{4}{15} = x$

10. $18 = z + 1$

11. $x - 20 = -15$

12. $x - 10 = -11$

13. $y + 3.4 = -2.5$

14. $y + 1.6 = -3.7$

15. $x + 3.6 = 2.4$

16. $x + 2.7 = 3.8$

17. $5x = 45$

18. $9x = 108$

19. $32 = 4y$

20. $51 = 17y$

21. $\dfrac{3x}{4} = 15$

22. $\dfrac{5x}{7} = 65$

23. $4 = \dfrac{2y}{5}$

24. $46 = \dfrac{46y}{5}$

25. $4x - 3x = 10 - 12$

26. $7x - 6x = 13 + 15$

27. $7x - 8x = 13 - 25$

28. $10n - 11n = 20 - 14$

29. $3n - 2n + 6 = 14$

30. $7n - 6n + 13 = 22$

31. $1.7y + 1.3y = 6.3$

32. $2.5y + 7.5y = 4.2$

33. $\dfrac{3}{4}x = \dfrac{5}{3}$

34. $\dfrac{5}{6}x = \dfrac{5}{3}$

35. $7.5x = -99.75$

36. $-14 = 0.7x$

37. $1.5y - 0.5y + 6.7 = -5.3$

38. $2.6y - 1.6y - 5.1 = -2.9$

39. $10x - 9x - \dfrac{1}{2} = -\dfrac{9}{10}$

40. $6x - 5x + \dfrac{3}{4} = -\dfrac{1}{12}$

41. $1.4x - 0.4x + 2.7 = -1.3$

42. $3.5y - 2.5y - 6.3 = -1.0 - 2.5$

43. $6.1x - 5.1x + 1.3 = 2.5 + 1.7$

44. $1.5n - 0.5n + 2.6 = 2.5 - 3.6$

45. $6.2 = -3.5 + 7n - 6n$

46. $-7.2 = 1.3n - 0.3n - 1.0$

47. $-8.2 - 1.5 = 2.6y - 1.6y + 3.5$

48. $-9.2 - 1.7 = 2.7n - 1.7n - 5.4$

49. $1.7x = -5.1 - 1.7$

50. $3.2x = 2.8 - 9.2$

Problems Related to Geometry
The perimeter, P, of a triangle is equal to the sum of the lengths of the sides a, b, and c.
($P = a + b + c$)

51. One side of a triangle is 11.5 inches long. A second side is 8.75 inches long. If the perimeter is 24 inches, find the length of the third side.

52. Two sides of a triangle measure 43.2 cm and 26.1 cm. Find the length of the third side if the perimeter is 98 cm.

53. The perimeter of a square is 4 times the length of a side. ($P = 4s$) Find the length of a side if the perimeter of a square is $64\frac{1}{2}$ meters.

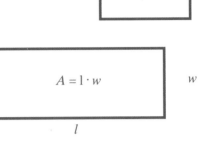

54. The area of a rectangle is found by multiplying the length times the width. ($A = lw$) Find the width of a rectangle with area 75 m² and length 10 m.

$A = l \cdot w$

55. The volume of a rectangular box of cereal is the product of its length, width, and height. ($V = lwh$) Find the width of a cereal box with volume 1260 cm³ if its length is 14 cm and its height is 20 cm.

14 cm

20 cm

The sum of the measures of the three angles of a triangle is 180° $\left(\alpha + \beta + \gamma = 180°\right)$. **(Note:** α, β, and γ **are the Greek lowercase letters alpha, beta, and gamma, respectively.)**

56. A triangle with two equal angles is called an **isosceles** triangle. If two angles of an isosceles triangle both measure 45°, is the triangle a **right** triangle? (Does the third angle measure 90°?) Draw a sketch of such a triangle.

57. If all three angles of a triangle measure less than 90°, then the triangle is called an **acute** triangle. Is a triangle with two angles of measure 55° and 20° an acute triangle? What is the measure of the third angle? (**Note:** If a triangle is not acute and not right, then the measure of one of the angles is more than 90°, and it is called an **obtuse** triangle.)

Problems Related to Profit
The profit, *P*, is equal to the revenue, *R*, minus the cost, *C*. (*P* = *R* − *C*)

58. Find the revenue (income) of a company that shows a profit of $3.2 million and costs of $1.8 million.

59. A suit cost the store $675 and the owner said that he wants to make a profit of $180. What price should he mark on the suit?

60. An automobile is advertised for a price of $22,500, but when the customer went to buy the car with the options that she wanted, the total price came to $25,300 (plus tax). What was the cost of the options?

Problems Related to Distance, Rate, and Time
The distance traveled, *d*, is equal to the product of the rate, *r*, and the time *t*. ($d = rt$)

61. How long will a truck driver take to travel 350 miles if he averages 50 mph?

62. What is the average rate of speed of a biker who bikes 21.92 miles in 68.5 min?

63. How long will it take a train traveling at 40 mph to go 140 miles?

Calculator Problems

Use a TI-84 Plus graphing calculator to help in solving the following equations.

64. $y + 32.861 = -17.892$ **65.** $x - 41.625 = 59.354$

66. $17.61x - 16.61x + 27.059 = 9.845$ **67.** $14.83y - 8.65 - 13.83y = 17.437 + 1.0$

68. $2.637x = 648.702$ **69.** $-0.3057y = 316.7052$

70. $-x = 145.6 + 17.89 - 10.32$ **71.** $-y = 143.5 + 178.462 - 200$

Hawkes Learning Systems: Introductory Algebra

Solving Linear Equations Using Addition and Subtraction
Applications of Linear Equations: Addition and Subtraction
Solving Linear Equations Using Multiplication and Division
Applications of Linear Equations: Multiplication and Division

3.2 Solving Linear Equations: *ax + b = c*

After completing this section, you will be able to:

*Solve equations of the form **ax** + **b** = **c**.*

In Section 3.1 the equations to be solved were in one of two forms: $x + b = c$ or $ax = c$. In the first type the coefficient of the variable, *x*, was always +1 and we had to add or subtract constants to solve the equation. (We used the Addition Principle.) In the second type, we had to multiply both sides by the reciprocal of the coefficient of the variable (or divide both sides by the coefficient itself). Now, we will apply both of these techniques in solving the same equation. That is, in the form $ax + b = c$ the coefficient *a* may be a number other than 1.

The **General Procedure for Solving Linear Equations** is now a combination of those stated in Section 3.1.

Procedure for Solving Linear Equations that Simplify to the Form $ax+b=c$

1. *Combine like terms on both sides of the equation.*
2. *Use the Addition Principle of Equality and add the opposite of the constant term to both sides.*
3. *Use the Multiplication (or Division) Principle of Equality to multiply both sides by the reciprocal of the coefficient of the variable (or divide both sides by the coefficient itself). The coefficient of the variable will become +1.*
4. *Check your answer by substituting it into the original equation.*

Example 1: Solving Linear Equations

Solve each of the following equations.

a. $5x + 3 - 2x = -18$

Solution: $5x + 3 - 2x = -18$	Write the equation.
$3x + 3 = -18$	Combine like terms.
$3x + 3 - 3 = -18 - 3$	Add -3 to both sides.
$3x = -21$	Simplify.
$\dfrac{3x}{3} = \dfrac{-21}{3}$	Divide both sides by 3.
$x = -7$	Simplify.

Check: $5x + 3 - 2x = -18$	
$5(-7) + 3 - 2(-7) \overset{?}{=} -18$	Substitute $x = -7$.
$-35 + 3 + 14 \overset{?}{=} -18$	Simplify.
$-18 = -18$	True statement

b. $-26 = 2y - 14 - 4y$

Solution: $-26 = 2y - 14 - 4y$	Write the equation.
$-26 = -2y - 14$	Combine like terms.
$-26 + 14 = -2y - 14 + 14$	Add 14 to both sides.
$-12 = -2y$	Simplify.
$\dfrac{-12}{-2} = \dfrac{-2y}{-2}$	Divide both sides by -2.
$6 = y$	Simplify.

Check: $-26 = 2y - 14 - 4y$	
$-26 = 2(6) - 14 - 4(6)$	Substitute $y = 6$.
$-26 \overset{?}{=} 12 - 14 - 24$	Simplify.
$-26 = -26$	True statement

Example 2: Solving Linear Equations

Solve each of the following equations.

a. $16.53 - 18.2z - 7.43 = 0$

Solution:

$16.53 - 18.2z - 7.43 = 0$	Write the equation.
$100(16.53 - 18.2z - 7.43) = 100(0)$	Multiply both sides by 100. (This results in integer coefficients.)
$1653 - 1820z - 743 = 0$	Simplify.
$910 - 1820z = 0$	Combine like terms.
$910 - 910 - 1820z = 0 - 910$	Subtract 910 from both sides.
$-1820z = -910$	Simplify.
$\dfrac{-1820z}{-1820} = \dfrac{-910}{-1820}$	Divide both sides by the coefficient -1820.
$z = 0.5 \left(\text{or } z = \dfrac{1}{2} \right)$	Simplify.

Check:

$16.53 - 18.2z - 7.43 = 0$	
$16.53 - 18.2(0.5) - 7.43 \overset{?}{=} 0$	Substitute $z = 0.5$.
$16.53 - 9.10 - 7.43 = 0$	Simplify.
$0 = 0$	True statement

b. $\dfrac{1}{2}x + \dfrac{3}{4}x + \dfrac{7}{2} - \dfrac{2}{3}x = 0$

Solution:

$\dfrac{1}{2}x + \dfrac{3}{4}x + \dfrac{7}{2} - \dfrac{2}{3}x = 0$	Write the equation.
$12\left(\dfrac{1}{2}x + \dfrac{3}{4}x + \dfrac{7}{2} - \dfrac{2}{3}x\right) = 12(0)$	Multiply both sides by 12. (the LCM of the denominators.)
$12\left(\dfrac{1}{2}x\right) + 12\left(\dfrac{3}{4}x\right) + 12\left(\dfrac{7}{2}\right) - 12\left(\dfrac{2}{3}x\right) = 12(0)$	Apply the distributive property.
$6x + 9x + 42 - 8x = 0$	Simplify.
$7x + 42 = 0$	Simplify.
$7x + 42 - 42 = 0 - 42$	Add -42 to both sides.
$7x = -42$	Simplify.
$\dfrac{7x}{7} = \dfrac{-42}{7}$	Divide both sides by 7.
$x = -6$	Simplify. Checking will show that -6 is the solution.

NOTES

The techniques illustrated in Example 2a and 2b could have been used to solve equations with decimals or fractions in Section 3.1. That is,
1. either work with the decimals or fractions as they are, or
2. multiply both sides in such a way that constants and coefficients are integers.

NOTES

ABOUT CHECKING: Checking can be quite time-consuming and need not be done for every problem. This is particularly important on exams. You should check only if you have time after the entire exam is completed.

Example 3: Geometry

The perimeter, P, of a rectangle is the sum of twice the length, l, and twice the width, w. $\left(P = 2l + 2w \right)$

$$P = 2l + 2w$$

w

l

The perimeter of a football field is 290 yards. If the length is 100 yards, what is the width?

Solution:

$$P = 2l + 2w$$

$290 = 2 \cdot 100 + 2w$	Substitute 290 for P and 100 for l.
$290 = 200 + 2w$	Simplify.
$290 - 200 = 200 + 2w - 200$	Subtract 200 from both sides.
$90 = 2w$	Simplify.
$\dfrac{90}{2} = \dfrac{2w}{2}$	Divide both sides by 2.
$45 = w$	

The width of the field is 45 yards.

Practice Problems

Solve the following linear equations.

1. $x + 14 - 8x = -7$

2. $2.1 = 3.1y - 6.4y - 1.2$

3. $n - \dfrac{2n}{3} - \dfrac{1}{2} = \dfrac{1}{6}$

4. $\dfrac{3x}{14} + \dfrac{1}{2} - \dfrac{x}{7} = 0$

3.2 Exercises

Solve each of the following linear (first-degree) equations.

1. $3x + 11 = 2$

2. $3x + 10 = -5$

3. $5x - 4 = 6$

4. $4y - 8 = -12$

5. $-5x + 2.9 = 3.5$

6. $3x + 2.7 = -2.7$

7. $5y - 3y + 2 = 2$

8. $6y + 8y - 7 = -7$

9. $x - 4x + 25 = 31$

10. $3y + 9y - 13 = 11$

11. $-20 = 7y - 3y + 4$

12. $-20 = 5y + y + 16$

13. $4n - 10n + 35 = 1 - 2$

14. $-5n - 3n + 2 = 34$

15. $3n - 15 - n = 1$

16. $2n + 12 + n = 0$

17. $5.4x - 0.2x = 0$

18. $0 = 5.1x + 0.3x$

19. $\dfrac{5}{8}x - \dfrac{1}{4}x + \dfrac{1}{2} = \dfrac{3}{10}$

20. $\dfrac{1}{2}x + \dfrac{3}{4}x - \dfrac{5}{3} = \dfrac{5}{6}$

21. $\dfrac{y}{2} + \dfrac{1}{5} = 3$

22. $\dfrac{y}{3} - \dfrac{2}{3} = 7$

23. $\dfrac{7}{8} = \dfrac{3}{4}x - \dfrac{5}{8}$

24. $\dfrac{1}{10} = \dfrac{4}{5}x + \dfrac{3}{10}$

25. $\dfrac{y}{7} + \dfrac{y}{28} + \dfrac{1}{2} = \dfrac{3}{4}$

26. $\dfrac{5y}{6} - \dfrac{7y}{8} - \dfrac{1}{12} = \dfrac{1}{3}$

27. $x + 1.2x + 6.9 = -3.0$

28. $3x - 0.75x - 1.72 = 3.23$

29. $10 = x - 0.5x + 32$

30. $33 = y + 3 - 0.4y$

31. $2.5x + 0.5x - 3.5 = 2.5$

32. $4.7 - 0.5x - 0.3x = -0.1$

33. $6.4 + 1.2x + 0.3x = 0.4$

34. $5.2 - 1.3x - 1.5x = -0.4$

35. $-12.13 = 2.42y + 0.6y - 13.64$

36. $-7.01 = 1.75x + 3.05x - 8.45$

37. $-0.4x + x + 17.2 = 18.1$

38. $y - 0.75y + 13.76 = 14.66$

39. $0 = 17.3x - 15.02x - 0.456$

40. $0 = 20.5x - 16.35x + 0.1245$

Answers to Practice Problems: 1. $x = 3$ **2.** $y = -1$ **3.** $n = 2$ **4.** $x = -7$

For Exercises 41 and 42, use the formula discussed in Example 3: $P = 2l + 2w$.

41. One hundred eighty-four ft of fencing is needed to enclose a rectangular-shaped garden plot. If the plot is 28 ft wide, what is the length?

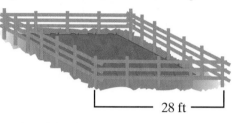

28 ft

42. A rectangular shaped parking lot is to have a perimeter of 450 yards. If the width must be 90 yards because of a building code, what will the length need to be?

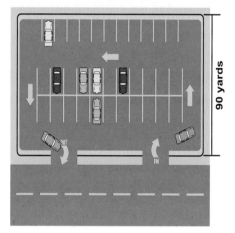

90 yards

When purchasing an item on an installment plan, you find the total cost, *C*, by multiplying the monthly payment, *p*, by the number of months, *t*, and adding the product to the down payment, *d*. ($C = pt + d$)

43. A refrigerator costs $857.60 if purchased on the installment plan. If the monthly payments are $42.50 and the down payment is $92.60, how long will it take to pay for the refrigerator?

$$$$

44. A used automobile will cost $3250 if purchased on an installment plan. If the monthly payments are $115 for 24 months, what will be the down payment?

Calculator Problems

Use a TI-84 Plus graphing calculator to help in solving the linear equations in Exercises 45 – 48.

45. $0.15x + 5.23x - 17.815 = 15.003$

46. $15.97y - 12.34y + 16.95 = 8.601$

47. $13.45x - 20x - 17.36 = -24.696$

48. $26.75y - 30y + 23.28 = 4.4625$

Writing and Thinking About Mathematics

49. Discuss, briefly, how you would apply Pólya's four-step problem-solving process to a problem that you faced today. (For example, what route to take to school, what time to spend studying algebra, what movie to see, etc.)
   a. What was the problem?
   b. What was the plan?
   c. How did you carry out the plan?
   d. Did your solution make sense? Could you have solved the problem in a different way?

Hawkes Learning Systems: Introductory Algebra

Solving Linear Equations

3.3   More Linear Equations: $ax + b = dx + c$

After completing this section, you will be able to:

Solve equations of the form $ax + b = dx + c$.

Now we are ready to solve linear equations of the most general form $ax + b = dx + c$ where constants and variables may be on both sides. There may also be parentheses or other symbols of inclusion. **Remember that the objective is to get the variable on one side of the equation by itself with a coefficient of +1.**

General Procedure for Solving Linear Equations that Simplify to the Form $ax+b = dx+c$

1. *Simplify each side of the equation by removing any grouping symbols and combining like terms on both sides of the equation.*

2. *Use the Addition Principle of Equality and add the opposite of the constant term and/or variable term to both sides so that variables are on one side and constants are on the other side.*

3. *Use the Multiplication (or Division) Principle of Equality to multiply both sides by the reciprocal of the coefficient of the variable (or divide both sides by the coefficient itself). The coefficient of the variable will become +1.*

4. *Check your answer by substituting it into the original equation.*

Example 1: Variables on Both Sides

Solve the following equations.

a. $5x + 3 = 2x - 18$ **b.** $4x + 1 - x = 2x - 13 + 5$ **c.** $6y + 2.5 = 7y - 3.6$

Solutions:

a.

$5x + 3 = 2x - 18$	Write the equation.
$5x + 3 - 3 = 2x - 18 - 3$	Add -3 to both sides.
$5x = 2x - 21$	Simplify.
$5x - 2x = 2x - 21 - 2x$	Add $-2x$ to both sides.
$3x = -21$	Simplify.
$\dfrac{\cancel{3}x}{\cancel{3}} = \dfrac{-21}{3}$	Divide both sides by 3. (Or, multiply by $\dfrac{1}{3}$.)
$x = -7$	Simplify.

Check:

$5x + 3 = 2x - 18$	
$5(-7) + 3 \stackrel{?}{=} 2(-7) - 18$	Substitute $x = -7$.
$-35 + 3 \stackrel{?}{=} -14 - 18$	Simplify.
$-32 = -32$	True statement

b.

$4x + 1 - x = 2x - 13 + 5$	Write the equation.
$3x + 1 = 2x - 8$	Combine like terms on each side.
$3x + 1 + (-1) = 2x - 8 + (-1)$	Add -1, the opposite of $+1$, to both sides.
$3x + 1 - 1 = 2x - 8 - 1$	
$3x = 2x - 9$	Simplify.
$3x + (-2x) = 2x - 9 + (-2x)$	Add $-2x$, the opposite of $+2x$, to both sides.
$x = -9$	Simplify.

Check:

$4x + 1 - x = 2x - 13 + 5$	
$4(-9) + 1 - (-9) \stackrel{?}{=} 2(-9) - 13 + 5$	Substitute $x = -9$.
$-36 + 1 + 9 \stackrel{?}{=} -18 - 13 + 5$	Simplify.
$-26 = -26$	True statement

Continued on next page...

c. $6y + 2.5 = 7y - 3.6$ Write the equation.

$6y + 2.5 + 3.6 = 7y - 3.6 + 3.6$ Add 3.6, the opposite of -3.6, to both sides.

$6y + 6.1 = 7y$ Simplify.

$6y + 6.1 - 6y = 7y - 6y$ Add $-6y$, the opposite of $6y$, to both sides.

$6.1 = y$ Simplify.

Check: $6y + 2.5 = 7y - 3.6$

$6(6.1) + 2.5 \overset{?}{=} 7(6.1) - 3.6$ Substitute $y = 6.1$.

$36.6 + 2.5 \overset{?}{=} 42.7 - 3.6$ Simplify.

$39.1 = 39.1$ True statement

Example 2: Parentheses

Solve the following equations.

a. $2(y-7) = 4(y+1) - 26$ **b.** $-2(5x+13) - 2 = -6(3x-2) - 41$ **c.** $\frac{1}{3}x + \frac{1}{6} = \frac{2}{5}x - \frac{7}{10}$

Solutions:

a. $2(y-7) = 4(y+1) - 26$ Write the equation.

$2y - 14 = 4y + 4 - 26$ Use the distributive property.

$2y - 14 = 4y - 22$ Combine like terms.

$2y - 14 + 22 = 4y - 22 + 22$ Add 22 to both sides. Here we will put the variables on the right side to get a positive coefficient of y.

$2y + 8 = 4y$ Simplify.

$-2y + 2y + 8 = -2y + 4y$ Add $-2y$ to both sides.

$8 = 2y$ Simplify.

$\dfrac{8}{2} = \dfrac{2y}{2}$ Divide both sides by 2.

$4 = y$ Simplify.

b. $-2(5x+13)-2 = -6(3x-2)-41$ Write the equation.

$-10x-26-2 = -18x+12-41$ Use the distributive property.
Be careful with the signs.

$-10x-28 = -18x-29$ Simplify.

$-10x-28+18x = -18x-29+18x$ Add $18x$ to both sides.

$8x-28 = -29$ Simplify.

$8x-28+28 = -29+28$ Add 28 to both sides.

$8x = -1$ Simplify.

$\dfrac{8x}{8} = \dfrac{-1}{8}$ Divide both sides by 8.

$x = -\dfrac{1}{8}$ Simplify.

c. $\dfrac{1}{3}x + \dfrac{1}{6} = \dfrac{2}{5}x - \dfrac{7}{10}$ Write the equation.

$30\left(\dfrac{1}{3}x + \dfrac{1}{6}\right) = 30\left(\dfrac{2}{5}x - \dfrac{7}{10}\right)$ Multiply both sides by 30.

$30\left(\dfrac{1}{3}x\right) + 30\left(\dfrac{1}{6}\right) = 30\left(\dfrac{2}{5}x\right) - 30\left(\dfrac{7}{10}\right)$ Apply the distributive property.

$10x+5 = 12x-21$ Simplify.

$10x+5-5 = 12x-21-5$ Subtract 5 from both sides.

$10x = 12x-26$ Simplify.

$10x-12x = 12x-26-12x$ Add $-12x$ to both sides.

$-2x = -26$ Simplify.

$\dfrac{-2x}{-2} = \dfrac{-26}{-2}$ Divide both sides by -2.

$x = 13$ Simplify.

Example 3: Application

A student bought a calculator and a textbook for a total of $200.80 (including tax). If the textbook cost $20.50 more than the calculator, what was the cost of each item?

Solution: Let x = cost of the calculator.
Then $x + 20.50$ = cost of the textbook.
The equation to be solved is:

$$\underbrace{x + 20.50}_{\substack{\text{cost of}\\\text{textbook}}} + \underbrace{x}_{\substack{\text{cost of}\\\text{calculator}}} = \underbrace{200.80}_{\substack{\text{total}\\\text{spent}}}$$

$$2x + 20.50 = 200.80$$
$$2x + 20.50 - 20.50 = 200.80 - 20.50$$
$$2x = 180.3$$
$$x = 90.15 \qquad \text{cost of calculator}$$
$$x + 20.50 = 110.65 \qquad \text{cost of textbook}$$

The calculator costs $90.15 and the textbook costs $90.15 + $20.50 = $110.65, with tax included in each price.

Practice Problems

Solve the following linear equations.

1. $x + 14 - 6x = 2x - 7$

2. $6.4x + 2.1 = 3.1x - 1.2$

3. $5 - (y - 3) = 14 - 4(y + 2)$

4. $\dfrac{2x}{3} - \dfrac{1}{2} = x + \dfrac{1}{6}$

5. $\dfrac{3}{14}n + \dfrac{1}{4} = \dfrac{1}{7}n - \dfrac{1}{4}$

Answers to Practice Problems: 1. 3 **2.** –1 **3.** $-\dfrac{2}{3}$ **4.** –2 **5.** –7

3.3 Exercises

Solve each of the following linear equations.

1. $3x + 2 = x - 8$ 2. $5x + 1 = 2x - 5$ 3. $4n - 3 = n + 6$

4. $6y + 3 = y - 7$ 5. $3y + 18 = 7y - 6$ 6. $2y + 5 = 8y + 10$

7. $14n = 3n$ 8. $1.6x = 0.8x$ 9. $13x + 5 = 2x + 5$

10. $6y - 2.1 = y - 2.1$ 11. $2(z + 1) = 3z + 3$ 12. $6x - 3 = 3(x + 2)$

13. $16y + 23y - 5 = 14y$ 14. $5x - 2x + 4 = 3x + x - 1$

15. $6.5 + 1.2x = 0.5 - 0.3x$ 16. $x - 0.1x + 0.9 = 0.2(x + 1)$

17. $0.25 + 3x + 6.5 = 0.75x$ 18. $0.9y + 3 = 0.4y + 1.5$

19. $\dfrac{2}{3}x + 1 = \dfrac{1}{3}x - 6$ 20. $\dfrac{4}{5}n + 2 = \dfrac{2}{5}n - 4$

21. $\dfrac{y}{5} + \dfrac{3}{4} = \dfrac{y}{2} + \dfrac{3}{4}$ 22. $\dfrac{3}{8}\left(y - \dfrac{1}{2} \right) = \dfrac{1}{8}\left(y + \dfrac{1}{2} \right)$

23. $\dfrac{1}{2}(x + 1) = \dfrac{1}{3}(x - 1)$ 24. $\dfrac{2x}{3} + \dfrac{x}{3} = -\dfrac{3}{4} + \dfrac{x}{2}$

25. $\dfrac{5n}{6} + \dfrac{1}{9} = \dfrac{3n}{2}$ 26. $\dfrac{3}{4}x + \dfrac{1}{5}x = \dfrac{1}{2}x - \dfrac{3}{10}$

27. $\dfrac{1}{6}n + \dfrac{7}{15} - \dfrac{2}{5}n = 0$ 28. $x + \dfrac{2}{3}x - 2x = \dfrac{x}{6} - \dfrac{1}{8}$

29. $3x + \dfrac{1}{2}x - \dfrac{2}{5}x = \dfrac{x}{10} + \dfrac{7}{20}$ 30. $7(2x - 1) = 5(x + 6) - 13$

31. $5 - 3(2x + 1) = 4(x - 5) + 6$ 32. $-2(y + 5) - 4 = 6(y - 2) + 2$

33. $8 + 4(2x - 3) = 5 - (x + 3)$ 34. $8(3x + 5) - 9 = 9(x - 2) + 14$

35. $4.7 - 0.3x = 0.5x - 0.1$ 36. $5.8 - 0.1x = 0.2x - 0.2$

37. $0.2(x + 3) = 0.1(x - 5)$ 38. $0.4(x + 3) = 0.3(x - 6)$

39. $0.6x - 22.9 = 1.5x - 18.4$ 40. $0.1y + 3.8 = 5.72 - 0.3y$

41. $0.3x + 0.1(x - 15) = 0.116$ 42. $-0.5(x + 8.5) = -0.3x - 10.3$

43. $0.12n + 0.25n - 5.895 = 4.3n$ 44. $0.15n + 32n - 21.0005 = 10.5n$

45. $0.7(x + 14.1) = 0.3(x + 32.9)$ 46. $0.8(x - 6.21) = 0.2(x - 24.84)$

47. The area of a triangle is found by multiplying $\frac{1}{2}$ times the base times the height $\left(A = \frac{1}{2}bh \right)$. Find the length of the base of a triangle with area 180 in.^{2} and height 10 in.

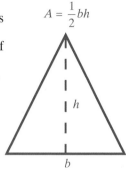

$A = \frac{1}{2}bh$

48. A sail is in the shape of a triangle. Find the height of the sail if its base is 20 ft and its area is 300 ft^2.

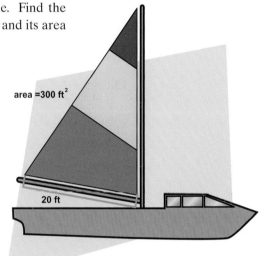

area =300 ft^2

20 ft

The area, A, of a trapezoid is $\frac{1}{2}$ the height times the sum of the two parallel sides, b and c. (The parallel sides are sometimes called bases.) $\left[A = \frac{1}{2}h(b+c) \right]$

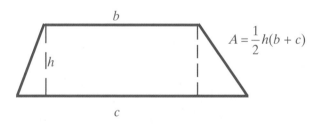

b

h

$A = \frac{1}{2}h(b + c)$

c

49. The area of a trapezoid is 108 cm^2. The height is 12 cm and the length of one of the parallel sides is 8 cm. Find the length of the second parallel side.

50. The lengths of the bases of a trapezoid are 10 in. and 15 in. Find the height if the area is 225 in.^{2}

51. The height of a trapezoid is 10 ft and its area is 250 ft² If one of the bases is 2 ft more than 3 times the other base, what are the lengths of the bases?

52. The total cost of a computer flash drive and a color printer was $225.50, including tax. If the cost of the flash drive was $170.70 less than the printer, what was the cost of each item?

53. If 14.56 plus twice a number equals the sum of the number and 1.46, what is the number?

54. If six times a number is increased by 15.35, the result is 12.5 less than the number. What is the number?

55. If the rental on a car is a fixed price per day plus $0.28 per mile, what was the fixed price per day if the total paid was $140 and the car was driven for 250 miles in two days?

56. Find the value of x in the figure if the area of the shaded portion is 95 inches².

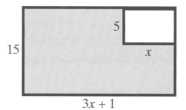

[**Note:** The area of a rectangle is found by multiplying the length times the width. $(A = lw)$]

Calculator Problems

Use a TI-84 Plus graphing calculator to help in solving the linear equations in Exercises 57 – 60.

57. $0.17x - 23.0138 = 1.35x + 36.234$

58. $48.512 - 1.63x = 2.58x + 87.63553$

59. $0.32(x + 14.1) = 2.47x + 2.2188$

60. $1.6(9.3 + 2x) = 0.2(3x + 133.94)$

Hawkes Learning Systems: Introductory Algebra

More Linear Equations $ax + b = dx + c$

Solving Linear Inequalities and Applications

After completing this section, you will be able to:

1. *Understand intervals of real numbers.*

2. *Solve linear inequalities.*

3. *Graph the solutions for linear inequalities.*

4. *Solve applications by using linear inequalities.*

Intervals of Real Numbers

Real numbers, graphs of real numbers on real number lines, and inequalities with the symbols $<, >, \leq,$ and $\geq$ were discussed in Section 1.1. In this section, we will expand our work with inequalities to include the concept of **intervals of real numbers**. Remember that dots are used to indicate the graphs of individual real numbers on a real number line.

For example, the set of numbers $\left\{ -\dfrac{3}{2}, -1, 0, 2, \sqrt{10} \right\}$ is graphed in Figure 3.1.

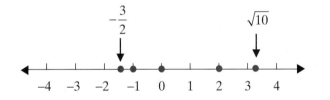

Figure 3.1

Irrational numbers, such as π, $\sqrt{2}$, $-\sqrt{7}$, and $\sqrt{10}$ have been discussed to some extent and will be discussed in more detail in Chapter 9. What is important here is that irrational numbers do indeed correspond to points on a real number line. In fact, we can make the following statement concerning real numbers and points on a line.

One-To-One Correspondence

> *There is a one-to-one correspondence between the real numbers and the points on a line.* *That is, each point on a number line corresponds to one real number, and each real number corresponds to one point on a number line.*

Now, when an inequality such as $x < 8$ is given, the implication is that x can be **any real number less than 8**. That is, the **solution** to the inequality is the set of all real numbers less than 8. To indicate this solution graphically, an open circle is drawn around 8 on a number line and the line is shaded to the left of 8 as shown in Figure 3.2.

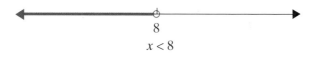

8

$x < 8$

Figure 3.2

Suppose that a and b are two real numbers and that $a < b$. The set of all real numbers between a and b is called an **interval of real numbers**. Intervals of real numbers are commonly used in relationship to ideas such as the length of time to heal a wound, the shelf life of chemicals, the stopping distances for a car traveling at certain speeds, and the distances that a camera can be expected to take clear pictures. As a specific example, a bicyclist may regularly ride from 1 to 3 hours for exercise. This interval of time can be represented in the following way with two inequalities:

$$1 \le t \le 3 \quad \text{where } t \text{ represents the time in hours.}$$

Intervals are classified and graphed as indicated in the box that follows. You should keep in mind the following facts as you consider the boxed information.

1. x is understood to represent real numbers.
2. Open dots at endpoints a and b indicate that these points **are not** included in the graph.
3. Solid dots at endpoints a and b indicate that these points **are** included in the graph.

Intervals of Real Numbers

Name	Symbolic Representation	Graph
Open Interval	$a < x < b$	
Closed Interval	$a \leq x \leq b$	
Half-Open Interval	$\begin{cases} a < x \leq b \\ a \leq x < b \end{cases}$	
Open Interval	$\begin{cases} x > a \\ x < a \end{cases}$	
Half-Open Interval	$\begin{cases} x \geq a \\ x \leq a \end{cases}$	

Example 1: Graph

a. Represent the following graph using algebraic notation, and tell what kind of interval it is.

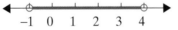

Solution: $-1 < x < 4$ is an open interval.

b. Represent the following graph using algebraic notation, and tell what kind of interval it is.

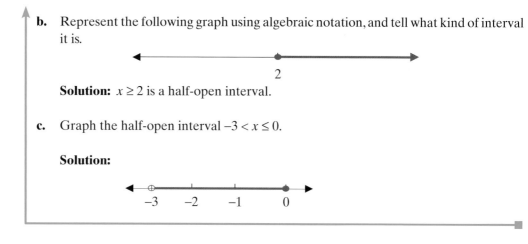

Solution: $x \geq 2$ is a half-open interval.

c. Graph the half-open interval $-3 < x \leq 0$.

Solution:

Solving Linear Inequalities

Inequalities of the forms

$$ax + b < c \quad \text{and} \quad ax + b \leq c$$
$$ax + b > c \quad \text{and} \quad ax + b \geq c$$
$$c < ax + b < d \quad \text{and} \quad c \leq ax + b \leq d$$

are called **linear inequalities** or **first-degree inequalities**.

The solutions to linear inequalities are intervals of real numbers and the methods for solving linear inequalities are similar to those used to solve linear equations. There is only one important exception:

> **Multiplying or dividing both sides of an inequality by a negative number causes the "sense" of the inequality to be reversed.**

By the sense of the inequality, we mean "less than" or "greater than." Consider the following examples.

We know that $4 < 20$.

Add 5 to both sides	**Multiply both sides by 3**	**Add −10 to both sides**
$4 < 20$	$4 < 20$	$4 < 20$
$4 + 5 \ ? \ 20 + 5$	$3 \cdot 4 \ ? \ 3 \cdot 20$	$4 + (-10) \ ? \ 20 + (-10)$
$9 < 25$	$12 < 60$	$-6 < 10$

In the cases just illustrated involving addition or multiplication by a positive number, the sense of the inequality stayed the same, in these cases, $<$. Now we will see that multiplying or dividing each side by a negative number will reverse the sense of the inequality, from $<$ to $>$, or from $>$ to $<$.

Multiply both sides by −3	**Divide both sides by −2**
$4 < 20$	$4 < 20$
$-3 \cdot 4 \ ? \ -3 \cdot 20$	$\dfrac{4}{-2} \ ? \ \dfrac{20}{-2}$
$-12 > -60$	$-2 > -10$

In each of these two cases, the sense of the inequality is changed from $<$ to $>$. While two examples do not prove a rule to be true, this particular rule is true and is included in the following rules for solving inequalities.

Rules for Solving Linear (or First Degree) Inequalities

1. *Simplify each side of the inequality by applying the distributive property to remove any parentheses (if they are present) and combining like terms.*

2. *Add constants and/or variables to or subtract them from both sides of the inequality so that variables are on one side and constants on the other.*

3. *Divide both sides by the coefficient (or multiply them by the reciprocal) of the variable, and **reverse the sense of the inequality if this coefficient is negative**.*

4. *A quick (and generally satisfactory) check is to select any one number in your solution and substitute it into the original inequality.*

As with solving equations, the object of solving an inequality is to find equivalent inequalities of simpler form that have the same solution set. We want the variable with a coefficient of +1 on one side of the inequality and any constants on the other side. One key difference between solving linear equations and solving linear inequalities is that linear equations have only one solution while linear inequalities generally have an infinite number of solutions.

> **NOTES**
>
> **Special Note About Reading Inequalities**
>
> Inequalities can be read from left to right or from right to left. When reading an inequality with a variable, **read the variable first**.
>
> For example,
>
> | | | |
> |---:|:---|:---|
> | | $4 < x$ | should be read from right to left as "x is greater than 4" |
> | while | $x > 4$ | should be read from left to right as "x is greater than 4" |
> | | $-2 < x < 5.1$ | should be read as follows: "x is greater than -2 and x is less than 5.1" |
>
> Reading the variable first will help in understanding the inequalities and drawing the correct related graphs.

Example 2: Solving Linear Inequalities

Solve the following linear inequalities and graph the solutions.

a. $5x + 4 \leq -1$

Solution:

$5x + 4 \leq -1$	Write the inequality.
$5x + 4 - 4 \leq -1 - 4$	Add -4 to both sides.
$5x \leq -5$	Simplify.
$\dfrac{5x}{5} \leq \dfrac{-5}{5}$	Divide both sides by 5.
$x \leq -1$	Simplify.

(**Note:** The shaded dot means that -1 is included.)

Check: As a quick check, pick any number less than or equal to -1, say, -4, and substitute it into the original inequality. If a false statement results, then you have made a mistake.

$$5(-4) + 4 \leq -1$$
$$-16 \leq -1 \qquad \text{True statement}$$

Continued on next page...

b. $x + 1 > 3 - 2x$

Solution:

$x + 1 > 3 - 2x$	Write the inequality.
$x + 1 + 2x > 3 - 2x + 2x$	Add $2x$ to both sides.
$3x + 1 > 3$	Simplify.
$3x + 1 - 1 > 3 - 1$	Add -1 both sides.
$3x > 2$	Simplify.
$\dfrac{3x}{3} > \dfrac{2}{3}$	Divide both sides by 3.
$x > \dfrac{2}{3}$	Simplify.

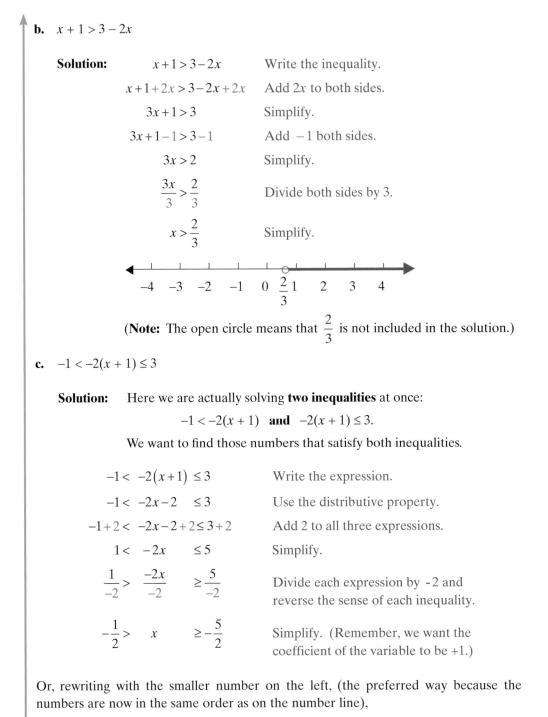

(**Note:** The open circle means that $\dfrac{2}{3}$ is not included in the solution.)

c. $-1 < -2(x + 1) \le 3$

Solution: Here we are actually solving **two inequalities** at once:

$$-1 < -2(x + 1) \quad \textbf{and} \quad -2(x + 1) \le 3.$$

We want to find those numbers that satisfy both inequalities.

$-1 < -2(x+1) \le 3$	Write the expression.
$-1 < -2x - 2 \le 3$	Use the distributive property.
$-1 + 2 < -2x - 2 + 2 \le 3 + 2$	Add 2 to all three expressions.
$1 < -2x \le 5$	Simplify.
$\dfrac{1}{-2} > \dfrac{-2x}{-2} \ge \dfrac{5}{-2}$	Divide each expression by -2 and reverse the sense of each inequality.
$-\dfrac{1}{2} > x \ge -\dfrac{5}{2}$	Simplify. (Remember, we want the coefficient of the variable to be $+1$.)

Or, rewriting with the smaller number on the left, (the preferred way because the numbers are now in the same order as on the number line),

$$-\frac{5}{2} \le x < -\frac{1}{2}$$

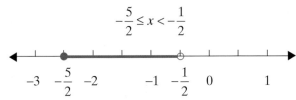

Check: For a quick check, substitute $x = -2$.

$$-1 < -2(-2+1) \le 3$$

$$-1 < -2(-1) \le 3$$

$$-1 < 2 \le 3 \qquad \text{True statement}$$

While a check does not guarantee there are no errors, it is a good indicator.

Applications of Linear Inequalities

We are generally familiar with the use of equations to solve word problems. In the following two examples, we show that solving inequalities can be related to real-world problems.

Example 3: Applications

a. A physics student has grades of 85, 98, 93, and 90 on four examinations. If he must average 90 or better to receive an A for the course, what score can he receive on the final exam and earn an A? (Assume that the final exam counts the same as the other exams.)

Solution: Let x = score on final exam.

The average is found by adding the scores and dividing by 5.

$$\frac{85 + 98 + 93 + 90 + x}{5} \ge 90$$

$$\frac{366 + x}{5} \ge 90$$

$$5\left(\frac{366 + x}{5}\right) \ge 5 \cdot 90$$

$$366 + x \ge 450$$

$$x \ge 450 - 366$$

$$x \ge 84$$

If the student scores 84 or more on the final exam, he will average 90 or more and receive an A in physics.

Continued on next page...

b. Ellen is going to buy 30 stamps, some 23-cent and some 41-cent. If she has $8.34, what is the maximum number of 41-cent stamps she can buy?

Solution: Let x = number of 41-cent stamps,
then $30 - x$ = number of 23-cent stamps.

Ellen cannot spend more than $8.34.

$$0.41x + 0.23(30 - x) \leq 8.34$$
$$0.41x + 6.90 - 0.23x \leq 8.34$$
$$0.18x + 6.90 \leq 8.34$$
$$0.18x \leq 8.34 - 6.90$$
$$0.18x \leq 1.44$$
$$x \leq 8$$

Ellen can buy at most eight 41-cent stamps if she buys a total of 30 stamps.

Practice Problems

Solve the inequalities and graph the solutions.

1. $7 + x < 3$

2. $\dfrac{x}{2} + 1 \geq \dfrac{x}{3}$

3. $-5 \leq 2x + 1 < 9$

Answers to Practice Problems: 1. $x < -4$ **2.** $x \geq -6$ **3.** $-3 \leq x < 4$

3.4 Exercises

In Exercises 1 – 10, determine whether each of the inequalities is true or false. If the inequality is false, rewrite it in a form that is true using the same numbers or expressions.

1. $3 < -3$

2. $-14 > 4$

3. $\left|-5\right| > -5$

4. $\left|-7\right| \le 7$

5. $-3.4 > -3.5$

6. $\dfrac{1}{2} > \dfrac{1}{3}$

7. $-6 \le 0$

8. $\dfrac{5}{8} \ge \dfrac{7}{16}$

9. $\left|-4\right| \le \left|4\right|$

10. $-12.5 < -10.2$

In Exercises 11 – 20, graph the set of real numbers that satisfies each of the inequalities and tell what type of interval it represents.

11. $-2 \le x < 3$

12. $-4 \le x \le -1$

13. $x \le -4$

14. $x > 6$

15. $-2 < x \le 1$

16. $x \ge -5$

17. $1 \le x < \dfrac{10}{3}$

18. $x < \dfrac{13}{4}$

19. $x > \dfrac{1}{3}$

20. $3 \le x \le 10$

Solve the inequalities in Exercises 21 – 50 and graph the solutions.

21. $4x + 5 \ge -6$

22. $2x + 3 > -8$

23. $3x + 2 > 2x - 1$

24. $3y + 2 \le y + 8$

25. $y - 6.1 \le 4.2 - y$

26. $4x - 2.3 < 6x + 6.5$

27. $2.9x - 5.7 > -3 - 0.1x$

28. $3y - 21.9 \ge 11.04 - 3.1y$

29. $5y + 6 < 2y - 2$

30. $4 - 2x < 5 + x$

31. $4 + x > 1 - x$

32. $x - 6 > 3x + 5$

33. $\dfrac{x}{4} + 1 \le 5 - \dfrac{x}{4}$

34. $\dfrac{x}{2} - 1 \le \dfrac{5x}{2} - 3$

35. $\dfrac{x}{3} - 2 > 1 - \dfrac{x}{3}$

36. $\dfrac{5x}{3} + 2 > \dfrac{x}{3} - 1$

37. $-(x + 5) \le 2x + 4$

38. $-3(2x - 5) \le 3(x - 1)$

39. $x - (2x + 5) \ge 7 - (4 - x) + 10$

40. $x - 3(4 - x) + 5 \ge -2(3 - 2x) - x$

41. $-5 < 1 - x < 3$

42. $2 < 2x - 4 \le 8$

43. $1 < 3x - 2 < 4$

44. $-5 < 4x + 1 < 7$

45. $-5 \le 5x + 3 < 9$

46. $-1 < 3 - 2x < 6$

47. $\dfrac{3}{4} \le \dfrac{1}{2}x + 6 \le \dfrac{15}{2}$

48. $\dfrac{1}{3} \le \dfrac{5}{6}x - 2 < \dfrac{14}{15}$

49. $-\dfrac{1}{2} < -\dfrac{2}{3}x + 5 \le \dfrac{5}{9}$

50. $0 \le -\dfrac{5}{8}x + 1 < \dfrac{9}{20}$

51. To receive a B grade, a student must average 80 or more but less than 90. If John received a B in the course and had five grades of 94, 86, 78, 91, and 87 before taking the final exam, what were the possible grades for his final if there were 100 points possible?

1st test	94
2nd test	86
HW avg	78
quiz avg	91
last test	87
final exam	?

52. The range for a C grade is 70 or more but less than 80. Before taking the final exam, Clyde had grades of 59, 68, 76, 84, and 69. If the final exam is counted as two tests, what is the minimum grade he could make on the final to receive a C? If there were 100 points possible, could he receive a B grade if an average of at least 80 points was required?

53. The temperature of a mixture in a chemistry experiment varied from 15°C to 65°C. What is the temperature expressed in degrees Fahrenheit?

$$\left[C = \frac{5}{9}(F - 32) \right]$$

54. The temperature at Braver Lake ranged from a low of 23°F to a high of 59°F. What is the equivalent range of temperatures in degrees Celsius?

$$\left[F = \frac{9}{5}C + 32 \right]$$

55. The sum of four times a number and 21 is greater than 45 and less than 73. What are the possible values for the number?

56. In order for Chuck to receive a B in his mathematics class, he must have a total of at least 400 points. If he has scores of 72, 68, 85, and 89, what scores can he make on the final and receive a B? (The maximum possible score on the final is 100.)

57. In Exercise 56, if the final exam counted twice, could Chuck receive a grade of A in the class if it takes 540 points for an A? Assume that the maximum possible score on the final is 100 points.

58. The Concert Hall has 400 seats. For a concert, the admission will be $6.00 for adults and $3.50 for students. If the expense of producing the concert is $1825, what is the least number of adult tickets that must be sold to realize a profit if all seats are taken? (Hint: Let x = number of adult tickets and $400 - x$ = number of student tickets.)

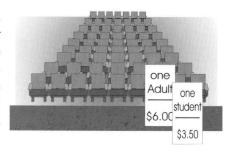

59. The Pep Club is selling candied apples to raise money. The price per apple is $2.50 until Friday, when they will sell for $2.00. If the Pep Club sells 200 apples, what is the minimum number they must sell at $2.50 each in order to raise at least $475.00?

60. Better-Car Rental charges $15 per day plus $0.25 per mile and Best-Car Rental charges $25 per day plus $0.15 per mile. How many miles can be driven in one day with a Better car and still have a bill less than what a Best car would cost?

Hawkes Learning Systems: Introductory Algebra

Solving Linear Inequalities

3.5   Working with Formulas

After completing this section, you will be able to:

1. Evaluate formulas for given values of the variables.

2. Solve formulas for specified variables in terms of the other variables.

3. Use formulas to solve a variety of applications.

Formulas are general rules or principles stated mathematically. There are many formulas in such fields of study as business, economics, medicine, physics, and chemistry as well as mathematics. Some of these formulas and their meanings are shown here.

NOTES

SPECIAL COMMENT: Be sure to use the letters just as they are given in the formulas. In mathematics, there is little or no flexibility between capital and small letters as they are used in formulas. In general, capital letters have special meanings that are different from corresponding small letters. For example, capital **A** may mean the area of a triangle and small **a** may mean the length of one side, two completely different ideas.

Listed here are a few formulas with a description of the meaning of each formula. There are of course many more formulas. Some you will see in the exercises here and more (related to geometry) in Section 3.8.

Formula	Meaning
1. $I = Prt$	The simple interest (I) earned by investing money is equal to the product of the principal (P) times the rate of interest (r) times the time (t) in one year or part of a year.
2. $C = \dfrac{5}{9}(F - 32)$	Temperature in degrees Celsius (C) equals $\dfrac{5}{9}$ times the difference between the Fahrenheit temperature (F) and 32.
3. $d = rt$	The distance traveled (d) equals the product of the rate of speed (r) and the time (t).
4. $P = 2l + 2w$	The perimeter (P) of a rectangle is equal to twice the length (l) plus twice the width (w).

5. $L = 2\pi rh$ The lateral surface area, L, (top and bottom not included) of a cylinder is equal to 2π times the radius (r) of the base times the height (h).

6. $F = ma$ In physics, the force, F, acting on an object is equal to its mass (m) times its acceleration (a).

7. $\alpha + \beta + \gamma = 180°$ The sum of the angles (α, β, and γ) of a triangle is 180°. (**Note:** α, β, and γ are the Greek lowercase letters alpha, beta, and gamma, respectively.)

8. $P = a + b + c$ The perimeter (P) of a triangle is equal to the sum of the lengths of the three sides (a, b, and c).

Evaluating Formulas

If you know values for all but one variable in a formula, you can substitute those values and find the value of the unknown variable by using the techniques for solving equations discussed in this chapter. This section presents a variety of formulas from real-life situations, with complete descriptions of the meanings of these formulas. Working with these formulas will help you become familiar with a wide range of applications of algebra and provide practice in solving equations.

Example 1: Evaluating Formulas

a. A **note** is a loan for a period of 1 year or less, and the interest earned (or paid) is called **simple interest**. A note involves only one payment at the end of the term of the note and includes both principal and interest. The formula for calculating simple interest is:

$$I = Prt$$

where I = **Interest** (earned or paid)
          P = **Principal** (the amount invested or borrowed)
          r = **rate** of interest (stated as an annual or yearly rate)
          t = **time** (one year or part of a year)

Note: The rate of interest is usually given in percent form and converted to decimal or fraction form for calculations. For the purpose of calculations, we will use 360 days in one year and 30 days in a month. Before the use of computers, this was common practice in business and banking.

Maribel loaned $5000 to a friend for 6 months at an interest rate of 8%. How much will her friend pay her at the end of the 6 months?

Continued on next page...

Solution: Here, $P = \$5000$,

$$r = 8\% = 0.08,$$

$$t = 6 \text{ months} = \frac{6}{12} \text{ year} = \frac{1}{2} \text{ year}.$$

Find the interest by substituting in the formula $I = Prt$ and evaluating.

$$I = 5000 \cdot \overset{0.04}{\cancel{0.08}} \cdot \frac{1}{\cancel{2}}$$

$$= 5000 \cdot 0.04$$

$$= \$200.00$$

The interest is $200 and the amount to be paid at the end of 6 months is Principal + Interest = $5000 + $200 = $5200.

b. Given the formula $C = \dfrac{5}{9}(F - 32)$, first find C if $F = 212°$ (212 degrees Fahrenheit) and then find F if $C = 20°$ (20 degrees Celsius).

Solution: $F = 212°$, so substitute 212 for F in the formula.

$$C = \frac{5}{9}(212 - 32)$$

$$-\frac{5}{9}(180)$$

$$= 100$$

That is, 212°F is the same as 100°C. Water will boil at 212°F at sea level. This means that if the temperature is measured in degrees Celsius instead of degrees Fahrenheit, water will boil at 100°C at sea level.

$C = 20°$, so substitute 20 for C in the formula.

$$20 = \frac{5}{9}(F - 32) \qquad \text{Now solve for } F.$$

$$\frac{9}{5} \cdot 20 = \frac{9}{5} \cdot \frac{5}{9}(F - 32) \qquad \text{Multiply both sides by } \frac{9}{5}.$$

$$36 = F - 32 \qquad \text{Simplify.}$$

$$68 = F \qquad \text{Add 32 to both sides.}$$

That is, a temperature of 20°C is the same as a comfortable spring day temperature of 68°F.

c. The lifting force, F, exerted on an airplane wing is found by multiplying some constant, k, by the area, A, of the wing's surface and by the square of the plane's velocity, v. The formula is $F = kAv^2$. Find the force on a plane's wing if the area of the wing is 120 ft², k is $\dfrac{4}{3}$, and the plane is traveling 80 miles per hour during take off.

Solution: We know that $k = \dfrac{4}{3}$, $A = 120$, and $v = 80$. Substitution gives

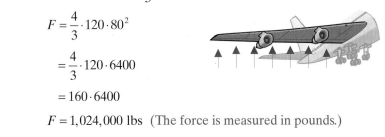

$$F = \frac{4}{3} \cdot 120 \cdot 80^2$$

$$= \frac{4}{3} \cdot 120 \cdot 6400$$

$$= 160 \cdot 6400$$

$$F = 1{,}024{,}000 \text{ lbs} \quad \text{(The force is measured in pounds.)}$$

Solving Formulas for Different Variables

We say that the formula $d = rt$ is "solved for" d in terms of r and t. Similarly, the formula $A = \dfrac{1}{2}bh$ is solved for A in terms of b and h, and the formula $P = S - C$ (profit is equal to selling price minus cost) is solved for P in terms of S and C. Many times we want to use a certain formula in another form. We want the formula "solved for" some variable other than the one given in terms of the remaining variables. **Treat the variables just as you would constants in solving linear equations**. Study the following examples carefully.

Example 2: Solving for Different Variables

a. Given $d = rt$, solve for t in terms of d and r. We want to represent the time in terms of distance and rate. We will use this concept later in word problems.

Solution:	$d = rt$	Treat r and d as if they were constants.
	$\dfrac{d}{r} = \dfrac{rt}{r}$	Divide both sides by r.
	$\dfrac{d}{r} = t$	Simplify.

Continued on next page...

b. Given $P = a + b + c$, solve for a in terms of $P, b,$ and c. This would be a convenient form for the case in which we know the perimeter and two sides of a triangle and want to find the third side.

Solution:

$$P = a + b + c \qquad \text{Treat } P, b, \text{ and } c \text{ as if they were constants.}$$
$$P - b - c = a + b + c - b - c \qquad \text{Add } -b - c \text{ to both sides.}$$
$$P - b - c = a \qquad \text{Simplify.}$$

c. Given $C = \dfrac{5}{9}(F - 32)$ as in Example 1b, solve for F in terms of C. This would give a formula for finding Fahrenheit temperature given a Celsius temperature value.

Solution:

$$C = \frac{5}{9}(F - 32) \qquad \text{Treat } C \text{ as a constant.}$$

$$\frac{9}{5} \cdot C = \frac{9}{5} \cdot \frac{5}{9}(F - 32) \qquad \text{Multiply both sides by } \frac{9}{5}.$$

$$\frac{9}{5}C = F - 32 \qquad \text{Simplify.}$$

$$\frac{9}{5}C + 32 = F \qquad \text{Add 32 to both sides.}$$

Thus,

$$F = \frac{9}{5}C + 32 \text{ is solved for } F \text{ and } C = \frac{5}{9}(F - 32) \text{ is solved for } C.$$

These are two forms of the same formula.

d. Given the equation $2x + 4y = 10$, **(i.)** solve first for x in terms of y, and then **(ii.)** solve for y in terms of x. This equation is typical of the algebraic equations that we will discuss in Chapter 7.

Solution: i. Solving for x yields

$$2x + 4y = 10 \qquad \text{Treat } 4y \text{ as a constant.}$$
$$2x + 4y - 4y = 10 - 4y \qquad \text{Subtract } 4y \text{ from both sides. (This is the same}$$
$$\text{as adding } -4y.)$$
$$2x = 10 - 4y$$
$$\frac{2x}{2} = \frac{10 - 4y}{2} \qquad \text{Divide both sides by 2.}$$
$$x = \frac{10}{2} - \frac{4y}{2} \qquad \text{Simplify.}$$
$$x = 5 - 2y$$

ii. Solving for y yields

$$2x + 4y = 10 \qquad \text{Treat } 2x \text{ as a constant.}$$

$$2x + 4y - 2x = 10 - 2x \qquad \text{Subtract } 2x \text{ from both sides.}$$

$$4y = 10 - 2x \qquad \text{Simplify.}$$

$$\frac{4y}{4} = \frac{10 - 2x}{4} \qquad \text{Divide both sides by 4.}$$

$$y = \frac{10}{4} - \frac{2x}{4} \qquad \text{Simplify.}$$

$$y = \frac{5}{2} - \frac{x}{2}$$

or we can write

$$y = \frac{5-x}{2} \quad \text{or} \quad y = -\frac{1}{2}x + \frac{5}{2}. \qquad \text{All forms are correct.}$$

e. Given $3x - y = 15$, solve for y in terms of x.

Solution: Solving for y gives

$$3x - y = 15$$

$$-y = 15 - 3x \qquad \text{Subtract } 3x \text{ from both sides.}$$

$$-1(-y) = -1(15 - 3x) \qquad \text{Multiply both sides by } -1 \text{ (or divide both sides by } -1).$$

$$y = -15 + 3x \qquad \text{Simplify using the distributive property.}$$

$$\text{or} \quad y = 3x - 15$$

f. Given $V = \dfrac{k}{P}$, solve for P in terms of V and k.

Solution: $\qquad V = \dfrac{k}{P}$

$$V \cdot P = k \qquad \text{Multiply both sides by } P.$$

$$\frac{V \cdot P}{V} = \frac{k}{V} \qquad \text{Divide both sides by } V.$$

$$P = \frac{k}{V}$$

Practice Problems

1. $2x - y = 5$; *solve for y.*

2. $2x - y = 5$; *solve for x.*

3. $A = \dfrac{1}{2}bh$; *solve for h.*

4. $L = 2\pi rh$; *solve for r.*

5. $P = 2l + 2w$; *solve for w.*

3.5 Exercises

Simple Interest

For Exercises 1 – 5, refer to Example 1(a) for information concerning simple interest and the related formula $I = Prt$. (**Note:** For ease of calculation, assume 360 days in a year and 30 days in each month.)

1. You want to borrow $4000 at 12% for only 90 days. How much interest would you pay?

2. A savings account of $3500 is left for 9 months and draws simple interest at a rate of 7%.
   a. How much interest is earned?
   b. What is the balance in the account at the end of the 9 months?

3. For how many days must you leave $1000 in a savings account at 5.5% to earn $11.00 in interest?

4. What is the rate of interest charged if a 3-month loan of $2500 is paid off with $2562.50?

5. What principal would you need to invest to earn $450 in simple interest in 6 months if the rate of interest was 9%?

In Exercises 6 – 20, read the descriptive information carefully and then substitute the values given in the problem for the corresponding variables in the formulas. Evaluate the resulting expression for the unknown variable.

Velocity

If an object is shot upward with an initial velocity, v_0, in feet per second, the velocity, v, in feet per second is given by the formula $v = v_0 - 32t$, where t is time in seconds.

6. Find the velocity at the end of 3 seconds if the initial velocity is 144 feet per second.

7. Find the initial velocity of an object if the velocity after 4 seconds is 48 feet per second.

Answers to Practice Problems: 1. $y = 2x - 5$ **2.** $x = \dfrac{y+5}{2}$ **3.** $h = \dfrac{2A}{b}$ **4.** $r = \dfrac{L}{2\pi h}$ **5.** $w = \dfrac{P-2l}{2}$

8. An object projected upward with an initial velocity of 106 feet per second has a velocity of 42 feet per second. How many seconds have passed?

Medicine

In nursing, one procedure for determining the dosage for a child is

$$\text{child's dosage} = \frac{\text{age of child in years}}{\text{age of child} + 12} \cdot \text{adult dosage}$$

9. If the adult dosage of a drug is 20 milliliters, how much should a 3-year-old child receive?

10. If the adult dosage of a drug is 340 milligrams, how much should a 5-year-old child receive?

Investment

The amount of money due from investing P dollars is given by the formula $A = P + Prt$, where r is the rate expressed as a decimal and t is the time in years.

11. Find the amount due if $1000 is invested at 6% for 2 years.

12. How long will it take an investment of $600, at an annual rate of 5%, to be worth $750?

Carpentry

The number, N, of rafters in a roof or studs in a wall can be found by the formula $N = \dfrac{l}{d} + 1$, where l is the length of the roof or wall and d is the center-to-center distance from one rafter or stud to the next. Note that l and d must be in the same units.

13. How many rafters will be needed to build a roof 26 ft long if they are placed 2 ft on center?

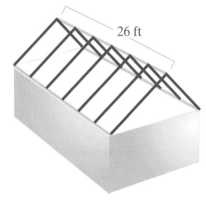

26 ft

14. A wall has studs placed 16 in. on center. If the wall is 20 ft long, how many studs are in the wall?

⊢⋯ 16 in. ⋯⊦⋯ 16 in. ⋯⊣

15. How long is a wall if it requires 22 studs placed 16 in. on center?

Cost

The total cost, C, of producing x items can be found by the formula $C = ax + k$, where a is the cost per item and k is the fixed costs (rent, utilities, and so on).

16. Find the cost of producing 30 items if each costs $15 and the fixed costs are $580.

17. It costs $1097.50 to produce 80 dolls per week. If each doll costs $9.50 to produce, find the fixed costs.

18. It costs a company $3.60 to produce a calculator. Last week the total costs were $1308. If the fixed costs are $480 weekly, how many calculators were produced last week?

Depreciation

Many items decrease in value as time passes. This decrease in value is called **depreciation**. One type of depreciation is called **linear depreciation**. The value, V, of an item after t years is given by $V = C - Crt$, where C is the original cost and r is the rate of depreciation expressed as a decimal.

19. If you buy a car for $6000 and depreciate it linearly at a rate of 10% per year, what will be its value after 6 years?

20. A contractor buys a 4 year-old piece of heavy equipment valued at $20,000. If the original cost of this equipment was $25,000, find the rate of depreciation.

Solve for the indicated variable in Exercises 21 – 55.

21. $P = a + b + c$; solve for b.   **22.** $P = 3s$; solve for s.

23. $F = ma$; solve for m.   **24.** $C = \pi d$; solve for d.

25. $A = lw$; solve for w.   **26.** $P = R - C$; solve for C.

27. $R = np$; solve for n.   **28.** $v = k + gt$; solve for k.

29. $I = A - P$; solve for P.   **30.** $L = 2\pi rh$; solve for h.

31. $A = \dfrac{m+n}{2}$; solve for m.   **32.** $P = a + 2b$; solve for a.

33. $I = Prt$; solve for t.   **34.** $R = \dfrac{E}{I}$; solve for E.

35. $P = a + 2b$; solve for b.   **36.** $c^2 = a^2 + b^2$; solve for b^2.

37. $\alpha + \beta + \gamma = 180°$; solve for β.   **38.** $y = mx + b$; solve for x.

39. $V = lwh$; solve for h.

40. $v = -gt + v_0$; solve for t.

41. $A = \dfrac{1}{2}bh$; solve for b.

42. $R = \dfrac{E}{I}$; solve for I.

43. $A = \pi r^2$; solve for π.

44. $A = \dfrac{R}{2L}$; solve for L.

45. $K = \dfrac{mv^2}{2g}$; solve for g.

46. $x + 4y = 4$; solve for y.

47. $2x + 3y = 6$; solve for y.

48. $3x - y = 14$; solve for y.

49. $5x + 2y = 11$; solve for x.

50. $-2x + 2y = 5$; solve for x.

51. $A = \dfrac{1}{2}h(b+c)$; solve for b.

52. $A = \dfrac{1}{2}h(b+c)$; solve for h.

53. $R = \dfrac{3(x-12)}{8}$; solve for x.

54. $-2x - 5 = -3(x + y)$; solve for x.

55. $3y - 2 = x + 4y + 10$; solve for y.

Make up a formula for each of the following situations.

56. Each ticket for a concert costs $\$t$ per person and parking costs $\$9.00$. What is the total cost per car, C, if there are n people in a car?

57. ABC Car Rental charges $\$25$ per day plus $\$0.12$ per mile. What would you pay per day for renting a car from ABC if you were to drive the car x miles in one day?

58. Top-Of-The-Line computer company knows that the cost (labor and materials) of producing a computer is $\$325$ per computer per week and the fixed overhead costs (lighting, rent, etc.) are $\$5400$ per week. What are the company's weekly costs of producing n computers per week?

59. If T-O-T-L (see Exercise 58) sells its computers for $\$683$ each, what is its profit per week if it sells the same number, n, that it produces? (Remember that profit is equal to revenue minus costs, or $P = R - C$.)

Writing and Thinking About Mathematics

60. The formula $z = \dfrac{x - \bar{x}}{s}$ is used extensively in statistics. In this formula, x represents one of a set of numbers, $\bar{x}$ represents the average (or mean) of those numbers in the set, and s represents a value called the standard deviation of the numbers. (The standard deviation is a positive number and is a measure of how "spread out" the numbers are.) The values for z are called z-scores, and they measure the number of standard deviation units a number x is from the mean $\bar{x}$.

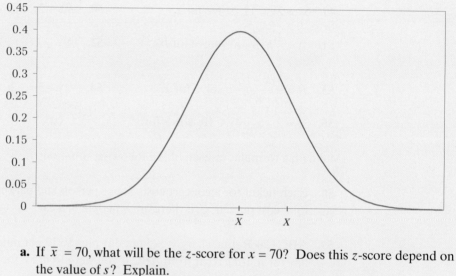

a. If $\bar{x} = 70$, what will be the z-score for $x = 70$? Does this z-score depend on the value of s? Explain.

b. For what values of x will the corresponding z-scores be negative?

c. Calculate your z-score on each of the last two scores in this class. (Your instructor will give you the mean and standard deviation for each test.) What do these scores tell you about your performance on the two exams?

61. Suppose that, for a particular set of exam scores, $\bar{x} = 72$ and $s = 6$. Find the z-score that corresponds to score of

    a. 78   **b.** 66   **c.** 81   **d.** 60

Hawkes Learning Systems: Introductory Algebra

Solving Formulas
Working with Formulas

3.6

Applications: Number Problems and Consecutive Integers

Objectives

After completing this section, you will be able to:

Solve word problems involving translating number phrases and consecutive integers.

Pólya's four-step process stated in Section 3.1 is outlined here for easy reference and to emphasize the importance of organization in problem solving.

Pólya's Four-Step Process for Solving Problems

1. *Understand the problem. (Read the problem carefully and be sure that you understand all the terms used.)*
2. *Devise a plan. (Set up an equation or a table or chart relating the information.)*
3. *Carry out the plan. (Perform any operations indicated in Step 2.)*
4. *Look back over the results. (Ask yourself if the answer seems reasonable and if you could solve similar problems in the future.)*

In this section we will concentrate on two types of problems:

1. Number problems involving translating phrases, such as those we discussed in Section 2.6.
2. Problems related to the concept of consecutive integers.

Number Problems

In Section 2.6, we discussed translating English phrases into algebraic expressions. Now we will use those skills to read number problems and translate the sentences and phrases in the problem into a related equation. The solution of this equation will be the solution to the problem.

As you learn more abstract mathematical ideas, you will find that you will use these ideas and the related processes to solve a variety of everyday problems as well as problems in specialized fields of study. Generally, you may not even be aware of the fact that you are using your mathematical talents. However, these skills and ideas will be part of your thinking and problem solving techniques for the rest of your life.

Example 1: Number Problems

a. Three times the sum of a number and 5 is equal to twice the number plus 5. Find the number.

Solution: Let x = the unknown number.

3 times the sum of a number and 5	is equal to	twice the number plus 5
$3(x + 5)$	$=$	$2x + 5$

$$3x + 15 = 2x + 5$$
$$3x + 15 - 2x = 2x + 5 - 2x$$
$$x + 15 = 5$$
$$x + 15 - 15 = 5 - 15$$
$$x = -10$$

The number is −10.

b. If a number is decreased by 36 and the result is 76 less than twice the number, what is the number?

Solution: Let n = the unknown number.

a number decreased by 36	the result is	76 less than twice the number
$n - 36$	$=$	$2n - 76$

$$n - 36 - n = 2n - 76 - n$$
$$-36 = n - 76$$
$$-36 + 76 = n - 76 + 76$$
$$40 = n$$

The number is 40.

c. One integer is 4 more than three times a second integer. Their sum is 24. What are the two integers?

Solution: Let n = the second integer.
Then $3n + 4$ = the first integer.

$$(\text{First}) + (\text{Second}) = 24$$
$$(3n + 4) + n = 24 \qquad \text{Their sum is 24.}$$
$$4n + 4 = 24$$
$$4n + 4 - 4 = 24 - 4$$
$$4n = 20$$
$$\frac{4n}{4} = \frac{20}{4}$$
$$n = 5$$
$$3n + 4 = 19$$

The two integers are 5 and 19.

d. Joe pays \$300 per month to rent an apartment. If this is $\dfrac{2}{5}$ of his monthly income, what is his monthly income?

Solution: Let x = Joe's monthly income, then $\dfrac{2}{5}(\text{Monthly income}) = \text{Rent}$.

$$\frac{2}{5}x = 300$$
$$\frac{5}{2} \cdot \frac{2}{5}x = \frac{5}{2} \cdot \frac{300}{1}$$
$$x = 750$$

Joe's monthly income is \$750.

Consecutive Integers

Remember that the set of **integers** consists of the whole numbers and their opposites:

$$\{ \dots, -4, -3, -2, -1, 0, 1, 2, 3, 4, \dots \}.$$

Even integers are integers that are divisible by 2. The even integers are

$$E = \{ \dots, -6, -4, -2, 0, 2, 4, 6, \dots \}.$$

Odd integers are integers that are not even. If an odd integer is divided by 2 the remainder will be 1. The odd integers are

$$O = \{ \dots, -5, -3, -1, 1, 3, 5, \dots \}.$$

In this discussion we will be dealing only with integers. Therefore, if you get a result that has a fraction or decimal number (not an integer), you will know that an error has been made and you should correct some part of your work.

The following terms and the ways of representing the integers must be understood before attempting the problems.

Consecutive Integers

Integers are **consecutive** *if each is 1 more than the previous integer. Three consecutive integers can be represented as*

$$n, \ n+1, \ \text{and} \ \ n+2.$$

Consecutive Odd Integers

Odd integers are **consecutive** *if each is 2 more than the previous odd integer. Three consecutive odd integers can be represented as*

$$n, \ n+2, \ \text{and} \ \ n+4$$

where n is an **odd** *integer.*

Consecutive Even Integers

Even integers are **consecutive** *if each is 2 more than the previous even integer. Three consecutive even integers can be represented as*

$$n, \ n+2, \ \text{and} \ \ n+4$$

where n is an **even** *integer.*

Note that consecutive even and consecutive odd integers are represented in the same way:

$$n, \ n+2, \ \text{and} \ \ n+4.$$

The value of the first integer, n, determines whether the remaining integers are odd or even. For example,

	n is odd		**n is even**
If	$n = 11$	**or**	If $n = 36$
then	$n + 2 = 13$		then $n + 2 = 38$
and	$n + 4 = 15$		and $n + 4 = 40$

Example 2: Consecutive Integers

a. Three consecutive odd integers are such that their sum is −3. What are the integers?

Solution: Let n = the first odd integer
$n + 2$ = the second odd integer
$n + 4$ = the third odd integer

Set up and solve the related equation.

$$
\begin{aligned}
\text{(First)} + \text{(Second)} + \text{(Third)} &= -3 \\
n + (n + 2) + (n + 4) &= -3 \\
3n + 6 &= -3 \\
3n &= -9 \\
n &= -3 \\
n + 2 &= -1 \\
n + 4 &= 1
\end{aligned}
$$

The three consecutive odd integers are −3, −1, and 1.

Check: $(-3) + (-1) + (1) = -3$

b. Find three consecutive integers such that the sum of the first and third is 76 less than three times the second.

Solution: Let n = the first integer
$n + 1$ = the second integer
$n + 2$ = the third integer

Set up and solve the related equation.

$$
\begin{aligned}
\text{(First)} + \text{(Third)} &= 3\text{(Second)} - 76 \\
n + (n + 2) &= 3(n + 1) - 76 \\
2n + 2 &= 3n + 3 - 76 \\
2n + 2 &= 3n - 73 \\
2n + 2 + 73 - 2n &= 3n - 73 + 73 - 2n \\
75 &= n \\
76 &= n + 1 \\
77 &= n + 2
\end{aligned}
$$

The three consecutive integers are 75, 76, and 77.

Check: $75 + 77 = 152$ and $3(76) - 76 = 228 - 76 = 152$

Continued on next page...

255

c. Find three consecutive even integers such that three times the first is 10 more than the sum of the second and third.

Solution: Let n = the first even integer
$n + 2$ = the second even integer
$n + 4$ = the third even integer

Set up and solve the related equation.

$$3(\text{First}) = (\text{Second}) + (\text{Third}) + 10$$
$$3n = (n + 2) + (n + 4) + 10$$
$$3n = 2n + 16$$
$$3n - 2n = 2n + 16 - 2n$$
$$n = 16$$
$$n + 2 = 18$$
$$n + 4 = 20$$

The three even integers are 16, 18, and 20.
(Checking shows that $3 \cdot 16$ is 10 more than $18 + 20$.)

3.6 Exercises

Read each problem carefully, translate the various phrases into algebraic expressions, set up an equation, and solve the equation.

1. Five less than a number is equal to 13 decreased by the number. Find the number.

2. Three less than twice a number is equal to the number. What is the number?

3. Thirty-six is 4 more than twice a certain number. Find the number.

4. Fifteen decreased by twice a number is 27. Find the number.

5. Seven times a certain number is equal to the sum of twice the number and 35. What is the number?

6. The difference between twice a number and 3 is equal to 6 decreased by the number. Find the number.

7. Fourteen more than three times a number is equal to 6 decreased by the number. Find the number.

8. Two added to the quotient of a number and 7 is equal to –3. What is the number?

9. The quotient of twice a number and 5 is equal to the number increased by 6. What is the number?

10. Three times the sum of a number and 4 is equal to –9. Find the number.

11. Four times the difference between a number and 5 is equal to the number increased by 4. What is the number?

12. When 17 is added to six times a number, the result is equal to 1 plus twice the number. What is the number?

13. If the sum of twice a number and 5 is divided by 11, the result is equal to the difference between 4 and the number. Find the number.

14. Twice a number increased by three times the number is equal to 4 times the sum of the number and 3. Find the number.

15. If 21 is subtracted from a number and the result is 8 times the number, what is the number?

16. Twice the difference between a number and 10 is equal to 6 times the number plus 16. What is the number?

17. The perimeter (distance around) of a triangle is 24 centimeters. If two sides are equal and the length of the third side is 4 centimeters, what is the length of each of the other two sides?(Reminder: The formula for the perimeter of a triangle is $P = a + b + c$.)

18. The length of a wire is bent to form a triangle with two sides equal. If the wire is 30 inches long and the two equal sides are each 9 inches long, what is the length of the third side?

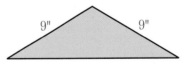

19. Two sides of a triangle have the same length and the third side is 3 cm less than the sum of the other two sides. If the perimeter of the triangle is 45 cm, what are the lengths of the three sides?

20. The perimeter of a rectangular shaped swimming pool is 172 meters. If the width is 16 meters less than the length, how long are the width and length of the pool?

21. A classic car is now selling for $1500 more than three times its original price. If the selling price is now $12,000, what was the car's original price?

22. A real estate agent says that the current value of a home is $90,000 more than twice its value when it was new. If the current value is $310,000, what was the value of the home when it was new? The home is 25 years old.

23. The sum of two consecutive odd integers is 60. What are the integers?

24. Find three consecutive integers whose sum is 69.

25. Find three consecutive integers whose sum is 93.

26. The sum of four consecutive integers is 74. What are the integers?

27. 171 minus the first of three consecutive integers is equal to the sum of the second and third. What are the integers?

28. If the first of three consecutive integers is subtracted from 120, the result is the sum of the second and third. What are the integers?

29. Four consecutive integers are such that if 3 times the first is subtracted from 208, the result is 50 less than the sum of the other three. What are the integers?

30. Find two consecutive integers such that twice the first plus three times the second equals 83.

31. Find three consecutive even integers such that the first plus twice the second is 54 less than four times the third.

32. Find three consecutive even integers such that if the first is subtracted from the sum of the second and third, the result is 66.

33. Find three consecutive odd integers such that 4 times the first is 44 more than the sum of the second and third.

34. Find three consecutive even integers such that their sum is 168 more than the second.

35. Find four consecutive integers whose sum is 90.

36. The perimeter of a rectangular parking lot is 410 yards. If the width is 75 yards, what is the length of the parking lot?

37. Twenty-four plus a number is equal to twice the number plus three times the same number. What is the number?

38. The width of a rectangular-shaped room is $\frac{2}{3}$ of the length. If the room is 18 ft wide, find the length.

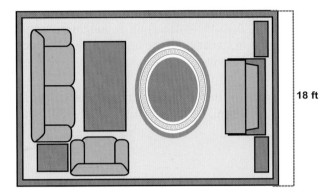

18 ft

39. Joe Johnson decided to buy a lot and build a house on the lot. He knew that the cost of constructing the house was going to be $25,000 more than the cost of the lot. He told a friend that the total cost was going to be $275,000. As a test to see if his friend remembered the algebra they had together in school, he challenged his friend to calculate what he paid for the lot and what he was going to pay for the house. What was the cost of the lot and the cost of the house?

40. A 29 foot board is cut into three pieces at a sawmill. The second piece is 2 feet longer than the first and the third piece is 4 feet longer than the second. What is the length of each of the three pieces?

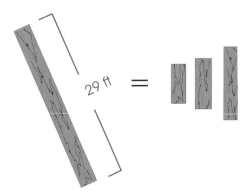

29 ft =

41. Lucinda bought two boxes of golf balls. She gave the pro-shop clerk a 50-dollar bill and received $10.50 in change. What was the cost of one box of golf balls? (Tax was included.)

42. A mathematics student bought a graphing calculator and a textbook for a course in statistics. If the text cost $49.50 more than the calculator, and the total cost for both was $125.74, what was the cost of each item?

43. The three sides of a triangle are x, $3x - 1$, and $2x + 5$ (as shown in the figure). If the perimeter of the triangle is 64 inches, what is the length of each side?

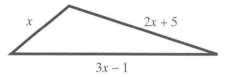

44. Find three consecutive odd integers such that the sum of twice the first and three times the second is 7 more than twice the third.

45. Find three consecutive even integers such that the sum of three times the first and twice the third is twenty less than six times the second.

Make up your own word problem that might use the given equation in its solution. Be creative! Then solve the equation and check to see that the answer is reasonable.

46. $n + (n + 1) = 33$

47. $3(n + 1) = n + 53$

48. $n + (n + 4) = 3(n + 2)$

49. $x - \dfrac{5}{6}x = \dfrac{5}{3}$

50. $\dfrac{x}{2} + \dfrac{1}{3} = \dfrac{3x}{4}$

Hawkes Learning Systems: Introductory Algebra

Applications: Number Problems and Consecutive Integers

3.7 Applications: Percent Problems (Discount, Taxes, Commission, Profit, and Others)

Objectives

After completing this section, you will be able to:

1. Change decimals to percents.

2. Change percents to decimals.

3. Find percents of numbers.

4. Find fractional parts of numbers.

5. Solve word problems involving decimals, fractions, and percents.

Our daily lives are filled with decimal numbers and percents: stock market reports, batting averages, won-lost records, salary raises, measures of pollution, taxes, interest on savings, home loans, discounts on clothes and cars, and on and on. Since decimal numbers and percents play such a prominent role in everyday life, we need to understand how to operate with them and apply them correctly in a variety of practical situations.

A review of percent and the relationship between percent and decimals and fractions is discussed in the Review Chapter. If you have not already, you may need to spend some time reviewing these concepts.

Briefly:

The word percent comes from the Latin *per centum,* meaning "per hundred." So, **percent means hundredths**. Thus,

$$72\%, \frac{72}{100}, \text{ and } 0.72 \text{ all have the same meaning.}$$

Change a decimal to percent:

$0.036 = 3.6\%$ Move the decimal point two places to the right and add the % sign.

Change a percent to a decimal:

$56\% = 0.56$ Move the decimal point two places to the left and drop the % sign.

Change a percent to a fraction (or mixed number):

$$30\% = \frac{30}{100} = \frac{3}{10}$$ Write the % in hundredths form and reduce.

$$125\% = 1.25 = 1\frac{25}{100} = 1\frac{1}{4}$$ Write the % in decimal form and change to mixed number form.

The Basic Percent Formula $R \cdot B = A$

Now, consider the statement

"15% of 80 is 12."

This statement has three numbers in it. In general, solving a percent problem involves knowing two of these numbers and trying to find the third. That is, **there are three basic types of percent problems**. We can break down the sentence in the following way.

Sentence →	15%	of	80	is	12
	↓	↓	↓	↓	↓
Equation →	0.15	·	80	=	12
	↑	↑	↓	↑	↓
	percent written as decimal	"of" changed to times		"is" changed to =	
	↓	↓	↓	↓	↓
Basic Formula →	Rate	·	Base	=	Amount

The terms that we have just discussed are explained in detail in the following box.

The Basic Formula $R \cdot B = A$

R = **RATE** or percent (as a decimal or fraction)

B = **BASE** (number we are finding the percent of)

A = **AMOUNT** or percentage (a part of the base)

"**of**" means to multiply.

"**is**" means equal (=).

The relationship among R, B, and A is given in the formula

$$R \cdot B = A \quad (or\ A = R \cdot B)$$

Even though there are just three basic types of percent problems, many people have difficulty deciding whether to multiply or divide in a particular problem. Using the formula $R \cdot B = A$ helps to avoid these difficulties. If the values of any two of the quantities in the formula are known, they can be substituted in the formula and then the missing number can be found by solving the equation. **The process of solving the equation for the unknown quantity determines whether multiplication or division is needed.**

The following examples illustrate how to substitute into the formula and how to solve the resulting equations.

Example 1: Percent of a Number

a. What is 72% of 800?

> **Solution:** $R = 0.72$ and $B = 800$ and A is unknown.
>
> $$R \ \cdot \ B \ = A$$
> $$\downarrow \quad \downarrow \quad \downarrow$$
> $$0.72 \cdot 800 = A \qquad \text{Here, simply multiply to find } A.$$
> $$576 = A$$
>
> So, **576** is 72% of 800.

b. 57% of what number is 163.191?

> **Solution:** $R = 0.57$ and B is unknown and A is 163.191.
>
> $$R \cdot B = A$$
> $$\downarrow \quad \downarrow \quad \downarrow$$
> $$0.57 \cdot B = 163.191$$
> $$\frac{0.57 \cdot B}{0.57} = \frac{163.191}{0.57} \qquad \text{Now, divide both sides by 0.57 to find } B.$$
> $$B = 286.3$$
>
> So, 57% of **286.3** is 163.191.

Continued on next page...

c. What percent of 180 is 45?

> **Solution:** R is the unknown and B is 180 and A is 45.
>
> $$R \cdot B = A$$
> $$\downarrow \quad \downarrow \quad \downarrow$$
> $$R \cdot 180 = 45$$
> $$\frac{R \cdot 180}{180} = \frac{45}{180} \qquad \text{Now, divide both sides by 180 to find } R.$$
> $$R = 0.25$$
> $$R = 25\% \qquad \text{Write the decimal in percent form.}$$
>
> So, **25%** of 180 is 45.

Applications with Percent

To sell goods that have been in stock for some time or simply to attract new customers, retailers and manufacturers sometimes offer a **discount**, a reduction in the selling price usually stated as a percent of the original price. The new, reduced price is called the **sale price**.

Sales tax is a tax charged on goods sold by retailers, and it is assessed by states and cities for income to operate various services. The **rate of sales tax** (a percent) varies by location.

Example 2: Applications

A bicycle was purchased at a discount of 25% of its original price of $1600. If 6% sales tax was added to the purchase price, what was the total paid for the bicycle?

Solution: There are two ways to approach this problem. One way is to find the discount and then subtract this amount from $1600. Another way is to subtract 25% from 100% to get 75% and then find 75% of $1600. In either case, the sales tax must be calculated on the sale price and added to the sale price.

Here, we will calculate the discount and then subtract from the original price.

$$R \quad \cdot \quad B \quad = \quad A$$
$$\downarrow \quad \quad \downarrow \quad \quad \downarrow$$
$$0.25 \cdot 1600 = 400 \qquad \text{Discount}$$

Original Price – Discount = $1600 – $400 = $1200   Sale Price

Sales Tax = 6% of Sale Price = 0.06 · $1200 = $72

Total Paid = Sale Price + Sales Tax = $1200 + $72 = $1272

The total paid for the bicycle, including sales tax, was $1272.

A **commission** is a fee paid to an agent or salesperson for a service. Commissions are usually a percent of a negotiated contract (as to a real estate agent) or a percent of sales.

Example 3: Commission

A saleswoman earns a salary of $1200 a month plus a commission of 8% on whatever she sells after she has sold $8000 in furniture. What did she earn the month she sold $25,000 worth of furniture?

Solution: First subtract $8000 from $25,000 to find the amount on which the commission is based.

$25,000 – $8000 = $17,000

Now, find the amount of the commission.

$$R \quad \cdot \quad B \quad = \quad A$$
$$\downarrow \qquad \downarrow \qquad \downarrow$$

0.08 · $17,000 = $1360 commission

Now, add the commission to her salary to find what she earned that month.

$1200 + $1360 = $2560 earned for the month

Percent of profit for an investment is the ratio (a fraction) that compares the money made to the money invested. If you make two investments of different amounts of money, then the amount of money you make on each investment is not a fair comparison. In comparing such investments, the investment with the greater percent of profit is considered the better investment. **To find the percent of profit, form the fraction (or ratio) of profit divided by the investment and change the fraction to a percent**.

Example 4: Percent of Profit

Calculate the percent of profit for both a and b and tell which is the better investment.
a. $300 profit on an investment of $2400.
b. $500 profit on an investment of $5000.

Solution: Set up ratios and find the corresponding percents.

a. $\dfrac{\$300 \text{ profit}}{\$2400 \text{ invested}} = \dfrac{300 \cdot 1}{300 \cdot 8} = \dfrac{1}{8} = 0.125 = 12.5\%$ percent of profit

b. $\dfrac{\$500 \text{ profit}}{\$5000 \text{ invested}} = \dfrac{500 \cdot 1}{500 \cdot 10} = \dfrac{1}{10} = 0.1 = 10\%$ percent of profit

Clearly, $500 is more than $300, but 12.5% is greater than 10% and investment **a.** is the better investment.

Practice Problems

1. *Write 0.3 as a percent.*

2. *Write 6.4% as a decimal.*

3. *Write $\dfrac{1}{5}$ as a percent.*

4. *Find 12% of 200.*

5. *Find the percent of profit if $400 is made on an investment of $1500.*

3.7 Exercises

Write each of the following numbers in percent form.

1. 0.91   **2.** 0.625   **3.** 1.37   **4.** 0.0075

5. $\dfrac{3}{8}$ **6.** $\dfrac{87}{100}$ **7.** $1\dfrac{1}{2}$ **8.** $\dfrac{2}{3}$

Write each of the following percents in decimal form.

9. 69%   **10.** 7.5%   **11.** 11.3%   **12.** 162%
13. 0.5%   **14.** 235%   **15.** 82%   **16.** 31.4%

Write each of the following percents in fraction form (reduced).

17. 35%   **18.** 72%   **19.** 130%   **20.** 40%

Answers to Practice Problems: 1. 30% **2.** 0.064 **3.** 20% **4.** 24 **5.** 26.67%

21. Heather's monthly income is $2500. Using the given circle graph answer the following questions.

    a. What percentage of her income does Heather spend on rent each month?

    b. What percentage of her income is spent on food and entertaiment each month?

    c. What percentage of her income does Heather save each month?

Monthly Expenses

Entertainment

$300

Utilities $300

Savings $300

Food $400

Rent $800

$300

Other $100

Car expenses

22. Worldwide there were 29,534 registered earthquakes in 2006. Using the given graph answer the following questions. (Round answers to the nearest tenth of a percent.)

    a. What percentage of earthquakes in 2006 were less than a magnitude of 3.0?

    b. What percentage of earthquakes were of magnitude 3.0 - 3.9?

    c. What percentage of earthquakes were 3.0 and above?

Magnitude of Earthquakes in 2006

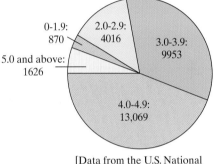

0-1.9: 870   2.0-2.9: 4016

3.0-3.9: 9953

5.0 and above: 1626

4.0-4.9: 13,069

[Data from the U.S. National Earthquake Information Center]

23. In 2006, the U.S. adult population was estimated as 217 million. (Round each answer to the nearest tenth.)

    a. What percentage of the U.S. adult population was married in 2006?

    b. What percentage of the adult population was either widowed or divorced?

Marital Status of the U.S. Population, 2006
(Data is in millions)

Divorced 23.2

Widowed 13.8

Never married 42.9

Living with a Partner 12.7

Married 124.4

[Data from the Centers for Disease Control & Prevention.]

24. In July, Daddy Fat's Record Shop sold 3900 CD's.

    a. R&B and Rap made up what percentage of their sales?

    b. What percentage of their sales was not Rock nor Country CD's?

Daddy Fat's CD Sales for July

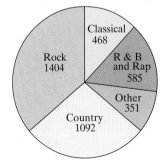

Rock 1404

Classical 468

R & B and Rap 585

Other 351

Country 1092

Find the missing rate, base, or amount.

25. 81% of 76 is _____.
26. 102% of 87 is _____.
27. _____% of 150 is 60.
28. _____ % of 160 is 240.
29. 3% of _____ is 65.4
30. 20% of _____ is 45.
31. What percent of 32 is 40?
32. 1250 is 50% of what number?
33. 100 is 125% of what number?
34. Find 24% of 244.

35. What is the percent of profit if
    a. $400 is made on an investment of $5000?
    b. $350 is made on an investment of $3500?
    c. Which was the better investment?

36. What is the percent of profit if
    a. $400 is made on an investment of $2000?
    b. $510 is made on an investment of $3000?
    c. Which was the better investment?

37. Carlos bought 100 shares of CISCO stock at $45 per share. One year later he sold the stock for $5490.
    a. What was the amount of his profit?
    b. What was his percent of profit?

38. Patrick took his parents to dinner and the bill was $70 plus 6% tax. If he left a tip of 15% of the total (food plus tax), what was the total cost of the dinner?

39. A calculator salesman's monthly income is $1500 salary plus a commission of 7% of his sales over $5000. What was his income the month that he sold $11,470 in calculators?

40. A computer programmer was told that he would be given a bonus of 5% of any money his programs could save the company. How much would he have to save the company to earn a bonus of $1000?

41. The property tax on a home was $2550. What was the tax rate if the home was valued at $170,000?

42. In one season a basketball player missed 14% of her free throws. How many free throws did she make if she attempted 150 free throws?

43. A student missed 6 problems on a statistics exam and received a grade of 85%. If all the problems were of equal value, how many problems were on the test?

44. At a department store sale, sheets and pillowcases were discounted 25%. Sheets were originally marked $30.00 and pillowcases were originally marked $12.00.
  a. What was the sale price of the sheets?
  b. What was the sale price of the pillowcases?

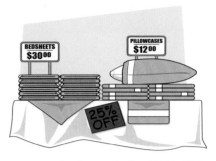

45. The Golf Pro Shop had a set of golf clubs that were marked on sale for $560. This was a discount of 20% off the original selling price.
  a. What was the original selling price?
  b. If the clubs cost the shop $420, what was the percent of profit based on cost?

46. If sales tax is figured at 7.25%, how much tax will be added to the total purchase price of three textbooks, priced at $35.00, $55.00, and $70.00? What will be the total amount paid for the books?

47. The dealer's discount to the buyer of a new motorcycle was $499.40. The manufacturer gave a rebate of $1000 on this particular model.
  a. What was the original price if the dealer discount was 7% of the original price?
  b. What would a customer pay for the motorcycle if taxes were 5% of the final selling price (after the rebate) and license fees were $642.60?

48. Data cartridges for computer backup were on sale for $27.00 (a package of two cartridges). This price was a discount of 10% off the original price.
  a. Was the original price more than $27 or less than $27?
  b. What was the original price?

49. A man weighed 200 pounds. He lost 20 pounds in 3 months. Then he gained back 20 pounds 2 months later.
  a. What percent of his weight did he lose in the first 3 months?
  b. What percent of his weight did he gain back?
  c. The loss and the gain are the same amount, but the two percents are different. Explain.

50. A car dealer bought a used car for $2500. He marked up the price so that he would make a profit of 25% of his cost.
  a. What was the selling price?
  b. If the customer paid 8% of the selling price in taxes and fees, what was the total cost of the car to the customer? The car was 6 years old.

Writing and Thinking About Mathematics

51. A man and his wife wanted to sell their house and contacted a realtor. The realtor said that the price for the services (listing, advertising, selling, and paperwork) was 6% of the selling price. The realtor asked the couple how much cash they wanted after the realtor fee was deducted from the selling price. They said $141,000 and that therefore, the asking price should be 106% of $141,000 or a total of $149,460. The realtor said that this was the wrong figure and that the percent used should be 94% and not 106%.

   a. Explain why 106% is the wrong percent and $149,460 is not the right selling price.

   b. Explain why and how 94% is the correct percent and just what the selling price should be for the couple to receive $141,000 after the realtor's fee is deducted.

Collaborative Learning Exercise

With the class separated into teams of two to four students, each team is to analyze the following problem and decide how to answer the related questions. Then each team leader is to present the team's answers and related ideas to the class for general discussion.

52. Sam works at a picture framing store and gets a salary of $500 per month plus a commission of 3% on sales he makes over $2000. Maria works at the same store, but she has decided to work on a straight commission of 8% of her sales.

   a. At what amount of sales will Sam and Maria make the same amount of money?

   b. Up to that point, who would be making more?

   c. After that point, who would be making more? Explain briefly. (If you were offered a job at that store, which method of payment would you choose?)

Hawkes Learning Systems: Introductory Algebra

Percents and Applications

<table>
<tr><td>**3.8**</td><td></td></tr>
</table>

Formulas in Geometry

After completing this section, you will be able to:

Recognize and use appropriate geometric formulas for computations.

A **formula** is an equation that represents a general relationship between two or more quantities or measurements. Several variables may appear in a formula. And, as was illustrated by the variety of formulas presented in Section 3.5, formulas are useful in mathematics, economics, chemistry, medicine, physics, and many other fields of study. In this section, we will discuss some formulas related to simple geometric figures such as rectangles, circles, and triangles.

Perimeter

The **perimeter** of a geometric figure is the total distance around the figure. Perimeters are measured in units of length such as inches, feet, yards, miles, centimeters, and meters. The formulas for the perimeters of various geometric figures are shown in Figure 3.3.

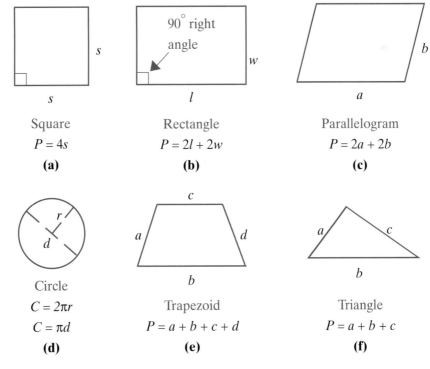

Square
$P = 4s$
(a)

Rectangle
$P = 2l + 2w$
(b)

Parallelogram
$P = 2a + 2b$
(c)

Circle
$C = 2\pi r$
$C = \pi d$
(d)

Trapezoid
$P = a + b + c + d$
(e)

Triangle
$P = a + b + c$
(f)

Figure 3.3

Terminology and Notation Related to Circles

Term	Notation	Definition
1. *Circumference*	*C*	*The perimeter of a circle is called its **circumference**.*
2. *Radius*	*r*	*The distance from the center of a circle to any point on the circle is called its **radius**.*
3. *Diameter*	*d*	*The distance from one point on a circle to another point on the circle measured through its center is called its **diameter**.*
		*(**Note:** The diameter is twice the radius: **d = 2r**.)*
4. *Pi*	π	*π is a Greek letter used for the constant 3.1415926535...*
		(π is an irrational number.)

For the calculations involving π in this text we will round off to two decimal places and use $\pi = 3.14$. Remember, however, that π is an irrational number and is an infinite nonrepeating decimal. Historically, π was discovered by mathematicians trying to find a relationship between the circumference and diameter of any circle. **The result is that π is the ratio of the circumference to the diameter of any circle, a rather amazing fact.**

Example 1: Perimeter

a. Find the perimeter of a rectangle with a length of 10 feet and a width of 8 feet.

Solution: **i.** Sketch the figure.

ii. $P = 2l + 2w$

$P = 2 \cdot 10 + 2 \cdot 8$

$P = 20 + 16 = 36$

The perimeter is 36 feet.

$w = 8$ ft

$l = 10$ ft

b. Find the circumference of a circle with a diameter of 3 centimeters.

Solution 1: **i.** Sketch the figure.
ii. $C = \pi d$
$C = 3.14(\,3\,)$
$= 9.42$ cm

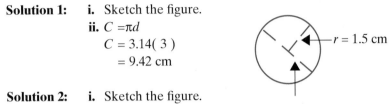

Solution 2: **i.** Sketch the figure.
ii. $C = 2\pi r$
$C = 2(\,3.14\,)(\,1.5\,)$
$= 9.42$ cm

$d = 3$ cm

Note that the result is the same with either formula. The circumference is approximately 9.42 centimeters.

c. Find the perimeter of the triangle with sides labeled as in the figure.

Solution: $P = a + b + c$
$P = 3 + 6.2 + 8.1$
$= 17.3$ in.

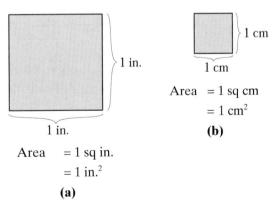

The perimeter is 17.3 inches.

Area

Area is a measure of the interior, or enclosure, of a surface. Area is measured in square units such as square feet, square inches, square meters, or square miles. For example, the area enclosed by a square of 1 inch on each side is 1 sq in. (or 1 in.^{2}) [Figure 3.4(a)], while the area enclosed by a square of 1 centimeter on each side is 1 sq cm (or 1 cm^2) [Figure 3.4(b)].

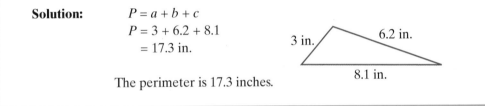

Figure 3.4

The formulas for finding the areas of several geometric figures are shown in Figure 3.5.

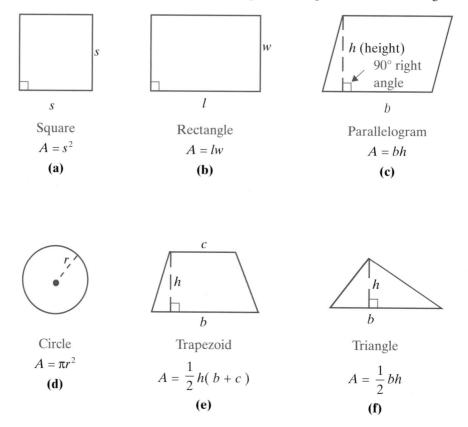

Square
$A = s^2$
(a)

Rectangle
$A = lw$
(b)

Parallelogram
$A = bh$
(c)

Circle
$A = \pi r^2$
(d)

Trapezoid
$A = \dfrac{1}{2}h(b + c)$
(e)

Triangle
$A = \dfrac{1}{2}bh$
(f)

Figure 3.5

Example 2: Area

a. Find the area of a triangle with a height of 3 centimeters and a base of 4 centimeters.

Solution: i. Sketch the figure.

ii. $A = \dfrac{1}{2} \cdot b \cdot h$

$A = \dfrac{1}{2} \cdot 4 \cdot 3$

$= 2 \cdot 3 = 6$

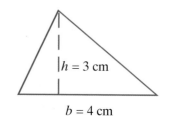

The area is 6 square centimeters.

b. Find the area of a circle with a radius of 6 inches. (Use $\pi = 3.14$.)

Solution: i. Sketch the figure.

ii. $A = \pi r^2$

$A = 3.14(\,6\,)^2$

$\quad = 3.14(\,36\,)$

$\quad = 113.04$

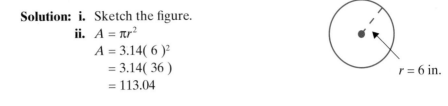

$r = 6$ in.

The area is approximately 113.04 square inches.

Volume

Volume is a measure of the space enclosed by a three-dimensional figure. Volume is measured in cubic units, such as cubic inches, cubic centimeters, and cubic feet. For example, the volume enclosed by a cube of 1 inch on each edge is 1 cu in. (or 1 in.^{3}) [Figure 3.6(a)], while the volume enclosed by a cube of 1 centimeter on each edge is 1 cu cm (or 1 cm^3) [Figure 3.6(b)].

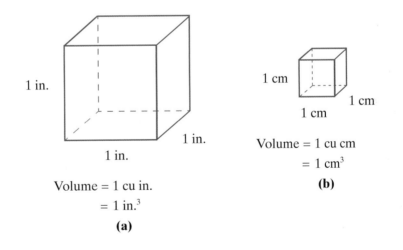

1 in.

1 in.

1 in.

Volume = 1 cu in.

$= 1$ in.^{3}

(a)

1 cm

1 cm

1 cm

Volume = 1 cu cm

$= 1$ cm^3

(b)

Figure 3.6

The formulas for finding the volume of several geometric figures are shown in Figure 3.7.

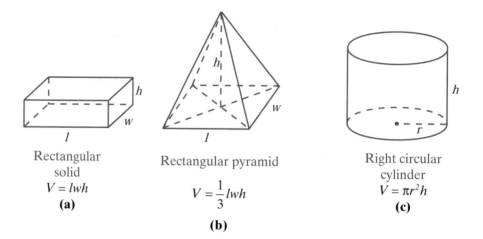

Rectangular
solid
$V = lwh$
(a)

Rectangular pyramid

$V = \dfrac{1}{3}lwh$

(b)

Right circular
cylinder
$V = \pi r^2 h$
(c)

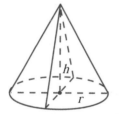

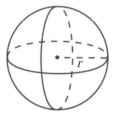

Right circular cone

$V = \dfrac{1}{3}\pi r^2 h$

(d)

Sphere

$V = \dfrac{4}{3}\pi r^3$

(e)

Figure 3.7

Example 3: Volume

a. Find the volume of a right circular cylinder with a radius of 2 centimeters and a height of 5 centimeters.

Solution: **i.** Sketch the figure.

ii. $V = \pi r^2 h$

$$= 3.14\left(2^2\right) \cdot 5$$

$$= 3.14(4) \cdot 5$$

$$= 3.14(20)$$

$$= 62.80 \text{ cm}^3$$

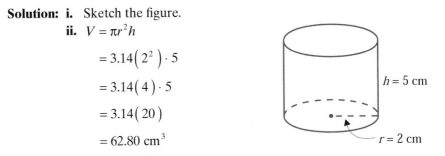

$h = 5$ cm

$r = 2$ cm

The volume is approximately 62.80 cubic centimeters.

b. Find the volume of a sphere with a radius of 4 feet.

Solution: **i.** Sketch the figure.

ii. $V = \dfrac{4}{3}\pi r^3$

$$V = \frac{4}{3}(3.14) \cdot 4^3$$

$$= \frac{4(3.14) \cdot 64}{3}$$

$$= \frac{803.84}{3} \text{ ft}^3 \text{ or about } 267.95 \text{ ft}^3$$

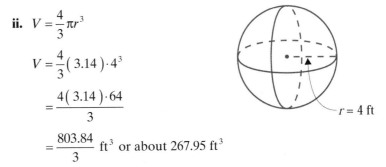

$r = 4$ ft

The volume is approximately 267.95 cubic feet.

Practice Problems

1. Find the area of a square with sides 4 cm long.

2. Find the area of a circle with a diameter 6 in.

3. Find the perimeter of a rectangle 3.5 m long and 1.6 m wide.

4. Find the volume of a rectangular solid with a length of 16 ft, a width of 10 ft, and a height of 1.4ft.

Answers to Practice Problems: 1. 16 cm^2 **2.** 28.26 in.^2 (or exactly $9\pi \text{ in.}^2$) **3.** 10.2 m **4.** 224 ft^3

3.8 Exercises

In Exercises 1 – 15, select the answer from the right-hand column that correctly matches the statement.

_____ 1. The formula for the perimeter of a square

a. $V = \dfrac{1}{3}lwh$

_____ 2. The formula for the circumference of a circle

b. $A = \dfrac{1}{2}h(b+c)$

_____ 3. The formula for the perimeter of a triangle

c. $A = s^2$

_____ 4. The formula for the area of a rectangle

d. $A = \dfrac{1}{2}bh$

_____ 5. The formula for the area of a square

e. $P = 4s$

_____ 6. The formula for the area of a trapezoid

f. $A = lw$

_____ 7. The formula for the area of a triangle

g. $P = 2l + 2w$

_____ 8. The formula for the area of a parallelogram

h. $V = \dfrac{1}{3}\pi r^2 h$

_____ 9. The formula for the perimeter of a rectangle

i. $V = \dfrac{4}{3}\pi r^3$

_____ 10. The formula for the volume of a rectangular pyramid

j. $C = 2\pi r$

_____ 11. The formula for the volume of a rectangular solid

k. $P = a + b + c$

_____ 12. The formula for the area of a circle

l. $V = lwh$

_____ 13. The formula for the volume of a right circular cylinder

m. $V = \pi r^2 h$

_____ 14. The formula for the volume of a sphere

n. $A = \pi r^2$

_____ 15. The formula for the volume of a right circular cone

o. $A = bh$

Find (a) the perimeter and (b) the area of each figure in Exercises 16 – 21.

16. $P =$ _____ $A =$ _____

18 mm

18 mm

17. $P =$ _____ $A =$ _____

15 cm

20 cm

18. $P =$ _____ $A =$ _____

6 in. 7 in.

12 in.

19. $P =$ _____ $A =$ _____

13 mm 20 mm

12 mm

21 mm

20. $P =$ _____ $A =$ _____

8 cm

5 cm 4 cm 5 cm

14 cm

21. $C =$ _____ $A =$ _____

5 in.

Find the volume of each figure in Exercise 22 – 26. (Use $\pi = 3.14$ and round to the nearest hundredth.)

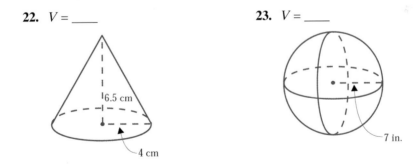

22. $V =$ _____

6.5 cm

4 cm

23. $V =$ _____

7 in.

24. $V =$ ____

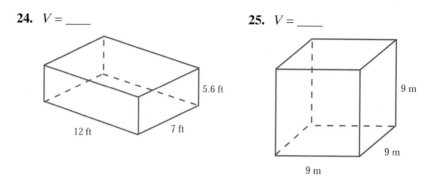

5.6 ft

12 ft   7 ft

25. $V =$ ____

9 m

9 m

9 m

26. $V =$ ____

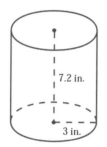

7.2 in.

3 in.

Solve the problems in Exercises 27 – 38. (Use $\pi = 3.14$)

27. What is the perimeter of a rectangle with length of 17 in. and width of 11 in.?

28. Find the area of a square with sides that are 9 ft long.

29. The base of a triangle is 14 cm and the height is 9 cm. Find the area.

30. Find the circumference of a circle with a radius of 8 m.

31. The radius of the base of a cylindrical tank is 14 ft If the tank is 10 ft high, find the volume.

32. The sides of a triangle are 6.2 m, 8.6 m, and 9.4 m. Find the perimeter.

33. What is the volume of a cube with edges of 5 ft?

34. A rectangular garden plot is 60 ft long and 42 ft wide. Find the area.

35. A parallelogram has a base of 20 cm and a height of 13.6 cm. Find the area.

36. A rectangular box is 18 in. long, 10.3 in. wide, and 8 in. high. Find the volume.

37. The diameter of the base of a right circular cone is 15 in. If the height of the cone is 9 in., find the volume.

38. Find the volume of a sphere whose radius is 10 cm. (Round to the nearest hundredth.)

Find the perimeter and area for each figure in Exercises 39 and 40. (Use $\pi = 3.14$)

39. $P =$ _____ $A =$ _____

40. $P =$ _____ $A =$ _____

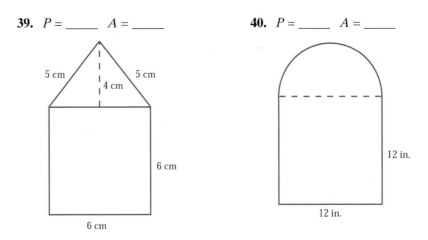

Calculator Problems

In Exercises 41 – 45, find the answers accurate to two decimal places.

41. Find the area of a rectangle that is 16.54 m long and 12.82 m wide.

42. The radius of a circle is 8.32 in. Find the circumference. (Use $\pi = 3.14$.)

43. Find the area of a trapezoid with bases of 22.36 in. and 17.48 in. and a height of 6.53 in.

44. Find the area of a triangle with a base of 63.52 cm and a height of 41.78 cm.

45. The radius of a sphere is 114.8 cm. Find the volume. (Use $\pi = 3.14$.)

Hawkes Learning Systems: Introductory Algebra

Formulas in Geometry

Chapter 3 Index of Key Ideas and Terms

Section 3.1 Solving Linear Equations: $x + b = c$ and $ax = c$

Linear Equations   pages 200 - 201

If a, b, and c are **constants** and $a \neq 0$, then a **linear equation in x**
(or **first-degree equation in x**) is an equation that can be written
in the form $ax + b = c$.

Addition Principle of Equality   page 201

If the same algebraic expression is added to both sides of an
equation, the new equation has the same solutions as the original
equation. Symbolically, if A, B, and C are algebraic expressions,
then the equations

$$A = B \quad \text{and} \quad A + C = B + C$$

have the same solutions.

Equivalent Equations   page 201

Equations with the same solutions are said to be **equivalent**.

Multiplication (or Division) Principle of Equality   page 204

If both sides of an equation are multiplied by (or divided by) the
same nonzero constant, the new equation has the same solutions
as the original equation. Symbolically, if A and B are algebraic
expressions and C is any nonzero constant, then the equations

$$A = B$$

$$\text{and } AC = AB \qquad \text{where } C \neq 0$$

$$\text{and } \frac{A}{C} = \frac{B}{C} \qquad \text{where } C \neq 0$$

have the same solutions.

Procedure for Solving Applications   pages 206 - 209

George Pólya's four-step process as an approach to problem solving:
1. Understand the problem
2. Devise a plan.
3. Carry out the plan.
4. Look back over the results.

Section 3.2 Solving Linear Equations: $ax + b = c$

Procedure for Solving Linear Equations that Simplify to the Form $ax + b = c$ page 213
1. Combine like terms on both sides of the equation
2. Use the Addition Principle of Equality and add the opposite of the constant term to both sides.
3. Use the Multiplication (or Division) Principle of Equality to multiply both sides by the reciprocal of the coefficient of the variable (or divide both sides by the coefficient itself), The coefficient of the variable will become +1.
4. Check your answer by substituting it into the original equation.

Section 3.3 More Linear Equations: $ax + b = dx + c$

General Procedure for Solving Linear Equations that Simplify to the   page 220
Form $ax + b = dx + c$
1. Simplify each side of the equation by removing any grouping symbols and combining like terms. (This may also involve multiplying both sides by the LCM of the denominators or by a power of 10 to remove decimals. In this manner we will be dealing only with **integer** constants and coefficients.)
2. Use the Addition Principle of Equality to add the opposites of constants and/or variables so that variables are on one side and constants are on the other side.
3. Use the Multiplication Principle of Equality to multiply both sides by the reciprocal of the coefficient (or to divide by the coefficient).
4. Check your answer by substituting it into the original equation.

Section 3.4 Solving Linear Inequalities and Applications

Intervals of Real Numbers   pages 228 - 230
There is a one-to-one correspondence between the real numbers and the points on a line. That is, each point on a number line corresponds to one real number, and each real number corresponds to one point on a number line.
Types of Intervals
Open Interval: $a < x < b$, $x < a$, $x > a$
Closed Interval: $a \leq x \leq b$
Half-Open Interval: $a < x \leq b$, $a \leq x < b$, $x \geq a$, $x \leq a$

Continued on next page...

Section 3.4   Solving Linear Inequalities   (continued)

Linear (or First-Degree) Inequalities   pages 231 - 233
    Rules for Solving Linear Inequalities   page 232
        1. Simplify each side of the inequality by removing any
           parentheses and combining like terms.
        2. Add constants and/or variables to or subtract them from
           both sides of the inequality so that variables are on one side
           and constants are on the other.
        3. Divide both sides by the coefficient (or multiply them by
           the reciprocal) of the variable and reverse the sense of the
           inequality if this coefficient is negative.
        4. A quick (and generally satisfactory) check is to select any
           one number in your solution and substitute it into the
           original inequality.

Section 3.5   Working with Formulas

Formulas   pages 240 - 241

Section 3.6   Applications: Number Problems and Consecutive Integers

Number Problems   pages 251 - 253

Consecutive Integers   pages 253 - 255
    Consecutive integers are two integers that differ by 1.
    Two consecutive integers can be represented as n and $n + 1$.

Consecutive Even Integers   pages 253 - 255
    Consecutive even integers are two even integers that differ by 2.
    Two consecutive even integers can be represented as n and $n + 2$
    where n is even.

Consecutive Odd Integers   pages 253 - 255
    Consecutive odd integers are two odd integers that differ by 2.
    Two consecutive odd integers can be represented as n and $n + 2$
    where n is odd.

Section 3.7 Applications: Percent Problems (Discount, Taxes, Commission, Profit, and Others)

Percent   pages 261 - 262

To Change a Decimal to a Percent: page 261
Step 1: Move the decimal point two places to the right.
(This is the same as multiplying by 100.)
Step 2: Add the % symbol.

To Change a Percent to a Decimal: page 261
Step 1: Move the decimal point two places to the left.
Step 2: Delete the % symbol.

To Change a Percent to a Fraction (or Mixed Number):   page 262
Step 1: Write the percent as a fraction with denominator 100 and
delete the % symbol.
Step 2: Reduce the fraction (or change it to a mixed number if you
prefer with the fraction part reduced).

The Formula for Solving Percent Problems   pages 262 - 263
$R \cdot B = A$ (or $A = R \cdot B$)
R = **RATE** or percent (as a decimal or fraction)
B = **BASE** (number we are finding the percent of)
A = **AMOUNT** or percentage (a part of the base)
"of" means to multiply.
"is" means equal (=).

Percent of Profit   pages 265 - 266
To find the **percent of profit**, form the fraction (or ratio) of
profit divided by the investment and change the fraction to a
percent.

Section 3.8 Formulas in Geometry

Formulas in Geometry   pages 271 - 276
    Perimeter   pages 271 - 272

        Square: $P = 4s$

        Rectangle: $P = 2l + 2w$

        Parallelogram: $P = 2a + 2b$

        Circle: $C = 2\pi r$ or $C = \pi d$

        Trapezoid: $P = a + b + c + d$

        Triangle: $P = a + b + c$

    Area   pages 274 - 274

        Square: $A = s^2$

        Rectangle: $A = lw$

        Parallelogram: $A = bh$

        Circle: $A = \pi r^2$

        Trapezoid: $A = \frac{1}{2}h(b+c)$

        Triangle: $A = \frac{1}{2}bh$

    Volume   pages 275 - 276

        Rectangular Solid: $V = lwh$

        Rectangular Pyramid: $V = \frac{1}{3}lwh$

        Right Circular Cylinder: $V = \pi r^2 h$

        Right Circular Cone: $V = \frac{1}{3}\pi r^2 h$

        Sphere: $V = \frac{4}{3}\pi r^3$

Hawkes Learning Systems: Introductory Algebra

For a review of the topics and problems from Chapter 3, look at the following lessons from *Hawkes Learning Systems: Introductory Algebra*

Solving Linear Equations Using Addition and Subtraction
Applications of Linear Equations: Addition and Subtraction
Solving Linear Equations Using Multiplication and Division
Applications of Linear Equations: Multiplication and Division
Solving Linear Equations
More Linear Equations: $ax + b = dx + c$
Solving Linear Inequalities
Solving Formulas
Working with Formulas
Applications: Number Problems and Consecutive Integers
Percents and Applications

Chapter 3 Review

3.1 Solving Linear Equations: $x + b = c$ and $ax = c$

Solve each of the following linear equations.

1. $x - 6 = 1$ **2.** $x - 4 = 10$ **3.** $y + 5 = -6$ **4.** $y + 7 = -1$

5. $0 = n + 8$ **6.** $0 = n + 13$ **7.** $4 = x + 9$ **8.** $13 = x + 17$

9. $6n = 60$ **10.** $7n = 42$ **11.** $\dfrac{2}{3}y = -12$ **12.** $\dfrac{5}{6}y = -25$

13. $2.9x - 1.9x = 7 + 11$ **14.** $3.2y - 2.2y = 1.3 - 0.9$ **15.** $-13.5 = 4.5x$

3.2 Solving Linear Equations: $ax + b = c$

Solve each of the following linear equations.

16. $2x + 3 = -7$ **17.** $3x + 5 = -19$ **18.** $2y + 2y - 6 = 30$

19. $3y + 10 + 2y = -25$ **20.** $20 = 5x - 3x - 4$ **21.** $-30 = 6x - 4x + 20$

22. $0.9y + 0.3y - 5 = 2.2$ **23.** $0.5y + 0.6y + 1.7 = -1.6$ **24.** $\dfrac{1}{5}n + \dfrac{2}{5}n - \dfrac{1}{3} = \dfrac{4}{5}$

25. $\dfrac{3}{8}n + \dfrac{1}{4}n + \dfrac{1}{2} = \dfrac{3}{4}$ **26.** $\dfrac{2}{3}x - \dfrac{1}{2}x + \dfrac{5}{6} = 2$ **27.** $\dfrac{3}{4}x - \dfrac{7}{12}x - \dfrac{2}{3} = 4$

28. $4.78 - 0.3x + 0.5x = 2.31$ **29.** $5.32 + 0.4x - 0.6x = 7.1$

30. $8.62 = 0.45x - 0.15x + 5.02$

3.3 More Linear Equations: $ax + b = dx + c$

Solve the following linear equations.

31. $8y - 2.1 = y + 5.04$ **32.** $6y - 4.5 = y + 4.5$ **33.** $2.4n = 3.6n$

34. $8n = -1.3n$ **35.** $2(x + 3) = 3(x - 7)$ **36.** $4(x - 2) = 3(x + 6)$

37. $\dfrac{1}{3}n + 6 = \dfrac{1}{4}n + 5$ **38.** $\dfrac{2}{5}n + 7 = \dfrac{1}{5}n + 4$

39. $6(2x - 1) = 4(x + 3) + 14$ **40.** $8 + 4(x - 4) = 5 - (x - 14)$

41. $-2(y + 2) - 4 = 6 - (y + 12)$ **42.** $7 - 3(x + 5) = -2(x + 3) - 15$

43. $\dfrac{5x}{6} + \dfrac{1}{18} = \dfrac{x}{3}$ **44.** $\dfrac{x}{6} + \dfrac{1}{15} - \dfrac{2x}{3} = 0$

45. $2(y + 8.3) + 1.2 = y - 16.6 - 1.2$

3.4 Solving Linear Inequalities and Applications

In Exercises 46 – 50, graph the set of real numbers that satisfies each of the inequalities and tell what type of interval it represents.

46. $-3 < x < 1$ **47.** $-3 \leq x \leq 4$ **48.** $-1 \leq x < \dfrac{3}{4}$

49. $x < \dfrac{2}{3}$ **50.** $x \geq 7.5$

Solve the following inequalities and graph the solutions.

51. $4x + 6 \geq 10$ **52.** $3x - 7 \leq 14$ **53.** $5y + 8 < 2y + 5$

54. $y - 6 > 3y + 4$ **55.** $\dfrac{x}{3} - 1 > 2 - \dfrac{x}{4}$ **56.** $\dfrac{1}{2} + \dfrac{x}{5} < 1 - \dfrac{x}{5}$

57. $1 \leq 3x - 5 \leq 10$ **58.** $-4 < 5x + 1 < 16$ **59.** $0 < -\dfrac{3}{4}x + 1 \leq 4$

60. $0 < -\dfrac{1}{2}x + 3 \leq 2$

3.5 Working with Formulas

Solve for the indicated variable in Exercises 61 – 70.

61. $L = 2\pi rh$; solve for π.   **62.** $P = R - C$; solve for R.

63. $A = \dfrac{1}{2}bh$; solve for b.   **64.** $\alpha + \beta + \gamma = 180$; solve for α.

65. $v = v_0 - gt$; solve for g.   **66.** $K = \dfrac{mv^2}{2g}$; solve for m.

67. $3x + y = 6$; solve for y.   **68.** $6x - y = 3$; solve for y.

69. $5x - 2y = 10$; solve for x.   **70.** $-3x + 4y = -7$; solve for x.

71. Given the formula $C = \dfrac{5}{9}(F - 32)$, find the value of C if $F = 32°$.

72. Using the formula $d = rt$, find Karl's average rate of speed if he drove 190 miles in 4 hours.

73. If the perimeter of a rectangular soccer field is planned to be 150 meters and because of building restrictions the width has to be 35 meters, what is the planned length of the field? $(P = 2l + 2w)$

74. Volume and pressure of gas in a container are known to be related by the formula $V = \dfrac{k}{P}$ where k is a constant that depends on the type of gas involved. Find P if $k = 3.2$ and $V = 20$ cubic centimeters.

75. Given the equation $4x + y = 10$, find the value of y that corresponds to a value of 3 for x.

3.6 Applications: Number Problems and Consecutive Integers

76. Forty-two is 4 more than twice a certain number. What is the number?

77. The difference between three times a number and 7 is equal to 17 decreased by the number. Find the number.

78. The quotient of twice a number and 9 is seven less than the number. What is the number?

79. Twice the difference between a number and 8 is equal to 5 times the number plus 11. What is the number?

80. Find three consecutive integers whose sum is 75.

81. Find three consecutive odd integers whose sum is 279.

82. Three consecutive even integers are such that if the second is added to twice the first the result is 18 more than the third. Find these integers.

83. Find four consecutive integers whose sum is 190.

84. Find three consecutive odd integers whose sum is –81.

85. A real estate agent says that the current value of a home is $150,000 more than twice its value when it was new. If the current value is $640,000, what was the value of the home when it was new?

3.7 Applications: Percent Problems (Discount, Taxes, Commission, Profit, and Others)

86. 92% of 85 is _____.   **87.** _____ % of 150 is 105.

88. The property tax on a home was $4440. What was the tax rate if the home was valued at $370,000?

89. In one season a basketball player made 85% of her free throws. How many free throws did she miss if she attempted 120 free throws?

90. At a department store sale shoes were discounted 33%. What was the sale price of shoes that were originally marked $90.00?

91. If sales tax is figured at 8.25%, how much tax will be added to a purchase of $260? What will be the total amount paid?

92. A golden retriever weighed 80 pounds and was put on a diet so she lost 5 pounds. What percent of her weight did she lose?

The formula $I = Prt$ is used to calculate simple interest (I), where P is the principal invested, r is the interest rate, and t is the time (in years or a part of a year). In these exercises (for ease in calculations) assume 360 days in a year and 30 days in a month.

93. The interest earned on an investment of $3240 at 10% was $81. What was the time that the money was invested?

94. A savings account pays an annual interest rate of 5%. How much must be invested to make $400 in interest in 9 months?

95. If you have a credit card debt of $2000 and the simple interest for one month is $40, what is the interest rate you are paying?

3.8 Formulas in Geometry

96. Write the formula for the perimeter of a parallelogram.

97. Write the formula for the area of a circle.

98. Write the formula for the volume of a sphere.

99. Find the perimeter of the triangle with sides labeled as in the figure.

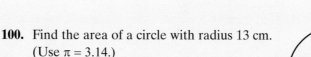

9.6 in.   5.7 in.   4.9 in.

100. Find the area of a circle with radius 13 cm. (Use $\pi = 3.14$.)

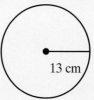

13 cm

101. What is the volume of a cube with sides 3 feet long?

102. A flower garden is in the shape of a parallelogram with a base of 30 meters and a height of 15.6 meters. What is the area of the flower garden?

103. One size of pizza served at Joe's Pizza Parlor is 10 in. in diameter. What is the area of this particular (circular) size pizza? (Use $\pi = 3.14$.)

104. A can of beans has a diameter of 8.2 cm and a height of 13 cm. What is the volume of the can of beans (to the nearest tenth of a cubic centimeter)?

105. Find the perimeter and area for the figure shown.

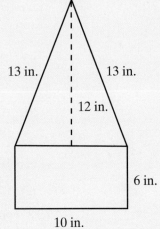

13 in.

13 in.

12 in.

6 in.

10 in.

Chapter 3 Test

Solve the equations in Exercises 1 – 6.

1. $\dfrac{5}{3}x + 1 = -4$

2. $4x - 5 - x = 2x + 5 - x$

3. $8(3 + x) = -4(2x - 6)$

4. $\dfrac{3}{2}x + \dfrac{1}{2}x = -3 + \dfrac{3}{4}$

5. $0.7x + 2 = 0.4x + 8$

6. $4(2x - 1) + 3 = 2(x - 4) - 5$

Find the missing number or percent in Exercises 7 and 8.

7. 62% of 180 is _____.

8. 48 is _____% of 150.

9. What simple interest would be earned if $5000 is invested at 9% for 9 months?

10. Which is a better investment, an investment of $6000 that earns a profit of $360 or an investment of $10,000 that earns a profit of $500? Explain your answer in terms of percent.

11. The triangle shown here indicates that the sides can be represented as $x, x + 1,$ and $2x - 1.$ What is the length of each side, if the perimeter is 12 meters?

$x + 1$ $2x - 1$

x

In Exercises 12 – 13, solve each formula for the indicated variable.

12. $N = mrt + p$; solve for m.

13. $5x + 3y - 7 = 0$; solve for y.

14. Given the formula $C = \dfrac{5}{9}(F - 32)$ that relates Fahrenheit and Celsius temperatures, find the Celsius temperature equal to $-4°F$.

15. Given the equation $3x - 2y = 18,$ find the value of x that corresponds to the value of 4 for y.

16. Find (a) the perimeter and (b) the area of the figure shown here.

17. Find the volume of a sphere with diameter 20 cm. (Round to the nearest hundreth.)

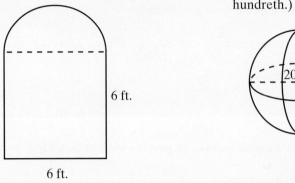

6 ft.

6 ft.

Solve each of the inequalities in Exercises 18 – 21 and graph each solution set on a real number line.

18. $3x + 2 \le -8$

19. $-\dfrac{3}{4}x - 6 > 9$

20. $2(x - 7) < 4(2x + 3)$

21. $-2 \le 4x + 7 \le 3$

In Exercises 22 – 26, set up an equation for each word problem and solve.

22. One number is 5 more than twice another number. Their sum is −22. Find the numbers.

23. Find two consecutive integers such that twice the first added to three times the second is equal to 83.

24. A rectangle is 37 inches long and has a perimeter of 122 inches.
a. Find the width of the rectangle.
b. Find the area of the rectangle.

25. A man's suit was on sale for $547.50.
a. If this price was a discount of 25% from the original price, what was the original price?
b. What would the suit cost a buyer if sales tax was 6% and the alterations were $25?

26. Find three consecutive odd integers such that three times the second is equal to 27 more than the sum of the first and the third.

27. a. Find the volume (in cubic feet) of concrete needed for a sidewalk that is 6 feet by 100 feet by 4 inches. (**Note:** 4 in. = $\frac{1}{3}$ ft)

b. Find the number of cubic yards (to the nearest tenth) in the sidewalk. (**Note:** 12 in. = 1 ft and 27 ft^3 = 1 yd.^{3})

28. If a cone has a height of 6 in. and a circular base with a radius of 4 in., what is its volume?

Cumulative Review: Chapters 1 – 3

Determine whether each statement in Exercises 1 – 3 is true or false. If a statement is false, rewrite it in a form that is a true statement. (There may be more than one way to correct a statement.)

1. $|-5| \leq 5$ **2.** $2^3 - 8 > 8 - 3 \cdot 4$ **3.** $\dfrac{3}{4} \geq |-1|$

Exercises 4 and 5: On a number line, graph the integers that satisfy the stated inequality.

4. $|x| \leq 6$ **5.** $|x| = x$

6. Explain, in your own words, why division by 0 is undefined.

7. Find the LCM for each set of numbers and variables.
     a. $\{25, 35, 40, 50\}$ **b.** $\{6x^2, 15xy^2, 20xy\}$

8. Find the average of each set of numbers.
     a. $\{32, 56, 70, 86, 91\}$ **b.** $\{13.62, 20.15, 26.24, 33.19\}$

Exercises 9 – 13: Use the rules for order of operations to evaluate each expression.

9. $5^2 \cdot 3^2 - 24 \div 2 \cdot 3$ **10.** $-36 \div (-2)^2 + 20 - 2(16 - 17)$

11. $(7 - 10)[49 \div (-7) + 20 \cdot 3 - 5 \cdot 15 - (-10)] - 1$

12. $\left(\dfrac{2}{3}\right)^2 \div \dfrac{5}{18} - \dfrac{3}{8}$ **13.** $\left(\dfrac{9}{10} + \dfrac{2}{15}\right) \div \left(\dfrac{1}{2} - \dfrac{2}{3}\right)$

14. Use a calculator to find the value of the expression: $(15)^3 - 160 + 13(-5)$

Exercises 15 – 18: (a) Simplify each expression by combining like terms. (b) Evaluate the simplified form of each expression for $x = -2$ and $y = 3$.

15. $-4(y + 3) + 2y$ **16.** $3(x + 4) - 5 + x$

17. $2(x^2 + 4x) - (x^2 - 5x)$ **18.** $\dfrac{3(5x - x)}{4} - 2x - 7$

Exercises 19 – 22: Write an algebraic expression described by each of the phrases.

19. The difference between 9 and twice a number

20. Three times the sum of a number and 10

21. The number of hours in x days and 5 hours

22. The number of points scored by a football team on T touchdowns (6 points each), E extra points (1 point each), and F field goals (3 points each).

Exercises 23 – 30: Solve the following equations.

23. $9x - 8 = 4x - 13$ **24.** $6 - (x - 2) = 5$ **25.** $10 + 4x = 5x - 3(x + 4)$

26. $4.4 + 0.6x = 1.2 - 0.2x$ **27.** $-2(5x + 12) - 5 = -6(3x - 2) - 10$

28. $\dfrac{3}{4}y - \dfrac{1}{2} = \dfrac{5}{8}y + \dfrac{1}{10}$ **29.** $\dfrac{1}{2} + \dfrac{1}{5}y = \dfrac{2}{15}y + 1$ **30.** $-\dfrac{3}{4}y + \dfrac{3}{8}y = \dfrac{1}{6}y$

Exercises 31 – 33: Solve for the indicated variable.

31. $h = vt - 16t^2$; solve for v.   **32.** $A = P + Prt$; solve for r.
33. $5x - 3y = 14$; solve for y.

Exercises 34 – 36: Find the missing number.

34. 75% of _____ is 102.   **35.** 20.24 is _____ % of 92.   **36.** 25.2% of 65 is _____ .

Exercises 37 – 40: Solve each of the inequalities and graph each solution set on a real number line.

37. $2x + 5 - 3 \le 6$ **38.** $-3(7 - 2x) \ge 2 + 3x - 10$

39. $-\dfrac{1}{2}x + \dfrac{7}{8} < \dfrac{2}{3}x + \dfrac{1}{2}$ **40.** $-6.2 \le 2x - 4 \le 10.8$

41. Find (a) the perimeter and (b) area of the figure shown here. (Use $\pi = 3.14$.)

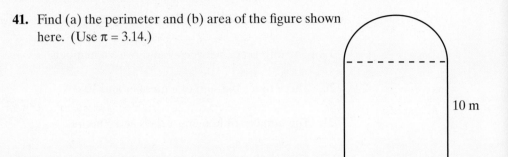

10 m

10 m

42. Find the volume of the circular cylinder with dimensions as shown in the figure. (Use $\pi = 3.14$.)

15 in.

5 in.

Exercises 43 – 54: Set up an equation or inequality for each problem and solve for the unknown quantity.

43. Your neighbor has a used (red) pickup for sale and it is perfect for you. However, you have no money. With your new job you would be able to buy it in 6 months, but you know that the truck will certainly be sold by then. Your uncle has agreed to loan you $2000 for 6 months at 8% interest and you have agreed to the terms. How much will you pay your uncle in interest for the 6 months loan?

44. Computers are on sale at a 30% discount. If you know that you will pay 6.5% in sales taxes, what will you pay for a computer that is priced at $680?

45. If twice a certain number is increased by 3, the result is 8 less than three times the number. Find the number.

46. Find three consecutive even integers such that the sum of the first and twice the second is equal to 14 more than twice the third.

47. Find four consecutive integers such that the sum of the first, second, and fourth integers is 2 less than the third.

48. ALPHA Truck Rental charges \$35 a day plus \$0.55 per mile and BETA Truck Rental charges \$45 a day but only \$0.25 per mile. For how many miles can you drive an ALPHA truck in one day and keep the cost below the cost of a BETA truck?

49. Jay works at a furniture store for a salary of \$800 per month plus a commission of 6% on his sales over \$5000. Kay works at the same store but works only on a straight commission of 10%. In one particular month, Jay and Kay sold the same value in furniture and made the same amount of money. What was the dollar value of furniture sold by each? What amount of money did each make for that month?

50. Find the height of a cone that has a volume of 175π in.^{3} and a radius of 7 in.

51. The range for a grade of B is an average of 75 or more but less than 85 in an English class. If your grades on the first four exams were 73, 65, 77, and 74, what possible grades can you get on the final exam to earn a grade of B? (Assume that 100 is the maximum number of points on the final exam.)

52. What principal would you need to invest to earn \$450 in interest in 6 months if the rate of interest was 9%?

53. Which is the better investment: (a) a profit of \$800 on an investment of \$10,000 or (b) a profit of \$1200 on an investment of \$15,000? Explain in terms of percents.

54. An artist has a piece of wire 30 inches long and wants to bend it into the shape of a triangle so that the sides are of length $x, 2x + 2$, and $2x + 3$. This triangle will be a right triangle and the two shortest sides will be perpendicular to each other.
a. What will be the length of each side of the triangle?
b. What will be the area of the triangle?

Graphing Linear Equations and Inequalities in Two Variables

Did You Know?

In Chapter 4 you will be introduced to the idea of a graph of an algebraic equation. A graph is simply a picture of an algebraic relationship. This topic is more formally called **analytic geometry**. It is a combination of algebra (the equation) and geometry (the picture).

The idea of combining algebra and geometry was not thought of until René Descartes wrote his famous *Discourse on the Method of Reasoning* in 1637. The third appendix in this book, "La Geometrie," made Descartes' system of analytic geometry known to the world. In fact, you will find that analytic geometry is sometimes called **Cartesian geometry**.

René Descartes is perhaps better known as a philosopher than as a mathematician; he is often referred to as the father of modern philosophy. His method of reasoning was to apply the same logical structure to philosophy that had been developed in mathematics, especially geometry.

In the Middle Ages, the highest forms of knowledge were believed to be mathematics, philosophy, and theology. Many famous people in history who have reputations as poets, artists, philosophers, and theologians were also creative mathematicians. Almost every royal court had mathematicians whose work reflected glory on the royal sponsor who paid the mathematician for his research and court presence.

Descartes, in fact, died in 1650 after accepting a position at the court of the young warrior-queen, Christina of Sweden. Apparently, the frail French philosopher-mathematician, who spent his mornings in bed doing mathematics, could not stand the climate of Sweden and the hardships imposed by Christina in her demand that Descartes tutor her in mathematics each morning at 5 o'clock in an unheated castle library.

4.1 The Cartesian Coordinate System and Reading Graphs

4.2 Graphing Linear Equations in Two Variables: $Ax + By = C$

4.3 The Slope-Intercept Form: $y = mx + b$

4.4 The Point-Slope Form: $y - y_1 = m(x - x_1)$

4.5 Introduction to Functions and Function Notation

4.6 Graphing Linear Inequalities: $y < mx + b$

"Divide each problem that you examine into as many parts as you can and as you need to solve them more easily."

René Descartes (1596 – 1650)

The Cartesian Coordinate System and Reading Graphs

After completing this section, you will be able to:

1. Name ordered pairs of real numbers corresponding to points on a graph.

2. Graph points corresponding to ordered pairs in the Cartesian coordinate system.

3. Find ordered pairs of real numbers that satisfy a given equation.

4. Read points on a given graph of a straight line.

René Descartes (1596 – 1650), a famous French mathematician, developed a system for solving geometric problems using algebra. This system is called the **Cartesian coordinate system** in his honor. Descartes based his system on a relationship between points in a plane and **ordered pairs** of real numbers. This section begins by relating algebraic formulas with ordered pairs and then shows how these ideas can be related to geometry.

Equations in Two Variables

Equations such as $d = 60t$, $I = 0.05P$, and $y = 2x + 3$ represent relationships between pairs of variables. For example, in the first equation, if $t = 3$, then $d = 60 \cdot 3 = 180$. With the understanding that t is first and d is second, we can represent $t = 3$ and $d = 180$ in the form of an ordered pair $(3, 180)$. In general, if t is the first number and d is the second number, then solutions to the equation $d = 60t$ can be written in the form of ordered pairs (t, d). Thus we see that $(180, 3)$ is different from $(3, 180)$. **The order of the numbers in an ordered pair is critical**.

We say that $(3, 180)$ **is a solution of** (or **satisfies**) the equation $d = 60t$. Similarly, $(5, 300)$ represents $t = 5$ and $d = 300$ and satisfies the equation $d = 60t$. In the same way, $(100, 5)$ satisfies $I = 0.05P$ where $P = 100$ and $I = 0.05 \cdot 100 = 5$. In this equation, solutions are ordered pairs in the form (P, I).

For the equation $y = 2x + 3$, ordered pairs are in the form (x, y), and $(2, 7)$ satisfies the equation. If $x = 2$, then substituting in the equation gives $y = 2 \cdot 2 + 3 = 7$. In the ordered pair (x, y), x is called the **first coordinate** (or **first component**) and y is called the **second coordinate** (or **second component**). To find ordered pairs that satisfy an equation in two variables, we can **choose any value** for one variable and find the corresponding value for the other variable by substituting into the equation. For example, for the equation $y = 2x + 3$:

Choices for x:	**Substitution:**	**Ordered Pairs:**
$x = 1$	$y = 2 \cdot 1 + 3 = 5$	$(1, 5)$
$x = -2$	$y = 2(-2) + 3 = -1$	$(-2, -1)$
$x = \dfrac{1}{2}$	$y = 2 \cdot \dfrac{1}{2} + 3 = 4$	$\left(\dfrac{1}{2}, 4\right)$

All the ordered pairs $(1, 5)$, $(-2, -1)$, and $\left(\dfrac{1}{2}, 4\right)$ satisfy the equation $y = 2x + 3$. **There is an infinite number of such ordered pairs. Any real number could have been chosen for x and the corresponding value for y calculated.**

Since the equation $y = 2x + 3$ is solved for y, we say that the value of y "depends" on the choice of x. Thus, in an ordered pair of the form (x, y), the second coordinate, y, is called the **dependent variable** and the first coordinate, x, is called the **independent variable**.

In the following table the first variable, in each case, is the independent variable and the second variable is the dependent variable. Corresponding ordered pairs would be of the form (t, d), (P, I), and (x, y). The choices for the values of the independent variables are arbitrary. There are an infinite number of other values that could have just as easily been chosen.

$d = 60t$		$I = 0.05P$		$y = 2x + 3$	
t	d	P	I	x	y
5	$60 \cdot 5 = 300$	100	$0.05\,(100) = 5$	-2	$2(-2) + 3 = -1$
10	$60 \cdot 10 = 600$	200	$0.05\,(200) = 10$	-1	$2(-1) + 3 = 1$
12	$60 \cdot 12 = 720$	500	$0.05\,(500) = 25$	0	$2(0) + 3 = 3$
15	$60 \cdot 15 = 900$	1000	$0.05\,(1000) = 50$	3	$2(3) + 3 = 9$

Graphing Ordered Pairs

The Cartesian coordinate system relates algebraic equations and ordered pairs to geometry. In this system, two number lines intersect at right angles and separate the plane into four **quadrants**. The **origin**, designated by the ordered pair $(0, 0)$, is the point of intersection of the two lines. The horizontal number line is called the **horizontal axis** or **x-axis**. The vertical number line is called the **vertical axis** or **y-axis**. Points that lie on either axis are not in any quadrant. They are simply on an axis (Figure 4.1).

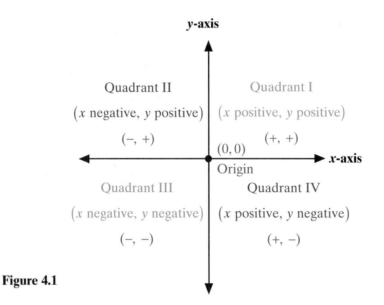

Figure 4.1

The following important relationship between ordered pairs of real numbers and points in a plane is the cornerstone of the Cartesian coordinate system.

There is a one-to-one correspondence between points in a plane and ordered pairs of real numbers.

In other words, for each point there is one and only one corresponding ordered pair of real numbers, and for each ordered pair of real numbers there is one and only one corresponding point.

The **graphs of the points** A $(2, 1)$, B $(-2, 3)$, C $(-3, -2)$, D $(1, -2)$, and E $(3, 0)$ are shown in Figure 4.2. [**Note**: An ordered pair of real numbers and the corresponding point on the graph are frequently used to refer to each other. Thus the ordered pair $(2, 1)$ and the point $(2, 1)$ are interchangeable ideas.]

POINT	QUADRANT
$A(2, 1)$	I
$B(-2, 3)$	II
$C(-3, -2)$	III
$D(1, -2)$	IV
$E(3, 0)$	x-axis

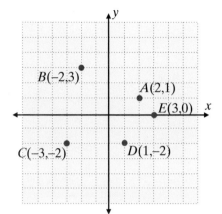

Figure 4.2

NOTES

Important Note: Unless otherwise stated, in this text, assume that the grid lines (as shown here in Figure 4.2) are one unit apart.

Example 1: Graphing the Ordered Pairs

Graph the sets of ordered pairs.

a. $\{(-2, 1), (-1, -4), (0, 2), (1, 3), (2, -3)\}$
(**Note:** The listing of ordered pairs within the braces can be in any order.)

Solution: To locate each point, **start at the origin**, and:
For $(-2, 1)$, move 2 units left and 1 unit up.
For $(-1, -4)$, move 1 unit left and 4 units down.
For $(0, 2)$, move no units left or right and 2 units up.
For $(1, 3)$, move 1 unit right and 3 units up.
For $(2, -3)$, move 2 units right and 3 units down.

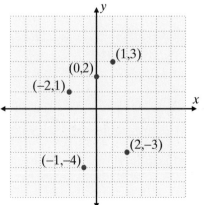

Continued on next page...

b. $\{(-1, 3), (0, 1), (1, -1), (2, -3), (3, -5)\}$

Solution: To locate points; start at the origin, move left or right for the *x*-coordinate and up or down for the *y*-coordinate.
For $(-1, 3)$, move 1 unit left and 3 units up.
For $(0, 1)$, move no units left or right and 1 unit up.
For $(1, -1)$, move 1 unit right and 1 unit down.
For $(2, -3)$, move 2 units right and 3 units down.
For $(3, -5)$, move 3 units right and 5 units down.

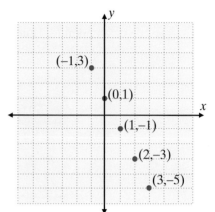

The points (ordered pairs) in Example 1b can be shown to satisfy the equation $y = -2x + 1$. For example, using $x = -1$ in the equation yields,

$$y = -2(-1) + 1 = 2 + 1 = 3$$

and the ordered pair $(-1, 3)$ satisfies the equation. Similarly, letting $y = 1$ gives,

$$1 = -2x + 1$$
$$0 = -2x$$
$$0 = x$$

and the ordered pair $(0, 1)$ satisfies the equation.

We can write all the ordered pairs in Example 1b in table form.

x	−2x + 1 = y
−1	$-2(-1) + 1 = 3$
0	$-2(0) + 1 = 1$
1	$-2(1) + 1 = -1$
2	$-2(2) + 1 = -3$
3	$-2(3) + 1 = -5$

Example 2: Determining Ordered Pairs

a. Determine which, if any, of the ordered pairs $(0, -2)$, $\left(\dfrac{2}{3}, 0\right)$, and $(2, 5)$ satisfy the equation $y = 3x - 2$.

Solution: We will substitute 0, $\dfrac{2}{3}$, and 2 for x and see if the corresponding y-values match those in the given ordered pairs.

$$x = 0: \qquad y = 3(\,0\,) - 2 = -2 \qquad \text{so, } (0, -2) \text{ satisfies the equation.}$$

$$x = \frac{2}{3}: \qquad y = 3\left(\frac{2}{3}\right) - 2 = 0 \qquad \text{so, } \left(\frac{2}{3}, 0\right) \text{ satisfies the equation.}$$

$$x = 2: \qquad y = 3(\,2\,) - 2 = 4 \qquad \text{so, } (2, 4) \text{ satisfies the equation.}$$

The point $(2, 5)$ does not satisfy the equation $y = 3x - 2$ because, as just illustrated, $y = 4$ when $x = 2$.

b. Determine the missing coordinate in each of the following ordered pairs so that the point will satisfy the equation $2x + 3y = 12$:

$$(0, \quad), (3, \quad), (\quad, 0), (\quad, -2).$$

Solution:

The missing value can be found by substituting the given value for x (or for y) into the equation and solving for the other variable.

For $(0, \quad)$, let $x = 0$:
$$2(\,0\,) + 3y = 12$$
$$3y = 12$$
$$y = 4$$

The ordered pair is $(0, 4)$.

For $(3, \quad)$, let $x = 3$:
$$2(\,3\,) + 3y = 12$$
$$6 + 3y = 12$$
$$3y = 6$$
$$y = 2$$

The ordered pair is $(3, 2)$.

For $(\quad, 0)$, let $y = 0$:
$$2x + 3(\,0\,) = 12$$
$$2x = 12$$
$$x = 6$$

The ordered pair is $(6, 0)$.

For $(\quad, -2)$, let $y = -2$:
$$2x + 3(-2\,) = 12$$
$$2x - 6 = 12$$
$$2x = 18$$
$$x = 9$$

The ordered pair is $(9, -2)$.

Continued on next page...

c. Complete the table below so that each ordered pair will satisfy the equation $y = 1 - 2x$.

x	y
0	
	3
$\dfrac{1}{2}$	
5	

Solution: Substituting each given value for x and y into the equation $y = 1 - 2x$ gives the following table of ordered pairs.

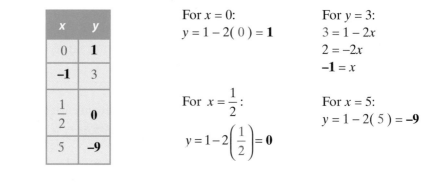

x	y
0	**1**
−1	3
$\dfrac{1}{2}$	**0**
5	**−9**

For $x = 0$:
$y = 1 - 2(\,0\,) = \mathbf{1}$

For $y = 3$:
$3 = 1 - 2x$
$2 = -2x$
$\mathbf{-1} = x$

For $x = \dfrac{1}{2}$:
$y = 1 - 2\left(\dfrac{1}{2}\right) = \mathbf{0}$

For $x = 5$:
$y = 1 - 2(\,5\,) = \mathbf{-9}$

NOTES
Although this discussion is related to ordered pairs of real numbers, most of the examples use ordered pairs of **integers**. This is because ordered pairs of integers are relatively easy to locate on a graph and relatively easy to read from a graph. Ordered pairs with fractions, decimals, or radicals (irrational numbers) must be located by estimating the positions of the points. The precise coordinates intended for such points can be difficult or impossible to read because large dots must be used so the points can be seen. **Even with these difficulties, you should understand that we are discussing ordered pairs of real numbers and that points with fractions, decimals, and radicals as coordinates do exist and should be plotted by estimating their positions**.

Example 3: Reading Points on a Graph

The graphs of two straight lines are given. Each line has an infinite number of points. Use the grid to help you locate (or estimate) three points on each line.

a. 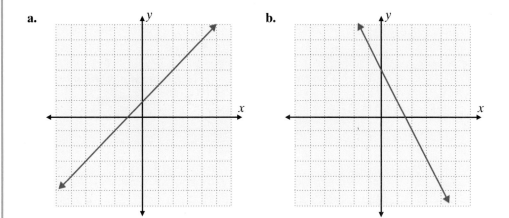 **b.**

Solutions:

a. Three points on this graph are $(-2, -1), (1, 2),$ and $(3, 4)$. (Of course there is no one answer to this type of question. Use your own judgment.)

b. Three points on this graph are $(0, 3), (2, -1),$ and $(1, 1)$. (You may estimate with fractions. For example, one point appears to be $\left(\frac{1}{2}, 2\right)$.)

1. Determine which ordered pairs satisfy the equation $3x + y = 14$.

    a. $(5, -1)$ **b.** $(4, 2)$ **c.** $(-1, 17)$

2. Given $3x + y = 5$, find the missing coordinate of each ordered pair so that it will satisfy the equation.

    a. $(0, \ )$ **b.** $\left(\dfrac{1}{3}, \ \right)$ **c.** $(\ , 2)$

3. Complete the table so that each ordered pair will satisfy the equation $y = \dfrac{2}{3}x + 1$.

x	y
0	
	-2
-3	
6	

4.1 Exercises

List the sets of ordered pairs corresponding to the graphs in Exercises 1 - 10. Assume that the grid lines are marked one unit apart. (**Note:** There is no particular order to be followed in listing ordered pairs.)

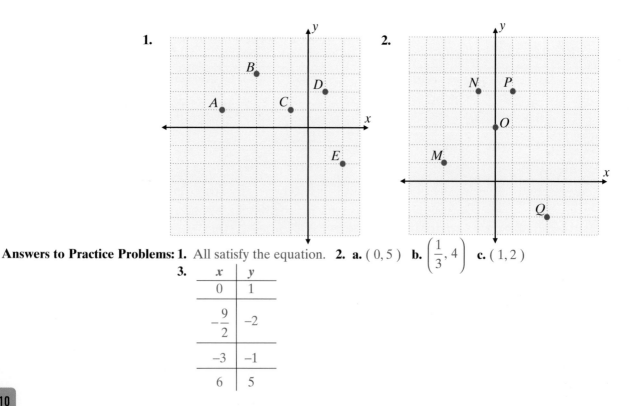

1.

2.

Answers to Practice Problems: 1. All satisfy the equation. **2. a.** $(0, 5)$ **b.** $\left(\dfrac{1}{3}, 4\right)$ **c.** $(1, 2)$

3.

x	y
0	1
$-\dfrac{9}{2}$	-2
-3	-1
6	5

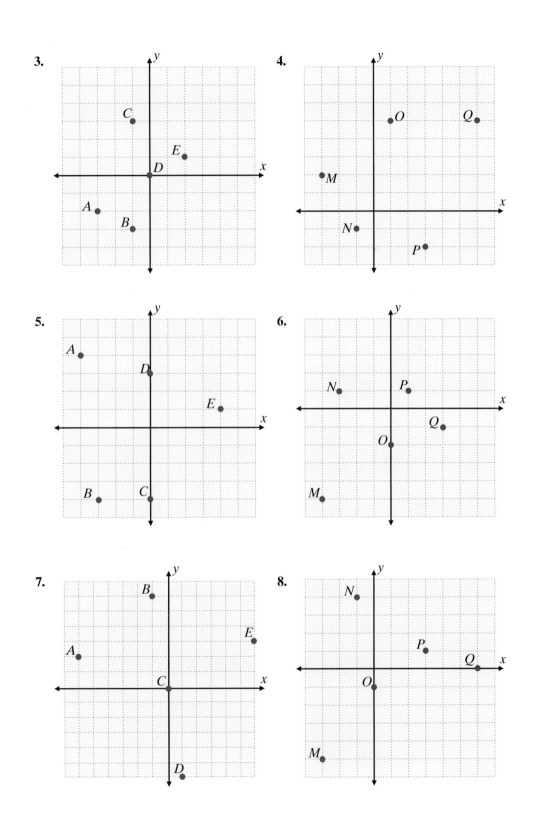

3.

4.

5.

6.

7.

8.

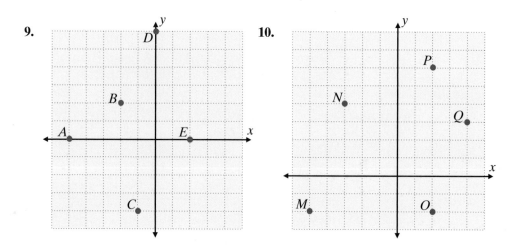

Graph the sets of ordered pairs and label the points in Exercises 11 – 24.

11. $\{A(4,-1), B(3,2), C(0,5), D(1,-1), E(1,4)\}$

12. $\{A(-1,-1), B(-3,-2), C(1,3), D(0,0), E(2,5)\}$

13. $\{A(1,2), B(0,2), C(-1,2), D(2,2), E(-3,2)\}$

14. $\{A(1,0), B(3,0), C(-2,1), D(-1,1), E(0,0)\}$

15. $\{A(-1,4), B(0,-3), C(2,-1), D(4,1)\}$

16. $\{A(-1,-1), B(0,1), C(1,3), D(2,5), E(3,10)\}$

17. $\{A(4,1), B(0,-3), C(1,-2), D(2,-1)\}$

18. $\{A(0,1), B(1,0), C(2,-1), D(3,-2), E(4,-3)\}$

19. $\{A(1,4), B(-1,-2), C(0,1), D(2,7), E(-2,-5)\}$

20. $\left\{A(1,-3), B\left(-4, \frac{3}{4}\right), C\left(2, -2\frac{1}{2}\right), D\left(\frac{1}{2}, 4\right)\right\}$

21. $\left\{A(0,0), B\left(-1, \frac{7}{4}\right), C\left(3, -\frac{1}{2}\right)\right\}$

22. $\left\{A\left(\frac{3}{4}, \frac{1}{2}\right), B\left(2, -\frac{5}{4}\right), C\left(\frac{1}{3}, -2\right), D\left(-\frac{5}{3}, 2\right)\right\}$

23. $\{A(1.6, -2), B(3, 2.5), C(-1, 1.5), D(0, -2.3)\}$

24. $\{A(-2, 2), B(-3, 1.6), C(3, 0.5), D(1.4, 0)\}$

Determine which of the given ordered pairs satisfy the equation in Exercises 25 – 30.

25. $2x - y = 4$
    a. $(1, 1)$
    b. $(2, 0)$
    c. $(1, -2)$
    d. $(3, 2)$

26. $x + 2y = -1$
    a. $(1, -1)$
    b. $(1, 0)$
    c. $(2, 1)$
    d. $(3, -2)$

27. $4x + y = 5$
    a. $\left(\dfrac{3}{4}, 2\right)$
    b. $(4, 0)$
    c. $(1, 1)$
    d. $(0, 3)$

28. $2x - 3y = 7$
    a. $(1, 3)$
    b. $\left(\dfrac{1}{2}, -2\right)$
    c. $\left(\dfrac{7}{2}, 0\right)$
    d. $(2, 1)$

29. $2x + 5y = 8$
    a. $(4, 0)$
    b. $(2, 1)$
    c. $(1, 1.2)$
    d. $(1.5, 1)$

30. $3x + 4y = 10$
    a. $(-2, 3)$
    b. $(0, 2.5)$
    c. $(4, -2)$
    d. $(1.2, 1.6)$

Determine the missing coordinate in each of the ordered pairs so that it will satisfy the equation given in Exercises 31 – 40.

31. $x - y = 4$
    $(0, \ ), (2, \ ), (\ , 0), (\ , -3)$

32. $x + y = 7$
    $(0, \ ), (-1, \ ), (\ , 0), (\ , 3)$

33. $x + 2y = 6$
    $(0, \ ), (2, \ ), (\ , 0), (\ , 4)$

34. $3x + y = 9$
    $(0, \ ), (4, \ ), (\ , 0), (\ , 3)$

35. $4x - y = 8$
    $(0, \ ), (1, \ ), (\ , 0), (\ , 4)$

36. $x - 2y = 2$
    $(0, \ ), (4, \ ), (\ , 0), (\ , 3)$

37. $2x + 3y = 6$
    $(0, \ ), (-1, \ ), (\ , 0), (\ , -2)$

38. $5x + 3y = 15$
    $(0, \ ), (2, \ ), (\ , 0), (\ , 4)$

39. $3x - 4y = 7$
    $(0, \ ), (1, \ ), (\ , 0), \left(\ , \dfrac{1}{2}\right)$

40. $2x + 5y = 6$
    $(0, \ ), \left(\dfrac{1}{2}, \ \right), (\ , 0), (\ , 2)$

Complete the tables in Exercises 41 – 55 so that each ordered pair will satisfy the given equation. Graph the resulting sets of ordered pairs.

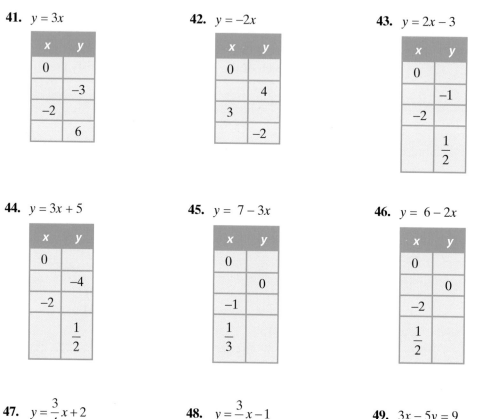

41. $y = 3x$

x	y
0	
	−3
−2	
	6

42. $y = -2x$

x	y
0	
	4
3	
	−2

43. $y = 2x - 3$

x	y
0	
	−1
−2	
	$\frac{1}{2}$

44. $y = 3x + 5$

x	y
0	
	−4
−2	
	$\frac{1}{2}$

45. $y = 7 - 3x$

x	y
0	
	0
−1	
$\frac{1}{3}$	

46. $y = 6 - 2x$

x	y
0	
	0
−2	
$\frac{1}{2}$	

47. $y = \frac{3}{4}x + 2$

x	y
0	
	5
−4	
	$\frac{5}{4}$

48. $y = \frac{3}{2}x - 1$

x	y
0	
	2
−2	
	$-\frac{5}{2}$

49. $3x - 5y = 9$

x	y
0	
	0
−2	
	−1

50. $4x + 3y = 6$

x	y
0	
	0
3	
	−1

51. $5x - 2y = 10$

x	y
0	
	0
−2	
	−1

52. $3x - 2y = 12$

x	y
	0
0	
	−3
6	

53. $x - y = 1.5$

x	y
0	
1.5	
	2.3
	−2.5

54. $2x + 3.2y = 6.4$

x	y
0	
3.2	
	0.8
	−0.2

55. $3x + y = -2.4$

x	y
	0
0	
	0.6
1.6	

In Exercises 56 – 65, the graph of a straight line is shown. Each line has an infinite number of points. List any three points on each line. (There is no one correct answer.)

56.

57.

58.

59.

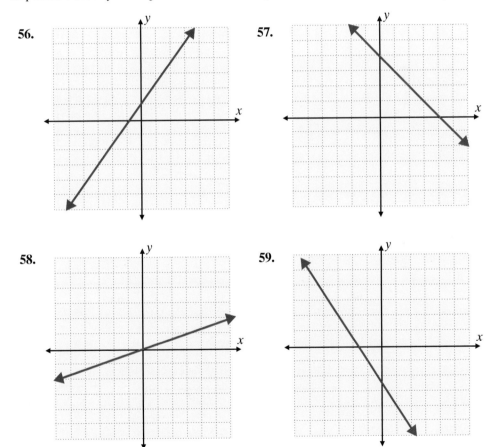

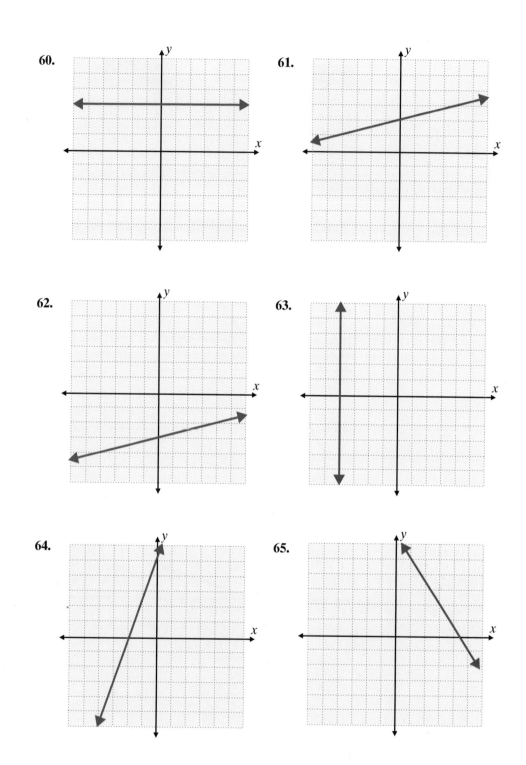

60.

61.

62.

63.

64.

65.

66. Given the equation $I = 0.08P$, where I is the interest earned on a principal P at the rate of 8%:

a. Make a table of ordered pairs for the values of P and I if P has the values $1000, $2000, $3000, $4000, and $5000.

b. Graph the points corresponding to the ordered pairs.

P	I
1000	
2000	
3000	
4000	
5000	

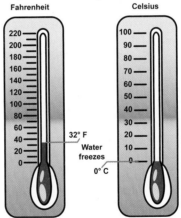

67. Given the equation $F = \dfrac{9}{5}C + 32$ where C is temperature in degrees Celsius and F is the corresponding temperature in degrees Fahrenheit:

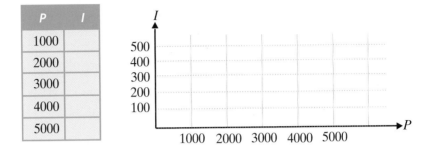

a. Make a table of ordered pairs for the values of C and F if C has the values $-20°, -10°, -5°, 0°, 5°, 10°,$ and $15°$.

b. Graph the points corresponding to the ordered pairs.

C	F
−20	
−10	
−5	
0	
5	
10	
15	

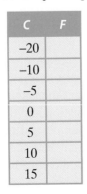
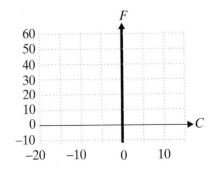

317

68. Given the equation $d = 16t^2$, where d is the distance an object falls in feet and t is the time in seconds that the object falls:

   a. Make a table of ordered pairs for the values of t and d with the values of $1, 2$, $3.5, 4, 4.5$, and 5 for t seconds.

   b. Graph the points corresponding to the ordered pairs.

   c. These points do not lie on a straight line. What feature of the equation might indicate to you that the graph is not a straight line?

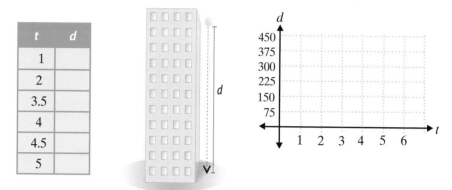

t	d
1	
2	
3.5	
4	
4.5	
5	

69. Given the equation $V = 9h$, where V is the volume (in cubic centimeters) of a box with a variable height h in centimeters and a fixed base of area $9\ \text{cm}^2$.

   a. Make a table of ordered pairs for the values of h and V with h as the values $2\ \text{cm}, 3\ \text{cm}, 5\ \text{cm}, 8\ \text{cm}, 9\ \text{cm}$, and $10\ \text{cm}$.

   b. Graph the points corresponding to the ordered pairs.

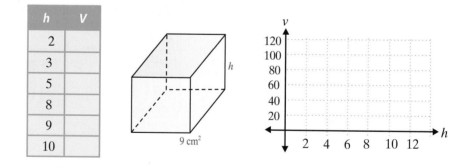

h	V
2	
3	
5	
8	
9	
10	

Writing and Thinking About Mathematics

In statistics, data is sometimes given in the form of ordered pairs where each ordered pair represents two pieces of information about one person. For example, ordered pairs might represent the height and weight of a person or the person's number of years of education and that person's annual income. The ordered pairs are plotted on a graph and the graph is called a **scatter diagram** (or **scatter plot**). Such scatter diagrams are used to see if there is any pattern to the data and, if there is, then the diagram is used to predict the value for one of the variables if the value of the other is known. For example, if you know that a person's height is 5 ft. 6 in., then his or her weight might be predicted from information indicated in a scatter diagram that has several points of known information about height and weight.

70. **a.** The following table of values indicates the number of push-ups and the number of sit-ups that ten students did in a physical education class. Plot these points in a scatter diagram.

Person	#1	#2	#3	#4	#5	#6	#7	#8	#9	#10
x (push-ups)	20	15	25	23	35	30	42	40	25	35
y (sit-ups)	25	20	20	30	32	36	40	45	18	40

b. Does there seem to be a pattern in the relationship between push-ups and sit-ups? What is this pattern?

c. Using the scatter diagram in Part a, predict the number of sit-ups that a student might be able to do if he or she has just done each of the following numbers of push-ups: 22, 32, 35, and 45. (**Note:** In each case, there is no one correct answer. The answers are only estimates based on the diagram.)

71. Ask ten friends or fellow students what their height and weight is. Organize the data in table form and then plot the corresponding scatter diagram. Knowing your own height, does the pattern indicated in the scatter diagram seem to predict your weight?

72. Ask ten friends or fellow students what their height and age is. Organize the data in table form and then plot the corresponding scatter diagram. Knowing your own height, does the pattern indicated in the scatter diagram seem to predict your age? Do you think that all scatter diagrams can be used to predict information related to the two variables graphed? Explain.

Hawkes Learning Systems: Introductory Algebra

Introduction to the Cartesian Coordinate System

4.2

Graphing Linear Equations in Two Variables: $Ax + By = C$

After completing this section, you will be able to:

1. Plot points that satisfy a linear equation and draw the graph of the corresponding line.

2. Recognize the standard form of a linear equation in two variables: $Ax + By = C$

*3. Find the **x-intercept** and **y-intercept** of a line and graph the corresponding line.*

4. Recognize equations of the form $x = a$ and $y = b$ as the forms for vertical and horizontal lines, respectively.

The Standard Form: $Ax + By = C$

In Section 4.1, we discussed ordered pairs and graphed a few points (ordered pairs) that satisfied particular equations. Now suppose we want to graph all the points that satisfy the equation $y = 2x + 3$. The fact is, there are an infinite number of such points. In Figure 4.3, we have graphed five points to try to find a pattern.

x	$2x + 3 = y$	(x, y)
0	$2(0) + 3 = 3$	$(0, 3)$
-1	$2(-1) + 3 = 1$	$(-1, 1)$
$\dfrac{1}{2}$	$2\left(\dfrac{1}{2}\right) + 3 = 4$	$\left(\dfrac{1}{2}, 4\right)$
1	$2(1) + 3 = 5$	$(1, 5)$
-2	$2(-2) + 3 = -1$	$(-2, -1)$

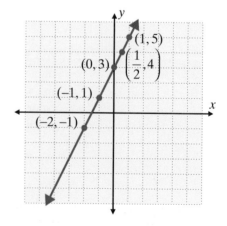

Figure 4.3

The five points in Figure 4.3 appear to lie on a straight line. They in fact do lie on a straight line, and any ordered pair that satisfies the equation $y = 2x + 3$ will also lie on that same line.

What determines whether or not the points that satisfy an equation will lie on a straight line? The points that satisfy any equation of the form

$$Ax + By = C \quad \text{(Standard Form)}$$

will lie on a straight line. The equation is called a **linear equation** and is considered the **standard form** for the equation of a line. We can write the equation $y = 2x + 3$ in the standard form as

$$-2x + y = 3 \quad \text{or} \quad 2x - y = -3.$$

NOTES

Note that in the standard form $Ax + By = C$, A and B may be positive, negative, or 0, but A and B cannot **both** be 0.

Since we now know that the graph will be a straight line, only two points are necessary to determine the entire graph (two points determine a line). The choice of the two points depends on the choice of any two values of x or any two values of y. A third point is sometimes chosen as insurance against a mistake and to help place the graph of the line in the right position. Remember that there are an infinite number of points on a line, so there are an infinite number of possible choices for x (or y). The only restriction is that the ordered pair must satisfy the equation.

Example 1: Graph by Plotting Points

a. Draw the graph of the linear equation $x + 3y = 6$.

Solution:

For $x = 0$:	For $x = 3$:
$0 + 3y = 6$	$3 + 3y = 6$
$y = 2$	$3y = 3$
	$y = 1$
$(0, 2)$	$(3, 1)$

Two points on the graph are $(0, 2)$ and $(3, 1)$. You may have chosen two other values for x and calculated the corresponding y-values. You would still get the same straight line as the graph. **For accuracy in graphing, avoid choosing two points close together**.

Continued on next page...

b. Draw the graph of the linear equation $2x - 5y = 10$.

Solution: For $x = 5$:	For $x = 0$:	For $x = -5$:
$2 \cdot 5 - 5y = 10$	$2 \cdot 0 - 5y = 10$	$2(-5) - 5y = 10$
$10 - 5y = 10$	$0 - 5y = 10$	$-10 - 5y = 10$
$-5y = 0$	$y = -2$	$-5y = 20$
$y = 0$		$y = -4$
$(5, 0)$	$(0, -2)$	$(-5, -4)$

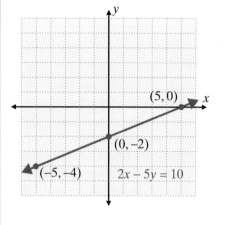

Graphing three points is a good idea, if only to be sure the graph is in the right position and no error has been made in the calculations for the other points.

y-intercept and x-intercept

While the choice of the values for x or y can be arbitrary, letting $x = 0$ will locate the point on the graph where the line crosses the y-axis. This point is called the **y-intercept**. The **x-intercept** is the point found by letting $y = 0$. These two points are generally easy to locate and are frequently used as the two points for drawing the graph of a linear equation.

Intercepts

1. To find the **y-intercept** *(where the line crosses the y-axis), substitute x = 0 and solve for y.*

2. To find the **x-intercept** *(where the line crosses the x-axis), substitute y = 0 and solve for x.*

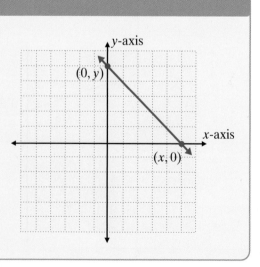

Example 2: Graph Using the *y*- and *x*- Intercepts

Graph the following linear equations by locating the *y*-intercepts and the *x*-intercepts.

a. $3y + 2x = 6$

Solution: $x = 0 \rightarrow y = 2$
$y = 0 \rightarrow x = 3$

$(0, 2)$ is the *y*-intercept
$(3, 0)$ is the *x*-intercept

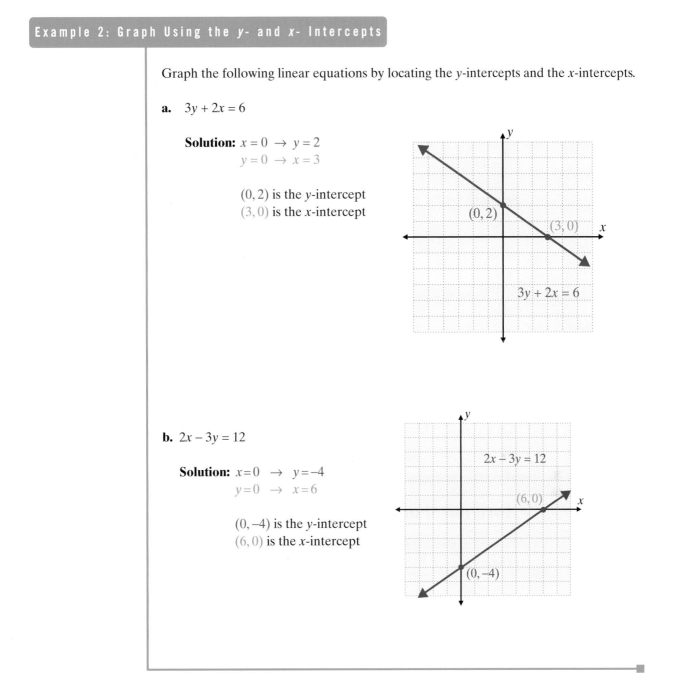

b. $2x - 3y = 12$

Solution: $x = 0 \rightarrow y = -4$
$y = 0 \rightarrow x = 6$

$(0, -4)$ is the *y*-intercept
$(6, 0)$ is the *x*-intercept

With the understanding that the *x*-coordinate of the **y-intercept** is always 0, the corresponding *y*-coordinate is also called the **y-intercept**. Thus, in Examples 2a and 2b just given, we will say that the *y*-intercept is 2 and –4, respectively rather than giving the ordered pairs (0, 2) and (0, –4). Similarly, the ***x*-intercept** is 3 in Example 2a, and 6 in Example 2b.

Lines that Contain the Origin

If the line goes through the origin, both the x-intercept and y-intercept will be 0. In this case, some other point must be used.

Example 3: Lines that Contain the Origin

Graph the linear equation $y = 3x$.

Solution: Locate two points on the graph.

$$x = 0 \rightarrow y = 0$$
$$x = 2 \rightarrow y = 6$$

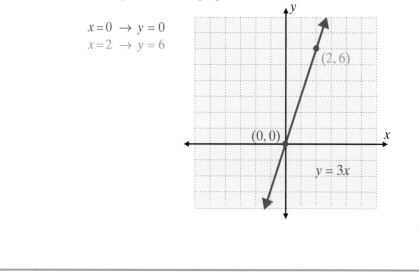

Horizontal and Vertical Lines

Now consider an equation in the form $0x + By = C$ where the coefficient of x is 0. For example,

$$0x + y = 3 \qquad \text{or just} \qquad y = 3.$$

Every value chosen for x will be multiplied by 0 and the corresponding y-value will be 3. In effect, x-values have no influence on the y-value. In table form three such points are:

x	y
−2	3
0	3
5	3

Regardless of what value is chosen for x, the corresponding y-value is 3. Thus, the graph of the equation $y = 3$ is a **horizontal line** (see Figure 4.4).

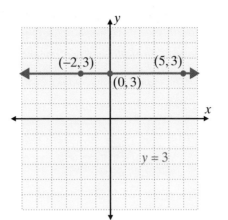

Figure 4.4

Next, consider an equation in the form $Ax + 0y = C$ where the coefficient of y is 0. For example,

$$x + 0y = 2 \qquad \text{or just} \qquad x = 2.$$

Every value chosen for y will be multiplied by 0 and the corresponding x-value will be 2. In effect the y-values have no influence on the value of x. In table form three such points are:

x	y
2	4
2	0
2	−3

Regardless of what is chosen for y, the corresponding x-value is 2. Thus, the graph of the equation $x = 2$ is a **vertical line** (see Figure 4.5).

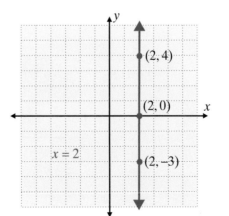

Figure 4.5

Horizontal and Vertical Lines

We can make the following two statements about horizontal and vertical lines:

> **1.** *The graph of any equation of the form $y = b$ is a **horizontal line**.*
>
> **2.** *The graph of any equation of the form $x = a$ is a **vertical line**.*

Example 4: Horizontal and Vertical Lines

Graph each of the following linear equations.

a. $2y = 5$

> **Solution:** Solving for y:
>
> $$2y = 5$$
> $$y = \frac{5}{2}$$

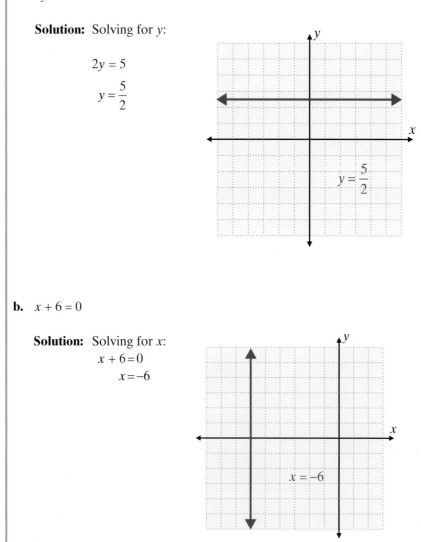

$$y = \frac{5}{2}$$

b. $x + 6 = 0$

> **Solution:** Solving for x:
> $$x + 6 = 0$$
> $$x = -6$$

$$x = -6$$

Using a TI-84 Plus Calculator to Graph Straight Lines

By following the steps outlined here you will be able to see the graphs of all straight lines that are not vertical. The calculator will not graph vertical lines using the methods described here.

To have the calculator graph a nonvertical straight line, you must first solve the equation for *y*. For example,

given the equation: $2x + y = 3$

solving for *y* gives: $y = -2x + 3$ (or $y = 3 - 2x$)

Step 1: Press the **MODE** key and set all the highlighted keys as shown in the diagram. If the mode is not as shown, press the down arrow until you reach the desired line and press **ENTER** .

It is particularly important that **Func** is highlighted. This stands for function. Function concepts and notation will be introduced later in this chapter.

Step 2: Press the **ZOOM** key and then press 6 to get the **ZStandard** window for your graphs. This will give scales from −10 to 10 for both the *x*-axis and the *y*-axis. If you later decide that you would like some other scales on either axis, press **WINDOW** and set the window values to the appropriate values.

Step 3: Press the ⬭ Y= key in the upper left corner of the keyboard. Here you will see the following window.

Step 4: Enter the expression for *y* here after $Y_1 =$. The letter *x* is entered by pressing the X,T,θ,n key located just below the MODE key. Also note that the negative sign (−) is next to the ENTER key.

The screen should appear as follows:

Note that the notation on the screen $Y_1 =, Y_2 =, Y_3 =$, and so forth, allows you to graph several equations at once. We will discuss and use this feature later.

Step 5: Press the GRAPH key in the upper right corner of the keyboard. Your graph should appear as follows:

To clear the screen you may press CLEAR or 2ND QUIT.

4.2 Exercises

Determine which of the ordered pairs, if any, in each table satisfy the given equation. (**Hint**: Substitute the values into the equation to see if you get a true statement or not.)

1. $x + y = -12$

x	y
6	6
0	12
−14	2
−1	−11

2. $x - y = 10$

x	y
2	−8
−2	−10
0	10
5	5

3. $y = -2x + 8$

x	y
2	4
0	8
4	0
3	−2

4. $y = 4x - 6$

x	y
2	−2
−2	2
1.5	0
0	6

5. $2x + y = 5$

x	y
0	5
1	3
2.5	0
−3	1

6. $3x + 2y = -6$

x	y
0	−2
2	0
−3	1
1	2

For Exercises 7 – 12, use your knowledge of *x*-intercepts and *y*-intercepts to match each of the following equations with its graph.

7. $4x + 3y = 12$ **8.** $4x - 3y = 12$

9. $x + 2y = 8$ **10.** $-x + 2y = 8$

11. $x + 4y = 0$ **12.** $5x - y = 10$

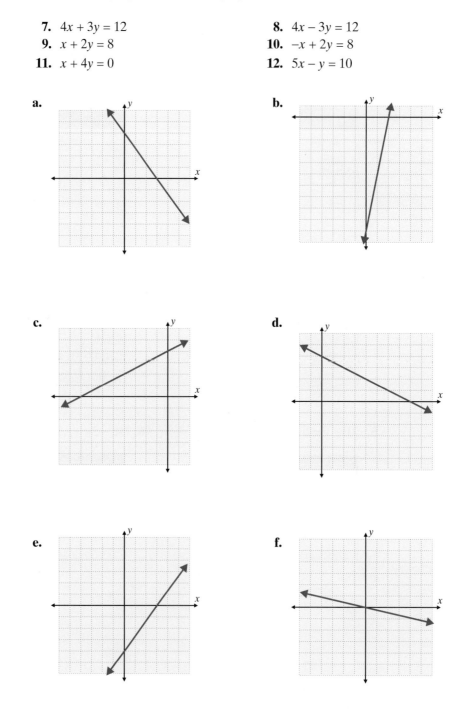

In Exercises 13 – 46, first tell (just by looking at the equation) whether the graph will be a vertical line, a horizontal line, or neither. Then, graph the line corresponding to each linear equation by locating any two (or three) ordered pairs (points) that satisfy the equation.

13. $y = 2x$

14. $y = 3x$

15. $y = -x$

16. $y = -5x$

17. $y = x - 4$

18. $y = x + 3$

19. $y = x + 2$

20. $y = x - 6$

21. $y = 4 - x$

22. $y = 8 - x$

23. $y = 2x - 1$

24. $y = 5 - 2x$

25. $3y = 12$

26. $10 - 2y = 0$

27. $2x - 8 = 0$

28. $9 - 4x = 0$

29. $x - 2y = 4$

30. $x + 3y = 5$

31. $2x + y = 0$

32. $2y = x$

33. $5y = 0$

34. $2y - 3 = 0$

35. $4x = 0$

36. $-\dfrac{3}{4}x - 1 = 0$

37. $2x + 3y = 7$

38. $4x + 3y = 11$

39. $3y - 2x = 4$

40. $3x - 2y = 6$

41. $5x + 2y = 9$

42. $2x - 7y = -14$

43. $4x + 2y = -10$

44. $y = \dfrac{1}{2}x + 1$

45. $y = \dfrac{1}{3}x - 3$

46. $\dfrac{2}{3}x + y = 4$

Graph the linear equations in Exercises 47 – 61 by locating the *y*-intercept and the *x*-intercept.

47. $x + y = 4$

48. $x - 2y = 6$

49. $3x - 2y = 6$

50. $3x - 4y = 12$

51. $5x + 2y = 10$

52. $3x + 7y = -21$

53. $2x - y = 9$

54. $4x + y = 7$

55. $x + 3y = 5$

56. $x - 6y = 3$

57. $\dfrac{1}{2}x - y = 4$

58. $\dfrac{2}{3}x - 3y = 4$

59. $\dfrac{1}{2}x - \dfrac{3}{4}y = 6$

60. $5x + 3y = 7$

61. $2x + 3y = 5$

In Exercises 62 – 71, solve each of the equations for y and use your graphing calculator to graph each line. Estimate the y-intercept and the x-intercept by looking at the graph. (Note: You may need to adjust the ⬤WINDOW.)

62. $3x + y = -6$

63. $2x + y = 14$

64. $y = -4x + 10$

65. $x - 3y = 9$

66. $x - 2y = 20$

67. $x + y = 5$

68. $2y + x = 0$

69. $3y - 15 = 0$

70. $3x - y = 4$

71. $5x - 2y = 10$

Writing and Thinking About Mathematics

Each of the following equations is not linear and the corresponding graph is not a straight line. Make a table and find several points, some where x is positive and some where x is negative, to determine the nature of the corresponding graph. After you have plotted your points and analyzed the nature of the graph, enter the expression for y in your graphing calculator to verify that you have a reasonably accurate graph.

72. $y = \dfrac{4}{x}$

73. $y = x^2$

74. $y = -x^2$

75. $y = x^2 - 5$

76. $y = x^3$

Hawkes Learning Systems: Introductory Algebra

Graphing Linear Equations by Plotting Points

The concept of slope also relates to situations that involve ideas other than construction. For example, miles per hour in traveling and gallons of water per day used at your home, can be illustrated graphically and the **rate of change** is the slope of the line in the graph. (See Figure 4.10.)

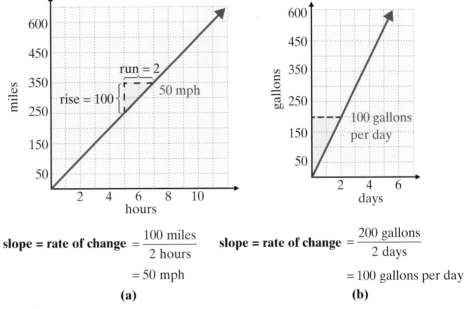

$$\text{slope} = \text{rate of change} = \frac{100 \text{ miles}}{2 \text{ hours}}$$

$$= 50 \text{ mph}$$

(a)

$$\text{slope} = \text{rate of change} = \frac{200 \text{ gallons}}{2 \text{ days}}$$

$$= 100 \text{ gallons per day}$$

(b)

Figure 4.10

Calculating the Slope: $m = \dfrac{y_2 - y_1}{x_2 - x_1}$

To Calculate the Slope of a Line

1. Locate *any two* points on the line.
2. Find the difference in the y-values and divide by the difference in the x-values.
3. This ratio (of rise divided by run) is the slope of the line.

Regardless of the two points used on the line, the right triangles formed (as illustrated in Figure 4.7) will be similar triangles. A fact of geometry is that **corresponding sides of similar triangles are proportional**. That is, the ratio of rise to run, which indicates the slope of the line, is the same for every one of these right triangles.

For example, consider the line $y = 3x - 1$, and find any two points on the line. Two such points can be found by letting $x = -1$ and $x = 3$.

For $x = -1$:	For $x = 3$:
$y = 3(-1) - 1$	$y = 3(3) - 1$
$y = -3 - 1$	$y = 9 - 1$
$y = -4$	$y = 8$

Now, using $P_1(-1, -4)$ and $P_2(3, 8)$, the coordinates of P_3 are $(3, -4)$, as shown in Figure 4.11(b). P_3 has the same x-coordinate as P_2 and the same y-coordinate as P_1.

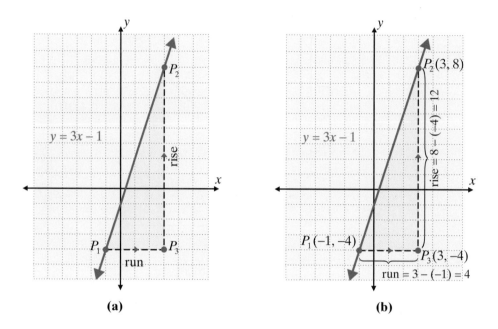

(a)   (b)

Figure 4.11

NOTES   In the notation P_1, 1 is called a **subscript** and P_1 is read "P sub 1". Similarly, P_2 is read "P sub 2" and P_3 is read "P sub 3."

For the line $y = 3x - 1$ and using the points $(-1, -4)$ and $(3, 8)$,

$$\textbf{slope} = \frac{\text{rise}}{\text{run}} = \frac{\text{difference in } y\text{-values}}{\text{difference in } x\text{-values}} = \frac{8 - (-4)}{3 - (-1)} = \frac{12}{4} = 3$$

Now find the slope of the line $y = 3x + 2$. Locate two points on the line and calculate the ratio. For example,

For $x = 0$:	For $x = 1$:
$y = 3 \cdot 0 + 2 = 2$	$y = 3 \cdot 1 + 2 = 5$

Thus, $P_1(0, 2)$ and $P_2(1, 5)$ are as shown in Figure 4.12. The point $P_3(1, 2)$ is shown to help illustrate the rise and the run.

$$\text{slope} = \frac{\text{rise}}{\text{run}} = \frac{5-2}{1-0} = \frac{3}{1} = 3$$

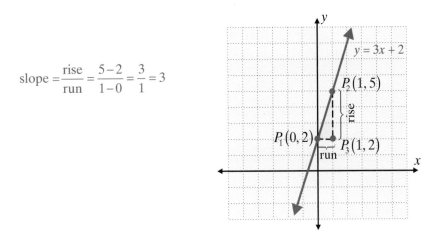

Figure 4.12

Notice that the two lines $y = 3x - 1$ and $y = 3x + 2$ have the same slope, 3. This means that the lines are **parallel**. They never intersect (cross each other). Figure 4.13 shows three such lines.

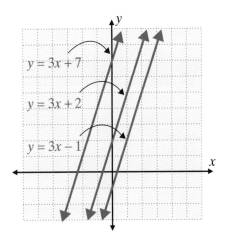

Figure 4.13

NOTES   **All lines with the same slope are parallel.**

By using subscript notation, we can develop a formula for the slope of any line.

Slope

Let $P_1(x_1, y_1)$ and $P_2(x_2, y_2)$ be two points on a line. Then $P_3(x_2, y_1)$ is at the right angle shown in Figure 4.14, and the **slope** can be calculated as follows.

$$slope = \frac{rise}{run} = \frac{y_2 - y_1}{x_2 - x_1}$$

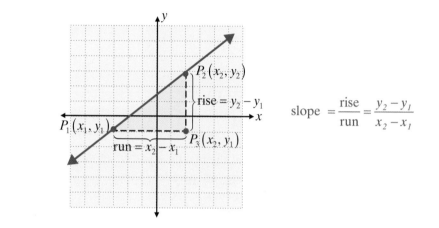

Figure 4.14

Example 1: Finding the Slope of a Line

a. Using the formula for slope, find the slope of the line $2x + 3y = 6$.

Solution: Let $x_1 = 0$:

$$2 \cdot 0 + 3y_1 = 6$$
$$3y_1 = 6$$
$$y_1 = 2$$

Let $x_2 = -3$:

$$2(-3) + 3y_2 = 6$$
$$-6 + 3y_2 = 6$$
$$3y_2 = 12$$
$$y_2 = 4$$

$$(x_1, y_1) = (0, 2) \quad \text{and} \quad (x_2, y_2) = (-3, 4)$$

$$slope = \frac{y_2 - y_1}{x_2 - x_1} = \frac{4 - 2}{-3 - 0} = \frac{2}{-3} = -\frac{2}{3}$$

b. Suppose that the order of the points in Example 1a is changed. That is, $(x_1, y_1) = (-3, 4)$ and $(x_2, y_2) = (0, 2)$. Will this make a difference in the slope?

Solution: Slope $= \dfrac{y_2 - y_1}{x_2 - x_1} = \dfrac{2 - 4}{0 - (-3)} = \dfrac{-2}{3} = -\dfrac{2}{3}$

Thus, **the slope is the same even if the order of the points is reversed**.

As demonstrated in Examples 1a and 1b, changing the order of the points does not make a difference in the value of the slope. Both the numerator and the denominator change signs so the fraction has the same value. In Example 1a, $\dfrac{2}{-3} = -\dfrac{2}{3}$ and in Example 1b, $\dfrac{-2}{3} = -\dfrac{2}{3}$. The important part of the procedure is that **the coordinates must be subtracted in the same order in both the numerator and the denominator**.

In general,

$$slope = \frac{y_2 - y_1}{x_2 - x_1} = \frac{y_1 - y_2}{x_1 - x_2}$$

In Example 1a, the **negative slope** for the line $2x + 3y = 6$ **means that the line slants** (or slopes) **downward to the right** (See Figure 4.15).

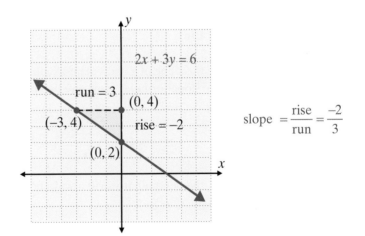

Figure 4.15

NOTES

Lines with **positive slope** go up as we move along the line from left to right.

Lines with **negative slope** go down as we move along the line from left to right.

Slopes of Horizontal and Vertical Lines

In Section 4.2, we discussed the graphs of horizontal lines ($y = b$) and vertical lines ($x = a$), but the slopes of lines of these types were not discussed.

To find the slope of a horizontal line, such as $y = 3$, find two points on the line and substitute into the slope formula. Note that any two points on the line will have the same y-coordinate; namely, 3. Two such points are $(-2, 3)$ and $(5, 3)$. Using these two points in the formula for slope gives:

$$\text{slope} = \frac{y_2 - y_1}{x_2 - x_1} = \frac{3 - 3}{5 - (-2)} = \frac{0}{7} = 0$$

In fact, the numerator will always be 0 because the y-values will all be 3 regardless of the x-values. (See Figure 4.16.)

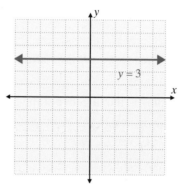

Figure 4.16

To find the slope of a vertical line, such as $x = 4$, find two points on the line and substitute into the slope formula. Now any two points on the line will have the same x-coordinate; namely, 4. Two such points are $(4, 1)$ and $(4, 6)$. The slope formula gives:

$$\text{slope} = \frac{y_2 - y_1}{x_2 - x_1} = \frac{6 - 1}{4 - 4} = \frac{5}{0}, \text{ which is } \textbf{undefined}.$$

In fact, the denominator will always be 0 because the x-values will all be 4 regardless of the y-values. (See Figure 4.17.)

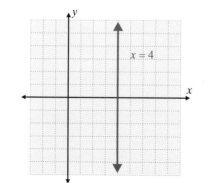

Figure 4.17

Horizontal and Vertical Lines

The following two general statements are true for horizontal and vertical lines:

1. **For horizontal lines (of the form $y = b$), the slope is 0.**

2. **For vertical lines (of the form $x = a$), the slope is undefined.**

Example 2: Slopes of Vertical and Horizontal Lines

a. Find the equation and slope of the horizontal line through the point $(-2, 6)$.

Solution: The equation is $y = 6$ and the slope is 0.

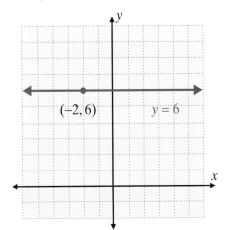

b. Find the equation and slope of the vertical line through the point $(5, 2)$.

Solution: The equation is $x = 5$ and the slope is undefined.

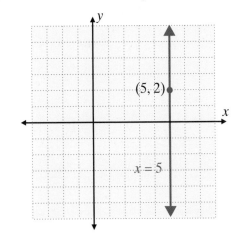

> **NOTES**
>
> **Special Note About Slopes of Lines**
> All lines that slant downward to the right have negative slopes.
> All lines that slant upward to the right have positive slopes.
> All horizontal lines have slope 0.
> All vertical lines have slope undefined.

The Slope-Intercept Form: $y = mx + b$

In the beginning of this section, we graphed two lines, $y = 3x - 1$ and $y = 3x + 2$.   In both cases we found that the slope was 3.   In Example 1a, we found that the line $2x + 3y = 6$ has a slope of $-\dfrac{2}{3}$.   If we solve this equation for y, we get the following results:

$$2x + 3y = 6$$

$$3y = -2x + 6$$

$$y = \frac{-2x}{3} + \frac{6}{3}$$

$$y = -\frac{2}{3}x + 2$$

Observing the relationship between slopes and the equations just discussed, we see that when the equation is solved for y the coefficient of x is the slope in each case.   This observation leads to the following discussion which states that if the equation is solved for y, the coefficient of x will always be the slope of the line.

For $y = mx + b$, m is the slope

Statement: *Given an equation in the form $y = mx + b$, then m is the slope.*

Proof: *Suppose that the equation is solved for y and $y = mx + b$. Let (x_1, y_1) and (x_2, y_2) be two points on the line where $x_1 \neq x_2$. Then $y_1 = mx_1 + b$ and $y_2 = mx_2 + b$ so the slope can be calculated as follows:*

$$slope = \frac{y_2 - y_1}{x_2 - x_1} = \frac{(mx_2 + b) - (mx_1 + b)}{(x_2 - x_1)}$$

$$= \frac{mx_2 + b - mx_1 - b}{(x_2 - x_1)}$$

$$= \frac{mx_2 - mx_1}{(x_2 - x_1)}$$

$$= \frac{m(x_2 - x_1)}{(x_2 - x_1)} = m$$

Therefore, for an equation in the form $y = mx + b$, the slope of the line is m.

For the line $y = mx + b$, the point where $x = 0$ is the point where the line will cross the y-axis. This point is called the **y-intercept**. By letting $x = 0$, we get

$$y = mx + b$$
$$y = m \cdot 0 + b$$
$$y = b$$

Thus, the point $(0, b)$ is the y-intercept. As discussed in Section 4.2, we generally say that b is the y-intercept. These concepts of slope and y-intercept lead to the following definition.

Slope - Intercept Form

$y = mx + b$ *is called the **slope-intercept form** for the equation of a line.* m *is the **slope** and*

b *is the **y-intercept**.*

The following examples show how to find the slope and y-intercept of a line by solving the equation for y and writing the result in the slope-intercept form. With this form, we will see that the lines can be easily graphed.

Example 3: Using the Form $y = mx + b$

a. Find the slope and y-intercept of $2x + 3y = 3$ and graph the line.

Solution: $2x + 3y = 3$ (Solve for y.)

$$3y = -2x + 3$$

$$y = \frac{-2x + 3}{3}$$

$$y = \frac{-2x}{3} + \frac{3}{3}$$

$$y = -\frac{2}{3}x + 1$$

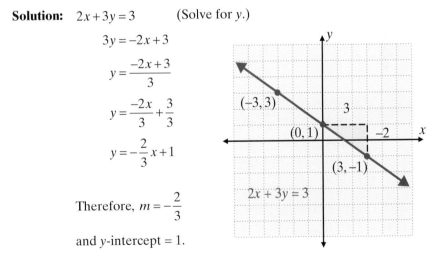

Therefore, $m = -\frac{2}{3}$

and y-intercept = 1.

As shown in the figure, if we first "run" 3 units to the right and "rise" 2 units down **from the y-intercept (0, 1)**, we locate another point $(3, -1)$. The line can be drawn through these two points.

Continued on next page...

Note: We could also first "rise" 2 units up and "run" 3 units to the left from $(0, 1)$ and locate another point on the graph, $(-3, 3)$.

That is, we have $\text{slope} = -\dfrac{2}{3} = \dfrac{-2}{3} = \dfrac{2}{-3}$

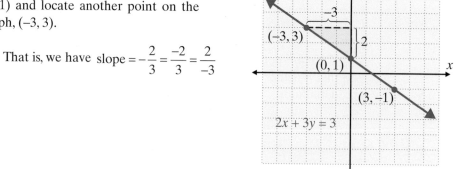

b. Find the slope and y-intercept of $x - 2y = 6$ and graph the line.

Solution: $x - 2y = 6$

$$-2y = -x + 6$$

$$y = \frac{-x + 6}{-2}$$

$$y = \frac{-x}{-2} + \frac{6}{-2}$$

$$y = \frac{1}{2}x - 3$$

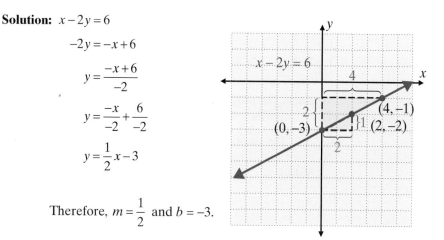

Therefore, $m = \dfrac{1}{2}$ and $b = -3$.

Because $\boldsymbol{m} = \dfrac{1}{2}$ and the y-intercept is $(0, -3)$, we can locate another point by moving first [from $(0, -3)$] a "run" of 2 units right and a "rise" of 1 unit up. This gives the point $(2, -2)$ and we can draw the line through these two points.

Or, we can first move a "rise" of 2 units up and a "run" of 4 units right and locate the point $(4, -1)$. [Note: The ratio of the rise to run must be 1 to 2, which is the slope of the line: $\boldsymbol{m} = \dfrac{\text{rise}}{\text{run}} = \dfrac{1}{2} = \dfrac{2}{4}$.]

c. Find the slope and *y*-intercept of $-4x + 2y = 7$ and graph the line.

Solution:

$$-4x + 2y = 7$$

$$2y = 4x + 7$$

$$y = \frac{4x + 7}{2}$$

$$y = \frac{4x}{2} + \frac{7}{2}$$

$$y = 2x + \frac{7}{2}$$

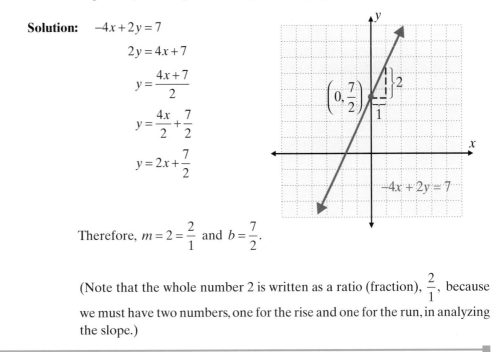

Therefore, $m = 2 = \frac{2}{1}$ and $b = \frac{7}{2}$.

(Note that the whole number 2 is written as a ratio (fraction), $\frac{2}{1}$, because we must have two numbers, one for the rise and one for the run, in analyzing the slope.)

In your study of Examples 3a, 3b, and 3c, note that in 3a the slope is negative and the line slants downward to the right. In both 3b and 3c the slopes are positive and the lines slant upward to the right.

Example 4: Finding the Equation of a Line

Find the equation of the line with *y*-intercept $(0, -2)$ and slope $m = \frac{3}{4}$.

Solution: In this case, we can substitute directly into the slope-intercept form $y = mx + b$ with $b = -2$ and $m = \frac{3}{4}$.

$$y = \frac{3}{4}x + (-2) \quad \text{or} \quad y = \frac{3}{4}x - 2$$

In standard form, $3x - 4y = 8$. Note that $\frac{3}{4}x - y = 2$ is also in standard form. For convenience, we multiply by 4 so that the coefficients are integers. Both forms are correct and either form is an acceptable answer.

To quickly graph a horizontal line, any two points with the y-value indicated by the equation can be used. To graph a vertical line, any two points with the x-value indicated by the equation can be used.

Example 5: Graphing Horizontal and Vertical Lines

a. Graph the horizontal line $y = 2$.

Solution: Two points with y-value 2 are $(-4, 2)$ and $(3, 2)$. The line can be drawn through these two points.

Note that the slope $m = 0$ and the y-intercept is $(0, 2)$.

b. Graph the vertical line $x = -3$.

Solution: Two points with x-value -3 are $(-3, 5)$ and $(-3, -1)$. The line can be drawn through these two points.

Note that the slope m is undefined and there is no y-intercept.

Practice Problems

1. Find the slope of the line determined by the point $(1, 5)$ and $(2, -2)$.
2. Find the equation of the line through the point $(0, 3)$ with slope $-\dfrac{1}{2}$.
3. Find the slope, m, and y-intercept, b, for the equation $2x + 3y = 6$.
4. Write an equation for the horizontal line through the point $(-2, 3)$.
5. Write an equation for the vertical line through the point $(-1, -2)$.

4.3 Exercises

Find the slope of the line determined by each pair of points in Exercises 1 – 20.

1. $(1, 4), (2, -1)$

2. $(3, 1), (5, 0)$

3. $(4, 7), (-3, -1)$

4. $(-6, 2), (1, 3)$

5. $(-3, 8), (5, 9)$

6. $(0, 0), (-6, -4)$

7. $(1, 5), (4, 8)$

8. $(2, 6), (6, 2)$

9. $(0, 0), (-5, 3)$

10. $(0, 0), (9, -3)$

11. $(-2, 0), (5, 1)$

12. $(-3, -2), (0, 4)$

13. $(-4, 0), (4, 4)$

14. $(-1, 5), (4, -5)$

15. $(-3, 4), (6, -14)$

16. $(-2, -2), (6, -18)$

17. $(4, 2), \left(-1, \dfrac{1}{2}\right)$

18. $\left(\dfrac{3}{4}, 2\right), \left(1, \dfrac{3}{2}\right)$

19. $\left(\dfrac{7}{2}, 3\right), \left(\dfrac{1}{2}, -\dfrac{3}{4}\right)$

20. $\left(-2, \dfrac{4}{5}\right), \left(\dfrac{3}{2}, \dfrac{1}{10}\right)$

Write the equation and draw the graph of the line passing through the given y-intercept with the given slope in Exercises 21 – 40.

21. $(0, 0), \ m = \dfrac{2}{3}$

22. $(0, 1), \ m = \dfrac{1}{5}$

23. $(0, -3), \ m = -\dfrac{3}{4}$

24. $(0, 6), m = 0$

25. $(0, 3), \ m = -\dfrac{5}{3}$

26. $(0, 2), \ m = 4$

27. $(0, -1), \ m = 2$

28. $(0, 4), \ m = -\dfrac{3}{5}$

29. $(0, -7), m = 0$

Answers to Practice Problems: **1.** $m = -7$ **2.** $y = -\dfrac{1}{2}x + 3$ **3.** $m = -\dfrac{2}{3}$ and $b = 2$ **4.** $y = 3$ **5.** $x = -1$

30. $(0, 5)$, $m = -3$

31. $(0, -4)$, $m = \dfrac{3}{2}$

32. $(0, 0.5)$, $m = 0$

33. $(0, 5)$, $m = 1$

34. $(0, 6)$, $m = -1$

35. $(0, -6)$, $m = \dfrac{2}{5}$

36. $(0, -2)$, $m = 0$

37. $(0, -2)$, $m = \dfrac{4}{3}$

38. $(0, -5)$, $m = -\dfrac{1}{4}$

39. $(0, 0)$, $m = -\dfrac{1}{3}$

40. $(0, -1.6)$, $m = 0$

Find the slope, m, and the y-intercept, b, for each of the equations in Exercises 41 – 60. Then graph the line.

41. $y = 2x - 3$

42. $y = 3x + 4$

43. $y = 2 - x$

44. $y = -\dfrac{2}{3}x + 1$

45. $x + y = 5$

46. $2x - y = 3$

47. $3x - y = 4$

48. $4x + y = 0$

49. $x - 3y = 9$

50. $x + 4y = -8$

51. $3x + 2y = 6$

52. $2x - 5y = 10$

53. $4x - 3y = -3$

54. $7x + 2y = 4$

55. $6x + 5y = -15$

56. $3x + 8y = -16$

57. $x + 4y = 6$

58. $2x + 3y = 8$

59. $-3x + 6y = 4$

60. $5x - 2y = 3$

In Exercises 61 – 70, find two points on each of the lines and graph the line. State the slope of each line.

61. $y + 5 = 0$

62. $y - 4 = 0$

63. $x = -1$

64. $x - 5 = 0$

65. $2y = -16$

66. $3x + 2 = 0$

67. $5y - 12 = 0$

68. $3y - 5 = 0$

69. $1.5x = -3$

70. $-2x + 6 = 0$

Writing and Thinking About Mathematics

71. The slope of a road is called a "grade." A steep grade is cause for truck drivers to have slow speed limits in mountains. What do you think that a "grade of 12%" means? Draw a picture of a right triangle that would indicate a grade of 12%.

72. Ramps for persons in wheelchairs or otherwise handicapped are now built into most buildings and walkways. (If ramps are not present in a building, then there must be elevators.) What do you think that the slope of a ramp should be for handicapped access? Look in your library or contact your local building permit office to find the recommended slope for such ramps.

Hawkes Learning Systems: Introductory Algebra

The Slope-Intercept Form: $y = mx + b$

Point-Slope Form: $y - y_1 = m(x - x_1)$

After completing this section, you will be able to:

1. *Graph lines given the slope and one point on each line.*

2. *Write the equations of lines given the slope and one point on each line.*

3. *Write the equations of lines given two points on each line.*

4. *Write linear equations from information given in applied problems.*

Graphing a Line Given a Point and the Slope

Lines represented by equations in the standard form $Ax + By = C$ and in the slope-intercept form $y = mx + b$ have been discussed in Sections 4.2 and 4.3. In Section 4.3, we discussed how to graph a line by using the y-intercept point $(0, b)$, and the slope by moving horizontally and then vertically (or by moving vertically and then horizontally) from the y-intercept.

Now suppose that you are given the slope of a line and a point on the line. This point may or may not be the y-intercept. The graph of the line can be drawn by using the same technique discussed when using the y-intercept and slope. Consider the following example.

Example 1: Graph a Line Given a Point and a Slope

a. Graph the line with slope $m = \dfrac{3}{4}$ which passes through the point $(4, 5)$.

Solution: Start from the point $(4, 5)$ and locate another point on the line using the slope as $\dfrac{\text{rise}}{\text{run}} = \dfrac{3}{4}$. We can proceed in either of the following ways:

1. Move 3 units up (rise) and 4 units right (run), or
2. Move 4 units right (run) and 3 units up (rise).

Either way we arrive at the same point $(8, 8)$.
(Note: any numbers in the ratio of 3 to 4 can be used for the moves, such as 6 to 8 or 9 to 12.)

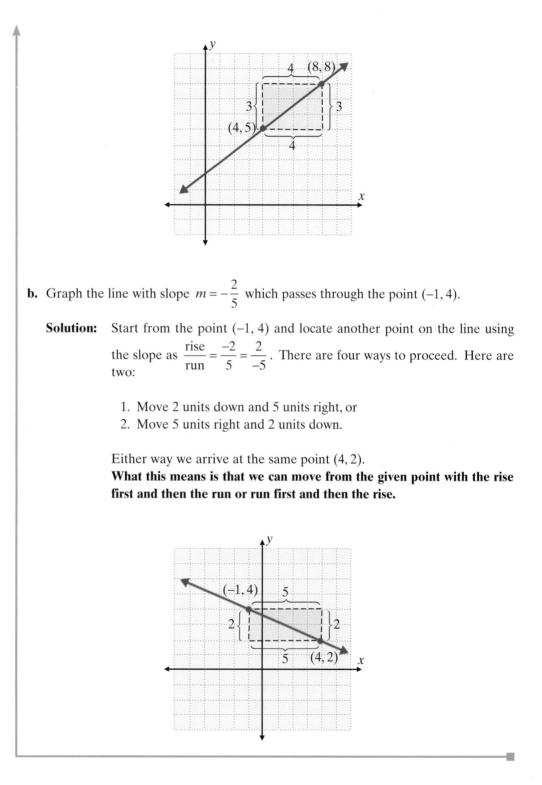

b. Graph the line with slope $m = -\dfrac{2}{5}$ which passes through the point $(-1, 4)$.

Solution: Start from the point $(-1, 4)$ and locate another point on the line using the slope as $\dfrac{\text{rise}}{\text{run}} = \dfrac{-2}{5} = \dfrac{2}{-5}$. There are four ways to proceed. Here are two:

1. Move 2 units down and 5 units right, or
2. Move 5 units right and 2 units down.

Either way we arrive at the same point $(4, 2)$.
What this means is that we can move from the given point with the rise first and then the run or run first and then the rise.

Note that in Example 1a the slope is positive and, as expected, the line slants upward to the right. In Example 1b, the slope is negative and the line slants downward to the right.

The Point-Slope Form: $y - y_1 = m(x - x_1)$

Now consider finding the equation of the line given a point on the line and the slope. For example, if the given slope is $m = \dfrac{1}{2}$ and a point on the line is (–3, 1) [denoted as (x_1, y_1)], how would you determine the equation of the line that satisfies these conditions? You could proceed as follows:

1. Suppose that (x, y) is any other point on the line.
2. Now use the fact that you know the slope and the formula for the slope:

   the formula for the slope gives $m = \dfrac{y-1}{x-(-3)} = \dfrac{y-1}{x+3}$ and slope $= m = \dfrac{1}{2}$.

3. Setting these two expressions equal gives the desired equation:

   $$\dfrac{y-1}{x+3} = \dfrac{1}{2}$$

This is not the usual form so we multiply both sides by $x + 3$ and get

$$y - 1 = \dfrac{1}{2}(x + 3)$$

If we recognize that the point given was (–3, 1) = (x_1, y_1) and the slope given was $m = \dfrac{1}{2}$, this last equation is in the form

$$y - y_1 = m(x - x_1) \quad \text{called } \textbf{the point-slope form.}$$

Point-Slope Form

$y - y_1 = m(x - x_1)$ is called the **point-slope form** for the equation of a line. m is the **slope**, and (x_1, y_1) is a given **point** on the line.

Example 2: Using the Point-Slope Form

a. Find an equation of the line that has slope $\dfrac{2}{3}$ and passes through the point $(1, 5)$.

Solution: Substitute into the point-slope form $y - y_1 = m(x - x_1)$.

$$y - 5 = \dfrac{2}{3}(x - 1) \qquad \text{The Point-Slope Form}$$

This same equation can be written in the slope-intercept form and in standard form.

$$y - 5 = \dfrac{2}{3}x - \dfrac{2}{3}$$

$$y = \dfrac{2}{3}x - \dfrac{2}{3} + 5 \qquad\qquad -\dfrac{2}{3} + 5 = -\dfrac{2}{3} + \dfrac{15}{3} = \dfrac{13}{3}$$

$$y = \dfrac{2}{3}x + \dfrac{13}{3} \qquad\qquad \text{Slope-intercept form}$$

Or multiplying each term by 3 gives

$$y - 5 = \frac{2}{3}(x - 1)$$

$$3(y - 5) = 3 \cdot \frac{2}{3}(x - 1)$$

$$3y - 15 = 2x - 2$$

$$\left. \begin{array}{l} -2x + 3y = 13 \\ \text{or} \quad 2x - 3y = -13 \end{array} \right\} \qquad \text{Standard form}$$

b. Find three forms for the equation of the line with slope $-\dfrac{5}{2}$ that contains the point $(6, -1)$.

Solution: Use the point-slope form: $y - y_1 = m(x - x_1)$.

$$y - (-1) = -\frac{5}{2}(x - 6)$$

$$y + 1 = -\frac{5}{2}(x - 6) \qquad \text{Point-slope form}$$

$$y + 1 = -\frac{5}{2}x + 15$$

$$y = -\frac{5}{2}x + 14 \qquad \text{Slope-intercept form}$$

$$2y = -5x + 28 \qquad \text{Multiply by 2.}$$

$$5x + 2y = 28 \qquad \text{Standard form}$$

You must understand that all three forms for the equation of a line are acceptable and correct. Your answer may be in a different form from that in the back of the text, but it should be algebraically equivalent.

Finding the Equations of Lines Given Two Points

Given two points that lie on a line, the equation of the line can be found by proceeding as follows:

Finding the Equation of a Line

If two points are given,

1. Use the formula $m = \dfrac{y_2 - y_1}{x_2 - x_1}$ *to find the slope.*

2. Use this slope, m, and either point in the point-slope formula $y - y_1 = m(x - x_1)$.

The following examples illustrate this common situation.

Example 3: Using Two Points to Find the Equation of Line

a. Find an equation of the line passing through the two points $(-1, -3)$ and $(5, -2)$.

Solution: Find m using the formula for slope.

$$m = \frac{y_2 - y_1}{x_2 - x_1} = \frac{-2 - (-3)}{5 - (-1)} = \frac{-2 + 3}{5 + 1} = \frac{1}{6}$$

Using the point-slope form [either point $(-1, -3)$ or point $(5, -2)$ will give the same result] yields

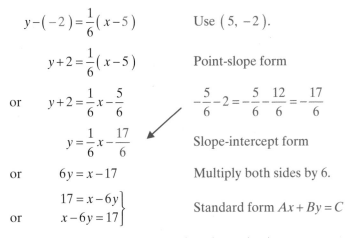

$$y - (-2) = \frac{1}{6}(x - 5) \qquad \text{Use } (5, -2).$$

$$y + 2 = \frac{1}{6}(x - 5) \qquad \text{Point-slope form}$$

$$\text{or} \quad y + 2 = \frac{1}{6}x - \frac{5}{6} \qquad\qquad -\frac{5}{6} - 2 = -\frac{5}{6} - \frac{12}{6} = -\frac{17}{6}$$

$$y = \frac{1}{6}x - \frac{17}{6} \qquad \text{Slope-intercept form}$$

$$\text{or} \quad 6y = x - 17 \qquad \text{Multiply both sides by 6.}$$

$$\left.\begin{array}{l} 17 = x - 6y \\ \text{or} \quad x - 6y = 17 \end{array}\right\} \quad \text{Standard form } Ax + By = C$$

b. Find an equation of the line that contains the points $(-2, 3)$ and $(5, 1)$.

Solution: $m = \frac{y_2 - y_1}{x_2 - x_1} = \frac{1 - 3}{5 - (-2)} = \frac{-2}{7} = -\frac{2}{7}$

Using $(-2, 3)$ in the point-slope form gives

$$y - 3 = -\frac{2}{7}\left[x - (-2)\right] \quad \text{Use } (-2, 3).$$

$$y - 3 = -\frac{2}{7}(x + 2) \qquad \text{Point-slope form}$$

$$y - 3 = -\frac{2}{7}x - \frac{4}{7}$$

$$y = -\frac{2}{7}x - \frac{4}{7} + 3 \qquad\qquad -\frac{4}{7} + 3 = -\frac{4}{7} + \frac{21}{7} = \frac{17}{7}$$

$$y = -\frac{2}{7}x + \frac{17}{7} \qquad \text{Slope-intercept form}$$

Or, using $(5, 1)$ in the point-slope form gives

$$y - 1 = -\frac{2}{7}(x - 5) \qquad \text{Point-slope form}$$

$$y - 1 = -\frac{2}{7}x + \frac{10}{7}$$

$$y = -\frac{2}{7}x + \frac{10}{7} + 1$$

$$y = -\frac{2}{7}x + \frac{17}{7} \qquad \text{Slope-intercept form}$$

Thus we see that using either point yields the same equation. The equation may be written in standard form $2x + 7y = 17$.

Parallel and Perpendicular Lines

Parallel lines are lines that never intersect (cross each other) and these lines have the same slope. However, if lines are **perpendicular to each other**, then their slopes are **negative (opposite) reciprocals of each other** and the lines will intersect at one point. For example, if a line has slope $-\frac{2}{3}$ then any line parallel to this line will have slope $-\frac{2}{3}$. However any line perpendicular to this line will have slope $+\frac{3}{2}$, the negative (or opposite) reciprocal of $-\frac{2}{3}$.

Example 4: Finding the Equations of Parallel Lines

Find the equation of the line through the point $(2, -4)$ and parallel to the line $y = 3x - 1$.

Solution:

Because we know that the line $y = 3x - 1$ has slope $m = 3$, we can use the point $(2, -4)$ as (x_1, y_1) and go directly to the form $y - y_1 = m(x - x_1)$. This gives the desired equation:

$$y - (-4) = 3(x - 2) \qquad \text{Point - slope form}$$

or, simplifying

$$y + 4 = 3x - 6$$

$$\left.\begin{array}{l} 10 = 3x - y \\ \text{or} \quad 3x - y = 10 \end{array}\right\} \qquad \text{The Standard Form}$$

Continued on next page...

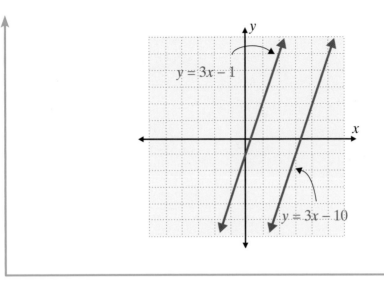

Example 5: Finding the Equations of Perpendicular Lines

Find the equation of the line perpendicular to the line $3x + y = 7$ and passing through the point $(5, -2)$.

Solution:

For the equation $3x + y = 7$, we solve for y and get $y = -3x + 7$.

This line has slope $m = -3$. Therefore, any line perpendicular to this line will have slope $m = -\left(-\dfrac{1}{3}\right) = \dfrac{1}{3}$.

Now using the point slope form gives:

$$y - (-2) = \frac{1}{3}(x - 5) \text{ or } y + 2 = \frac{1}{3}(x - 5)$$

Simplifying gives:

$$3(y + 2) = x - 5$$
$$3y + 6 = x - 5$$
$$\left.\begin{array}{c} 11 = x - 3y \\ \text{or} \quad x - 3y = 11 \end{array}\right\} \quad \text{The Standard Form}$$

Thus the line $x - 3y = 11$ passes through the point $(5, -2)$ and is perpendicular to the line $3x + y = 7$. Both lines are graphed.

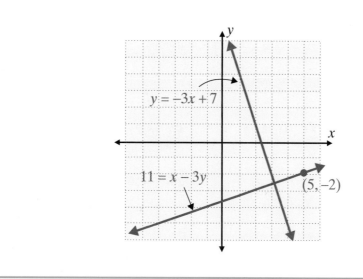

Practice Problems

1. *Write an equation in standard form for the line with slope $\dfrac{5}{8}$ that contains the point $(2, -2)$.*

2. *Write an equation in standard form for the line passing through the two points $(6, 2)$ and $(-3, 1)$.*

3. *Write an equation in slope-intercept form for the line passing through the point $(-5, 2)$ and parallel to the line $3x + y = 8$. (Remember, parallel lines have the same slope.)*

4.4 Exercises

In Exercises 1 – 12, write an equation (in slope-intercept form or standard form) for the line passing through the given point with the given slope.

1. $(0, 3), m = 2$ **2.** $(1, 4), m = -1$ **3.** $(-1, 3), m = -\dfrac{2}{5}$

4. $(-2, -5), m = \dfrac{7}{2}$ **5.** $(-3, 2), m = \dfrac{3}{4}$ **6.** $(6, -1), m = \dfrac{5}{3}$

7. $(3, -1), m = 0$ **8.** $(4, 6), m = 0$ **9.** $(2, -4),$ undefined slope

10. $(-3, 2),$ undefined slope   **11.** $\left(1, \dfrac{2}{3}\right), m = \dfrac{3}{2}$ **12.** $\left(\dfrac{1}{2}, -2\right), m = \dfrac{4}{3}$

Answers to Practice Problems: 1. $5x - 8y = 26$ **2.** $x - 9y = -12$ **3.** $y = -3x - 13$

In Exercises 13 – 24, write an equation (in slope-intercept form or standard form) for the line passing through the given points.

13. $(-2, 3), (1, 2)$

14. $(-5, 1), (2, 0)$

15. $(3, 4), (6, 2)$

16. $(-4, -4), (3, 1)$

17. $(8, 2), (-1, 2)$

18. $\left(\frac{1}{2}, -1\right), (3, -1)$

19. $(-5, 2), (1, 4)$

20. $(-2, 6), (3, 1)$

21. $(-2, 4), (-2, 1)$

22. $(3, 0), (3, -3)$

23. $(-4, 2), \left(1, \frac{1}{2}\right)$

24. $(0, 2), \left(1, \frac{3}{4}\right)$

25. Write an equation for the horizontal line through point $(-2, 5)$.

26. Write an equation for the line parallel to the x-axis through point $(6, -2)$.

27. Write an equation for the line parallel to the y-axis through point $(-1, -3)$.

28. Write an equation for the vertical line through point $(-1, -1)$.

29. Write an equation for the horizontal line through point $(-1, -6)$.

30. Write an equation for the line parallel to the line $3x - y = 4$ through the origin. Graph both lines.

31. Write an equation for the line parallel to the line $2x = 5$ through point $(0, 3)$. Graph both lines.

32. Write an equation for the line through point $\left(\frac{3}{2}, 1\right)$ and parallel to the line $4y - 6 = 0$. Graph both lines.

33. Write an equation for the line that contains the point $(2, 5)$ and is perpendicular to the line $y = \frac{3}{4}x + 2$. Graph both lines.

34. Write an equation for the line that contains the point $(-1, 3)$ and is perpendicular to the line $y = -\frac{1}{2}x - 10$. Graph both lines.

35. Write an equation for the line that contains the point $(7, -2)$ and is perpendicular to the line $2x + 3y = 6$. Graph both lines.

36. Write an equation for the line that contains the point $(-4, -6)$ and is perpendicular to the line $3x - y = 10$. Graph both lines.

37. The two lines $y = 5x + 3$ and $y = \dfrac{1}{5}x + 4$ are (parallel, perpendicular, or neither). Explain your answer.

38. The two lines $y = \dfrac{4}{3}x + 2$ and $4x - 3y = 12$ are (parallel, perpendicular, or neither). Explain your answer.

39. Find an equation for the line parallel to $2x - y = 7$ having the same y-intercept as $x - 3y = 6$.

40. Find an equation for the line parallel to $3x - 2y = 4$ having the same y-intercept as $5x + 4y = 12$.

41. The rental rate schedule for a car rental is described as "$30 per day plus 15¢ per mile driven." Write a linear equation for the daily cost of renting a car for x miles. $(C = mx + b)$

42. For each job, a print shop charges $5.00 plus $0.05 per page printed.
   a. Write a linear equation for the cost of printing x pages. $(C = mx + b)$
   b. Graph the corresponding part of the line for $x \geq 0$.
   c. Interpret the meaning of the slope of the line.
   d. Explain the fact that the line makes sense only for positive integer values of x.

43. A salesperson's weekly income depends on the amount of her sales. Her weekly income is a salary of $200 plus 9% of her weekly sales.
   a. Write a linear equation for her weekly income for sales of x dollars. $(I = mx + b)$
   b. Graph the corresponding part of the line for $x \geq 0$.
   c. Interpret the meaning of the slope of the line.

44. A manufacturer has determined that the cost, C, of producing x units is given by a linear equation $(C = mx + b)$.
   a. If 12 units cost $752 to produce and 15 units cost $800 to produce, find a linear equation for the cost of producing x units.
   b. What is the slope of the line?
   c. Interpret the meaning of the slope of the line.
   d. Explain the fact that the line makes sense only for positive integer values of x.

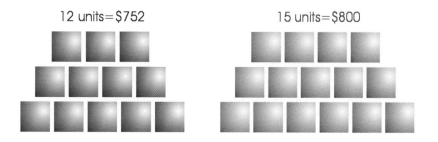
12 units=$752   15 units=$800

45. A local amusement park found that if the admission was $7, the attendance was about 1000 customers per day. When the price of admission was dropped to $6, attendance increased to about 1200 customers per day.

    a. Write a linear equation for the attendance in terms of the price, p.
       ($A = mp + b$)

    b. What is the slope of the line?

    c. Interpret the meaning of the slope of the line.

    d. Explain the fact that the line makes sense only for positive values of p.

46. A department store manager has determined that if the price for a necktie is $35, the store will sell 100 neckties per month. However, only 50 neckties per month are sold if the price is raised to $45.

    a. Write a linear equation for the number of neckties sold per month in terms of the price, p. ($N = mp + b$)

    b. What is the slope of the line?

    c. Interpret the meaning of the slope of the line.

    d. Explain the fact that the line makes sense only for positive values of p.

In Exercises 47 – 51, (a) use the given information to find the equation of each line and (b) use a graphing calculator to graph the line.

47. contains the two points $(-1, 2)$ and $(3, 1)$

48. has slope $\dfrac{5}{2}$ and y-intercept $(0, -1)$

49. has slope $-\dfrac{1}{2}$ and passes through the origin

50. contains the two points $(-2, 5)$ and $(3, 5)$

51. contains the point $(-3, -4)$ and has slope $\dfrac{3}{2}$

Hawkes Learning Systems: Introductoy Algebra

The Point-Slope Form: $y - y_1 = m(x - x_1)$

4.5 Introduction to Functions and Function Notation

After completing this section, you will be able to:

1. *Determine whether a relation is or is not a function.*

2. *Find the domain and range of a relation or a function.*

3. *Use the vertical line test to determine whether a graph is or is not the graph of a function.*

4. *Use function notation.*

The cost of producing computers is a **function** of the number of computers produced; the rate of your heart beat is a function of your anxiety level; your grade in this class is a function of the time you spend studying. The concept of a function is present in many aspects of our daily lives. In fact, the concept of a function is one of the most useful and important in all of mathematics. With this concept and the related topics, we are able to develop calculus and perform engineering feats such as building bridges, building airplanes, and sending satellites into orbit. Here we will simply introduce the idea of a function and how to use function notation.

Introduction to Functions

In Section 4.1 we discussed ordered pairs of real numbers and their graphs in the Cartesian coordinate system. Every set of ordered pairs of real numbers can be classified as a **relation**.

Relation

*A **relation** is a set of ordered pairs.*

*The **domain**, **D**, of a relation is the set of all first coordinates in the relation.*

*The **range**, **R**, of a relation is the set of all second coordinates in the relation.*

In graphing relations, the x-axis is sometimes called the **domain axis** and the y-axis is called the **range axis**. In real life applications the variables may not be x and y. For example, the distance, d, an object falls depends on the time, t, that it falls and this relationship can be represented in the form $d = 16t^2$. Thus, in graphing such a relation, the horizontal axis (domain axis) would be labeled the t-axis and the vertical axis (range axis) would be labeled the d-axis.

Example 1: Finding the Domain and Range

Find the domain and range for each of the following relations.

a. $r = \{(5, 7), (6, 2), (6, 3), (-2, -1)\}$

Solution: $D = \{-2, 5, 6\}$
$R = \{-1, 2, 3, 7\}$

Note that 6 is listed only once in the domain even though it appears in two ordered pairs. The numbers may be listed in any order in both the domain and range. We have listed them in order here for convenience only.

b. $f = \{(-1, 1), (1, 5), (3, 0), (4, 0)\}$

Solution: $D = \{-1, 1, 3, 4\}$
$R = \{0, 1, 5\}$

The relation $f = \{(-1, 1), (1, 5), (3, 0), (4, 0)\}$ in Example 1b meets a particular condition in that **each first coordinate has only one corresponding second coordinate**. Such a relation is called a **function**. Notice that r in Example 1a is **not** a function because the first coordinate 6 has more than one corresponding second coordinate. That is, 6 appears as a first coordinate more than once in the listing of the ordered pairs. The fact that 0 appears more than once as a second coordinate in the relation f is not a factor in determining whether or not f is a function.

Also, you should be aware of the fact that most functions have an infinite number of ordered pairs and therefore the domains and ranges cannot be listed as illustrated in these examples. These ideas will be studied in greater detail in later courses in mathematics.

Function

> A **function** is a relation in which each domain element has exactly one corresponding range element.
>
> OR
>
> A **function** is a relation in which each domain element occurs only once.

Example 2: Determining Functions

Determine whether each of the following relations is or is not a function.

a. $r = \{(2, 3), (1, 6), (2, 5), (0, -1), (-3, 0)\}$

Solution: r is **not** a function. The first coordinate 2 appears more than once.

b. $s = \{(0, 0), (1, 1), (2, 4), (3, 9)\}$

Solution: s is a function. Each first coordinate has only one corresponding second coordinate.

c. $t = \{(1, 5), (3, 5), (-2, 5), (-1, 5), (-4, 5)\}$

Solution: t is a function. Each first coordinate appears only once. The fact that the second coordinates are all the same has no relevance in the definition of a function.

The Vertical Line Test

If one point in the graph of a relation is directly above or below another point in the graph, then these points have the same first coordinate (or the same x-coordinate). Such a relation would **not** be a function. Therefore, we can tell whether or not a graph represents a function by using the following **vertical line test**.

Vertical Line Test

> *If any vertical line intersects a graph of a relation in more than one point, then the relation graphed is not a function.*

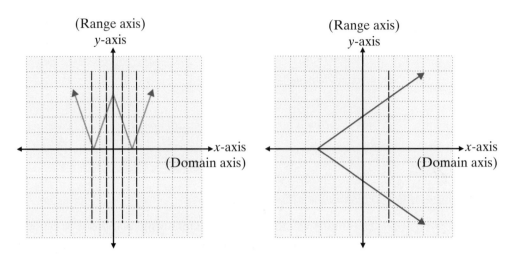

This graph represents a function. Several vertical lines are drawn to illustrate that vertical lines do not intersect the graph in more than one point. Each *x*-value has only one corresponding *y*-value.

(a)

This graph is not a function because the vertical line drawn intersects the graph in more than one point. Thus, for at least one *x*-value, there is more than one corresponding *y*-value.

(b)

Figure 4.18

Example 3: Vertical Line Test

Use the vertical line test to determine whether each of the following graphs does or does not represent a function.

a.

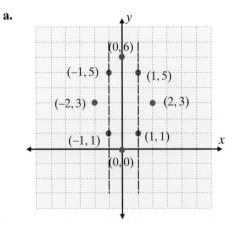

Solution: Not a function. At least one vertical line intersects the graph in more than one point. Listing the ordered pairs shows that several *x*-coordinates are repeated:

$$\{(-2, 3), (-1, 1), (-1, 4), (0, 0), (0, 5), (1, 1), (1, 4), (2, 3)\}$$

b.

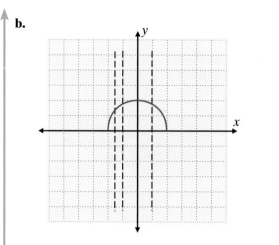

Solution: A function. No vertical line will intersect the graph in more than one point. Several vertical lines are drawn to illustrate this fact. Furthermore, the domain and range can be determined by investigating the graph. We see that the domain and range consist of an infinite number of real values and are intervals of real numbers:

Domain: $-2 \le x \le 2$

Range: $0 \le y \le 2$

c.

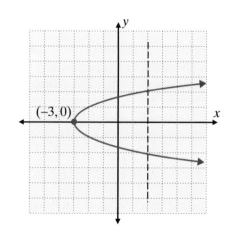

Solution: Not a function. A vertical line intersects the graph in more than one point.

Continued on next page...

d.

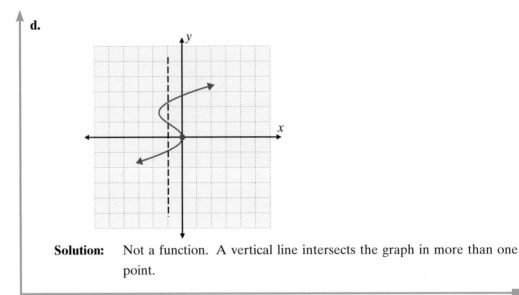

Solution: Not a function. A vertical line intersects the graph in more than one
point.

Function Notation

We have used ordered pair notation (x, y) to represent points on the graphs of relations
and functions. As the vertical line test will verify, equations of the form

$$y = mx + b$$

where the equation is solved for y, represent linear functions. The use of x and y in this
manner is standard in mathematics and is particularly useful when graphing because we
are familiar with the concepts of the x-axis and y-axis.

Another notation, called **function notation**, is more convenient for indicating calculations
of values of a function and for indicating operations performed with functions. In function
notation,
            instead of writing y,   write $f(x)$,   read "f of x."

The letter f is the name of the function. Any letter will do. The letters f, g, h, F, G, and
H are commonly used. We have used r, s, and t in previous examples.

Suppose that $y = 3x - 2$ is a given linear equation. Since the equation is solved for y, it
represents a linear function and we can replace y with the notation $f(x)$ as follows:

$$y = 3x - 2 \quad \text{can be written in the form} \quad f(x) = 3x - 2.$$

That is, the notation $f(x)$ represents the y-value for some corresponding x-value.

Now, in function notation, $f(5)$ means to replace x with 5 in the function,

$$f(5) = 3 \cdot 5 - 2 = 15 - 2 = 13$$

Thus the ordered pair $(5, 13)$ is the same as $(5, f(5))$.

Example 4: Function Evaluation

For the function $g(x) = 4x + 3$, find

a. $g(2)$ **b.** $g(-3)$ **c.** $g(0)$

Solution: **a.** $g(2) = 4 \cdot 2 + 3 = 11$
b. $g(-3) = 4(-3) + 3 = -9$
c. $g(0) = 4 \cdot 0 + 3 = 3$

Not all functions are linear functions just as not all graphs are straight lines. The function notation is valid for a wide variety of types of functions. The following examples illustrate the use of function notation with functions other than linear functions.

Example 5: Nonlinear Function Evaluation

For the function $f(x) = x^2 - 2x + 1$, find
a. $f(-2)$ **b.** $f(0)$ **c.** $f(4)$

Solution: **a.** $f(-2) = (-2)^2 - 2(-2) + 1 = 4 + 4 + 1 = 9$

b. $f(0) = 0^2 - 2(0) + 1 = 0 - 0 + 1 = 1$

c. $f(4) = 4^2 - 2 \cdot 4 + 1 = 16 - 8 + 1 = 9$

Example 6: Nonlinear Function Evaluation

For the function $h(x) = 2x^3 - 5x$, find
a. $h(-1)$ **b.** $h(3)$ **c.** $h(-4)$

Solution: **a.** $h(-1) = 2(-1)^3 - 5(-1) = 2(-1) + 5 = -2 + 5 = 3$

b. $h(3) = 2(3)^3 - 5(3) = 54 - 15 = 39$

c. $h(-4) = 2(-4)^3 - 5(-4) = 2(-64) + 20 = -128 + 20 = -108$

Example 7: Nonlinear Function Evaluation

Given the function $f(x) = x^2 - 10$ with restricted domain $D = \{-1, 0, 2, 3\}$, write the function as a set of ordered pairs.

Solution: For each x-value in D, find the corresponding y-value by substituting into $f(x)$.

$$f(-1) = (-1)^2 - 10 = 1 - 10 = -9$$

$$f(0) = 0^2 - 10 = 0 - 10 = -10$$

$$f(2) = 2^2 - 10 = 4 - 10 = -6$$

$$f(3) = 3^2 - 10 = 9 - 10 = -1$$

So the function can be written as the following set of ordered pairs:

$$\{(-1, -9), (0, -10), (2, -6), (3, -1)\}$$

Practice Problems

1. *State the domain and range of the relation { (1, 2), (3, 4), (5, 6) }. Is the relation a function?*

2. *Write the function as a set of ordered pairs given $f(x) = 2x - 5$ and $D = \{ -4, 0, 2, 3 \}$.*

4.5 Exercises

In Exercises 1 – 10, (a) list the sets of ordered pairs corresponding to the points, (b) give the domain and range, and (c) state whether the relation is or is not a function.

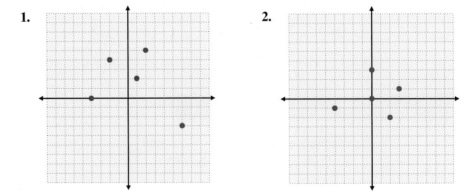

1.

2.

Answers to Practice Problems: 1. $D = \{1, 3, 5\}$ and $R = \{2, 4, 6\}$. It is a function. **2.** $\{(-4, -13), (0, -5), (2, -1), (3, 1)\}$

368

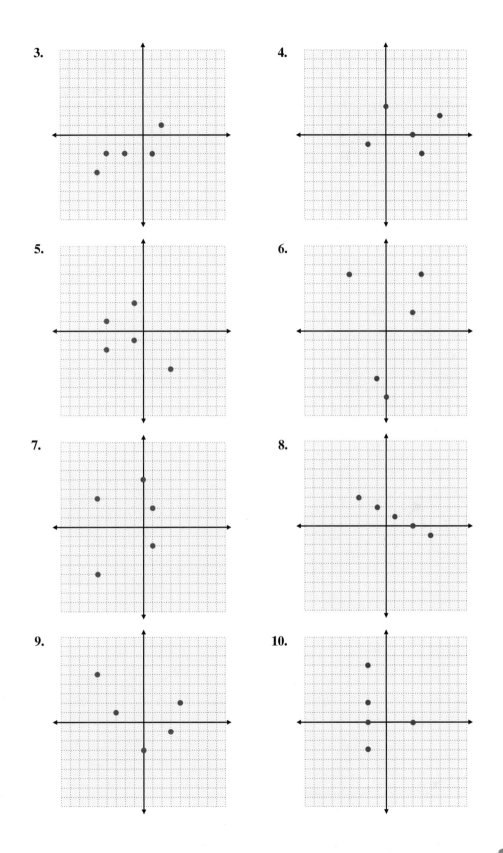

In Exercises 11 – 20, graph each set of ordered pairs and state whether the relation is or is not a function. If the relation is not a function, tell why not.

11. $\{(0,0), (1,6), (4,-2), (-3,5), (2,-1)\}$

12. $\{(1,-5), (2,-3), (-1,-3), (0,2), (4,3)\}$

13. $\{(-4,4), (3,-2), (1,0), (2,-3)\}$

14. $\{(-3,-3), (0,1), (-2,1), (3,1)\}$

15. $\{(0,2), (-1,1), (2,4), (3,5), (-3,5)\}$

16. $\{(-1,4), (-1,2), (-1,0), (-1,6), (-1,-2)\}$

17. $\{(0,0), (-2,5), (2,0.5), (4,-4.6), (5,0.2)\}$

18. $\{(2,-4), (2,-2), (0.5,-1), (0.5,3)\}$

19. $\{(-2,6), (1,-5), (3,1), (-4,3), (-3,2)\}$

20. $\{(5,-1), (6,-1), (-2,0), (5,0), (0.3,0.6)\}$

Use the vertical line test to determine whether or not each of the graphs in Exercises 21 – 32 is a function. If the graph represents a function, state its domain and range.

21.   **22.**

23.   **24.**

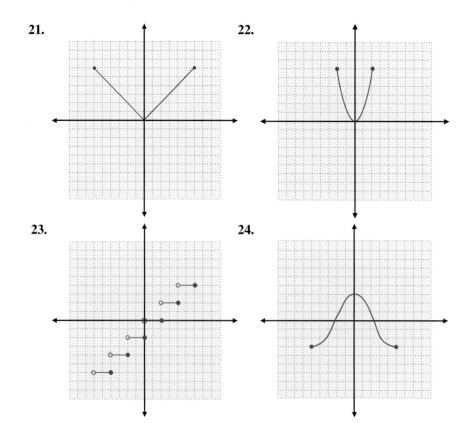

25.

26.

27.

28.

29.

30.

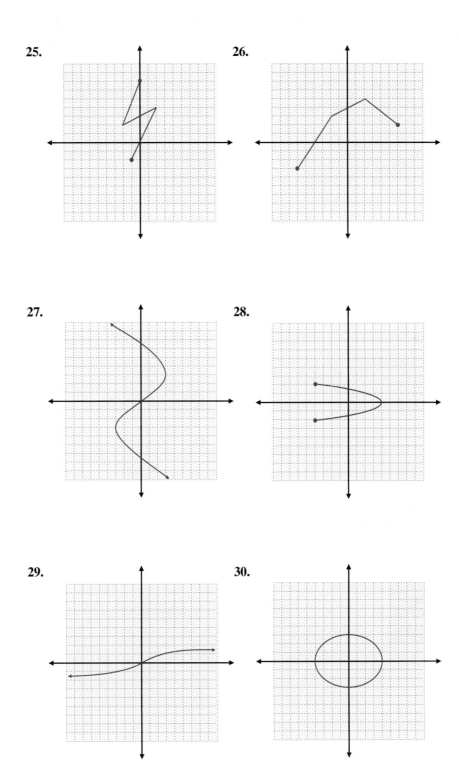

31.   **32.**

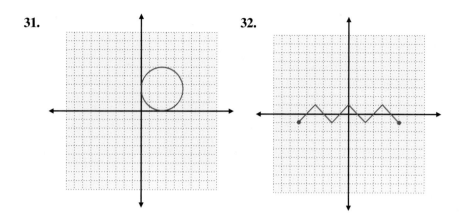

For Exercises 33 – 45, assume that each function has the same domain, $D = \{-2, -1, 0, 1, 5\}$ and find the corresponding functional value (y-value) for each x-value in the domain.

33. $f(x) = 2x + 1$ **34.** $f(x) = 3x - 5$ **35.** $g(x) = -4x + 2$

36. $g(x) = -x - 7$ **37.** $h(x) = x^2$ **38.** $h(x) = x^3$

39. $F(x) = x^2 - 4x + 4$ **40.** $G(x) = x^2 + 5x + 6$ **41.** $H(x) = x^3 - 8x$

42. $C(x) = 4x^2 + 3x - 7$ **43.** $P(x) = 0.5x^2 - 2x + 3.5$ **44.** $R(x) = 1.5x^2 + 4x - 9$

45. $f(x) = x^3 - 5x^2 + 7x - 2$

Hawkes Learning Systems: Introductory Algebra

Introduction to Functions and Function Notation

4.6 Graphing Linear Inequalities: $y < mx + b$

Graphing Linear Inequalities: $y < mx + b$

Concepts Related to Straight Lines

In summary, we have discussed the following concepts related to straight lines:

$$m = \frac{y_2 - y_1}{x_2 - x_1} \qquad \text{the slope of a line through two points}$$

$$y = mx + b \qquad \text{the slope-intercept form}$$

$$y - y_1 = m(x - x_1) \qquad \text{the point-slope form}$$

$$x = a \qquad \text{represents a vertical line}$$

$$y = b \qquad \text{represents a horizontal line}$$

Parallel lines have the same slope.

Perpendicular lines have slopes that are negative (or opposite) reciprocals of each other.

Graphing calculators can be used to graph straight lines that are not vertical.

In this section, we will develop techniques of analyzing and graphing **linear inequalities**. We need the following terminology:

Half-plane	A straight line separates a plane into two **half-planes**. The points on one side of the line are in one of the half-planes, and the points on the other side of the line are in the other half-plane.
Boundary line	The line itself is called the **boundary line**.
Closed half-plane	If the boundary line is included, then the half-plane is said to be **closed**.
Open half-plane	If the boundary line is not included, then the half-plane is said to be **open**.

Figure 4.19 illustrates an **open half-plane** and a **closed half-plane** with the line $5x - 3y = 15$ as the boundary line in both cases.

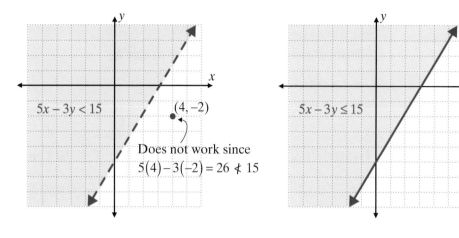

The points on the line $5x - 3y = 15$ are not included so the line is dashed.
The half-plane is open.
(a)

The points on the line $5x - 3y = 15$ are included so the line is solid.
The half-plane is closed.
(b)

Figure 4.19

Two Methods for Graphing Linear Inequalities

First, graph the boundary line (dashed if the inequality is < or >, solid if the inequality is ≤ or ≥).

Method 1

    a. *Test any one point obviously on one side of the line.*

    b. *If the test-point satisfies the inequality, shade the half-plane on that side of the line. Otherwise, shade the other half-plane.*

Method 2

    a. *Solve the inequality for y (assuming that the line is not vertical).*

    b. *If the solution shows y < or y ≤, then shade the half-plane below the line.*

    c. *If the solution shows y > or y ≥, then shade the half-plane above the line.*

*(**Note**: If the boundary line is vertical, then it is of the form x = a and Method 1 should be used.)*

Example 1: Graphing Linear Inequalities

Graph the following inequalities.

a. Graph the half-plane that satisfies the inequality $2x + y \le 6$.

Solution: Method 1 is used in this example.

Step 1: Graph the line $2x + y = 6$ as a solid line.

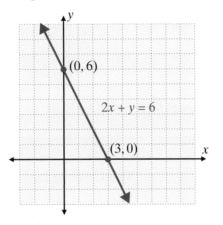

Step 2: Test any point on one side of the line. In this example, we have chosen $(0, 0)$.

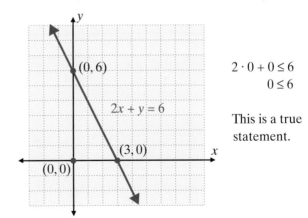

$2 \cdot 0 + 0 \le 6$
$0 \le 6$

This is a true statement.

Continued on next page...

Step 3: Shade the points on the same side as the point $(0,0)$.

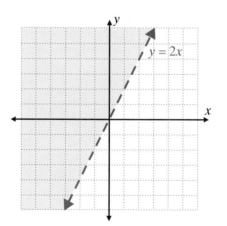

b. Graph the solution set to the inequality $y > 2x$.

Solution: Since the inequality is already solved for y, Method 2 is easy to apply.

Step 1: Graph the line $y = 2x$ as a dashed line.

Step 2: By Method 2, the graph consists of those points above the line. Shade the half-plane above the line.

c. Graph the half-plane that satisfies the inequality $y > 1$.

Solution: Again, the inequality is already solved for y and Method 2 is used.

Step 1: Graph the horizontal line $y = 1$ as a dashed line.

Step 2: By Method 2, shade the half-plane above the line.

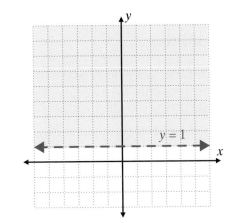

d. Graph the solution set to the inequality $x \le 0$.

> **Solution:** The boundary line is a vertical line and Method 1 is used.
>
> **Step 1:** Graph the line $x = 0$ as a solid line.
> Note that this is the y-axis.
>
> **Step 2:** Test the point $(2, 1)$.
>
> $2 \le 0$
>
> This statement is false.
>
> **Step 3:** Shade the half-plane on the opposite side of the line as $(2, 1)$.
> This half-plane consists of the points with x-coordinate 0 or negative.

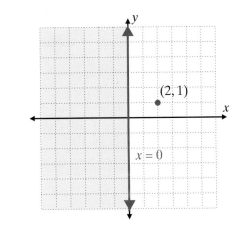

Using a TI-84 Plus Graphing Calculator to Graph Linear Inequalities

The first step in using the TI-84 Plus (or any other graphing calculator) to graph a linear inequality is to solve the inequality for y. This is necessary because this is the way that the boundary line equation can be graphed as a function. Thus Method 2 for graphing the correct half-plane is appropriate.

Note that when you press the 〔 Y= 〕 key on the calculator, a slash (\ ) appears to the left of the Y expression as in $\backslash Y_1 =$. This slash is actually a command to the calculator to graph the corresponding function as a solid line or curve. If you move the cursor to position over the slash and hit 〔ENTER〕, you will find the following options appearing:

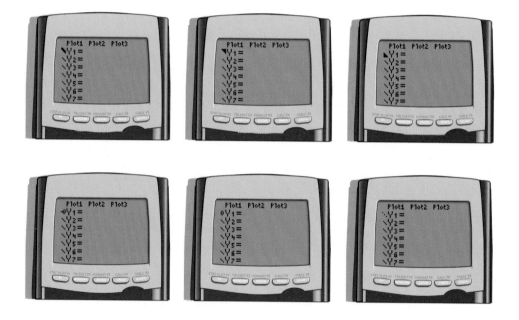

If the slash (which is actually four dots if you look closely) becomes a set of three dots, then the corresponding graph of the function will be dotted. To shade above the boundary choose ▜. To shade below the boundary choose ▙. The calculator is not good for determining whether the boundary curve is included or not. The following examples illustrate two situations.

Example 2: Graphing Using a TI-84 Plus

a.   Graph the linear inequality $2x + y \geq 3$.

Solution: **Step 1:** Solving the inequality for *y* gives: $y \geq -2x + 3$.

Step 2: Press the key $\boxed{\text{Y=}}$ and enter the function: $\backslash Y_1 = -2X + 3$.

Step 3: Go to the \ and hit $\boxed{\text{ENTER}}$ two times so that the display appears as follows:

Step 4: Press $\boxed{\text{GRAPH}}$ and (assuming that the window includes part of the line) the following graph should appear on the display.

Continued on next page ...

b. Graph the linear inequality $-3x + 2y < -4$

Solution: **Step 1:** Solving the inequality for y gives: $y < \dfrac{3}{2}x - 2$.

Step 2: Press the key 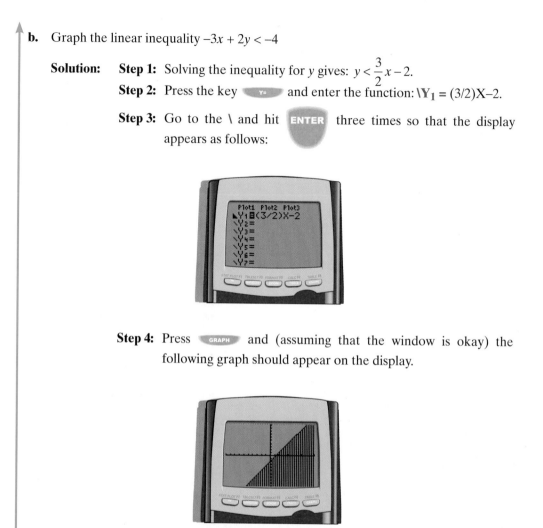 and enter the function: $\backslash Y_1 = (3/2)X{-}2$.

Step 3: Go to the $\backslash$ and hit **ENTER** three times so that the display appears as follows:

Step 4: Press **GRAPH** and (assuming that the window is okay) the following graph should appear on the display.

Practice Problems

1. *Which of the following points satisfy the inequality $x + y < 3$?*

    a. $(2, 1)$ *b.* $\left(\dfrac{1}{2}, 3\right)$ *c.* $(0, 5)$ *d.* $(-5, 2)$

2. *Which of the following points satisfy the inequality $x - 2y \geq 0$?*

    a. $(2, 1)$ *b.* $(1, 3)$ *c.* $(4, 2)$ *d.* $(3, 1)$

3. *Which of the following points satisfy the inequality $x < 3$?*

    a. $(1, 0)$ *b.* $(0, 1)$ *c.* $(4, -1)$ *d.* $(2, 3)$

Answers to Practice Problems: 1. (d) **2.** (a), (c), (d) **3.** (a), (b), (d)

4.6 Exercises

Graph each of the linear inequalities in Exercises 1 – 30.

1. $y > 3x$

2. $x - y > 0$

3. $y \leq 2x - 1$

4. $y < 4 - x$

5. $y > -7$

6. $2y - x \geq 2$

7. $2x - 3 \leq 0$

8. $4x + y < 2$

9. $5x - 2y \geq 4$

10. $3y - 8 \leq 2$

11. $y < -\dfrac{1}{4}x$

12. $2y - 3x \leq 4$

13. $2x - y < 1$

14. $4x + 3y \leq 6$

15. $2x - 3y \geq -3$

16. $2y + 5x \geq 0$

17. $2x + 5y \geq 10$

18. $4x + 2y \leq 10$

19. $3x < 4 - y$

20. $x < 2y + 3$

21. $2x - 3y < 9$

22. $x \leq 4 - 3y$

23. $5x - 2y \geq 4$

24. $3x + 6 \geq 2y$

25. $4y \geq 3x - 8$

26. $\dfrac{1}{2}x + y < 3$

27. $2x + \dfrac{1}{3}y < 2$

28. $2x - \dfrac{1}{2}y > -1$

29. $\dfrac{1}{2}x + \dfrac{2}{3}y \geq \dfrac{5}{6}$

30. $\dfrac{1}{4}x - \dfrac{1}{2}y \leq \dfrac{3}{4}$

Use your graphing calculator to graph each of the linear inequalities in Exercises 31 – 40.

31. $y > 2x$

32. $x - y \leq 4$

33. $x + 2y > 7$

34. $3x - 2y \geq 10$

35. $2x - y \leq 6$

36. $y \geq -2$

37. $x - 3y \geq 5$

38. $2x + 8y \geq 0$

39. $2x + 4y < 9$

40. $4x - 3y > 15$

Hawkes Learning Systems: Introductory Algebra

Graphing Linear Inequalities

Chapter 4 Index of Key Ideas and Terms

Section 4.1 The Cartesian Coordinate System and Reading Graphs

Cartesian Coordinate System — page 302
    Ordered Pairs — page 302
        First coordinate (or first component)
        Second coordinate (or second component)
    Quadrants — page 304
    Origin — page 304
    Axes — page 304
        Horizontal Axis (or x-axis) — page 304
        Vertical Axis (or y-axis) — page 304

One-To-One Correspondence — page 304
    There is a **one-to-one correspondence** between points in a plane and ordered pairs of real numbers.

Graphs of Points — page 309

Section 4.2 Graphing Linear Equations in Two Variables: $Ax + By = C$

Linear Equations With Two Variables — pages 320 - 322
    Standard Form: $Ax + By = C$ — pages 320 - 322
    (A and B cannot both be 0.)

Intercepts — pages 322 - 323
    y-intercept (point where a line crosses the y-axis)
    x-intercept (point where a line crosses the x-axis)

Horizontal and Vertical Lines — pages 324 - 326
    1. The graph of any equation of the form $y = b$ is a horizontal line.
    2. The graph of any equation of the form $x = a$ is a vertical line.

Section 4.3 The Slope-Intercept Form: $y = mx + b$

Slope of a Line — pages 333 - 335
    Formula for Slope: — pages 334 - 338

$$\text{slope} = \frac{\text{rise}}{\text{run}} = \frac{y_2 - y_1}{x_2 - x_1} = \frac{y_1 - y_2}{x_1 - x_2}$$

Continued on next page...

Section 4.3   The Slope-Intercept Form: $y = mx + b$ (continued)

To Calculate the Slope of a Line   pages 333 - 335
     1. Locate any two points on the line.
     2. Find the difference in the y-values and divide by the difference in the x-values.
     3. This ratio (of rise divided by run) is the slope of the line.

Parallel Lines   page 337
     All lines with the same slope are **parallel**.

Slope Related to Graphs of Lines   pages 335 - 341
     1. All lines with **negative slope** slant downard as we go from left to right
     2. All lines with **positive slope** slant upward as we go from left to right.
     3. For **horizontal lines** ($y = b$), the slope is 0.
     4. For **vertical lines** ($x = a$), the slope is undefined.

Slope-Intercept Form   pages 342 - 343
     $y = mx + b$ is called **slope-intercept form** for the equation of a line.
     m is the slope and b is the y-intercept.

Section 4.4   The Point-Slope Form: $y - y_1 = m(x - x_1)$

Point-Slope Form   pages 352 - 353
     $y - y_1 = m(x - x_1)$ is called the **point-slope form** for the equation of a line.
     m is the slope and (x_1, y_1) is a given point on the line.

Finding the Equation of a Line Given Two Points   pages 353 - 355
     If two points are given,

     1. Use the formula $m = \dfrac{y_2 - y_1}{x_2 - x_1}$ to find the slope.

     2. Use this slope, m, and either point in the point-slope formula $y - y_1 = m(x - x_1)$.

Parallel Lines   pages 355 - 356
     If two lines are **parallel,** then they never intersect and they have the same slope.

Perpendicular Lines   pages 355 - 357
     If two lines are **perpendicular**, then their slopes are negative (opposite) reciprocals of each other.

Section 4.5   Introduction to Functions and Function Notation

Relation   page 361

A **relation** is a set of ordered pairs.
The **domain**, D, of a relation is the set of all first coordinates
in the relation.
The **range**, R, of a relation is the set of all second coordinates
in the relation.

Function   pages 361 - 363

A **function** is a relation in which each domain element has
exactly one corresponding range element. (or equivalently)
A **function** is a relation in which each domain element occurs
only once.

Vertical Line Test   pages 363 - 366

If any vertical line intersects a graph of a relation in
more than one point, then the relation is not a function.

Function Notation: $f(x)$ page 366

The notation $f(x)$ is read "f of x."
The notation $f(x)$ represents the y-value for some corresponding
x-value.

Section 4.6   Graphing Linear Inequalities: $y < mx + b$

Concepts related to Straight Lines   page 373

Linear Inequalities   page 373
     Half-Plane   page 373
     Boundary Line   page 373
     Closed Half-Plane   pages 373 - 374
     Open Half-Plane   pages 373 - 374

Two Methods for Graphing Linear Inequalities   pages 374 - 377

Using a Calculator to Graph Linear Inequalities   pages 378 - 380

Hawkes Learning Systems: Introductory Algebra

For a review of the topics and problems from Chapter 4, look at the following lessons from *Hawkes Learning Systems: Introductory Algebra*

Introduction to the Cartesian Coordinate System
Graphing Linear Equations by Plotting Points
Graphing Linear Equations in Two Variables: $Ax + By = C$
The Slope-Intercept Form: $y = mx + b$
The Point-Slope Form: $y - y_1 = m(x - x_1)$
Introduction to Functions and Function Notation
Graphing Linear Inequalities

Chapter 4 Review

4.1 The Cartesian Coordinate System and Reading Graphs

List the sets of ordered pairs corresponding to the graphs in Exercises 1 – 4.

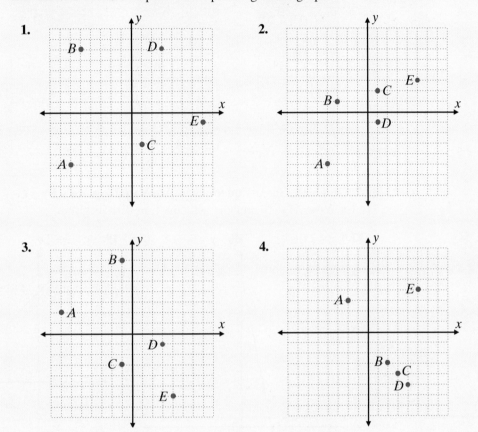

1.

2.

3.

4.

In Exercises 5 – 8, graph the sets of ordered pairs and label the points.

5. $\{A(-3,1), B(-2,-2), C(0,-5), D(1,4)\}$ **6.** $\{E(-3,-4), F(0,0), G(5,-1), H(1.5,-3)\}$

7. $\{M(6,2), N(3,-1), O(0,-2), P(-4,-5)\}$ **8.** $\{Q(-4,-2), R(-4,0), S(-4,3), T(-4,5)\}$

In Exercises 9 – 12, determine which, if any, of the given ordered pairs satisfy the equation.

9. $3x - y = -2$

   a. $(1, 1)$

   b. $(-1, 1)$

   c. $(2, 2)$

   d. $(-2, 2)$

10. $x + 2y = 3$

   a. $(-3, 1)$

   b. $(2, 1)$

   c. $\left(2, \dfrac{1}{2}\right)$

   d. $(1, -2)$

11. $y = 4x$

   a. $(1, 4)$

   b. $(-1, 4)$

   c. $\left(\dfrac{1}{4}, 1\right)$

   d. $(-3, -12)$

12. $2x - 3y = 5$

   a. $(1, -1)$

   b. $\left(2, -\dfrac{1}{3}\right)$

   c. $(2.5, 0)$

   d. $(0, -2)$

Complete the tables in Exercises 13 – 16 so that each order pair will satisfy the given equation. Graph the resulting sets of ordered pairs.

13. $y = 5x$

x	y
0	
	−1
−2	
	5

14. $y = 6 - 3x$

x	y
0	
	0
−1	
	$\frac{1}{3}$

15. $y = 8 - 2x$

x	y
0	
	0
3	
	$-\frac{1}{2}$

16. $3x + y = -5$

x	y
0	
	−1
−3	
	−0.5

4.2 Graphing Linear Equations in Two Variables: $Ax + By = C$

In Exercises 17 – 20, match each equation with its graph.

17. $y = 2x + 4$ **18.** $y = 6 - x$ **19.** $y = 3 - 3x$ **20.** $-5x + y = 10$

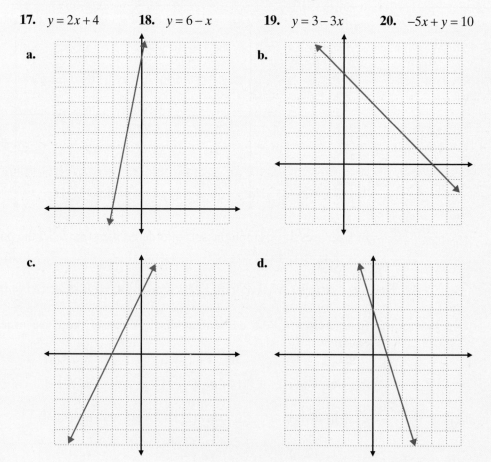

In Exercises 21 – 24, locate the *x*-intercepts and the *y*-intercepts. Do not graph the lines.

21. $5x + y = 1.5$ **22.** $3x - y = 7.5$ **23.** $x + 2y = 3$ **24.** $6x - 2y = 7$

In Exercises 25 – 30, first tell (just by looking and using your understanding of lines) whether the graph will be a vertical line, a horizontal line, or neither. Then graph the line by locating any two (or more) points that satisfy the equation.

25. $y = -\dfrac{1}{2}x + 4$ **26.** $y = \dfrac{2}{3}x$ **27.** $x = 7$

28. $y = -3.5$ **29.** $-10 = y$ **30.** $5x + y = 10$

4.3 The Slope-Intercept Form: $y = mx + b$
In Exercises 31 – 38, find the slope of the line determined by each pair of points.

31. $(1, 5), (2, -2)$ **32.** $(-4, 1), (3, -2)$ **33.** $(-5, -2), (1, 3)$

34. $(-6, 1), (-1, 6)$ **35.** $(0, 4), (8, 0)$ **36.** $(0, 3), (5, 13)$

37. $\left(\dfrac{3}{4}, \dfrac{1}{2}\right), \left(\dfrac{1}{4}, -\dfrac{1}{6}\right)$ **38.** $(2.6, 1.8), (3.6, 2.3)$

In Exercises 39 – 42, write the equation and draw the graph of the line passing through the given *y*-intercept with the given slope.

39. $(0, -1), m = \dfrac{5}{8}$ **40.** $(0, -4), m = -1$ **41.** $(0, 3), m = 0$ **42.** $(0, 5), m = \dfrac{2}{3}$

For each of the equations in Exercises 43 – 50, find the slope, *m*, and the *y*-intercept, *b*. Then graph each line.

43. $y = 2x - 6$ **44.** $y = -3x + 4$ **45.** $x + y = 1.5$ **46.** $x - 2y = 10$

47. $x - y = 8$ **48.** $2x - y = 12$ **49.** $x + 4y = 2$ **50.** $y = 2 - 3x$

4.4 The Point-Slope Form: $y - y_1 = m(x - x_1)$

In Exercises 51 – 56, write an equation for the line passing through the given point with the given slope.

51. $(1, 5)$, $m = 3$ **52.** $(-1, 4)$, $m = -1$ **53.** $(-3, 2)$, $m = \dfrac{1}{5}$

54. $(2, -4)$, $m = \dfrac{2}{3}$ **55.** $(5, 6)$, $m = $ undefined   **56.** $(-2, -7)$, $m = 0$

In Exercises 57 – 62, write an equation for the line passing through the given points.

57. $(1, 8), (2, 3)$ **58.** $(2, 4), (5, -1)$ **59.** $(0, -2), (5, 0)$

60. $(-4, 1), (0, 8)$ **61.** $(-5, -2), (3, 1)$ **62.** $(-5, 6), (4, -3)$

63. Write an equation for the horizontal line through point $(-3, 4)$.

64. Write an equation for the line parallel to the x-axis through point $(-2, 5)$.

65. Write an equation for the vertical line through point $(5, -1)$.

66. Write an equation for the horizontal line through point $(0, -2)$.

67. Write an equation for the line through point $\left(\dfrac{5}{2}, 2\right)$ and parallel to the line $x - 4y = 3$. Graph both lines.

68. Write an equation for the line through point $(-6, -1)$ and parallel to the line $3x + y = -7$. Graph both lines.

69. Write an equation for the line perpendicular to the line $y = 2x - 3$ and passing through point $(0, 1)$. Graph both lines.

70. Write an equation for the line perpendicular to the line $x + 2y = 10$ and passing through point $(-5, 0)$. Graph both lines.

4.5 Introduction to Functions and Function Notation

In Exercises 71 – 74, graph each set of ordered pairs and state whether the relation is or is not a function. If the relation is not a function, tell why not.

71. $\{(0, 2), (-1, 3), (0, 5), (4, 2), (3, -2)\}$ **72.** $\{(0, 0), (-2, 6), (0.5, 2), (-2, -1), (4, 0)\}$

73. $\{(-5, 0), (-2, 1), (0, 2), (1, 2), (4, 0)\}$ **74.** $\{(-6, 1), (-4, 0), (-2, -1), (0, -2), (2, -4)\}$

In Exercises 75 – 80, use the vertical line test to determine whether or not each of the graphs is a function. If the graph represents a function, state its domain and range.

75. **76.**

77. **78.**

79. **80.**

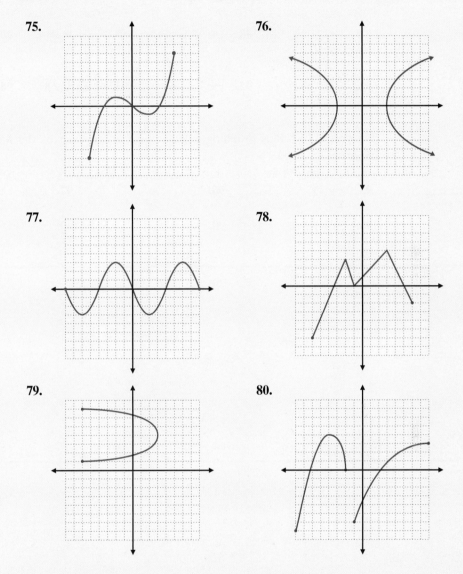

For Exercises 81 – 86, assume that each function has the domain $D = \{-3, -1, 0, 2, 3\}$ and find the corresponding functional value (y-value) for each x-value in the domain.

81. $f(x) = 3x - 10$ **82.** $g(x) = 6 - 4x$ **83.** $h(x) = x^2 + 4$

84. $f(x) = x^2 - 8x + 7$ **85.** $F(x) = 3x^2 + 4x + 1$ **86.** $C(x) = x^2 + 4x$

4.6 Graphing Linear Inequalities: $y < mx + b$

Graph each of the linear inequalities in Exercises 87 – 94.

87. $y \le 2x + 1$ 88. $y < -3$ 89. $y > -4$ 90. $y \ge 3x$

91. $y \ge \dfrac{1}{4}x - 2$ 92. $y < \dfrac{2}{3}x + 6$ 93. $\dfrac{2}{5}x + y > 5$ 94. $2x + 6 > 2y$

Use your TI-84 Plus graphing calculator to graph each of the linear inequalities in Exercises 95 – 100.

95. $y > 4x$ 96. $x - y \le 6$ 97. $2x + 4y > 0$

98. $2x + y \le 3$ 99. $y \ge 4$ 100. $x - 3y < 9$

Chapter 4 Test

1. Which of the following points lie on the line determined by the equation $x - 4y = 7$?

   a. $(2, -2)$ **b.** $(-1, -2)$ **c.** $\left(5, -\dfrac{1}{2}\right)$ **d.** $(0, 7)$

2. List the ordered pairs corresponding to the points on the graph.

   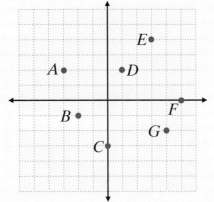

3. Graph the following set of ordered pairs:

   $$\left\{ L(0, 2), M(4, -1), N(-3, 2), O(-1, -5), P\left(2, \frac{3}{2}\right) \right\}.$$

4. Graph the line $2x - y = 4$.

5. Graph the equation $3x + 4y = 9$ by locating and labeling the x-intercept and y-intercept.

6. Graph the line passing through the point $(1, -5)$ with the slope $m = \dfrac{5}{2}$.

7. For the line $3x + 5y = -15$, determine the slope, m and the y-intercept, b. Then graph the line.

8. What is the slope of the line that contains the two points $(4, -3)$ and $\left(2, \dfrac{1}{2}\right)$?

9. Write an equation for the line passing through the point $(-3, 2)$ with the slope $m = \dfrac{1}{4}$.

10. Write an equation for the line passing through the points $(3, 1)$ and $(-2, 6)$. What is the slope of this line?

11. Write an equation for the line parallel to the x-axis through the point $(3, -2)$. What is the slope of this line?

12. A carpet cleaning machine can be rented for a fixed cost of $9.00 plus $4.00 per hour.

    a. Write a linear equation representing the cost, C, in terms of hours used, t.

    b. Explain why this equation only makes sense if t is positve.

    c. Explain the meaning of the slope of this line.

13. Determine whether the two lines $2x + 3y = 6$ and $y = -\dfrac{2}{3}x + 1$ are or are not parallel. Explain.

14. Write an equation for the line that passes through the point $(1, -6)$ and is parallel to the line $y - 2x = 4$. Graph both lines.

15. Write an equation for the line through $(4, 2)$ that is perpendicular to the line $x + y = 3$. Graph both lines.

16. State the domain and range of the relation $\{(0, 6), (-1, 4), (2, -3), (5, 6)\}$.

17. Given the relation $r = \{(2, -5), (3, -4), (2, 0), (1, -5)\}$. Is this relation a function? Explain.

18. For the function $f(x) = 3x - 4$, find   **19.** For the function $h(x) = 3x^2 - 5$, find

    a. $f(-1)$ **b.** $f(6)$ **a.** $h(-3)$ **b.** $h(2)$

Use the vertical line test to determine whether each graph in Exercises 20 – 23 does or does not represent a function. If the graph represents a function, state its domain and range.

20.   **21.**

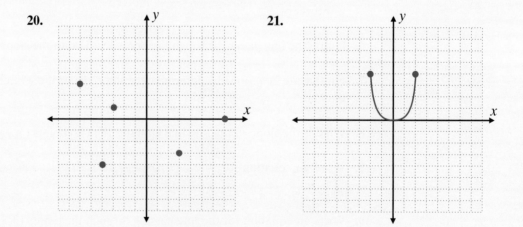

22. 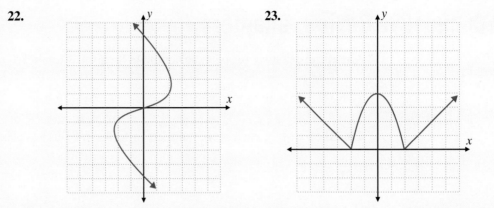 **23.**

24. Graph the half-plane that satisfies the inequality $y > -2x$. Is this half-plane open or closed?

25. Graph the half-plane $x \le 5$. Is this half-plane open or closed?

Cumulative Review: Chapters 1 – 4

Use the distributive property to complete the expression in Exercises 1 and 2.

1. $3x + 45 = 3(\quad)$

2. $6x + 16 = 2(\quad)$

Find the LCM for each set of numbers or terms in Exercises 3 and 4.

3. $\{12, 15, 54\}$

4. $\{6a^2, 24ab^2, 30ab, 40a^2b\}$

5. Find the average of the set of numbers: $\{25, 35, 42\}$

6. Find the average of the set of numbers: $\{3.6, 8.9, 14.7, 25.3\}$

7. Fernando scored 80, 88, and 82 on three exams in statistics. What was his average score on these exams (to the nearest tenth)?

8. Fran was flying her new airplane and descended from 12,000 feet to 8000 feet. What was her change in altitude?

In Exercises 9 – 12, determine whether the inequality is true or false. If the inequality is false, rewrite it in a form that is true using the same numbers or expressions.

9. $-15 > 5$

10. $|-6| \geq 6$

11. $\dfrac{7}{8} \leq \dfrac{7}{10}$

12. $-10 < 0$

Solve the equations in Exercises 13 – 16.

13. $7(4 - x) = 3(x + 4)$

14. $-2(5x + 1) + 2x = 4(x + 1)$

15. $\dfrac{2}{3}x + \dfrac{1}{2} = \dfrac{3}{4}$

16. $1.5x - 3.7 = 3.6x + 2.6$

In Exercises 17 – 20, solve for the indicated variable.

17. $C = \pi d$; solve for d.

18. $3x + 5y = 10$, solve for y.

19. $P = 2l + 2w$, solve for w.

20. $V = \dfrac{1}{3}\pi r^2 h$, solve for h.

In Exercises 21 – 24, solve the inequality for x and graph the solution set on a real number line.

21. $5x + 3 \geq 2x - 15$ **22.** $-16 < 3x + 5 < 17$

23. $\dfrac{1}{2}x + 5 \leq \dfrac{2}{3}x - 7$ **24.** $-15 < \dfrac{3}{4}x - 9 \leq \dfrac{1}{3}$

On a number line, graph the integers that satisfy each inequality in Exercises 25 and 26.

25. $|x| < 5$ **26.** $|x| > 6$

For each relation represented in Exercises 27 – 30, state the domain and range and state whether or not the relation represents a function.

27.

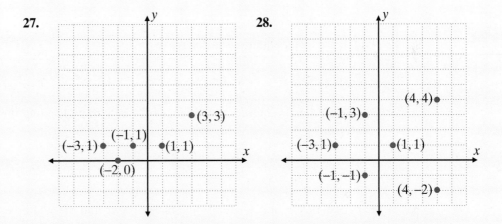

28.

29.

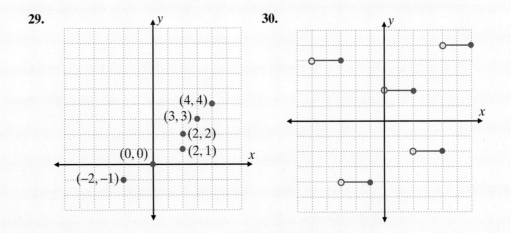

30.

In Exercises 31 and 32, (a) graph each of the relations, (b) state the domain and range, and (c) state whether the relation is or is not a function.

31. $\{(0,4), (2,2), (3,1), (-2,6), (0,-3)\}$ **32.** $\left\{\left(2, \frac{1}{2}\right), (-1, 3), \left(\frac{2}{3}, -2\right), (1, 1)\right\}$

Determine which of the given points lie on the line determined by the equations in Exercises 33 and 34.

33. $4x - y = 7$

    a. $(2, 1)$ **b.** $(3, 5)$ **c.** $(1, -3)$ **d.** $(0, 7)$

34. $2x + 5y = 6$

    a. $\left(\frac{1}{2}, 1\right)$ **b.** $(3, 0)$ **c.** $\left(0, \frac{6}{5}\right)$ **d.** $(-2, 2)$

Graph the linear equations in Exercises 35 – 38.

35. $y = -4x$ **36.** $x + 2y = 4$ **37.** $x + 4 = 0$ **38.** $2y - 6 = 0$

Graph the linear equations in Exercises 39 and 40 by locating the x-intercept and the y-intercept.

39. $2x + y = 4$ **40.** $5x - y = 1$

For Exercises 41 and 42, determine the slope, m, and the y-intercept, b, and then graph the line.

41. $y - 2x = 3$ **42.** $2x + 5y = 10$

Find the equation of the line determined by each pair of points in Exercises 43 and 44.

43. $(2, 4), (-1, 6)$ **44.** $(6, -2), (3, 2)$

In Exercises 45 – 48, graph the line and then write an equation (in slope-intercept form or standard form) for it.

45. $(6, -1), m = 0$ **46.** $\left(3, \frac{5}{2}\right), m = -\frac{1}{4}$

47. $(-3, 4)$, slope undefined   **48.** $(4, -5), m = \frac{3}{2}$

49. Write an equation for the line parallel to $2x + y = 5$ passing through $(1, -2)$.

50. Write an equation for the line parallel to $x - 2y = 5$ having the same y-intercept as $5x + 3y = 9$.

51. Write an equation for the line through $(0, 5)$ that is perpendicular to $3x + y = 9$. Graph both lines.

52. Write an equation for the line through $(-1, 6)$ that is perpendicular to $y = 3 - x$. Graph both lines.

53. Graph the half-plane $y \geq -2x + 3$. Is this half-plane open or closed?

54. Graph the half-plane $-3x + y < 6.3$. Is this half-plane open or closed?

55. For the function $f(x) = -3x + 4$, find **a.** $f(5)$ **b.** $f(-6)$

56. For the function $F(x) = 2x^2 - 5x + 1$, find **a.** $F(2)$ **b.** $F(-5)$

57. Find three consecutive even integers such that the sum of the second and third is equal to three times the first decreased by 14.

58. The length of a rectangle is 9 cm more than its width. The perimeter is 82 cm. Find the dimensions of the rectangle.

59. Twice the difference between a number and 16 is equal to four times the number increased by 6. What is the number?

60. To receive a B grade, a student must average 80 or more but less than 90. If Izumi received a B in the course and had four scores of 93, 77, 90, and 86 before taking the final exam, what were her possible scores for the final? (Assume that the maximum score on the final was 100.)

61. Acme Car Rental charges $0.25 for each mile driven plus $15 a day. Zenith Car Rental charges $50 a day with no extra charge for mileage. How many miles a day can you drive an Acme car and still keep the cost less than a Zenith car?

62. Find three consecutive odd integers such that the sum of the first and three times the second is 28 more than twice the third.

Systems of Linear Equations

Did You Know?

The subject of systems of linear equations received a great deal of attention in nineteenth-century mathematics. However, problems of this type are very old in the history of mathematics. The solution of simultaneous systems of equations was well known in China and Japan. The great Japanese mathematician Seki Shinsuku Kowa (1642-1708) wrote a book on the subject in 1683 that was well in advance of European work on the subject.

Seki Kowa is probably the most distinguished of all Japanese mathematicians. Born into a samurai family, he showed great mathematical talent at an early age. He is credited with the independent development of calculus and is sometimes referred to as "the Newton of Japan." There is a traditional story that Seki Kowa made a journey to the Buddhist shrines at Nara. There, ancient Chinese mathematical manuscripts were preserved, and Seki is supposed to have spent three years learning the contents of the manuscripts that previously no one had been able to understand. As was the custom, much of Seki Kowa's work was done through the solution of very intricate problems. In 1907, the Emperor of Japan presented a posthumous award to the memory of Seki Kowa, who did so much to awaken interest in scientific and mathematical research in Japan.

Native Japanese mathematics (the **wasan**) flourished until the nineteenth century when western mathematics and notation were completely adopted. In the 1940's, solution of large systems of equations became a part of the new branch of mathematics called **operations research**. Operations research is concerned with deciding how to best design and operate man-machine systems, usually under conditions requiring the allocation of scarce resources. Although this new science initially had only military applications, recent applications have been in the areas of business, industry, transportation, meteorology, and ecology. Computers now make it possible to solve extremely large systems of linear equations that are used to model the system being studied. Although computers do much of the work involved in solving systems of equations, it is necessary for you to understand the principles involved by studying small systems involving two unknowns as presented in Chapter 5.

5.1 Systems of Equations: Solutions by Graphing

5.2 Systems of Equations: Solutions by Substitution

5.3 Systems of Equations: Solutions by Addition

5.4 Applications: Distance-Rate-Time, Number Problems, Amounts and Costs

5.5 Applications: Interest and Mixture

5.6 Graphing Systems of Linear Inequalities

"The advancement and perfection of mathematics are intimately connected with the prosperity of the state."

Napoleon Bonaparte (1769 - 1821)

M any applications involve two (or more) quantities and, by using two (or more) variables, we can form linear equations using the information given. Such a set of equations is called a system, and in this chapter we will develop techniques for solving systems of linear equations.

Graphing systems of equations in two variables is helpful in visualizing the relationship between two equations. However, this approach is somewhat limited since numbers might be quite large or solutions might involve fractions that must be estimated on the graph. Therefore, other algebraic techniques are necessary. We will discuss two of these: substitution and addition. These algebraic techniques are also necessary when we try to solve more than two equations with more than two variables. These ideas and graphing in three dimensions will be left to later courses.

5.1   Systems of Equations: Solutions by Graphing

Objectives

After completing this section, you will be able to:

1. Determine if given points lie on both lines in specified systems of equations.

2. Determine, by graphing, if systems of linear equations are consistent, inconsistent, or dependent.

3. Estimate, by graphing, the coordinates of the intersections of consistent systems of linear equations.

Systems of Linear Equations

In Chapter 4, straight lines and their characteristics (slopes and intercepts) were analyzed in terms of linear equations. Many applications, as we will see in Sections 5.4 and 5.5, involve solving pairs of linear equations. Such pairs are said to form a **system of linear equations** (or a set of **simultaneous linear equations**). For example,

$$\begin{cases} x - 2y = 0 \\ 3x + y = 7 \end{cases} \qquad \text{and} \qquad \begin{cases} y = -x + 4 \\ y = 2x + 1 \end{cases}$$

are two systems of linear equations.

The **solution of a system** of linear equations is the set of ordered pairs (or points) that satisfy **both** equations. To determine whether or not a particular ordered pair is a solution to a system, substitute the values for x and y in **both** equations. If the results for both equations are true statements, then the ordered pair is a solution to the system.

Example 1: Solution of a System

Show that (2, 1) is a solution to the system $\begin{cases} x - 2y = 0 \\ 3x + y = 7 \end{cases}$.

Solution: Substitute $x = 2$ and $y = 1$ into **both** equations.
In the first equation:

$$2 - 2(1) \stackrel{?}{=} 0$$

$$2 - 2 = 0 \quad \text{a true statement}$$

In the second equation:

$$3(2) + 1 \stackrel{?}{=} 7$$

$$6 + 1 = 7 \quad \text{a true statement}$$

Because (2, 1) satisfies both equations, **(2, 1) is a solution to the system**.

Example 2: Not a Solution of a System

Show that (0, 4) is not a solution to the system $\begin{cases} y = -x + 4 \\ y = 2x + 1 \end{cases}$.

Solution: Substitute $x = 0$ and $y = 4$ into **both** equations.
In the first equation:

$$4 \stackrel{?}{=} -(0) + 4$$

$$4 = 0 + 4 \quad \text{a true statement}$$

In the second equation:

$$4 \stackrel{?}{=} 2(0) + 1$$

$$4 = 0 + 1 \quad \text{a false statement}$$

Because (0, 4) does not satisfy both equations, **(0, 4) is NOT a solution to the system**.

Solutions by Graphing

In this chapter, we will discuss three methods for solving systems of linear equations. The first method is to **solve by graphing**. In this method, both equations are graphed and the point of intersection (if there is one) is the solution to the system. Geometrically there are three possibilities for the graphs of two lines relative to each other. What do you think that these three possibilities are? The following discussion analyzes the three situations.

A Consistent System:

In Example 2, we showed that $(0, 4)$ is not a solution to the system $\begin{cases} y = -x + 4 \\ y = 2x + 1 \end{cases}$.

However, by graphing both equations as shown in Figure 5.1, we see that the two lines appear to **intersect** at the point $(1, 3)$.

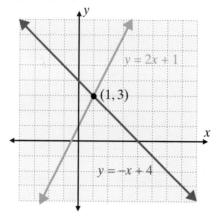

Figure 5.1

We can check that $(1, 3)$ satisfies both equations by substituting $x = 1$ and $y = 3$ into **both equations**.

$$y = 2x + 1$$
$$3 \overset{?}{=} 2 \cdot 1 + 1$$
$$3 = 3$$

$$y = -x + 4$$
$$3 \overset{?}{=} -1 + 4$$
$$3 = 3$$

Since both substitutions give true statements, the lines do intersect at the point $(1, 3)$, the system is said to be **consistent**, and **(1, 3) is the solution to the system**.

An Inconsistent System:

The graphs in Figure 5.2 indicate that the lines are parallel and do not intersect. The system is said to be **inconsistent**, and the system has **no solution**.

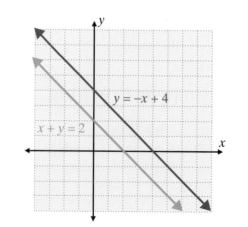

Figure 5.2

We can show that the lines in Figure 5.2 are **parallel** by writing both equations in the slope-intercept form and noting that they have the same slope and different y-intercepts. Solving for y in both equations gives the following:

first equation: $y = -x + 4$ with $m = -1$ and y-intercept 4
second equation: $y = -x + 2$ with $m = -1$ and y-intercept 2

Thus the lines are **parallel** and do not intersect, and there is **no solution** to the system.

A Dependent System:

For the system $\begin{cases} y = -x + 4 \\ 2y + 2x = 8 \end{cases}$, the graphs of both lines are shown in Figure 5.3.

Why is there just one line?

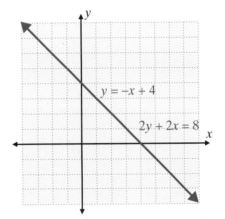

Figure 5.3

There is just one line in Figure 5.3 because both equations represent the same line. The lines not only intersect, they are the same line. That is, they **coincide**. Any point that satisfies one equation will also satisfy the other, and the system has an infinite number of solutions. Putting both equations in the slope-intercept form shows that they are identical.

Solving $2y + 2x = 8$ for y gives:

$$2y + 2x = 8$$
$$2y = -2x + 8$$
$$\frac{2y}{2} = \frac{-2x}{2} + \frac{8}{2}$$
$$y = -x + 4$$

This is the same as the first equation: $y = -x + 4$.

In this case we say that the system is **dependent** and the **solution is the set of all points of the form $(x, -x + 4)$**. In general, in a dependent system there is an infinite number of points in the solution and the points are of the form (x, y) where one of the equations is solved for y and this expression is substituted for y. For equations in the form $y = mx + b$, the solutions can be written in the form $(x, mx + b)$ for all real values of x.

The following table and graphs summarize the basic ideas and terminology related to systems of linear equations.

Solutions By Graphing

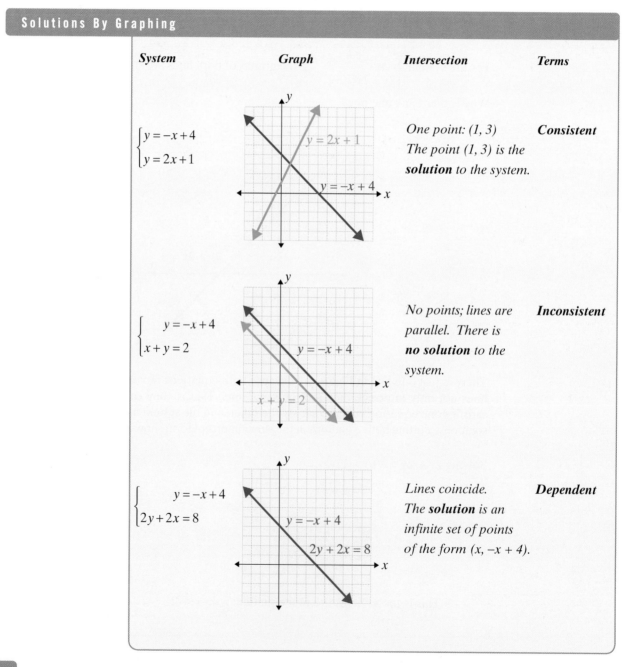

System	Graph	Intersection	Terms
$\begin{cases} y = -x + 4 \\ y = 2x + 1 \end{cases}$	$y = 2x + 1$ $y = -x + 4$	One point: $(1, 3)$ The point $(1, 3)$ is the **solution** to the system.	**Consistent**
$\begin{cases} y = -x + 4 \\ x + y = 2 \end{cases}$	$y = -x + 4$ $x + y = 2$	No points; lines are parallel. There is **no solution** to the system.	**Inconsistent**
$\begin{cases} y = -x + 4 \\ 2y + 2x = 8 \end{cases}$	$y = -x + 4$ $2y + 2x = 8$	Lines coincide. The **solution** is an infinite set of points of the form $(x, -x + 4)$.	**Dependent**

NOTES

The graphing method of solving a system of linear equations shows the geometric relationship between the two lines. However, determining the exact point of intersection on a graph can be difficult or impossible. In particular, when the solution of a system involves fractions or roots, the point of intersection can only be estimated. Algebraic methods, which give exact solutions, will be discussed in Sections 5.2 and 5.3.

Example 3: Solving by Graphing

Solve the following systems graphically and state whether the systems are consistent, inconsistent, or dependent.

a. $\begin{cases} y - x = 0 \\ 2x + y = 3 \end{cases}$
b. $\begin{cases} x - 2y = 1 \\ x + 3y = 0 \end{cases}$
c. $\begin{cases} y = 3x \\ y - 3x = -4 \end{cases}$
d. $\begin{cases} x + 2y = 6 \\ y = -\dfrac{1}{2}x + 3 \end{cases}$

Solutions:

a. The system is consistent.

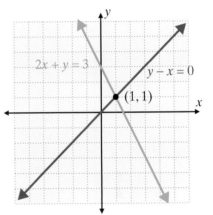

The point of intersection is $(1, 1)$.
That is, $(1,1)$ is the solution of the system.

Check: Substitute $x = 1$ and $y = 1$
into **both equations**.
For $y - x = 0$, we have $1 - 1 = 0$.
For $2x + y = 3$,
we have $2 \cdot 1 + 1 = 3$.

b. The system is consistent, but the point of intersection can be only estimated. This example points out the main weakness in solving a system graphically. In such cases, any reasonable estimate will be acceptable. For example, if you estimated

$\left(\dfrac{1}{2}, -\dfrac{1}{4} \right)$ or $\left(\dfrac{3}{4}, -\dfrac{1}{3} \right)$, or some such point,

your answer is acceptable. The actual point of intersection (and the solution to

the system) is $\left(\dfrac{3}{5}, -\dfrac{1}{5} \right)$. We will be able

to locate this point precisely using the techniques of the next section.

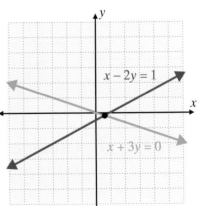

Continued on next page...

407

c. The system is inconsistent. The lines are parallel with the same slope, 3, and there are no points of intersection.

There is no solution to the system.

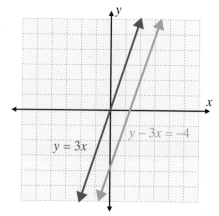

d. The system is dependent. All points that lie on one line also lie on the other line. For example, $(4, 1)$ is a point on the line $x + 2y = 6$ since $4 + 2(\,1\,) = 6$. The point $(4, 1)$ is also on the line $y = -\dfrac{1}{2}x + 3$ since $1 = -\dfrac{1}{2}(4) + 3$. Therefore, the solution can be stated as all of the points that satisfy the equation $x + 2y = 6$.

The solution is the set of all points of the form $\left(x, -\dfrac{1}{2}x + 3\right)$, an infinite set of solutions.

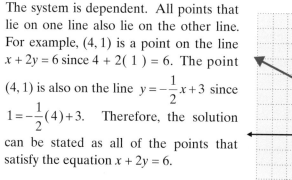

Remember:
1. To use the graphing method, graph the lines as accurately as you can.
2. Be sure to check your solution by substituting back into both of the original equations. (Of course, fractional estimates may not check exactly.)

Using a TI-84 Plus Graphing Calculator to Solve a System of Linear Equations

A graphing calculator can be used to locate (or estimate) the point of intersection of two lines (and therefore the solution to the system.)

Example 4: Using a Graphing Calculator

Consider the system $\begin{cases} 2x + y = 8 \\ x - y = 1 \end{cases}$ and proceed as follows:

Step 1: Solve each equation for y. For this system $\begin{cases} y = -2x + 8 \\ y = x - 1 \end{cases}$.

Step 2: Press 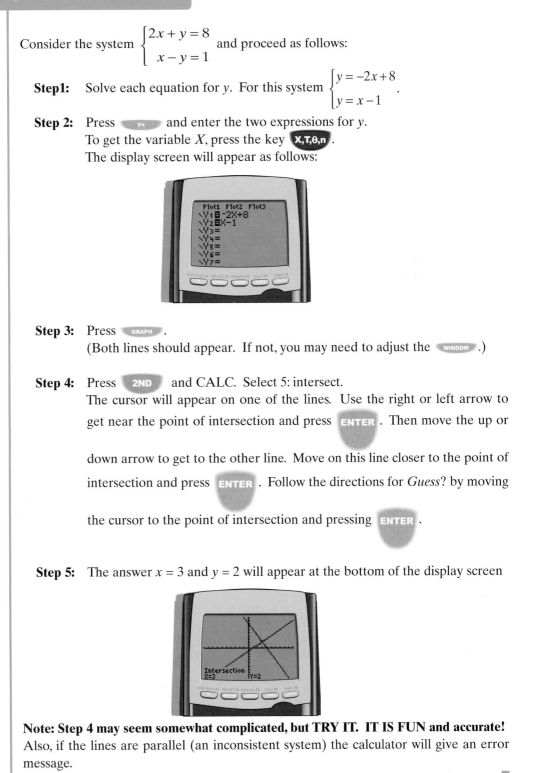 and enter the two expressions for y. To get the variable X, press the key **X,T,θ,n**. The display screen will appear as follows:

Step 3: Press **GRAPH**.
(Both lines should appear. If not, you may need to adjust the **WINDOW**.)

Step 4: Press **2ND** and CALC. Select 5: intersect. The cursor will appear on one of the lines. Use the right or left arrow to get near the point of intersection and press **ENTER**. Then move the up or down arrow to get to the other line. Move on this line closer to the point of intersection and press **ENTER**. Follow the directions for *Guess?* by moving the cursor to the point of intersection and pressing **ENTER**.

Step 5: The answer $x = 3$ and $y = 2$ will appear at the bottom of the display screen

Note: Step 4 may seem somewhat complicated, but TRY IT. IT IS FUN and accurate!
Also, if the lines are parallel (an inconsistent system) the calculator will give an error message.

5.1 Exercises

In Exercises 1 – 4, determine which of the given points, if any, lie on both of the lines in the systems of equations by substituting each point into both equations.

1. $\begin{cases} x - y = 6 \\ 2x + y = 0 \end{cases}$

    a. $(1, -2)$
    b. $(4, -2)$
    c. $(2, -4)$
    d. $(-1, 2)$

2. $\begin{cases} x + 3y = 5 \\ 3y = 4 - x \end{cases}$

    a. $(2, 1)$
    b. $(2, -2)$
    c. $(-1, 2)$
    d. $(4, 0)$

3. $\begin{cases} 2x + 4y - 6 = 0 \\ 3x + 6y - 9 = 0 \end{cases}$

    a. $(1, 1)$
    b. $(2, 0)$
    c. $\left(0, \dfrac{3}{2}\right)$
    d. $(-1, 3)$

4. $\begin{cases} 5x - 2y - 5 = 0 \\ 5x = -3y \end{cases}$

    a. $(1, 0)$
    b. $\left(\dfrac{3}{5}, -1\right)$
    c. $(0, 0)$
    d. $(1, 4)$

In Exercises 5 – 8, the graphs of the lines represented by each system of equations are given. Determine the solution of the system by looking at the graph. Check your solution by substituting into both equations.

5. $\begin{cases} x + 2y = 4 \\ x - y = -2 \end{cases}$
  6. $\begin{cases} x + 2y = 1 \\ 2x + y = -1 \end{cases}$
  7. $\begin{cases} 2x - y = 6 \\ 3x + y = 14 \end{cases}$
  8. $\begin{cases} x - y = 4 \\ 3x - y = 6 \end{cases}$

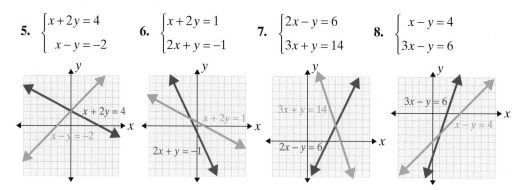

In Exercises 9 – 12, show that each system of equations is inconsistent by determining the slope of each line and the y-intercept. (That is, show that the lines are parallel and do not intersect.)

9. $\begin{cases} 2x + y = 3 \\ 4x + 2y = 5 \end{cases}$
  10. $\begin{cases} y = \dfrac{1}{2}x + 3 \\ x - 2y = 1 \end{cases}$
  11. $\begin{cases} 3x - y = 8 \\ x - \dfrac{1}{3}y = 2 \end{cases}$
  12. $\begin{cases} 3x - 5y = 1 \\ 6x - 10y = 4 \end{cases}$

Solve the following systems graphically and state whether the systems are consistent, inconsistent, or dependent.

13. $\begin{cases} 2x - y = 4 \\ 3x + y = 6 \end{cases}$

14. $\begin{cases} x + y - 5 = 0 \\ x - 4y = 5 \end{cases}$

15. $\begin{cases} 3x - y = 6 \\ y = 3x \end{cases}$

16. $\begin{cases} x - y = 5 \\ x = -3 \end{cases}$

17. $\begin{cases} x + 2y = 8 \\ 3x - 2y = 0 \end{cases}$

18. $\begin{cases} 5x - 4y = 5 \\ 8y = 10x - 10 \end{cases}$

19. $\begin{cases} 4x - 2y = 10 \\ -6x + 3y = -15 \end{cases}$

20. $\begin{cases} x = 5 \\ y = -1 \end{cases}$

21. $\begin{cases} y = 2x + 5 \\ 4x - 2y = 7 \end{cases}$

22. $\begin{cases} \dfrac{1}{2}x + 2y = 7 \\ 2x = 4 - 8y \end{cases}$

23. $\begin{cases} 4x + 3y + 7 = 0 \\ 5x - 2y + 3 = 0 \end{cases}$

24. $\begin{cases} 2x + 3y = 4 \\ 4x - y = 1 \end{cases}$

25. $\begin{cases} 7x - 2y = 1 \\ y = 3 \end{cases}$

26. $\begin{cases} 2x - 5y = 6 \\ y = \dfrac{2}{5}x + 1 \end{cases}$

27. $\begin{cases} y = 4x - 3 \\ x = 2y - 8 \end{cases}$

28. $\begin{cases} y = \dfrac{1}{2}x + 2 \\ x - 2y + 4 = 0 \end{cases}$

29. $\begin{cases} y = 7 \\ x = 8 \end{cases}$

30. $\begin{cases} 2x + 3y = 5 \\ 3x - 2y = 1 \end{cases}$

31. $\begin{cases} \dfrac{2}{3}x + y = 2 \\ x - 4y = 3 \end{cases}$

32. $\begin{cases} x + y = 8 \\ 5y = 2x + 5 \end{cases}$

33. $\begin{cases} x - 2y = 4 \\ x = 4 \end{cases}$

34. $\begin{cases} y = 2x \\ 2x + y = 4 \end{cases}$

35. $\begin{cases} x - y = 4 \\ 2y = 2x - 4 \end{cases}$

36. $\begin{cases} y = x + 1 \\ y + x = -5 \end{cases}$

37. $\begin{cases} x + y = 4 \\ 2x - 3y = 3 \end{cases}$

38. $\begin{cases} x = 1.5 \\ x - 3y = 9 \end{cases}$

39. $\begin{cases} 2x + y = 0 \\ 4x + y = -2 \end{cases}$

40. $\begin{cases} 4x + y = 6 \\ 2x + \dfrac{1}{2}y = 3 \end{cases}$

41. $\begin{cases} \dfrac{1}{2}x + \dfrac{1}{3}y = \dfrac{1}{6} \\ \dfrac{1}{4}x + \dfrac{1}{4}y = 0 \end{cases}$

42. $\begin{cases} \dfrac{1}{4}x - y = \dfrac{13}{4} \\ \dfrac{1}{3}x + \dfrac{1}{6}y = -\dfrac{1}{6} \end{cases}$

For Exercises 43 – 46, a word problem is stated with equations given that represent a mathematical model for the problem. Solve the system graphically.

43. The sum of two numbers is 25 and their difference is 15. What are the two numbers?
Let x = one number and y = the other number.

The corresponding modeling system is $\begin{cases} x + y = 25 \\ x - y = 15 \end{cases}$.

44. The perimeter of a rectangle is 50 meters and the length is 5 meters longer than the width. Find the dimensions of the rectangle.
Let x = the length and y = the width.

The corresponding modeling system is $\begin{cases} 2x + 2y = 50 \\ x - y = 5 \end{cases}$.

perimeter = 50 m

45. Ten gallons of a salt solution consists of 30% salt. It is the result of mixing a 50% solution with a 25% solution. How many gallons of each of the mixing solutions was used?
Let x = the number of gallons of the 50% solution
and y = the number of gallons of the 25% solution.

The corresponding modeling system is $\begin{cases} x + y = 10 \\ 0.50x + 0.25y = 0.30(10) \end{cases}$.

46. A student bought a calculator and a textbook for a course in algebra. He told his friend that the total cost was $170 (without tax) and that the calculator cost $20 more than twice the cost of the textbook. What was the cost of each item?
Let c = the cost of the calculator
and t = the cost of the textbook.
The corresponding modeling system is $\begin{cases} c + t = 170 \\ c = 2t + 20 \end{cases}$.

In Exercises 47 – 54, use a TI-84 Plus graphing calculator to solve each system of equations. Remember to solve each equation for y and follow the steps outlined in Example 4. Use the `2ND`, CALC, intersect sequence of steps to locate the solution point on the display screen.

47. $\begin{cases} x + 2y = 9 \\ x - 2y = -7 \end{cases}$ **48.** $\begin{cases} x - 3y = 0 \\ 2x + y = 7 \end{cases}$ **49.** $\begin{cases} y = 2 \\ 2x - 3y = -3 \end{cases}$ **50.** $\begin{cases} 2x - 3y = 0 \\ 3x + 3y = \dfrac{5}{2} \end{cases}$

51. $\begin{cases} y = -3 \\ 2x + y = 0 \end{cases}$ **52.** $\begin{cases} 2x - 3y = 1.25 \\ x + 2y = 5 \end{cases}$ **53.** $\begin{cases} x + y = 3.5 \\ -2x + 5y = 7.7 \end{cases}$ **54.** $\begin{cases} 4x + y = -0.5 \\ x + 2y = -8 \end{cases}$

Hawkes Learning Systems: Introductory Algebra

Solving Systems of Linear Equations by Graphing

5.2   Systems of Equations: Solutions by Substitution

After completing this section, you will be able to:

Solve systems of linear equations by substitution.

As we discussed in Section 5.1, solving systems of linear equations by graphing is somewhat limited in accuracy. The graphs must be precisely drawn and even then the points of intersection (if there are any) can be difficult to estimate accurately. In this section, we will develop an algebraic method called the **method of substitution**. This method involves **solving one of the equations for one of the variables** and then **substituting the resulting expression into the other equation**. The objective is to get one equation in one variable. For example, consider the system

$$\begin{cases} y = -2x + 5 \\ x + 2y = 1 \end{cases}$$

How would you substitute? Since the first equation is already solved for y, a reasonable substitution would be to put $-2x + 5$ for y in the second equation. Try this and see what happens.

First equation already solved for y: $y = -2x + 5$

Substitute into second equation: $x + 2y = 1$

$$x + 2(-2x + 5) = 1$$

We now have one equation in only one variable, namely x. **The problem has been reduced from one of solving two equations in two variables to solving one equation in one variable.** Solve this equation for x. Then find the corresponding y-value by substituting this x-value into **either of the two original equations**.

$$x + 2(-2x + 5) = 1$$
$$x - 4x + 10 = 1$$
$$-3x = -9$$
$$x = 3$$

Substituting $x = 3$ into $y = -2x + 5$ gives

$$y = -2 \cdot 3 + 5$$
$$= -6 + 5$$
$$= -1.$$

Thus the system is consistent and the solution to the system is $x = 3$ and $y = -1$, or the point $(3, -1)$.

Substitution is not the only algebraic technique for solving a system of linear equations. It does work in all cases but generally is used only when one of the equations is easily solved for one variable.

In the following examples, note how the results in Example 1c indicate that the system is inconsistent and the results in Example 1d indicate that the system is dependent.

Example 1: Solve by Substitution

Solve the following systems of linear equations by using the technique of substitution.

a. $\begin{cases} x = -5 \\ y = 2x + 9 \end{cases}$
b. $\begin{cases} y = \dfrac{5}{6}x + 2 \\ \dfrac{1}{6}x + y = 8 \end{cases}$
c. $\begin{cases} 3x + y = 1 \\ 6x + 2y = 3 \end{cases}$
d. $\begin{cases} x - 2y = 1 \\ 3x - 6y = 3 \end{cases}$

Solutions:

a. $\begin{cases} x = -5 \\ y = 2x + 9 \end{cases}$

The first equation is already solved for x. Substituting -5 for x in the second equation gives:

$$y = 2(-5) + 9 = -10 + 9 = -1$$

The system is **consistent** and the **solution** is $x = -5$ and $y = -1$, or $(-5, -1)$.

Continued on next page...

b. $\begin{cases} y = \dfrac{5}{6}x + 2 \\ \dfrac{1}{6}x + y = 8 \end{cases}$

The first equation is already solved for y. Substituting $\dfrac{5}{6}x + 2$ for y in the second equation gives

$$\frac{1}{6}x + \left(\frac{5}{6}x + 2\right) = 8$$

$$\mathbf{6} \cdot \frac{1}{6}x + \mathbf{6} \cdot \left(\frac{5}{6}x + 2\right) = \mathbf{6} \cdot 8 \qquad \text{Multiply both sides by } \mathbf{6}, \text{ the LCD.}$$

$$x + 5x + 12 = 48$$

$$6x + 12 = 48$$

$$6x = 36$$

$$x = 6$$

Substituting 6 for x in the first equation gives the corresponding value for y:

$$y = \frac{5}{6} \cdot 6 + 2 = 5 + 2 = 7$$

The system is **consistent** and the **solution** is $(6, 7)$.

c. $\begin{cases} 3x + y = 1 \\ 6x + 2y = 3 \end{cases}$

Solving the first equation for y gives $y = 1 - 3x$. Substituting $1 - 3x$ for y in the second equation gives:

$$6x + 2\left(1 - 3x\right) = 3$$

$$6x + 2 - 6x = 3$$

$$2 = 3$$

This last equation ($2 = 3$) is never true. This tells us that the system is **inconsistent** and there is **no solution**. Graphically, the lines are parallel and there is no intersection.

d. $\begin{cases} x - 2y = 1 \\ 3x - 6y = 3 \end{cases}$

Solving the first equation for x gives $x = 1 + 2y$. Substituting $1 + 2y$ for x in the second equation gives:

$$3(1 + 2y) - 6y = 3$$
$$3 + 6y - 6y = 3$$
$$3 = 3$$

This last equation ($3 = 3$) is always true. This tells us that the system is **dependent** and the **solutions** are of the form $(1 + 2y, y)$ for all values of y (Or, solving one of the equations for y we have $\left(x, \dfrac{1}{2}x - \dfrac{1}{2} \right)$ for all values of x.)

NOTES

Note that there are two forms for the solutions of a dependent system, as illustrated in example 1d. In one form we can solve for x as illustrated with $(1 + 2y, y)$ or we can solve for y as illustrated with $\left(x, \dfrac{1}{2}x - \dfrac{1}{2} \right)$.

Practice Problems

Solve the following systems using the technique of substitution.

1. $\begin{cases} x + y = 3 \\ y = 2x \end{cases}$ 　　 **2.** $\begin{cases} y = 3x - 1 \\ 2x + y = 4 \end{cases}$ 　　 **3.** $\begin{cases} x + 2y = -1 \\ x - 4y = -4 \end{cases}$

5.2 Exercises

Solve the systems of linear equations in Exercises 1 – 42 by using the technique of substitution. If the system is inconsistent or dependent, say so in your answer.

1. $\begin{cases} x + y = 6 \\ y = 2x \end{cases}$ 　　 **2.** $\begin{cases} 5x + 2y = 21 \\ x = y \end{cases}$ 　　 **3.** $\begin{cases} x - 7 = 3y \\ y = 2x - 4 \end{cases}$

4. $\begin{cases} y = 3x + 4 \\ 2y = 3x + 5 \end{cases}$ 　　 **5.** $\begin{cases} x = 3y \\ 3y - 2x = 6 \end{cases}$ 　　 **6.** $\begin{cases} 4x = y \\ 4x - y = 7 \end{cases}$

Answer to Practice Problems: **1.** $x = 1, y = 2$ **2.** $x = 1, y = 2$ **3.** $x = -2, \ y = \dfrac{1}{2}$

7. $\begin{cases} x - 5y + 1 = 0 \\ \quad\quad x = 7 - 3y \end{cases}$

8. $\begin{cases} 2x + 5y = 15 \\ \quad\quad x = y - 3 \end{cases}$

9. $\begin{cases} 7x + y = 9 \\ \quad\quad y = 4 - 7x \end{cases}$

10. $\begin{cases} 3y + 5x = 5 \\ \quad\quad y = 3 - 2x \end{cases}$

11. $\begin{cases} 3x - y = 7 \\ \quad x + y = 5 \end{cases}$

12. $\begin{cases} 4x - 2y = 5 \\ \quad\quad y = 2x + 3 \end{cases}$

13. $\begin{cases} 6x + 3y = 9 \\ \quad\quad y = 3 - 2x \end{cases}$

14. $\begin{cases} 15x + 5y = 20 \\ \quad\quad y = -3x + 4 \end{cases}$

15. $\begin{cases} \quad x - y = 5 \\ 2x + 3y = 0 \end{cases}$

16. $\begin{cases} \quad 4x = 8 \\ 3x + y = 8 \end{cases}$

17. $\begin{cases} \quad x + y = 8 \\ 3x + 2y = 8 \end{cases}$

18. $\begin{cases} \quad\quad y = 2x - 5 \\ 2x + y = -3 \end{cases}$

19. $\begin{cases} 2x + 3y = 5 \\ \quad x - 6y = 0 \end{cases}$

20. $\begin{cases} \quad\quad 2y = 5 \\ 3x - 4y = -4 \end{cases}$

21. $\begin{cases} x + 5y = 1 \\ x - 3y = 5 \end{cases}$

22. $\begin{cases} 3x + 8y = -2 \\ \quad x + 2y = -1 \end{cases}$

23. $\begin{cases} 4x - 4y = 9 \\ 3x + y = 2 \end{cases}$

24. $\begin{cases} 5x + 2y = -10 \\ \quad 7x = 4 - y \end{cases}$

25. $\begin{cases} \quad x - 2y = -4 \\ 3x + y = -5 \end{cases}$

26. $\begin{cases} \quad x + 4y = 3 \\ 3x - 4y = 7 \end{cases}$

27. $\begin{cases} 3x - y = -1 \\ 7x - 4y = 0 \end{cases}$

28. $\begin{cases} \quad x + 5y = -1 \\ 2x + 7y = 1 \end{cases}$

29. $\begin{cases} \quad x + 3y = 5 \\ 3x + 2y = 7 \end{cases}$

30. $\begin{cases} 3x - 4y - 39 = 0 \\ 2x - y - 13 = 0 \end{cases}$

31. $\begin{cases} \dfrac{x}{3} + \dfrac{y}{5} = 1 \\ x + 6y = 12 \end{cases}$

32. $\begin{cases} \dfrac{x}{5} + \dfrac{y}{4} - 3 = 0 \\ \dfrac{x}{10} - \dfrac{y}{2} + 1 = 0 \end{cases}$

33. $\begin{cases} \quad 6x - y = 15 \\ 0.2x + 0.5y = 2.1 \end{cases}$

34. $\begin{cases} \quad x + 2y = 3 \\ 0.4x + y = 0.6 \end{cases}$

35. $\begin{cases} 0.2x - 0.1y = 0 \\ \quad\quad y = x + 10 \end{cases}$

36. $\begin{cases} 0.1x - 0.2y = 1.4 \\ \quad 3x + y = 14 \end{cases}$

37. $\begin{cases} 3x - 2y = 5 \\ \quad\quad y = \dfrac{3}{2}x + 2 \end{cases}$

38. $\begin{cases} \quad\quad x = 2y - 7.5 \\ 2x + 4y = -15 \end{cases}$

39. $\begin{cases} \dfrac{1}{2}x + \dfrac{1}{3}y = 4 \\ 3x + 2y = 24 \end{cases}$

40. $\begin{cases} \dfrac{1}{3}x + \dfrac{1}{7}y = 2 \\ 7x + 3y = 42 \end{cases}$

41. $\begin{cases} \dfrac{1}{4}x - \dfrac{3}{2}y = 5 \\ -x + 6y = 20 \end{cases}$

42. $\begin{cases} \dfrac{-4}{3}x + 2y = 7 \\ \dfrac{8}{3}x - 4y = -5 \end{cases}$

For Exercises 43 – 46, a word problem is stated with equations given that represent a mathematical model for the problem. Solve the system by using the method of substitution. (**Note:** These exercises were also given in Section 5.1. Check to see that you arrived at the same answers by both methods.)

43. The sum of two numbers is 25 and their difference is 15. What are the two numbers?
Let x = one number and y = the other number.

The corresponding modeling system is $\begin{cases} x+y=25 \\ x-y=15 \end{cases}$.

44. The perimeter of a rectangle is 50 meters and the length is 5 meters longer than the width. Find the dimensions of the rectangle.
Let x = the length and y = the width.

The corresponding modeling system is $\begin{cases} 2x+2y=50 \\ x-y=5 \end{cases}$.

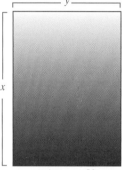

perimeter = 50 m

45. Ten gallons of a salt solution consists of 30% salt. It is the result of mixing a 50% solution with a 25% solution. How many gallons of each of the mixing solutions were used?
Let x = the number of gallons of the 50% solution and y = the number of gallons of the 25% solution.

The corresponding modeling system is $\begin{cases} x+y=10 \\ 0.50x+0.25y=0.30(10) \end{cases}$.

46. A student bought a calculator and a textbook for a course in algebra. He told his friend that the total cost was $170 (without tax) and that the calculator cost $20 more than twice the cost of the textbook. What was the cost of each item?
Let c = the cost of the calculator and t = the cost of the textbook.

The corresponding modeling system is $\begin{cases} c+t=170 \\ c=2t+20 \end{cases}$.

Hawkes Learning Systems: Introductory Algebra

Solving Systems of Linear Equations By Substitution

Systems of Equations: Solutions by Addition

After completing this section, you will be able to:

1. *Solve systems of linear equations by addition.*

2. *Use systems of equations to find the equation of a line through two given points.*

We have discussed two methods for solving systems of linear equations:

1. by **graphing**
2. by **substitution**

We know that solutions by graphing are not necessarily exact; and, in some cases, the method of substitution can lead to complicated algebraic steps. In this section, we consider a third method:

3: **addition** (or **method of elimination**)

A Consistent System

Consider solving the following system where both equations are written in standard form:

$$\begin{cases} x - 2y = -9 \\ x + 2y = 11 \end{cases}.$$

In the **method of addition**, we write one equation under the other so that like terms are aligned vertically. (Note that, in this example, the coefficients of y are opposites, namely -2 and $+2$.)

Then add like terms as follows:

$$x - 2y = -9$$

$$\underline{x + 2y = 11}$$

$$2x \quad = 2 \qquad \text{The } y \text{ terms are eliminated because the coefficients are opposites.}$$

$$x \quad = 1$$

Now substitute $x = 1$ into either of the original equations and solve for y:

$$1 - 2y = -9$$

$$-2y = -10$$

$$y = 5$$

Therefore, the system is **consistent** and the **solution** to the system is $x = 1$ and $y = 5$ or $(1, 5)$.

A Consistent System

To make two coefficients opposites (either for x or for y), we can multiply each equation by some non-zero constant. Then, when we add like terms, at least one of the variables will be eliminated. For example, consider the following system:

$$\begin{cases} 4x - 3y = 1 \\ 3x - 2y = 4 \end{cases}$$

Eliminating y-terms
Multiplying **all terms** in the first equation by 2 and **all terms** in the second equation by -3 will give opposite y-coefficients(-6 and $+6$). Adding like terms eliminates y as follows:

$$\begin{cases} [2] \quad 4x - 3y = 1 \rightarrow \quad 8x - 6y = \quad 2 \\ [-3] \ 3x - 2y = 4 \rightarrow \underline{-9x + 6y = -12} \end{cases}$$

$$-x \quad\quad = -10$$
$$x \quad\quad = \ 10$$

Substitute $x = 10$ into one of the original equations to find the corresponding y-value.

$$4x - 3y = 1$$
$$4 \cdot 10 - 3y = 1$$
$$40 - 3y = 1$$
$$-3y = -39$$
$$y = 13$$

Thus the solution is $x = 10$ and $y = 13$ or $(10, 13)$.

Eliminating x-terms
Multiplying all terms in the first equation by -3 and all terms in the second equation by 4 will give opposite x-coefficients (-12 and $+12$). Adding like terms eliminates x as follows:

$$\begin{cases} [-3] \quad 4x - 3y = 1 \rightarrow -12x + 9y = -3 \\ [4] \quad\ 3x - 2y = 4 \rightarrow \underline{\ 12x - 8y = 16} \end{cases}$$

$$y = 13$$

Substitute $y = 13$ into one of the original equations to find the corresponding x-value.

$$4x - 3y = 1$$
$$4x - 3 \cdot 13 = 1$$
$$4x - 39 = 1$$
$$4x = 40$$
$$x = 10$$

We have the same solution as before: $(10, 13)$.

In either case, we see that this system is **consistent** and the solution is $(10, 13)$.

What if both the x- and y-terms are eliminated? In this situation, the system is either inconsistent or dependent. We solve one of each type.

An Inconsistent System

For the system

$$\begin{cases} 2x - y = 6 \\ 4x - 2y = 1 \end{cases}$$

multiplying the terms in the first equation by -2 and then adding gives

$$\begin{cases} [-2]\ 2x - y = 6 & \rightarrow & -4x + 2y = -12 \\ \quad\ 4x - 2y = 1 & \rightarrow & \underline{\ 4x - 2y = \quad 1} \\ & & 0 = -11 \end{cases}$$

Since the equation $0 = -11$ is not true, the system is **inconsistent**. There is **no solution**.

A Dependent System

For the system

$$\begin{cases} 2x - 2y = 1 \\ 3x - 3y = \dfrac{3}{2} \end{cases}$$

multiplying the terms in the first equation by 3 and the terms in the second equation by -2 and then adding gives

$$\begin{cases} [3]\ 2x - 2y = 1 & \rightarrow & 6x - 6y = \quad 3 \\ [-2]3x - 3y = \dfrac{3}{2} & \rightarrow & \underline{-6x + 6y = -3} \\ & & 0 = \quad 0 \end{cases}$$

Since the equation $0 = 0$ is always true, the system is **dependent**.

The solution consists of all points that satisfy the equation $2x - 2y = 1$ and, by solving for y, can be written in the general form $\left(x, x - \dfrac{1}{2}\right)$.

Solve each of the following systems by using the method of addition.

a. $\begin{cases} 5x + 3y = -3 \\ 2x - 7y = 7 \end{cases}$ **b.** $\begin{cases} y = 4x - 2 \\ 8x - 2y = 4 \end{cases}$ **c.** $\begin{cases} x + 0.4y = 3.08 \\ 0.1x - y = 0.1 \end{cases}$

Solutions:

a. $\begin{cases} 5x + 3y = -3 \\ 2x - 7y = 7 \end{cases}$

$$\begin{cases} [-2] \ 5x + 3y = -3 \rightarrow -10x - 6y = 6 \\ [5] \ \ \ 2x - 7y = 7 \ \rightarrow \ \underline{10x - 35y = 35} \\ -41y = 41 \\ y = -1 \end{cases}$$

Substitute $y = -1$ into one of the original equations.

$$5x + 3y = -3$$
$$5x + 3(-1) = -3$$
$$5x - 3 = -3$$
$$5x = 0$$
$$x = 0$$

The solution is $x = 0$ and $y = -1$, or $(0, -1)$.

b. $\begin{cases} y = 4x - 2 \\ 8x - 2y = 4 \end{cases}$

Rearranging so that both equations are in standard form gives

$$\begin{cases} -4x + y = -2 \\ 8x - 2y = 4 \end{cases}$$

Continued on next page...

Then

$$\begin{cases} & -4x+ \; y=-2 \; \rightarrow \; -4x+y=-2 \\ \left[\dfrac{1}{2}\right] & 8x-2y= \; 4 \; \rightarrow \; \underline{4x-y= \; 2} \end{cases}$$

$$0=0$$

The system is dependent. The solution is the set of all points that satisfy the equation $y = -2 + 4x$ (or the equation $8x - 2y = 4$).

c. $\begin{cases} x+0.4y = 3.08 \\ 0.1x - y = 0.1 \end{cases}$

$$\begin{cases} & x+0.4y = 3.08 \; \rightarrow \; 1.0x+ \; 0.4y= 3.08 \\ [-10] & 0.1x- \; y = 0.1 \; \rightarrow \underline{-1.0x+10.0y = -1.0} \end{cases}$$

$$10.4y = 2.08$$

$$y = 0.2$$

Substitute $y = 0.2$ into one of the original equations.

$$x+0.4y = 3.08$$

$$x+0.4(0.2) = 3.08$$

$$x+0.08 = 3.08$$

$$x = 3$$

The solution is $x = 3$ and $y = 0.2$, or $(3, 0.2)$.

Example 2: Using Systems to Find the Equation of a Line

Using the formula $y = mx + b$, find the equation of the line determined by the two points $(3, 5)$ and $(-6, 2)$.

Solution: Write two equations in m and b by substituting the coordinates of the points for x and y.

$$\begin{cases} & 5= \; 3m+b \; \rightarrow \; 5 = 3m+b \\ [-1] & 2 = -6m+b \rightarrow \underline{-2 = 6m - b} \end{cases}$$

$$3 = 9m$$

$$\frac{1}{3} = \; m$$

Substitute $m = \dfrac{1}{3}$ into one of the original equations.

$$5 = 3m + b$$

$$5 = 3 \cdot \frac{1}{3} + b$$

$$5 = 1 + b$$

$$4 = b$$

The equation is $y = \dfrac{1}{3}x + 4$.

Summary of the Method of Addition

1. *Rewrite (if necessary) both equations in the standard form $Ax + By = C$.*

2. *Multiply (if necessary) all terms in one (or both) equations so that the coefficients of one of the variables are opposites.*

3. *Add the like terms of the equations so that one of the variables is eliminated and solve the resulting equation.*

   (If both variables are eliminated and the constant is not 0, the system is inconsistent.)

   (If both variables are eliminated and the constant is 0, the system is dependent.)

4. *Substitute the solution from Step 3 back into either of the two original equations and solve for the other variable.*

5. *Check the solutions **in both** of the original equations.*

Now that you know three methods for solving a system of linear equations (graphing, substitution, and addition), which method should you use? Consider the following guidelines in making your decision.

Guidelines for Deciding which Method to Use in Solving a System of Linear Equations

1. *The graphing method is helpful in "seeing" the geometric relationship between the lines and finding approximate solutions. A calculator can be very helpful here.*

2. *Both the substitution method and the addition method give exact solutions.*

3. *The substitution method may be reasonable and efficient if one of the coefficients of one of the variables is 1.*

4. *In general, the method of addition will prove to be most efficient.*

Practice Problems

Solve the following systems by using the method of addition.

1. $\begin{cases} 2x + 2y = 4 \\ x - y = -3 \end{cases}$

2. $\begin{cases} 3x + 4y = 12 \\ \dfrac{1}{3}x - 8y = -5 \end{cases}$

3. $\begin{cases} 0.02x + 0.06y = 1.48 \\ 0.03x - 0.02y = 0.02 \end{cases}$

5.3 Exercises

In Exercises 1 – 42, solve the system by using either the substitution method or the addition method (whichever seems better to you). If the system is inconsistent or dependent, say so in your answer.

1. $\begin{cases} 2x - y = 7 \\ x + y = 2 \end{cases}$

2. $\begin{cases} x + 3y = 9 \\ x - 7y = -1 \end{cases}$

3. $\begin{cases} 3x + 2y = 0 \\ 5x - 2y = 8 \end{cases}$

4. $\begin{cases} 4x - y = 7 \\ 4x + y = -3 \end{cases}$

5. $\begin{cases} 2x + 2y = 5 \\ x + y = 3 \end{cases}$

6. $\begin{cases} y = 2x + 14 \\ x = 14 - 3y \end{cases}$

7. $\begin{cases} x = 11 + 2y \\ 2x - 3y = 17 \end{cases}$

8. $\begin{cases} 6x - 3y = 6 \\ y = 2x - 2 \end{cases}$

9. $\begin{cases} x - 2y = 4 \\ y = \dfrac{1}{2}x - 2 \end{cases}$

10. $\begin{cases} 2x + y = 3 \\ 4x + 2y = 7 \end{cases}$

11. $\begin{cases} x = 3y + 4 \\ y = 6 - 2x \end{cases}$

12. $\begin{cases} 8x - y = 29 \\ 2x + y = 11 \end{cases}$

13. $\begin{cases} 7x - y = 16 \\ 2y = 2 - 3x \end{cases}$

14. $\begin{cases} 3x + y = -10 \\ 2y - 1 = x \end{cases}$

15. $\begin{cases} 4x - 2y = 8 \\ 2x - y = 4 \end{cases}$

16. $\begin{cases} 3x + 3y = 18 \\ 4x + 2y = 32 \end{cases}$

17. $\begin{cases} 3x + 2y = 4 \\ x + 5y = -3 \end{cases}$

18. $\begin{cases} x + 2y = 0 \\ 2x = 4y \end{cases}$

19. $\begin{cases} \dfrac{1}{2}x + y = -4 \\ 3x - 4y = 6 \end{cases}$

20. $\begin{cases} x + y = 1 \\ x - \dfrac{1}{3}y = \dfrac{11}{3} \end{cases}$

21. $\begin{cases} 4x + 3y = 2 \\ 3x + 2y = 3 \end{cases}$

22. $\begin{cases} x - 3y = 4 \\ 3x - 9y = 10 \end{cases}$

23. $\begin{cases} 5x - 2y = 17 \\ 2x - 3y = 9 \end{cases}$

24. $\begin{cases} \dfrac{1}{2}x + 2y = 9 \\ 2x - 3y = 14 \end{cases}$

Answer to Practice Problems: 1. $x = -\dfrac{1}{2}, y = \dfrac{5}{2}$ **2.** $x = 3, y = \dfrac{3}{4}$ **3.** $x = 14, y = 20$

25. $\begin{cases} 3x + 2y = 14 \\ 7x + 3y = 26 \end{cases}$ **26.** $\begin{cases} 4x + 3y = 28 \\ 5x + 2y = 35 \end{cases}$ **27.** $\begin{cases} 2x + 7y = 2 \\ 5x + 3y = -24 \end{cases}$

28. $\begin{cases} 7x - 6y = -1 \\ 5x + 2y = 37 \end{cases}$ **29.** $\begin{cases} 3x + 3y = 9 \\ x + y = 3 \end{cases}$ **30.** $\begin{cases} 9x + 2y = -42 \\ 5x - 6y = -2 \end{cases}$

31. $\begin{cases} 10x + 4y = 7 \\ 5x + 2y = 15 \end{cases}$ **32.** $\begin{cases} 6x - 5y = -40 \\ 8x - 7y = -54 \end{cases}$ **33.** $\begin{cases} \dfrac{3}{4}x - \dfrac{1}{2}y = 2 \\ \dfrac{1}{3}x - \dfrac{7}{6}y = 1 \end{cases}$

34. $\begin{cases} x + y = 12 \\ 0.05x + 0.25y = 1.6 \end{cases}$ **35.** $\begin{cases} x + 0.5y = 8 \\ 0.1x + 0.01y = 0.64 \end{cases}$ **36.** $\begin{cases} 0.5x - 0.3y = 7 \\ 0.3x - 0.4y = 2 \end{cases}$

37. $\begin{cases} 0.6x + 0.5y = 5.9 \\ 0.8x + 0.4y = 6 \end{cases}$ **38.** $\begin{cases} 2.5x + 1.8y = 7 \\ 3.5x - 2.7y = 4 \end{cases}$ **39.** $\begin{cases} \dfrac{2}{3}x + \dfrac{1}{2}y = \dfrac{2}{3} \\ 3x + 2y = \dfrac{17}{6} \end{cases}$

40. $\begin{cases} \dfrac{3}{4}x - \dfrac{1}{4}y = \dfrac{3}{8} \\ \dfrac{1}{2}x + \dfrac{1}{2}y = \dfrac{3}{4} \end{cases}$ **41.** $\begin{cases} \dfrac{1}{6}x - \dfrac{1}{12}y = -\dfrac{13}{6} \\ \dfrac{1}{5}x + \dfrac{1}{4}y = 2 \end{cases}$ **42.** $\begin{cases} \dfrac{5}{3}x - \dfrac{2}{3}y = -\dfrac{29}{30} \\ 2x + 5y = 0 \end{cases}$

In Exercises 43 – 48, write an equation for the line determined by the two given points using the formula $y = mx + b$ to set up a system of equations with m and b as the unknowns.

43. $(2, 3), (1, -2)$ **44.** $(4, 7), (-3, 2)$ **45.** $(-4, 1), (5, 2)$

46. $(0, 6), (-3, -3)$ **47.** $(3, -4), (7, 7)$ **48.** $(1, -3), (5, -3)$

Calculator Problems

In Exercises 49 – 54 use a TI-84 Plus graphing calculator to find the solutions to each system of equations by graphing. Remember to solve each equation for y and follow the steps outlined in Example 4 of Section 5.1. Use the 2ND , CALC, intersect sequence of steps to locate the solution point on the display screen. You may need to adjust the WINDOW for some problems.

49. $\begin{cases} 0.9x + 1.3y = 1.4 \\ 1.2x - 0.7y = 4.3 \end{cases}$ **50.** $\begin{cases} 1.8x + 2.0y = 4.4 \\ 1.2x + 1.2y = -3.6 \end{cases}$ **51.** $\begin{cases} 1.4x + 3.5y = 7.28 \\ 2.4x - 2.1y = -0.48 \end{cases}$

52. $\begin{cases} 2.2x + 1.5y = 7.69 \\ 4.0x - 0.8y = 7.28 \end{cases}$ **53.** $\begin{cases} 1.3x + 4.1y = 9.282 \\ 0.7x - 1.6y = 4.503 \end{cases}$ **54.** $\begin{cases} 0.09x + 0.17y = 0.6198 \\ 2.10x - 0.90y = 1.6140 \end{cases}$

For Exercises 55 – 58, a word problem is stated with equations given that represent a mathematical model for the problem. Solve the system by using either the method of substitution or the method of addition.

55. Georgia has $10,000 to invest and she is going to put the money into two accounts. One of the accounts will pay 6% interest and the other will pay 10%. How much will she put in each account if she knows that the interest from the 10% account will exceed the interest from the 6% account by $40?

Let x = amount in 10% account and y = amount in 6% account.

Then the system that models the problem is $\begin{cases} x + y = 10,000 \\ 0.10x - 0.06y = 40 \end{cases}$

56. Money is invested at two rates of interest. One rate is 8% and the other is 6%. If there is $1000 more invested at 8% than at 6%, find the amount invested at each rate if the annual interest from both investments is $640.

Let x = amount invested at 8% and y = amount invested at 6%.

Then the system that models the problem is $\begin{cases} x = y + 1000 \\ 0.08x + 0.06y = 640 \end{cases}$

57. How many liters each of a 30% acid solution and a 40% acid solution must be used to produce 100 liters of a 36% acid solution?

Let x = amount of 30% solution and y = amount of 40% solution.

Then the system that models the problem is $\begin{cases} x + y = 100 \\ 0.30x + 0.40y = 0.36(100) \end{cases}$

58. Two cars leave Denver at the same time traveling in opposite directions. One travels at an average speed of 55 mph and the other at 65 mph. In how many hours will they be 420 miles apart?

Let x = time of travel for first car and y = time of travel for second car.

Then the system that models the problem is $\begin{cases} x = y \\ 55x + 65y = 420 \end{cases}$

65 mph

Denver

55 mph

Solving Systems of Linear Equations by Addition

| 5.4 | # Applications: Distance-Rate-Time, Number Problems, Amounts and Costs |

Objectives

After completing this section, you will be able to:

1. *Solve applied problems related to distance, rate, and time by using systems of linear equations.*

2. *Solve applied problems related to numbers by using systems of linear equations.*

3. *Solve applied problems related to amounts and costs by using systems of linear equations.*

Systems of equations occur in many practical situations such as supply and demand in business, velocity and acceleration in engineering, money and interest in investments, and mixture in physics and chemistry. Many of the problems in the exercises for Sections 5.1, 5.2, and 5.3 illustrated these ideas, and corresponding systems of equations that served to solve the problems were given. The equations were given in those problems to help lead you to understand how the variables and equations are set up in solving problems with systems of equations. As you study the applications in Sections 5.4 and 5.5, you may want to refer to some of those exercises, as well as the examples, as guides in solving the new applications.

In this section, we will study applications related to distance-rate-time concepts ($d = rt$) as well as "fun type" reasoning problems involving such topics as coins and people's ages. **Remember that the emphasis in all applications in a course in introductory algebra is to develop your reasoning skills and to teach you how to transfer English phrases into abstract algebraic expressions.**

Example 1: Distance-Rate-Time (Rates Unknown)

A small plane flew 300 miles in 2 hours flying with the wind. Then on the return trip, flying against the wind, it traveled only 200 miles in 2 hours. What were the wind velocity and the speed of the plane? (**Note:** The "speed of the plane" means how fast the plane would be flying with no wind.)

Solution: Let s = speed of plane
and w = wind velocity

Continued on next page...

When flying with the wind, the plane's actual rate will increase to $s + w$. When flying against the wind, the plane's actual rate will decrease to $s - w$. (Note: If the wind had been strong enough, the plane could actually have been flying backward, away from its destination.)

	Rate	×	Time	=	Distance
With the wind	$s + w$		2		$2(s + w)$
Against the wind	$s - w$		2		$2(s - w)$

The system of linear equations is

with the wind $\longrightarrow$ $\begin{cases} 2(s+w) = 300 \rightarrow 2s + 2w = 300 \\ 2(s-w) = 200 \rightarrow \underline{2s - 2w = 200} \end{cases}$

against the wind $\longrightarrow$

$$4s \quad = 500$$
$$s \quad = 125$$

Substitute $s = 125$ into one of the original equations.

$$2(125 + w) = 300$$
$$125 + w = 150$$
$$w = 25$$

The speed of the plane was 125 mph, and the wind velocity was 25 mph.

Example 2:   Distance-Rate-Time (Times Unknown)

Two buses leave Eureka traveling in opposite directions. One leaves at noon and the second leaves at 1 P.M. The first one travels at an average speed of 55 mph and the second one at an average speed of 59 mph. At what time will the buses be 226 miles apart?

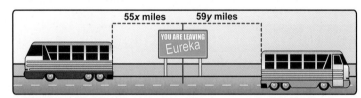

Solution: Let x = time of travel for first bus
and y = time of travel for second bus

	Rate	×	Time	=	Distance
Bus 1	55		x		$55x$
Bus 2	59		y		$59y$

The system of linear equations is

$$\begin{cases} x = y+1 & \text{The first bus travels 1 hour longer than the second.} \\ 55x + 59y = 226 & \text{The sum of the distances is 226 miles.} \end{cases}$$

Substitution gives

$$55(y+1) + 59y = 226$$
$$55y + 55 + 59y = 226$$
$$114y = 171$$
$$y = 1.5$$

This gives $x = 1.5 + 1 = 2.5$.

Thus the first bus travels 2.5 hours and the second bus travels 1.5 hours. Therefore, the buses will be 226 miles apart at 2:30 P.M.

Example 3: Number Problem

The sum of two numbers is 80 and their difference is 10. What are the two numbers?

Solution: Let x = one number
and y = the other number.

The system of linear equations is

$$\begin{cases} x + y = 80 & \text{The sum is 80.} \\ x - y = 10 & \text{The difference is 10.} \end{cases}$$

Solving by addition gives

$$\begin{aligned} x + y &= 80 \\ \underline{x - y} &= \underline{10} \\ 2x &= 90 \\ x &= 45 \end{aligned}$$

Substituting 45 for x in the first equation gives: $45 + y = 80$
$$y = 80 - 45 = 35$$

The two numbers are 45 and 35.

Check: $45 + 35 = 80$ and $45 - 35 = 10$

Example 4: Counting Coins

Mike has $1.05 worth of change in nickels and quarters. If he has twice as many nickels as quarters, how many of each type of coin does he have?

Solution: We use two equations – one relating the number of coins and the other relating the value of the coins. The value of each nickel is 5 cents and the value of each quarter is 25 cents.

Let n = number of nickels
and q = number of quarters.

Coins	# of coins	Value	Total value
Nickel	n	0.05	$0.05n$
Quarters	q	0.25	$0.25q$

The system of linear equations is

$$\begin{cases} n = 2q \\ 0.05n + 0.25q = 1.05 \end{cases}$$

   This equation relates the number of coins.
   This equation relates the values of the coins.

Note carefully that in the first equation q is multiplied by 2 because the number of nickels is twice the number of quarters. Therefore, n is bigger.

The first equation is already solved for n, so substitution into the second equation gives

$$0.05(2q) + 0.25q = 1.05$$
$$0.10q + 0.25q = 1.05$$
$$10q + 25q = 105 \qquad \text{Multiply the equation by 100.}$$
$$35q = 105$$
$$q = 3$$

$$n = 2q = 6$$

Mike has 3 quarters and 6 nickels.

Example 5: Calculating Age

Pat is 6 years older than her sister Sue. In 3 years, she will be twice as old as Sue. How old is each girl now?

Solution: We use two equations – one relating their ages now and the other relating their ages in 3 years.

Let P = Pat's age now
and S = Sue's age now
Then the system of linear equations is

$$\begin{cases} P - S = 6 & \text{The difference in their \textbf{current} ages is 6.} \\ P + 3 = 2(S+3) & \textbf{In 3 years} \text{ each age is increased by 3.} \end{cases}$$

Rewrite the second equation in standard form and solve by addition.

$$P + 3 = 2(S+3)$$
$$P + 3 = 2S + 6$$
$$P - 2S = 3$$

$$\begin{cases} P - S = 6 \rightarrow \quad P - \ \ S = \ \ 6 \\ [-1]P - 2S = 3 \rightarrow \underline{-P + 2S = -3} \end{cases}$$
$$\qquad\qquad\qquad\qquad S = 3$$

Substitute $S = 3$ into one of the original equations.

$$P - 3 = 6$$
$$P = 6 + 3$$
$$P = 9$$

Pat is 9 years old; Sue is 3 years old.

Example 6: Amounts and Costs

Three hot dogs and two orders of French fries cost $10.30. Four hot dogs and four orders of fries cost $15.60. What is the cost of a hot dog? What is the cost of an order of fries?

Solution: Let x = cost of one hot dog
and y = cost of one order of fries.
Then the system of linear equations is

$$\begin{cases} 3x + 2y = 10.30 & \text{Three hot dogs and two orders of French fries cost \$10.30.} \\ 4x + 4y = 15.60 & \text{Four hot dogs and four orders of fries cost \$15.60.} \end{cases}$$

Both equations are in standard form. Solve using the addition method.

$$\begin{cases} [-2] \; 3x + 2y = 10.30 \rightarrow & -6x - 4y = -20.60 \\ 4x + 4y = 15.60 \rightarrow & \underline{4x + 4y = 15.60} \end{cases}$$

$$\begin{aligned} -2x &= -5.00 \\ x &= 2.50 \end{aligned}$$

Substitute $x = 2.50$ into one of the original equations.

$$3(2.50) + 2y = 10.30$$
$$7.50 + 2y = 10.30$$
$$2y = 2.80$$
$$y = 1.40$$

One hot dog costs $2.50 and one order of fries costs $1.40.

NOTES You should consider making tables similar to those illustrated in Examples 1, 2, and 4 when working with applications. These tables can help you organize the information in a more understandable form.

5.4 Exercises

Solve each problem by setting up a system of two equations in two unknowns and solving the system.

1. The sum of two numbers is 56. Their difference is 10. Find the numbers.

2. The sum of two numbers is 40. The sum of twice the larger and 4 times the smaller is 108. Find the numbers.

3. The sum of two numbers is 36. Three times the smaller plus twice the larger is 87. Find the two numbers.

4. The difference between two numbers is 17. Four times the smaller is equal to 7 more than the larger. What are the numbers?

5. Ken makes a 4-mile motorboat trip downstream in 20 minutes $\left(\frac{1}{3}\text{hr.}\right)$. The return trip takes 30 minutes $\left(\frac{1}{2}\text{hr.}\right)$. Find the rate of the boat in still water and the rate of the current.

6. Mr. McKelvey finds that flying with the wind he can travel 1188 miles in 6 hours. However, when flying against the wind, he travels only $\frac{2}{3}$ of the distance in the same amount of time. Find the speed of the plane in still air and the wind speed.

7. Randy made a business trip of 190 miles. He averaged 52 mph for the first part of the trip and 56 mph for the second part. If the total trip took $3\frac{1}{2}$ hours, how long did he travel at each rate?

8. Marian drove to a resort 335 miles from her home. She averaged 60 mph for the first part of her trip and 55 mph for the second part. If her total driving time was $5\frac{3}{4}$ hours, how long did she travel at each rate?

9. Mr. Green traveled to a city 200 miles from his home to attend a meeting. Due to car trouble, his average speed returning was 10 mph less than his speed going. If the outbound trip was 4 hours and the return trip was 5 hours, at what rate of speed did he travel to the city?

10. Two trains leave Kansas City at the same time. One train travels east and the other travels west. The speed of the westbound train is 5 mph greater than the speed of the eastbound train. After 6 hours, they are 510 miles a part. Find the rate of each train. Assume the trains travel in a straight line in opposite directions.

11. Steve travels 4 times as fast as Fred. Traveling in opposite directions, they are 105 miles apart after 3 hours. Find their rates of travel.

12. Sue travels 5 mph less than twice as fast as June. Starting at the same point and traveling in the same direction, they are 80 miles apart after 4 hours. Find their speeds.

13. Mary and Linda live 324 miles apart. They start at the same time and travel toward each other. Mary's speed is 8 mph greater than Linda's. If they meet in 3 hours, find their speeds.

14. Two planes leave from points 1860 miles apart at the same time and travel toward each other (at slightly different altitudes, of course). If their rates are 220 mph and 400 mph, how soon will they meet?

15. A jogger runs into the countryside at a rate of 10 mph. He returns along the same route at 6 mph. If the total trip took 1 hour, 36 minutes, how far did he jog?

16. A cyclist traveled to her destination at an average rate of 15 mph. By traveling 3 mph faster, she took 30 minutes less to return. What distance did she travel each way?

17. An airliner's average speed is $3\frac{1}{2}$ times the average speed of a private plane. Two hours after traveling in the same direction they are 580 miles apart. What is the average speed of each plane? (**Hint:** Since they are traveling in the same direction, the distance between them will be the difference of their distances.)

18. Sonja has some nickels and dimes. If she has 30 coins worth a total of $2.00, how many of each type of coin does she have?

19. Louis has a total of 27 coins consisting of quarters and dimes. The total value of the coins is $5.40. How many of each type of coin does he have?

20. Your friend challenges you to figure out how many dimes and quarters are in a cash register. He tells you that there are 65 coins and that their value is $11.90. How many dimes and how many quarters are in the register?

21. A bag contains pennies and nickels only. If there are 182 coins in all and their value is $3.90, how many pennies and how many nickels are in the bag?

22. The length of a rectangle is 10 meters more than one-half the width. If the perimeter is 44 meters, what are the length and width?

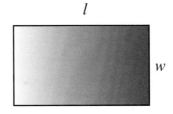

23. The length of a rectangle is 1 meter less than twice the width. If each side is increased by 4 meters, the perimeter will be 116 meters. Find the length and the width of the original rectangle.

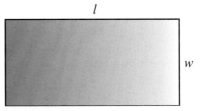

24. The line $y = mx + b$ passes through the two points $(1, 3)$ and $(5, 1)$. Find the equation of the line. (**Hint:** Substitute the values for x and y in the equation and solve the resulting system of equations for m and b.)

25. The line $y = mx + b$ passes through the two points $(-2, -1)$ and $(6, -7)$. Find the equation of the line. (**Hint:** Substitute the values for x and y in the equation and solve the resulting system of equations for m and b.)

26. Jill is 8 years older than her brother Curt. Four years from now, Jill will be twice as old as Curt. How old is each at the present time?

27. Two years ago, Anna was half as old as Beth. Eight years from now, she will be two-thirds as old as Beth. How old are they now?

28. A Christmas charity party sold tickets for $45.00 for adults and $25.00 for children. The total number of tickets sold was 320 and the total for the ticket sales was $13,000. How many adult and how many children's tickets were sold?

29. Tickets for the local high school basketball game were priced at $3.50 for adults and $2.50 for students. If the income for one game was $9550 and the attendance was 3500, how many adults and how many students attended that game?

30. The width of a rectangle is $\dfrac{3}{4}$ of its length. If the perimeter of the rectangle is 140 feet, what are the dimensions of the rectangle?

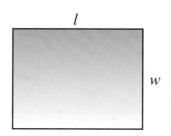

31. A farmer has 260 meters of fencing to build a rectangular corral. He wants the length to be 3 times as long as the width. What dimensions should he make his corral?

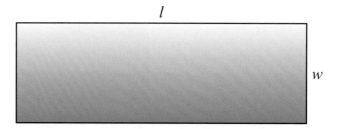

32. Joan went to a book sale on campus and bought paperback books for $0.25 each and hardback books for $1.75 each. If she bought a total of 15 books for $11.25, how many of each type of book did she buy?

33. Admission to the baseball game is $2.00 for general admission and $3.50 for reserved seats. The receipts were $36,250 for 12,500 paid admissions. How many of each ticket, general and reserved, were sold?

34. A men's clothing store sells two styles of sports jackets, one selling for $95 and one selling for $120. Last month, the store sold 40 jackets, with receipts totaling $4250. How many of each style did the store sell?

35. Seventy children and 160 adults attended a movie theater. The total receipts were $620. One adult ticket and 2 children's tickets cost $7. Find the price of each type of ticket.

36. Morton took some old newspapers and aluminum cans to the recycling center. Their total weight was 180 pounds. He received 1.5¢ per pound for the newspapers and 30¢ per pound for the cans. The total received was $14.10. How many pounds of each did Morton have?

37. Frank bought 2 shirts and 1 pair of slacks for a total of $55. If he had bought 1 shirt and 2 pairs of slacks, he would have paid $68. What was the price of each shirt and each pair of slacks?

38. Four hamburgers and three orders of french fries cost $5.15. Three hamburgers and five orders of fries cost $5.10. What would one hamburger and one order of fries cost?

39. A small manufacturer produces two kinds of radios, model X and model Y. Model X takes 4 hours to produce and costs $8 each to make. Model Y takes 3 hours to produce and costs $7 each to make. If the manufacturer decides to allot a total of 58 hours and $126 each week, how many of each model will be produced?

40. A furniture shop refinishes chairs. Employees use two methods to refinish a chair. Method I takes 1 hour and the material costs $3. Method II takes $1\frac{1}{2}$ hours and the material costs $1.50. Last week, they took 36 hours and spent $60 refinishing chairs. How many did they refinish with each method?

Writing and Thinking About Mathematics

41. A two digit number can be written as ab, where a and b are the digits. We do **not** mean that the digits are multiplied, but the value of the number is $10a + b$. For example, the two digit number 34 has a value of $10 \cdot 3 + 4$. Set up and solve a system of equations for the following problem:

The sum of the digits of a two digit number is 13. If the digits are reversed, then the value of the number is increased by 45. What is the number?

Hawkes Learning Systems: Introductory Algebra

Applications: Distance-Rate-Time, Number Problems, Amounts and Costs

Applications: Interest and Mixture

After completing this section, you will be able to:

1. *Solve applied problems related to interest by using systems of linear equations.*

2. *Solve applied problems related to mixture by using systems of linear equations.*

In this section we will study two more types of applications that can be solved by using systems of linear equations; interest on money invested and mixture. These applications can be "wordy" and you will need to read carefully and analyze the information thoroughly to be able to translate it into a system involving two variables.

Interest

People in business and banking know several formulas for calculating interest. The formula used depends on the method of payment (monthly or yearly) and the type of interest (simple or compound). Also, penalties for late payments and even penalties for early payments might be involved. In any case, standard notation is the following:

$P \longrightarrow$ the principal (amount of money invested or borrowed)
$r \longrightarrow$ the rate of interest (an annual rate)
$t \longrightarrow$ the time (in years or part of a year)
$I \longrightarrow$ the interest (paid or earned)

In this section, we will use only the basic formula for simple interest

$$I = Prt$$

with interest calculated on an annual basis. In this special case, we have $t = 1$ and the formula becomes

$$I = Pr$$

Example 1: Interest

James has two investment accounts, one pays 6% interest and the other pays 10% interest. He has $1000 more in the 10% account than he has in the 6% account. The interest from the 10% account exceeds the interest from the 6% account by $260 each year. How much does he have in each account?

Solution:

Careful reading indicates two types of information:
1. He has two accounts.
2. He earns two amounts of interest.

Let x = amount (principal) invested at 6%
       y = amount (principal) invested at 10%,

then $0.06x$ = interest earned on first account
       $0.10y$ = interest earned on second account.

Now set up two equations.

$$\begin{cases} y = x + 1000 \\ 0.10y - 0.06x = 260 \end{cases}$$

y is larger than x by $1000.

Interest from the 10% account exceeds (is larger than) interest from the 6% account by $260.

Because the first equation is already solved for y, we use the substitution method and substitute for y in the second equation:

$$0.10(x + 1000) - 0.06x = 260$$
$$10(x + 1000) - 6x = 26,000 \qquad \text{Multiply by 100.}$$
$$10x + 10,000 - 6x = 26,000$$
$$4x = 16,000$$
$$x = 4000$$

Substitute $x = 4000$ into one of the original equations to find y.

$$y = x + 1000 = 4000 + 1000 = 5000$$

James has $4000 invested at 6% and $5000 invested at 10%.

Example 2: Interest

Lila has $7000 to invest. She decides to separate her funds into two accounts. One yields interest at the rate of 7% and the other at 12%. If she wants a total annual income from both accounts to be $690, how should she split the money? (**Note:** The higher interest account is considered more risky. Otherwise, she would put the entire $7000 into that account.)

Solution:

Again, careful reading indicates two types of information:
1. She has two accounts.
2. She earns two amounts of interest.

Let x = amount (principal) invested at 7%
       y = amount (principal) invested at 12%,

Then $0.07x$ = interest earned on first account
        $0.12y$ = interest earned on second account.

Now set up two equations.

$$\begin{cases} x + y = 7000 \\ 0.07x + 0.12y = 690 \end{cases}$$

The sum of both accounts is $7000.

The total interest from both accounts is $690.

Both equations are in standard form. Solve by addition. Multiply the first equation by -7 and the second by 100 to get opposite coefficients for x as follows:

$$\begin{cases} [-7] \quad\quad x + \quad y = 7000 \;\rightarrow\; -7x - 7y = -49,000 \\ [100] \quad 0.07x + 0.12y = 690 \;\rightarrow\; \underline{\;\; 7x + 12y = \;\; 69,000} \end{cases}$$

$$5y = 20,000$$
$$y = 4000$$

Substitute $y = 4000$ into one of the original equations.

$$x + 4000 = 7000$$
$$x = 3000$$

She should invest $3000 at 7% and $4000 at 12%.

Mixture

Problems involving mixtures occur in physics and chemistry and in such places as candy stores or coffee shops. Two or more items of a different percentage of concentration of a chemical such as salt, chlorine, or antifreeze are to be mixed; or two or more types of coffee are to be mixed to form a final mixture that satisfies certain conditions of percentage of concentration.

The basic plan is to write an equation that deals with only one part of the mixture (such as the salt in the mixture). The following examples explain how this can be accomplished.

Example 3: Mixture

How many ounces each of a 10% salt solution and a 15% salt solution must be used to produce 50 ounces of a 12% salt solution?

Solution: Let x = amount of 10% solution
and y = amount of 15% solution.

	Amount of solution	× Percent of salt	= Amount of salt
10% solution	x	0.10	$0.10x$
15% solution	y	0.15	$0.15y$
12% solution	50	0.12	$0.12(50)$

Continued on next page...

Then the system of linear equations is

$$\begin{cases} x + y = 50 & \text{The sum of the two amounts must be 50 ounces.} \\ 0.10x + 0.15y = 0.12(50) & \text{The sum of the amount of salt from each solution} \\ & \text{equals the total amount of salt in the final solution.} \end{cases}$$

Multiplying the first equation by −10 and the second by 100 gives

$$\begin{cases} [-10] & x + y = 50 \quad \rightarrow \quad -10x - 10y = -500 \\ [100] & 0.10x + 0.15y = 0.12(50) \rightarrow \quad \underline{10x + 15y = 600} \end{cases}$$

$$5y = 100$$
$$y = 20$$

Substitute $y = 20$ into one of the original equations.

$$x + 20 = 50$$
$$x = 30$$

Use 30 ounces of the 10% solution and 20 ounces of the 15% solution.

Example 4: Mixture

How many gallons of a 20% acid solution should be mixed with a 30% acid solution to produce 100 gallons of a 23% solution?

Solution: Let x = amount of 20% solution
and y = amount of 30% solution.

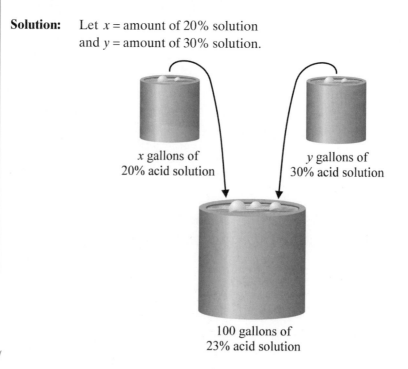

x gallons of
20% acid solution

y gallons of
30% acid solution

100 gallons of
23% acid solution

	Amount of solution ×	Percent of acid =	Amount of acid
20% solution	x	0.20	$0.20x$
30% solution	y	0.30	$0.30y$
23% solution	100	0.23	$0.23(100)$

Then the system of linear equations is

$$\begin{cases} x + y = 100 & \text{The sum of the two amounts must be 100 gallons.} \\ 0.20x + 0.30y = 0.23(100) & \text{The sum of the amount of acid from each solution} \\ & \text{equals the total amount of acid in the final solution.} \end{cases}$$

Multiplying the first equation by –20 and the second by 100 gives,

$$\begin{cases} [-20] & x + & y = 100 & \to -20x - 20y = -2000 \\ [100] & 0.20x + 0.30y = 0.23(100) & \to & \underline{20x + 30y = 2300} \end{cases}$$

$$10y = 300$$
$$y = 30$$

Substitute $y = 30$ into one of the original equations.

$$x + 30 = 100$$
$$x = 70$$

Seventy gallons of the 20% solution should be added to 30 gallons of the 30% solution. This will produce 100 gallons of a 23% solution.

5.5 Exercises

Solve each problem by setting up a system of two equations in two unknowns and solving the system.

1. Carmen invested $9000, part in a 6% passbook account and the rest in a 10% certificate account. If her annual interest was $680, how much did she invest at each rate?

2. Mr. Brown has $12,000 invested. Part is invested at 6% and the remainder at 8%. If the interest from the 6% investment exceeds the interest from the 8% investment by $230, how much is invested at each rate?

3. Ten thousand dollars is invested, part at 5.5% and part at 6%. The interest from the 5.5% investment exceeds the interest from the 6% investment by $251. How much is invested at each rate?

4. On two investments totaling $9500, Bill lost 3% on one and earned 6% on the other. If his net annual receipts were $282, how much was each investment?

5. Marsha has money in two savings accounts. One rate is 8% and the other is 10%. If she has $200 more in the 10% account, how much is invested at 8% if the total interest is $101?

6. Money is invested at two rates. One rate is 9% and the other is 13%. If there is $700 more invested at 9%, find the amount invested at each rate if the annual interest is $239.

7. Frank has half of his investments in stock paying an 11% dividend and the other half in a debentured stock paying 13% interest. If his total annual interest is $840, how much does he have invested?

8. Betty invested some of her money at 12% interest. She invested $300 more than twice that amount at 10%. How much is invested at each rate if her income is $318 annually?

9. GFA invested some money in a development yielding 24% and $9000 less in a development yielding 18%. If the first investment produces $2820 more per year than the second, how much is invested in each development?

10. Judy invests a certain amount of money at 7% annual interest and three times that amount at 8%. If her annual income is $232.50, how much does she have invested at each rate?

11. Norman has a certain amount of money invested at 5% annual interest and $500 more than twice that amount invested in bonds yielding 7%. His total income from interest is $187. How much does he have invested at each rate?

12. A total of $6000 is invested, part at 8% and the remainder at 12%. How much is invested at each rate if the annual interest is $620?

13. Ms. Merriman has $12,000 invested. Part is invested at 9% and the remainder at 11%. If the interest from the 9% investment exceeds the interest from the 11% investment by $380, how much is invested at each rate?

14. Eight thousand dollars is invested, part at 15% and the remainder at 12%. If the annual income from the 15% investment exceeds the income from the 12% investment by $66, how much is invested at each rate?

15. A metallurgist has one alloy containing 20% copper and another containing 70% copper. How many pounds of each alloy must he use to make 50 pounds of a third alloy containing 50% copper?

16. A manufacturer has received an order for 24 tons of a 60% copper alloy. His stock contains only alloys of 80% copper and 50% copper. How much of each will he need to fill the order?

17. A tobacco shop wants 50 ounces of tobacco that is 24% rare Turkish blend. How much each of a 30% Turkish blend and a 20% Turkish blend will be needed?

The Tobacco Shop

18. How many liters each of a 40% acid solution and a 55% acid solution must be used to produce 60 liters of a 45% acid solution?

19. A dairy man wants to mix a 35% protein supplement and a standard 15% protein ration to make 1800 pounds of a high-grade 20% protein ration. How many pounds of each should he use?

20. To meet the government's specifications, an alloy must be 65% aluminum. How many pounds each of a 70% aluminum alloy and a 54% aluminum alloy will be needed to produce 640 pounds of the 65% aluminum alloy?

21. A meat market has ground beef that is 40% fat and extra lean ground beef that is only 15% fat. How many pounds of each will be needed to obtain 50 pounds of lean ground beef that is 25% fat?

22. George decides to mix grades of gasoline in his truck. He puts in 8 gallons of regular and 12 gallons of premium for a total cost of $55.80. If premium gasoline costs $0.15 more per gallon than regular, what was the price of each grade of gasoline?

23. How many grams of pure acid (100% acid) and how many grams of a 40% solution should be mixed together to get a total of 30 grams of a 60% solution?

24. Pure dark coffee beans are to be mixed with a mixture that is 60% of these dark beans. How much of each (pure dark and 60% dark beans) should be used to get a mixture of 50 pounds that contains 70% of the dark beans?

25. Salt is to be added to a 4% salt solution. How many ounces of salt and how many ounces of the 4% solution should be mixed together to get 60 ounces of a 20% salt solution?

Hawkes Learning Systems: Introductory Algebra

Applications: Interest and Mixture

5.6 Graphing Systems of Linear Inequalities

After completing this section, you will be able to:

Solve systems of linear inequalities graphically.

In some branches of mathematics, in particular a topic called (interestingly enough) game theory, the solution to a very sophisticated problem can involve the set of points that satisfy a system of several **linear inequalities**. In business these ideas relate to problems such as minimizing the cost of shipping goods from several warehouses to distribution outlets. In this section we will consider graphing the solution sets to only two inequalities. We will leave the problem solving techniques to another course.

In Section 4.6 we graphed linear inequalities of the form $y < mx + b$ (or $y \leq mx + b$ or $y > mx + b$ or $y \geq mx + b$). These graphs are called **half-planes**. The line $y = mx + b$ is called the **boundary line** and the half-planes are **open** (the boundary line is not included) or **closed** (the boundary line is included).

In this section, we will develop techniques for graphing (and therefore solving) **systems of two linear inequalities**. The **solution set** (if there are any solutions) to such a system of linear inequalities consists of the points in the **intersection** of two half-planes and portions of boundary lines indicated by the inequalities. The following procedure may be used to solve a system of linear inequalities.

To Solve a System of Two Linear Inequalities

1. *Graph both half-planes.*
2. *Shade the region that is common to both of these half-planes.*
   *(This region is called the **intersection** of the two half-planes.)*
3. *To check, pick one test-point in the intersection and verify that it satisfies both inequalities.*

*(**Note:** If there is no intersection, then the system is inconsistent and has no solution.)*

Example 1: Solving a System of Linear Inequalities

Graph the points that satisfy the system of linear inequalities $\begin{cases} x \leq 2 \\ y \geq -x+1 \end{cases}$.

Solution: **Step 1:** For $x \leq 2$, the points are to the left of and on the line $x = 2$.

Step 2: For $y \geq -x+1$, the points are above and on the line $y = -x + 1$.

Step 3: Shade only the region with points that satisfy both inequalities. In this case, we test the point $(0, 3)$.

$0 \leq 2$ A true statement

$3 \geq -0+1$ A true statement

Example 2: Solving a System of Linear Inequalities

Solve the system of linear inequalities graphically: $\begin{cases} 2x+y \leq 6 \\ x+y < 4 \end{cases}$.

Solution: **Step 1:** Solve each inequality for y: $\begin{cases} y \leq -2x+6 \\ y < -x+4 \end{cases}$.

Step 2: For $y \leq -2x+6$, the points are below and on the line $y = -2x + 6$.

Step 3: For $y < -x + 4$, the points are below but not on the line $y = -x + 4$.

Step 4: Shade only the region with points that satisfy both inequalities. Note that the line $y = -x + 4$ is dashed. In this case, we test the point $(0, 0)$.

$2 \cdot 0 + 0 \leq 6$ A true statement

$0 + 0 < 4$ A true statement

Using a TI-84 Plus Graphing Calculator to Graph Systems of Linear Inequalities

To graph (and therefore solve) a system of linear inequalities (with no vertical line) with a TI-84 Plus graphing calculator, proceed as follows:

1. Solve each inequality for y.
2. Press the ▬ Y= ▬ key and enter the two expressions for Y_1 and Y_2.

3. Move to the left of Y_1 and Y_2 and press **ENTER** until the desired graphing symbol appears.
4. Press ▬ GRAPH ▬ and the desired region will be graphed as a cross-hatched area. (You may need to reset the ▬ WINDOW ▬ so that both regions appear.)

Example 3 illustrates how this can be done.

Example 3: Using a Graphing Calculator to Solve a System of Linear Inequalities

Use a TI-84 Plus graphing calculator to graph the following system of linear inequalities:

$$\begin{cases} 2x + y < 4 \\ 2x - y \le 0 \end{cases}$$

Solution: **Step 1:** Solve each inequality for y: $\begin{cases} y < -2x + 4 \\ y \ge 2x \end{cases}$.

(**Note:** Solving $2x - y \le 0$ for y can be written as $2x \le y$ and then as $y \ge 2x$.)

Steps 2 and 3: Press the ▬ Y= ▬ key and enter both functions and the corresponding symbols as they appear here:

Continued on next page...

Step 4: Press 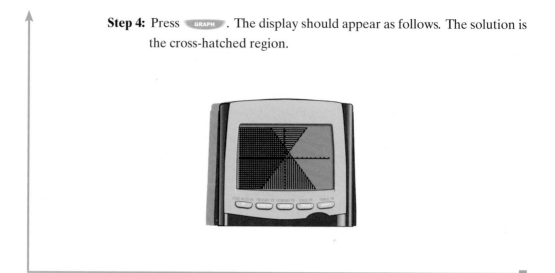 GRAPH . The display should appear as follows. The solution is the cross-hatched region.

Exercises 5.6

In Exercises 1 – 20, solve the systems of two linear inequalities graphically.

1. $\begin{cases} y > 2 \\ x \geq -3 \end{cases}$ **2.** $\begin{cases} 2x+5 < 0 \\ y \geq 2 \end{cases}$ **3.** $\begin{cases} x < 3 \\ y > -x+2 \end{cases}$ **4.** $\begin{cases} y \leq -5 \\ y \geq x-5 \end{cases}$

5. $\begin{cases} x \leq 3 \\ 2x+y > 7 \end{cases}$ **6.** $\begin{cases} 2x-y > 4 \\ y < -1 \end{cases}$ **7.** $\begin{cases} x-3y \leq 3 \\ x < 5 \end{cases}$ **8.** $\begin{cases} 3x-2y \geq 8 \\ y \geq 0 \end{cases}$

9. $\begin{cases} x-y \geq 0 \\ 3x-2y \geq 4 \end{cases}$ **10.** $\begin{cases} y \geq x-2 \\ x+y \geq -2 \end{cases}$ **11.** $\begin{cases} 3x+y \leq 10 \\ 5x-y \geq 6 \end{cases}$ **12.** $\begin{cases} y \geq 2x-5 \\ 3x+2y > -3 \end{cases}$

13. $\begin{cases} 3x+4y \geq -7 \\ y < 2x+1 \end{cases}$ **14.** $\begin{cases} 2x-3y \geq 0 \\ 8x-3y < 36 \end{cases}$ **15.** $\begin{cases} x+y < 4 \\ 2x-3y < 3 \end{cases}$ **16.** $\begin{cases} 2x+3y < 12 \\ 3x+2y > 13 \end{cases}$

17. $\begin{cases} x+y \geq 0 \\ x-2y \geq 6 \end{cases}$ **18.** $\begin{cases} y \geq 2x+3 \\ y \leq x-2 \end{cases}$ **19.** $\begin{cases} x+3y \leq 9 \\ x-y \geq 5 \end{cases}$ **20.** $\begin{cases} x-y \geq -2 \\ x+2y < -1 \end{cases}$

Use a graphing calculator to solve the systems of linear inequalities in Exercises 21 – 30.

21. $\begin{cases} y \geq 0 \\ 3x-5y \leq 10 \end{cases}$ **22.** $\begin{cases} 3x+2y \leq 15 \\ 2x+5y \geq 10 \end{cases}$ **23.** $\begin{cases} 4x-3y \geq 6 \\ 3x-y \leq 3 \end{cases}$ **24.** $\begin{cases} y \leq 0 \\ 3x+y \leq 11 \end{cases}$

25. $\begin{cases} 3x-4y \geq -6 \\ 3x+2y \leq 12 \end{cases}$ **26.** $\begin{cases} 3y \leq 2x+2 \\ x+2y \leq 11 \end{cases}$ **27.** $\begin{cases} x+y \leq 8 \\ 3x-2y \geq -6 \end{cases}$ **28.** $\begin{cases} x+y \leq 7 \\ 2x-y \leq 8 \end{cases}$

29. $\begin{cases} y \leq x \\ y < 2x+1 \end{cases}$ **30.** $\begin{cases} x-y \geq -2 \\ 4x-y < 16 \end{cases}$

Writing and Thinking About Mathematics

31. The material in the text and the exercises do not discuss cases in solving a system of two linear inequalities in which the boundary lines are parallel. Describe, in your own words, what you think the solutions in such cases might be. After you have done this, solve the following systems graphically and discuss how these cases relate to your previous analysis.

a. $\begin{cases} y \leq 2x - 5 \\ y \geq 2x + 3 \end{cases}$ **b.** $\begin{cases} y \leq -x + 2 \\ y \geq -x - 1 \end{cases}$ **c.** $\begin{cases} y \leq \dfrac{1}{2}x + 3 \\ y \geq \dfrac{1}{2}x - 3 \end{cases}$

Hawkes Learning Systems: Introductory Algebra

Systems of Linear Inequalities

Chapter 5 Index of Key Ideas and Terms

Section 5.1   Systems of Equations: Solutions by Graphing

System of Linear Equations   page 402

A pair of linear equations considered together is called a **system of linear equations** (or a **set of simultaneous equations**).

Solutions by Graphing   pages 403 - 408

The two lines related to a system of equations can be graphed.

    a. If the two lines intersect in a single point, the point is the   page 404
    solution to the system. The system is said to be **consistent**.

    b. If the two lines are parallel (do not intersect), the system is   pages 404 - 405
    said to be **inconsistent** and there are no solutions.

    c. If the two lines coincide (are the same line), the system is said   pages 405 - 406
    to be **dependent** and there are an infinite number of solutions.

Section 5.2   Systems of Equations: Solutions by Substitution

Solutions by Substitution   pages 414 - 417

    1. Solve one of the equations for one of the variables.
    2. Substitute the resulting expression into the other equation.
    3. Solve the new equation for the one variable.
    4. Substitute the result back into one of the original equations to find
    the value of the other variable.

Section 5.3   Systems of Equations: Solutions by Addition

Solutions by Addition   pages 420 - 425

    1. Write the equations in standard form.
    2. Arrange the equations vertically so that like terms are aligned.
    3. Multiply one or both of the equations by a constant so that the
    coefficients of one of the variables are opposites.
    4. Add the like terms.
    5. Solve the resulting equation.
    6. Substitute the results back into one of the original equations to
    find the value of the other variable.

Continued on next page...

Section 5.3 Systems of Equations: Solutions by Addition (continued)

Guidelines for Deciding which Method to Use in Solving a System of Linear Equations   page 426
    1. The graphing method is helpful in "seeing" the geometric relationship between the lines and finding approximate solutions. A calculator can be very helpful here.
    2. Both the substitution method and the addition method give exact solutions.
    3. The substitution method may be reasonable and efficient if one of the coefficients of one of the variables is 1.
    4. In general, the method of addition will prove to be most efficient.

Section 5.4 Applications: Distance-Rate-Time, Number Problems, Amounts and Costs

Applications   pages 429 - 434
    Distance-Rate-Time, Number Problems, Amounts and Costs

Section 5.5 Applications: Interest and Mixture

Applications   pages 440 - 445
    Interest and Mixture

Section 5.6 Graphing Systems of Linear Inequalities

To Solve a System of Two Linear Inequalities   page 449
    1. Graph both half-planes.
    2. Shade the region that is common to both of these half-planes. (This region is called the intersection of the two half-planes.)
    3. To check, pick one test-point in the intersection and verify that it satisfies both inequalities.
    (Note: If there is no intersection, then the system is inconsistent and has no solution.)

Hawkes Learning Systems: Introductory Algebra

For a review of the topics and problems from Chapter 5, look at the following lessons from *Hawkes Learning Systems: Introductory Algebra*

Solving Systems of Linear Equations by Graphing
Solving Systems of Linear Equations By Substitution
Solving Systems of Linear Equations By Addition
Applications: Distance-Rate-Time, Number Problems, Amounts and Costs
Applications: Interest and Mixture
Systems of Linear Inequalities

Chapter 5 Review

5.1 Systems of Equations: Solutions by Graphing

In Exercises 1 – 4, show that each system of equations is inconsistent by determining the slope of each line and the y-intercept. (That is, show that the lines are parallel and do not intersect.)

1. $\begin{cases} y = 2x \\ -2x + y = 5 \end{cases}$ **2.** $\begin{cases} 3x + y = 4 \\ 6x + 2y = -1 \end{cases}$ **3.** $\begin{cases} 2x - y = -7 \\ x - \dfrac{1}{2}y = 2 \end{cases}$ **4.** $\begin{cases} 5x - 3y = -6 \\ 10x - 6y = 3 \end{cases}$

Solve the following systems graphically and state whether the systems are consistent, inconsistent, or dependent.

5. $\begin{cases} y = -2x + 5 \\ y = x + 2 \end{cases}$ **6.** $\begin{cases} y = -\dfrac{1}{2}x \\ 3x - 2y = 8 \end{cases}$ **7.** $\begin{cases} x + 4y = 6 \\ y = -\dfrac{1}{4}x + \dfrac{3}{2} \end{cases}$ **8.** $\begin{cases} 5x + y = 4 \\ 10x + 2y = 8 \end{cases}$

9. $\begin{cases} -3x + y = 7 \\ -6x + 2y = 9 \end{cases}$ **10.** $\begin{cases} y = -x + 10 \\ x + y = 4 \end{cases}$ **11.** $\begin{cases} x + y = 3 \\ 2x + y = 2 \end{cases}$ **12.** $\begin{cases} x + 2y = 12 \\ 3x - y = -6 \end{cases}$

In Exercises 13 – 16, use a TI-84 Plus graphing calculator to find the solutions to each of the systems of linear equations.

13. $\begin{cases} 2x + 3y = 10.5 \\ x - y = -1 \end{cases}$ **14.** $\begin{cases} x - 3y = 2.5 \\ 2x + y = -5.5 \end{cases}$ **15.** $\begin{cases} y = 7 + x \\ x + 3y = 21 \end{cases}$ **16.** $\begin{cases} x - 5y = -1.5 \\ x - y = 4.3 \end{cases}$

5.2 Systems of Equations: Solutions by Substitution

Solve the systems of linear equations in Exercises 17 – 24 by using the method of substitution. If the system is inconsistent or dependent, say so in your answer.

17. $\begin{cases} y = 2x \\ x - y = -1 \end{cases}$ **18.** $\begin{cases} y = 4x + 1 \\ 8x - 2y = -2 \end{cases}$ **19.** $\begin{cases} 3x - 2y = 4 \\ y = \dfrac{3}{2}x + 6 \end{cases}$ **20.** $\begin{cases} y = -2x + 5 \\ 4x + 2y = -1 \end{cases}$

21. $\begin{cases} 5x - y = 6 \\ 2y = 10x - 12 \end{cases}$ **22.** $\begin{cases} y = -\dfrac{1}{8}x \\ x - 4y = 4 \end{cases}$ **23.** $\begin{cases} 2x - y = 2 \\ 2x + y = -2 \end{cases}$ **24.** $\begin{cases} x + \dfrac{1}{2}y = 8 \\ y = x + 1 \end{cases}$

5.3 Systems of Equations: Solutions by Addition

In Exercises 25 – 32, solve the system of linear equations by using either the substitution method or the addition method (whichever seems better to you). If the system is inconsistent or dependent, say so in your answer.

25. $\begin{cases} x + y = 1 \\ x - y = 3 \end{cases}$
  26. $\begin{cases} 3x - 2y = 5 \\ x + 3y = 17 \end{cases}$
  27. $\begin{cases} x + 2y = 1 \\ 2x - 3y = 0 \end{cases}$
  28. $\begin{cases} 5x + y = 1 \\ 3x + y = 3 \end{cases}$

29. $\begin{cases} -3x + y = 6 \\ 6x - 2y = -12 \end{cases}$
  30. $\begin{cases} 3x + 4y = 8 \\ y = -\dfrac{3}{4}x + 4 \end{cases}$
  31. $\begin{cases} y = -\dfrac{1}{2}x + 10 \\ x + 2y = 6 \end{cases}$
  32. $\begin{cases} y = \dfrac{1}{3}x + 10 \\ x - 3y = -30 \end{cases}$

5.4 Applications: Distance-Rate-Time, Number Problems, Amounts and Costs

33. Alice is 8 years older than her brother John. In 5 years she will be twice as old as John. How old are Alice and John now?

34. Jorge has $2.10 worth of change in quarters and dimes. If he has eight times as many dimes as quarters how many of each type of coin does he have?

35. The sum of two numbers is –12. One number is 3 more than twice the other.

36. The difference between two numbers is 2. Three times the smaller is equal to 8 more than the larger. What are the numbers?

37. Karl decided to hike in the mountains a total distance of 14.4 miles (on a trail he knew). He averaged 2 mph for the first part of the hike and 3 mph for the second part. If the total hike took 6 hours, how many hours did he hike at each rate?

38. The width of a rectangle is 5 meters less than the length. If the perimeter of the rectangle is 80 meters, what are the length and width?

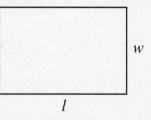

39. A golf pro shop sells two particular types of shirts, the first for $110 each and the other for $65 each. Last month the shop sold 50 of these shirts for a total of $4600. How many of each type of shirt did they sell?

40. The line $y = mx + b$ passes through the two points (2, 5) and (–1, –1). Find the equation of the line.

5.5 Applications: Interest and Mixture

41. Mr. Smith has $20,000 invested, part at 6% and the remainder at 8%. How much does he have invested at each rate if the annual interest is $1300?

42. Mary Jane invests a certain amount of money at 8% and twice as much at 5%. If her annual income from the two investments is $720, how much does she have invested at each rate?

43. If you invest half of your savings at 7% and the other half at 10% and you make $1275 annually, how much do you have invested at each rate?

44. A financial advisor advises Jennifer to invest her money equally in stocks and bonds. If the return on bonds is 4.5% and the return on the stocks is anticipated to be 7%, how much should she invest in each type of investment to make $2530 each year?

45. How many gallons each of a 25% salt solution and a 40% salt solution must be used to produce 60 gallons of a 35% solution?

46. A butcher decides to mix hamburger so it can be sold at a competitive price. If he mixes some that is 22% fat with some that is 10% fat, how many pounds of each type should he mix to get 80 pounds of hamburger that is 16% fat?

47. How many ounces of pure acid and how many ounces of a 10% acid solution should be mixed to make a total of 50 ounces of a 64% acid solution?

48. A manufacturer decided to mix two aluminum alloys, one is 20% alloy and the other is 60% alloy. How many tons of each type should he use to get a total of 100 tons of 28% aluminum alloy?

5.6 Graphing Systems of Linear Inequalities

In Exercises 49 – 54, solve the systems of two linear inequalities graphically.

49. $\begin{cases} y > 2x - 3 \\ y < x + 2 \end{cases}$ **50.** $\begin{cases} 2x + y \geq 12 \\ y \leq x + 3 \end{cases}$ **51.** $\begin{cases} y \geq x - 2 \\ y \leq 2x + 3 \end{cases}$

52. $\begin{cases} y < -x \\ y < 2x \end{cases}$ **53.** $\begin{cases} x \geq -1 \\ y \geq -1 \end{cases}$ **54.** $\begin{cases} x < 2 \\ y \geq -2x - 3 \end{cases}$

Use a TI-84 Plus graphing calculator to solve the systems of linear inequalities in Exercises 55 and 56.

55. $\begin{cases} y \leq 0 \\ y \leq 2x + 2 \end{cases}$ **56.** $\begin{cases} 2x + y > 10 \\ x - y > 0 \end{cases}$

Chapter 5 Test

In Exercises 1 and 2, determine which of the points satisfy the given system of linear equations.

1. $\begin{cases} 3x - 7y = 5 \\ 5x - 2y = -11 \end{cases}$

2. $\begin{cases} x - 2y = 7 \\ 2x - 3y = 5 \end{cases}$

a. $(1, 8)$

b. $(4, 1)$

c. $(-3, -2)$

d. $(13, 10)$

a. $(0, 3)$

b. $(7, 0)$

c. $(1, -1)$

d. $(-11, -9)$

For Exercises 3 – 12 solve as directed. In each exercise state whether the system is consistent, inconsistent, or dependent.

Solve the systems in Exercises 3 and 4 by graphing.

3. $\begin{cases} y = 2 - 5x \\ x - y = 6 \end{cases}$

4. $\begin{cases} x - y = 3 \\ 2x + 3y = 11 \end{cases}$

Solve the systems in Exercises 5 and 6 by using substitution.

5. $\begin{cases} 5x - 2y = 0 \\ y = 3x + 4 \end{cases}$

6. $\begin{cases} x = \dfrac{1}{3}y - 4 \\ 2x + \dfrac{3}{2}y = 5 \end{cases}$

Solve the systems in Exercises 7 and 8 by using the method of additon.

7. $\begin{cases} -2x + 3y = 6 \\ 4x + y = 1 \end{cases}$

8. $\begin{cases} 3x + 4y = 10 \\ x + 6y = 1 \end{cases}$

Solve the systems in Exercises 9 – 12 by using any method.

9. $\begin{cases} x + y = 2 \\ y = -2x - 1 \end{cases}$

10. $\begin{cases} 6x + 2y - 8 = 0 \\ y = -3x \end{cases}$

11. $\begin{cases} 7x + 5y = -9 \\ 6x + 2y = 6 \end{cases}$

12. $\begin{cases} x + 3y = -2 \\ 2x = -6y - 4 \end{cases}$

13. Solve the following system of linear inequalities graphically: $\begin{cases} x \geq -3 \\ y \geq 3x \end{cases}$

14. Determine the values of a and b such that the straight line $ax + by = 11$ passes through the two points $(1, -3)$ and $(2, 5)$.

15. Pete's boat can travel 48 miles upstream in 4 hours. The return trip takes 3 hours. Find the speed of the boat in still water and the speed of the current.

16. Eight pencils and two pens cost $2.22. Three pens and four pencils cost $2.69. What is the price of each pen and each pencil?

17. Gary has two investments yielding a total annual interest of $185.60. The amount invested at 8% is $320 less than twice the amount invested at 6%. How much is invested at each rate?

18. A metallurgist needs 2000 pounds of an alloy that is 80% copper. In stock, he has only alloys of 83% copper and 68% copper. How many pounds of each must be used?

19. The perimeter of a rectangle is 60 inches and the length is 4 inches longer than the width. Find the dimensions of the rectangle.

20. Sonia has a bag of coins with only nickels and quarters. She wants you to figure out how many of each type of coin she has and tells you that she has 105 coins and that the value of the coins is $17.25. Tell her that you know algebra and determine how many nickels and how many quarters she has.

Cumulative Review: Chapters 1 – 5

Use the rules for order of operations to evaluate the expressions in Exercises 1 – 4.

1. $-20 + 15 \div (-5) \cdot 2^3 - 11^2$

2. $24 \div 4 \cdot 6 - 36 \cdot 2 \div 3^2$

3. $\dfrac{3}{4} \div \dfrac{5}{8} - \dfrac{2}{3} \cdot \dfrac{6}{5} + \left(\dfrac{1}{5}\right)^2$

4. $3\left(4^2 - 16\right) + 5\left(9 - 3^2\right) - \left(6^2 - 9 \cdot 4\right)$

5. What is 110% of 50?

6. Find $\dfrac{7}{8}$ of 10,000.

Solve each of the equations in Exercises 7 – 10.

7. $2(x - 7) + 14 = -3x + 1$

8. $\dfrac{5x}{6} - \dfrac{2}{3} = \dfrac{x}{2} + \dfrac{1}{4}$

9. $2.3y - 1.6 = 3(1.2y + 2.5)$

10. $\dfrac{a}{5} + \dfrac{a}{7} = \dfrac{a}{2} - \dfrac{5}{14}$

In Exercises 11 and 12, solve the inequality and graph the solution set on a real number line.

11. $3x - 14 \geq 25$

12. $0 \leq 2x + 10 < 1.6$

In Exercises 13 and 14, solve for the indicated variable.

13. $v = k + gt$; solve for t.

14. $3x - 4y = 6$; solve for y.

15. For the equation $2x + 4y = 11$, determine the slope, m, and the y-intercept, b. Then use a graphing calculator to graph the line.

16. Write an equation for the line parallel to the line $x - 3y = -1$ and passing through the point $(1, 2)$. Use a graphing calculator to graph both lines.

17. Determine which of the points lie on both of the lines in the given system of equations.
$$\begin{cases} x - 2y = 6 \\ y = \dfrac{1}{2}x - 3 \end{cases}$$

    a. $(0, 6)$ **b.** $(6, 0)$ **c.** $(2, -2)$ **d.** $\left(3, -\dfrac{3}{2}\right)$

In Exercises 18 and 19, use a TI-84 Plus graphing calculator to locate the solutions to the systems of linear equations.

18. $\begin{cases} x = y + 3 \\ x + y = 10 \end{cases}$

19. $\begin{cases} x + y = 1 \\ 3x - y = 0 \end{cases}$

For Exercises 20 – 35 solve as directed. In each exercise state whether the system is consistent, inconsistent, or dependent.

Solve the systems in Exercises 20 – 23 by graphing.

20. $\begin{cases} 2x+3y=4 \\ 3x-y=6 \end{cases}$ **21.** $\begin{cases} 5x+2y=3 \\ y=4 \end{cases}$ **22.** $\begin{cases} 3x+y=3 \\ x-3y=4 \end{cases}$ **23.** $\begin{cases} y=4x-6 \\ 8x-2y=-4 \end{cases}$

Solve the systems in Exercises 24 – 27 by using the method of substitution.

24. $\begin{cases} x+y=-4 \\ 2x+7y=2 \end{cases}$ **25.** $\begin{cases} x=2y \\ y=\dfrac{1}{2}x+9 \end{cases}$ **26.** $\begin{cases} 4x+3y=8 \\ x+\dfrac{3}{4}y=2 \end{cases}$ **27.** $\begin{cases} 2x+y=0 \\ 7x+6y=-10 \end{cases}$

Solve the systems in Exercises 28 – 31 by using the method of addition.

28. $\begin{cases} 2x+y=7 \\ 2x-y=1 \end{cases}$ **29.** $\begin{cases} 3x-2y=9 \\ x-2y=11 \end{cases}$ **30.** $\begin{cases} 2x+4y=9 \\ 3x+6y=8 \end{cases}$ **31.** $\begin{cases} x+5y=10 \\ y=2-\dfrac{1}{5}x \end{cases}$

Solve the systems in Exercises 32 – 35 by using any method.

32. $\begin{cases} 3y-x=7 \\ x-2y=-2 \end{cases}$ **33.** $\begin{cases} x-\dfrac{2}{5}y=\dfrac{4}{5} \\ \dfrac{3}{4}x+\dfrac{3}{4}y=\dfrac{5}{4} \end{cases}$ **34.** $\begin{cases} x+3y=7 \\ 5x-2y=1 \end{cases}$ **35.** $\begin{cases} 3x-5y=17 \\ x+2y=4 \end{cases}$

In Exercises 36 and 37, write an equation for the line determined by the two given points. Use the formula $y = mx + b$ to set up a system of equations with m and b as the unknowns.

36. $(3,-1),(2,6)$ **37.** $(-2,5),(4,-3)$

38. a. Two angles are **complementary** if the sum of their measures is 90°. Find two complementary angles such that one is 10° more than three times the other.

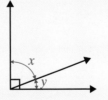

b. Two angles are **supplementary** if the sum of their measures is 180°. Find two supplementary angles such that one is 30° less than one-half of the other.

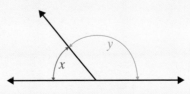

39. Bernice's secretary bought 15 stamps, some 41¢ and some 58¢. If he spent $7.00, how many of each kind did he buy?

40. A boat can travel 24 miles downstream in 2 hours. The return trip takes 3 hours. Find the speed of the boat in still water and the speed of the current.

41. The perimeter of a rectangle is 50 yards and the width is only 3 yards less than the length. What are the dimensions of the rectangle?

42. A company manufactures two kinds of dresses, Model A and Model B. Each Model A dress takes 4 hours to produce and each costs $18. Each Model B takes 2 hours to produce and each costs $7. If during a week there were 52 hours of production and costs of $198, how many of each model were produced?

43. Two trains leave Kansas City at the same time. One train travels east and the other travels west. The speed of the westbound train is 5 mph greater than the speed of the eastbound train. After 6 hours, they are 510 miles apart. Find the rate of each train. (Assume that the trains travel in a straight line in opposite directions.)

44. In a biology experiment, Owen observed that the bacteria count in a culture was approximately 100 at the end of 1 hour. At the end of 3 hours, the bacteria count was about 2000. Write a linear equation describing the bacteria count, N, in terms of the time, t.

45. List the set of ordered pairs corresponding to the points on the graph. Give the domain and range, and indicate if the relation is or is not a function.

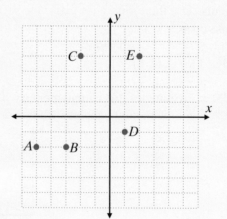

In Exercises 46 and 47, use the vertical line test to determine whether each of the graphs does or does not represent a function.

46.

47.

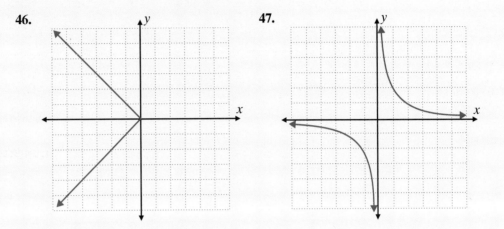

48. Given that $F(x) = 3x - 7$, find each of the following:

    a. $F(-5)$ **b.** $F(10)$ **c.** $F(3.5)$

49. Given that $f(x) = 2x^2 + 3x - 10$, find each of the following:

    a. $f(0)$ **b.** $f(-3)$ **c.** $f(6)$

In Exercises 50 – 52, solve the systems of linear inequalities graphically.

50. $\begin{cases} x > 2 \\ y < -x + 5 \end{cases}$ **51.** $\begin{cases} y \geq 1 \\ y \leq 2x + 5 \end{cases}$ **52.** $\begin{cases} 2x + 5y < 31 \\ x - 2y < -9 \end{cases}$

In Exercises 53 and 54, use a TI-84 Plus graphing calculator to solve each system of linear inequalities.

53. $\begin{cases} y \leq 4x + 6 \\ y \leq -x + 2 \end{cases}$ **54.** $\begin{cases} 2x - 3y < 12 \\ x + y < 6 \end{cases}$

Exponents and Polynomials

Did You Know?

One of the most difficult problems for students in beginning algebra is to become comfortable with the idea that letters or symbols can be manipulated just like numbers in arithmetic. These symbols may be the cause of "math anxiety." A great deal of publicity has recently been given to the concept that a large number of people suffer from math anxiety, a painful uneasiness caused by mathematical symbols or problem-solving situations. Persons affected by math anxiety find it almost impossible to learn mathematics, or they may be able to learn but be unable to apply their knowledge or do well on tests. Persons suffering from math anxiety often develop math avoidance, so they avoid careers, majors, or classes that will require mathematics courses or skills. The sociologist Lucy Sells has determined that mathematics is a critical filter in the job market. Persons who lack quantitative skills are channeled into high unemployment, low-paying, non-technical areas.

What causes math anxiety? Researchers are investigating the following hypotheses:

1. A lack of skills that leads to a lack of confidence and, therefore, to anxiety;
2. An attitude that mathematics is not useful or significant to society;
3. Career goals that seem to preclude mathematics;
4. A self-concept that differs radically from the stereotype of a mathematician;
5. Perceptions that parents, peers, or teachers have low expectations for the person in mathematics;
6. Social conditioning to avoid mathematics.

We hope that you are finding your present experience with algebra successful and that the skills you are acquiring now will enable you to approach mathematical problems with confidence.

6.1 Exponents

6.2 Exponents and Scientific Notation

6.3 Introduction to Polynomials

6.4 Addition and Subtraction with Polynomials

6.5 Multiplication with Polynomials

6.6 Special Products of Binomials

6.7 Division with Polynomials

"Mathematics is the queen of the sciences and arithmetic the queen of mathematics."

Karl F. Gauss (1777 – 1855)

In Chapter 6, you will learn the rules of exponents and how exponents can be used to simplify very large and very small numbers. Astronomers and chemists are very familiar with these ideas and the notation used is appropriately called scientific notation. This notation is also used in scientific calculators. (Multiply 5,000,000 by 5,000,000 on your calculator and read the results.)

Polynomials and operations with polynomials (the topics in Chapters 6, and 7) appear at almost every level of mathematics from elementary and intermediate algebra through statistics, calculus, and beyond. Be aware that the knowledge and skills you learn about polynomials will be needed again and again in any mathematics courses you take in the future.

6.1   Exponents

Objectives

After completing this section, you will be able to:

1. Simplify expressions by using properties of integer exponents.

2. Recognize which property of exponents is used to simplify an expression.

The Product Rule

In Section R.1, an **exponent** was defined as a number that tells how many times a number (called the **base**) is used in multiplication. This definition is limited because it is valid only if the exponents are positive integers. In this section, we will develop four properties of exponents that will help in simplifying algebraic expressions and expand your understanding of exponents to include variable bases, negative exponents, and the exponent 0. (In later sections you will study fractional exponents.)

From Section R.1, we know that

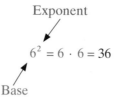

$$6^2 = 6 \cdot 6 = 36$$

and

$$6^3 = 6 \cdot 6 \cdot 6 = 216.$$

Also, the base may be a variable so that

$$x^3 = x \cdot x \cdot x \quad \text{and} \quad x^5 = x \cdot x \cdot x \cdot x \cdot x.$$

Now, to find the products of expressions such as $6^2 \cdot 6^3$ or $x^3 \cdot x^5$ and to simplify these products, we can write down all the factors as follows:

$$6^2 \cdot 6^3 = (6 \cdot 6) \cdot (6 \cdot 6 \cdot 6) = 6^5$$

and

$$x^3 \cdot x^5 = (x \cdot x \cdot x) \cdot (x \cdot x \cdot x \cdot x \cdot x) = x^8.$$

With these examples in mind, what do you think would be a simplified form for the product $3^4 \cdot 3^3$? You were right if you thought 3^7. That is, $3^4 \cdot 3^3 = 3^7$. Notice that in each case, **the base stays the same**.

The preceding discussion, along with the basic concept of whole-number exponents, leads to the following **Product Rule for Exponents**.

The Product Rule for Exponents

If a is a nonzero real number and m and n are integers, then

$$\boldsymbol{a^m \cdot a^n = a^{m+n}}$$

In words, to multiply two powers with the same base, keep the base and add the exponents.

NOTES Remember (see Section R.1) about the **exponent 1**. If a variable or constant has no exponent written, the exponent is understood to be 1.

For example,

$$y = y^1$$

and

$$7 = 7^1.$$

In general, for any real number a,

$$\boldsymbol{a = a^1}.$$

Example 1: The Product Rule for Exponents

Use the product rule for exponents to simplify the following expressions.

a. $x^2 \cdot x^4$

   Solution: $x^2 \cdot x^4 = x^{2+4} = x^6$

b. $y \cdot y^6$

   Solution: $y \cdot y^6 = y^1 \cdot y^6 = y^{1+6} = y^7$

c. $4^2 \cdot 4$

   Solution: $4^2 \cdot 4 = 4^{2+1} = 4^3 = 64$ Note that the base stays 4.

   That is, the bases are not multiplied.

d. $2^3 \cdot 2^2$

   Solution: $2^3 \cdot 2^2 = 2^{3+2} = 2^5 = 32$ Note that the base stays 2.

   That is, the bases are not multiplied.

e. $(-2)^4(-2)^3$

   Solution: $(-2)^4(-2)^3 = (-2)^{4+3} = (-2)^7 = -128$

To multiply terms that have numerical coefficients and variables with exponents, we need to be aware that the **coefficients are multiplied** as usual and the **exponents are added** by using the product rule. Example 2 illustrates these concepts.

Example 2: Multiplying Terms with Coefficients

Use the product rule for exponents when simplifying the following expressions.

a. $2y^2 \cdot 3y^9$

   Solution: $2y^2 \cdot 3y^9 = 2 \cdot 3 \cdot y^2 \cdot y^9$

   $= 6y^{2+9}$ Coefficients 2 and 3 are multiplied

   $= 6y^{11}$ and exponents 2 and 9 are added.

b. $\left(-3x^3\right)\left(-4x^3\right)$

Solution:

$$\left(-3x^3\right)\left(-4x^3\right) = (-3)(-4) \cdot x^3 \cdot x^3 \quad \text{Coefficients } -3 \text{ and } -4 \text{ are multiplied}$$

$$= 12x^{3+3} \qquad \text{and exponents 3 and 3 are added.}$$

$$= 12x^6$$

c. $\left(-6ab^2\right)\left(8ab^3\right)$

Solution: $\left(-6ab^2\right)\left(8ab^3\right) = -6 \cdot 8 \cdot a^1 \cdot a^1 \cdot b^2 \cdot b^3 \quad$ Coefficients -6 and 8

$$= -48 \cdot a^{1+1} \cdot b^{2+3} \qquad \text{are multiplied}$$

$$= -48a^2b^5 \qquad \text{and exponents on each}$$

variable are added.

The Exponent 0

The Product Rule is stated for m and n as **integer** exponents. This means that the rule is also valid for 0 and for negative exponents. As an aid for understanding 0 as an exponent, consider the following patterns of exponents for powers of 2, 3, and 5.

Powers of 2	Powers of 3	Powers of 5
$2^5 = 32$	$3^5 = 243$	$5^5 = 3125$
$2^4 = 16$	$3^4 = 81$	$5^4 = 625$
$2^3 = 8$	$3^3 = 27$	$5^3 = 125$
$2^2 = 4$	$3^2 = 9$	$5^2 = 25$
$2^1 = 2$	$3^1 = 3$	$5^1 = 5$
$2^0 = ?$	$3^0 = ?$	$5^0 = ?$

Do you notice that the patterns indicate that the exponent 0 gives the same value for the last number in each column? That is, $2^0 = 1$, $3^0 = 1$, and $5^0 = 1$.

Another approach to understanding 0 as an exponent involves the product rule. Remember that 1 is the multiplicative identity (i.e. $1 \cdot a = a$). Consider the following analysis:

$$2^0 \cdot 2^3 = 2^{0+3} = 2^3 \qquad \text{Using the product rule.}$$

and

$$1 \cdot 2^3 = 2^3.$$

So,

$$2^0 \cdot 2^3 = 1 \cdot 2^3 \quad \text{and} \quad 2^0 = 1.$$

Similarly,

$$7^0 \cdot 7^2 = 7^{0+2} = 7^2$$

and

$$1 \cdot 7^2 = 7^2.$$

So,

$$7^0 \cdot 7^2 = 1 \cdot 7^2 \quad \text{and} \quad 7^0 = 1.$$

This discussion leads directly to the **rule for 0 as an exponent**.

The Exponent 0

If a is a nonzero real number, then

$$a^0 = 1.$$

The expression 0^0 is undefined.

Example 3: The Exponent 0

Simplify the following expressions using the rule for 0 as an exponent.

a. 10^0

Solution: $10^0 = 1$

b. $x^0 \cdot x^3$

Solution: $x^0 \cdot x^3 = x^{0+3} = x^3$ or $x^0 \cdot x^3 = 1 \cdot x^3 = x^3$

c. $(-6)^0$

Solution: $(-6)^0 = 1$

The Quotient Rule

Now consider a fraction in which the numerator and denominator are powers with the same base, such as $\dfrac{5^4}{5^2}$ or $\dfrac{x^5}{x^2}$. We can write,

$$\frac{5^4}{5^2} = \frac{\not{5} \cdot \not{5} \cdot 5 \cdot 5}{\not{5} \cdot \not{5} \cdot 1} = \frac{5^2}{1} = 25 \quad \text{or} \quad \frac{5^4}{5^2} = 5^{4-2} = 5^2 = 25$$

and

$$\frac{x^5}{x^2} = \frac{\not{x} \cdot \not{x} \cdot x \cdot x \cdot x}{\not{x} \cdot \not{x} \cdot 1} = \frac{x^3}{1} = x^3 \quad \text{or} \quad \frac{x^5}{x^2} = x^{5-2} = x^3$$

In fractions, as just illustrated, the exponents can be subtracted. Again, the base remains the same. We now have the following **Quotient Rule for Exponents**.

The Quotient Rule for Exponents

If a is a nonzero real number and m and n are integers, then

$$\frac{a^m}{a^n} = a^{m-n}$$

In words, to divide two powers with the same base, keep the base and subtract the exponents. (Subtract the denominator exponent from the numerator exponent.)

Example 4: The Quotient Rule for Exponents

Use the quotient rule for exponents to simplify the following expressions.

a. $\dfrac{x^6}{x}$

  Solution: $\dfrac{x^6}{x} = x^{6-1} = x^5$

b. $\dfrac{y^8}{y^2}$

  Solution: $\dfrac{y^8}{y^2} = y^{8-2} = y^6$

Continued on next page...

c. $\dfrac{x^2}{x^2}$

Solution: $\dfrac{x^2}{x^2} = x^{2-2} = x^0 = 1$

Note how this example shows another way to justify the idea that $a^0 = 1$. Since the numerator and denominator are the same and not 0, it makes sense that the fraction is equal to 1.

In division with terms that have numerical coefficients the **coefficients are divided** as usual and any **exponents are subtracted** by using the quotient rule. These ideas are illustrated in Example 5.

Example 5: Dividing Terms with Coefficients

Use the Quotient Rule for exponents when simplifying the following expressions.

a. $\dfrac{15x^{15}}{3x^3}$

Solution: $\dfrac{15x^{15}}{3x^3} = \dfrac{15}{3} \cdot \dfrac{x^{15}}{x^3}$

$= 5 \cdot x^{15-3}$ Note that the coefficients are divided

$= 5x^{12}$ and the exponents are subtracted.

b. $\dfrac{20x^{10}y^6}{2x^2y^3}$

Solution: $\dfrac{20x^{10}y^6}{2x^2y^3} = \dfrac{20}{2} \cdot \dfrac{x^{10}}{x^2} \cdot \dfrac{y^6}{y^3}$

$= 10 \cdot x^{10-2} \cdot y^{6-3}$ Again, note that the coefficients

$= 10x^8y^3$ are divided and the exponents

are subtracted.

Negative Exponents

The Quotient Rule for exponents leads directly to the development of an understanding of negative exponents. In Examples 4 and 5, for each base, the exponent in the numerator was larger than or equal to the exponent in the denominator. Therefore, when the exponents were subtracted using the Quotient Rule, the result was either a positive exponent or the exponent 0. But, what if the larger exponent is in the denominator and we still apply the Quotient Rule?

For example, applying the quotient rule to $\dfrac{4^3}{4^5}$ gives

$$\frac{4^3}{4^5} = 4^{3-5} = 4^{-2} \text{ which results in a negative exponent.}$$

But, simply reducing $\dfrac{4^3}{4^5}$ gives

$$\frac{4^3}{4^5} = \frac{\cancel{4} \cdot \cancel{4} \cdot \cancel{4}}{\cancel{4} \cdot \cancel{4} \cdot \cancel{4} \cdot 4 \cdot 4} = \frac{1}{4 \cdot 4} = \frac{1}{4^2}.$$

This means that $4^{-2} = \dfrac{1}{4^2}$.

Similar discussions will show that $2^{-1} = \dfrac{1}{2}$, $5^{-2} = \dfrac{1}{5^2}$, and $6^{-3} = \dfrac{1}{6^3}$.

The **Rule for Negative Exponents** follows.

Rule for Negative Exponents

If a is a nonzero real number and n is an integer, then

$$a^{-n} = \frac{1}{a^n}$$

In words, a negative exponent indicates the reciprocal of the expression.

Example 6: Negative Exponents

Use the rule for negative exponents to simplify each expression so that it contains only positive exponents.

a. 5^{-1}

Solution: $\quad 5^{-1} = \dfrac{1}{5^1} = \dfrac{1}{5} \qquad\qquad$ Using the rule for negative exponents.

b. x^{-3}

Solution: $\quad x^{-3} = \dfrac{1}{x^3} \qquad\qquad$ Using the rule for negative exponents.

Continued on next page...

c. $x^{-9} \cdot x^7$

Solution: Here we use the product rule first and then the rule for negative exponents.

$$x^{-9} \cdot x^7 = x^{-9+7} = x^{-2} = \frac{1}{x^2}$$

Each of the expressions in Example 7 is simplified by using the appropriate rules for exponents. Study each example carefully. In each case, **the expression is considered simplified if each base appears only once and each base has only positive exponents.**

(**Note:** There is nothing wrong with negative exponents. In fact, negative exponents are preferred in later courses. However, so that all answers are the same, in this course we will consider expressions to be simplified if they have only positive exponents.)

Example 7: Combining the Rules of Exponents

Use the rule for negative exponents to simplify each expression so that it contains only positive exponents.

a. $2^{-5} \cdot 2^8$

Solution: $2^{-5} \cdot 2^8 = 2^{-5+8} = 2^3 = 8$ Using the product rule with positive and negative exponents.

b. $\dfrac{x^6}{x^{-1}}$

Solution: $\dfrac{x^6}{x^{-1}} = x^{6-(-1)} = x^{6+1} = x^7$ Using the quotient rule with positive and negative exponents.

c. $\dfrac{10^{-5}}{10^{-2}}$

Solution: $\dfrac{10^{-5}}{10^{-2}} = 10^{-5-(-2)}$ Using the quotient rule with positive and negative exponents.

$= 10^{-5+2}$ Using the rule for negative

$= 10^{-3}$ exponents.

$= \dfrac{1}{10^3}$ or $\dfrac{1}{1000}$

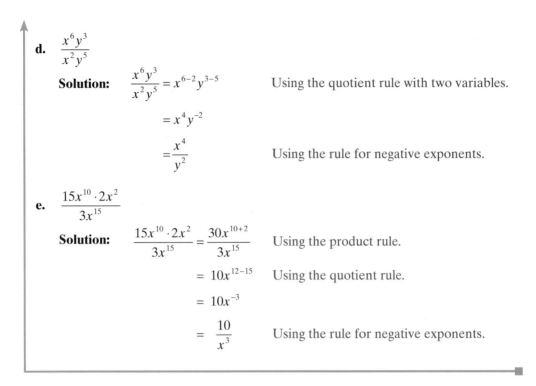

d. $\dfrac{x^6 y^3}{x^2 y^5}$

    Solution: $\dfrac{x^6 y^3}{x^2 y^5} = x^{6-2} y^{3-5}$ Using the quotient rule with two variables.

                     $= x^4 y^{-2}$

                     $= \dfrac{x^4}{y^2}$ Using the rule for negative exponents.

e. $\dfrac{15 x^{10} \cdot 2 x^2}{3 x^{15}}$

    Solution: $\dfrac{15 x^{10} \cdot 2 x^2}{3 x^{15}} = \dfrac{30 x^{10+2}}{3 x^{15}}$ Using the product rule.

                       $= 10 x^{12-15}$ Using the quotient rule.

                       $= 10 x^{-3}$

                       $= \dfrac{10}{x^3}$ Using the rule for negative exponents.

NOTES

Special Note about Using the Quotient Rule:
Regardless of the size of the exponents or whether they are positive or negative, the following single subtraction rule can be used with the quotient rule.

(numerator exponent – denominator exponent)

This subtraction will always lead to the correct answer.

Summary of the Rules for Exponents

For any nonzero real number a and integers m and n:

1. The Exponent 1: $a = a^1$

2. The Exponent 0: $a^0 = 1$

3. The Product Rule: $a^m \cdot a^n = a^{m+n}$

4. The Quotient Rule: $\dfrac{a^m}{a^n} = a^{m-n}$

5. Negative Exponents: $a^{-n} = \dfrac{1}{a^n}$

Practice Problems

Simplify each expression.

1. $2^3 \cdot 2^4$

2. $\dfrac{2^3}{2^4}$

3. $\dfrac{x^7 \cdot x^{-3}}{x^{-2}}$

4. $\dfrac{10^{-8} \cdot 10^2}{10^{-7}}$

5. $\dfrac{14x^{-3}y^2}{2x^{-3}y^{-2}}$

6. $\left(9x^4\right)^0$

6.1 Exercises

Simplify each expression and tell which rule (or rules) for exponents you used. The final form of the expressions with variables should contain only positive exponents. Assume that all variables represent nonzero numbers.

1. $3^2 \cdot 3$

2. $7^2 \cdot 7^3$

3. $8^3 \cdot 8^0$

4. $5^0 \cdot 5^2$

5. 3^{-1}

6. 4^{-2}

7. $(-5)^{-2}$

8. $(-6)^{-3}$

9. $(-2)^4 \cdot (-2)^0$

10. $3 \cdot 2^3$

11. $(-4)^3 \cdot (-4)^0$

12. $6 \cdot 3^2$

13. $-4 \cdot 5^3$

14. $-2 \cdot 3^3$

15. $3 \cdot 2^{-3}$

16. $4 \cdot 3^{-2}$

17. $-3 \cdot 5^{-2}$

18. $-5 \cdot 2^{-2}$

19. $x^2 \cdot x^3$

20. $x^3 \cdot x$

21. $y^2 \cdot y^0$

22. $y^3 \cdot y^8$

23. x^{-3}

24. y^{-2}

25. $2x^{-1}$

26. $5y^{-4}$

27. $-8y^{-2}$

28. $-10x^{-3}$

29. $5x^6y^{-4}$

30. x^0y^{-2}

31. $3x^0 + y^0$

32. $5y^0 - 3x^0$

33. $\dfrac{7^3}{7}$

34. $\dfrac{9^5}{9^2}$

35. $\dfrac{10^3}{10^4}$

36. $\dfrac{10}{10^5}$

Answers to Practice Problems: 1. $2^7 = 128$ **2.** $\dfrac{1}{2}$ **3.** x^6 **4.** 10 **5.** $7y^4$ **6.** 1

37. $\dfrac{2^3}{2^6}$ **38.** $\dfrac{5^7}{5^4}$ **39.** $\dfrac{x^4}{x^2}$ **40.** $\dfrac{x^6}{x^3}$

41. $\dfrac{x^3}{x}$ **42.** $\dfrac{y^7}{y^2}$ **43.** $\dfrac{x^7}{x^3}$ **44.** $\dfrac{x^8}{x^3}$

45. $\dfrac{x^{-2}}{x^2}$ **46.** $\dfrac{x^{-3}}{x}$ **47.** $\dfrac{x^4}{x^{-2}}$ **48.** $\dfrac{x^5}{x^{-1}}$

49. $\dfrac{x^{-3}}{x^{-5}}$ **50.** $\dfrac{x^{-4}}{x^{-1}}$ **51.** $\dfrac{y^{-2}}{y^{-4}}$ **52.** $\dfrac{y^3}{y^{-3}}$

53. $3x^3 \cdot x^0$ **54.** $3y \cdot y^4$ **55.** $(5x^2)(2x^2)$ **56.** $(3x^2)(3x)$

57. $(4x^3)(9x^0)$ **58.** $(5x^2)(3x^4)$ **59.** $(-2x^2)(7x^3)$ **60.** $(3y^3)(-6y^2)$

61. $(-4x^5)(3x)$ **62.** $(6y^4)(5y^5)$ **63.** $\dfrac{8y^3}{2y^2}$ **64.** $\dfrac{12x^4}{3x}$

65. $\dfrac{9y^5}{3y^3}$ **66.** $\dfrac{-10x^5}{2x}$ **67.** $\dfrac{-8y^4}{4y^2}$ **68.** $\dfrac{12x^6}{-3x^3}$

69. $\dfrac{21x^4}{-3x^2}$ **70.** $\dfrac{10 \cdot 10^3}{10^{-3}}$ **71.** $\dfrac{10^4 \cdot 10^{-3}}{10^{-2}}$ **72.** $\dfrac{10 \cdot 10^{-1}}{10^2}$

73. $(9x^2)^0$ **74.** $(9x^2y^3)(-2x^3y^4)$ **75.** $(-3xy)(-5x^2y^{-3})$ **76.** $\dfrac{-8x^2y^4}{4x^3y^2}$

77. $\dfrac{-8x^{-2}y^4}{4x^2y^{-2}}$ **78.** $(-2x^{-3}y^5)^0$ **79.** $(3a^2b^4)(4ab^5c)$ **80.** $(-6a^3b^{-4})(4a^{-2}b^8)$

81. $\dfrac{36a^5b^0c}{-9a^{-5}b^{-3}}$

82. $\dfrac{25y^6 \cdot 3y^{-2}}{15xy^4}$

Calculator Problems

Use a TI-84 Plus calculator to evaluate each expression.

83. $(2.16)^0$

84. $(-5.06)^2$

85. $(1.6)^{-2}$

86. $(6.4)^5 \cdot (2.3)^2$

87. $(-14.8)^2 \cdot (21.3)^2$

Hawkes Learning Systems: Introductory Algebra

Simplifying Integer Exponent I

6.2 Exponents and Scientific Notation

Objectives

After completing this section, you will be able to:

1. Simplify powers of expressions by using the properties of integer exponents.

2. Write a decimal number in scientific notation.

3. Operate with decimal numbers by using scientific notation.

The summary of the rules for exponents given in Section 6.1 is repeated here for easy reference.

Summary of the Rules for Exponents

For any nonzero real number a and integers m and n:

1. The Exponent 1: $a = a^1$

2. The Exponent 0: $a^0 = 1$

3. The Product Rule: $a^m \cdot a^n = a^{m+n}$

4. The Quotient Rule: $\dfrac{a^m}{a^n} = a^{m-n}$

5. Negative Exponents: $a^{-n} = \dfrac{1}{a^n}$

Power Rule

Now, consider what happens when a power is raised to a power. For example, to simplify the expressions $(x^2)^3$ and $(2^5)^2$, we can write

$$\left(x^2\right)^3 = x^2 \cdot x^2 \cdot x^2 = x^{2+2+2} = x^6$$

and

$$\left(2^5\right)^2 = 2^5 \cdot 2^5 = 2^{5+5} = 2^{10}$$

However, this technique can be quite time-consuming when the exponent is large such as in $(3y^3)^{17}$. The **Power Rule for Exponents** gives a convenient way to handle powers raised to powers.

Power Rule for Exponents

If a is a nonzero real number and m and n are integers, then

$$\left(a^m\right)^n = a^{mn}.$$

In words, the value of a power raised to a power can be found by multiplying the exponents and keeping the base.

Example 1: Power Rule for Exponents

Simplify each expression using the power rule for exponents.

a. $\left(x^2\right)^4$

   Solution: $\left(x^2\right)^4 = x^{2 \cdot 4} = x^8$

b. $\left(x^5\right)^{-2}$

   Solution: $\left(x^5\right)^{-2} = x^{5(-2)} = x^{-10} = \dfrac{1}{x^{10}}$

   or $\left(x^5\right)^{-2} = \dfrac{1}{\left(x^5\right)^2} = \dfrac{1}{x^{5 \cdot 2}} = \dfrac{1}{x^{10}}$

c. $\left(y^{-7}\right)^2$

   Solution: $\left(y^{-7}\right)^2 = y^{(-7)2} = y^{-14} = \dfrac{1}{y^{14}}$

d. $\left(2^3\right)^{-2}$

   Solution: $\left(2^3\right)^{-2} = 2^{3(-2)} = 2^{-6} = \dfrac{1}{2^6}$ $\left(\text{Evaluating gives } \dfrac{1}{2^6} = \dfrac{1}{64}.\right)$

   Another approach, because we have a numerical base, would be

   $\left(2^3\right)^{-2} = (8)^{-2} = \dfrac{1}{8^2}.$ $\left(\text{Evaluating gives } \dfrac{1}{8^2} = \dfrac{1}{64}.\right)$

   We see that while the base and exponent may be different, the value is the same.

Rule for Power of a Product

If the base of an exponent is a product, we will see that each factor in the product can be raised to the power indicated by the exponent. For example, $(10x)^3$ indicates that the product of 10 and x is to be raised to the 3rd power and $(-2x^2y)^5$ indicates that the product of -2, x^2, and y is to be raised to the 5th power. We can simplify these expressions as follows:

$$\left(10x\right)^3 = 10x \cdot 10x \cdot 10x = 10 \cdot 10 \cdot 10 \cdot x \cdot x \cdot x = 10^3 \cdot x^3 = 1000x^3$$

and

$$\left(-2x^2y\right)^5 = \left(-2x^2y\right) \cdot \left(-2x^2y\right) \cdot \left(-2x^2y\right) \cdot \left(-2x^2y\right) \cdot \left(-2x^2y\right)$$
$$= (-2)(-2)(-2)(-2)(-2) \cdot x^2 \cdot x^2 \cdot x^2 \cdot x^2 \cdot x^2 \cdot y \cdot y \cdot y \cdot y \cdot y$$
$$= (-2)^5 \cdot (x^2)^5 \cdot y^5$$
$$= -32x^{10}y^5.$$

We can simplify expressions such as these in a much easier fashion by using the following **Power of a Product Rule for Exponents**.

Power of a Product

If a and b are nonzero real numbers and n is an integer then

$$(ab)^n = a^n b^n.$$

In words, a power of a product is found by raising each factor to that power.

Example 2: Rule for Power of a Product

Simplify each expression using the rule for power of a product.

a. $(5x)^2$

Solution: $(5x)^2 = 5^2 \cdot x^2 = 25x^2$

b. $(xy)^3$

Solution: $(xy)^3 = x^3 \cdot y^3 = x^3y^3$

Continued on next page...

c. $(-7ab)^2$

Solution: $(-7ab)^2 = (-7)^2 a^2 b^2 = 49a^2b^2$

d. $(ab)^{-5}$

Solution: $(ab)^{-5} = a^{-5} \cdot b^{-5} = \dfrac{1}{a^5} \cdot \dfrac{1}{b^5} = \dfrac{1}{a^5 b^5}$

or, using the rule of negative exponents first and then the rule for the power of a product,

$$(ab)^{-5} = \dfrac{1}{(ab)^5} = \dfrac{1}{a^5 b^5}$$

e. $(x^2 y^{-3})^4$

Solution: $(x^2 y^{-3})^4 = (x^2)^4 \cdot (y^{-3})^4 = x^8 \cdot y^{-12} = x^8 \cdot \dfrac{1}{y^{12}} = \dfrac{x^8}{y^{12}}$

NOTES

Special Note about Negative Numbers and Exponents:

In an expression such as $-x^2$, we know that -1 is understood to be the coefficient of x^2. That is,

$$-x^2 = -1 \cdot x^2.$$

The same is true for expressions with numbers such as -7^2. That is,

$$-7^2 = -1 \cdot 7^2 = -1 \cdot 49 = -49.$$

We see that the exponent refers to 7 and **not** to -7. For the exponent to refer to -7 as the base, -7 **must be in parentheses** as follows:

$$(-7)^2 = (-7) \cdot (-7) = +49.$$

As another example,

$$-2^0 = -1 \cdot 2^0 = -1 \cdot 1 = -1 \text{ and } (-2)^0 = 1.$$

Rule for Power of a Quotient

If the base of an exponent is a quotient (in fraction form), we will see that both the numerator and denominator can be raised to the power indicated by the exponent. For example, in the expression $\left(\dfrac{2}{x}\right)^3$ the quotient (or fraction) $\dfrac{2}{x}$ is raised to the 3rd power.

We can simplify this expression as follows:

$$\left(\dfrac{2}{x}\right)^3 = \dfrac{2}{x} \cdot \dfrac{2}{x} \cdot \dfrac{2}{x} = \dfrac{2 \cdot 2 \cdot 2}{x \cdot x \cdot x} = \dfrac{2^3}{x^3} = \dfrac{8}{x^3}.$$

Or, we can simplify the expression in a much easier manner by applying the following **Rule for the Power of a Quotient**.

Rule for the Power of a Quotient

If a and b are nonzero real numbers and n is an integer, then

$$\left(\frac{a}{b}\right)^n = \frac{a^n}{b^n}.$$

In words, a power of a quotient (in fraction form) is found by raising both the numerator and the denominator to that power.

Example 3: Rule for the Power of a Quotient

Simplify each expression using the rule for the power of a quotient.

a. $\left(\dfrac{y}{x}\right)^5$

Solution: $\left(\dfrac{y}{x}\right)^5 = \dfrac{y^5}{x^5}$

b. $\left(\dfrac{2}{a}\right)^4$

Solution: $\left(\dfrac{2}{a}\right)^4 = \dfrac{2^4}{a^4} = \dfrac{16}{a^4}$

Using Combinations of Rules for Exponents

There may be more than one way to apply the various rules for exponents. As illustrated in the following examples, if you apply the rules correctly (even in a different sequence), the answer will be the same in every case.

Example 4: Applying Combinations of Rules for Exponents

Simplify each expression by using the appropriate rules for exponents.

a. $\left(\dfrac{-2x}{y^2}\right)^3$

Solution: $\left(\dfrac{-2x}{y^2}\right)^3 = \dfrac{\left(-2x\right)^3}{\left(y^2\right)^3} = \dfrac{\left(-2\right)^3 x^3}{y^6} = \dfrac{-8x^3}{y^6}$

Continued on next page...

b. $\left(\dfrac{3a^2 b}{a^3 b^2}\right)^2$

Solution:

Option 1: Simplify inside the parentheses first

$$\left(\frac{3a^2 b}{a^3 b^2}\right)^2 = \left(3a^{2-3}b^{1-2}\right)^2 = \left(3a^{-1}b^{-1}\right)^2 = 3^2 a^{-2} b^{-2} = \frac{9}{a^2 b^2}$$

Option 2: Apply the Power Rule first

$$\left(\frac{3a^2 b}{a^3 b^2}\right)^2 = \frac{3^2 a^{2\cdot2} b^2}{a^{3\cdot2} b^{2\cdot2}} = \frac{9a^4 b^2}{a^6 b^4} = 9a^{4-6} b^{2-4} = 9a^{-2} b^{-2} = \frac{9}{a^2 b^2}$$

Note that the answer is the same even though the rules were applied in a different order.

Another general approach with fractions involving negative exponents is to note that

$$\left(\frac{a}{b}\right)^{-n} = \frac{a^{-n}}{b^{-n}} = \frac{b^n}{a^n} = \left(\frac{b}{a}\right)^n.$$

In effect, there are two basic shortcuts with negative exponents and fractions:
1. Taking the reciprocal of a fraction changes the sign of any exponent on the fraction.
2. Moving any term from numerator to denominator or vice versa, changes the sign of the corresponding exponent.

Example 5:   Two Approaches with Fractional Expressions and Negative Exponents

Simplify: $\left(\dfrac{x^3}{y^5}\right)^{-4}$

Solution:

Option 1: Use the ideas of reciprocals first:

$$\left(\frac{x^3}{y^5}\right)^{-4} = \left(\frac{y^5}{x^3}\right)^4 = \frac{y^{5\cdot4}}{x^{3\cdot4}} = \frac{y^{20}}{x^{12}}$$

Option 2: Apply the power rule first:

$$\left(\frac{x^3}{y^5}\right)^{-4} = \frac{x^{3(-4)}}{y^{5(-4)}} = \frac{x^{-12}}{y^{-20}} = \frac{y^{20}}{x^{12}}$$

Example 6: A More Complex Example

This example involves the application of a variety of steps. Study it carefully and see if you can get the same result by following a different sequence of steps.

Simplify: $\left(\dfrac{2x^2y^3}{3xy^{-2}}\right)^{-2}\left(\dfrac{4x^2y^{-1}}{3x^{-5}y^3}\right)^{-1}$

Solution:

$$\left(\frac{2x^2y^3}{3xy^{-2}}\right)^{-2}\left(\frac{4x^2y^{-1}}{3x^{-5}y^3}\right)^{-1} = \frac{2^{-2}\cdot x^{2(-2)}\cdot y^{3(-2)}}{3^{-2}\cdot x^{-2}y^{-2(-2)}}\cdot\frac{4^{-1}\cdot x^{2(-1)}\cdot y^{-1(-1)}}{3^{-1}\cdot x^{-5(-1)}\cdot y^{3(-1)}}$$

$$= \frac{3^2}{2^2}\cdot\frac{x^{-4}}{x^{-2}}\cdot\frac{y^{-6}}{y^4}\cdot\frac{3}{4}\cdot\frac{x^{-2}}{x^5}\cdot\frac{y^1}{y^{-3}}$$

$$= \frac{9\cdot3\cdot x^{-4-2}\,y^{-6+1}}{4\cdot4\cdot x^{-2+5}\,y^{4-3}} = \frac{27x^{-6}y^{-5}}{16x^3y^1}$$

$$= \frac{27x^{-6-3}\,y^{-5-1}}{16} = \frac{27x^{-9}y^{-6}}{16} = \frac{27}{16x^9y^6}$$

Now a complete summary of the rules for exponents shows eight rules.

Summary of the Rules for Exponents

If a and b are nonzero real numbers and m and n are integers:

1. The Exponent 1: $a = a^1$

2. The Exponent 0: $a^0 = 1$

3. The Product Rule: $a^m \cdot a^n = a^{m+n}$

4. The Quotient Rule: $\dfrac{a^m}{a^n} = a^{m-n}$

5. Negative Exponents: $a^{-n} = \dfrac{1}{a^n}$

6. Power Rule: $(a^m)^n = a^{mn}$

7. Power of a Product: $(ab)^n = a^n b^n$

8. Power of a Quotient: $\left(\dfrac{a}{b}\right)^n = \dfrac{a^n}{b^n}$

Scientific Notation and Calculators

A basic application of integer exponents occurs in scientific disciplines, such as astronomy and biology, when very large and very small numbers are involved. For example, the distance from the earth to the sun is approximately 93,000,000 miles, and the approximate radius of a carbon atom is 0.000 000 007 7 centimeters.

In **scientific notation** (an option in all scientific and graphing calculators), **decimal numbers are written as the product of a number greater than or equal to 1 and less than 10, and an integer power of 10.** In scientific notation there is just one digit to the left of the decimal point. For example,

$$250,000 = 2.5 \times 10^5 \quad \text{and} \quad 0.000\,000\,345 = 3.45 \times 10^{-7}.$$

The exponent tells how many places the decimal point is to be moved and in what direction. If the exponent is positive, the decimal point is moved to the right:

$$5.6 \times 10^4 = 56,000. \qquad \text{4 places right}$$

A negative exponent indicates that the decimal point should move to the left:

$$4.9 \times 10^{-3} = 0.004.9 \qquad \text{3 places left}$$

Scientific Notation

*If N is a decimal number, then in **scientific notation***

$$N = a \times 10^n \quad \text{where } 1 \le a < 10 \text{ and n is an integer.}$$

Example 7: Decimals in Scientific Notation

Write the following decimal numbers in scientific notation.

    a. 8,720,000   **b.** 0.000 000 376

Solutions:

    a. $8,720,000 = 8.72 \times 10^6$ 8.72 is between 1 and 10

To check, move the decimal point 6 places to the right and get the original number:

$$8.72 \times 10^6 = 8.720000. = 8,720,000.$$
$$1\;2\;3\;4\;5\;6$$

    b. $0.000\,000\,376 = 3.76 \times 10^{-7}$ 3.76 is between 1 and 10

To check, move the decimal point 7 places to the left and get the original number:

$$3.76 \times 10^{-7} = 0.0000003.76 = 0.000000376.$$
$$\phantom{3.76 \times 10^{-7} = 0.000}7\;6\;5\;4\;3\;2\;1$$

Example 8: Properties of Exponents

Simplify the following expressions by first writing the decimal numbers in scientific notation and then using the properties of exponents.

a. $\dfrac{0.085 \times 41,000}{0.00017}$

Solution: $\dfrac{0.085 \times 41,000}{0.00017} = \dfrac{8.5 \times 10^{-2} \times 4.1 \times 10^{4}}{1.7 \times 10^{-4}}$

$$= \dfrac{\overset{5}{\cancel{8.5}} \times 4.1}{\cancel{1.7}} \times \dfrac{10^{-2} \times 10^{4}}{10^{-4}} = 20.5 \times \dfrac{10^{2}}{10^{-4}}$$

$$= 2.05 \times 10^{1} \times \dfrac{10^{2}}{10^{-4}} = 2.05 \times 10^{1+2-(-4)}$$

$$= 2.05 \times 10^{7}$$

b. $\dfrac{11,100 \times 0.064}{8,000,000 \times 370}$

Solution: $\dfrac{11,100 \times 0.064}{8,000,000 \times 370} = \dfrac{1.11 \times 10^{4} \times 6.4 \times 10^{-2}}{8.0 \times 10^{6} \times 3.7 \times 10^{2}}$

$$= \dfrac{\overset{0.3}{\cancel{1.11}} \times \overset{0.8}{\cancel{6.4}}}{\cancel{8.0} \times \cancel{3.7}} \times \dfrac{10^{2}}{10^{8}}$$

$$= 0.24 \times 10^{2-8} = 2.4 \times 10^{-1} \times 10^{-6}$$

$$= 2.4 \times 10^{-1-6} = 2.4 \times 10^{-7}$$

c. Light travels approximately 3×10^{8} meters per second. How many meters per minute does light travel?

Solution: Since there are 60 seconds in one minute, multiply by 60.

$$3 \times 10^{8} \times 60 = 180 \times 10^{8} = 1.8 \times 10^{2} \times 10^{8} = 1.8 \times 10^{10}$$

Thus light travels 1.8×10^{10} meters per minute.

Example 9:   Scientific Notation and Calculators

a.   Use a TI-84 Plus graphing calculator to evaluate the expression $\dfrac{0.0042 \cdot 3,000,000}{0.21}$.
Leave the answer in scientific notation.

(**Note:** The caret key  is used to indicate an exponent. Also, you can press
the **MODE** key and select SCI on the first line to have all decimal calculations in
scientific notation.)

Solution: With a TI-84 Plus calculator (set in scientific notation mode) the display
should appear as shown below:

b.   A light-year is the distance light travels in one year. Use a TI-84 Plus calculator to
find the length of a light-year in scientific notation if light travels 186,000 miles per
second.

Solution: 60 sec = 1 minute

60 min = 1 hour

24 hr = 1 day

365 days = 1 year

Multiplication gives the following display on your calculator.

(Note that the **E** in the display indicates an exponent with base 10.)

Thus a light-year is 5,865,696,000,000 miles.

Practice Problems

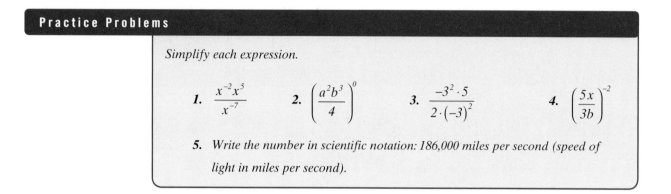

Simplify each expression.

1. $\dfrac{x^{-2}x^{5}}{x^{-7}}$

2. $\left(\dfrac{a^{2}b^{3}}{4}\right)^{0}$

3. $\dfrac{-3^{2}\cdot 5}{2\cdot(-3)^{2}}$

4. $\left(\dfrac{5x}{3b}\right)^{-2}$

5. Write the number in scientific notation: 186,000 miles per second (speed of light in miles per second).

6.2 Exercises

Use the rules for exponents to simplify each of the expressions in Exercises 1 – 44. Assume that all variables represent nonzero real numbers.

1. -3^{4}

2. -5^{2}

3. $(-3)^{4}$

4. -20^{2}

5. $(-10)^{6}$

6. $(-4)^{6}$

7. $(6x^{3})^{2}$

8. $(-3x^{4})^{2}$

9. $4(-3x^{2})^{3}$

10. $7(y^{-2})^{4}$

11. $-3(7xy^{2})^{0}$

12. $5(x^{2}y^{-1})$

13. $-2(3x^{5}y^{-2})^{-3}$

14. $\left(\dfrac{3x}{y}\right)^{3}$

15. $\left(\dfrac{-4x}{y^{2}}\right)^{2}$

16. $\left(\dfrac{6m^{3}}{n^{5}}\right)^{0}$

17. $\left(\dfrac{3x^{2}}{y^{3}}\right)^{2}$

18. $\left(\dfrac{-2x^{2}}{y^{-2}}\right)^{2}$

19. $\left(\dfrac{x}{y}\right)^{-2}$

20. $\left(\dfrac{2a}{b}\right)^{-1}$

21. $\left(\dfrac{2x}{y^{5}}\right)^{-2}$

22. $\left(\dfrac{3x}{y^{-2}}\right)^{-1}$

23. $\left(\dfrac{4a^{2}}{b^{-3}}\right)^{-3}$

24. $\left(\dfrac{-3}{xy^{2}}\right)^{-3}$

25. $\left(\dfrac{5xy^{3}}{y}\right)^{2}$

26. $\left(\dfrac{m^{2}n^{3}}{mn}\right)^{2}$

27. $\left(\dfrac{2ab^{3}}{b^{2}}\right)^{4}$

28. $\left(\dfrac{-7^{2}x^{2}y}{y^{3}}\right)^{-1}$

29. $\left(\dfrac{2ab^{4}}{b^{2}}\right)^{-3}$

Answers to Practice Problems: **1.** x^{10} **2.** 1 **3.** $\dfrac{-5}{2}$ **4.** $\dfrac{9b^{2}}{25x^{2}}$ **5.** 1.86×10^{5}

30. $\left(\dfrac{5x^3 y}{y^2}\right)^2$ **31.** $\left(\dfrac{2x^2 y}{y^3}\right)^{-4}$ **32.** $\left(\dfrac{x^3 y^{-1}}{y^2}\right)^2$ **33.** $\left(\dfrac{2a^2 b^{-1}}{b^2}\right)^3$

34. $\left(\dfrac{6y^5}{x^2 y^{-2}}\right)^2$ **35.** $\dfrac{\left(7x^{-2}y\right)^2}{\left(xy^{-1}\right)^2}$ **36.** $\dfrac{-5x^6 y^7}{3x^{-6}y^2}$ **37.** $\dfrac{\left(3x^2 y^{-1}\right)^{-2}}{\left(6x^{-1}y\right)^{-3}}$

38. $\dfrac{\left(2x^{-3}\right)^{-3}}{\left(5y^{-2}\right)^{-2}}$ **39.** $\dfrac{\left(4x^{-2}\right)\left(6x^5\right)}{\left(9y\right)\left(2y^{-1}\right)}$ **40.** $\dfrac{\left(5x^2\right)\left(3x^{-1}\right)^2}{\left(25y^3\right)\left(6y^{-2}\right)}$ **41.** $\left(\dfrac{3xy^3}{4x^2 y^{-3}}\right)^{-1}\left(\dfrac{2x^3 y^{-1}}{9x^{-3}y^{-1}}\right)^2$

42. $\left(\dfrac{5a^4 b^{-2}}{6a^{-4}b^3}\right)^{-2}\left(\dfrac{5a^3 b^4}{2^{-2}a^{-2}b^{-2}}\right)^3$ **43.** $\left(\dfrac{6x^{-4}yz^{-2}}{4^{-1}x^{-4}y^3 z^{-2}}\right)^{-1}\left(\dfrac{2^{-2}xyz^{-3}}{12x^2 y^2 z^{-1}}\right)^{-2}$

44. $\left(\dfrac{3^{-5}a^5 b^3 c^{-1}}{3^{-2}abc}\right)^{-2}\left(\dfrac{7^{-1}a^{-4}bc^2}{7^{-2}a^{-3}bc^{-2}}\right)^{-2}$

In Exercises 45 – 50, write the numbers in scientific notation.

45. 86,000   **46.** 927,000   **47.** 0.0362

48. 0.0061   **49.** 18,300,000   **50.** 376,000,000

In Exercises 51 – 56, write the numbers in decimal form.

51. 4.2×10^{-2} **52.** 8.35×10^{-3} **53.** 7.56×10^6

54. 6.132×10^{-5} **55.** 8.515×10^8 **56.** 9.374×10^7

In Exercises 57 – 70, first write each of the numbers in scientific notation. Then perform the indicated operations and leave your answer in scientific notation.

57. $300 \cdot 0.00015$

58. $0.000024 \cdot 40{,}000$

59. $0.0003 \cdot 0.0000025$

60. $0.00005 \cdot 0.00013$

61. $\dfrac{3900}{0.003}$

62. $\dfrac{4800}{12{,}000}$

63. $\dfrac{125}{50{,}000}$

64. $\dfrac{0.0046}{230}$

65. $\dfrac{0.02 \cdot 3900}{0.013}$

66. $\dfrac{0.0084 \cdot 0.003}{0.21 \cdot 60}$

67. $\dfrac{0.005 \cdot 650 \cdot 3.3}{0.0011 \cdot 2500}$

68. $\dfrac{5.4 \cdot 0.003 \cdot 50}{15 \cdot 0.0027 \cdot 200}$

69. $\dfrac{\left(1.4 \times 10^{-2}\right)(922)}{\left(3.5 \times 10^{3}\right)\left(2.0 \times 10^{6}\right)}$

70. $\dfrac{(4300)\left(3.0 \times 10^{2}\right)}{\left(1.5 \times 10^{-3}\right)\left(860 \times 10^{-2}\right)}$

71. One light-year is approximately 9.46×10^{15} meters. The distance to a certain star is 4.3 light-years. How many meters is this?

72. The mass of a hydrogen atom is approximately 0.00000000000000000000000167 grams. Write this number in scientific notation.

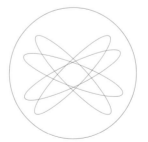

73. The weight of an atom is expressed in atomic weight units (amu), where 1 amu = 0.000000000000000000000000000016605 kilograms. Express the weight in scientific notation.

74. The distance light travels in one year is called a light-year. A light-year is approximately 5,866,000,000,000 miles. Write this number in scientific notation. How far does light travel in 2.3 years?

75. Light travels approximately 3×10^{10} centimeters per second. How many centimeters would this be per minute? Per hour? Express your answers in scientific notation.

76. The mass of the earth is about 5,980,000,000,000,000,000,000,000,000 grams. Write this number in scientific notation.

77. An atom of gold weighs approximately 3.25×10^{-22} grams. What would be the weight of 2000 atoms of gold? Express your answer in scientific notation.

78. An ounce of gold contains 5×10^{22} atoms. All the gold ever taken out of the earth is estimated to be 3.0×10^{31} atoms. How many ounces of gold is this?

79. There are approximately 6×10^{13} cells in an adult human body. Express this number in decimal form.

80. Write the number 1.0×10^{10} in decimal form. About how many years are in 1.0×10^{10} minutes?

In Exercises 81 – 88, use your calculator (set in scientific notation mode) to evaluate each expression. Leave the answer in scientific notation.

81. $8900 \times 90,000$

82. $0.0081 \div 9000$

83. $400 \times 175,000 + 5000 \times 3000$

84. $7000 \times 6000 + 200 \times 450,000$

85. $\dfrac{5.6 \cdot 0.003 \cdot 5000}{15 \cdot 0.0028 \cdot 20}$

86. $\dfrac{0.0006 \cdot 660 \cdot 40.4}{0.00011 \cdot 3600}$

87. $\dfrac{\left(1.8 \times 10^{-3}\right)(932)}{\left(4.5 \times 10^{3}\right)\left(2.0 \times 10^{-6}\right)}$

88. $\dfrac{(86,000)\left(3.0 \times 10^{4}\right)}{\left(4.3 \times 10^{-2}\right)\left(1.5 \times 10^{-3}\right)}$

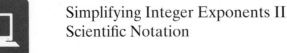

Hawkes Learning Systems: Introductory Algebra

Simplifying Integer Exponents II
Scientific Notation

| 6.3 | Introduction to Polynomials |

After completing this section, you will be able to:

1. Define a polynomial.

2. Classify a polynomial as a monomial, binomial, trinomial, or a polynomial with more than three terms.

3. Evaluate a polynomial for given values of the variable.

Definition of a Polynomial

A **term** is an expression that involves only multiplication and/or division with constants and/or variables. Remember that a number written next to a variable indicates multiplication, and the number is called the **numerical coefficient** (or **coefficient**) of the variable. For example,

$$3x, \; -5y^2, \; 17, \text{ and } \frac{x}{y}$$

are all algebraic terms. In the term $3x$, 3 is the coefficient of x. In the term $-5y^2$, -5 is the coefficient of y^2 and 1 is the coefficient of the term $\frac{x}{y}$. A term that consists of only a number, such as 17, is called a **constant** or a **constant term**.

*A **monomial in x** is a term of the form*

$$kx^n$$

*where **k** is a real number and **n** is a whole number.*

***n** is called the **degree** of the term, and **k** is called the **coefficient**.*

A monomial may have more than one variable, and **the degree of such a monomial is the sum of the degrees of its variables**. For example,

$$4x^2y^3 \text{ is a } 5^{\text{th}} \text{ degree monomial in } x \text{ and } y.$$

However, in this chapter, only monomials of one variable (note that any variable may be used in place of x) will be discussed.

Monomials may have fractional or negative coefficients; however, monomials may **not** have fractional or negative exponents. These facts are part of the definition since in the expression kx^n, k (the coefficient) can be any real number, but n (the exponent) must be a whole number.

$$\text{Expressions that } \textbf{are not} \text{ monomials: } 3\sqrt{x}, \ -15x^{\frac{2}{3}}, \ 4a^{-2}$$

$$\text{Expressions that } \textbf{are} \text{ monomials: } 17, \ 3x, \ 5y^2, \ \frac{2}{7}a^4, \ \pi x^2, \ \sqrt{2}x^3$$

Since $x^0 = 1$, a nonzero constant can be multiplied by x^0 without changing its value. Thus we say that a **nonzero constant is a monomial of degree 0**. For example,

$$17 = 17x^0 \quad \text{and} \quad -6 = -6x^0$$

which means that the constants 17 and –6 are monomials of degree 0. However, for the special number 0, we can write

$$0 = 0x^2 = 0x^5 = 0x^{13}$$

and we say that **the constant 0 is a monomial of no degree**.

A **polynomial** is a monomial or the algebraic sum of monomials. Examples of polynomials are

$$3x, y + 5, 4x^2 - 7x + 1, \text{ and } a^{10} + 5a^3 - 2a^2 + 6.$$

Polynomial

*A **polynomial** is a monomial or the indicated sum or difference of monomials.*

*The **degree of a polynomial** is the largest of the degrees of its terms.*

*The coefficient of the term of the largest degree is called the **leading coefficient**.*

Special Terminology for Some Polynomials

		Examples
Monomial:	*polynomial with one term*	$-2x^3$ *and* $4a^5$
Binomial:	*polynomial with two terms*	$3x + 5$ *and* $a^2 + 3$
Trinomial:	*polynomial with three terms*	$x^2 + 6x - 7$ *and* $a^3 - 8a^2 + 12a$

Examples of polynomials are

$$-1.4x^5 \text{ a fifth-degree monomial in } x \text{ (leading coefficient is } -1.4\text{)},$$

$$4z^3 - 7.5z^2 - 5z \text{ a third-degree trinomial in } z \text{ (leading coefficient is 4)},$$

$$\frac{3}{4}y^4 - 2y^3 + 4y - 6 \text{ a fourth-degree polynomial in } y \text{ (leading coefficient is } \frac{3}{4}\text{)}.$$

In each of these examples, the terms have been written so that the exponents on the variables decrease in order from left to right. We say that the terms are written in **descending order**. If the exponents on the terms increase in order from left to right, we say that the terms are written in **ascending order**. **As a general rule, for consistency and style in operating with polynomials, the polynomials in this chapter will be written in descending order.**

Example 1: Simplifying Polynomials

Simplify each of the following polynomials by combining like terms. Write the polynomial in descending order and state the degree and type of the polynomial.

a. $5x^3 + 7x^3$

Solution: $5x^3 + 7x^3 = (5 + 7)x^3 = 12x^3$ Third-degree monomial

b. $5x^3 + 7x^3 - 2x$

Solution: $5x^3 + 7x^3 - 2x = 12x^3 - 2x$ Third-degree binomial

c. $\frac{1}{2}y + 3y - \frac{2}{3}y^2 - 7$

Solution: $\frac{1}{2}y + 3y - \frac{2}{3}y^2 - 7 = -\frac{2}{3}y^2 + \frac{7}{2}y - 7$ Second-degree trinomial

d. $x^2 + 8x - 15 - x^2$

Solution: $x^2 + 8x - 15 - x^2 = 8x - 15$ First-degree binomial

e. $-3y^4 + 2y^2 + y^{-1}$

Solution: This expression is not a polynomial since y has a negative exponent.

Evaluating Polynomials

To evaluate a polynomial for a given value of the variable, substitute the value for the variable wherever it occurs in the polynomial and follow the rules for order of operations. A convenient notation for evaluating polynomials is the function notation, $p(x)$ [read "p of x"] discussed in Section 4.5. For example,

$$\text{if} \quad p(x) = x^2 - 4x + 13,$$

$$\text{then} \quad p(5) = 5^2 - 4 \cdot 5 + 13 = 25 - 20 + 13 = 18.$$

Example 2: Evaluating Polynomials

a. Evaluate $p(x) = 4x^2 + 5x - 15$ for $x = 3$.

Solution: For $p(x) = 4x^2 + 5x - 15$,

$$\begin{aligned}
p(3) &= 4 \cdot 3^2 + 5 \cdot 3 - 15 \qquad \text{Substitute 3 for } x.\\
&= 4 \cdot 9 + 15 - 15\\
&= 36 + 15 - 15\\
&= 36
\end{aligned}$$

b. Evaluate $p(y) = 5y^3 + y^2 - 3y + 8$ for $y = -2$.

Solution: For $p(y) = 5y^3 + y^2 - 3y + 8$,

$$\begin{aligned}
p(-2) &= 5(-2)^3 + (-2)^2 - 3(-2) + 8 \quad \text{Note the use of}\\
&= 5(-8) + 4 - 3(-2) + 8 \qquad \text{parentheses around } -2.\\
&= -40 + 4 + 6 + 8\\
&= -22
\end{aligned}$$

Practice Problems

Combine like terms and state the degree and type of the polynomial.

1. $8x^3 - 3x^2 - x^3 + 5 + 3x^2$ *2.* $5y^4 + y^4 - 3y^2 + 2y^2 + 4$

3. *For the polynomial* $p(x) = x^2 - 5x - 5$, *find (a)* $p(3)$ *and (b)* $p(-1)$.

Answers to Practice Problems: 1. $7x^3 + 5$; third degree binomial   **2.** $6y^4 - y^2 + 4$; fourth degree trinomial   **3. a.** -11 **b.** 1

6.3 Exercises

In Exercises 1 – 10, identify the expression as a monomial, binomial, trinomial, or not a polynomial.

1. $3x^4$ **2.** $5y^2 - 2y + 1$ **3.** $-2x^{-2}$ **4.** $8x^3 - 7$

5. $14a^7 - 2a - 6$ **6.** $17x^{\frac{2}{3}} + 5x^2$ **7.** $6a^3 + 5a^2 - a^{-3}$ **8.** $-3y^4 + 2y^2 - 9$

9. $\frac{1}{2}x^3 - \frac{2}{5}x$ **10.** $\frac{5}{8}x^5 + \frac{2}{3}x^4$

Simplify the polynomials in Exercises 11 – 30. Write the polynomial in descending order and state the degree and type of the simplified polynomial. For each polynomial, state the leading coefficient.

11. $y + 3y$ **12.** $4x^2 - x + x^2$ **13.** $x^3 + 3x^2 - 2x$

14. $3x^2 - 8x + 8x$ **15.** $x^4 - 4x^2 + 2x^2 - x^4$ **16.** $2 - 6y + 5y - 2$

17. $-x^3 + 6x + x^3 - 6x$ **18.** $11x^2 - 3x + 2 - 7x^2$ **19.** $6a^5 + 2a^2 - 7a^3 - 3a^2$

20. $2x^2 - 3x^2 + 2 - 4x^2 - 2 + 5x^2$ **21.** $4y - 8y^2 + 2y^3 + 8y^2$

22. $2x + 9 - x + 1 - 2x$ **23.** $5y^2 + 3 - 2y^2 + 1 - 3y^2$

24. $13x^2 - 6x - 9x^2 - 4x$ **25.** $7x^3 + 3x^2 - 2x + x - 5x^3 + 1$

26. $-3y^5 + 7y - 2y^3 - 5 + 4y^2 + y^2$ **27.** $x^4 + 3x^4 - 2x + 5x - 10 - x^2 + x$

28. $a^3 + 2a^2 - 6a + 3a^3 + 2a^2 + 7a + 3$ **29.** $2x + 4x^2 + 6x + 9x^3$

30. $15y - y^3 + 2y^2 - 10y^2 + 2y - 16$

In Exercises 31 – 40, evaluate the given polynomials as indicated.

31. Given $p(x) = x^2 + 14x - 3$, find $p(-1)$.

32. Given $p(y) = y^3 - 5y^2 + 6y + 2$, find $p(2)$.

33. Given $p(x) = 3x^3 - 9x^2 - 10x - 11$, find $p(3)$.

34. Given $p(x) = -5x^2 - 8x + 7$, find $p(-3)$.

35. Given $p(x) = 8x^4 + 2x^3 - 6x^2 - 7$, find $p(-2)$.

36. Given $p(a) = a^3 + 4a^2 + a + 2$, find $p(-5)$.

37. Given $p(a) = 2a^4 + 3a^2 - 8a$, find $p(-1)$.

38. Given $p(y) = -4y^3 + 5y^2 + 12y - 1$, find $p(-10)$.

39. Given $p(x) = x^5 - x^3 + x - 2$, find $p(-2)$.

40. Given $p(x) = 3x^6 - 2x^5 + x^4 - x^3 - 3x^2 + 2x - 1$, find $p(1)$.

In Exercises 41 – 46, a polynomial is given. Rewrite the polynomial by substituting for the variable as indicated with the function notation.

41. Given $p(x) = 3x^4 + 5x^3 - 8x^2 - 9x$, find $p(a)$.

42. Given $p(x) = 6x^5 + 5x^2 - 10x + 3$, find $p(c)$.

43. Given $f(x) = 3x + 5$, find $f(a+2)$.

44. Given $f(x) = -4x + 6$, find $f(a-2)$.

45. Given $g(x) = 5x - 10$, find $g(2a+7)$.

46. Given $g(x) = -4x - 8$, find $g(3a+1)$.

First-degree polynomials are also called **linear polynomials**, second-degree polynomials are called **quadratic polynomials**, and third-degree polynomials are called **cubic polynomials**. The related functions are called linear functions, quadratic functions, and cubic functions, respectively.

47. Use a graphing calculator to graph the following linear functions.

    a. $p(x) = 2x + 3$ **b.** $p(x) = -3x + 1$ **c.** $p(x) = \dfrac{1}{2}x$

48. Use a graphing calculator to graph the following quadratic functions.

    a. $p(x) = x^2$ **b.** $p(x) = x^2 + 6x + 9$ **c.** $p(x) = -x^2 + 2$

49. Use a graphing calculator to graph the following cubic functions.

    a. $p(x) = x^3$ **b.** $p(x) = x^3 - 4x$ **c.** $p(x) = x^3 + 2x^2 - 5$

50. Make up a few of your own linear, quadratic, and cubic functions and graph these functions with your calculator. Using the results from Exercises 47, 48, and 49, and your own functions, describe in your own words:

    a. the nature of graphs of linear functions.

    b. the nature of graphs of quadratic functions.

    c. the nature of graphs of cubic functions.

Hawkes Learning Systems: Introductory Algebra

Identifying Polynomials
Evaluating Polynomials

Addition and Subtraction with Polynomials

After completing this section, you will be able to:

1. Add polynomials.

2. Subtract polynomials.

3. Simplify expressions by removing grouping symbols and combining like terms.

Addition with Polynomials

The **sum** of two or more polynomials is found by combining **like terms**. Remember that like terms (or similar terms) are constants or terms that contain the same variables raised to the same powers. (See Section 2.1.) For example,

$3x^2$, $-10x^2$, and $1.4x^2$ are all **like terms**. Each has the same variable raised to the same power.

$5a^3$ and $-7a$ are **not** like terms. Each has the same variable, but the powers are different.

When adding polynomials, the polynomials may be written horizontally or vertically. For example,

$$(x^2 - 5x + 3) + (2x^2 - 8x - 4) + (3x^3 + x^2 - 5)$$
$$= 3x^3 + (x^2 + 2x^2 + x^2) + (-5x - 8x) + (3 - 4 - 5)$$
$$= 3x^3 + 4x^2 - 13x - 6.$$

If the polynomials are written in a vertical format, we align like terms, one beneath the other, in a column format and combine like terms in each column.

$$
\begin{array}{r}
x^2 \;-\; 5x \;+\; 3 \\
2x^2 \;-\; 8x \;-\; 4 \\
3x^3 + x^2 \qquad -\; 5 \\
\hline
3x^3 + 4x^2 - 13x \;-\; 6
\end{array}
$$

Example 1: Addition with Polynomials

a. Add as indicated: $(5x^3 - 8x^2 + 12x + 13) + (-2x^2 - 8) + (4x^3 - 5x + 14)$

Solution: $(5x^3 - 8x^2 + 12x + 13) + (-2x^2 - 8) + (4x^3 - 5x + 14)$

$$= (5x^3 + 4x^3) + (-8x^2 - 2x^2) + (12x - 5x) + (13 - 8 + 14)$$

$$= 9x^3 - 10x^2 + 7x + 19$$

b. Find the sum: $(x^3 - x^2 + 5x) + (4x^3 + 5x^2 - 8x + 9)$

Solution:

$$\begin{array}{r} x^3 - x^2 + 5x \\ 4x^3 + 5x^2 - 8x + 9 \\ \hline 5x^3 + 4x^2 - 3x + 9 \end{array}$$

Subtraction with Polynomials

If a negative sign is written in front of a polynomial in parentheses, the meaning is the opposite of the entire polynomial. The opposite can be found by changing the sign of every term in the polynomial.

$$-(2x^2 + 3x - 7) = -2x^2 - 3x + 7$$

We can also think of the opposite of a polynomial as -1 times the polynomial, then applying the distibutive property as follows:

$$\begin{aligned} -(2x^2 + 3x - 7) &= -1(2x^2 + 3x - 7) \\ &= -1(2x^2) - 1(3x) - 1(-7) \\ &= -2x^2 - 3x + 7. \end{aligned}$$

The result is the same with either approach. So the **difference** between two polynomials can be found by changing the sign of each term of the second polynomial and then combining like terms.

$$\begin{aligned} (5x^2 - 3x - 7) - (2x^2 + 5x - 8) &= 5x^2 - 3x - 7 - 2x^2 - 5x + 8 \\ &= 5x^2 - 2x^2 - 3x - 5x - 7 + 8 \\ &= 3x^2 - 8x + 1 \end{aligned}$$

If the polynomials are written in a vertical format, one beneath the other, we change the signs of the terms of the polynomial being subtracted and then combine like terms.

Subtract:

$$5x^2 - 3x - 7 \qquad\qquad 5x^2 - 3x - 7$$
$$\underline{-\,(2x^2 + 5x - 8)} \longrightarrow \underline{-\,2x^2 - 5x + 8}$$
$$3x^2 - 8x + 1$$

Example 2: Subtraction with Polynomials

a. Subtract as indicated: $(9x^4 - 22x^3 + 3x^2 + 10) - (5x^4 - 2x^3 - 5x^2 + x)$

Solution: $(9x^4 - 22x^3 + 3x^2 + 10) - (5x^4 - 2x^3 - 5x^2 + x)$
$$= 9x^4 - 22x^3 + 3x^2 + 10 - 5x^4 + 2x^3 + 5x^2 - x$$
$$= 9x^4 - 5x^4 - 22x^3 + 2x^3 + 3x^2 + 5x^2 - x + 10$$
$$= 4x^4 - 20x^3 + 8x^2 - x + 10$$

b. Find the difference:
$$8x^3 + 5x^2 - 14$$
$$\underline{-(-2x^3 + x^2 + 6x)}$$

Solution:

$$8x^3 + 5x^2 - 14 \qquad\qquad 8x^3 + 5x^2 + 0x - 14$$
$$\underline{-\,(-2x^3 + x^2 + 6x)} \longrightarrow \underline{2x^3 - x^2 - 6x + 0}$$
$$10x^3 + 4x^2 - 6x - 14$$

Write in 0's for missing powers to help with alignment of like terms.

Simplifying Algebraic Expressions

If an algebraic expression contains more than one pair of grouping symbols, such as parentheses ( ), brackets [ ], or braces { }, simplify by working to remove the innermost pair of symbols first. **Apply the rules for order of operations** just as if the variables were numbers and proceed to combine like terms.

Example 3: Simplifying Algebraic Expressions

Simplify each of the following expressions.
a. $5x - [\, 2x + 3\,(4-x) + 1\,] - 9$

Solution: $5x - [\, 2x + 3(4-x) + 1\,] - 9$
$$= 5x - [\, 2x + 12 - 3x + 1\,] - 9$$
$$= 5x - [\, -x + 13\,] - 9$$
$$= 5x + x - 13 - 9$$
$$= 6x - 22$$

Work with the parentheses first since they are included inside the brackets.

b. $10 - x + 2 [x + 3 (x - 5) + 7]$

Solution: $10 - x + 2 [x + 3(x - 5) + 7]$ Work with the parentheses first since
$$= 10 - x + 2 [x + 3x - 15 + 7]$$ they are included inside the brackets.
$$= 10 - x + 2 [4x - 8]$$
$$= 10 - x + 8x - 16$$
$$= 7x - 6$$

Practice Problems

1. *Add:* $(15x + 4) + (3x^2 - 9x - 5)$

2. *Subtract:* $(- 5x^3 - 3x + 4) - (3x^3 - x^2 + 4x - 7)$

3. *Simplify:* $2 - [3a - (4 - 7a) + 2a]$

6.4 Exercises

Find the indicated sum in Exercises 1 – 20.

1. $(2x^2 + 5x - 1) + (x^2 + 2x + 3)$ **2.** $(x^2 + 2x - 3) + (x^2 + 5)$

3. $(x^2 + 7x - 7) + (x^2 + 4x)$ **4.** $(x^2 + 3x - 8) + (3x^2 - 2x + 4)$

5. $(2x^2 - x - 1) + (x^2 + x + 1)$ **6.** $(3x^2 + 5x - 4) + (2x^2 + x - 6)$

7. $(-2x^2 - 3x + 9) + (3x^2 - 2x + 8)$ **8.** $(x^2 + 6x - 7) + (3x^2 + x - 1)$

9. $(-4x^2 + 2x - 1) + (3x^2 - x + 2) + (x - 8)$ **10.** $(8x^2 + 5x + 2) + (-3x^2 + 9x - 4)$

11. $(x^2 + 2x - 1) + (3x^2 - x + 2) + (x - 8)$

12. $(x^3 + 2x - 9) + (x^2 - 5x + 2) + (x^3 - 4x^2 + 1)$

13. $x^2 + 4x - 4$ **14.** $x^3 + 3x^2 + x$ **15.** $7x^3 + 5x^2 + x - 6$
$$-2x^2 + 3x + 1$$ $$-2x^3 - x^2 + 2x - 4$$ $$-3x^2 + 4x + 11$$
$$-3x^3 - x^2 - 5x + 2$$

Answers to Practice Problems: 1. $3x^2 + 6x - 1$ **2.** $- 8x^3 + x^2 - 7x + 11$ **3.** $-12a + 6$

16. $5x^2 - 3x + 11$
$\underline{-2x^2 + x - 6}$

17. $x^3 + 3x^2 - 4$
$7x^2 + 2x + 1$
$\underline{x^3 + x^2 - 6x}$

18. $x^3 + 2x^2 - 5$
$-2x^3 + x - 9$
$\underline{x^3 - 2x^2 + 14}$

19. $2x^2 + 4x - 3$
$\underline{3x^2 - 9x + 2}$

20. $x^3 + 5x^2 + 7x - 3$
$4x^2 + 3x - 9$
$\underline{4x^3 + 2x^2 - 2}$

Find the indicated difference in Exercises 21 – 40.

21. $(2x^2 + 4x + 8) - (x^2 + 3x + 2)$

22. $(3x^2 + 7x - 6) - (x^2 + 2x + 5)$

23. $(x^4 + 8x^3 - 2x^2 - 5) - (2x^4 + 10x^3 - 2x^2 + 11)$

24. $(-3x^4 + 2x^3 - 7x^2 + 6x + 12) - (x^4 + 9x^3 + 4x^2 + x - 1)$

25. $(x^2 - 9x + 2) - (4x^2 - 3x + 4)$

26. $(2x^2 - x - 10) - (-x^2 + 3x - 2)$

27. $(7x^2 + 4x - 9) - (-2x^2 + x - 9)$

28. $(6x^2 + 11x + 2) - (4x^2 - 2x - 7)$

29. $(3x^4 - 2x^3 - 8x - 1) - (5x^3 - 3x^2 - 3x - 10)$

30. $(x^5 + 6x^3 - 3x^2 - 5) - (2x^5 + 8x^3 + 5x + 17)$

31. $(9x^2 + 6x - 5) - (13x^2 + 6)$

32. $(8x^2 + 9) - (4x^2 - 3x - 2)$

33. $(x^3 + 4x^2 - 7) - (3x^3 + x^2 + 2x + 1)$

34. $14x^2 - 6x + 9$
$\underline{-\left(8x^2 + x - 9\right)}$

35. $x^3 + 6x^2 - 3$
$\underline{-\left(-x^3 + 2x^2 - 3x + 7\right)}$

36. $9x^2 - 3x + 2$
$\underline{-\left(4x^2 - 5x - 1\right)}$

37. $5x^4 + 8x^2 + 11$
$\underline{-\left(-3x^4 + 2x^2 - 4\right)}$

38. $11x^2 + 5x - 13$
$\underline{-\left(-3x^2 + 5x + 2\right)}$

39. $3x^3 + 9x - 17$
$\underline{-\left(x^3 + 5x^2 - 2x - 6\right)}$

40. $-3x^4 + 10x^3 - 8x^2 - 7x - 6$

$-\left(2x^4 + x^3 + 5x + 6\right)$

Simplify each of the algebraic expressions in Exercises 41 – 55, and write the polynomials in descending order.

41. $5x + 2(x - 3) - (3x + 7)$

42. $-4(x - 6) - (8x + 2)$

43. $11 + [\, 3x - 2(1 + 5x)\,]$

44. $2x + [\, 9x - 4(3x + 2) - 7\,]$

45. $8x - [\, 2x + 4(x - 3) - 5\,]$

46. $17 - [\, -3x + 6(2x - 3) + 9\,]$

47. $3x^3 - [\, 5 - 7(x^2 + 2) - 6x^2\,]$

48. $10x^3 - [\, 8 - 5(3 - 2x^2) - 7x^2\,]$

49. $(2x^2 + 4) - [\, -8 + 2(7 - 3x^2) + x\,]$

50. $-[\, 6x^2 - 3(4 + 2x) + 9\,] - (x^2 + 5)$

51. $2[\, 3x + (x - 8) - (2x + 5)\,] - (x - 7)$

52. $-3[\, -x + (10 - 3x) - (8 - 3x)\,] + (2x - 1)$

53. $(x^2 - 1) + x[\, 4 + (3 - x)\,]$

54. $x\,(x - 5) + [\, 6x - x(4 - x)\,]$

55. $x(2x + 1) - [\, 5x - x(2x + 3)\,]$

56. Subtract $2x^2 - 4x$ from $7x^3 + 5x$.

57. Subtract $3x^2 - 4x + 2$ from the sum of $4x^2 + x - 1$ and $6x - 5$.

58. Subtract $-2x^2 + 6x + 12$ from the sum of $2x^2 + 3x - 4$ and $x^2 - 13x + 2$.

59. Add $5x^3 - 8x + 1$ to the difference between $2x^3 + 14x - 3$ and $x^2 + 6x + 5$.

60. Find the sum of $10x - 2(3x + 5)$ and $3(x - 4) + 16$.

Hawkes Learning Systems: Introductory Algebra

Addition and Subtraction with Polynomials

<table>
<tr><td>**6.5**</td><td></td></tr>
</table>

6.5 | Multiplication with Polynomials

After completing this section, you will be able to:

Multiply polynomials.

Up to this point, we have multiplied terms such as $5x^2 \cdot 3x^4 = 15x^6$ by using the product rule for exponents. Also, we have applied the distributive property to expressions such as $5(2x + 3) = 10x + 15$.

Now we will use both the product rule for exponents and the distributive property to multiply polynomials. We discuss three cases here:

    a. the product of a monomial with a polynomial of two or more terms,
    b. the product of two binomials, and
    c. the product of a binomial with a polynomial of more than two terms.

Multiplying a Polynomial by a Monomial

Using the distributive property $a(b + c) = ab + ac$ with multiplication indicated on the left, we can find the product of a monomial with a polynomial of two or more terms as follows:

$$5x(2x + 3) = 5x \cdot 2x + 5x \cdot 3 = 10x^2 + 15x$$
$$3x^2(4x - 1) = 3x^2 \cdot 4x + 3x^2(-1) = 12x^3 - 3x^2$$
$$-4a^5(a^2 - 8a + 5) = -4a^5 \cdot a^2 - 4a^5(-8a) - 4a^5(5) = -4a^7 + 32a^6 - 20a^5.$$

Multiplying Two Polynomials

Now suppose that we want to multiply two binomials, say, $(x + 3)(x + 7)$. We will apply the distributive property in the following way with multiplication indicated on the right of the parentheses.

Compare $(x + 3)(x + 7)$ to

$$(a + b)c = ac + bc.$$

Think of $(x + 7)$ as taking the place of c. Thus,

takes the form $(x + 3)(x + 7)$ = $x(x + 7)$ + $3(x + 7)$.

Completing the products on the right, using the distributive property twice again, gives

$$(x + 3)(x + 7) = x(x + 7) + 3(x + 7)$$
$$= x \cdot x + x \cdot 7 + 3 \cdot x + 3 \cdot 7$$
$$= x^2 + 7x + 3x + 21$$
$$= x^2 + 10x + 21.$$

In the same manner,

$$(x + 2)(3x + 4) = x(3x + 4) + 2(3x + 4)$$
$$= x \cdot 3x + x \cdot 4 + 2 \cdot 3x + 2 \cdot 4$$
$$= 3x^2 + 4x + 6x + 8$$
$$= 3x^2 + 10x + 8.$$

Similarly,

$$(2x - 1)(x^2 + x - 5) = 2x(x^2 + x - 5) - 1(x^2 + x - 5)$$
$$= 2x \cdot x^2 + 2x \cdot x + 2x(-5) - 1 \cdot x^2 - 1 \cdot x - 1(-5)$$
$$= 2x^3 + 2x^2 - 10x - x^2 - x + 5$$
$$= 2x^3 + x^2 - 11x + 5.$$

One quick way to check if your products are correct is to substitute some convenient number for x into the original two factors and into the product. Choose any nonzero number that you like. The values of both expressions should be the same. For example, let $x = 1$. Then,

$$(x + 2)(3x + 4) = (1 + 2)(3 \cdot 1 + 4) = (3)(7) = 21$$

and

$$3x^2 + 10x + 8 = 3 \cdot 1^2 + 10 \cdot 1 + 8 = 3 + 10 + 8 = 21.$$

The product $(x + 2)(3x + 4) = 3x^2 + 10x + 8$ seems to be correct. We could double check by letting $x = 5$.

$$(x + 2)(3x + 4) = (5 + 2)(3 \cdot 5 + 4) = (7)(19) = 133$$

and

$$3x^2 + 10x + 8 = 3 \cdot 5^2 + 10 \cdot 5 + 8 = 75 + 50 + 8 = 133.$$

Convinced? This is just a quick check, however, and is not foolproof unless you try this process with more values of x than indicated by the degree of the product.

The product of two polynomials can also be found by writing one polynomial under the other. **The distributive property is applied by multiplying each term of one polynomial by each term of the other.** Consider the product $(2x^2 + 3x - 4)(3x + 7)$. Now, writing one polynomial under the other and applying the distributive property, we obtain

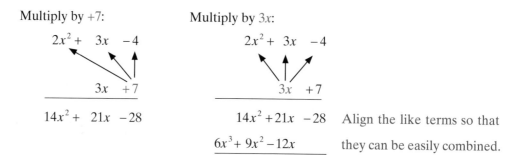

Multiply by $+7$:

$$2x^2 + \quad 3x \quad -4$$
$$\underline{\qquad\qquad 3x \quad +7}$$
$$14x^2 + \quad 21x \quad -28$$

Multiply by $3x$:

$$2x^2 + \quad 3x \quad -4$$
$$\underline{\qquad\qquad 3x \quad +7}$$
$$14x^2 + 21x \quad -28 \qquad \text{Align the like terms so that}$$
$$\underline{6x^3 + 9x^2 - 12x} \qquad\qquad \text{they can be easily combined.}$$

Finally, combine like terms:

$$2x^2 \quad + 3x \quad - 4$$
$$\underline{\qquad\qquad\qquad 3x \quad + 7}$$
$$14x^2 \quad + 21x \quad -28$$
$$\underline{6x^3 + \quad 9x^2 \quad -12x}$$
$$6x^3 + 23x^2 \quad + 9x \quad -28 \qquad \text{Combine like terms.}$$

Example 1: Multiply

Find each product.

a. $-4x\left(x^2 - 3x + 12\right)$

Solution: $-4x\left(x^2 - 3x + 12\right) = -4x \cdot x^2 - 4x(-3x) - 4x \cdot 12$
$$= -4x^3 + 12x^2 - 48x$$

b. $(2x - 4)(5x + 3)$

Solution: $(2x - 4)(5x + 3) = 2x(5x + 3) - 4(5x + 3)$
$$= 2x \cdot 5x + 2x \cdot 3 - 4 \cdot 5x - 4 \cdot 3$$
$$= 10x^2 + 6x - 20x - 12$$
$$= 10x^2 - 14x - 12$$

c. $7y^2 - 3y + 2$

$\qquad \underline{2y + 3}$

Solution: $\qquad 7y^2 - 3y + 2$

$\qquad \underline{\qquad\qquad 2y + 3}$

$\qquad\qquad 21y^2 - 9y + 6$ $\qquad$ Multiply by 3.

$\qquad \underline{14y^3 - 6y^2 + 4y}$ $\qquad$ Multiply by $2y$.

$\qquad 14y^3 + 15y^2 - 5y + 6$ $\qquad$ Combine like terms.

d. $x^2 + 3x - 1$

$\qquad \underline{x^2 - 3x + 1}$

Solution: $\qquad x^2 + 3x - 1$

$\qquad \underline{\qquad\qquad x^2 - 3x + 1}$

$\qquad\qquad x^2 + 3x - 1$ $\qquad$ Multiply by 1.

$\qquad -3x^3 - 9x^2 + 3x$ $\qquad$ Multiply by $-3x$.

$\qquad \underline{x^4 + 3x^3 - x^2}$ $\qquad$ Multiply by x^2.

$\qquad x^4 \qquad - 9x^2 + 6x - 1$ $\qquad$ Combine like terms.

e. $(x - 5)(x + 2)(x - 1)$

Solution: First multiply $(x - 5)(x + 2)$, then multiply this result by $(x - 1)$.

$$(x - 5)(x + 2) = x(x + 2) - 5(x + 2)$$
$$= x^2 + 2x - 5x - 10$$
$$= x^2 - 3x - 10$$

$$(x^2 - 3x - 10)(x - 1) = (x^2 - 3x - 10)x + (x^2 - 3x - 10)(-1)$$
$$= x^3 - 3x^2 - 10x - x^2 + 3x + 10$$
$$= x^3 - 4x^2 - 7x + 10$$

Practice Problems

Find each product.

1. $2x(3x^2 + x - 1)$

2. $(x + 3)(x - 7)$

3. $(x - 1)(x^2 + x - 4)$

4. $(x + 1)(x - 2)(x + 2)$

6.5 Exercises

Multiply as indicated in Exercises 1 – 40 and simplify if possible.

1. $-3x^2(2x^3 + 5x)$

2. $4x^5(x^2 - 3x + 1)$

3. $5x^2(-4x^2 + 6)$

4. $9x^3(2x^3 - x^2 + 5x)$

5. $-1(y^5 - 8y + 2)$

6. $-7(2y^4 + 3y^2 + 1)$

7. $-4x^3(x^5 - 2x^4 + 3x)$

8. $a^2(a^5 + 2a^4 - 5a + 1)$

9. $7t^3(-5t^2 + 2t + 1)$

10. $5x^3(5x^2 - x + 2)$

11. $-x(x^3 + 5x - 4)$

12. $-2x^4(x^3 - x^2 + 2x)$

13. $3x(2x + 1) - 2(2x + 1)$

14. $x(3x + 4) + 7(3x + 4)$

15. $3a(3a - 5) + 5(3a - 5)$

16. $6x(x - 1) + 5(x - 1)$

17. $5x(-2x + 7) - 2(-2x + 7)$

18. $y(y^2 + 1) - 1(y^2 + 1)$

19. $x(x^2 + 3x + 2) + 2(x^2 + 3x + 2)$

20. $4x(x^2 - x + 1) + 3(x^2 - x + 1)$

21. $(x + 4)(x - 3)$

22. $(x + 7)(x - 5)$

23. $(a + 6)(a - 8)$

24. $(x + 2)(x - 4)$

25. $(x - 2)(x - 1)$

26. $(x - 7)(x - 8)$

27. $3(t + 4)(t - 5)$

28. $-4(x + 6)(x - 7)$

29. $x(x + 3)(x + 8)$

Answers to Practice Problems: 1. $6x^3 + 2x^2 - 2x$ **2.** $x^2 - 4x - 21$ **3.** $x^3 - 5x + 4$ **4.** $x^3 + x^2 - 4x - 4$

30. $t(t-4)(t-7)$ **31.** $(2x+1)(x-4)$ **32.** $(3x-1)(x+4)$

33. $(6x-1)(x+3)$ **34.** $(3t+5)(3t-5)$ **35.** $(2x+3)(2x-3)$

36. $(8x+15)(x+1)$ **37.** $(4x+1)(4x+1)$ **38.** $(5x-2)(5x-2)$

39. $(y+3)(y^2-y+4)$ **40.** $(2x+1)(x^2-7x+2)$

In Exercises 41 – 45, multiply as indicated and simplify.

41. $\begin{array}{r} 3x+7 \\ x-5 \\ \hline \end{array}$ **42.** $\begin{array}{r} x^2+3x+1 \\ 5x-9 \\ \hline \end{array}$ **43.** $\begin{array}{r} 8x^2+3x-2 \\ -2x+7 \\ \hline \end{array}$

44. $\begin{array}{r} 2x^2+3x+5 \\ x^2+2x-3 \\ \hline \end{array}$ **45.** $\begin{array}{r} 6x^2-x+8 \\ 2x^2+5x+6 \\ \hline \end{array}$

Find the product in Exercises 46 – 70 and simplify if possible. Check by letting the variable equal 2.

46. $(3x-4)(x+2)$ **47.** $(t+6)(4t-7)$ **48.** $(2x+5)(x-1)$

49. $(5a-3)(a+4)$ **50.** $(2x+1)(3x-8)$ **51.** $(x-2)(3x+8)$

52. $(7x+1)(x-2)$ **53.** $(3x+7)(2x-5)$ **54.** $(2x+3)(2x+3)$

55. $(5y+2)(5y+2)$ **56.** $(x+3)(x^2-4)$ **57.** $(y^2+2)(y-4)$

58. $(2x+7)(2x-7)$ **59.** $(3x-4)(3x+4)$ **60.** $(x+1)(x^2-x+1)$

61. $(x-2)(x^2+2x+4)$ **62.** $(7a-2)(7a-2)$ **63.** $(5a-6)(5a-6)$

64. $(2x+3)(x^2-x-1)$ **65.** $(3x+1)(x^2-x+9)$ **66.** $(x+1)(x+2)(x+3)$

67. $(t-1)(t-2)(t-3)$ **68.** $(a^2+a-1)(a^2-a+1)$ **69.** $(y^2+y+2)(y^2+y-2)$

70. $(t^2+3t+2)^2$

Simplify the expressions in Exercises 71 – 75.

71. $(x-3)(x+5)-(x+3)(x+2)$ **72.** $(y-2)(y-4)+(y-1)(y+1)$

73. $(2a+1)(a-5)+(a-4)(a-4)$ **74.** $(2t+3)(2t+3)-(t-2)(t-2)$

75. $(y+6)(y-6)+(y+5)(y-5)$

Hawkes Learning Systems: Introductory Algebra

Multiplying a Polynomial by a Monomial
Multiplying Two Polynomials

<table>
<tr><td>**6.6**</td><td></td></tr>
</table>

Special Products of Binomials

Objectives

After completing this section, you will be able to:

1. *Multiply binomials using the FOIL method.*

2. *Multiply binomials, finding products that are the difference of squares.*

3. *Square binomials, finding products that are perfect square trinomials.*

4. *Identify the difference of two squares and perfect square trinomials.*

In Section 6.5, we emphasized the use of the distributive property in finding products of polynomials. This approach is correct and serves to provide a solid understanding of multiplication. However, there is a more efficient way to **multiply binomials** which is used by most students. After some practice, you can probably find these products mentally.

The FOIL Method

Consider finding the product $(2x+3)(5x+2)$. Using the distributive property gives:

$$(2x+3)(5x+2) = 2x(5x+2)+3(5x+2)$$
$$= 2x \cdot 5x + 2x \cdot 2 + 3 \cdot 5x + 3 \cdot 2$$

If we look carefully at both binomials, we see that

$2x$ and $5x$ are the **F**irst two terms in each binomial,
$2x$ and 2 are the **O**utside terms in the expression,
3 and $5x$ are the **I**nside terms in the expression,
3 and 2 are the **L**ast two terms in each binomial.

Thus we can write the product more directly as follows:

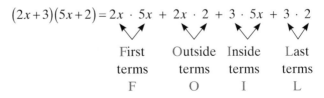

$$(2x+3)(5x+2) = 2x \ \cdot \ 5x \ + \ 2x \ \cdot \ 2 \ + \ 3 \ \cdot \ 5x \ + \ 3 \ \cdot \ 2$$

First terms	Outside terms	Inside terms	Last terms
F	O	I	L

This analysis leads to the mnemonic device called the **FOIL method** of mentally multiplying **two binomials. Remember that this FOIL approach is valid only for two binomials**.

The following layout shows how the FOIL method can be applied mentally.

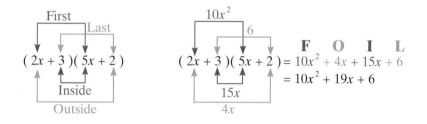

The same layout is shown again to help you get a mental image of the process.

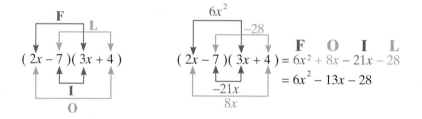

Example 1: FOIL Method

Use the FOIL method to find each product of binomials.

a. $(2x + 1)(4x + 3)$

  Solution:

$$(2x + 1)(4x + 3) = 8x^2 + 6x + 4x + 3 = 8x^2 + 10x + 3$$

b. $(x + 6)(x - 6)$

  Solution:

$$(x + 6)(x - 6) = x^2 - 6x + 6x - 36 = x^2 - 36$$

The Difference of Two Squares: $(x + a)(x - a) = x^2 - a^2$

In Example 1b, the middle terms, $-6x$ and $+6x$, are opposites of each other and their sum is 0. Therefore, the resulting product has only two terms.

$$(x + 6)(x - 6) = x^2 - 36$$

The simplified product is in the form of a difference and both terms are squares. Thus we have the following special case called the **difference of two squares**.

Difference of Two Squares

$$(x + a)(x - a) = x^2 - a^2$$

When the two binomials are in the form of the sum and difference of the same two terms, the product will always be the difference of the squares of the terms. In such a case, we can write the answer directly with no calculations. You should memorize the following squares of the positive integers from 1 to 20. The squares of integers are called **perfect squares**.

Perfect Squares From 1 to 400

1, 4, 9, 16, 25, 36, 49, 64, 81, 100, 121, 144, 169, 196, 225, 256, 289, 324, 361, 400

Example 2: Difference of Two Squares

Find each product.
a. $(x + 5)(x - 5)$

Solution: The two binomials represent the sum and difference of x and 5. So, the product is the difference of their squares.
$$(x + 5)(x - 5) = x^2 - 5^2 = x^2 - 25$$

b. $(8x + 3)(8x - 3)$

Solution: $(8x + 3)(8x - 3) = (8x)^2 - (3)^2 = 64x^2 - 9$

c. $(x^3 + 2)(x^3 - 2)$

Solution: $(x^3 + 2)(x^3 - 2) = (x^3)^2 - 2^2 = x^6 - 4$ **Note:** $(x^3)^2 = x^3 \cdot x^3 = x^{3+3} = x^6$

Perfect Square Trinomials: $(x + a)^2 = x^2 + 2ax + a^2$
$(x - a)^2 = x^2 - 2ax + a^2$

We now want to consider the case where the two binomials being multiplied are the same. That is, we want to consider the **square of a binomial**. As the following discussion shows, there is a pattern that, after some practice, allows us to go directly to the product,

$$(x + 5)^2 = (x + 5)(x + 5) = x^2 + 5x + 5x + 25$$
$$= x^2 + 2 \cdot 5x + 25$$
$$= x^2 + 10x + 25$$

$$(x + 7)^2 = (x + 7)(x + 7) = x^2 + 2 \cdot 7x + 49$$
$$= x^2 + 14x + 49$$

$$(x + 10)^2 = (x + 10)(x + 10) = x^2 + 20x + 100.$$

Perfect Square Trinomial: $x^2 + 2ax + a^2$

$$(x + a)^2 = (x + a)(x + a) = x^2 + 2ax + a^2$$

The simplified product $x^2 + 2ax + a^2$ is called a **perfect square trinomial** because it is the result of squaring a binomial.

One interesting device for remembering the result of squaring a binomial is the square shown in Figure 6.1 where the total area is the sum of the shaded areas.

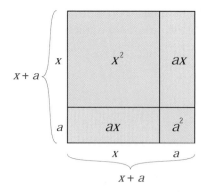

For the area of the square:

$$(x + a)^2 = x^2 + ax + ax + a^2$$
$$= x^2 + 2ax + a^2.$$

Figure 6.1

Another perfect square trinomial results if an expression of the form $(x - a)$ is squared. The sign between x and a is $-$ instead of $+$.

$$(x - 5)^2 = (x - 5)(x - 5) = x^2 - 5x - 5x + 25$$
$$= x^2 - 2 \cdot 5x + 25$$
$$= x^2 - 10x + 25$$

$$(x - 7)^2 = (x - 7)(x - 7) = x^2 - 2 \cdot 7x + 49 = x^2 - 14x + 49$$

$$(x - 10)^2 = (x - 10)(x - 10) = x^2 - 20x + 100$$

In general, we have the following **perfect square trinomial** formula.

Perfect Square Trinomial: $x^2 - 2ax + a^2$

$$(x - a)^2 = (x - a)(x - a) = x^2 - 2ax + a^2$$

Example 3: Perfect Square Trinomials

Find the following products.

a. $(2x + 3)^2$

Solution: The pattern for squaring a binomial gives
$$(2x + 3)^2 = (2x)^2 + 2 \cdot 3 \cdot 2x + (3)^2 \quad \textbf{Note: } (2x)^2 = 2x \cdot 2x = 4x^2$$
$$= 4x^2 + 12x + 9$$

b. $(5x - 1)^2$

Solution: $(5x - 1)^2 = (5x)^2 - 2(1)(5x) + (1)^2$
$$= 25x^2 - 10x + 1$$

c. $(9 - x)^2$

Solution: $(9 - x)^2 = (9)^2 - 2(9)(x) + x^2$
$$= 81 - 18x + x^2$$

d. $(y^3 + 1)^2$

Solution: $(y^3 + 1)^2 = (y^3)^2 + 2(1)(y^3) + 1^2 \quad \textbf{Note: } (y^3)^2 = y^3 \cdot y^3 = y^{3+3} = y^6$
$$= y^6 + 2y^3 + 1$$

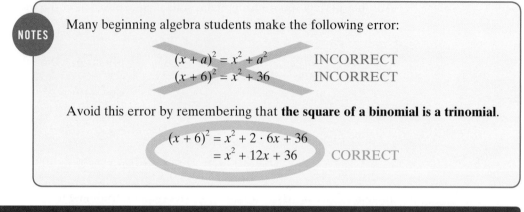

NOTES Many beginning algebra students make the following error:

$$(x + a)^2 = x^2 + a^2 \qquad \text{INCORRECT}$$
$$(x + 6)^2 = x^2 + 36 \qquad \text{INCORRECT}$$

Avoid this error by remembering that **the square of a binomial is a trinomial**.

$$(x + 6)^2 = x^2 + 2 \cdot 6x + 36$$
$$= x^2 + 12x + 36 \qquad \text{CORRECT}$$

Practice Problems

Find the indicated products.

1. $(x + 10)(x - 10)$ **2.** $(x + 3)^2$ **3.** $(2x - 1)(x + 3)$

4. $(2x - 5)^2$ **5.** $(x^2 + 4)(x^2 - 3)$

6.6 Exercises

Find the product in Exercises 1 – 30, and identify those that are the difference of two squares or perfect square trinomials.

1. $(x + 3)(x - 3)$ **2.** $(x - 7)^2$ **3.** $(x - 5)^2$

4. $(x + 4)(x + 4)$ **5.** $(x - 6)(x + 6)$ **6.** $(x + 9)(x - 9)$

7. $(x + 8)(x + 8)$ **8.** $(x + 12)(x - 12)$ **9.** $(2x + 3)(x - 1)$

10. $(3x + 1)(2x + 5)$ **11.** $(3x - 4)^2$ **12.** $(5x + 2)(5x - 2)$

13. $(2x + 1)(2x - 1)$ **14.** $(3x + 1)^2$ **15.** $(3x - 2)(3x - 2)$

16. $(4x + 5)(4x - 5)$ **17.** $(3 + x)^2$ **18.** $(8 - x)(8 - x)$

Answers to Practice Problems: 1. $x^2 - 100$ **2.** $x^2 + 6x + 9$ **3.** $2x^2 + 5x - 3$ **4.** $4x^2 - 20x + 25$
    5. $x^4 + x^2 - 12$

19. $(5-x)(5-x)$ **20.** $(11-x)(11+x)$ **21.** $(5x-9)(5x+9)$

22. $(4-x)^2$ **23.** $(2x+7)(2x+7)$ **24.** $(3x+2)^2$

25. $(9x+2)(9x-2)$ **26.** $(6x+5)(6x-5)$ **27.** $(5x^2+2)(2x^2-3)$

28. $(4x^2+7)(2x^2+1)$ **29.** $(1+7x)^2$ **30.** $(2-5x)^2$

Write the indicated products in Exercises 31 – 58.

31. $(x+2)(5x+1)$ **32.** $(7x-2)(x-3)$ **33.** $(4x-3)(x+4)$

34. $(x+11)(x-8)$ **35.** $(3x-7)(x-6)$ **36.** $(x+7)(2x+9)$

37. $(5+x)(5+x)$ **38.** $(3-x)(6-x)$ **39.** $(x^2+1)(x^2-1)$

40. $(x^2+5)(x^2-5)$ **41.** $(x^2+3)(x^2+3)$ **42.** $(x^2-4)^2$

43. $(x^3-2)^2$ **44.** $(x^3+8)(x^3-8)$ **45.** $(x^2-6)(x^2+9)$

46. $(x^2+3)(x^2-5)$ **47.** $\left(x+\dfrac{2}{3}\right)\left(x-\dfrac{2}{3}\right)$ **48.** $\left(x-\dfrac{1}{2}\right)\left(x+\dfrac{1}{2}\right)$

49. $\left(x+\dfrac{3}{4}\right)\left(x-\dfrac{3}{4}\right)$ **50.** $\left(x+\dfrac{3}{8}\right)\left(x-\dfrac{3}{8}\right)$ **51.** $\left(x+\dfrac{3}{5}\right)\left(x+\dfrac{3}{5}\right)$

52. $\left(x+\dfrac{4}{3}\right)\left(x+\dfrac{4}{3}\right)$ **53.** $\left(x-\dfrac{5}{6}\right)^2$ **54.** $\left(x-\dfrac{2}{7}\right)^2$

55. $\left(x+\dfrac{1}{4}\right)\left(x-\dfrac{1}{2}\right)$ **56.** $\left(x-\dfrac{1}{5}\right)\left(x+\dfrac{2}{3}\right)$ **57.** $\left(x+\dfrac{1}{3}\right)\left(x+\dfrac{1}{2}\right)$

58. $\left(x-\dfrac{4}{5}\right)\left(x-\dfrac{3}{10}\right)$

Calculator Problems

Use a TI-84 Plus calculator as an aid in multiplying the binomials in Exercises 59 – 72.

59. $(x + 1.4)(x - 1.4)$

60. $(x - 2.1)(x + 2.1)$

61. $(x - 2.5)^2$

62. $(x + 1.7)^2$

63. $(x + 2.15)(x - 2.15)$

64. $(x + 1.36)(x - 1.36)$

65. $(x + 1.24)^2$

66. $(x - 1.45)^2$

67. $(1.42x + 9.6)^2$

68. $(0.46x - 0.71)^2$

69. $(11.4x + 3.5)(11.4x - 3.5)$

70. $(2.5x + 11.4)(1.3x - 16.9)$

71. $(12.6x - 6.8)(7.4x + 15.3)$

72. $(3.4x + 6)(3.4x - 6)$

73. A square is 20 inches on each side. A square x inches on each side is cut from each corner of the square.
   a. Represent the area of the remaining portion of the square in the form of a polynomial function $A(x)$.
   b. Represent the perimeter of the remaining portion of the square in the form of a polynomial function $P(x)$.

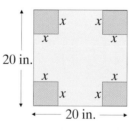

74. In the case of binomial probabilities, if x is the probability of success in one trial of an event, then the expression $f(x) = 15x^4(1 - x)^2$ is the probability of 4 successes in 6 trials where $0 \le x \le 1$.
   a. Represent the expression $f(x)$ as a single polynomial.
   b. If a fair coin is tossed, the probability of heads occurring is $\dfrac{1}{2}$. That is, $x = \dfrac{1}{2}$. Find the probability of 4 heads occurring in 6 tosses.

75. A rectangle has sides ($x + 3$) ft and ($x + 5$) ft. If a square x feet on a side is cut from the rectangle, represent the remaining area (light shaded area in the figure shown) in the form of a polynomial function $A(x)$.

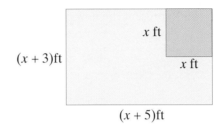

76. A pool, 20 meters by 50 meters, is surrounded by a concrete deck that is x meters wide.
a. Represent the area covered by the deck and the pool in the form of a polynomial function.
b. Represent the area covered by the deck only in the form of a polynomial function.

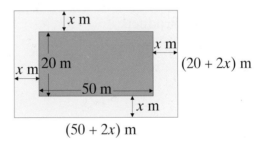

77. A rectangular piece of cardboard that is 10 inches by 15 inches has squares of length x inches on a side cut from each corner. (Assume that $0 < x < 5$.)
a. Represent the remaining area in the form of a polynomial function $A(x)$.
b. Represent the perimeter of the remaining figure in the form of a polynomial function $P(x)$.
c. If the flaps of the figure are folded up, an open box is formed. Represent the volume of this box in the form of a polynomial function $V(x)$.

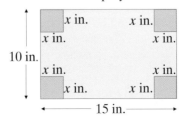

78. A rectangular room is 8 feet by 12 feet. The length and width are both increased by x feet.
  a. Represent the new area of the room in the form of a polynomial function $A(x)$.
  b. Represent the new perimeter of the room in the form of a polynomial function $P(x)$.

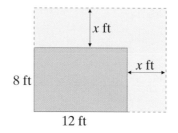

Hawkes Learning Systems: Introductory Algebra

The FOIL Method
Special Products

6.7 Division with Polynomials

Objectives

After completing this section, you will be able to:

1. Divide polynomials by monomials.

2. Divide polynomials by other polynomials using long division.

Fractions, such as $\dfrac{135}{8}$ and $\dfrac{7}{8}$, in which the numerator and denominator are integers are called rational numbers. Fractions in which the numerator and denominator are polynomials are called **rational expressions**. (No denominator can be 0).

$$\text{Rational numbers: } \frac{2}{3}, \frac{-5}{16}, \frac{22}{7}, \frac{0}{6}, \frac{3}{4}, -\frac{3}{4}$$

$$\text{Rational expressions: } \frac{x}{x^2+1}, \frac{x^2+5x+6}{x^2+7x+12}, \frac{x^2+2x+1}{x}, \text{ and } \frac{1}{5x}$$

In this section, we want to treat a rational expression as an indicated division problem. From this basis, there are two situations to consider:

1. the denominator (divisor) is a monomial, or
2. the denominator (divisor) is not a monomial.

Division by a Monomial

We know from arithmetic that the sum of fractions with the same denominator can be written as a single fraction by adding the numerators and using the common denominator. For example,

$$\frac{3}{a}+\frac{2b}{a}+\frac{5c}{a}=\frac{3+2b+5c}{a}$$

If, instead of adding the fractions, we want to divide the numerator by the denominator (with a monomial in the denominator), we divide each term in the numerator by the monomial denominator and simplify each fraction.

$$\frac{3x^3+6x^2+9x}{3x}=\frac{3x^3}{3x}+\frac{6x^2}{3x}+\frac{9x}{3x}=x^2+2x+3$$

Similarly,

$$\frac{x^2 + 2x + 1}{x} = \frac{x^2}{x} + \frac{2x}{x} + \frac{1}{x} = x + 2 + \frac{1}{x}$$

and

$$\frac{3xy + 6xy^2 - 18}{2y} = \frac{3xy}{2y} + \frac{6xy^2}{2y} - \frac{18}{2y} = \frac{3x}{2} + 3xy - \frac{9}{y}.$$

Example 1: Division by a Monomial

Divide each polynomial by the monomial denominator by writing a sum of fractions. Reduce each fraction, if possible.

a. $\dfrac{8x^2 - 14x + 1}{2}$

Solution: $\dfrac{8x^2 - 14x + 1}{2} = \dfrac{8x^2}{2} - \dfrac{14x}{2} + \dfrac{1}{2} = 4x^2 - 7x + \dfrac{1}{2}$

b. $\dfrac{12x^3 + 3x^2 - 9x}{3x^2}$

Solution: $\dfrac{12x^3 + 3x^2 - 9x}{3x^2} = \dfrac{12x^3}{3x^2} + \dfrac{3x^2}{3x^2} - \dfrac{9x}{3x^2}$

$$= 4x + 1 - \frac{3}{x}$$

c. $\dfrac{10x^2y + 25xy + 3y^2}{5xy^2}$

Solution: $\dfrac{10x^2y + 25xy + 3y^2}{5xy^2} = \dfrac{10x^2y}{5xy^2} + \dfrac{25xy}{5xy^2} + \dfrac{3y^2}{5xy^2}$

$$= \frac{2x}{y} + \frac{5}{y} + \frac{3}{5x}$$

The Division Algorithm

In arithmetic, the process (or series of steps) that we follow in dividing two numbers is called the **division algorithm** (or **long division**). By this division algorithm, we can find $135 \div 8$ as follows:

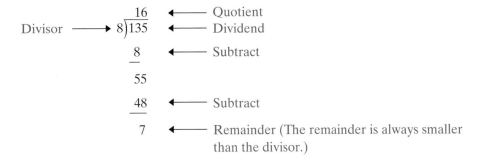

Check: $8 \cdot 16 + 7 = 128 + 7 = 135$ (Multiply the divisor times the quotient and add the remainder. The result should be the original dividend.)

We can also write the division in fraction form and the remainder over the divisor, giving a mixed number.

$$135 \div 8 = \frac{135}{8} = 16 + \frac{7}{8} = 16\frac{7}{8}$$

In algebra, the division algorithm with polynomials is quite similar. If we divide one polynomial by another, the quotient will be another polynomial with a remainder. Symbolically, if we have $P \div D$, then

$$P = Q \cdot D + R$$

or

$$\frac{P}{D} = Q + \frac{R}{D}$$

where
P is the dividend,
D is the divisor, and
R is the remainder.

The remainder must be of smaller degree than the divisor. If the remainder is 0, then the divisor and quotient are factors of the dividend.

In the following two examples, the **division algorithm** is illustrated in a step-by-step form with first-degree divisors. Each step is explained in detail and should be studied carefully.

Example 2: The Division Algorithm

a. $\dfrac{3x^2 - 5x + 2}{x - 4}$ or $\left(3x^2 - 5x + 2\right) \div \left(x - 4\right)$

Solution:

	Steps	**Explanation**
Step 1:	$x - 4\overline{)3x^2 - 5x + 2}$	Write both polynomials in order of descending powers. **If any powers are missing, fill in with 0's.**
Step 2:	$\begin{array}{r} 3x \\ x-4\overline{)\,3x^2 -\;\; 5x + 2} \end{array}$	Mentally divide $3x^2$ by x: $\dfrac{3x^2}{x} = 3x$. Write $3x$ above $3x^2$.
Step 3:	$\begin{array}{r} 3x \\ x-4\overline{)\,3x^2 -\;\; 5x + 2} \\ -(3x^2 - 12x) \end{array}$	Multiply $3x$ times $(x - 4)$ and write the terms under the like terms in the dividend. Use a '–' sign to indicate that the product is to be subtracted.
Step 4:	$\begin{array}{r} 3x \\ x-4\overline{)\,3x^2 -\;\; 5x + 2} \\ -3x^2 + 12x \\ \hline + 7x \end{array}$	Subtract $3x^2 - 12x$ by changing signs and adding.
Step 5:	$\begin{array}{r} 3x \\ x-4\overline{)\,3x^2 -\;\; 5x + 2} \\ -3x^2 + 12x \\ \hline + 7x + 2 \end{array}$	Bring down the +2.
Step 6:	$\begin{array}{r} 3x +\;\; 7 \\ x-4\overline{)\,3x^2 -\;\; 5x + 2} \\ -3x^2 + 12x \\ \hline +\;\; 7x + 2 \end{array}$	Mentally divide $+7x$ by x: $\dfrac{+7x}{x} = +7$. Write +7 in the quotient.

Step 7:

$$\begin{array}{r} 3x \;+\; 7 \\ x-4\overline{\smash{\big)}\,3x^2 - 5x + 2} \\ \underline{-3x^2 + 12x} \\ +7x+\;2 \\ \underline{-\left(+7x-28\right)} \end{array}$$

Multiply +7 times ($x - 4$) and write the terms under the like terms in the expression $+7x + 2$. Use a '–' sign to indicate that the product is to be subtracted.

Step 8:

$$\begin{array}{r} 3x \;+\; 7 \\ x-4\overline{\smash{\big)}\,3x^2 - 5x + 2} \\ \underline{-3x^2 + 12x} \\ +\;7x+\;2 \\ \underline{-\;7x+28} \\ 30 \end{array}$$

Subtract $7x - 28$ by changing signs and adding.

Thus the quotient is $3x + 7$ and the remainder is 30.

The answer can also be written in the form $Q + \dfrac{R}{D}$ as $3x + 7 + \dfrac{30}{x-4}$.

Step 9: **Check:** Multiply the divisor and quotient, and add the remainder. The result should be the dividend.

$$\begin{aligned} Q \cdot D + R &= (\,3x + 7\,)(\,x - 4\,) + 30 \\ &= 3x^2 - 12x + 7x - 28 + 30 \\ &= 3x^2 - 5x + 2 \end{aligned}$$

b. $\dfrac{4x^3 - 7x - 5}{2x+1}$ or $\left(4x^3 - 7x - 5\right) \div \left(2x+1\right)$

This example will be done in fewer steps. Note that $0x^2$ is inserted so that like terms will be aligned.

Solution:

Steps	Explanation

Step 1: $2x+1\overline{\smash{\big)}\,4x^3 + 0x^2 - 7x - 5}$ x^2 is a missing power, so $0x^2$ is supplied.

Continued on next page...

Step 2:

$$\begin{array}{r} 2x^2 \\ 2x+1\overline{)\,4x^3+0x^2-7x-5} \\ -(4x^3+2x^2) \end{array}$$

$\dfrac{4x^3}{2x}=2x^2$ and
$2x^2\,(\,2x\,+1\,)=4x^3+2x^2.$

Step 3:

$$\begin{array}{r} 2x^2 \\ 2x+1\overline{)\,4x^3+0x^2-7x-5} \\ \underline{-4x^3-2x^2} \\ -2x^2-7x \end{array}$$

Subtract $4x^3+2x^2$ and bring down $-7x.$

Step 4:

$$\begin{array}{r} 2x^2-x \\ 2x+1\overline{)\,4x^3+0x^2-7x-5} \\ \underline{-4x^3-2x^2} \\ -2x^2-7x \\ \underline{-(-2x^2-x)} \end{array}$$

$-\dfrac{2x^2}{2x}=-x.$ Multiply $-x$ times $(\,2x+1\,).$

Step 5:

$$\begin{array}{r} 2x^2-x-3 \\ 2x+1\overline{)\,4x^3+0x^2-7x-5} \\ \underline{-4x^3-2x^2} \\ -2x^2-7x \\ \underline{+2x^2+x} \\ -6x-5 \\ \underline{+6x+3} \\ -2 \end{array}$$

Continue dividing, using the same procedure. The remainder is $-2.$

The answer can also be written in the form

$$2x^2-x-3+\frac{-2}{2x+1} \quad \text{or} \quad 2x^2-x-3-\frac{2}{2x+1}$$

Step 6:   **Check:** $(2x+1)(2x^2-x-3)-2=4x^3-2x^2-6x+2x^2-x-3-2$

$$=4x^3-7x-5$$

The division algorithm can be used to divide two polynomials whenever the degree of the dividend, P (the numerator), is greater than or equal to the degree of the divisor, D (the denominator). The remainder, R, must be of smaller degree than the divisor, D.

Example 3: Long Division

Use the division algorithm (or long division) to divide:

$$\left(x^3 - 3x^2 - 5x - 8\right) \div \left(x^2 + 2x + 3\right)$$

Solution:

$$
\begin{array}{r}
x - 5 \\
x^2 + 2x + 3 \overline{)x^3 - 3x^2 - 5x - 8} \quad \text{Dividend} \\
-\left(x^3 + 2x^2 + 3x\right) \\
\hline
-5x^2 - 8x - 8 \\
-\left(-5x^2 - 10x - 15\right) \\
\hline
2x + 7 \quad \text{Remainder is first-degree, which is}
\end{array}
$$

a smaller degree than the divisor.

$$Q + \frac{R}{D} = x - 5 + \frac{2x + 7}{x^2 + 2x + 3}$$

Check: $\left(x^2 + 2x + 3\right)\left(x - 5\right) + \left(2x + 7\right) = x^3 + 2x^2 + 3x - 5x^2 - 10x - 15 + 2x + 7$

$$= x^3 - 3x^2 - 5x - 8$$

Practice Problems

1. *Express the quotient as a sum of fractions:* $\dfrac{4x^2 + 6x + 1}{2x}$.

Use the division algorithm to divide.

2. $(3x^2 + 8x - 4) \div (x + 2)$

3. $(x^3 + 4x^2 - 5) \div (x^2 + x - 1)$

Answers to Practice Problems: 1. $2x + 3 + \dfrac{1}{2x}$ **2.** $3x + 2 - \dfrac{8}{x + 2}$ **3.** $x + 3 + \dfrac{-2x - 2}{x^2 + x - 1}$

6.7 Exercises

Express each quotient in Exercises 1 – 20 as a sum of fractions and simplify if possible.

1. $\dfrac{4x^2 + 8x + 3}{4}$ **2.** $\dfrac{6x^2 - 10x + 1}{2}$ **3.** $\dfrac{10x^2 - 15x - 3}{5}$ **4.** $\dfrac{9x^2 - 12x + 5}{3}$

5. $\dfrac{2x^2 + 5x}{x}$ **6.** $\dfrac{8x^2 - 7x}{x}$ **7.** $\dfrac{x^2 + 6x - 3}{x}$ **8.** $\dfrac{-2x^2 - 3x + 8}{x}$

9. $\dfrac{4x^2 + 6x - 3}{2x}$ **10.** $\dfrac{3x^3 - 2x^2 + x}{x^2}$ **11.** $\dfrac{6x^3 - 9x^2 - 3x}{3x^2}$

12. $\dfrac{5x^2y - 10xy^2 - 3y}{5xy}$ **13.** $\dfrac{7x^2y^2 + 21xy^3 - 11y^3}{7xy^2}$ **14.** $\dfrac{12x^3y + 6x^2y^2 - 3xy}{6x^2y}$

15. $\dfrac{3x^3y - 8xy^2 - 4y^3}{4xy}$ **16.** $\dfrac{2x^3y^2 - 6x^2y^3 + 15xy}{3xy^2}$ **17.** $\dfrac{5x^2y^2 - 8xy^2 + 16xy}{8xy^2}$

18. $\dfrac{3x^3y - 14x^2y - 7xy}{7x^2y}$ **19.** $\dfrac{8x^3y^2 - 9x^2y^3 + 5xy}{9xy^2}$ **20.** $\dfrac{24x^2y^2 + 12xy - 6xy}{3xy}$

Divide in Exercises 21 – 60 by using the long division procedure. Write your answers in the form $Q + \dfrac{R}{D}$. Check each answer by showing that $P = Q \cdot D + R$.

21. $278 \div 23$ **22.** $326 \div 64$

23. $(x^2 + 3x + 2) \div (x + 2)$ **24.** $(x^2 + x - 6) \div (x + 3)$

25. $(y^2 + 8y + 15) \div (y + 4)$ **26.** $(a^2 - 2a - 15) \div (a - 2)$

27 $(x^2 - 7x - 18) \div (x + 5)$ **28.** $(y^2 - y - 42) \div (y + 4)$

29. $(4a^2 - 21a + 2) \div (a - 6)$ **30.** $(5y^2 + 14y - 7) \div (y + 5)$

31. $(8x^2 + 10x - 4) \div (2x + 3)$

32. $(8c^2 + 2c - 14) \div (2c + 3)$

33. $(6x^2 + x - 4) \div (2x - 1)$

34. $(10m^2 - m - 6) \div (5m - 3)$

35. $(x^2 - 6) \div (x + 2)$

36. $(x^2 + 3x) \div (x + 5)$

37. $(2x^3 - x^2 - 13x - 6) \div (x - 3)$

38. $(y^3 - 9y^2 + 26y - 24) \div (y - 2)$

39. $(3t^3 + 10t^2 + 3t + 2) \div (3t + 1)$

40. $(12a^3 - 3a^2 + 4a + 1) \div (4a - 1)$

41. $(2x^3 + x^2 - 6) \div (x + 4)$

42. $(x^3 + 2x^2 - 5) \div (x - 5)$

43. $(x^3 - 8) \div (x - 2)$

44. $(x^3 + 27) \div (x + 3)$

45. $(2x^3 - x + 3) \div (x^2 - 2)$

46. $(3x^3 + 11x + 1) \div (x^2 + 3)$

47. $(x^3 + 7x^2 + x - 2) \div (x^2 - x + 1)$

48. $(x^3 + 3x^2 + 1) \div (x^2 + 2x + 3)$

49. $\dfrac{8x^3 - 27}{2x - 3}$

50. $\dfrac{27a^3 - 64}{3a - 4}$

51. $\dfrac{4x^2 - x + 5}{x - 1}$

52. $\dfrac{5x^3 - 4x^2 + 81}{x + 3}$

53. $\dfrac{3x^3 - 10x^2 - 3x - 20}{x - 4}$

54. $\dfrac{x^4 + 2x^2 - 5}{x^2 + 1}$

55. $\dfrac{2a^3 + 3a^2 + 6}{a^2 + 2}$

56. $\dfrac{27y^3 + 64}{3y + 4}$

57. $\dfrac{8a^3 + 125}{2a + 5}$

58. $\dfrac{a^3 - 125}{a - 5}$

59. $\dfrac{x^4 - 1}{x - 1}$

60. $\dfrac{x^4 + 1}{x + 1}$

Hawkes Learning Systems: Introductory Algebra

Division by a Monomial
The Division Algorithm

Chapter 6 Index of Key Ideas and Terms

Section 6.1   Exponents

Summary of the Rules for Exponents   page 477
    1. The Exponent 1: $a = a^1$ page 469
    2. The Product Rule: $a^m \cdot a^n = a^{m+n}$ pages 469 - 471
    3. The Exponent 0: $a^0 = 1$ $(a \neq 0)$ pages 471 - 472
    4. The Quotient Rule: $\dfrac{a^m}{a^n} = a^{m-n}$ pages 473 - 474

    5. Negative Exponents: $a^{-n} = \dfrac{1}{a^n}$ pages 474 - 476

Section 6.2   Exponents and Scientific Notation

Summary of the Rules for Exponents   pages 481, 487
    1. The Exponent 1: $a = a^1$ pages 481, 487

    2. The Exponent 0: $a^0 = 1$ $(a \neq 0)$ pages 481, 487

    3. The Product Rule: $a^m \cdot a^n = a^{m+n}$ pages 481, 487

    4. Power Rule: $(a^m)^n = a^{mn}$ pages 481 - 482

    5. Power of a Product: $(ab)^n = a^n b^n$ pages 483 - 484

    6. The Quotient Rule: $\dfrac{a^m}{a^n} = a^{m-n}$ pages 481, 487

    7. Negative Exponents: $a^{-n} = \dfrac{1}{a^n}$ pages 482, 486

    8. Power of a Quotient: $\left(\dfrac{a}{b}\right)^n = \dfrac{a^n}{b^n}$ pages 484 - 485

Scientific Notation and Calculator   pages 488 - 490
    If N is a decimal number, then in **scientific notation**
       $N = a \times 10^n$ where $1 \leq a < 10$ and n is an integer.

Section 6.3 Introduction to Polynomials

Monomial

pages 495 - 496

A **monomial in x** is a term of the form kx^n where k is a real number and n is a whole number. n is called the **degree** of the term, and k is called the **coefficient**.

Polynomial

pages 496 - 497

A **polynomial** is a monomial or the indicated sum or difference of monomials.

The **degree of a polynomial** is the largest of the degrees of its terms.

The coefficient of the term of the largest degree is called the **leading coefficient**.

Special Terminology for some Polynomials

page 496

Monomial: a polynomial with one term
Binomial: a polynomial with two terms
Trinomial: a polynomial with three terms

Evaluation of Polynomials

page 498

A polynomial can be treated as a function and the notation $p(x)$ can be used.

Section 6.4 Addition and Subtraction with Polynomials

Addition with Polynomials

pages 502 - 503

Subtraction with Polynomials

pages 503 - 504

Simplifying Algebraic Expressions

pages 504 - 505

Section 6.5 Multiplication with Polynomials

Multiplication with Polynomials

pages 508 - 511

Section 6.6   Special Products of Binomials

The FOIL Method   pages 515 - 516

The Difference of Two Squares   page 517
$$(x + a)(x - a) = x^2 - a^2$$

Perfect Square Trinomials   pages 518 - 520
$$(x + a)^2 = (x + a)(x + a) = x^2 + 2ax + a^2$$
$$(x - a)^2 = (x - a)(x - a) = x^2 - 2ax + a^2$$

Section 6.7   Division with Polynomials

Division with Polynomials
    Division by a Monomial   pages 525 - 526
    The Division Algorithm   pages 527 - 531

Hawkes Learning Systems: Introductory Algebra

For a review of the topics and problems from Chapter 6, look at the following lessons from *Hawkes Learning Systems: Introductory Algebra*

Simplifying Integer Exponents I
Simplifying Integer Exponents II
Scientific Notation
Identifying Polynomials
Evaluating Polynomials
Addition and Subtraction with Polynomials
Multiplying a Polynomial By a Monomial
Multiplying Two Polynomials
The FOIL Method
Special Products
Division by a Monomial
The Division Algorithm

Chapter 6 Review

Section 6.1 Exponents
Simplify each expression in Exercises 1 – 15 by using rules for exponents. The final form of the expressions with variables should contain only positive exponents. Assume that all variables represent nonzero numbers.

1. $5^2 \cdot 5$ **2.** $4^3 \cdot 4^0$ **3.** $-3 \cdot 2^{-3}$ **4.** $5 \cdot 3^{-2}$

5. y^{-3} **6.** $x^3 \cdot x^4$ **7.** $y^{-2} \cdot y^0$ **8.** $4x^{-1}$

9. $\left(6x^2\right)\left(-2x^2\right)$ **10.** $\dfrac{16y^3}{8y^2}$ **11.** $\dfrac{-12x^7}{2x^3}$ **12.** $\left(-3a^3\right)^0$

13. $\left(3a^3b^2c\right)\left(-2ab^2c\right)$ **14.** $\dfrac{36a^6b^0}{12a^{-1}b^3}$ **15.** $\dfrac{35y^5 \cdot 2y^{-2}}{14xy^4}$

Section 6.2 Exponents and Scientific Notation
Use the rules for exponents to simplify each of the expressions in Exercises 16 – 29. Assume that all variables represent nonzero real numbers.

16. $\left(5x^3\right)^2$ **17.** $\left(-2x^4\right)^3$ **18.** $17\left(x^{-3}\right)^2$ **19.** $\left(\dfrac{y}{x}\right)^{-1}$

20. $\left(\dfrac{3x}{y^{-2}}\right)^2$ **21.** $\left(\dfrac{4a}{b}\right)^{-2}$ **22.** $\left(\dfrac{3x^3y}{y^3}\right)^2$ **23.** $\left(\dfrac{m^3n^2}{mn^{-1}}\right)^3$

24. $\dfrac{\left(5x^{-2}\right)\left(-2x^6\right)}{\left(y^2\right)^{-3}}$ **25.** $\dfrac{\left(6x^{-2}\right)\left(5x^{-1}\right)}{\left(3y\right)\left(2y^{-1}\right)}$

26. $\left(\dfrac{3xy^3}{4xy^{-2}}\right)^{-2}\left(\dfrac{6x^3y^{-1}}{8x^{-4}y^3}\right)^{-1}$ **27.** $\left(\dfrac{15a^3b^{-1}}{20ab^2}\right)^{-1}\left(\dfrac{-3a^2b}{a^{-4}b^{-2}}\right)^{-2}$

28. $\left(\dfrac{-2xy^3}{12x^{-1}y^3}\right)^2\left(\dfrac{6x^{-4}y^2}{y^{-3}}\right)^2$ **29.** $\left(\dfrac{3a^4bc^{-1}}{7ab^{-1}c^2}\right)^{-2}\left(\dfrac{9abc^{-1}}{7b^2c^2}\right)^2$

In Exercises 30 – 33, write the numbers in scientific notation.

30. 97,000   **31.** 29,300,000   **32.** 0.0562   **33.** 0.0075

In Exercises 34 – 37, write the numbers in decimal form.

34. 5.3×10^{-2} **35.** 7.24×10^{-4} **36.** 8.23×10^5 **37.** 9.485×10^7

In Exercises 38 – 41, first write each of the numbers in scientific notation. Then perform the indicated operations and leave your answer in scientific notation.

38. $\dfrac{125}{100,000}$

39. $\dfrac{0.0058}{290}$

40. $\dfrac{2.7 \cdot 0.002 \cdot 25}{54 \cdot 0.0005}$

41. $\dfrac{\left(1.5 \times 10^{-3}\right)(822)}{\left(4.11 \times 10^{3}\right)(300)}$

Section 6.3 Introduction to Polynomials

Simplify the polynomials in Exercises 42 – 49. Write the polynomial in descending order and state the degree and type of the simplified polynomial.

42. $x + 5x$

43. $5x^2 - x + x^2$

44. $x^3 - 3x^2 + 5x$

45. $12x^3 - 8x + 4x - 7x^3$

46. $6x^2 + 4 - 2x^2 + 1 - 4x^2$

47. $11x^3 - 6x^2 - 9x^3 - 5 + 6x^3$

48. $a^3 - 2a^2 - 8a + 3a^3 - 2a^2 + 6a + 9$

49. $3a + 4a^2 + 6a - 10a^3$

In Exercises 50 – 56, evaluate the given polynomial as indicated.

50. Given $p(x) = x^2 - 13x + 5$, find $p(-1)$.

51. Given $p(x) = 3x^2 + 4x - 6$, find $p(2)$.

52. Given $p(y) = 3y^3 - 3y^2 + 10$, find $p(4)$.

53. Given $p(y) = y^2 - 8y + 13$, find $p(-5)$.

54. Given $p(x) = x^2 + 6x + 9$, find $p(a)$.

55. Given $p(x) = -3x^3 - 10x + 15$, find $p(c)$.

56. Given $p(x) = x^4 + 3x^3 - x^2 - 20x + 5$, find $p(1)$.

Section 6.4 Addition and Subtraction with Polynomials

Find the indicated sum in Exercises 57 – 64.

57. $\left(3x^2 + 6x - 1\right) + \left(x^2 + 3x + 4\right)$

58. $\left(-5x^2 + 2x - 3\right) + \left(4x^2 - x + 6\right)$

59. $\left(x^2 + 7x + 12\right) + (2x - 8) + \left(x^2 + 10x\right)$

60. $\left(-8x^2 + 5x + 2\right) + \left(3x^2 - 9x + 5\right) + (x - 11)$

61. $\begin{aligned}2x^2 - 4x + 4 \\ \underline{-x^2 - 3x + 4}\end{aligned}$

62. $\begin{aligned}x^3 + 4x^2 + x \\ \underline{-3x^3 - x^2 + 2x + 7}\end{aligned}$

63. $\begin{aligned}x^3 + 4x^2 - 6 \\ 8x^2 + 3x - 1 \\ \underline{x^3 + x^2 - 4x}\end{aligned}$

64. $\begin{aligned}4x^3 - 9x^2 + 7x - 12 \\ \underline{2x^3 - 2x^2 + 2x - 15}\end{aligned}$

Find the indicated difference in Exercises 65 – 70.

65. $\left(3x^2 + 4x - 4\right) - \left(x^2 + 5x + 6\right)$

66. $\left(x^4 + 7x^3 - x + 13\right) - \left(-x^4 + 6x^2 - x + 3\right)$

67. $\left(5x^2 - x - 10\right) - \left(-x^2 - 10\right)$

68. $\left(7x^2 - 14\right) - \left(4x^2 - 2x - 12\right)$

69. $\quad 13x^2 - 6x + 7$

$\quad\;\; -\left(9x^2 + \;\; x + 5\right)$

70. $\quad 11x^3 \qquad\; + 10x - 20$

$\quad\;\; -\left(\;\; x^3 + 5x^2 - 10x - 22\right)$

Simplify each of the expressions in Exercises 71 – 76 and write the polynomials in descending order.

71. $6x + 2\left(x - 7\right) - \left(3x + 4\right)$

72. $3x + \left[8x - 4\left(3x + 1\right) - 9\right]$

73. $4x^3 - \left[7 - 5\left(2x - 3\right) + 10x^3\right]$

74. $\left(x^2 + 4\right) - \left[-6 + 2\left(3 - 2x^2\right) + x\right]$

75. $2\left[4x + 5\left(x - 8\right) - \left(2x + 3\right)\right] + \left(3x - 5\right)$

76. $-\left[5x^2 + 7\left(x^2 - 30x + 2\right) - x^2 - 14\right]$

77. Subtract $3x^2 - 2x + 4$ from $8x^2 - 10x + 3$.

78. Subtract $4x^2 - 8x + 1$ from the sum of $2x^2 - 5x + 2$ and $x^2 - 14x + 11$.

Section 6.5 Multiplication with Polynomials

Multiply as indicated in Exercises 79 – 90 and simplify if possible.

79. $-2x^2\left(3x^3 + 4x\right)$

80. $7a^3\left(-5a^2 + 3a - 1\right)$

81. $3a\left(-2a + 6\right) + 5\left(-2a + 6\right)$

82. $x\left(x^2 + 5x + 4\right) - 2\left(x^2 + 5x + 4\right)$

83. $x\left(x - 5\right)\left(x - 6\right)$

84. $-4\left(x + 3\right)\left(x - 5\right)$

85. $\left(2y + 1\right)\left(y^2 - 3y + 2\right)$

86. $\left(x + 2\right)\left(x + 3\right)\left(x + 4\right)$

87. $\quad 4x + 7$

$\quad\;\; \underline{\;\; x - 6}$

88. $\quad y^2 + 4y + 1$

$\quad\;\; \underline{\;\; 3y - 8}$

89. $\quad x^2 + 3x + 5$

$\quad\;\; \underline{x^2 + 3x - 2}$

90. $\quad 2a^2 - \;\; a - 3$

$\quad\;\; \underline{a^2 + 6a + 5}$

Section 6.6 Special Products of Binomials

Find the product in Exercises 91 – 102 and identify those that are the difference of two squares or perfect square trinomials.

91. $\left(2x + 3\right)\left(x + 1\right)$

92. $\left(3x + 5\right)\left(2x + 5\right)$

93. $\left(x + 13\right)\left(x - 13\right)$

94. $\left(y + 8\right)\left(y - 8\right)$

95. $\left(2x + 9\right)^2$

96. $\left(3x - 1\right)^2$

97. $\left(y^2 + 6\right)\left(y^2 - 4\right)$

98. $\left(3x - 7\right)\left(2x + 11\right)$

99. $\left(x^3 - 10\right)\left(x^3 + 10\right)$

100. $\left(t+\dfrac{1}{4}\right)\left(t-\dfrac{1}{4}\right)$ **101.** $\left(y+\dfrac{2}{3}\right)\left(y-\dfrac{1}{2}\right)$ **102.** $\left(7-3x\right)^2$

Section 6.7 Division with Polynomials

Express each quotient in Exercises 103 – 110 as a sum of fractions and simplify if possible.

103. $\dfrac{9x^2+6x+1}{3}$ **104.** $\dfrac{-2x^2-8x-3}{-2}$ **105.** $\dfrac{9x^3-12x^2+3x}{x^2}$

106. $\dfrac{4x^2y-8xy^2+y^3}{2xy}$ **107.** $\dfrac{5a^3-10a^2+4a}{5a}$ **108.** $\dfrac{4x^2y^2+16xy^2-8y^2}{8y^2}$

109. $\dfrac{12a^2b^2+6ab-3ab}{3ab}$ **110.** $\dfrac{7x^3y^2-8x^2y^3-5xy}{4xy^2}$

Divide in Exercises 111 – 122 by using the long division procedure. Write your answers in the form $Q+\dfrac{R}{D}$.

111. $\left(a^2-2a-15\right)\div\left(a-5\right)$ **112.** $\left(x^2+9x+20\right)\div\left(x+4\right)$

113. $\left(y^2-y-56\right)\div\left(y-8\right)$ **114.** $\left(x^2+7x-18\right)\div\left(x+9\right)$

115. $\left(4x^2-21x+3\right)\div\left(x-2\right)$ **116.** $\left(8y^2+10y+5\right)\div\left(2y+3\right)$

117. $\left(6x^2+x-6\right)\div\left(2x-1\right)$ **118.** $\left(x^2-9\right)\div\left(x+4\right)$

119. $\left(x^3-1\right)\div\left(x-1\right)$ **120.** $\left(8x^3+27\right)\div\left(2x+3\right)$

121. $\left(y^3+5y^2\right)\div\left(y+5\right)$ **122.** $\left(64x^3-125\right)\div\left(4x-5\right)$

Chapter 6 Test

Use the rules for exponents to simplify each expression in Exercises 1 – 6. Each answer should have only positive exponents.

1. $(5a^2b^5)(-2a^3b^{-5})$ 　　 **2.** $(-7x^4y^{-3})^0$ 　　 **3.** $\dfrac{(-8x^2y^{-3})^2}{16xy}$

4. $\dfrac{3x^{-2}}{9x^{-3}y^2}$ 　　 **5.** $\left(\dfrac{4xy^2}{x^3}\right)^{-1}$ 　　 **6.** $\left(\dfrac{2x^0y^3}{x^{-1}y}\right)^2$

7. Write each of the following numbers in decimal notation.

   a. 1.35×10^5 　　　　　 **b.** 2.7×10^{-6}

8. Perform the indicated operations by writing each number in scientific notation and write the result in scientific notation.

   a. $250 \cdot 500,000$ 　　　 **b.** $\dfrac{65 \cdot 0.012}{1500}$

In Exercises 9 – 11, simplify the polynomials and state the degree and type of the polynomial.

9. $3x + 4x^2 - x^3 + 4x^2 + x^3$ 　　　 **10.** $2x^2 + 3x - x^3 + x^2 - 1$

11. $17 - 14 + 4x^5 - 3x + 2x^4 + x^5 - 8x$

12. For the polynomial $p(x) = 2x^3 - 5x^2 + 7x + 10$, find (a) $p(2)$ and (b) $p(-3)$.

Simplify each expression in Exercises 13 and 14.

13. $7x + [2x - 3(4x + 1) + 5]$ 　　　 **14.** $12x - 2[5 - (7x + 1) + 3x]$

In Exercises 15 – 26, perform the indicated operations and simplify each expression. Tell which, if any, answers are the difference of two squares and which, if any, are perfect square trinomials.

15. $(5x^3 - 2x + 7) + (-x^2 + 8x - 2)$ 　　 **16.** $(4x^2 + 3x - 1) - (6x^2 + 2x + 5)$

17. $(x^4 + 3x^2 + 9) - (-6x^4 - 11x^2 + 5)$ 　 **18.** $(3x^3 - 2x^2) + (4x^3 - 7x + 1)$

19. $5x^2(3x^5 - 4x^4 + 3x^3 - 8x^2 - 2)$ 　　 **20.** $(7x + 3)(7x - 3)$

21. $(4x + 1)^2$ 　　　　　　　　　 **22.** $(6x - 5)^2$

23. $(2x + 5)(6x - 3)$

24. $3x(x - 7)(2x - 9)$

25. $(3x + 1)(3x - 1) - (2x + 3)(x - 5)$

26. $\begin{array}{r} 2x^3 - 3x - 7 \\ \times \quad 5x + 2 \\ \hline \end{array}$

In Exercises 27 and 28, express each quotient as a sum of fractions and simplify.

27. $\dfrac{4x^3 + 3x^2 - 6x}{2x^2}$

28. $\dfrac{5a^2b + 6a^2b^2 + 3ab^3}{3a^2b}$

In Exercises 29 and 30, divide by using the division algorithm and write the answers in the form $Q + \dfrac{R}{D}$.

29. $(2x^2 - 9x - 20) \div (2x + 3)$

30. $\dfrac{x^3 - 8x^2 + 3x + 15}{x^2 + x - 3}$

31. Subtract $5x^2 - 3x + 4$ from the sum of $4x^2 + 2x + 1$ and $3x^2 - 8x - 10$.

32. Divide the product of $2y + 7$ and $3y - 6$ by $y - 3$ and write the answer in the form $Q + \dfrac{R}{D}$.

Cumulative Review: Chapters 1 – 6

Use the distributive property to complete the expression in Exercises 1 and 2.

1. $3x + 45 = 3($ $)$ **2.** $6x + 16 = 2($ $)$

Find the LCM for each set of numbers or terms in Exercises 3 and 4.

3. $12, 15, 54$ **4.** $6a^2, 24ab^3, 30ab, 40a^2b^2$

For each pair of numbers in Exercises 5 – 8, find two factors of the first number whose sum is the second number.

5. $32, 18$ **6.** $36, -13$ **7.** $-56, 10$ **8.** $-48, -8$

In Exercises 9 – 12, simplify each polynomial and state its degree and type. Write the values for each of the coefficients $a_n, a_{n-1}, \ldots, a_1, a_0$.

9. $6x^2 + 10x - (x^2 - 6x)$ **10.** $4(2x - 3) - 3(x + 2)$

11. $4x - [(6x + 7) - (3x + 4)] - 6$ **12.** $3x^2 - 8(x - 5) + x^4 - 2x^3 + x^2 - 2x$

Solve the equations in Exercises 13 – 16.

13. $7(4 - x) = 3(x + 4)$ **14.** $-2(5x + 1) + 2x = 4(x + 1)$

15. $\dfrac{2}{3}x + \dfrac{1}{2} = \dfrac{3}{4}$ **16.** $1.5x - 3.7 = 3.6x + 2.6$

Solve each formula in Exercises 17 and 18 for the indicated variable.

17. $C = \pi d$; solve for d.   **18.** $3x + 5y = 10$; solve for y.

In Exercises 19 and 20, solve the inequality and graph the solution set on a real number line.

19. $5x + 3 \geq 2x - 15$ **20.** $-16 < 3x + 5 < 17$

21. Write an equation for the line that has slope $m = \dfrac{2}{3}$ and contains the point $(-1, 2)$. Graph the line.

22. Write an equation for the line passing through the two points $(-5, 1)$ and $(3, -4)$. Graph the line.

23. Write an equation for the line that is parallel to the line $2x + 5y = 3$ and passes through the point $(0, 7)$. Graph both lines.

24. Determine whether the lines $2x + 6y = 5$ and $x + 3y = 9$ are (parallel, perpendicular, or neither). Explain your reasoning.

25. Determine whether the lines $x - 2y = 4$ and $y = -2x + 8$ are (parallel, perpendicular, or neither). Explain your reasoning.

26. For the function $f(x) = 3x^2 - 4x + 2$, find (a) $f(-2)$ and (b) $f(5)$.

Graph the following inequalities and tell what type of half-plane is graphed.

27. $3x + y < 10$

28. $4x + 2y \geq 9$

Use a graphing calculator to graph and solve the systems of equations in Exercises 29 and 30. You may need to estimate the answers.

29. $\begin{cases} 2x + 3y = 18 \\ 3x - 2y = -12 \end{cases}$

30. $\begin{cases} y = \dfrac{1}{3}x + 5 \\ y = -2x + 1 \end{cases}$

Solve each system of equations in Exercises 31 and 32 by using an algebraic method.

31. $\begin{cases} 3x - 4y = -25 \\ 2x + y = 9 \end{cases}$

32. $\begin{cases} x + 8y = -22 \\ 3x - y = -9 \end{cases}$

Solve each system of inequalities in Exercises 33 and 34.

33. $\begin{cases} 2x + y < 6 \\ -x + 2y > 4 \end{cases}$

34. $\begin{cases} x \leq 3 \\ y \geq -3x + 1 \end{cases}$

Simplify each expression in Exercises 35 – 44 so that it has no exponents or only positive exponents.

35. $\dfrac{4x^3}{2x^{-2}x^4}$

36. $(4x^2 y)^3$

37. $(7x^5 y^{-2})^2$

38. $\left(\dfrac{6x^2}{y^5}\right)^2$

39. $(a^{-3}b^2)^{-2}$

40. $\left(\dfrac{3xy^4}{x^2}\right)^{-1}$

41. $\left(\dfrac{8x^{-3}y^2}{xy^{-1}}\right)^0$

42. $\left(\dfrac{3^{-1}x^3 y^{-1}}{x^{-1}y^2}\right)^2$

43. $\left(\dfrac{5x^{-2}y^3}{12x^2 y}\right)^{-1}\left(\dfrac{10x^2 y^2}{3x^{-1}y^{-1}}\right)^2$

44. $\left(\dfrac{22a^3 b^2 c}{6a^{-3}bc^2}\right)^{-1}\left(\dfrac{14ab^2 c}{7a^{-2}b^2 c^3}\right)^3$

45. Write each number in decimal notation.

    a. 2.8×10^{-7} **b.** 3.51×10^{4}

In Exercises 46 – 48, first write each of the numbers in scientific notation. Then perform the indicated operations and leave your answer in scientific notation.

46. $0.0015 \cdot 4200$ **47.** $\dfrac{840}{0.00021}$ **48.** $\dfrac{0.005 \cdot 77}{0.011 \cdot 3500}$

In Exercises 49 – 52, (a) simplify each polynomial, (b) state its degree and type, (c) find $p(3)$ and (d) find $p(-5)$.

49. $p(x) = -x^2 + 5x + 2x^2 - x$ **50.** $p(x) = 9x - x^3 + 3x^2 - x + x^3$

51. $p(x) = -x^4 + 4x^3 - 3x + x^2 - x^3 + x$ **52.** $p(x) = 8x - 7x^2 + x^2 - x^3 - 6x - 4$

Perform the indicated operations in Exercises 53 – 73 and simplify each expression.

53. $(-2x^2 - 11x + 1) + (x^3 + 3x - 7) + (2x - 1)$

54. $(4x^2 + 2x - 7) - (5x^2 + x - 2)$

55. $(x^3 + 4x^2 - x) - (-2x^3 + 6x + 3)$

56. $(6x^2 + x - 10) - (x^3 - x^2 + x - 4)$

57. $(x^2 + 2x + 6) + (5x^2 - x - 2) - (8x + 3)$

58. $(2x^2 - 5x - 7) - (3x^2 - 4x + 1) + (x^2 - 9)$

59. $-3x(x^2 - 4x + 1)$ **60.** $5x^3(x^2 + 2x)$ **61.** $(x + 6)(x - 6)$

62. $(x + 4)(x - 3)$ **63.** $(3x + 7)(3x + 7)$ **64.** $(2x - 1)(2x + 1)$

65. $(x^2 + 5)(x^2 - 5)$ **66.** $(x^2 - 2)(x^2 - 2)$ **67.** $-1(x^2 - 5x + 2)$

68. $(x - 4)^2$ **69.** $(2x - 9)(x + 4)$ **70.** $(x - 6)(5x + 3)$

71. $(3x - 4)(2x + 3)$ **72.** $(3x + 8)(3x - 8)$ **73.** $(4x + 1)(x^2 - x)$

In Exercises 74 – 76, express each quotient as a sum of fractions and simplify if possible.

74. $\dfrac{8x^2 - 14x + 6}{2x}$ **75.** $\dfrac{13x^3y + 10x^2y^2 + 5xy^3}{5x^2y}$

76. $\dfrac{x^2y^2 - 21x^2y^3 + 7xy - 28x^2y^4}{7x^2y^2}$

In Exercises 77 – 79, divide by using the division algorithm and express each answer in the form $Q + \dfrac{R}{D}$.

77. $\dfrac{x^2 + 7x - 18}{x + 9}$ **78.** $\dfrac{4x^2 + 5x - 2}{4x - 3}$ **79.** $\dfrac{2x^3 + 5x^2 + 7}{x + 3}$

80. Find three consecutive even integers such that the sum of the second and third is equal to three times the first decreased by 14.

81. The length of a rectangle is 9 centimeters more than its width. The perimeter is 82 centimeters. Find the dimensions of the rectangle.

82. Twice the difference between a number and 16 is equal to four times the number increased by 6. What is the number?

83. An isosceles triangle has two equal sides. The third side is 4 feet more than the length of the two equal sides. If each side is increased by 3 feet in length, the perimeter of the triangle will be 127 feet. What were the lengths of each side of the original triangle?

84. Determine the values of a and b such that the straight line $ax + by = 9$ passes through the two points $(-1, -3)$ and $(1, -1.5)$.

85. Sylvester has two investments that total $100,000. One investment earns interest at 6% and the other investment earns interest at 8%. If the total amount of interest from the two investments is $6700 in one year, how much money does he have invested at each rate?

86. A chemist needs 8 ounces of a mixture that is 15% iodine. He has a 10% mixture and a 30% mixture in stock. How many ounces of each mixture must he use to get the mixture he wants?

Factoring Polynomials and Solving Quadratic Equations

Did You Know?

You have probably noticed by now that almost every algebraic skill somehow relates to equation solving and applied problems. The emphasis on equation solving has always been a part of classical algebra.

In Italy during the Renaissance, it was the custom for one mathematician to challenge another mathematician to an equation-solving contest. A large amount of money, often in gold, was supplied by patrons or sponsoring cities as the prize. At that time, it was important not to publish equation-solving methods, since mathematicians could earn large amounts of money if they could solve problems that their competitors could not. Equation-solving techniques were passed down from a mathematician to an apprentice, but they were never shared.

A Venetian mathematician, Niccolo Fontana (1500? – 1557) known as Tartaglia, "the stammerer," discovered how to solve third-degree or cubic equations. At that time, everyone could solve first and second-degree equations and special kinds of equations of higher degree. Tartaglia easily won equation-solving contests simply by giving his opponents third-degree equations to solve.

Tartaglia planned to keep his method secret, but after receiving a pledge of secrecy, he gave his method to Girolamo Cardano (1501 – 1576). Cardano broke his promise by publishing one of the first successful Latin algebra texts, Ars Magna, "The Great Art." He included not only Tartaglia's solution to the third-degree equations but also a pupil's (Ferrari) discovery of the general solution to fourth-degree equations. Until recently, Cardano received credit for discovering both methods.

It was not until 300 years later that it was shown that there are no general algebraic methods for solving fifth- or higher-degree equations. As you can see, a great deal of time and energy has gone into developing the methods of equation solving that you are learning.

7.1   **Greatest Common Factor and Factoring by Grouping**

7.2   **Factoring Trinomials: $x^2 + bx + c$**

7.3   **More on Factoring Trinomials: $ax^2 + bx + c$**

7.4   **Factoring Special Products: Difference of Two Squares and Perfect Square Trinomials**

7.5   **Solving Quadratic Equations by Factoring**

7.6   **Applications of Quadratic Equations**

7.7   **Additional Applications of Quadratic Equations**

"Algebra is the intellectual instrument which has been created for rendering clear the quantitative aspect of the world."

Alfred North Whitehead (1861-1947)

Factoring is the reverse of multiplication. That is, to factor polynomials, you need to remember how you multiplied them. In this way, the concept of factoring is built on your previous knowledge and skills with multiplication. For example, if you are given a product, such as $x^2 - a^2$ (the difference of two squares), you must recall the factors of $(x + a)$ and $(x - a)$ from your work in multiplying polynomials in Chapter 6.

Studying mathematics is a building process with each topic dependent on previous topics with a few new ideas added each time. The equations and applications in Chapter 7 involve many of the concepts studied earlier, yet you will find them a step higher and more interesting and more challenging.

7.1   Greatest Common Factor and Factoring by Grouping

Objectives

After completing this section, you will be able to:

1. Find the greatest common factor of a set of terms.

2. Factor polynomials by finding the greatest common monomial factor.

3. Factor polynomials by grouping.

In Chapter 6 we used the distributive property and the FOIL method to multiply polynomials and other algebraic expressions. The result of multiplication is called the **product** and the numbers or expressions being multiplied are called **factors** of the product. In this chapter we will study the reverse of this process, which is called **factoring**. That is, given a product, we want to find the factors. For example,

Multiplying Polynomials   *Factoring Polynomials*

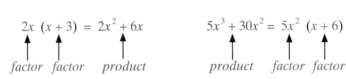

$$2x\,(x + 3) = 2x^2 + 6x \qquad 5x^3 + 30x^2 = 5x^2\,(x + 6)$$

factor factor   product   product   factor factor

Greatest Common Factor

The **greatest common factor (GCF)** of two or more integers is the largest integer that is a factor (or divisor) of all of the integers. For example, the GCF of 30 and 40 is 10. Note that 5 is also a common factor of 30 and 40, but 5 is not the **greatest** common factor. The number 10 is the largest number that will divide into both 30 and 40.

One way of finding the GCF is to use the prime factorization of each number. For example, to find the GCF for 36 and 60, we can write

$$36 = 4 \cdot 9 = \mathbf{2} \cdot \mathbf{2} \cdot \mathbf{3} \cdot 3,$$
$$60 = 4 \cdot 15 = \mathbf{2} \cdot \mathbf{2} \cdot \mathbf{3} \cdot 5.$$

The common factors are 2, 2, and 3 and their product is the GCF:

$$\text{GCF} = 2 \cdot 2 \cdot 3 = 12.$$

Writing the prime factorizations using exponents gives

$$36 = 2^2 \cdot 3^2 \quad \text{and} \quad 60 = 2^2 \cdot 3 \cdot 5.$$

We can see that the GCF is the product of the greatest power of each factor that is common to both numbers. That is,

$$\text{GCF} = 2^2 \cdot 3 = 12.$$

This procedure can be used to find the GCF for any set of integers or algebraic terms with integer exponents.

Procedure for Finding the GCF of a Set of Terms

1. *Find the prime factorization of all integers and integer coefficients.*
2. *List all factors common to all terms, including variables.*
3. *Choose the greatest power of each factor common to all terms.*
4. *Multiply these powers to find the GCF.*

   (***Note****: If there is no common prime factor or variable, then the GCF is 1.*)

Find the GCF for each of the following sets of algebraic terms.

a.   30, 45, 75

Solution: Find the prime factorization of each number:
$$30 = 2 \cdot 3 \cdot 5, \quad 45 = 3^2 \cdot 5, \quad \text{and} \quad 75 = 3 \cdot 5^2$$

The common factors are 3 and 5 and the greatest power of each common to all numbers is 1.

Thus, GCF $= 3^1 \cdot 5^1 = 15$.

b. $20x^4y,\ 15x^3y,\ 10x^5y^2$

Solution: Writing each integer coefficient in prime factored form gives:
$$20x^4y \;=\; 2^2 \cdot 5 \cdot x^4 \cdot y,$$
$$15x^3y \;=\; 3 \cdot 5 \cdot x^3 \cdot y,$$
$$10x^5y^2 \;=\; 2 \cdot 5 \cdot x^5 \cdot y^2.$$

The common factors are $5, x$, and y and after finding the greatest power of each common to all three terms, we have $5^1, x^3$, and y^1.

Thus, GCF $= 5^1 \cdot x^3 \cdot y^1 = 5x^3y$.

Common Monomial Factors

Now consider the polynomial $3n + 15$. We want to write this polynomial as a product of two factors. Since
$$3n + 15 = 3 \cdot n + 3 \cdot 5$$

we see that 3 is a common factor of the two terms in the polynomial. By using the distributive property, we can write
$$3n + 15 = 3 \cdot n + 3 \cdot 5 = 3(n + 5)$$

In this way, the polynomial $3n + 15$ has been **factored** into the product of 3 and $(n + 5)$.

For a more general approach to finding factors of a polynomial, we can use the quotient property of exponents:
$$\frac{a^m}{a^n} = a^{m-n}$$

This property is used when dividing terms. For example,

$$\frac{35x^8}{5x^2} = 7x^6 \quad \text{and} \quad \frac{16a^5}{-8a} = -2a^4.$$

To divide a polynomial by a monomial (a procedure that we discussed in Chapter 6) each term in the polynomial is divided by the monomial. For example,

$$\frac{8x^3 - 14x^2 + 10x}{2x} = \frac{8x^3}{2x} - \frac{14x^2}{2x} + \frac{10x}{2x} = 4x^2 - 7x + 5.$$

Note: **With practice, this division can be done mentally**.

This concept of dividing each term by a monomial is part of finding a monomial factor of a polynomial. Finding the greatest common monomial factor (GCF) of a polynomial means to **find the GCF of the terms of the polynomial**. This monomial will be one factor, and the sum of the various quotients found by dividing each term by the GCF will be the other factor. Thus, using the quotients just found and the fact that $2x$ is the GCF of the terms,

$$8x^3 - 14x^2 + 10x = 2x(4x^2 - 7x + 5).$$

We say that $2x$ is **factored out** and $2x$ and $(4x^2 - 7x + 5)$ are the factors of $8x^3 - 14x^2 + 10x$.

Now factor $24x^6 - 12x^4 - 18x^3$. The GCF is $6x^3$ and is factored out as follows:

$$24x^6 - 12x^4 - 18x^3 = 6x^3 \cdot 4x^3 + 6x^3(-2x) + 6x^3(-3)$$
$$= 6x^3(4x^3 - 2x - 3).$$

By definition, the GCF of a polynomial will have a positive coefficient. **However, if the leading coefficient is negative, we may choose to factor out the negative of the GCF (or** $-1 \cdot$ **GCF).** This technique will leave a positive coefficient for the first term of the other polynomial factor. For example,

the GCF for $-10a^4b + 15a^4$ is $5a^4$ and we can factor as follows:

$$-10a^4b + 15a^4 = 5a^4(-2b + 3)$$

or $\quad -10a^4b + 15a^4 = -5a^4(2b - 3).$

Both answers are correct.

Factoring out a monomial is essentially using the distributive property in a sort of reverse sense: $ab + ac = a(b + c)$ where a is the greatest common factor of the two terms ab and ac.

Example 2: Finding the GCF of a Polynomial

Factor each polynomial by factoring out the greatest common monomial factor.

a. $6n + 30$

    Solution: $6n + 30 = 6 \cdot n + 6 \cdot 5 = 6(n + 5)$

b. $x^3 + x$

    Solution: $x^3 + x = x \cdot x^2 + x(+1) = x(x^2 + 1)$ **Note:** $+1$ is the coefficient of x.

c. $5x^3 - 15x^2$

    Solution: $5x^3 - 15x^2 = 5x^2 \cdot x + 5x^2 (-3) = 5x^2(x - 3)$

d. $2x^4 - 3x^2 + 2$

    Solution: $2x^4 - 3x^2 + 2$ No common monomial factor other than 1.

e. $-4a^5 + 2a^3 - 6a^2$

    Solution: The GCF is $2a^2$ and we can factor as follows:

$$-4a^5 + 2a^3 - 6a^2 = 2a^2\left(-2a^3 + a - 3\right)$$

However, the leading coefficient is negative and we can also factor as follows:

$$-4a^5 + 2a^3 - 6a^2 = -2a^2\left(2a^3 - a + 3\right)$$

Both answers are correct. However, we will see later that having a positive leading coefficient for the polynomial in parentheses may make that polynomial easier to factor.

A polynomial may be in more than one variable. For example, $5x^2y + 10xy^2$ is in the two variables x and y. Thus the GCF may have more than one variable.

$$5x^2y + 10xy^2 = 5xy \cdot x + 5xy \cdot 2y$$
$$= 5xy(x + 2y)$$

Similarly,

$$4xy^3 - 2x^2y^2 + 8xy^2 = 2xy^2 \cdot 2y + 2xy^2(-x) + 2xy^2 \cdot 4$$
$$= 2xy^2(2y - x + 4)$$

Example 3: Finding the GCF of a Polynomial

Factor each polynomial by finding the GCF.

a. $4ax^3 + 4ax$

Solution: $4ax^3 + 4ax = 4ax(x^2 + 1)$ Note that $4ax = 1 \cdot 4ax$.

b. $3x^2y^2 - 6xy^2$

Solution: $3x^2y^2 - 6xy^2 = 3xy^2(x - 2)$

c. $-14by^3 - 7b^2y + 21by^2$

Solution: $-14by^3 - 7b^2y + 21by^2 = -7by\left(2y^2 + b - 3y\right)$ Note that by factoring out a negative term, in this case $-7by$, the leading coefficient in parentheses is positive.

d. $13a^4b^5 - 3a^2b^9 - 4a^6b^3$

Solution: $13a^4b^5 - 3a^2b^9 - 4a^6b^3 = a^2b^3(13a^2b^2 - 3b^6 - 4a^4)$

Factoring by Grouping

Consider the expression

$$y(x + 4) + 2(x + 4)$$

as the sum of two "terms," $y(x + 4)$ and $2(x + 4)$. Each of these "terms" has the common **binomial** factor $(x + 4)$. Factoring out this common binomial factor by using the distributive property gives,

$$y(x + 4) + 2(x + 4) = (x + 4)(y + 2)$$

Similarly,

$$3(x - 2) - a(x - 2) = (x - 2)(3 - a)$$

Now consider the product

$$(x + 3)(y + 5) = (x + 3)y + (x + 3)5$$
$$= xy + 3y + 5x + 15$$

which has four terms and no like terms. Yet the product has two factors, namely $(x + 3)$ and $(y + 5)$. Factoring polynomials with four or more terms can sometimes be accomplished by grouping the terms and using the distributive property, as in the above discussion and the following examples. Keep in mind that the common factor can be a binomial or other polynomial.

Example 4: Factoring Out a Common Binomial Factor

Factor each polynomial.

a. $3x^2(5x + 1) - 2(5x + 1)$

Solution: $3x^2(5x + 1) - 2(5x + 1) = (5x + 1)(3x^2 - 2)$

b. $7a(2x - 3) + (2x - 3)$

Solution: $7a(2x-3)+(2x-3) = 7a(2x-3)+1\cdot(2x-3)$ **Note:** 1 is the understood coefficient of $(2x - 3)$

$$= (2x-3)(7a+1)$$

In the examples just discussed, the common binomial factor was in parentheses. However, many times the expression to be factored is in a form with four or more terms. For example, by multiplying in Example 4a, we get the following expression:

$$3x^2(5x + 1) - 2(5x + 1) = 15x^3 + 3x^2 - 10x - 2$$

The expression $(15x^3 + 3x^2 - 10x - 2)$ has four terms with no common monomial factor; yet, we know that it has the two binomial factors $(5x + 1)$ and $(3x^2 - 2)$. We can find the binomial factors by **grouping**. This means looking for common factors in each group and then looking for common binomial factors. The process is illustrated in Example 5.

Example 5: Factoring by Grouping

Factor each polynomial by grouping.

a. $xy + 5x + 3y + 15$

Solution:

$$xy+5x+3y+15 = (xy+5x)+(3y+15)$$ Group terms that have a common monomial factor.

$$= x(y+5)+3(y+5)$$ Using the distributive property.

$$= (y+5)(x+3)$$ **Note:** $y+5$ is a common binomial factor.

b. $x^2 - xy - 5x + 5y$

Solution: $x^2 - xy - 5x + 5y = (x^2 - xy) + (-5x + 5y)$

$$= x(x-y) + 5(-x+y)$$

This does not work because $(x-y) \neq (-x+y)$. However, these two expressions are **opposites**. Thus we can find a common factor by factoring -5 instead of $+5$ from the last two terms.

$$x^2 - xy - 5x + 5y = (x^2 - xy) + (-5x + 5y)$$
$$= x(x-y) - 5(x-y)$$
$$= (x-y)(x-5) \qquad \text{Success!}$$

c. $x^2 + ax + 3x + 3y$

Solution: $x^2 + ax + 3x + 3y = x(x+a) + 3(x+y)$
But $x + a \neq x + y$ and there is no common factor.
So $x^2 + ax + 3x + 3y$ is **not factorable**.

d. $xy + 5x + y + 5$

Solution: $xy + 5x + y + 5 = x(y+5) + (y+5)$ **Note**: 1 is the understood coefficient of $(y+5)$.

$$= (y+5)(x+1)$$

e. $5xy + 6uv - 3vy - 10ux$

Solution: In the expression $5xy + 6uv - 3vy - 10ux$ there is no common factor in the first two terms. However, the first and third terms have a common factor so we rearrange the terms as follows:

$$5xy + 6uv - 3vy - 10ux = 5xy - 3vy + 6uv - 10ux$$
$$= y(5x - 3v) + 2u(3v - 5x)$$

Now we see that $5x - 3v$ and $3v - 5x$ are opposites and we factor out $-2u$ from the last two terms. The result is as follows:

$$5xy + 6uv - 3vy - 10xu = 5xy - 3vy + 6uv - 10ux$$
$$= y(5x - 3v) - 2u(-3v + 5x)$$
$$= (5x - 3v)(y - 2u)$$

Note that $5x - 3v = -3v + 5x$.

Practice Problems

Factor each expression.

1. $2x - 16$

2. $-5x^2 - 5x$

3. $7ax^2 - 7ax$

4. $a^4b^5 + 2a^3b^2 - a^3b^3$

5. $9x^2y^2 + 12x^2y - 6x^3$

6. $6a^3(x + 3) - (x + 3)$

7. $5x + 35 - xy - 7y$

7.1 Exercises

Find the GCF for each set of terms in Exercises 1 – 14.

1. $10, 15, 20$

2. $25, 30, 75$

3. $16, 40, 56$

4. $30, 42, 54$

5. $9, 14, 22$

6. $44, 66, 88$

7. $30x^3, 40x^5$

8. $15y^4, 25y$

9. $26ab^2, 39a^2b, 52a^2b^2$

10. $8a^3, 16a^4, 20a^2$

11. $28c^2d^3, 14c^3d^2, 42cd^2$

12. $45x^2y^2z^2, 75xy^2z^3$

13. $36xy, 48xy, 60xy$

14. $21a^5b^4c^3, 28a^3b^4c^3, 35a^3b^4c^2$

Simplify the expressions in Exercises 15 – 22. Assume that no denominator is equal to 0.

15. $\dfrac{x^7}{x^3}$

16. $\dfrac{x^8}{x^3}$

17. $\dfrac{-8y^3}{2y^2}$

18. $\dfrac{12x^2}{2x}$

19. $\dfrac{9x^5}{3x^2}$

20. $\dfrac{-10x^5}{2x}$

21. $\dfrac{4x^3y^2}{2xy}$

22. $\dfrac{21x^4y^3}{-3xy^2}$

Complete the factoring of the polynomial as indicated in Exercises 23 – 30.

23. $3m + 27 = 3(\quad)$

24. $2x + 18 = 2(\quad)$

25. $5x^2 - 30x = 5x(\quad)$

26. $6y^3 - 24y^2 = 6y^2(\quad)$

Answers to Practice Problems: 1. $2(x - 8)$ **2.** $-5x(x + 1)$ **3.** $7ax(x - 1)$ **4.** $a^3b^2(ab^3 + 2 - b)$
5. $3x^2(3y^2 + 4y - 2x)$ **6.** $(6a^3 - 1)(x + 3)$ **7.** $(5 - y)(x + 7)$

27. $13ab^2 + 13ab = 13ab($ $)$ **28.** $8x^2y - 4xy = 4xy($ $)$

29. $-15xy^2 - 20x^2y - 5xy = -5xy($ $)$ **30.** $-9m^3 - 3m^2 - 6m = -3m($ $)$

Factor each of the polynomials in Exercises 31 – 50 by finding the greatest common monomial factor.

31. $11x - 121$ **32.** $14x + 21$ **33.** $16y^3 + 12y$

34. $4ax - 8ay$ **35.** $-8a - 16b$ **36.** $-3x^2 + 6x$

37. $-6ax + 9ay$ **38.** $10x^2y - 25xy$ **39.** $16x^4y - 14x^2y$

40. $18y^2z^2 - 2yz$ **41.** $-14x^2y^3 - 14x^2y$ **42.** $8y^2 - 32y + 8$

43. $5x^2 - 15x - 5$ **44.** $ad^2 + 10ad + 25a$ **45.** $8m^2x^3 - 12m^2y + 4m^2z$

46. $36t^2x^4 - 45t^2x^3 + 24t^2x^2$ **47.** $34x^4y^6 - 51x^3y^5 + 17x^5y^4$

48. $-56x^4z^3 - 98x^3z^4 - 35x^2z^5$ **49.** $15x^4y^2 + 24x^6y^6 - 32x^7y^3$

50. $-3x^2y^4 - 6x^3y^4 - 9x^2y^3$

In Exercises 51 – 60, factor each expression by factoring out the common binomial factor.

51. $7y^2(y + 3) + 2(y + 3)$ **52.** $6a(a - 7) - 5(a - 7)$ **53.** $3x(x - 4) + (x - 4)$

54. $2x^2(x + 5) + (x + 5)$ **55.** $4x^3(x - 2) - (x - 2)$ **56.** $9a(x + 1) - (x + 1)$

57. $10y(2y + 3) - 7(2y + 3)$ **58.** $a(x + 5) + b(x + 5)$

59. $a(x - 2) - b(x - 2)$ **60.** $3a(x - 10) + 5b(x - 10)$

Factor each of the polynomials in Exercises 61 – 84 by grouping. If a polynomial cannot be factored, write *not factorable*.

61. $bx + b + cx + c$

62. $3x + 3y + ax + ay$

63. $x^3 + 3x^2 + 6x + 18$

64. $2z^3 - 14z^2 + 3z - 21$

65. $x^2 - 4x + 6xy - 24y$

66. $3x + 3y - bx - by$

67. $5xy + yz - 20x - 4z$

68. $x - 3xy + 2z - 6zy$

69. $24y - 3yz + 2xz - 16x$

70. $10xy - 2y^2 + 7yz - 35xz$

71. $ax + 5ay + 3x + 15y$

72. $6ax + 12x + a + 2$

73. $4xy + 3x - 4y - 3$

74. $xy + x + y + 1$

75. $xy + x - y - 1$

76. $z^2 + 3 + az^2 + 3a$

77. $x^2 - 5 + x^2y + 5y$

78. $ab^2 + 6ab + b^2 + 12$

79. $7xy - 3y + 2x^2 - 3x$

80. $10a^2 - 5az + 2a + z$

81. $3xy - 4yu - 6xv + 8uv$

82. $xy + 5yv + 6xu + 30uv$

83. $6ac - 9ad + 2bc - 3bd$

84. $2ac - 3bc + 6ad - 9bd$

Hawkes Learning Systems: Introductory Algebra

Greatest Common Factor of Two or More Terms
Greatest Common Factor of a Polynomial
Factoring Expressions by Grouping

7.2

Factoring Trinomials: $x^2 + bx + c$

Objectives

After completing this section, you will be able to:

1. Factor trinomials with leading coefficient 1 (of the form $x^2 + bx + c$).

2. Factor out a common monomial factor and then factor the remaining factor, a trinomial with leading coefficient 1.

In Chapter 6 we learned to use the FOIL method to multiply two binomials. In many cases the simplified form of the product was a trinomial. In the next two sections, we will learn to factor trinomials by reversing the FOIL method or by using the grouping method. In this section we consider only methods for factoring trinomials in one variable with leading coefficient 1.

Factoring Trinomials with Leading Coefficient 1

Using the FOIL method to multiply $(x + 5)$ and $(x + 3)$, we find

$$(x + 5)(x + 3) = x^2 + 3x + 5x + 15 = x^2 + 8x + 15.$$

Note that the the coefficient 8 is the sum of 5 and 3 and the constant term 15 is the product of 5 and 3. That is,

$$15 = 5 \cdot 3 \text{ and } 8 = 5 + 3$$

$$x^2 + 8x + 15$$

$$\uparrow \quad \uparrow$$

$$5 + 3 \quad 5 \cdot 3$$

This relationship is true in general. Consider the following general product:

$$(x+a)(x+b) = x^2 + ax + bx + ab$$

$$= x^2 + (a+b)x + ab$$

$$\uparrow \qquad \uparrow$$

sum of   product of
constants   constants
a and b a and b

Now, given a trinomial with leading coefficient 1, we want to find the binomial factors, if any. Reversing the relationship between a and b, as shown previously, we can proceed as follows:

To factor a trinomial with leading coefficient 1, find two factors of the constant term whose sum is the coefficient of the middle term.

For example,

(a) To factor $x^2 + 9x + 14$, find two factors of $+14$ whose sum is $+9$.
Since $14 = 2 \cdot 7$ and $9 = 2 + 7$, we have $a = 2$ and $b = 7$ for

$$x^2 + 9x + 14 = (x + 2)(x + 7)$$

(b) To factor $x^2 - 6x + 5$, find two factors of $+5$ whose sum is -6.
Since, $5 = (-1)(-5)$ and $-6 = (-1) + (-5)$, we have $a = -1$ and $b = -5$ for

$$x^2 - 6x + 5 = (x - 1)(x - 5)$$

Note: Because of the commutative property of multiplication, the order of the factors does not matter. We could just as easily write $(x - 5)(x - 1)$.

To check that the factors are correct, multiply the factors to see that the outer and inner products add up to the middle term.

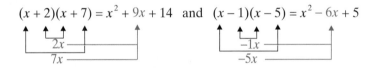

$$(x + 2)(x + 7) = x^2 + 9x + 14 \quad \text{and} \quad (x - 1)(x - 5) = x^2 - 6x + 5$$

Example 1: Factoring Trinomials with Leading Coefficient 1

Factor the following trinomials.

a. $x^2 + 8x + 12$

Solution: 12 has three pairs of positive integer factors as illustrated in the following table. Of these 3 pairs, only $2 + 6$ is equal to 8:

Factors of 12

1	12
2	6
3	4

$\rightarrow 2 + 6 = 8$ Thus $x^2 + 8x + 12 = (x + 2)(x + 6)$

Note: If the middle term had been $-8x$, then we would have wanted pairs of negative integer factors to find a sum of -8.

b. $y^2 - 8y - 20$

> **Solution:** We want a pair of integer factors of -20 whose sum is -8. In this case, one of the factors must be positive and the other negative.

Factors of -20

-1	20
1	-20
-2	10
2	-10 $\rightarrow 2 + (-10) = -8$
-4	5
4	-5

> We have listed all the pairs of integer factors of -20. You can see that 2 and -10 are the only two whose sum is -8. Thus listing all the pairs is not necessary. This stage is called the **trial-and-error stage**. That is, you can **try** different pairs (mentally or by making a list) until you find the correct pair. If such a pair does not exist, the polynomial is **not factorable**.
>
> In this case, we have
>
> $$y^2 - 8y - 20 = (y + 2)(y - 10)$$
>
> Note that by the commutative property of multiplication, the order of the factors does not matter. That is, we can also write
>
> $$y^2 - 8y - 20 = (y - 10)(y + 2)$$

To Factor Trinomials of the Form $x^2 + bx + c$

To factor $x^2 + bx + c$, if possible, find an integer pair of factors of c whose sum is b.

1. *If c is positive, then both factors must have the same sign.*

   a. *Both will be positive if b is positive.*

   b. *Both will be negative if b is negative.*

2. *If c is negative, then one factor must be positive and the other negative.*

Finding a Common Monomial Factor First

If a trinomial does not have a leading coefficient of 1, then we look for a common monomial factor. If there is a common monomial factor, factor out this common monomial factor and factor the remaining trinomial factor, if possible. (We will discuss factoring trinomials with leading coefficients other than 1 in the next section.) A polynomial is **completely factored** if none of its factors can be factored.

Example 2: Finding a Common Monomial Factor

Completely factor the following trinomials by first factoring out the GCF in the form of a common monomial factor.

a. $5x^3 - 15x^2 + 10x$

Solution: First factor out the GCF, $5x$.

$$5x^3 - 15x^2 + 10x = 5x(x^2 - 3x + 2) \quad \text{Factored, but not completely factored.}$$

Now factor the trinomial $x^2 - 3x + 2$. Look for factors of $+2$ that add up to -3. Since $(-1)(-2) = +2$ and $(-1) + (-2) = -3$, we have

$$5x^3 - 15x^2 + 10x = 5x(x^2 - 3x + 2)$$
$$= 5x(x - 1)(x - 2). \quad \text{Completely factored.}$$

b. $10y^5 - 20y^4 - 80y^3$

Solution: First factor out the GCF, $10y^3$.

$$10y^5 - 20y^4 - 80y^3 = 10y^3(y^2 - 2y - 8)$$

Now factor the trinomial $y^2 - 2y - 8$. Look for factors of -8 that add up to -2. Since $(-4)(+2) = -8$ and $(-4) + (+2) = -2$, we have

$$10y^5 - 20y^4 - 80y^3 = 10y^3(y^2 - 2y - 8) = 10y^3(y - 4)(y + 2).$$

NOTES

When factoring polynomials, always look for a common monomial factor first. Then, if there is one, remember to include this common monomial factor as part of the answer. Not all polynomials are factorable. For example, no matter what combinations are tried, $x^2 + 3x + 4$ does not have two binomial factors with integer coefficients. (There are no factors of $+4$ that will add to $+3$.) We say that the polynomial is **not factorable** (or **irreducible** or **prime**). **A polynomial is not factorable if it cannot be factored as the product of polynomials with integer coefficients**.

7.2 Exercises

In Exercises 1 – 10, list all pairs of integer factors for the given integer. Remember to include negative integers as well as positive integers.

1. 15   **2.** 12   **3.** 20   **4.** 30   **5.** –6

6. –7   **7.** 16   **8.** –18   **9.** –10   **10.** 25

In Exercises 11 – 20, find the pair of integers whose product is the first integer and whose sum is the second integer.

11. 12, 7   **12.** 25, 26   **13.** –14, –5   **14.** –30, –1   **15.** –8, 7

16. –40, –6   **17.** 36, –12   **18.** 16, –10   **19.** 20, –9   **20.** 4, –5

In Exercises 21 – 26, complete the factorization.

21. $x^2 + 6x + 5 = (x + 5)(\quad)$ **22.** $y^2 - 7y + 6 = (y - 1)(\quad)$

23. $p^2 - 9p - 10 = (p + 1)(\quad)$ **24.** $m^2 + 4m - 45 = (m - 5)(\quad)$

25. $a^2 + 12a + 36 = (a + 6)(\quad)$ **26.** $n^2 - 2n - 3 = (n - 3)(\quad)$

Factor the trinomials in Exercises 27 – 40. If the trinomial cannot be factored, write *not factorable*.

27. $x^2 - x - 12$ **28.** $x^2 + 7x + 12$ **29.** $y^2 + y - 30$

30. $y^2 - 3y + 2$ **31.** $a^2 + a + 2$ **32.** $m^2 + 3m - 1$

33. $x^2 + 3x + 5$ **34.** $x^2 - 8x + 16$ **35.** $x^2 + 3x - 18$

36. $y^2 + 12y + 35$ **37.** $y^2 - 14y + 24$ **38.** $a^2 + 10a + 25$

39. $x^2 - 6x - 27$ **40.** $x^2 + 5x - 36$

Completely factor the polynomials in Exercises 41 – 55.

41. $x^3 + 10x^2 + 21x$

42. $x^3 + 8x^2 + 15x$

43. $5x^2 - 5x - 60$

44. $6x^2 + 24x + 18$

45. $10y^2 - 10y - 60$

46. $7y^3 - 70y^2 + 168y$

47. $4p^4 + 36p^3 + 32p^2$

48. $15m^5 - 30m^4 + 15m^3$

49. $2x^4 - 14x^3 - 36x^2$

50. $3y^6 + 33y^5 + 90y^4$

51. $2x^2 - 2x - 72$

52. $3x^2 - 18x + 30$

53. $a^2 - 30a - 216$

54. $a^4 + 30a^3 + 81a^2$

55. $3y^5 - 21y^4 - 24y^3$

In Exercises 56 – 60, the polynomials are in more than one variable. Completely factor each polynomial.

56. $x^2 - 10xy + 16y^2$ **57.** $x^2 - 2xy - 3y^2$ **58.** $5a^2 + 10ab - 30b^2$

59. $20a^2 + 40ab + 20b^2$ **60.** $6x^2 - 12xy + 6y^2$

61. The area (in square inches) of the rectangle shown is given by the polynomial function $A(x) = 4x^2 + 20x$. If the width of the rectangle is $4x$ inches, what is the length?

$$A(x) = 4x^2 + 20x \qquad 4x$$

62. The area (in square feet) of the rectangle shown is given by the polynomial function $A(x) = x^2 + 11x + 24$. If the length of the rectangle is $(x + 8)$ feet, what is the width?

$$A(x) = x^2 + 11x + 24$$

$x + 8$

63. The volume of an open box is found by cutting equal squares (x units on a side) from a sheet of cardboard that is 10 inches by 40 inches. The function representing this volume is $V(x) = 4x^3 - 100x^2 + 400x$, where $0 < x < 5$. Factor this function and use the factors to explain, in your own words, how the function represents the volume.

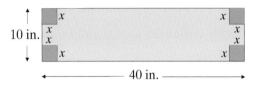

64. The area of a triangle is $\dfrac{1}{2}$ the product of its base and its height. If the area (in square meters) of the triangle shown is given by the function $A(x) = \dfrac{1}{2}x^2 + 24x$ find representations for the lengths of its base and its height.

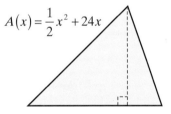

$$A(x) = \dfrac{1}{2}x^2 + 24x$$

Writing and Thinking About Mathematics

65. It is true that $2x^2 + 10x + 12 = (2x + 6)(x + 2) = (2x + 4)(x + 3)$. Explain how the trinomial can be factored in two ways. Is there some kind of error?

66. It is true that $5x^2 - 5x - 30 = (5x - 15)(x + 2)$. Explain why this is not the completely factored form of the trinomial.

Hawkes Learning Systems: Introductory Algebra

Factoring Trinomials: $x^2 + bx + c$

More on Factoring Trinomials: $ax^2 + bx + c$

Objectives

After completing this section, you will be able to:

1. Factor trinomials by using the ac-method.

2. Factor trinomials by using reverse FOIL (or trial-and-error method).

In this section we will investigate two methods for factoring trinomials in the general form

$$ax^2 + bx + c \quad \text{where the coefficients } a, b, \text{ and } c \text{ are integers.}$$

In this general form the coefficient a may be an integer other than 1. The two methods are related to two methods of factoring that we have already studied: factoring by grouping and the FOIL method. The **ac-method** involves factoring by grouping and the **trial-and-error method** is a reverse FOIL method.

The *ac*-Method (Grouping)

The *ac*-method is in reference to the coefficients a and c in the general form stated above:

$$ax^2 + bx + c \quad \text{where the coefficients } a, b, \text{ and } c \text{ are integers.}$$

The method is best explained by analyzing an example and explaining each step.

Consider the problem of factoring the trinomial

$$3x^2 + 17x + 10 \qquad \text{where } a = 3, b = 17, \text{ and } c = 10.$$

Analysis of Factoring by the ac-Method

$3x^2 + 17x + 10$	$ax^2 + bx + c$

Step 1: Multiply $3 \cdot 10 = 30$.

Multiply $a \cdot c$.

Step 2: Find two integers whose product is 30 and whose sum is 17. (In this case, $2 \cdot 15 = 30$ and $2 + 15 = 17$.)

Find two integers whose product is ac and whose sum is b. If this is not possible, then the trinomial is **not factorable**.

Step 3: Rewrite the middle term $(+17x)$ using $+2$ and $+15$ as coefficients.
$3x^2 + 17x + 10 = 3x^2 + 2x + 15x + 10$

Rewrite the middle term (bx) using the two numbers found in Step 2 as coefficients.

Step 4: Factor by grouping the first two terms and the last two terms.
$3x^2 + 2x + 15x + 10 = x(3x + 2) + 5(3x + 2)$

Factor by grouping the first two terms and the last two terms.

Step 5: Factor out the common binomial factor $(3x + 2)$. Thus,
$3x^2 + 17x + 10 = 3x^2 + 2x + 15x + 10$
$= x(3x + 2) + 5(3x + 2)$
$= (3x + 2)(x + 5)$

Factor out the common binomial factor to find two binomial factors of the trinomial $ax^2 + bx + c$.

Example 1: Using the ac-Method

a. Factor $4x^2 + 33x + 35$ by using the ac-method.

Solution: $a = 4, b = 33, c = 35$

Step 1: Find the product ac: $4 \cdot 35 = 140$.

Step 2: Find two integers whose product is 140 and whose sum is 33:

$$(+5)(+28) = 140 \quad \text{and} \quad (+5) + (+28) = +33$$

Continued on next page...

Note: this step may take some experimenting with factors. You might try prime factoring. For example

$$140 = 10 \cdot 14 = 2 \cdot 5 \cdot 2 \cdot 7.$$

With combinations of these prime factors we can write

Factors of 140		Sum	
1	140	$1 + 140 = 141$	
2	70	$2 + 70 = 72$	
4	35	$4 + 35 = 39$	
5	28	$5 + 28 = 33$	We can stop here!

5 and 28 are the desired coefficients.

Step 3: Rewrite $+33x$ as $+5x + 28x$, giving

$$4x^2 + 33x + 35 = 4x^2 + 5x + 28x + 35$$

Step 4: Factor by grouping:

$$4x^2 + 33x + 35 = 4x^2 + 5x + 28x + 35$$
$$= x(4x + 5) + 7(4x + 5)$$

Step 5: Factor out the common binomial factor $(4x + 5)$.

$$4x^2 + 33x + 35 = 4x^2 + 5x + 28x + 35$$
$$= x(4x + 5) + 7(4x + 5)$$
$$= (4x + 5)(x + 7)$$

Note that in Step 3 we could have written +33x as +28x + 5x. Try this to convince yourself that the result will be the same two factors.

b. Factor $12x^3 - 26x^2 + 12x$ by using the *ac*-method.

Solution: First factor out the greatest common factor $2x$:

$$12x^3 - 26x^2 + 12x = 2x(6x^2 - 13x + 6).$$

Now factor the trinomial $6x^2 - 13x + 6$ with $a = 6$, $b = -13$, and $c = 6$.

Step 1: Find the product *ac*: $6(6) = 36$.

Step 2: Find two integers whose product is 36 and whose sum is –13: (**Note:** This may take some time and experimentation. We do know that both numbers must be negative because the product is positive and the sum is negative.)

$$(-9)(-4) = +36 \text{ and } -9 + (-4) = -13$$

Steps 3 and 4:

Factor by grouping.

$$6x^2 - 13x + 6 = 6x^2 - 9x - 4x + 6$$
$$= 3x(2x - 3) - 2(2x - 3)$$

[**Note:** –2 is factored from the last two terms so that there will be a common binomial factor $(2x - 3)$.]

Step 5: Factor out the common binomial factor $(2x - 3)$:

$$6x^2 - 13x + 6 = 6x^2 - 9x - 4x + 6$$
$$= 3x(2x - 3) - 2(2x - 3)$$
$$= (2x - 3)(3x - 2)$$

Thus for the original expression,

$$12x^3 - 26x^2 + 12x = 2x(6x^2 - 13x + 6)$$
$$= 2x(2x - 3)(3x - 2)$$

Do not forget to write the common monomial factor, $2x$, in the answer.

c. Factor $4x^2 - 5x - 6$ by using the *ac*-method.

Solution: $a = 4, b = -5, c = -6$

Step 1: Find the product *ac*: $4(-6) = -24$.

Step 2: Find two integers whose product is –24 and whose sum is –5: (**Note:** We know that one number must be positive and the other negative because the product is negative.) In this example we have

$$(+3)(-8) = -24 \quad \text{and} \quad (+3) + (-8) = -5$$

Steps 3 and 4:

Factor by grouping.

$$4x^2 - 5x - 6 = 4x^2 + 3x - 8x - 6$$
$$= x(4x + 3) - 2(4x + 3)$$

Continued on next page...

Step 5: Factor out the common binomial factor $(4x + 3)$:

$$4x^2 - 5x - 6 = 4x^2 + 3x - 8x - 6$$
$$= x(4x + 3) - 2(4x + 3)$$
$$= (4x + 3)(x - 2).$$

NOTES

As illustrated in Example 1b, the first step in factoring should always be to factor out any common monomial factor. Also, if the leading coefficient is negative, factor out a negative monomial even if it is just -1. For example,

$$-12x^3 + 26x^2 - 12x = -2x(6x^2 - 13x + 6) = -2x(2x - 3)(3x - 2).$$

Having a positive leading coefficient for the trinomial in parentheses makes factoring this trinomial much easier.

Remember that the product of the factors must always equal the original polynomial. This is a good way to check your work.

The Trial-and-Error Method

The key to the trial-and-error method of factoring trinomials is the FOIL method of multiplication of two binomials that we studied in Section 6.6. The FOIL method is used in a reverse sense, since the product is given and the object is to find the factors. Reviewing the FOIL method of multiplication, we can find the product

$$(2x + 5)(3x + 1) = 6x^2 + 17x + 5.$$

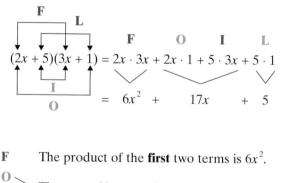

F   The product of the **first** two terms is $6x^2$.

O
$\Big\rangle$ The sum of **inner** and **outer** products is $17x$.
I

L   The product of the **last** two terms is 5.

We have already discussed in Section 7.2, the situation of factoring a trinomial with leading coefficient 1. For example, to factor $x^2 + 7x + 12$ we find factors of 12 whose sum is 7. This is a simpler form of the reverse FOIL method with $\mathbf{F} = x^2 = x \cdot x$ and $\mathbf{L} = 12$.

$$x^2 + 7x + 12 = (x \quad)(x \quad)$$

So the only problem is to find the factors of 12 whose sum is 7. The factors are 4 and 3 since $4 \cdot 3 = 12$ and $4 + 3 = 7$.

$$x^2 + 7x + 12 = (x + 3)(x + 4)$$
$$\mathbf{I} = 3x$$
$$\mathbf{O} = 4x$$

$3x + 4x = 7x$ Add to get the middle term.

Now consider the problem of factoring

$$6x^2 + 31x + 5$$

as a product of two binomials.

$$\mathbf{F} = 6x^2 \qquad \mathbf{L} = +5$$
$$6x^2 + 31x + 5 = (\qquad)(\qquad)$$

For $\mathbf{F} = 6x^2$, we know that $6x^2 = 3x \cdot 2x$ and $6x^2 = 6x \cdot x$.

For $\mathbf{L} = +5$, we know that $+5 = (+1)(+5)$ and $+5 = (-1)(-5)$.

Now we use various combinations for $\mathbf{F}$ and $\mathbf{L}$ in the **trial-and-error method** as follows:

1. List all the possible combinations of factors of $6x^2$ and $+5$ in their respective $\mathbf{F}$ and $\mathbf{L}$ positions.

2. Check the sum of the products in the $\mathbf{O}$ and $\mathbf{I}$ positions until you find the sum to be $31x$.

3. If none of these sums is $31x$, the trinomial is not factorable.

F
L

a. $(3x + 1)(2x + 5)$

b. $(3x + 5)(2x + 1)$

c. $(6x + 1)(x + 5)$

d. $(6x + 5)(x + 1)$

e. $(3x - 1)(2x - 5)$

f. $(3x - 5)(2x - 1)$

g. $(6x - 1)(x - 5)$

h. $(6x - 5)(x - 1)$

We really don't need to check these last four because the **O** and **I** sum would have to be negative, and we are looking for $+31x$. In this manner the trial-and-error method is more efficient than it first appears to be.

Now, investigating only the possibilities in the list with positive constants, we need to check the sums of the outer (**O**) and inner (**I**) products to find $+31x$.

a. $(3x + 1)(2x + 5)$; $15x + 2x = 17x$

$2x$
$15x$

b. $(3x + 5)(2x + 1)$; $3x + 10x = 13x$

$10x$
$3x$

c. $(6x + 1)(x + 5)$; $30x + x =$ ⟨$31x$⟩ ← We found $31x$!

$1x$
$30x$

The correct factors, $(6x + 1)$ and $(x + 5)$, have been found so we need not take the time to try the next product in the list $(6x + 5)(x + 1)$. Thus, even though the list of possibilities of factors may be long, the actual time may be quite short if the correct factors are found early in the trial-and-error method. So we have

$$6x^2 + 31x + 5 = (6x + 1)(x + 5).$$

Look at the constant term to determine what signs to use for the constants in the factors. The following guidelines will help limit the trial-and-error search.

NOTES
No matter which method you use (the *ac*-method or the trial-and-error method), factoring trinomials takes time. With practice you will become more efficient with either method. Make sure to be patient and observant.

1. *If the sign of the constant term is positive (+), the signs in both factors will be the same, either both positive or both negative.*

2. *If the sign of the constant term is negative (−), the signs in the factors will be different, one positive and one negative.*

Example 2: Using the Trial-and-Error Method

Factor by using the trial-and-error method.

a. Factor $2x^2 + 12x + 10$ completely.

Solution: First factor the common monomial, 2. Now, because the middle term is $+ 6x$ and the constant is 5, we know that the two factors of 5 must both be positive, $+5$ and $+1$.

First factor the common monomial factor, 2.

$$2x^2 + 12x + 10 = 2\left(x^2 + 6x + 5\right)$$
$$= 2(x+5)(x+1); \ x + 5x = 6x$$

$5x$

x

b. Factor $6x^2 - 31x + 5$.

Solution: Since the middle term is $-31x$ and the constant is $+5$, we know that the two factors of 5 must both be negative, -5 and -1. Also, from the previous discussion, $\mathbf{F} = 6x^2 = 6x \cdot x$.

$$6x^2 - 31x + 5 = (6x - 1)(x - 5); \ -30x - x = -31x$$

$-x$

$-30x$

c. $4x^2 - 4x - 15$

Solution: For **F**: $4x^2 = 4x \cdot x$ and $4x^2 = 2x \cdot 2x$.
For **L**: $-15 = -15 \cdot 1 = -1 \cdot 15$ and $-15 = -3 \cdot 5 = -5 \cdot 3$.

Trials: $(2x - 15)(2x + 1)$ $-30x + 2x = -28x$ is the wrong middle term.

$-30x$

$2x$

Continued on next page...

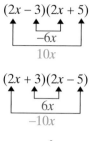

$(2x - 3)(2x + 5)$

$-6x$

$10x$

$-6x + 10x = +4x$ is the wrong middle term only because the sign is wrong. So just switch the signs and the factors will be right.

$(2x + 3)(2x - 5)$

$6x$

$-10x$

$+6x - 10x = -4x$ is the right middle term.

Thus $4x^2 - 4x - 15 = (2x + 3)(2x - 5)$.

There is no need to try all the possibilities with $(4x \quad)(x \quad)$.

NOTES

Reminder: To factor completely means to find factors of the polynomial, none of which are themselves factorable. Thus

$$2x^2 + 12x + 10 = (2x + 10)(x + 1)$$

is not factored completely since $2x + 10 = 2(x + 5)$.

We could write

$$2x^2 + 12x + 10 = (2x + 10)(x + 1)$$
$$= 2(x + 5)(x + 1).$$

This problem can be avoided by first factoring out the GCF (in this case, 2).

Example 3: Completely Factoring

Factor completely. Be sure to look first for the greatest common monomial factor.

a. $6x^3 - 8x^2 + 2x$

Solution: $6x^3 - 8x^2 + 2x = 2x(3x^2 - 4x + 1) = 2x(3x - 1)(x - 1)$

b. $-2x^2 - x + 6$

Solution: $-2x^2 - x + 6 = -1(2x^2 + x - 6) = -1(2x - 3)(x + 2)$

c. $6x^3 + 3x^2 + 3x$

Solution: $6x^3 + 3x^2 + 3x = 3x(2x^2 + x + 1)$

Now consider the trinomial:

$$2x^2 + x + 1 = (2x + ?)(x + ?)$$

There is no way to factor and get a middle term of $+x$ for the product. This trinomial is **not factorable**.

So we have

$$6x^3 + 3x^2 + 3x = 3x(2x^2 + x + 1)$$ Factored completely.

Practice Problems

Factor completely.

1. $3x^2 + 7x - 6$

2. $2x^2 + 6x - 8$

3. $3x^2 + 15x + 18$

4. $10x^2 - 41x - 18$

5. $x^2 + 11x + 28$

6. $-3x^3 + 6x^2 - 3x$

7.3 Exercises

In Exercises 1 – 70, factor completely, if possible. If a polynomial cannot be factored, write *not factorable*.

1. $x^2 + 5x + 6$

2. $x^2 - 6x + 8$

3. $2x^2 - 3x - 5$

4. $3x^2 - 4x - 7$

5. $6x^2 + 11x + 5$

6. $4x^2 - 11x + 6$

7. $-x^2 + 3x - 2$

8. $-x^2 - 5x - 6$

9. $x^2 - 3x - 10$

10. $x^2 - 11x + 10$

11. $-x^2 + 13x + 14$

12. $-x^2 + 12x - 36$

13. $x^2 + 8x + 64$

14. $x^2 + 2x + 3$

15. $-2x^3 + x^2 + x$

16. $-2y^3 - 3y^2 - y$

17. $4t^2 - 3t - 1$

18. $2x^2 - 3x - 2$

Answers to Practice Problems: 1. $(3x - 2)(x + 3)$ **2.** $2(x - 1)(x + 4)$ **3.** $3(x + 2)(x + 3)$
4. $(5x + 2)(2x - 9)$ **5.** $(x + 7)(x + 4)$ **6.** $-3x(x - 1)(x - 1)$

19. $5a^2 - a - 6$

20. $3a^2 + 4a + 1$

21. $7x^2 + 5x - 2$

22. $8x^2 - 10x - 3$

23. $4x^2 + 23x + 15$

24. $6x^2 + 23x + 21$

25. $x^2 + 6x - 16$

26. $9x^2 - 3x - 20$

27. $12x^2 - 38x + 20$

28. $12b^2 - 12b + 3$

29. $3x^2 - 7x + 2$

30. $7x^2 - 11x - 6$

31. $9x^2 - 6x + 1$

32. $4x^2 + 40x + 25$

33. $4x^2 + 4x + 1$

34. $6y^2 + 7y + 2$

35. $12y^2 - 7y - 12$

36. $x^2 - 46x + 45$

37. $3x^2 + 9x + 5$

38. $5a^2 - 7a + 2$

39. $8a^2b - 22ab + 12b$

40. $12m^3n - 146m^2n + 24mn$

41. $x^2 + x + 1$

42. $x^2 + 2x + 2$

43. $16x^2 - 8x + 1$

44. $3x^2 - 11x - 4$

45. $64x^2 - 48x + 9$

46. $9x^2 - 12x + 4$

47. $6x^2 + 2x - 20$

48. $12y^2 - 15y + 3$

49. $10x^2 + 35x + 30$

50. $24y^2 + 4y - 4$

51. $-18x^2 + 72x - 8$

52. $-45y^2 + 30y + 120$

53. $7x^4 - 5x^3 + 3x^2$

54. $12x^2 - 60x + 75$

55. $-12m^2 + 22m + 4$

56. $32y^2 + 50$

57. $6x^3 + 9x^2 - 6x$

58. $-5y^2 + 40y - 60$

59. $9x^3y^3 + 9x^2y^3 + 9xy^3$

60. $30a^3 + 51a^2 + 9a$

61. $12x^3 - 108x^2 + 243x$ **62.** $48x^2y - 354xy + 126y$ **63.** $48xy^3 - 100xy^2 + 48xy$

64. $24a^2x^2 + 72a^2x + 243x$ **65.** $21y^4 - 98y^3 + 56y^2$ **66.** $72a^3 - 306a^2 + 189a$

67. $ax + ay + 3x + 3y$ **68.** $2x^2 + 2y^2 + ax^2 + ay^2$ **69.** $5x^2 + 5y^2 + bx^2 + by^2$

70. $ax + bx - 4a - 4b$

Writing and Thinking About Mathematics

71. A ball is dropped from the top of a building that is 784 feet high. The height of the ball above ground level is given by the polynomial function $h(t) = -16t^2 + 784$ where t is measured in seconds.
   a. How high is the ball after 3 seconds? 5 seconds?
   b. How far has the ball traveled in 3 seconds? 5 seconds?
   c. When will the ball hit the ground? Explain your reasoning in terms of factors.

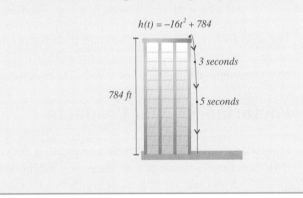

Hawkes Learning Systems: Introductory Algebra

Factoring Trinomials by Grouping
Factoring Trinomials by Trial and Error

Factoring Special Products: Difference of Two Squares and Perfect Square Trinonomials

Objectives

After completing this section, you will be able to:

1. Factor the differences of two squares.

2. Factor perfect square trinomials.

3. Complete the square of trinomials by determining the missing terms that make incomplete trinomials perfect squares.

In Chapter 6 we discussed the following three special products.

$$\textbf{I.} \quad (x+a)(x-a) = x^2 - a^2 \qquad \text{Difference of two squares}$$

$$\textbf{II.} \quad (x+a)^2 = x^2 + 2ax + a^2 \qquad \text{Perfect square trinomial}$$

$$\textbf{III.} \quad (x-a)^2 = x^2 - 2ax + a^2 \qquad \text{Perfect square trinomial}$$

The objective in this section is to learn to factor products of these types without referring to the *ac*-method or the trial-and-error method. That is, with practice you will learn to be able to recognize the "form" of the special product and go directly to the factors.

Factoring Special Products

Memorize the products and their names listed above. These products are like formulas. If you are familiar enough with them, you will be able see how some polynomials fit the forms, and you will be able to write the factors without using the techniques learned in Section 7.3. Obviously, many polynomials do not fit any of these forms and you will need to use the *ac*-method or the trial-and-error method to factor these polynomials. However, the difference of two squares and perfect square trinomials occur frequently enough to deserve special treatment.

Difference of Two Squares:

Consider the polynomial $x^2 - 9$. By recognizing this expression as the **difference of two squares** we can go directly to the factors:

$$x^2 - 9 = (x+3)(x-3).$$

Similarly, we have

$$25 - y^2 = (5+y)(5-y) \quad \text{and} \quad 36x^2 - 49 = (6x+7)(6x-7).$$

Remember to **look for a common monomial factor first**. For example,

$$5x^2y - 20y = 5y(x^2 - 4) = 5y(x+2)(x-2).$$

Perfect Square Trinomials:

In a **perfect square trinomial, both the first and last terms of the trinomial must be perfect squares**. If we think of the first term in the form of x^2 and the last term in the form of a^2, then the middle term must be of the form $2ax$ or $-2ax$. For example,

$$x^2 + 6x + 9 = (x+3)^2, \text{ here } 9 = 3^2 = a^2 \text{ and } 6 = 2 \cdot 3 = 2a$$

$$x^2 - 14x + 49 = (x-7)^2, \text{ here } 49 = 7^2 = a^2 \text{ and } -14 = -2 \cdot 7 = -2a$$

Again, remember to **look for a common monomial factor first**. For example,

$$2x^2y + 24xy + 72y = 2y(x^2 + 12x + 36) = 2y(x+6)^2.$$

Example 1: Factoring Special Products

Factor completely each of the following polynomials.

a. $y^2 - 10y + 25$

Solution: $y^2 - 10y + 25 = (y-5)^2$ Perfect square trinomial

b. $6a^2b - 6b$

Solution: $6a^2b - 6b = 6b(a^2 - 1)$ Factor out the common term.

$= 6b(a+1)(a-1)$ Difference of two squares

c. $3x^2 + 12 + 12x$

Solution: $3x^2 + 12 + 12x = 3x^2 + 12x + 12$ Rearrange the terms in order of descending powers.

$= 3(x^2 + 4x + 4)$ Factor out the common monomial factor.

$= 3(x+2)^2$ Perfect square trinomial

d. $a^6 - 100$ Even powers, such as a^6, can always be treated as squares: $a^6 = (a^3)^2$

Solution: $a^6 - 100 = (a^3 + 10)(a^3 - 10)$ Difference of two squares

Special Comment on the Sum of Two Squares:

The **sum of two squares** is an expression of the form $a^2 + b^2$ and is **not factorable**. For example, $x^2 + 4$ is the sum of two squares and is not factorable. There are no factors with integer coefficients whose product is $x^2 + 4$. To understand this situation, write

$$x^2 + 4 = x^2 + 0x + 4$$

and note that there are no factors of +4 that will add to 0.

Example 2: Using The Sum of Two Squares

Factor completely. Be sure to look first for the greatest common monomial factor.

a. $y^2 + 36$

    Solution: $y^2 + 36$ is the sum of two squares and is not factorable.

b. $4x^2 + 100$

    Solution: $4x^2 + 100 = 4\left(x^2 + 25\right)$ Factored completely

      We see that 4 is the greatest common monomial factor and $x^2 + 25$ is not factorable.

Summary of Procedures to Follow in Factoring Polynomials

1. *Look for a common monomial factor.*
2. *Check the number of terms:*
    a. *Two terms:*
      (1) *difference of two squares? - factorable*
      (2) *sum of two squares? - not factorable*
    b. *Three terms:*
      (1) *perfect square trinomial?*
      (2) *use trial-and-error method?*
      (3) *use the ac-method?*
    c. *Four terms:*
      (1) *group terms with a common factor*
3. *Check the possibility of factoring any of the factors.*

Completing the Square

Closely related to factoring special products is the procedure of **completing the square**. This procedure helps in understanding the nature of perfect square trinomials, and we will use it again in Chapter 10 in solving second-degree equations (called quadratic equations). The procedure involves adding a square term to a binomial so that the resulting trinomial is a perfect square trinomial, thus "completing the square." For example, we want

$$x^2 + 10x + \underline{\ \ \ } = (\qquad)^2 \quad \text{to be of the form} \quad x^2 + 2ax + a^2 = (x+a)^2.$$

The middle coefficient, 10, is twice the number that is to be squared. That is, $10 = 2a$. So taking half of this coefficient gives

$$\frac{1}{2}(10) = 5 = a$$

and squaring the result gives

$$5^2 = 25 = a^2$$

Thus, $x^2 + 10x + \underline{\ \ \ } = (\qquad)^2$ becomes $x^2 + 10x + \underline{25} = (x+5)^2$.

For $x^2 + 18x$, we have $\frac{1}{2}(18) = 9$ and $9^2 = 81$. So

$$x^2 + 18x + \underline{\ \ \ } = (\qquad)^2 \quad \text{becomes} \quad x^2 + 18x + \underline{81} = (x+9)^2$$

Example 3: Completing the Square

Complete the square as indicated. In Examples 3b and 3c, fractions are introduced into the process of completing the square to help in understanding that algebraic concepts need not be restricted to integers.

a. $y^2 + 20y + \underline{\ \ \ } = (\qquad)^2$

Solution: Since $\frac{1}{2}(20) = 10$ and $10^2 = 100$, then

$$y^2 + 20y + \underline{100} = (y+10)^2$$

b. $x^2 + 3x + \underline{\ \ \ } = (\qquad)^2$

Solution: In this example, $2a = 3$; and since 3 is an odd number, a will be a fraction, not an integer.

$$\text{Thus } \frac{1}{2}(3) = \frac{3}{2} \text{ and } \left(\frac{3}{2}\right)^2 = \frac{9}{4}$$

$$\text{This gives } x^2 + 3x + \frac{9}{\underline{4}} = \left(x + \frac{3}{2}\right)^2$$

Continued on next page...

c. $a^2 - 5a + \underline{\quad} = (\quad)^2$

Solution: We have $2a = -5$. Thus, $\frac{1}{2}(-5) = -\frac{5}{2}$ and $\left(-\frac{5}{2}\right)^2 = \frac{25}{4}$.

So, $a^2 - 5a + \underline{\frac{25}{4}} = \left(a - \frac{5}{2}\right)^2$

To help in understanding the relationships between the terms in a perfect square trinomial, we look at completing the square from a slightly different perspective. Suppose that the middle x-term is missing and we want to find this term so that the resulting trinomial is the square of a binomial. For example,

$$x^2 + \underline{\quad} + 36 = (\quad)^2$$

From the formulas,

$$(x+a)^2 = x^2 + \underline{2ax} + a^2 \text{ and } (x-a)^2 = x^2 - \underline{2ax} + a^2$$

we know that the middle (missing) term is to have coefficient $2a$ (or $-2a$) and, in this case,

$$a^2 = 36 = 6^2 \text{ and } 2a = 2 \cdot 6 = 12$$

So

$$x^2 + \underline{12x} + 36 = (x+6)^2$$

Similarly

$$x^2 - \underline{\quad} + 16 = (\quad)^2$$

where $a^2 = 16$ and $a = 4$. Thus $-2a = -2(4) = -8$, and we get $x^2 - \underline{8x} + 16 = (x-4)^2$.

Example 4: Finding the Missing Term

Find the missing term in the trinomial and complete the square as indicated.

a. $x^2 + \underline{\quad} + 100 = (\underline{\quad})^2$

Solution: Because $100 = 10^2$ and $2 \cdot 10 = 20$, we have $x^2 + \underline{20x} + 100 = (x+10)^2$.

b. $x^2 - \underline{} + \dfrac{9}{16} = ()^2$

Solution: Because $\dfrac{9}{16} = \left(\dfrac{3}{4}\right)^2$ and $-2 \cdot \dfrac{3}{4} = -\dfrac{3}{2}$,

we have $x^2 - \underline{\dfrac{3}{2}}x + \dfrac{9}{16} = \left(x - \dfrac{3}{4}\right)^2$

Practice Problems

Factor completely.

1. $x^2 - 16$ **2.** $25 - x^2$

3. $3y^2 + 6y + 3$ **4.** $y^{10} - 81$

Complete the square.

5. $x^2 + 6x + \underline{} = ()^2$ **6.** $x^2 + \underline{} + 49 = ()^2$

7.4 Exercises

Factor completely each of the polynomials in Exercises 1 – 50. If a polynomial cannot be factored, write *not factorable*.

1. $x^2 - 1$ **2.** $x^2 - 4$ **3.** $x^2 - 49$

4. $x^2 - 144$ **5.** $x^2 + 4x + 4$ **6.** $x^2 + 10x + 25$

7. $x^2 - 12x + 36$ **8.** $x^2 - 20x + 100$ **9.** $16x^2 - 9$

10. $4x^2 - 49$ **11.** $9x^2 - 1$ **12.** $36x^2 - 1$

13. $25 - 4x^2$ **14.** $16 - 9x^2$ **15.** $x^2 - 14x + 49$

16. $x^2 - 2x + 1$ **17.** $x^2 + 6x + 9$ **18.** $x^2 + 8x + 16$

Answers to Practice Problems: 1. $(x+4)(x-4)$ **2.** $(5+x)(5-x)$ **3.** $3(y+1)^2$ **4.** $\left(y^5+9\right)\left(y^5-9\right)$ **5.** $x^2 + 6x + \underline{9} = (x+3)^2$ **6.** $x^2 + \underline{14x} + 49 = (x+7)^2$

19. $3x^2 - 27y^2$

20. $27x^2 - 48y^2$

21. $x^3 - xy^2$

22. $12ax^2 - 75ay^2$

23. $x^2 - \dfrac{1}{4}$

24. $x^2 - \dfrac{4}{9}$

25. $x^2 - \dfrac{9}{16}$

26. $x^2 - \dfrac{25}{49}$

27. $x^4 - 1$

28. $x^4 - 16y^4$

29. $2x^2 - 32x + 128$

30. $3x^2 - 30x + 75$

31. $ay^2 + 2ay + a$

32. $ax^2 + 18ax + 81a$

33. $4x^2y^2 - 24xy^2 + 36y^2$

34. $2ax^2 - 44ax + 242a$

35. $16x^2 + 81$

36. $4y^2 + 49$

37. $x^2 + x + 1$

38. $y^2 + 4y + 5$

39. $y^2 + 2y + 2$

40. $x^2 + 4x + 1$

41. $16x^2 - 64$

42. $36y^2 - 81$

43. $ax^2 - 10ax + 25a$

44. $2y^2 - 16y + 32$

45. $147 - 3y^2$

46. $128 - 50x^2$

47. $x^2 + 5x + \dfrac{25}{4}$

48. $x^2 - \dfrac{4}{3}x + \dfrac{4}{9}$

49. $2x^2 - \dfrac{18}{25}$

50. $3y^2 - \dfrac{3}{4}$

In Exercises 51 – 66, complete the square by adding the correct missing term in the left equation, then factor as indicated.

51. $x^2 - 6x + \underline{} = \left(\right)^2$

52. $x^2 - 12x + \underline{} = \left(\right)^2$

53. $x^2 - 4x + \underline{} = \left(\right)^2$

54. $x^2 + 20x + \underline{} = \left(\right)^2$

55. $x^2 + \underline{} + 16 = \left(\right)^2$

56. $x^2 - \underline{} + 49 = \left(\right)^2$

57. $x^2 - \underline{} + 81 = \left(\right)^2$

58. $x^2 + \underline{} + 121 = \left(\right)^2$

59. $x^2 + x + \underline{\quad} = (\qquad)^2$

60. $x^2 - 7x + \underline{\quad} = (\qquad)^2$

61. $x^2 - 9x + \underline{\quad} = (\qquad)^2$

62. $x^2 + 3x + \underline{\quad} = (\qquad)^2$

63. $x^2 + \underline{\quad} + \dfrac{25}{4} = (\qquad)^2$

64. $x^2 - \underline{\quad} + \dfrac{121}{4} = (\qquad)^2$

65. $x^2 - \underline{\quad} + \dfrac{9}{4} = (\qquad)^2$

66. $x^2 + \underline{\quad} + \dfrac{81}{4} = (\qquad)^2$

Hawkes Learning Systems: Introductory Algebra

Special Factorizations - Squares

7.5   Solving Quadratic Equations by Factoring

Objectives

After completing this section, you will be able to:

Solve quadratic equations by factoring.

In Chapter 7 the emphasis has been on second-degree polynomials and functions. Such polynomials are particularly evident in physics with the path of thrown objects (or projectiles) affected by gravity, in mathematics involving area (area of a circle, area of a rectangle, square, and triangle), and in many situations in higher level mathematics. The methods of factoring we have studied (*ac*-method, trial-and-error method, difference of two squares, and perfect square trinomials) have been related, in general, to second-degree polynomials. Second-degree polynomials are called **quadratic polynomials** (or **quadratics**) and, as just discussed, they play a major role in many applications of mathematics. In fact, quadratic polynomials and techniques for solving quadratic equations are central topics in the first two courses in algebra.

Solving Quadratic Equations by Factoring

Quadratic Equations

Quadratic equations *are equations of the form*

$$ax^2 + bx + c = 0 \text{ where a, b, and c are constants and } a \neq 0$$

The form $ax^2 + bx + c = 0$ is called the **standard form** (or **general form**) of a **quadratic equation**. In the standard form, a quadratic polynomial is on one side of the equation and 0 is on the other side. For example,

$x^2 - 8x + 12 = 0$ is a quadratic equation in standard form.

While

$3x^2 - 2x = 27$ and $x^2 = 25x$ are quadratic equations, just not in standard form.

These last two equations can be manipulated algebraically so that 0 is on one side. In solving quadratic equations by factoring, having 0 on one side is necessary because of the **zero-factor property**.

Zero-Factor Property

If a product is 0, then at least one of the factors must be 0. That is, if a and b are real numbers, then

$$\textbf{if } a \cdot b = 0, \textbf{ then } a = 0 \textbf{ or } b = 0$$

Example 1: Solving Factored Quadratic Equations

Solve the following quadratic equation.

$$(x - 5)(2x - 7) = 0$$

Solution: Since the quadratic is already factored and the other side of the equation is 0, we use the zero-factor property and set each factor equal to 0. This process yields two linear equations, which can, in turn, be solved.

$$
\begin{array}{lll}
x - 5 = 0 & \text{or} & 2x - 7 = 0 \\
x = 5 & \text{or} & 2x = 7 \qquad \text{Add 7 to both sides.} \\
& & x = \dfrac{7}{2} \qquad \text{Divide both sides by 2.}
\end{array}
$$

Thus the two solutions to the original equation are $x = 5$ and $x = \dfrac{7}{2}$.

The solutions can be **checked** by substituting them one at a time for x in the equation. That is, there will be two "checks."

Substituting $x = 5$ gives

$$(x - 5)(2x - 7) = (5 - 5)(2 \cdot 5 - 7) = (0)(3) = 0$$

Substituting $x = \dfrac{7}{2}$ gives

$$(x - 5)(2x - 7) = \left(\frac{7}{2} - 5\right)\left(2 \cdot \frac{7}{2} - 7\right) = \left(-\frac{3}{2}\right)(7 - 7) = \left(-\frac{3}{2}\right)(0) = 0$$

Therefore, both 5 and $\dfrac{7}{2}$ are solutions to the original equation.

In Example 1 the polynomial was factored and the other side of the equation was 0. The equation was solved by setting each factor, in turn, equal to 0 and solving the resulting linear equations. These solutions tell us that the original equation has two solutions. **In general, a quadratic equation has two solutions**. In the special cases where the two factors are the same, there is only one solution and it is called a **double solution** (or **double root**). The following examples show how to solve quadratic equations by factoring. Remember that the equation must be in standard form with one side of the equation equal to 0. Study these examples carefully.

Example 2: Solving Quadratic Equations

Solve the following equations by writing the equation in standard form with one side 0 and factoring the polynomial. Then set each factor equal to 0 and solve. Checking is left as an exercise for the student.

a. $3x^2 = 6x$

Solution: $3x^2 = 6x$

$3x^2 - 6x = 0$ Write the equation in standard form with 0 on one side.

$3x(x-2) = 0$ Factor out the common monomial, $3x$.

$3x = 0$ or $x - 2 = 0$ Set each factor equal to 0.

$x = 0$ or $x = 2$ Solve each linear equation.

The solutions are 0 and 2.

b. $x^2 - 8x + 16 = 0$

Solution: $x^2 - 8x + 16 = 0$

$(x-4)^2 = 0$ The trinomial is a perfect square.

$x - 4 = 0$ or $x - 4 = 0$ Both factors are the same,

$x = 4$ $x = 4$ so there is only one distinct solution.

The only solution is 4, and it is called a **double solution** (or **double root**).

c. $4x^2 - 4x = 24$

Solution: $4x^2 - 4x = 24$

$4x^2 - 4x - 24 = 0$ One side must be 0.

$4\left(x^2 - x - 6\right) = 0$ 4 is a common monomial factor.

$4(x - 3)(x + 2) = 0$ The constant factor 4 can never be 0 and does not affect the solution.

$x - 3 = 0$ or $x + 2 = 0$

$x = 3$ $x = -2$

The solutions are 3 and –2.

d. $\dfrac{2x^2}{15} - \dfrac{x}{3} = -\dfrac{1}{5}$

Solution: $\dfrac{2x^2}{15} - \dfrac{x}{3} = -\dfrac{1}{5}$

$15 \cdot \dfrac{2x^2}{15} - \dfrac{x}{3} \cdot 15 = -\dfrac{1}{5} \cdot 15$ Multiply each term by 15, the LCM of the denominators, to get integer coefficients.

$2x^2 - 5x = -3$ Simplify.

$2x^2 - 5x + 3 = 0$ One side must be 0.

$(2x - 3)(x - 1) = 0$ Factor by the trial-and-error method or

$2x - 3 = 0$ or $x - 1 = 0$ the *ac*-method.

$2x = 3$ $x = 1$

$x = \dfrac{3}{2}$

The solutions are $\dfrac{3}{2}$ and 1.

e. $(x + 5)^2 = 36$

Solution: $(x + 5)^2 = 36$

$x^2 + 10x + 25 = 36$ Expand $(x + 5)^2$.

$x^2 + 10x - 11 = 0$ Add –36 to both sides so that one side is 0.

$(x + 11)(x - 1) = 0$ Factor.

$x + 11 = 0$ or $x - 1 = 0$

$x = -11$ $x = 1$

The solutions are –11 and 1.

Continued on next page...

f. $3x(x-1) = 2(5-x)$

Solution:

$$3x(x-1) = 2(5-x)$$

$$3x^2 - 3x = 10 - 2x \qquad \text{Use the distributive property.}$$

$$3x^2 - 3x + 2x - 10 = 0 \qquad \text{Arrange terms so that 0 is on one side.}$$

$$3x^2 - x - 10 = 0 \qquad \text{Simplify.}$$

$$(3x + 5)(x - 2) = 0 \qquad \text{Factor.}$$

$$3x + 5 = 0 \quad \text{or} \quad x - 2 = 0$$

$$x = -\frac{5}{3} \qquad\qquad x = 2$$

The solutions are $-\dfrac{5}{3}$ and 2.

In the next example, we show that factoring can be used to solve equations of degrees higher than second-degree. There will be more than two factors, but just as with quadratics, if the product is 0, then the solutions are found by setting each factor equal to 0.

Example 3: Solving Higher Degree Equations

Solve the following equation: $2x^3 - 4x^2 - 6x = 0$.

Solution:

$$2x^3 - 4x^2 - 6x = 0$$

$$2x(x^2 - 2x - 3) = 0 \qquad \text{Factor out the common monomial, } 2x.$$

$$2x(x - 3)(x + 1) = 0 \qquad \text{Factor the trinomial.}$$

$$2x = 0 \quad \text{or} \quad x - 3 = 0 \quad \text{or} \quad x + 1 = 0 \qquad \text{Set each factor equal to 0 and solve.}$$

$$x = 0 \qquad\qquad x = 3 \qquad\qquad x = -1$$

The solutions are 0, 3, and –1.

The procedures to solve a quadratic equation can be summarized as follows.

To Solve a Quadratic Equation by Factoring

1. *Add or subtract terms as necessary so that 0 is on one side of the equation and the equation is in the standard form* $ax^2 + bx + c = 0$ *where* **a**, **b**, *and* **c** *are real constants and* $a \neq 0$.

2. *Factor completely. (If there are any fractional coefficients, multiply each term by the least common denominator so that all coefficients will be integers.)*

3. *Set each nonconstant factor equal to 0 and solve each linear equation for the unknown.*

4. *Check each solution, one at a time, in the original equation.*

Special Comment: All of the quadratic equations in this section can be solved by factoring. That is, all of the quadratic polynomials are factorable. However, as we have seen in some of the previous sections, not all polynomials are factorable. In Chapter 10 we will develop techniques (other than factoring) for solving quadratic equations whether the quadratic polynomial is factorable or not.

NOTES

COMMON ERROR

A **common error** is to divide both sides of an equation by the variable x. This error can be illustrated by using the equation in Example 2.

$$3x^2 = 6x$$

$$\frac{3x^2}{x} = \frac{6x}{x}$$

$$3x = 6$$

$$x = 2$$

INCORRECT
Do not divide by x because you lose the solution $x = 0$.

Factoring is the method to use. By factoring, you will find all solutions as shown in the previous examples.

Practice Problems

Solve each equation by factoring.

1. $x^2 - 6x = 0$

2. $6x^2 - x - 1 = 0$

3. $(x - 2)^2 - 25 = 0$

4. $x^3 - 8x^2 + 16x = 0$

Answers to Practice Problems: 1. $x = 0, x = 6$ **2.** $x = \frac{1}{2}, x = -\frac{1}{3}$ **3.** $x = 7, x = -3$ **4.** $x = 0, x = 4$

7.5 Exercises

Solve the equations in Exercises 1 – 10 by setting each factor equal to 0 and solving the resulting linear equations.

1. $(x-3)(x-2)=0$ **2.** $(x+5)(x-2)=0$ **3.** $(2x-9)(x+2)=0$

4. $(x+7)(3x-4)=0$ **5.** $0=(x+3)(x+3)$ **6.** $0=(x+10)(x-10)$

7. $(x+5)(x+5)=0$ **8.** $(x+5)(x-5)=0$ **9.** $2x(x-2)=0$

10. $3x(x+3)=0$

In Exercises 11 – 72, solve the equations by factoring.

11. $x^2-3x=4$ **12.** $x^2+7x=-12$ **13.** $0=5x^2+15x$

14. $x^2-11x=-18$ **15.** $2x^2-24=2x$ **16.** $0=4x^2-12x$

17. $x^2+8=6x$ **18.** $x^2=x+30$ **19.** $2x^2+2x-24=0$

20. $9x^2+63x+90=0$ **21.** $0=2x^2-5x-3$ **22.** $0=2x^2-x-3$

23. $3x^2-4x-4=0$ **24.** $3x^2-8x+5=0$ **25.** $2x^2-7x-4=0$

26. $4x^2+8x+3=0$ **27.** $0=3x^2+2x-8$ **28.** $0=6x^2+7x+2$

29. $4x^2-12x+9=0$ **30.** $25x^2-60x+36=0$ **31.** $8x=5x^2$

32. $15x=3x^2$ **33.** $9x^2=36$ **34.** $4x^2-16=0$

35. $5x^2+10x+5=0$ **36.** $2x^2=4x+6$ **37.** $8x^2+32=32x$

38. $-6x^2+18x+24=0$ **39.** $\dfrac{x^2}{3}-2x+3=0$ **40.** $\dfrac{x^2}{9}=1$

41. $\dfrac{x^2}{5} - x - 10 = 0$

42. $\dfrac{2}{3}x^2 + 2x - \dfrac{20}{3} = 0$

43. $\dfrac{x^2}{8} + x + \dfrac{3}{2} = 0$

44. $\dfrac{x^2}{6} - \dfrac{1}{2}x - 3 = 0$

45. $x^2 - x + \dfrac{1}{4} = 0$

46. $x^2 - \dfrac{7}{6}x + \dfrac{1}{3} = 0$

47. $x^3 + 8x = 6x^2$

48. $x^3 = x^2 + 30x$

49. $6x^3 + 7x^2 = -2x$

50. $3x^3 = 8x - 2x^2$

51. $0 = x^2 - 100$

52. $0 = x^2 - 121$

53. $3x^2 - 75 = 0$

54. $5x^2 - 45 = 0$

55. $x^2 + 8x + 16 = 0$

56. $x^2 + 14x + 49 = 0$

57. $3x^2 = 18x - 27$

58. $5x^2 = 10x - 5$

59. $(x - 1)^2 = 4$

60. $(x - 3)^2 = 1$

61. $(x + 5)^2 = 9$

62. $(x + 4)^2 = 16$

63. $(x + 4)(x - 1) = 6$

64. $(x - 5)(x + 3) = 9$

65. $27 = (x + 2)(x - 4)$

66. $-1 = (x + 2)(x + 4)$

67. $x(x + 7) = 3(x + 4)$

68. $x(x + 9) = 6(x + 3)$

69. $3x(x + 1) = 2(x + 1)$

70. $2x(x - 1) = 3(x - 1)$

71. $x(2x + 1) = 6(x + 2)$

72. $3x(x + 3) = 2(2x - 1)$

Hawkes Learning Systems: Introductory Algebra

Solving Quadratic Equations By Factoring

7.6   Applications of Quadratic Equations

After completing this section, you will be able to:

Solve word problems by writing quadratic equations that can be factored and solved.

Whether or not word problems cause you difficulty depends a great deal on your personal experiences and general reasoning abilities. These abilities are developed over a long period of time. A problem that is easy for you, possibly because you have had experience in a particular situation, might be quite difficult for a friend and vice versa.

Most problems do not say specifically to add, subtract, multiply, or divide. You are to know from the nature of the problem what to do. You are to ask yourself, "What information is given? What am I trying to find? What tools, skills, and abilities do I need to use?"

Word problems should be approached in an orderly manner. You should have an "attack plan."

Attack Plan for Word Problems

1. *Read the problem carefully at least twice.*
2. *Decide what is asked for and assign a variable or variable expression to the unknown quantities.*
3. *Organize a chart, table, or diagram relating all the information provided.*
4. *Form an equation. (A formula of some type may be necessary.)*
5. *Solve the equation.*
6. *Check your solution with the wording of the problem to be sure it makes sense.*

Several types of problems lead to quadratic equations. The problems in this section are set up so that the equations can be solved by factoring. More general problems and approaches to solving quadratic equations are discussed in Chapter 10.

Example 1: Applications of Quadratic Functions

a. One number is four more than another and the sum of their squares is 296. What are the numbers?

Solution: Let x = smaller number. Then $x + 4$ = larger number.

$$x^2 + (x + 4)^2 = 296 \quad \text{Add the squares.}$$

$$x^2 + x^2 + 8x + 16 = 296$$

$$2x^2 + 8x - 280 = 0 \qquad \text{Write the equation in standard form.}$$

$$2(x^2 + 4x - 140) = 0 \qquad \text{Factor out the GCF.}$$

$$2(x + 14)(x - 10) = 0 \qquad \text{The constant factor 2 does not affect the solution.}$$

$$x + 14 = 0 \qquad \text{or} \qquad x - 10 = 0$$

$$x = -14 \qquad\qquad x = 10$$

$$x + 4 = -10 \qquad\qquad x + 4 = 14$$

There are two sets of answers to the problem: 10 and 14 or −14 and −10.

Check:

$$10^2 + 14^2 = 100 + 196 = 296$$

$$\text{and } (-14)^2 + (-10)^2 = 196 + 100 = 296$$

b. In an orange grove, there are 10 more trees in each row than there are rows. How many rows are there if there are 96 trees in the grove?

Solution: Let r = number of rows.
Then $r + 10$ = number of trees per row.
Set up the equation and solve.

$$r(r + 10) = 96$$

$$r^2 + 10r = 96$$

$$r^2 + 10r - 96 = 0$$

$$(r - 6)(r + 16) = 0$$

$$r - 6 = 0 \qquad \text{or} \qquad r + 16 = 0$$

$$r = 6 \qquad\qquad r = -16$$

$r + 10$ trees per row

r rows

There are 6 rows in the grove ($6 \cdot 16 = 96$ trees).

Note: While −16 is a solution to the equation, −16 does not fit the conditions of the problem and is discarded. You cannot have −16 rows.

Continued on next page...

c. A rectangle has an area of 135 square meters and perimeter of 48 meters. What are the dimensions of the rectangle?

Solution: The area of a rectangle is the product of its length and width ($A = lw$).
The perimeter of a rectangle is given by $P = 2l + 2w$.
Since the perimeter is 48 meters, then the length plus the width must be 24 meters (one-half of the perimeter).
Let w = width. Then $24 - w$ = length.
Set up the equation and solve.

$$w(24 - w) = 135 \qquad \text{Area} = lw = \text{length times width.}$$

$$24w - w^2 = 135$$

$$0 = w^2 - 24w + 135 \qquad \textbf{Note: } 0 \text{ can be on either side of}$$

$$0 = (w - 9)(w - 15) \qquad \text{the equation.}$$

$$w - 9 = 0 \quad \text{or} \quad w - 15 = 0$$

$$w = 9 \qquad\qquad w = 15 \;\Big\}\; \text{Note the solution because}$$

$$24 - 9 = 15 \qquad 24 - 15 = 9 \;\Big\}\; \text{generally length is greater than width.}$$

The width is 9 meters and the length is 15 meters ($9 \cdot 15 = 135$).

d. A man wants to build a block wall shaped like a rectangle along three sides of his property. If 180 feet of fencing is needed and the area of the lot is 4000 square feet, what are the dimensions of the lot?

Solution: Let x = one of two equal sides. Then $180 - 2x$ = third side.
Set up the equation and solve.

$$x(180 - 2x) = 4000$$

$$180x - 2x^2 = 4000$$

$$0 = 2x^2 - 180x + 4000$$

$$0 = 2(x^2 - 90x + 2000)$$

$$0 = 2(x - 50)(x - 40)$$

$$x - 50 = 0 \quad \text{or} \quad x - 40 = 0$$

$$x = 50 \qquad\qquad x = 40$$

$$180 - 2(50) = 80 \qquad 180 - 2(40) = 100$$

From this information there are two possible answers: the lot is 50 ft by 80 ft or the lot is 40 ft by 100 ft.

e. The sum of the squares of two consecutive positive odd integers is 202. Find the integers.

Solution: Let n = first odd integer. Then $n + 2$ = next consecutive odd integer. Set up the equation and solve.

$$n^2 + (n + 2)^2 = 202$$
$$n^2 + n^2 + 4n + 4 = 202$$
$$2n^2 + 4n - 198 = 0$$
$$2(n^2 + 2n - 99) = 0$$
$$2(n - 9)(n + 11) = 0$$

$n = 9$ or $n = -11$ Since the problem asked for

$n + 2 = 11$ positive integers, -11 is discarded.

The first integer is 9 and the next consecutive odd integer is 11.

Checking: $9^2 + 11^2 = 81 + 121 = 202$

7.6 Exercises

Determine a quadratic equation for each of the following problems in Exercises 1 – 44. Then solve the equation.

1. The square of an integer is equal to seven times the integer. Find the integer.

2. The square of an integer is equal to twice the integer. Find the integer.

3. The square of a positive integer is equal to the sum of the integer and 12. Find the integer.

4. If the square of a positive integer is added to three times the integer, the result is 28. Find the integer.

5. One number is seven more than another. Their product is 78. Find the numbers.

6. One positive number is three more than twice another. If the product is 27, find the numbers.

7. If the square of a positive integer is added to three times the number, the result is 54. Find the number.

8. One number is six more than another. The difference between their squares is 132. What are the numbers?

9. The difference between two positive integers is 8. If the smaller is added to the square of the larger, the sum is 124. Find the numbers.

10. One number is three less than twice another. The sum of their squares is 74. Find the numbers.

11. One number is five less than another. The sum of their squares is 97. Find the numbers.

12. Find a positive integer such that the product of the integer with a number three less than the integer is equal to the integer increased by 32.

13. The product of two consecutive positive integers is 72. Find the integers.

14. Find two consecutive integers whose product is 110.

15. Find two consecutive positive integers such that the sum of their squares is 85.

16. Find two consecutive positive integers such that the square of the second integer added to four times the first is equal to 41.

17. Find two consecutive positive integers such that the difference between their squares is 17.

18. The product of two consecutive odd integers is 63. Find the integers.

19. The product of two consecutive even integers is 120. Find the integers.

20. The product of two consecutive even integers is 168. Find the integers.

21. The length of a rectangle is twice the width. The area is 72 square inches. Find the length and width of the rectangle.

22. The length of a rectangle is three times the width. If the area is 147 square centimeters, find the length and width of the rectangle.

23. The length of a rectangular yard is 12 meters greater than the width. If the area of the yard is 85 square meters, find the length and width of the yard.

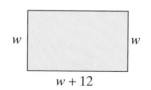

$w + 12$

24. The length of a rectangle is three centimeters greater than the width. The area is 108 square centimeters. Find the length and width of the rectangle.

25. The width of a rectangle is 4 feet less than the length. The area is 117 square feet. Find the length and width of the rectangle.

26. The height of a triangle is 4 feet less than the base. The area of the triangle is 16 square feet. Find the length of the base and height of the triangle.

27. The base of a triangle exceeds the height by 5 meters. If the area is 42 square meters, find the length of the base and the height of the triangle.

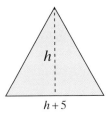

28. The base of a triangle is 15 inches greater than the height. If the area is 63 square inches, find the length of the base.

29. The base of a triangle is 6 feet less than the height. The area is 56 square feet. Find the length of the height.

30. The perimeter of a rectangle is 32 inches. The area of the rectangle is 48 square inches. Find the dimensions of the rectangle.

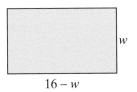

31. The area of a rectangle is 24 square centimeters. If the perimeter is 20 centimeters, find the length and width of the rectangle.

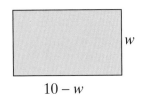

32. The perimeter of a rectangle is 40 meters and the area is 96 square meters. Find the dimensions of the rectangle.

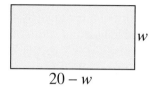

$$20 - w$$

33. An orchard has 140 orange trees. The number of rows exceeds the number of trees per row by 13. How many trees are there in each row?

34. One formation for a drill team is rectangular. The number of members in each row exceeds the number of rows by 3. If there is a total of 108 members in the formation, how many rows are there?

35. A theater can seat 144 people. The number of rows is 7 less than the number of seats in each row. How many rows of seats are there?

36. The length of a rectangle is 7 centimeters greater than the width. If 4 centimeters are added to both the length and width, the new area would be 98 square centimeters. Find the dimensions of the original rectangle.

37. The width of a rectangle is 5 meters less than the length. If 6 meters are added to both the length and width, the new area will be 300 square meters. Find the dimensions of the original rectangle.

38. Susan is going to fence a rectangular flower garden in her back yard. She has 50 feet of fencing and she plans to use the house as the fence on one side of the garden. If the area is 300 square feet, what are the dimensions of the flower garden?

39. A rancher is going to build a corral with 52 yards of fencing. He is planning to use the barn as one side of the corral. If the area is 320 square yards, what are the dimensions?

40. The area of a square is 81 square centimeters. How long is each side of the square?

41. The length of a rectangle is three times the width. If the area is 48 square feet, find the length and width of the rectangle.

42. The length of a rectangle is five times the width. If the area is 180 square inches, find the length and width of the rectangle.

43. One number is eight more than another. Their product is −16. What are the numbers?

44. One number is 10 more than another. If their product is −25, find the numbers.

Hawkes Learning Systems: Introductory Algebra

Applications of Quadratic Equations

Additional Applications of Quadratic Equations

Objectives

After completing this section, you will be able to:

Use given formulas to solve applied problems.

The following applications provide practice in using quadratic equations found in geometrical, engineering, physics, and business type settings. The formulas are provided with no discussion of how or why they fit the described situation. Further studies in other fields will show how some of these formulas are generated and how they are used.

You are to read the information and substitute the given data into the formula and then solve for the unknown quantity. Depending on the given information, finding the solution may or may not involve a quadratic equation. For example,

Body Mass Index

The **Body-Mass-Index** formula (used in a health and fitness program) for people is:

$$BMI = \frac{W \times 704.5}{H^2}$$

where W is the weight (in pounds) and H is the height (in inches) of a person.

a. Find the BMI for a person who weighs 172 pounds and is 5 ft 9 in. tall.
b. Find the BMI for a person who weighs 135 pounds and is 5 ft tall.

Solution (substituting in the numbers and using a calculator):

a. $BMI = \dfrac{172 \times 704.5}{69^2} = 25.45$ (Note: 5 ft 9 in. = 69 in.)

b. $BMI = \dfrac{135 \times 704.5}{60^2} = 26.42$ (Note: 5 ft = 60 in.)

7.7 Exercises

1. Calculate the BMI of a person who is 5 ft 7 in. tall and weighs 130 pounds.

2. Calculate the BMI of a person who is 6 ft 5 in. tall and weighs 260 pounds.

3. Find the height of a person who has a BMI of 24.37 and weighs 170 pounds (round to the nearest inch).

4. Find the height of a person who has a BMI of 27.15 and weighs 210 pounds (round to the nearest inch).

Load on a Wooden Beam

The safe load, L, of a horizontal wooden beam supported at both ends is expressed by the formula

$$L = \frac{kbd^2}{l}$$

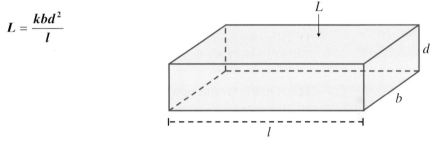

where L is expressed in pounds;

b is the breadth (or width) in inches;
d is the depth in inches;
l is the length in inches;
k, a constant, depends on the grade of the beam and is expressed in pounds per square inch.

5. What is the safe maximum load of a white pine beam 180 in. long, 3 in. wide, and 4 in. deep if $k = 3000$ lb per in.^{2}?

6. Find the constant k if a beam 144 in. long, 2 in. wide, and 6 in. deep supports a maximum load of 1100 lb.

7. A solid oak beam is required to support a load of 12,000 lbs. It can be no more than 8 in. deep and is 192 in. long. For this grade of oak, $k = 6000$ lb per in.^{2} How wide should the beam be?

8. The safe maximum load of a white pine beam 4 in. wide and 150 in. long is 2880 lb. For white pine, $k = 3000$ lb per in.^{2} Find the depth of the beam.

9. A Douglas fir beam is required to support a load of 20,000 lb. For Douglas fir, $k = 4800$ lb per in.^{2} If the beam is 6 in. wide and 144 in. long, what is the minimum depth for the beam to support the required load?

Height of a Projectile

The equation

$$h = -16t^2 + v_0t$$

gives the height h, in feet, that a body will be above the earth at time t, in seconds, if it is projected upward with an initial velocity v_0, in feet per second. (The initial velocity v_0 is the velocity when $t = 0$.)

10. Find the height of an object 3 seconds after it has been projected upward at a rate of 56 feet per second.

11. Find the height of an object 5 seconds after it has been projected upward at a rate of 120 feet per second.

12. A ball is thrown upward with a velocity of 144 feet per second. When will it strike the ground?

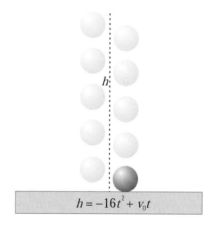

13. An object is projected upward at a rate of 160 feet per second. Find the time when it is 384 feet above the ground.

14. An object is projected upward at a rate of 96 feet per second. Find the time when it is 144 feet above the ground.

Electrical Power

The power output of a generator is given by the equation

$$P_0 = E_g I - r_g I^2$$

where P_0 is measured in kilowatts;
               E_g is measured in volts;
               r_g is measured in ohms;
               I is measured in amperes.

15. Find I if $P_0 = 120$ kilowatts, $E_g = 16$ volts, and $r_g = \dfrac{1}{2}$ ohm.

16. Find I if $P_0 = 180$ kilowatts, $E_g = 22$ volts, and $r_g = \dfrac{2}{3}$ ohm.

17. Find I if $P_0 = 210$ kilowatts, $E_g = 20$ volts, and $r_g = \dfrac{2}{5}$ ohm.

18. Find I if $P_0 = 240$ kilowatts, $E_g = 25$ volts, and $r_g = \dfrac{5}{8}$ ohm.

Consumer Demand

The demand for a product is the number of units of the product that consumers are willing to buy when the market price is p dollars. The consumers' total expenditure for the product is found by multiplying the price times the demand.

19. When fishing reels are priced at p dollars, local consumers will buy $36 - p$ fishing reels. What is the price if total sales were $320?

20. A manufacturer can sell $100 - 2p$ lamps at p dollars each. If the receipts from the lamps total $1200, what is the price of the lamps?

21. During the summer at a local market, consumers will buy $10 - 5p$ pounds of peaches at p dollars per pound. If someone buys $4.95 worth of peaches, what is the price per pound of the peaches?

22. Fans at the stadium will buy $2000 - 100p$ drinks for p dollars on a hot afternoon. If the total sales after the game was \$4375, what was the price per drink?

Consumer Demand

The demand for a certain commodity is given by

$$D = -20p^2 + ap + 1200$$

where p is the selling price and a is a constant

23. Find the selling price if 1403 units are sold and $a = 128$.

24. Find the selling price if 2268 units are sold and $a = 298$.

25. Find the selling price if 1120 units are sold and $D = -20p^2 + 60p + 1200$.

26. Find the selling price if 1860 units are sold and $D = -20p^2 + 232p + 1200$.

Volume of a Cylinder

The volume of a cylinder is given by the formula

$$V = \pi r^2 h$$

where V is the volume;
       $\pi = 3.14$ (3.14 is an approximation for π);
       r is the radius,
       h is the height.

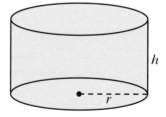

27. Find the volume of a cylinder with a radius of 6 in. and a height of 20 in.
28. Find the height of a cylinder if the volume is 282.6 in.³ and the radius in 3 in.
29. A cylinder has a height of 14 in. and a volume of 1099 in.³ Find the radius.
30. Find the radius of a cylinder whose volume is 2512 cm³ and whose height is 8 cm.

Hawkes Learning Systems: Introductory Algebra

Applications of Quadratic Equations

Chapter 7 Index of Key Ideas and Terms

Section 7.1   Greatest Common Factor and Factoring by Grouping

Greatest Common Factor (GCF)   pages 549 - 550

Procedure for Finding the GCF   page 549
   1. Find the prime factorization of all integer coefficients.
   2. List all factors to all terms, including variables.
   3. Choose the greatest power of each factor common to all terms.
   4. Multiply these powers to find the GCF.
      (**Note:** If there is no common prime factor or variables,
      then the GCF is 1.)

Factoring by Grouping   pages 553 - 555

Section 7.2   Factoring Trinomials: $x^2 + bx + c$

Factoring Trinomials with Leading Coefficient 1 ($x^2 + bx + c$)   pages 559 - 561
   Find factors of the constant term whose sum is the
   coefficient of the middle term.

Not Factorable (or Irreducible or Prime)   page 562

Section 7.3   More on Factoring Trinomials: $ax^2 + bx + c$

The *ac*-method (Grouping)   pages 566 - 570
   Step 1: Multiply $a \cdot c$.

   Step 2: Find two integers whose product is ac and whose sum is b.
            If this is not possible, then the trinomial is ***not factorable***.

   Step 3: Rewrite the middle term (bx) using the two numbers found
            in Step 2 as coefficients.

   Step 4: Factor by grouping the first two terms and the last two terms.

   Step 5: Factor out the common binomial factor to find two binomial
            factors of the trinomial $ax^2 + bx + c$.

Continued on next page...

Section 7.3 More on Factoring Trinomials: $ax^2 + bx + c$ (continued)

Trial-and-error method   pages 570 - 574

Guidelines for the Trial-and-Error Method   page 573

1. If the sign of the constant term is positive (+),
   the signs in both factors will be the same,
   either both positive or both negative.
2. If the sign of the constant term is negative (–),
   the signs in the factors will be different,
   one positive and one negative.

Section 7.4 Factoring Special Products: Difference of Two Squares and Perfect Square Trinomials

Factoring Special Products   pages 578 - 579

Difference of Two Squares   pages 578 - 579

$$x^2 - a^2 = (x+a)(x-a)$$

Perfect Square Trinomials   pages 578 - 579

$$(x+a)^2 = x^2 + 2ax + a^2$$

$$(x-a)^2 = x^2 - 2ax + a^2$$

General Guidelines for Factoring Polynomials   page 580

1. Always look for a common monomial factor
   first. If the leading coefficient is negative,
   factor out a negative monomial even if it is just –1.
2. Check the number of terms.
   a. Two terms:
      (1) difference of squares? - factorable
      (2) sum of squares? - not factorable
   b. Three terms:
      (1) perfect square trinomial?
      (2) use trial-and-error method?
      (3) use the *ac*-method?
   c. Four terms:
      (1) group terms with a common factor
          and factor out any common binomial factor.
3. Check the possibility of factoring any of the factors.

Completing the Square   pages 581 - 583

Add a square term to a binomial so that the resulting trinomial
is a perfect square trinomial.

Section 7.5 Solving Quadratic Equations by Factoring

Quadratic Equations   page 586
Standard Form: $ax^2 + bx + c = 0$ $(a \neq 0)$

Solve By Factoring   pages 586 - 591

Zero-factor Property   page 587
If a product is 0, then at least one of
the factors must be 0. That is, if a and b
are real numbers, then if $a \cdot b = 0$, then $a = 0$ or $b = 0$.

Procedure   page 591
1. Add or subtract terms as necessary so that 0 is on
    one side of the equation and the equation is in the
    standard form $ax^2 + bx + c = 0$ where a, b and c are
    real constants and $a \neq 0$.
2. Factor completely. (If there are any fractional coefficients,
    multiply each term by the least common denominator
    so that all coefficients will be integers.)
3. Set each nonconstant factor equal to 0 and solve each
    linear equation for the unknown.
4. Check each solution, one at a time, in the original equation.

Section 7.6 Applications of Quadratic Equations

Attack Plan for Word Problems   page 594
1. Read the problem carefully at least twice.
2. Decide what is asked for and assign a variable or
    variable expression to the unknown quantities.
3. Organize a chart, table, or diagram relating all
    the information provided.
4. Form an equation. (A formula of some type may be necessary.)
5. Solve the equation.
6. Check your solution with the wording of the problem to be sure
    it makes sense.

Hawkes Learning Systems: Introductory Algebra

For a review of the topics and problems from Chapter 7, look at the following lessons from *Hawkes Learning Systems: Introductory Algebra*

Greatest Common Factor of Two or More Terms
Greatest Common Factor of a Polynomial
Factoring Expressions by Grouping
Factoring Trinomials: $x^2 + bx + c$
Factoring Trinomials by Grouping
Factoring Trinomials by Trial and Error
Special Factorizations - Squares
Solving Quadratic Equations By Factoring
Applications of Quadratic Equations

Chapter 7 Review

The following guidelines for factoring polynomials are provided for easy reference.

General Guidelines for Factoring Polynomials

1. **Always look for a common monomial factor first.** *If the leading coefficient is negative, factor out a negative monomial even if it is just −1.*

2. **Check the number of terms.**

   a. **Two terms:**

      (1) difference of two squares? − factorable

      (2) sum of two squares? − not factorable

   b. **Three terms:**

      (1) perfect square trinomial?

      (2) use ac-method?

      Guidelines for the ac-method

      (a) Multiply a · c.

      (b) Find two integers whose product is ac and whose sum is b. If this is not possible, then the trinomial is not factorable.

      (c) Rewrite the middle terms (bx) using the two numbers found in Step (b) as coefficients.

      (d) Factor by grouping.

      (3) use trial-and-error method?

      Guidelines for the Trial-and-Error Method

      (a) If the sign of the constant term is positive (+), the signs in both factors will be the same, either both positive or both negative.

      (b) If the sign of the constant term is negative (−), the signs in the factors will be different, one positive and one negative.

   c. **Four terms:**

      (1) group terms with a common factor and factor out any common binomial factor.

3. **Check the possibility of factoring any of the factors.**

Section 7.1 Greatest Common Factor and Factoring by Grouping

Factor each of the polynomials in Exercises 1 – 8 by factoring out the greatest common monomial factor.

1. $11x - 22$

2. $3y + 24$

3. $-7a - 14b$

4. $-4y^2 + 28y$

5. $16x^3y - 12x^2y$

6. $am^2 + 11am + 25a$

7. $-7x^4t^3 + 98t^4x^3 + 35t^5x$

8. $-6x^2y^4 - 6x^3y^4 + 24x^2y^3$

In Exercises 9 – 12, factor each expression by factoring out the common binomial factor.

9. $8y(y + 5) + 3(y + 5)$

10. $3a(a + 7) - 2(a + 7)$

11. $a(x + 4) + b(x + 4)$

12. $2a(x - 10) + 3b(x - 10)$

Factor each of the polynomials in Exercises 13 – 20 by grouping. If a polynomial cannot be factored, write *not factorable*.

13. $ax - a + cx - c$

14. $4x + 4y + bx + by$

15. $x - 4xy + 2z - 8zy$

16. $z^2 + 5 + cz^2 + 5c$

17. $x^2 - x^2y + 6 + 6y$

18. $7ab - 3b + 2a^2 - 3a$

19. $10ab - 2b^2 + 7ac - 35bc$

20. $x - 3xy - z + 3zy$

Section 7.2 Factoring Trinomials: $x^2 + bx + c$

Completely factor each of the trinomials in Exercises 21 – 35. If the trinomial cannot be factored, write *not factorable*.

21. $m^2 + 7m + 6$

22. $a^2 - 4a + 3$

23. $x^2 + 11x + 18$

24. $y^2 + 8y + 15$

25. $n^2 - 8n + 12$

26. $x^2 - 10x + 24$

27. $x^2 + 3x - 10$

28. $y^2 + 13y + 36$

29. $y^2 + 2y + 24$

30. $c^2 + 3c + 10$

31. $x^2 + 12x + 35$

32. $x^2 + 17x + 72$

33. $a^2 - 5a - 50$

34. $x^2 - 4x - 21$

35. $x^2 + 29x + 100$

Section 7.3 More on Factoring Trinomials: $ax^2 + bx + c$

Completely factor each of the polynomials in Exercises 36 – 50. If the polynomial cannot be factored, write *not factorable*.

36. $2x^2 + 7x + 3$

37. $-2x^2 + 3x - 1$

38. $-12x^2 + 32x - 5$

39. $12x^2 + x - 6$

40. $6y^2 - 11y + 4$

41. $12y^2 - 11y + 4$

42. $63x^2 - 3x - 30$ **43.** $16x^2 + 12x - 70$ **44.** $24 + x - 3x^2$

45. $14 + 11x - 15x^2$ **46.** $2y^2 - 13y + 5$ **47.** $3y^2 + 10y + 6$

48. $200 + 20x - 4x^2$ **49.** $7y^2 + 14y - 168$ **50.** $18x^2 - 15x + 2$

Section 7.4 Factoring Special Products: Difference of Two Squares and Perfect Square Trinomials

Completely factor each of the polynomials in Exercises 51 – 65. If the polynomial cannot be factored, write *not factorable*.

51. $y^2 - 49$ **52.** $x^2 - 100$ **53.** $64 + 49x^2$

54. $36 + 25y^2$ **55.** $25x^2 - 36$ **56.** $2x^2 - 98$

57. $3x^2 - 147$ **58.** $64a^2 - 1$ **59.** $12x^2 - 60x - 75$

60. $21x^3 - 13x^2 - 2x$ **61.** $x^2 + 12x + 36$ **62.** $6y^2 + 120y + 600$

63. $5x^2 - 150x + 1125$ **64.** $7a^2 - 42a + 63$ **65.** $-2ax^2 - 44ax - 242a$

In Exercises 66 – 70, complete the square by adding the correct missing term on the left, then factor as indicated.

66. $y^2 + 22y + \underline{\qquad} = (\quad)^2$ **67.** $x^2 - 12x + \underline{\qquad} = (\quad)^2$

68. $x^2 - 7x + \underline{\qquad} = (\quad)^2$ **69.** $y^2 + \underline{\qquad} + 144 = (\quad)^2$

70. $x^2 - \underline{\qquad} + 49 = (\quad)^2$

Section 7.5 Solving Quadratic Equations by Factoring

In Exercises 71 – 85, solve the equations by factoring.

71. $3x^2 + 15x = 0$ **72.** $4x^2 - 24x = 0$ **73.** $3x^2 - 17x + 10 = 0$

74. $6y^2 + y - 35 = 0$ **75.** $16x^3 - 100x = 0$ **76.** $x^3 - 4x^2 - 12x = 0$

77. $25y^3 = 36y$ **78.** $x^3 = 81x$ **79.** $2a^2 + 28a + 98 = 0$

80. $3y^2 - 28y + 64 = 0$ **81.** $x^2 = x + 12$ **82.** $x^2 = 2x + 35$

83. $(x + 3)^2 = 9$ **84.** $(x - 5)^2 = 64$ **85.** $5x^2 - 65x = -200$

Section 7.6 Applications of Quadratic Equations

86. One positive number is three more than another and the sum of their squares is 269. What are the numbers?

87. A tract of new homes is arranged in rectangular fashion with 5 more houses on each street than there are streets. How many streets are there if there are 126 houses in the tract?

88. A rectangle has an area of 450 square yards and a perimeter of 90 yards. What are the dimensions of the rectangle?

89. The sum of the squares of two consecutive odd positive integers is 290. Find the integers.

90. Find two consecutive negative integers whose product is 156.

91. The base of a triangle is 12 inches greater than the height. If the area is 80 square inches, find the length of the base.

92. The length of a rectangle is 10 meters more than the width. If 3 meters are added to both the length and the width, the new area will be 299 square meters. Find the dimensions of the original rectangle.

93. The area of a square (in square inches) is numerically equal to its perimeter (in inches). What is the length of one side of the square?

94. One number is 20 more than another. Their product is –84. What are the numbers?

95. A small theater can seat 600 people. The number of rows is 10 less than the number of seats in each row. How many rows of seats are there? How many seats are in each row?

Section 7.7 Additional Applications of Quadratic Equations

96. The formula for the volume of a cylinder is $V = \pi r^2 h$. Use $\pi = 3.14$ to find the volume of a cylinder with a radius of 3 cm and a height of 15 cm.

97. The equation $h = -16t^2 + 56t$ (where t is measured in seconds and h in feet) gives the height of a projectile projected upward with an initial velocity of 56 ft per second. In how many seconds will the projectile be 24 feet above the ground? (**Hint**: This may occur two times.)

98. The equation $h = -16t^2 + 120t$ (where t is measured in seconds and h in feet) gives the height of a projectile projected upward with an initial velocity of 120 ft per second. In how many seconds will the projectile be 200 feet above the ground? (**Hint**: This may occur two times.)

99. The demand for a certain commodity is given by $D = -20p^2 + 500p + 2000$. Find the selling price (p), if 5000 units are sold. (**Hint**: This may occur two times.)

100. The demand for a certain commodity is given by $D = -20p^2 + 200p + 2000$. Find the selling price (p), if 2500 units are sold. (**Hint**: This may occur two times.)

Chapter 7 Test

Find the GCF for each of the sets of terms in Exercises 1 and 2.

1. $30, 75, 90$

2. $40x^2y, \ 48x^2y^3, \ 56x^2y^2$

In Exercises 3 and 4, simplify each expression. Assume that no denominator is 0.

3. $\dfrac{16x^4}{8x}$

4. $\dfrac{42x^3y^3}{-6x^3y}$

Factor completely, if possible, each of the polynomials in Exercises 5 – 16. If the polynomial cannot be factored, write *not factorable*.

5. $20x^3y^2 + 30x^3y + 10x^3$

6. $x^2 - 9x + 20$

7. $-x^2 - 14x - 49$

8. $6x^2 - 6$

9. $12x^2 + 2x - 10$

10. $3x^2 + x - 24$

11. $16x^2 - 25y^2$

12. $2x^3 - x^2 - 3x$

13. $6x^2 - 13x + 6$

14. $2xy - 3y + 14x - 21$

15. $4x^2 + 25$

16. $-3x^3 + 6x^2 - 6x$

In Exercises 17 and 18, complete the square by adding the correct missing term on the left, then factor as indicated.

17. $x^2 - 10x + \underline{\quad} = \left(\quad \right)^2$

18. $x^2 + 16x + \underline{\quad} = \left(\quad \right)^2$

Solve the equations in Exercises 19 – 25.

19. $(x + 2)(3x - 5) = 0$

20. $x^2 - 7x - 8 = 0$

21. $-3x^2 = 18x$

22. $\dfrac{2x^2}{5} - 6 = \dfrac{4x}{5}$

23. $0 = 4x^2 - 17x - 15$

24. $0 = 8x^2 - 2x - 15$

25. $(2x - 7)(x + 1) = 6x - 19$

26. One number is 10 less than five times another number. Their product is 120. Find the numbers.

27. The length of a rectangle is 7 centimeters less than twice the width. If the area of the rectangle is 165 square centimeters, find the length and width.

28. The product of two consecutive positive integers is 342. Find the two integers.

29. The difference between two positive numbers is 9. If the smaller is added to the square of the larger, the result is 147. Find the numbers.

30. A sheet of metal is in the shape of a rectangle with width 12 inches and length 20 inches. A slot (see the figure) of width x inches and length $x + 3$ inches is cut from the top of the rectangle.

   a. Write a polynomial function, $A(x)$, that represents the area of the remaining figure.

   b. Write a polynomial function, $P(x)$, that represents the perimeter of the remaining figure.

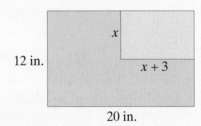

12 in.

x

$x + 3$

20 in.

Cumulative Review: Chapter 1 – 7

Find the LCM for each set of terms in Exercises 1 and 2.

1. $20, 12, 24$

2. $8x^2,\ 14x^2y,\ 21xy$

Perform the indicated operation in Exercises 3 – 6. Reduce all answers to lowest terms.

3. $\dfrac{7}{12} + \dfrac{9}{16}$

4. $\dfrac{11}{15a} - \dfrac{5}{12a}$

5. $\dfrac{6x}{25} \cdot \dfrac{5}{4x}$

6. $\dfrac{40}{92} \div \dfrac{2}{15x}$

Simplify each of the expressions in Exercises 7 – 10 so that it has only positive exponents.

7. $\dfrac{-24x^4y^2}{3x^2y}$

8. $\dfrac{36x^3y^5}{9xy^4}$

9. $\dfrac{-4x^2y}{12xy^{-3}}$

10. $\dfrac{21xy^2}{3x^{-1}y}$

11. Given the relation $r = \{(2,\ -3), (3,\ -2), (5,\ 0), (7.1,\ 3.2)\}$.
   a. What is the domain of the relation?
   b. What is the range of the relation?
   c. Is the relation a function? Explain.

12. For the function $f(x) = x^2 - 3x + 4$, find
   a. $f(-6)$
   b. $f(0)$
   c. $f\left(\dfrac{1}{2}\right)$

13. Find the equation of the line determined by the two points $(-5, 3)$ and $(2, -4)$. Graph the line.

14. Find the equation of the line parallel to the line $2x - y = 7$ and passing through the point $(-1, 6)$. Graph both lines.

15. Find the equation of the line parallel to the line $y = 3x - 7$ and passing through the point $(-1, 1)$. Graph both lines

16. Find the equation of the line perpendicular to the line $y = 3x - 7$ and passing through the point $(-1, 1)$. Graph both lines.

17. Find the equation of the line perpendicular to the line $2x - y = 8$ and passing through the point $(0, 4)$. Graph both lines.

In Exercises 18 and 19, use the vertical line test to determine whether each of the graphs does or does not represent a function.

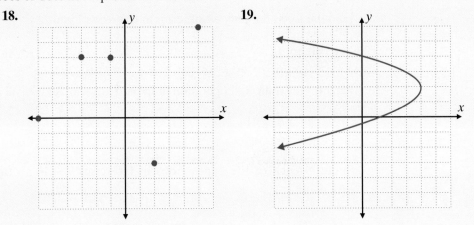

18.   **19.**

Solve the systems of equations in Exercises 20 – 22. Also state whether the system is consistent, inconsistent, or dependent.

20. $\begin{cases} 3x - y = 12 \\ x + 2y = -3 \end{cases}$ **21.** $\begin{cases} y = 4x - 1 \\ 5x + 2y = -2 \end{cases}$ **22.** $\begin{cases} y = 2x - 7 \\ 4x - 2y = 3 \end{cases}$

In Exercises 23 and 24, use a TI-84 Plus graphing calculator to find the solutions to each system of equations.

23. $\begin{cases} 5x - y = 5 \\ 3x + y = 5.4 \end{cases}$ **24.** $\begin{cases} -3x + y = -8.9 \\ 2x - y = 6.5 \end{cases}$

25. Use a TI-84 Plus graphing calculator to graph the two functions $f(x) = x$ and $g(x) = 20x^2 - 1$ and use the calculator to find the points of intersection.

Graph the inequalities in Exercises 26 – 28.

26. $y \le 4x - 5$ **27.** $y > 6x + 2$ **28.** $3x + 2y \ge 10$

Graph the systems of inequalities in Exercises 29 and 30.

29. $\begin{cases} 2x + y > 8 \\ x + 2y < 2 \end{cases}$ **30.** $\begin{cases} y \le 7 \\ x \le 3 \end{cases}$

Perform the indicated operations in Exercises 31 – 40 and simplify by combining like terms.

31. $2(4x + 3) + 5(x - 1)$ **32.** $2x(x + 5) - (x + 1)(x - 3)$

33. $\left(x^2 + 3x - 1\right) + \left(2x^2 - 8x + 4\right)$ **34.** $\left(2x^2 + 6x - 7\right) + \left(2x^2 - x - 1\right)$

35. $\left(2x^2 - 5x + 3\right) - \left(x^2 - 2x + 8\right)$ **36.** $(x + 1) - \left(4x^2 + 3x - 2\right)$

37. $(2x - 7)(x + 4)$ **38.** $-(x + 6)(3x - 1)$

39. $(x + 6)^2$ **40.** $(2x - 7)^2$

Express each quotient in Exercises 41 and 42 as a sum of fractions and simplify if possible.

41. $\dfrac{15x^3 - 10x^2 + 3x}{5x^2}$ **42.** $\dfrac{8x^2y^2 - 5xy^2 + 4xy}{4xy^2}$

In Exercises 43 and 44, divide by using long division and write the answers in the form $Q + \dfrac{R}{D}$.

43. $\left(3x^2 - 5x + 20\right) \div \left(x - 2\right)$ **44.** $\dfrac{x^3 - 4x^2 + 6x + 10}{x + 4}$

Factor each expression as completely as possible in Exercises 45 – 64.

45. $8x - 20$ **46.** $6x - 96$ **47.** $2x^2 - 15x + 18$

48. $6x^2 - x - 12$ **49.** $5x - 10$ **50.** $-12x^2 - 16x$

51. $16x^2y - 24xy$ **52.** $10x^4 - 25x^3 + 5x^2$ **53.** $4x^2 - 1$

54. $y^2 - 20y + 100$ **55.** $3x^2 - 48y^2$ **56.** $x^2 - 7x - 18$

57. $5x^2 + 40x + 80$ **58.** $x^2 + x + 3$ **59.** $25x^2 + 20x + 4$

60. $3x^2 + 5x + 2$ **61.** $2x^3 - 20x^2 + 50x$ **62.** $4x^3 + 100x$

63. $xy + 3x + 2y + 6$ **64.** $ax - 2a - 2b + bx$

Complete the square by adding the correct term in Exercises 65 – 69 so that the trinomials in each will factor as indicated.

65. $x^2 - 4x + \underline{\quad} = (\quad)^2$ **66.** $x^2 + 18x + \underline{\quad} = (\quad)^2$

67. $x^2 - 8x + \underline{\quad} = (\quad)^2$ **68.** $x^2 - \underline{\quad} + 25 = (\quad)^2$

69. $x^2 - \underline{\quad} + \dfrac{25}{4} = (\quad)^2$

Solve the equations in Exercises 70 – 84.

70. $(x - 7)(x + 1) = 0$ **71.** $x(3x + 5) = 0$ **72.** $4x^2 + 9x - 9 = 0$

73. $21x - 3x^2 = 0$ **74.** $0 = x^2 + 8x + 12$ **75.** $x^2 = 3x + 28$

76. $x^3 + 5x^2 - 6x = 0$ **77.** $\dfrac{1}{4}x^2 + x - 15 = 0$ **78.** $0 = 15 - 12x - 3x^2$

79. $x^3 + 14x^2 + 49x = 0$ **80.** $8x = 12x + 2x^2$ **81.** $6x^2 = 24x$

82. $2x(x - 2) = x + 25$ **83.** $(4x - 3)(x + 1) = 2$ **84.** $2x(x + 5)(x - 2) = 0$

85. A boat can travel 21 miles downstream in 3 hours. The return trip takes 7 hours. Find the speed of the boat in still water and the speed of the current.

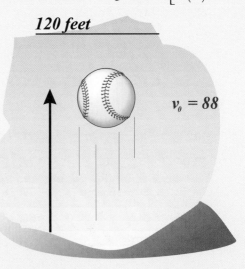

86. Karl invested a total of $10,000 in two separate accounts. One account paid 6% interest and the other paid 8% interest. If the annual income from both accounts was $650, how much did he have invested in each account?

87. The perimeter of a rectangle is 60 inches. The area of the rectangle is 221 square inches. Find the length and width of the rectangle. (**Hint**: After substituting and simplifying, you will have a quadratic equation.)

88. The difference between two positive numbers is 9. If the smaller is added to the square of the larger, the result is 147. Find the numbers.

89. Find two consecutive integers such that the sum of their squares is equal to 145.

90. A ball is thrown upward with an initial velocity of 88 feet per second $(v_0 = 88)$. When will the ball be 120 feet above the ground? $\left[h(t) = -16t^2 + v_0 t. \right]$

120 feet

$v_0 = 88$

Rational Expressions

Did You Know?

Pythagoras (c. 550 BC) was a Greek mathematician who founded a secret brotherhood whose objective was to investigate music, astronomy, and mathematics, especially geometry. The Pythagoreans believed that all physical phenomena were explainable in terms of "arithmos," the basic properties of whole numbers and their ratios.

In Chapter 8, you will be studying rational numbers (fractions), numbers written as the ratio of two integers. The Pythagorean Society attached mystical significance to these rational numbers. Their investigations of musical harmony revealed that a vibrating string must be divided exactly into halves, thirds, fourths, fifths, and so on, to produce tones in harmony with the string vibrating as a whole.

The recognition that the chords which are pleasing to the ear correspond to exact divisions of the vibrating string by whole numbers stimulated Pythagoras to propose that all physical properties could be described using rational numbers. The Pythagorean Society attempted to compute the orbits of the planets by relating them to the musical intervals. The Pythagoreans thought that as the planets moved through space they produced music, the music of the spheres.

Unfortunately, a scandalous idea soon developed within the Pythagorean Society. It became apparent that there were some numbers that could not be represented as the ratio of two whole numbers. This shocking idea came about when the Pythagoreans investigated their master's favorite theorem, the Pythagorean Theorem:

Given a right triangle, the length of the hypotenuse squared is equal to the sum of the lengths squared of the other two sides (as illustrated in the diagrams).

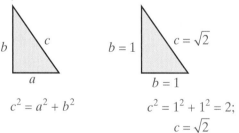

$$c^2 = a^2 + b^2$$

$$c^2 = 1^2 + 1^2 = 2;$$
$$c = \sqrt{2}$$

The hypotenuse of the second triangle is some number that, when squared, gives two. It can be proven (and it was by the Pythagoreans) that no such rational number exists. Imagine the amazement within the Brotherhood. Legend has it that the Pythagoreans attempted to keep the "irrational" numbers a secret. Hippaes, the Brother who first told an outsider about the new numbers, supposedly was expelled from the Society and punished by death for his unfaithfulness.

8.1   *Reducing Rational Expressions*

8.2   *Multiplication and Division with Rational Expressions*

8.3   *Addition and Subtraction with Rational Expressions*

8.4   *Complex Algebraic Fractions*

8.5   *Solving Proportions and Other Equations with Rational Expressions*

8.6   *Applications*

8.7   *Additional Applications: Variation*

"Numbers rule the Universe"

The Pythagoreans

R ational expressions are fractions. So, if you know how to work with fractions in arithmetic, you will find the same techniques used in Chapter 8. You will need to find a common denominator to add (or subtract) rational expressions, and division is accomplished by multiplying by the reciprocal of the divisor. The basic difference is that the numerators and denominators are polynomials and the skills of multiplying and factoring that you learned in Chapters 6 and 7 will be needed. If you keep in mind throughout Chapter 8 that you are following the same rules you learned for fractions in arithmetic, some of the difficult-looking expressions may seem considerably easier.

8.1   Reducing Rational Expressions

Objectives

After completing this section, you will be able to:

1. *Determine what values of the variable, if any, will make a rational expression undefined.*

2. *Find the numerical value of a rational expression.*

3. *Reduce rational expressions to lowest terms.*

Introduction to Rational Expressions

The term **rational number** is the technical name for a fraction in which both the numerator and denominator are integers. Similarly, the term **rational expression** is the technical name for a fraction in which both the numerator and denominator are polynomials.

Rational Expression

*A **rational expression** is an algebraic expression that can be written in the form*

$$\frac{P}{Q} \text{ where } P \text{ and } Q \text{ are polynomials and } Q \neq 0.$$

Examples of rational expressions are

$$\frac{5x^2}{7}, \quad \frac{m^2-9}{m^2+9}, \quad \text{and} \quad \frac{x^2+x-6}{x^3-2x^2}.$$

As with rational numbers, the denominators of rational expressions cannot be 0. If a numerical value is substituted for the variable in a rational expression and the denominator assumes a value of 0, we say that the expression is **undefined** for that value of the variable.

Remember, the denominator can never be 0. Division by 0 is undefined.

To Determine Which Values Make a Rational Expression Undefined

To determine which values, if any, of the variable will make a rational expression undefined:

1. *Set the denominator equal to 0.*
2. *Solve the resulting equation.*
3. *The solutions to this equation are the values that make the expression undefined.*

   *(These values are said to be **restrictions** on the variable.)*

   Note: *If the denominator is a **constant** or the denominator is **not factorable**, then there are no restrictions on the variable.*

Example 1: Undefined Values of Rational Expressions

Determine what values of the variable, if any, will make the rational expression undefined. (These values are called restrictions on the variable.)

a. $\dfrac{3}{2x-1}$

Solution: Set the denominator equal to 0: $\quad 2x-1=0$

Solve the equation: $\quad 2x = 1$

$$x = \frac{1}{2}$$

Thus the expression $\dfrac{3}{2x-1}$ is undefined for $x = \dfrac{1}{2}$. Any other real number may be substituted for x in the expression. We write $x \neq \dfrac{1}{2}$.

b. $\dfrac{m^2-4}{m^2-2m-3}$

Solution: Set the denominator equal to 0: $\quad m^2-2m-3=0$

Solve the equation: $\quad (m-3)(m+1)=0$

$$m-3=0 \quad \text{or} \quad m+1=0$$
$$m=3 \qquad\qquad m=-1$$

Thus, the expression $\dfrac{m^2-4}{m^2-2m-3}$ is undefined for $m = 3$ and $m = -1$.

We write $m \neq 3, -1$.

Continued on next page...

c. $\dfrac{x-5}{x^2+1}$

Solution: Set the denominator equal to 0: $x^2+1=0$

Solve the equation: $x^2+1=0$

$$x^2=-1$$

But there are no real numbers whose square is -1. Therefore, this equation has no real solutions and the denominator will not be 0. There are **no restrictions** on the variable in the real number system.

NOTES

Special Comment About the Numerator Being 0:
If the numerator of a fraction has a value of 0 and the denominator is not 0, then the fraction is defined and has a value of 0. For example, if, in Example 1c, $x=5$, then $\dfrac{x-5}{x^2+1}=\dfrac{5-5}{5^2+1}=\dfrac{0}{26}=0$. If both numerator and denominator are 0, then the fraction is undefined just as in the case where only the denominator is 0.

While we cannot substitute values for the variable that will make a denominator 0, other values may be substituted. The following examples illustrate how to find the numerical value of a rational expression by substituting a value for the variable.

Example 2: Evaluating Rational Expressions

Find the value of each rational expression for the given value of the variable.

a. $\dfrac{2x}{x^2+1}; x=2$

Solution: $\dfrac{2x}{x^2+1}=\dfrac{2\cdot(2)}{(2)^2+1}=\dfrac{4}{4+1}=\dfrac{4}{5}$

Note that any real number may be substituted for x since the denominator x^2+1 is never 0 for any real number.

b. $\dfrac{a-3}{a^2-5}; a=3$

Solution: $\dfrac{a-3}{a^2-5}=\dfrac{(3)-3}{(3)^2-5}=\dfrac{0}{9-5}=\dfrac{0}{4}=0$ Note that a numerator may be 0.

Example 2b illustrates the important fact about fractions stated in the notes above: **If the numerator is 0 and the denominator is not 0, then the value of the fraction is 0.**

Reducing Rational Expressions

The rules for operating with rational expressions are essentially the same as those for operating with fractions in arithmetic. That is, adding, subtracting, multiplying, and dividing with rational expressions are operations involving factoring and common denominators just as with fractions in arithmetic. The following rules for arithmetic with fractions were discussed in Chapter 2.

Summary of Arithmetic Rules for Rational Numbers (or Fractions)

*A **fraction** (or **rational number**) is a number that can be written in the form* $\dfrac{a}{b}$ *where a and b are integers and* **b ≠ 0**. *No denominator can be 0.*

The Fundamental Principle: $\dfrac{a}{b} = \dfrac{a \cdot k}{b \cdot k}$ *where b, k ≠ 0.*

The reciprocal *of* $\dfrac{a}{b}$ *is* $\dfrac{b}{a}$ *and* $\dfrac{a}{b} \cdot \dfrac{b}{a} = 1$ *where a, b ≠ 0.*

Multiplication: $\dfrac{a}{b} \cdot \dfrac{c}{d} = \dfrac{a \cdot c}{b \cdot d}$ *where b, d ≠ 0.*

Division: $\dfrac{a}{b} \div \dfrac{c}{d} = \dfrac{a}{b} \cdot \dfrac{d}{c}$ *where b, c, d ≠ 0.*

Addition: $\dfrac{a}{b} + \dfrac{c}{b} = \dfrac{a+c}{b}$ *where b ≠ 0.*

Subtraction: $\dfrac{a}{b} - \dfrac{c}{b} = \dfrac{a-c}{b}$ *where b ≠ 0.*

These same rules are valid for rational expressions. Each can be restated by replacing *a* and *b* with *P* and *Q* where *P* and *Q* represent polynomials. In particular, the Fundamental Principle can be restated as follows:

The Fundamental Principle of Rational Expressions

If $\dfrac{P}{Q}$ *is a rational expression and Q ≠ 0 and K is a polynomial and K ≠ 0, then*

$$\frac{P}{Q} = \frac{P \cdot K}{Q \cdot K}$$

The Fundamental Principle can be used to build a rational expression to higher terms or to reduce a rational expression. We will want to build to higher terms when adding and subtracting rational expressions in Section 8.3. At this time, we will discuss reducing rational expressions by factoring the numerators and denominators, just as we did with reducing rational numbers.

To reduce a rational expression, factor both the numerator and denominator and use the Fundamental Principle to "divide out" any common factors. A rational expression is reduced to lowest terms if the greatest common factor (GCF) of the numerator and denominator is 1.

For example, the following expressions are reduced to lowest terms.

$$\text{Fraction:} \quad \frac{15}{35} = \frac{\overset{1}{\cancel{5}} \cdot 3}{\underset{1}{\cancel{5}} \cdot 7} = 1 \cdot \frac{3}{7} = \frac{3}{7}$$

$$\text{Rational expression:} \quad \frac{x^2 + 5x + 6}{x^2 + 7x + 12} = \frac{\overset{1}{\cancel{(x+3)}}(x+2)}{\underset{1}{\cancel{(x+3)}}(x+4)} = 1 \cdot \frac{x+2}{x+4} = \frac{x+2}{x+4}$$

Remember that **no denominator can be 0 either before or after a rational expression is reduced**. Thus in the illustration here, $x + 3 \neq 0$ and $x + 4 \neq 0$ (or $x \neq -3$ and $x \neq -4$). Therefore, in the expression $\dfrac{x^2 + 5x + 6}{x^2 + 7x + 12}$, any real number, except -3 and -4, may be substituted for x. We will see that this idea is particularly important when solving equations.

NOTES

Common Error
The key word when reducing is **factor**. Many students incorrectly "divide out" terms or numbers that are not factors.

Consider the following incorrect statement that $2 = 4$:

$$2 = \frac{6}{3} = \frac{\cancel{3} + 3}{\cancel{3}} = \frac{1 + 3}{1} = \frac{4}{1} = 4 \quad \textbf{INCORRECT}$$

Obviously, incorrect reducing can lead to ridiculous results.

"Divide out" only common factors.

INCORRECT	INCORRECT
$\dfrac{5x}{x+5} = \dfrac{\cancel{5}x}{x+\cancel{5}} = \dfrac{x}{x+1}$	$\dfrac{5x}{x+5} = \dfrac{5\cancel{x}}{\cancel{x}+5} = \dfrac{5}{1+5}$
5 is not a factor of the denominator.	x is not a factor of the denominator.

CORRECT

$$\frac{5x + 15}{5x + 20} = \frac{\cancel{5}(x+3)}{\cancel{5}(x+4)} = \frac{x+3}{x+4}$$

5 is a common factor.

Example 3: Reducing Rational Expressions

Reduce each rational expression to lowest terms. State any restrictions on the variable. **Remember, no denominator can be 0.** This restriction applies to denominators **before and after** a rational expression is reduced.

a. $\dfrac{2x+4}{3x+6}$

Solution: $\dfrac{2x+4}{3x+6} = \dfrac{2\,\cancel{(x+2)}}{3\,\cancel{(x+2)}} = \dfrac{2}{3}\ (x \neq -2)$ Since $x+2 \neq 0, x \neq -2$.

b. $\dfrac{x^2-16}{x-4}$

Solution: $\dfrac{x^2-16}{(x-4)} = \dfrac{(x+4)\,\cancel{(x-4)}}{\cancel{(x-4)}} = x+4\ (x \neq 4)$ Since $x-4 \neq 0, x \neq 4$.

c. $\dfrac{a}{a^2-5a}$

Solution: $\dfrac{a}{a^2-5a} = \dfrac{\cancel{a}}{\cancel{a}(a-5)} = \dfrac{1}{a-5}\ (a \neq 0,5)$ Note that these restrictions are determined before reducing.

This example illustrates the importance of writing 1 in the numerator if all the factors divide out. Remember that 1 is an understood factor.

d. $\dfrac{3-y}{y-3}$

Solution: $\dfrac{3-y}{y-3} = \dfrac{-y+3}{y-3}$ First write both numerator and denominator in descending order. Now factor out -1 in the numerator.

$= \dfrac{-1(y-3)}{(y-3)}$

$= \dfrac{-1\,\cancel{(y-3)}}{\cancel{(y-3)}} = -1\ (y \neq 3)$

In Example 3d, the result was -1. The expression $3-y$ is the opposite of $y-3$ for any value of y, except 3. That is,

$$3-y = -y+3 = -1(y-3).$$

When nonzero opposites are divided, the quotient is always -1.

$$\dfrac{-8}{+8} = -1, \qquad \dfrac{14}{-14} = -1, \qquad \dfrac{x-5}{5-x} = \dfrac{\cancel{(x-5)}}{-1\,\cancel{(x-5)}} = \dfrac{1}{-1} = -1$$

Opposites

In general, for $b \neq a$, $(a-b)$ and $(b-a)$ are opposites and

$$\frac{a-b}{b-a} = \frac{a-b}{-1(a-b)} = -1$$

Now consider the problem of placement of the negative sign in a fraction. For example,

$$-\frac{6}{2} = -3, \quad \frac{-6}{2} = -3, \quad \text{and} \quad \frac{6}{-2} = -3.$$

Thus,

$$-\frac{6}{2} = \frac{-6}{2} = \frac{6}{-2}.$$

The following fact about negative fractions should be clear. Without changing the value of the fraction, the negative sign can be placed,

1. in front of the fraction, or
2. in the numerator, or
3. in the denominator.

The following general statements about the placement of negative signs can be made.

Fractions and Negative Signs

For integers a and b, $b \neq 0$, $-\dfrac{a}{b} = \dfrac{-a}{b} = \dfrac{a}{-b}$

For polynomials P and Q, $Q \neq 0$, $-\dfrac{P}{Q} = \dfrac{-P}{Q} = \dfrac{P}{-Q}$

At this stage of the discussion, we will assume that no denominator is 0. You should keep in mind that for any rational expression, there are certain restrictions on the variable; **but these restrictions are generally not stated, just understood**.

Example 4: Simplifying Rational Expressions

Simplify the following rational expressions.

a. $-\dfrac{-2x+6}{x^2-3x}$

Solution:
$$-\frac{-2x+6}{x^2-3x}=\frac{-(-2x+6)}{x^2-3x}=\frac{2x-6}{x^2-3x}=\frac{2\overset{1}{\cancel{(x-3)}}}{x\underset{1}{\cancel{(x-3)}}}=\frac{2}{x}$$

or

$$-\frac{-2x+6}{x^2-3x}=\frac{-(-2x+6)}{x^2-3x}=\frac{-(2)\overset{-1}{\cancel{(-x+3)}}}{x\underset{1}{\cancel{(x-3)}}}=\frac{2}{x}$$

or

$$-\frac{-2x+6}{x^2-3x}=\frac{-2x+6}{-(x^2-3x)}=\frac{-2\overset{1}{\cancel{(x-3)}}}{-1(x)\underset{1}{\cancel{(x-3)}}}=\frac{2}{x}$$

b. $-\dfrac{5-10x}{6x-3}$

Solution:
$$-\frac{5-10x}{6x-3}=-\frac{5\overset{-1}{\cancel{(1-2x)}}}{3\underset{1}{\cancel{(2x-1)}}}=-\frac{-5}{3}=\frac{5}{3}$$

or

$$-\frac{5-10x}{6x-3}=\frac{-(5-10x)}{6x-3}=\frac{-(5)\overset{-1}{\cancel{(1-2x)}}}{3\underset{1}{\cancel{(2x-1)}}}=\frac{5}{3}$$

or

$$-\frac{5-10x}{6x-3}=\frac{(5-10x)}{-(6x-3)}=\frac{(5)\overset{-1}{\cancel{(1-2x)}}}{-1(3)\underset{1}{\cancel{(2x-1)}}}=\frac{5}{3}$$

Practice Problems

Reduce each rational expression to lowest terms and state any restrictions on the variables.

1. $\dfrac{x^2}{x^2+4}$ **2.** $\dfrac{7x-7}{7x^2-7}$ **3.** $\dfrac{-4x+8}{x^2-4}$ **4.** $\dfrac{a^2-8a-9}{a^2-9a-10}$

8.1 Exercises

In Exercises 1 – 20, determine any restrictions on the variable. (These are the values of the variable that will make the rational expression undefined.)

1. $\dfrac{-3}{5x}$ **2.** $\dfrac{17}{4y}$ **3.** $\dfrac{5y^2}{3y-4}$

4. $\dfrac{20a}{a^2+2}$ **5.** $\dfrac{x+1}{x^2-4x}$ **6.** $\dfrac{2x+3}{x^2-7x-8}$

7. $\dfrac{y^2-2y+3}{y^2-3y-10}$ **8.** $\dfrac{x-1}{2x-5}$ **9.** $\dfrac{x^2+2x}{x^2+4}$

10. $\dfrac{5m^2+2m}{3m^2-m-2}$ **11.** $\dfrac{4-y}{y-4}$ **12.** $\dfrac{x-6}{6-x}$

13. $\dfrac{5x+20}{6x+24}$ **14.** $\dfrac{4x+3}{8x+6}$ **15.** $\dfrac{x^2}{x^2+5x}$

16. $\dfrac{y^2+7y}{y^2}$ **17.** $\dfrac{a^2-7a}{a^2-49}$ **18.** $\dfrac{a^2-a-6}{a^2-5a+6}$

19. $\dfrac{m^2+3}{m^2+9}$ **20.** $\dfrac{8x^2-8}{8x^2+8}$

Answers to Practice Problems: 1. $\dfrac{x^2}{x^2+4}$, no restriction **2.** $\dfrac{1}{x+1}$, $x \neq -1, 1$ **3.** $\dfrac{-4}{x+2}$, $x \neq -2, 2$
4. $\dfrac{a-9}{a-10}$, $x \neq -1, 10$

In Exercises 21 – 30, evaluate each rational expression for the given value of the variable.

21. $\dfrac{x-3}{3x^2}$; $x = 5$

22. $\dfrac{2x+1}{3x-2}$; $x = 1$

23. $\dfrac{5x^2}{x^2-4}$; $x = -3$

24. $\dfrac{3y-4}{y^2+25}$; $y = 3$

25. $\dfrac{2x^2+5x}{x^2-1}$; $x = 0$

26. $\dfrac{n^3}{n^2-5n+6}$; $n = -1$

27. $\dfrac{2m-7}{m^2+8m+12}$; $m = 2$

28. $\dfrac{-x+3}{x-3}$; $x = -10$

29. $-\dfrac{15-x}{x-15}$; $x = 1000$

30. $-\dfrac{16+x}{x^2-16}$; $x = 20$

In Exercises 31 – 64, reduce each rational expression to lowest terms and state any restrictions on the variable.

31. $\dfrac{4x-12}{4x}$

32. $\dfrac{3y+15}{3y}$

33. $\dfrac{x^2+2x}{2x}$

34. $\dfrac{x^3+5x^2}{5x^2}$

35. $\dfrac{2y+3}{4y+6}$

36. $\dfrac{x-2}{x^2-6x+8}$

37. $\dfrac{3x-6}{6x+3}$

38. $\dfrac{5x+20}{6x+24}$

39. $\dfrac{x}{x^2-4x}$

40. $\dfrac{4-x}{x-4}$

41. $\dfrac{7x-14}{2-x}$

42. $\dfrac{3x^2-3x}{2x-2}$

43. $\dfrac{5+3x}{3x+5}$

44. $\dfrac{4xy+y^2}{3y^2+2y}$

45. $\dfrac{4-4x^2}{4x^2-4}$

46. $\dfrac{4x-8}{(x-2)^2}$

47. $\dfrac{x^2+2x}{x^2+4x+4}$

48. $\dfrac{x^2-4}{2x+4}$

49. $\dfrac{x^2+7x+10}{x^2-25}$

50. $\dfrac{x^2-3x-10}{x^2-7x+10}$

51. $\dfrac{x^2-3x-18}{x^2+6x+9}$

52. $\dfrac{8x^2+6x-9}{16x^2-9}$

53. $\dfrac{x^2-5x+6}{8x-2x^3}$

54. $\dfrac{6x^2-11x+3}{4x^2-12x+9}$

55. $\dfrac{16x^2 + 40x + 25}{4x^2 + 13x + 10}$

56. $\dfrac{6x^2 - 11x + 4}{3x^2 - 7x + 4}$

57. $\dfrac{3x - 6}{8 - 4x}$

58. $\dfrac{5x - 10}{30 - 15x}$

59. $\dfrac{a^2 + 6a + 8}{a^2 + 4a + 4}$

60. $\dfrac{y^2 - 16}{2y - 8}$

61. $\dfrac{3y^2 - 3y - 18}{2y^2 - 10y + 12}$

62. $\dfrac{5x^2 + 5x - 10}{6x^2 + 18x + 12}$

63. $\dfrac{5y^2 + 17y + 6}{5y^2 + 16y + 3}$

64. $\dfrac{6x^2 + 11x - 2}{6x^2 + 17x - 3}$

Writing and Thinking About Mathematics

65. The following "proof" that $2 = 0$ contains an error. Discuss the error in your own words.

$$a = b$$
$$2a = 2b$$
$$2a - 2b = 0$$
$$2(a - b) = 0$$
$$\frac{2\cancel{(a-b)}}{\cancel{(a-b)}} = \frac{0}{a - b}$$
$$2 = 0$$

Hawkes Learning Systems: Introductory Algebra

Defining Rational Expressions
Reducing Rational Expressions

Multiplication and Division with Rational Expressions

After completing this section, you will be able to:

1. *Multiply rational expressions.*

2. *Divide rational expressions.*

Multiplying Rational Expressions

To multiply any two rational expressions, multiply the numerators and multiply the denominators, **keeping the expressions in factored form**. Then "divide out" any common factors. Assume that no denominator has a value of 0.

Multiplying Rational Expressions

If P, Q, R, and S are polynomials with Q, S ≠ 0, then

$$\frac{P}{Q} \cdot \frac{R}{S} = \frac{P \cdot R}{Q \cdot S}$$

Example 1: Multiply Rational Expressions

Multiply and reduce, if possible.

a. $\dfrac{x+3}{x} \cdot \dfrac{x-3}{x+5}$

Solution: $\dfrac{x+3}{x} \cdot \dfrac{x-3}{x+5} = \dfrac{(x+3)(x-3)}{x(x+5)} = \dfrac{x^2-9}{x(x+5)}$ In this case the numerator and denominator have no common factors.

or $\dfrac{x^2-9}{x^2+5x}$

b. $\dfrac{x+5}{7x} \cdot \dfrac{49x^2}{x^2-25}$

Solution: $\dfrac{x+5}{7x} \cdot \dfrac{49x^2}{x^2-25} = \dfrac{\overset{7x}{\cancel{49x^2}}\,\cancel{(x+5)}}{\cancel{7x}\,\cancel{(x+5)}(x-5)} = \dfrac{7x}{x-5}$

Continued on next page...

c. $\dfrac{x^2-7x+12}{2x+6}\cdot\dfrac{x^2-4}{x^2-2x-8}$

Solution: $\dfrac{x^2-7x+12}{2x+6}\cdot\dfrac{x^2-4}{x^2-2x-8}=\dfrac{(x-4)(x-3)(x+2)(x-2)}{2(x+3)(x-4)(x+2)}$

$$=\dfrac{(x-3)(x-2)}{2(x+3)}\ \text{or}\ \dfrac{x^2-5x+6}{2(x+3)}\ \text{or}\ \dfrac{x^2-5x+6}{2x+6}$$

As shown in Examples 1a and 1c, there may be more than one correct form for an answer. After a rational expression has been reduced, the numerator and denominator may be multiplied out or left in factored form. We will see in Section 8.3, that multiplying out the numerator and leaving only the denominator in factored form is useful for adding and subtracting.

Dividing Rational Expressions

To **divide** any two rational expressions, multiply by the **reciprocal** of the divisor.

Dividing Rational Expressions

If P, Q, R, and S are polynomials with Q, R, S ≠ 0, then

$$\dfrac{P}{Q}\div\dfrac{R}{S}=\dfrac{P}{Q}\cdot\dfrac{S}{R}.$$

Example 2: Dividing Rational Expressions

Divide and reduce, if possible.

a. $\dfrac{a^2-49}{12a^2}\div\dfrac{a^2+8a+7}{18a}$

Solution: $\dfrac{a^2-49}{12a^2}\div\dfrac{a^2+8a+7}{18a}=\dfrac{a^2-49}{12a^2}\cdot\dfrac{18a}{a^2+8a+7}$

$$=\dfrac{(a+7)(a-7)\cdot 6\cdot 3\cdot a}{6\cdot 2\cdot a^2\,(a+7)(a+1)}$$

$$=\dfrac{3(a-7)}{2a(a+1)}$$

$$=\dfrac{3a-21}{2a(a+1)}\quad\text{or}\quad\dfrac{3a-21}{2a^2+2a}$$

b. $\dfrac{2x^2+3x-2}{x^2+3x+2} \div \dfrac{1-2x}{x-2}$

Solution: $\dfrac{2x^2+3x-2}{x^2+3x+2} \div \dfrac{1-2x}{x-2} = \dfrac{2x^2+3x-2}{x^2+3x+2} \cdot \dfrac{x-2}{1-2x}$

$$= \dfrac{\overset{-1}{\cancel{(2x-1)}}\;\cancel{(x+2)}\,(x-2)}{(x+1)\,\cancel{(x+2)}\,\cancel{(1-2x)}}$$

$$= \dfrac{-1(x-2)}{x+1}$$

$$= \dfrac{-x+2}{x+1} \text{ or } \dfrac{2-x}{x+1}$$

c. $\dfrac{3x^2-4x-4}{x^2-4} \div \dfrac{3x^2-7x-6}{x^2+3x+2} \cdot \dfrac{3x+9}{x^2-9}$

Solution: $\dfrac{3x^2-4x-4}{x^2-4} \div \dfrac{3x^2-7x-6}{x^2+3x+2} \cdot \dfrac{3x+9}{x^2-9}$

$$= \dfrac{3x^2-4x-4}{x^2-4} \cdot \dfrac{x^2+3x+2}{3x^2-7x-6} \cdot \dfrac{3x+9}{x^2-9}$$

$$= \dfrac{\cancel{(3x+2)}\,\cancel{(x-2)}}{\cancel{(x+2)}\,\cancel{(x-2)}} \cdot \dfrac{\cancel{(x+2)}\,(x+1)}{\cancel{(3x+2)}\,(x-3)} \cdot \dfrac{3\cancel{(x+3)}}{\cancel{(x+3)}\,(x-3)}$$

$$= \dfrac{3(x+1)}{(x-3)(x-3)}$$

$$= \dfrac{3x+3}{(x-3)(x-3)} \quad \text{or} \quad \dfrac{3x+3}{x^2-6x+9}$$

Practice Problems

Perform the indicated operations and simplify. Assume that no denominators are 0.

1. $\dfrac{2y^2-16y}{6y^2+7y-3} \cdot \dfrac{2y^2+11y+12}{y^2-9y+8}$

2. $\dfrac{a-b}{b-a} \div \dfrac{a^2+2ab+b^2}{a^2+ab}$

3. $\dfrac{x^2+3x-4}{x^2-1} \div \dfrac{x^2+6x+8}{x+1}$

Answers to Practice Problems: 1. $\dfrac{2y^2+8y}{(3y-1)(y-1)}$ **2.** $\dfrac{-a}{a+b}$ **3.** $\dfrac{1}{x+2}$

8.2 Exercises

Perform the indicated operations in Exercises 1 – 40. Assume that no denominator has a value of 0.

1. $\dfrac{2x-4}{3x+6} \cdot \dfrac{3x}{x-2}$

2. $\dfrac{5x+20}{2x} \cdot \dfrac{4x}{2x+4}$

3. $\dfrac{4x^2}{x^2+3x} \cdot \dfrac{x^2-9}{2x-2}$

4. $\dfrac{x^2-x}{x-1} \cdot \dfrac{x+1}{x}$

5. $\dfrac{x^2-4}{x} \cdot \dfrac{3x}{x+2}$

6. $\dfrac{x-3}{15x} \div \dfrac{3x-9}{30x^2}$

7. $\dfrac{x^2-1}{5} \div \dfrac{x^2+2x+1}{10}$

8. $\dfrac{x-5}{3} \div \dfrac{x^2-25}{6x+30}$

9. $\dfrac{10x^2-5x}{6x^2+12x} \div \dfrac{2x-1}{x^2+2x}$

10. $\dfrac{x^2+2x-8}{4x} \div \dfrac{2x^2+5x+2}{3x^2}$

11. $\dfrac{x^2+x}{x^2+2x+1} \cdot \dfrac{x^2-x-2}{x^2-1}$

12. $\dfrac{x^2-x-6}{x^2-4} \cdot \dfrac{x^2-25}{x^2+2x-15}$

13. $\dfrac{x+3}{x^2+3x-4} \cdot \dfrac{x^2+x-2}{x+2}$

14. $\dfrac{x^2+5x+6}{2x+4} \cdot \dfrac{5x}{x+3}$

15. $\dfrac{6x^2-7x-3}{x^2-1} \cdot \dfrac{x-1}{2x-3}$

16. $\dfrac{x^2-9}{2x^2+7x+3} \cdot \dfrac{2x^2+11x+5}{x^2-3x}$

17. $\dfrac{x^2-8x+15}{x^2-9x+14} \cdot \dfrac{7-x}{x^2+4x-21}$

18. $\dfrac{x^2-16x+39}{6+x-x^2} \cdot \dfrac{4x+8}{x+1}$

19. $\dfrac{3x^2-7x+2}{1-9x^2} \cdot \dfrac{3x+1}{x-2}$

20. $\dfrac{16-x^2}{x^2+2x-8} \cdot \dfrac{4-x^2}{x^2-2x-8}$

21. $\dfrac{4x^2-1}{x^2-16} \div \dfrac{2x+1}{x^2-4x}$

22. $\dfrac{4x^2-13x+3}{16x^2-4x} \div \dfrac{x^2-6x+9}{8x^2}$

23. $\dfrac{x^2+x-6}{2x^2+6x} \div \dfrac{x^2-5x+6}{8x^2}$

24. $\dfrac{x^2-4}{x^2-5x+6} \div \dfrac{x^2+3x+2}{x^2-2x-3}$

25. $\dfrac{2x^2-5x-12}{x^2-10x+24} \div \dfrac{4x^2-9}{x^2-9x+18}$

26. $\dfrac{2x^2-7x+3}{x^2-3x+2} \div \dfrac{x^2-6x+9}{x^2-4x+3}$

27. $\dfrac{6x^2-x-2}{12x^2+5x-2} \div \dfrac{4x^2-1}{8x^2-6x+1}$

28. $\dfrac{8x^2+6x-9}{8x^2-26x+15} \div \dfrac{4x^2+12x+9}{16x^2+18x-9}$

29. $\dfrac{3x^2+11x+6}{4x^2+16x+7} \div \dfrac{9x^2+12x+4}{2x^2-x-28}$

30. $\dfrac{2x^2+5x-3}{5x^2+17x-12} \div \dfrac{2x^2+3x-2}{5x^2+7x-6}$

31. $\dfrac{x^2-16}{x^3-8x^2+16x} \cdot \dfrac{x^2}{x^2+4x} \cdot \dfrac{2x^2-2x}{x^2-2x+1}$

32. $\dfrac{10x^2+3x-1}{6x^2+x-2} \cdot \dfrac{2x^2-x}{2x^2-x-1} \cdot \dfrac{3x+2}{5x^2-x}$

33. $\dfrac{x^2-3x}{x^2+2x+1} \cdot \dfrac{x-3}{x^2-9} \cdot \dfrac{x^2-3x-18}{x-6}$

34. $\dfrac{x^2+2x-3}{2x^2+3x-2} \cdot \dfrac{3x+5}{2x-3} \div \dfrac{3x^2+2x-5}{2x^2+x-6}$

35. $\dfrac{2x^2+5x-3}{x^2+2x-3} \cdot \dfrac{x^2+3x-4}{4x^2-1} \div \dfrac{x^2+4x}{x^3+5x^2}$

36. $\dfrac{6x^2+7x-3}{2x^2+5x+3} \div \dfrac{x^2+5x+6}{4x^2+3x-1} \cdot \dfrac{x^2+2x-3}{4x^2+7x-2}$

37. $\dfrac{x^2+4x+3}{x^2+8x+7} \div \dfrac{x^2-7x-8}{35+12x+x^2} \cdot \dfrac{x^2+8x+15}{x^2-9x+8}$

38. $\dfrac{12x^2}{x^2-1} \cdot \dfrac{2x^2-x-3}{2x^2-5x+3} \div (6x+6)$

39. $\dfrac{2x^2-7x+3}{36x^2-1} \div \dfrac{6x^2+5x+1}{2x+1} \div \dfrac{2x^2-5x-3}{18x^2+3x-1}$

40. $\dfrac{12x^2-8x-15}{4x^2-8x+3} \cdot \dfrac{6x^2-13x+6}{4x^2+5x+1} \div \dfrac{18x^2+3x-10}{8x^2-2x-1}$

41. The area of a rectangle (in square feet) is represented by the polynomial function $A(x) = 4x^2 - 4x - 15$. If the length of the rectangle is $(2x + 3)$ feet, find a representation for the width.

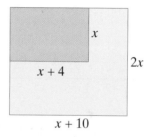

$$A(x) = 4x^2 - 4x - 15$$

$$2x + 3$$

42. A rectangle with dimensions x by $x + 4$ cm is cut from a rectangle with dimensions $2x$ by $x + 10$ centimeters. (a) Represent the remaining area as a polynomial function. (b) Find this area if $x = 8$ cm.

x

$x + 4$

$2x$

$x + 10$

Hawkes Learning Systems: Introductory Algebra

Multiplication and Division of Rational Expressions

8.3 Addition and Subtraction with Rational Expressions

After completing this section, you will be able to:

1. Add rational expressions.

2. Subtract rational expressions.

Adding Rational Expressions

To add rational expressions with a common denominator, proceed just as with fractions: add the numerators and keep the common denominator. For example,

$$\frac{5}{x+1} + \frac{6}{x+1} = \frac{5+6}{x+1} = \frac{11}{x+1}$$

Sometimes the sum can be reduced:

$$\frac{x^2}{x+1} + \frac{2x+1}{x+1} = \frac{x^2+2x+1}{x+1} = \frac{(x+1)^2}{x+1} = \frac{\cancel{(x+1)}(x+1)}{\cancel{(x+1)}} = x+1$$

$$\frac{2x^2-3}{x+3} + \frac{5x}{x+3} = \frac{2x^2+5x-3}{x+3} = \frac{(2x-1)(x+3)}{x+3} = \frac{(2x-1)\cancel{(x+3)}}{\cancel{(x+3)}} = 2x-1$$

Adding Rational Expressions

For polynomials P, Q, and R, with Q ≠ 0,

$$\frac{P}{Q} + \frac{R}{Q} = \frac{P+R}{Q}$$

Example 1: Adding Rational Expressions

Find the indicated sums and reduce if possible.

a. $\dfrac{x^2+5}{x+5}+\dfrac{6x}{x+5}$

Solution: $\dfrac{x^2+5}{x+5}+\dfrac{6x}{x+5}=\dfrac{x^2+5+6x}{x+5}=\dfrac{x^2+6x+5}{x+5}$

$$=\dfrac{\cancel{(x+5)}(x+1)}{\cancel{x+5}}=x+1$$

b. $\dfrac{3}{x^2+3x+2}+\dfrac{2x+1}{x^2+3x+2}$

Solution: $\dfrac{3}{x^2+3x+2}+\dfrac{2x+1}{x^2+3x+2}=\dfrac{3+2x+1}{x^2+3x+2}=\dfrac{2x+4}{(x+2)(x+1)}$

$$=\dfrac{2\cancel{(x+2)}}{\cancel{(x+2)}(x+1)}=\dfrac{2}{x+1}$$

The rational expressions added in Examples 1a and 1b had common denominators. To add expressions with different denominators, we need to find the least common multiple (LCM) of the denominators. The LCM was discussed in Section 2.2. We list the procedure here for polynomials and use this in finding the common denominator for rational expressions.

To Find the LCM for a Set of Polynomials

  1. Completely factor each polynomial (including prime factors for numerical factors).

  2. Form the product of all factors that appear, using each factor the most number of times it appears in any one polynomial.

Example 2: Using the LCM to Add Rational Expressions

Find the indicated sums by finding the LCM of the denominators and changing each expression to an equivalent expression with that LCM as the denominator.

a. $\dfrac{5}{x^2} + \dfrac{3}{x^2 - 4x}$

Solution: First find the LCM. Factoring, we have

$$\left.\begin{array}{l} x^2 = x^2 \\ x^2 - 4x = x(x-4) \end{array}\right\} \text{LCM} = x^2(x-4)$$

Thus $x^2(x-4)$ is the common denominator. Multiply the numerator and denominator of each fraction so that it has denominator $x^2(x-4)$.

$$\frac{5}{x^2} + \frac{3}{x^2 - 4x} = \frac{5(x-4)}{x^2(x-4)} + \frac{3 \cdot x}{x(x-4) \cdot x}$$

$$= \frac{5x - 20}{x^2(x-4)} + \frac{3x}{x^2(x-4)}$$

$$= \frac{5x - 20 + 3x}{x^2(x-4)} = \frac{8x - 20}{x^2(x-4)} \quad \left(\text{or } \frac{4(2x-5)}{x^2(x-4)}\right)$$

b. $\dfrac{5}{x^2 - 1} + \dfrac{4}{(x-1)^2}$

Solution:

$$\left.\begin{array}{l} x^2 - 1 = (x+1)(x-1) \\ (x-1)^2 = (x-1)^2 \end{array}\right\} \text{LCM} = (x+1)(x-1)^2$$

$$\frac{5}{x^2 - 1} + \frac{4}{(x-1)^2} = \frac{5(x-1)}{(x+1)(x-1)(x-1)} + \frac{4(x+1)}{(x-1)^2(x+1)}$$

$$= \frac{5x - 5 + 4x + 4}{(x-1)^2(x+1)} = \frac{9x - 1}{(x-1)^2(x+1)}$$

Subtracting Rational Expressions

To subtract one rational expression from another with the same denominator, simply subtract the numerators and use the common denominator. Since the numerators will be polynomials, **a good idea is to put both numerators in parentheses so that all changes in signs will be done correctly**.

Subtracting Rational Expressions

For polynomials P, Q, and R, with $Q \neq 0$

$$\frac{P}{Q} - \frac{R}{Q} = \frac{P-R}{Q}$$

Example 3: Subtraction

Find the indicated difference and reduce, if possible.

a. $\dfrac{3}{2x-4} - \dfrac{5}{2x-4}$

Solution: $\dfrac{3}{2x-4} - \dfrac{5}{2x-4} = \dfrac{3-5}{2x-4} = \dfrac{-2}{2x-4} = \dfrac{\overset{-1}{\cancel{-2}}}{\cancel{2}(x-2)} = \dfrac{-1}{x-2}$

b. $\dfrac{x+4}{x+3} - \dfrac{5x+1}{x+3}$

Solution: $\dfrac{x+4}{x+3} - \dfrac{5x+1}{x+3} = \dfrac{(x+4)-(5x+1)}{x+3} = \dfrac{x+4-5x-1}{x+3} = \dfrac{-4x+3}{x+3}$

NOTES

Common Error

Many beginning students make a mistake in subtracting fractions by not subtracting the entire numerator. They make a mistake similar to the following.

INCORRECT $\dfrac{7}{x+5} - \dfrac{2-x}{x+5} = \dfrac{7-2-x}{x+5}$

By using parentheses, you can avoid such mistakes.

CORRECT $\dfrac{7}{x+5} - \dfrac{2-x}{x+5} = \dfrac{7-(2-x)}{x+5} = \dfrac{7-2+x}{x+5} = \dfrac{5+x}{x+5} = \dfrac{x+5}{x+5} = 1$

As with addition, if the rational expressions do not have the same denominator, find the LCM of the denominators and change each fraction to an equivalent fraction with the LCM as denominator.

Example 4: Using the LCM to Subtract Rational Expressions

Subtract the following expression.

$$\dfrac{2x}{x^2-9} - \dfrac{1}{x^2+7x+12}$$

Solution:

$$\left. \begin{array}{l} x^2-9 = (x+3)(x-3) \\ x^2+7x+12 = (x+4)(x+3) \end{array} \right\} \text{LCM} = (x+3)(x-3)(x+4)$$

$$\dfrac{2x}{x^2-9} - \dfrac{1}{x^2+7x+12} = \dfrac{2x(x+4)}{(x+3)(x-3)(x+4)} + \dfrac{-1(x-3)}{(x+4)(x+3)(x-3)}$$

$$= \dfrac{2x^2+8x-x+3}{(x+3)(x-3)(x+4)} = \dfrac{2x^2+7x+3}{(x+3)(x-3)(x+4)}$$

$$= \dfrac{(2x+1)\,(x+3)}{(x+3)\,(x-3)(x+4)} = \dfrac{2x+1}{(x-3)(x+4)}$$

Whenever the denominators are opposites of each other or contain factors that are opposites of each other, it is important to use −1 as a factor or to multiply both the numerator and denominator by −1. This will simplify the work considerably.

643

Example 5: Simplifying Using −1 as a Factor

Simplify the following expression.

$$\frac{10}{x-2} + \frac{8}{2-x}$$

Solution: Note that $x - 2$ and $2 - x$ are opposites and $x - 2 = -1(2 - x)$.
So, if we multiply,

$$\frac{8}{2-x} \cdot \frac{(-1)}{(-1)} = \frac{-8}{x-2}$$

the fractions will have the same denominator and we do not have to find an LCM.

$$\frac{10}{x-2} + \frac{8}{2-x} = \frac{10}{x-2} + \frac{8}{2-x} \cdot \frac{(-1)}{(-1)} = \frac{10}{x-2} + \frac{-8}{x-2} = \frac{10-8}{x-2} = \frac{2}{x-2}$$

Do not try to reduce this last fraction, since 2 is **not a factor** of the denominator.

Practice Problems

Perform the indicated operations and reduce, if possible.

1. $\dfrac{5}{x-1} - \dfrac{4+x}{x-1}$

2. $\dfrac{5}{x+1} + \dfrac{10x}{x^2+4x+3}$

3. $\dfrac{1}{x^2+x} + \dfrac{4}{x^2} - \dfrac{2}{x^2-x}$

4. $\dfrac{x}{2x-1} - \dfrac{2}{1-2x}$

8.3 Exercises

Perform the indicated operations and reduce if possible in Exercises 1 – 56. Assume that no denominator has a value of 0.

1. $\dfrac{4x}{x+2} + \dfrac{8}{x+2}$

2. $\dfrac{x-1}{x+4} + \dfrac{x+9}{x+4}$

3. $\dfrac{3x-1}{2x-6} + \dfrac{x-11}{2x-6}$

4. $\dfrac{2x+5}{4(x+1)} + \dfrac{-x+3}{4x+4}$

5. $\dfrac{x^2}{x^2+2x+1} + \dfrac{-2x-3}{x^2+2x+1}$

Answers to Practice Problems: 1. -1 **2.** $\dfrac{15}{x+3}$ **3.** $\dfrac{3x^2-3x-4}{x^2(x+1)(x-1)}$ **4.** $\dfrac{x+2}{2x-1}$

6. $\dfrac{2x-1}{x^2-x-6}+\dfrac{1-x}{x^2-x-6}$

7. $\dfrac{-(3x+2)}{x^2-7x+6}+\dfrac{4x-4}{x^2-7x+6}$

8. $\dfrac{x+1}{x^2-2x-3}+\dfrac{-(3x-5)}{x^2-2x-3}$

9. $\dfrac{x^2+2}{x^2-4}+\dfrac{-(4x-2)}{x^2-4}$

10. $\dfrac{2x+5}{2x^2-x-1}+\dfrac{-(4x+2)}{2x^2-x-1}$

11. $\dfrac{3x+2}{4x-2}-\dfrac{x+2}{4x-2}$

12. $\dfrac{2x+1}{x^2-x-6}-\dfrac{x-1}{x^2-x-6}$

13. $\dfrac{2x^2-3x+1}{x^2-3x-4}-\dfrac{x^2+x+6}{x^2-3x-4}$

14. $\dfrac{x^2-x-2}{x^2-4}-\dfrac{x^2+x-2}{x^2-4}$

15. $\dfrac{3}{x-3}-\dfrac{2}{2x-6}$

16. $\dfrac{2x}{3x+6}+\dfrac{5}{2x+4}$

17. $\dfrac{8}{5x-10}-\dfrac{6}{3x-6}$

18. $\dfrac{7}{4x-20}-\dfrac{1}{3x-15}$

19. $\dfrac{2}{x}+\dfrac{1}{x+4}$

20. $\dfrac{8x+3}{x+1}+\dfrac{x-1}{x}$

21. $\dfrac{x}{x+4}+\dfrac{2}{x-4}$

22. $\dfrac{5}{x-2}+\dfrac{x}{x+3}$

23. $\dfrac{3}{2-x}+\dfrac{6}{x-2}$

24. $\dfrac{5}{2x-3}+\dfrac{2}{3-2x}$

25. $\dfrac{x}{5-x}+\dfrac{2x+3}{x-5}$

26. $\dfrac{x-1}{3x-1}+\dfrac{4}{x+2}$

27. $\dfrac{x}{x-1}-\dfrac{1}{x+2}$

28. $\dfrac{x+2}{x+3}+\dfrac{4}{3-x}$

29. $\dfrac{x+1}{x+4}+\dfrac{2x}{4-x}$

30. $\dfrac{8}{x}-\dfrac{x+1}{x-6}$

31. $\dfrac{8}{x-2}-\dfrac{4}{x+2}$

32. $\dfrac{8}{2x+2}+\dfrac{3}{3x+3}$

33. $\dfrac{x}{4x-8}-\dfrac{3x+1}{3x-6}$

34. $\dfrac{7x-1}{x^2-25}+\dfrac{4}{x+5}$

35. $\dfrac{2x+3}{x^2+4x-5}-\dfrac{4}{x-1}$

36. $\dfrac{2x-3}{x+1}-\dfrac{x+1}{2x-2}$

37. $\dfrac{3x}{6+x}-\dfrac{2x}{x^2-36}$

38. $\dfrac{x}{7+x} - \dfrac{7-13x}{x^2-49}$

39. $\dfrac{2x+1}{x-7} + \dfrac{3x}{x^2-8x+7}$

40. $\dfrac{x-4}{x-2} - \dfrac{x-7}{2x-10}$

41. $\dfrac{x+1}{x+2} - \dfrac{x+2}{2x+6}$

42. $\dfrac{3x-4}{x^2-x-20} - \dfrac{2}{x-5}$

43. $\dfrac{x+2}{x^2-16} - \dfrac{x+1}{2x-8}$

44. $\dfrac{7}{x-9} - \dfrac{x-1}{3x+6}$

45. $\dfrac{1}{x^2+x-2} - \dfrac{1}{x^2-1}$

46. $\dfrac{4}{x^2+x-6} + \dfrac{4}{x^2+5x+6}$

47. $\dfrac{2x}{x^2+x-12} + \dfrac{3x}{x^2-9}$

48. $\dfrac{x}{x^2+4x-21} + \dfrac{1-x}{x^2+8x+7}$

49. $\dfrac{x-3}{x^2+4x+4} - \dfrac{3x}{x^2+3x+2}$

50. $\dfrac{x-1}{2x^2+3x-2} + \dfrac{x}{2x^2-3x+1}$

51. $\dfrac{x}{x^2-16} - \dfrac{3x}{x^2+5x+4}$

52. $\dfrac{2x}{x^2-2x-15} - \dfrac{5}{x^2-6x+5}$

53. $\dfrac{x-2}{2x^2+5x-3} + \dfrac{2x-5}{2x^2-9x+4}$

54. $\dfrac{6}{x^2-4x-12} + \dfrac{x+1}{x^2-3x-18}$

55. $\dfrac{x}{x^2+3x-10} + \dfrac{3x}{x^2-4}$

56. $\dfrac{2x+3}{x^2-6x-7} + \dfrac{x-1}{x^2-5x-14}$

Hawkes Learning Systems: Introductory Algebra

Addition and Subtraction of Rational Expressions

8.4 Complex Algebraic Fractions

Objectives

After completing this section, you will be able to:

Simplify complex fractions and complex algebraic expressions.

Simplifying Complex Algebraic Fractions (First Method)

A fraction that contains various combinations of addition, subtraction, multiplication, and division with rational expressions is called a **complex algebraic fraction**. Examples of complex algebraic fractions are

$$\frac{\dfrac{1}{x}+\dfrac{1}{y}}{x+y} \quad \text{and} \quad \frac{\dfrac{1}{x+2}-\dfrac{1}{x}}{1+\dfrac{2}{x}}$$

An expression such as

$$\frac{4-x}{x+3}+\frac{x}{x+3}\div\frac{x}{x-3}$$

that involves rational expressions with more than one operation and is not written in fraction form is called a **complex algebraic expression**. The objective here is to develop techniques for simplifying complex algebraic fractions and expressions so that they are written in the form of a **single reduced rational expression**.

In a complex fraction such as $\dfrac{\dfrac{1}{x}+\dfrac{1}{y}}{x+y}$, the large fraction bar is a symbol of inclusion. The expression could be written as follows:

$$\frac{\dfrac{1}{x}+\dfrac{1}{y}}{x+y}=\left(\frac{1}{x}+\frac{1}{y}\right)\div(x+y).$$

Similarly,

$$\frac{\dfrac{1}{x+2}-\dfrac{1}{x}}{1+\dfrac{2}{x}}=\left(\frac{1}{x+2}-\frac{1}{x}\right)\div\left(1+\frac{2}{x}\right)$$

Thus a complex fraction indicates that the numerator is to be divided by the denominator. So a complex fraction can be simplified as follows.

Simplifying Complex Fractions

To simplify complex fractions:

1. *Simplify the numerator so that it is a single rational expression.*

2. *Simplify the denominator so that it is a single rational expression.*

3. *Divide the numerator by the denominator and reduce to lowest terms.*

This method is used to simplify the following two complex fractions. Study the examples closely so that you understand what happens at each step.

Example 1: First Method for Simplifying Complex Fractions

Simplify the following expression.

a. $\dfrac{\dfrac{1}{x}+\dfrac{1}{y}}{x+y}$

Solution: $\dfrac{\dfrac{1}{x}+\dfrac{1}{y}}{x+y} = \dfrac{\dfrac{1\cdot y}{x\cdot y}+\dfrac{1\cdot x}{y\cdot x}}{x+y} = \dfrac{\dfrac{y+x}{xy}}{\dfrac{x+y}{1}} = \dfrac{\cancel{y+x}}{xy}\cdot\dfrac{1}{\cancel{x+y}} = \dfrac{1}{xy}$

b. $\dfrac{\dfrac{1}{x+2}-\dfrac{1}{x}}{1+\dfrac{2}{x}}$

Solution: $\dfrac{\dfrac{1}{x+2}-\dfrac{1}{x}}{1+\dfrac{2}{x}} = \dfrac{\dfrac{1\cdot x}{(x+2)x}-\dfrac{1(x+2)}{x(x+2)}}{1\cdot\dfrac{x}{x}+\dfrac{2}{x}}$

$= \dfrac{\dfrac{x-(x+2)}{x(x+2)}}{\dfrac{x+2}{x}} = \dfrac{\dfrac{x-x-2}{x(x+2)}}{\dfrac{x+2}{x}}$

$= \dfrac{-2}{\cancel{x}(x+2)}\cdot\dfrac{\cancel{x}}{x+2} = \dfrac{-2}{(x+2)^2}$

Simplifying Complex Algebraic Fractions (Second Method)

A second method is to find the LCM of the denominators in the fractions in both the original numerator and the original denominator, and then multiply both the numerator and the denominator by this LCM.

Example 2: Second Method for Simplifying Complex Fractions

Simplify the following expression.

a. $\dfrac{\dfrac{1}{x}+\dfrac{1}{y}}{x+y}$

Solution: $\left.\begin{array}{c} x \\ y \\ 1 \end{array}\right\}$ LCM $= xy$

$$\frac{\dfrac{1}{x}+\dfrac{1}{y}}{x+y}=\frac{\left(\dfrac{1}{x}+\dfrac{1}{y}\right)xy}{\left(\dfrac{x+y}{1}\right)xy}=\frac{\dfrac{1}{x}\cdot xy+\dfrac{1}{y}\cdot xy}{(x+y)xy}=\frac{y+x}{(\cancel{x+y})xy}=\frac{1}{xy}$$

b. $\dfrac{\dfrac{1}{x+2}-\dfrac{1}{x}}{1+\dfrac{2}{x}}$

Solution: $\left.\begin{array}{c} x \\ x+2 \end{array}\right\}$ LCM $= x(x+2)$

$$\frac{\dfrac{1}{x+2}-\dfrac{1}{x}}{1+\dfrac{2}{x}}=\frac{\left(\dfrac{1}{x+2}-\dfrac{1}{x}\right)\cdot x(x+2)}{\left(1+\dfrac{2}{x}\right)\cdot x(x+2)}=\frac{\dfrac{1}{\cancel{x+2}}\cdot x(\cancel{x+2})-\dfrac{1}{\cancel{x}}\cdot \cancel{x}(x+2)}{1\cdot x(x+2)+\dfrac{2}{\cancel{x}}\cdot \cancel{x}(x+2)}$$

$$=\frac{x-(x+2)}{x(x+2)+2(x+2)}=\frac{x-x-2}{(x+2)(x+2)}=\frac{-2}{(x+2)^2}$$

Each of the techniques just described is valid. Sometimes one is easier to use than the other, but the choice is up to you.

Example 3: Simplifying Complex Fractions

Simplify the following complex algebraic fraction: $\dfrac{\dfrac{1}{x+y}-\dfrac{1}{x-y}}{\dfrac{2y}{x^2-y^2}}$.

Solution:

$$\frac{\dfrac{1}{x+y}-\dfrac{1}{x-y}}{\dfrac{2y}{x^2-y^2}}=\frac{\dfrac{1(x-y)}{(x+y)(x-y)}-\dfrac{1(x+y)}{(x-y)(x+y)}}{\dfrac{2y}{x^2-y^2}}=\frac{\dfrac{(x-y)-(x+y)}{(x-y)(x+y)}}{\dfrac{2y}{x^2-y^2}}$$

$$=\frac{x-y-x-y}{(x-y)(x+y)}\cdot\frac{x^2-y^2}{2y}=\frac{\overset{-1}{\cancel{-2y}}}{\cancel{(x-y)}\,\cancel{(x+y)}}\cdot\frac{\cancel{(x-y)}\,\cancel{(x+y)}}{\cancel{2y}}$$

$$=-1$$

Or use the technique of multiplying the numerator and the denominator by the LCM of the denominators of the various fractions.

$$\left.\begin{array}{l} x+y \\[4pt] x-y \\[4pt] x^2-y^2=(x+y)(x-y) \end{array}\right\}\ \text{LCM}=(x+y)(x-y)$$

$$\frac{\dfrac{1}{x+y}-\dfrac{1}{x-y}}{\dfrac{2y}{x^2-y^2}}=\frac{\left(\dfrac{1}{x+y}-\dfrac{1}{x-y}\right)(x+y)(x-y)}{\left(\dfrac{2y}{x^2-y^2}\right)(x+y)(x-y)}$$

$$=\frac{\dfrac{1}{\cancel{x+y}}\,\cancel{(x+y)}\,(x-y)-\dfrac{1}{\cancel{x-y}}\,(x+y)\,\cancel{(x-y)}}{\dfrac{2y}{\cancel{(x+y)}\,\cancel{(x-y)}}\,\cancel{(x+y)}\,\cancel{(x-y)}}$$

$$=\frac{(x-y)-(x+y)}{2y}=\frac{x-y-x-y}{2y}=\frac{-2y}{2y}=-1$$

Simplifying Complex Algebraic Expressions

A complex algebraic expression (as stated earlier) is an expression that involves rational expressions and more than one operation. In simplifying such expressions, the rules for order of operations apply. The objective is to simplify the expression so that it is written in the form of **a single reduced rational expression**.

Example 4: Simplifying Complex Algebraic Expressions

Simplify the following expression.

$$\frac{4-x}{x+3} + \frac{x}{x+3} \div \frac{x}{x-3}$$

Solution: In a complex algebraic expression such as

$$\frac{4-x}{x+3} + \frac{x}{x+3} \div \frac{x}{x-3}$$

the Rules for Order of Operations indicate that the division is to be done first.

$$\frac{4-x}{x+3} + \frac{x}{x+3} \div \frac{x}{x-3} = \frac{4-x}{x+3} + \frac{\cancel{x}}{x+3} \cdot \frac{x-3}{\cancel{x}} = \frac{4-x}{x+3} + \frac{x-3}{x+3}$$

$$= \frac{4-x+x-3}{x+3} = \frac{1}{x+3}$$

Practice Problems

Simplify the following expressions.

1. $\dfrac{\dfrac{1}{x}}{1+\dfrac{1}{x}}$

2. $\dfrac{1+\dfrac{3}{x-3}}{x-\dfrac{x^2}{x-3}}$

3. $\dfrac{5}{x} - \dfrac{3}{x-2} \div \dfrac{x}{x-2}$

Answers to Practice Problems: 1. $\dfrac{1}{x+1}$ **2.** $-\dfrac{1}{3}$ **3.** $\dfrac{2}{x}$

8.4 Exercises

Simplify the complex algebraic fractions in Exercises 1 – 32 so that each is a rational expression in reduced form.

1. $\dfrac{\dfrac{4}{5}}{\dfrac{7}{10}}$

2. $\dfrac{\dfrac{5}{6}}{\dfrac{2}{3}}$

3. $\dfrac{\dfrac{1}{2}+\dfrac{1}{3}}{\dfrac{5}{6}-\dfrac{1}{4}}$

4. $\dfrac{\dfrac{3}{4}-\dfrac{7}{8}}{\dfrac{1}{3}-\dfrac{1}{6}}$

5. $\dfrac{1+\dfrac{1}{3}}{2+\dfrac{2}{3}}$

6. $\dfrac{2+\dfrac{5}{7}}{3-\dfrac{2}{7}}$

7. $\dfrac{\dfrac{3x}{5}}{\dfrac{x}{5}}$

8. $\dfrac{\dfrac{x}{2y}}{\dfrac{3x}{y}}$

9. $\dfrac{\dfrac{4}{xy^2}}{\dfrac{6}{x^2y}}$

10. $\dfrac{\dfrac{x}{2y^2}}{\dfrac{5x^2}{6y}}$

11. $\dfrac{\dfrac{8x^2}{5y}}{\dfrac{x}{10y^2}}$

12. $\dfrac{\dfrac{15y^2}{2x^3}}{8y}$

13. $\dfrac{\dfrac{12x^3}{7y^4}}{\dfrac{3x^5}{2y}}$

14. $\dfrac{\dfrac{9x^2}{5y^3}}{\dfrac{3xy}{10}}$

15. $\dfrac{\dfrac{3x}{x+1}}{\dfrac{x-2}{x+1}}$

16. $\dfrac{\dfrac{x-3}{x+4}}{\dfrac{x-2}{x+4}}$

17. $\dfrac{\dfrac{x+3}{2x}}{\dfrac{2x-1}{4x^2}}$

18. $\dfrac{\dfrac{x-2}{6x}}{\dfrac{x+3}{3x^2}}$

19. $\dfrac{\dfrac{x^2-9}{x}}{x-3}$

20. $\dfrac{\dfrac{x^2-3x+2}{x-4}}{x-2}$

21. $\dfrac{\dfrac{x+2}{x-2}}{\dfrac{4x^2+8x}{x^2-4}}$

22. $\dfrac{\dfrac{x-1}{x+3}}{\dfrac{x+2}{x^2+2x-3}}$

23. $\dfrac{\dfrac{1}{x}-\dfrac{1}{3x}}{\dfrac{x+6}{x^2}}$

24. $\dfrac{\dfrac{1}{3}+\dfrac{1}{x}}{\dfrac{1}{2}-\dfrac{1}{x}}$

25. $\dfrac{1+\dfrac{1}{x}}{1-\dfrac{1}{x^2}}$

26. $\dfrac{\dfrac{3}{x}-\dfrac{6}{x^2}}{\dfrac{1}{x}-\dfrac{2}{x^2}}$

27. $\dfrac{\dfrac{1}{x}-\dfrac{1}{y}}{\dfrac{y}{x^2}-\dfrac{1}{y}}$

28. $\dfrac{\dfrac{x}{y}-\dfrac{1}{3}}{\dfrac{6}{y}-\dfrac{2}{x}}$

29. $\dfrac{2-\dfrac{4}{x}}{\dfrac{x^2-4}{x^2+x}}$

30. $\dfrac{\dfrac{1}{x}}{1-\dfrac{1}{x-2}}$

31. $\dfrac{x-\dfrac{2}{x+1}}{x+\dfrac{x-3}{x+1}}$

32. $\dfrac{x-\dfrac{2x-3}{x-2}}{2x-\dfrac{x+3}{x-2}}$

Write each of the expressions as a single fraction reduced to lowest terms in Exercises 33 – 40.

33. $\dfrac{1}{x+1}-\dfrac{3}{2x}\cdot\dfrac{4x}{x+1}$

34. $\dfrac{4}{x}-\dfrac{2}{x^2-2x}\cdot\dfrac{x-2}{5}$

35. $\left(\dfrac{8}{x}-\dfrac{3}{4x}\right)\div\dfrac{4x+5}{x}$

36. $\left(\dfrac{2}{x}+\dfrac{5}{x-3}\right)\cdot\dfrac{2x-6}{x}$

37. $\dfrac{x}{x-1}-\dfrac{3}{x-1}\cdot\dfrac{x+2}{x}$

38. $\dfrac{x+3}{x+2}+\dfrac{x}{x+2}\div\dfrac{x^2}{x-3}$

39. $\dfrac{x-1}{x+4}+\dfrac{x-6}{x^2+3x-4}\div\dfrac{x-4}{x-1}$

40. $\dfrac{x}{x+3}-\dfrac{3}{x-5}\cdot\dfrac{x^2-3x-10}{x-2}$

Writing and Thinking About Mathematics

41. Some complex fractions involve the sum (or difference) of complex fractions. Beginning with the "farthest" denominator, simplify each of the following expressions.

a. $1+\dfrac{1}{1+\dfrac{1}{1+\dfrac{1}{1+1}}}$

b. $2-\dfrac{1}{2-\dfrac{1}{2-\dfrac{1}{2-1}}}$

c. $x+\dfrac{1}{x+\dfrac{1}{x+\dfrac{1}{x+1}}}$

Hawkes Learning Systems: Introductory Algebra

Complex Algebraic Fractions

Solving Proportions and Other Equations with Rational Expressions

Objectives

After completing this section, you will be able to:

1. Solve proportions.

2. Solve word problems using proportions.

3. Solve equations involving rational expressions.

Proportions

A **ratio** is a comparison of two numbers by division. Ratios are written in the form

$$a:b \quad \text{or} \quad \frac{a}{b} \quad \text{or} \quad a \text{ to } b$$

For example, suppose the ratio of female to male faculty at the local high school is 3 to 2. We can also write this ratio in the form $3:2$ or in fraction form $\frac{3}{2}$. This ratio does not mean that there are only 5 teachers in the high school. There are many ratios (fractions) that reduce to $\frac{3}{2}$. If there are 35 teachers in the high school, then there are 21 women and 14 men teachers because $21 + 14 = 35$ and $\frac{21}{14} = \frac{3}{2}$.

Proportion

*A **proportion** is an equation stating that two ratios are equal.*

Proportions can involve only numbers as in $\frac{3}{4} = \frac{6}{8}$. However, proportions can also be used to find unknown quantities in word problems, and in such cases, will involve variables.

One method of solving proportions with variables is to "clear" the equation of fractions by first multiplying both sides of the equation by the LCM of the denominators. This method is illustrated in Example 1.

Example 1: Proportions

Solve the following proportions.

a. $\dfrac{4}{7x} = \dfrac{2}{35}$

Solution: $\dfrac{4}{7x} = \dfrac{2}{35}$ **Note :** $x \neq 0$.

$$35x \cdot \frac{4}{7x} = \frac{2}{35} \cdot 35x$$ Multiply both sides by the LCM $35x$.

$$5 \cdot 4 = 2 \cdot x$$

$$20 = 2x$$

$$10 = x$$

b. $\dfrac{4}{x-5} = \dfrac{2}{x+3}$

Solution: $\dfrac{4}{x-5} = \dfrac{2}{x+3}$ **Note :** $x \neq 5, -3$

$$(x-5)(x+3)\frac{4}{x-5} = \frac{2}{x+3}(x-5)(x+3)$$ Multiply both sides by

$$4(x+3) = 2(x-5)$$ the LCM $(x-5)(x+3)$.

$$4x + 12 = 2x - 10$$

$$4x + 12 - 2x - 12 = 2x - 10 - 2x - 12$$

$$2x = -22$$

$$x = -11$$

Proportions can be used to solve many everyday types of word problems. We must be very careful with the units in the setup of a proportion. One of the following conditions must be true.

1. **The numerators agree in type and the denominators agree in type**.
2. **The numerators correspond and the denominators correspond**.

Both arrangements are illustrated in the following example. Consider the situation where tires are on sale at 2 for $75, but you need 5 new tires. What would you pay for the 5 tires? Setting up a proportion gives one of the following types of equations:

Method 1 **Agree in Type**	**Method 2** **Correspond**

$$\frac{2 \text{ tires}}{\$75} = \frac{5 \text{ tires}}{\$x} \qquad \frac{\$75}{\$x} = \frac{2 \text{ tires}}{5 \text{ tires}}$$

$$75x \cdot \frac{2}{75} = 75x \cdot \frac{5}{x} \qquad 5x \cdot \frac{75}{x} = 5x \cdot \frac{2}{5}$$

$$2x = 375 \qquad\qquad 375 = 2x$$

$$x = 187.50 \qquad\qquad 187.50 = x$$

TIRE SALE
2 for $75

The cost of 5 tires would be $187.50.

Be careful to follow one of the patterns illustrated. Other arrangements will give the wrong answer. For example,

$$\frac{75 \text{ dollars}}{x \text{ dollars}} = \frac{5 \text{ tires}}{2 \text{ tires}} \text{ is \textbf{INCORRECT}.}$$

Looking at the numerators we see that $75 does not correspond to 5 tires.

Example 2: Proportion Word Problems

a. On an architect's scale drawing of a building, $\frac{1}{2}$ inch represents 12 feet. What does 3 inches represent?

Solution: Set up a proportion representing the information.

$$\frac{\frac{1}{2} \text{ inch}}{12 \text{ feet}} = \frac{3 \text{ inches}}{x \text{ feet}}$$

$$12x \cdot \frac{\frac{1}{2}}{12} = 12x \cdot \frac{3}{x} \qquad \text{Multiply both sides by the LCM } 12x.$$

$$\frac{1}{2}x = 36$$

$$2 \cdot \frac{1}{2}x = 2 \cdot 36$$

$$x = 72$$

Three inches represents 72 feet.

b. Making a statistical analysis, Mike finds 3 defective computer disks in a sample of 20 disks. If this ratio is consistent, how many bad disks does he expect to find in an order of 2400?

Solution: $\dfrac{3 \text{ defective disks}}{20 \text{ disks}} = \dfrac{x \text{ defective disks}}{2400 \text{ disks}}$ Set up a proportion representing the given information.

$$2400 \cdot \frac{3}{20} = 2400 \cdot \frac{x}{2400}$$

$$360 = x$$

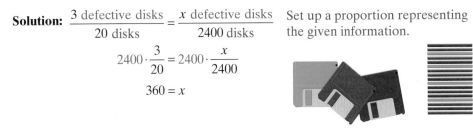

He expects to find 360 defective disks. (**Note:** He will probably return the order to the manufacturer.)

Equations Involving Rational Expressions

An equation such as

$$\frac{x+1}{3} + \frac{x}{2} = 1$$

which involves the sum of rational expressions is not a proportion. However, the method of finding the solution to the equation is similar in that we "clear" the fractions by multiplying both sides of the equation by the LCM of the denominators. In this example the LCM of the denominators is 6 and we can proceed as follows:

$$\frac{x+1}{3} + \frac{x}{2} = 1$$

$$6\left(\frac{x+1}{3} + \frac{x}{2}\right) = 1 \cdot 6 \quad \text{Multiply both sides by 6.}$$

$$\overset{2}{\cancel{6}}\left(\frac{x+1}{\cancel{3}}\right) + \overset{3}{\cancel{6}}\left(\frac{x}{\cancel{2}}\right) = 1 \cdot 6 \quad \text{Multiply all terms on both sides by 6.}$$

$$2x + 2 + 3x = 6 \qquad \text{This equation has only integer coefficients and constants.}$$

$$5x = 4$$

$$x = \frac{4}{5}$$

As we have seen, rational expressions may contain variables in either the numerator or denominator or both. In any case, a general approach to solving equations that contain rational expressions is as follows.

To Solve an Equation Containing Rational Expressions

1. Find the LCM of the denominators.

2. Multiply both sides of the equation by this LCM and simplify.

3. Solve the resulting equation. (This equation will have only polynomials on both sides.)

4. Check to see that no solution makes a denominator 0.

Note: Such solutions, if there are any, are called **extraneous solutions**.

Checking is particularly important when equations have rational expressions, because a solution to the new equation found by multiplying by the LCM may not be a solution of the original equation. In effect, multiplying by the LCM might be multiplication by 0. In such a case, a solution of the equation containing only polynomials may not be a solution of the original equation containing rational expressions. Such solutions are called **extraneous solutions** and occur because multiplication by a variable expression increases the degree of the equation.

Example 3: Solving Equations Involving Rational Expressions

First state any restrictions on the variable, then solve each of the following equations.

a. $\dfrac{3}{x}+\dfrac{1}{2}=\dfrac{7}{x}$

Solution:

$$\dfrac{3}{x}+\dfrac{1}{2}=\dfrac{7}{x} \qquad x \neq 0 \text{ and LCM} = 2x.$$

$$2x\left(\dfrac{3}{x}+\dfrac{1}{2}\right)=2x\left(\dfrac{7}{x}\right) \qquad \text{Multiply both sides by } 2x.$$

$$\overset{2}{2x}\left(\dfrac{3}{\cancel{x}}\right)+\overset{x}{2x}\left(\dfrac{1}{\cancel{2}}\right)=\overset{2}{2x}\left(\dfrac{7}{\cancel{x}}\right) \qquad \text{Multiply each fraction by } 2x.$$

$$6+x=14 \qquad \text{Simplify. This equation has only polynomials}$$

$$x=8 \qquad \text{on both sides.}$$

Since no denominator can be 0, $x \neq 0$. Since $8 \neq 0$, 8 is a possible solution. Checking, we have

$$\dfrac{3}{(8)}+\dfrac{1}{2}\overset{?}{=}\dfrac{7}{(8)}$$

$$\dfrac{3}{8}+\dfrac{4}{8}\overset{?}{=}\dfrac{7}{8}$$

$$\dfrac{7}{8}=\dfrac{7}{8}.$$

So, $x = 8$ is the solution.

b. $\dfrac{3}{x-2}+2=\dfrac{18}{x+1}$

Solution:

$$\dfrac{3}{x-2}+2=\dfrac{18}{x+1}$$

$x \neq 2,\ -1$ and
$\mathrm{LCM}=(x-2)(x-1).$

$$(x-2)(x+1)\left[\dfrac{3}{x-2}+2\right]=\dfrac{18}{x+1}(x-2)(x+1)$$

Multiply both sides by
$(x-2)(x-1).$

$$(x-2)(x+1)\left(\dfrac{3}{x-2}\right)+2(x-2)(x+1)=\left(\dfrac{18}{x+1}\right)(x-2)(x+1)$$

Multiply each term,
including 2, by
$(x-2)(x+1).$

$$3(x+1)+2(x^2-x-2)=18(x-2)$$

Both sides of the new
equation are
polynomials.

$$3x+3+2x^2-2x-4=18x-36$$

$$2x^2+x-1=18x-36$$

$$2x^2-17x+35=0$$

$$(2x-7)(x-5)=0$$

$$2x-7=0 \quad \text{or} \quad x-5=0$$

$$x=\dfrac{7}{2} \qquad x=5$$

There are two solutions: $\dfrac{7}{2}$ and 5. Both will check. (Note that neither of these is 2 or -1, the restricted values.)

c. $\dfrac{x}{x-2}+\dfrac{x-6}{x(x-2)}=\dfrac{5x}{x-2}-\dfrac{10}{x-2}$

Solution: $\dfrac{x}{x-2}+\dfrac{x-6}{x(x-2)}=\dfrac{5x}{x-2}-\dfrac{10}{x-2}$ $\mathrm{LCM}=x(x-2).$ Also, $x \neq 0,\ 2.$

$$\dfrac{x}{x-2}\cdot x(x-2)+\dfrac{x-6}{x(x-2)}\cdot x(x-2)=\dfrac{5x}{x-2}\cdot x(x-2)-\dfrac{10}{x-2}\cdot x(x-2)$$

Multiply each term by $x(x-2).$

Continued on next page...

659

$$x^2 + x - 6 = 5x^2 - 10x$$

$$0 = 4x^2 - 11x + 6$$

$$0 = (4x - 3)(x - 2)$$

$$4x - 3 = 0 \quad \text{or} \quad x - 2 = 0$$

$$4x = 3 \qquad\qquad x = 2$$

$$x = \frac{3}{4}$$

The only solution is $x = \dfrac{3}{4}$. Since $x \neq 2$, 2 is **not** a solution. No denominator can be 0.

d. $\dfrac{4}{x-4} - \dfrac{x}{x-4} = 1$

Solution: $\dfrac{4}{x-4} - \dfrac{x}{x-4} = 1$ $\qquad x \neq 4$ and LCM $= x - 4$.

$$(x-4)\left[\frac{4}{x-4} - \frac{x}{x-4}\right] = 1(x-4) \quad \text{Multiply both sides by } x - 4.$$

$$(x-4)\left(\frac{4}{x-4}\right) - (x-4)\left(\frac{x}{x-4}\right) = 1(x-4)$$

$$4 - x = x - 4$$

$$8 = 2x$$

$$4 = x$$

But $x \neq 4$ was the restriction because this value will give denominators of 0. Therefore, **this equation has no solution**.

Practice Problems

Solve the following equations.

1. $\dfrac{10}{x} = \dfrac{15}{48}$

2. $\dfrac{x-4}{7} = \dfrac{3}{5}$

3. $\dfrac{3}{5x+2} = \dfrac{2}{x-1}$

4. $\dfrac{3}{x} + \dfrac{2}{3} = \dfrac{23}{3x}$

5. $\dfrac{5}{8}x + 1 = x - \dfrac{1}{6}$

6. $\dfrac{x-5}{x} + \dfrac{7}{x} = x$

Answers to Practice Problems: 1. $x = 32$ **2.** $x = \dfrac{41}{5}$ **3.** $x = -1$ **4.** $x = 7$ **5.** $x = \dfrac{28}{9}$ **6.** $x = 2$ or $x = -1$

8.5 Exercises

State any restrictions on the variables and then solve the equations in Exercises 1 – 44.

1. $\dfrac{4}{5x} = \dfrac{2}{25}$

2. $\dfrac{2}{15} = \dfrac{8}{3x}$

3. $\dfrac{21}{x+4} = \dfrac{7}{8}$

4. $\dfrac{5}{x+3} = \dfrac{6}{11}$

5. $\dfrac{2x}{x-1} = \dfrac{3}{2}$

6. $\dfrac{x+7}{x-4} = \dfrac{5}{6}$

7. $\dfrac{10}{x} = \dfrac{5}{x-2}$

8. $\dfrac{8}{x-3} = \dfrac{12}{2x-3}$

9. $\dfrac{6}{x-4} = \dfrac{5}{x+7}$

10. $\dfrac{4}{x-6} = \dfrac{11}{5x+2}$

11. $\dfrac{4x}{3} - \dfrac{3}{4} = \dfrac{5x}{6}$

12. $\dfrac{5x}{8} - \dfrac{3}{4} = \dfrac{5}{16}$

13. $\dfrac{x-2}{3} - \dfrac{x-3}{5} = \dfrac{13}{15}$

14. $\dfrac{2+x}{4} - \dfrac{5x-2}{12} = \dfrac{8-2x}{5}$

15. $\dfrac{9}{3x} = \dfrac{1}{4} - \dfrac{1}{6x}$

16. $\dfrac{3}{8x} - \dfrac{7}{10} = \dfrac{1}{5x}$

17. $\dfrac{1}{x} - \dfrac{8}{21} = \dfrac{3}{7x}$

18. $\dfrac{3}{4x} = \dfrac{7}{8x} + \dfrac{2}{3}$

19. $\dfrac{5}{x-6} - 5 = \dfrac{2x}{x-6}$

20. $\dfrac{2x}{x-2} + 3 = \dfrac{9}{x-2}$

21. $3 + \dfrac{2x}{x-3} = \dfrac{4}{x-3}$

22. $\dfrac{x}{x+3} + \dfrac{1}{x+2} = 1$

23. $\dfrac{1}{3} + \dfrac{1}{x+4} = \dfrac{1}{x}$

24. $\dfrac{5}{x+3} - \dfrac{2}{x+1} = \dfrac{1}{4}$

25. $\dfrac{4}{x} - \dfrac{2}{x+1} = \dfrac{4}{3}$

26. $\dfrac{2}{x+2} + \dfrac{5}{x-2} = -3$

27. $\dfrac{6}{x+4} + \dfrac{2}{x-3} = -1$

28. $\dfrac{12}{x-4} = 6 + \dfrac{3x}{x-4}$

29. $\dfrac{6}{x-3} + 1 = \dfrac{2x}{x-3}$

30. $\dfrac{4}{x+2} + \dfrac{5}{x(x+2)} = \dfrac{3x+5}{x(x+2)}$

31. $\dfrac{x}{x+3} + \dfrac{1}{x+2} = 1$

32. $\dfrac{2x}{x-4} + \dfrac{2}{x+1} = 2$

33. $\dfrac{x}{x-4} - \dfrac{4}{2x-1} = 1$

34. $\dfrac{x}{x-2} - \dfrac{x+1}{x+4} = 1$

35. $\dfrac{2}{4x-1} + \dfrac{1}{x+1} = \dfrac{3}{x+1}$

36. $\dfrac{x-2}{x+4} - \dfrac{3}{2x+1} = \dfrac{x-7}{x+4}$

37. $\dfrac{x}{x-4} - \dfrac{12x}{x^2+x-20} = \dfrac{x-1}{x+5}$

38. $\dfrac{x-2}{x-3} + \dfrac{x-3}{x-2} = \dfrac{2x^2}{x^2-5x+6}$

39. $\dfrac{12}{x+1} + 1 = \dfrac{5}{x-1}$

40. $\dfrac{8}{x-4} - 3 = \dfrac{1}{x-10}$

41. $\dfrac{x}{x-3} + \dfrac{8}{x^2-2x-3} = \dfrac{-2}{x+1}$

42. $\dfrac{x}{x+4} + \dfrac{8x}{x^2-16} = \dfrac{1}{x-4}$

43. $\dfrac{x}{x+5} - \dfrac{5}{x^2+9x+20} = \dfrac{2}{x+4}$

44. $\dfrac{3}{x-1} - \dfrac{5}{x+4} = \dfrac{11}{14}$

45. The local sales tax is 6 cents for each dollar. What is the price of an item if the sales tax is 51 cents?

46. The property tax rate is $10.87 on every $100 of assessed valuation. Find the property tax on a house assessed at $38,000. (**Hint**: $38,000 = $380 \cdot 100$)

47. At the Bright-As-Day light bulb plant, 3 out of each 100 bulbs produced are defective. If the daily production is 4800 bulbs, how many are defective?

48. On a map, each inch represents $7\frac{1}{2}$ miles. What is the distance represented by 6 inches?

49. An elementary school has a ratio of 1 teacher for every 24 children. If the school presently has 21 teachers, how many students are enrolled?

50. A floor plan is drawn to scale in which 1 inch represents 4 feet. What size will the drawing be for a room that is 26 feet by 32 feet?

51. Pete is averaging 17 hits for every 50 times at bat. If he maintains this average, how many "at bats" will he need in order to get 204 hits?

52. Every 50 pounds of a certain alloy contains 6 pounds of copper. To have 27 pounds of copper, how many pounds of alloy will you need?

53. Instructions for Never-Ice Antifreeze state that 4 quarts of antifreeze are needed for every 10 quarts of radiator capacity. If Sal's car has a 22-quart radiator, how many quarts of antifreeze will it need?

54. The negative of a rectangular-shaped picture is $1\frac{1}{2}$ in. by 2 in. If the largest paper available is 11 in. long, what would be the maximum width of the enlargement?

55. A flagpole casts a 30-foot shadow at the same time as a 6-foot man casts a $4\frac{1}{2}$ foot shadow. How tall is the flagpole?

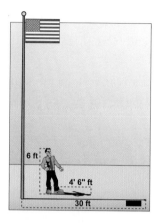

6 ft

4' 6" ft

30 ft

56. In a group of 30 people, there are 3 left-handed people. How many left-handed people would you expect to find in a group of 180 people?

Writing and Thinking About Mathematics

In simplifying rational expressions, the result is a rational expression. However, in solving equations that contain rational expressions, the goal is to find a value (or values) for the variable that will make the equation a true statement. Many students confuse these two ideas. To avoid confusing the techniques for adding and subtracting rational expressions with the techniques for solving equations, simplify the expression in part (a) and solve the equation in part (b). Explain, in your own words, the differences in your procedures.

57. a. $\dfrac{5}{x} + \dfrac{2}{x-1} + \dfrac{2x}{x-1}$ b. $\dfrac{5}{x} + \dfrac{2}{x-1} = \dfrac{2x}{x-1}$

58. a. $\dfrac{9}{x^2-9} + \dfrac{x}{3x+9} - \dfrac{1}{3}$ b. $\dfrac{9}{x^2-9} + \dfrac{x}{3x+9} = \dfrac{1}{3}$

Hawkes Learning Systems: Introductory Algebra

Solving Equations Involving Rational Expressions

| **8.6** | # Applications |

After completing this section, you will be able to:

1. Solve word problems using proportions.

2. Solve word problems related to distance, rate, and time.

3. Solve word problems related to the time it takes to complete a job.

In Section 8.5, we solved equations involving rational expressions (fractions) and some word problems using proportions. In this section, we continue to deal with rational expressions in solving a variety of applications involving numbers and proportions, time (using the relationship $d = rt$ in the form $t = \dfrac{d}{r}$), and work.

More on Proportions

Continuing our discussion of proportions from Section 8.5, we are interested in situations where a total amount is known and this total is to be separated into two parts according to some ratio. The key idea here is to recognize the relationship between the total and the two parts. In general,

$$
\begin{aligned}
&\text{If} &&T = \text{total amount}\\
&\text{and} &&x = \text{one part}\\
&\text{then } &&T - x = \text{second part.}
\end{aligned}
$$

For example, suppose that you have \$20 and you give \$5 to your brother and the remainder to your sister. What amount will you give to your sister? Answer: \$20 – \$5 = \$15.

Example 1: Solving Proportions

a. The sum of two integers is 104 and they are in the ratio of 8 to 5. Find the two numbers.

Solution: Here we know the total sum of the two numbers is 104. So let
$$n = \text{one number}$$
and $104 - n =$ second number.

Then set up a proportion using the ratio 8 to 5.
$$\frac{n}{104-n} = \frac{8}{5}$$

$$5(104-n) \cdot \frac{n}{104-n} = \frac{8}{5} \cdot 5(104-n)$$

$$5n = 8(104-n)$$

$$5n = 832 - 8n$$

$$\frac{13n}{13} = \frac{832}{13}$$

$$n = 64$$

$$104 - n = 40$$

The two numbers are 64 and 40.

(**Checking**, we get the sum $64 + 40 = 104$ and the ratio $\frac{64}{40} = \frac{8 \cdot 8}{8 \cdot 5} = \frac{8}{5}$.)

b. Suppose that an artist expects that for every 9 special brushes he orders, 7 will be good and 2 will be defective. If he orders 45 brushes, how many will he expect to be defective?

Solution: Let $n =$ number of defective brushes
and $45 - n =$ number of good brushes (Total $- n$).

Then, using the ratio of good to defective,
$$\frac{7 \text{ good}}{2 \text{ defective}} = \frac{45-n \text{ good}}{n \text{ defective}}$$

$$7n = 2(45-n)$$

$$7n = 90 - 2n$$

$$9n = 90$$

$$n = 10.$$

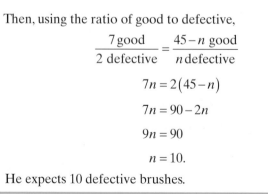

He expects 10 defective brushes.

Time $\left(d = rt \text{ or } t = \dfrac{d}{r} \right)$

Another application of rational expressions and proportions involves the formula $d = rt$ (distance equals rate times time). The same formula can be written in three forms:

$$d = rt, \quad r = \frac{d}{t}, \quad \text{and} \quad t = \frac{d}{r}.$$

The last formula, $t = \dfrac{d}{r}$, indicates that time can be represented as a ratio of distance divided by rate.

Example 2: Determining Rate Using Proportions

a. Janice can run 32 kilometers (about 20 miles) in the same amount of time that her twin sister, Jill, can run 24 kilometers. If Janice runs 2 kilometers per hour faster than Jill, how fast does each girl run?

Solution: Let r = Jill's rate
and $r + 2$ = Janice's rate (She runs faster than Jill.)

	Rate	Time $= \dfrac{\text{distance}}{\text{rate}}$	Distance
Jill	r	$\dfrac{24}{r}$	24
Janice	$r + 2$	$\dfrac{32}{r+2}$	32

Using the formula $t = \dfrac{d}{r}$ and the fact that **their times are equal**, we can write the equation:

$$\text{Jill's time} \} \rightarrow \frac{24}{r} = \frac{32}{r+2} \leftarrow \{ \text{Janice's time}$$

$$\cancel{r}\,(r+2) \cdot \frac{24}{\cancel{r}} = \frac{32}{\cancel{r+2}} \cdot r\,\cancel{(r+2)}$$

$$24(r+2) = 32r$$

$$24r + 48 = 32r$$

$$48 = 8r$$

$$6 = r$$

So, Jill's rate $= r = 6$ kph and Janice's rate $= r + 2 = 8$ kph.

Continued on next page...

b. A small plane can fly at 120 mph in still air. If it can fly 490 miles with a tailwind in the same time that it can fly 350 miles against a headwind, what is the speed of the wind? (**Note:** A tailwind is a wind that blows in the same direction that the plane is flying and, therefore increases the speed of the plane. A headwind is a wind that blows from the opposite direction the plane is flying and therefore decreases the speed of the plane.)

Solution: Let w = speed of the wind.

	Rate	Time = $\dfrac{\text{distance}}{\text{rate}}$	Distance
Tailwind	$120 + w$	$\dfrac{490}{120 + w}$	490
Headwind	$120 - w$	$\dfrac{350}{120 - w}$	350

Using the formula $t = \dfrac{d}{r}$ and the fact that the times are equal, the equation is

$$\text{time with the wind} \} \rightarrow \frac{490}{120 + w} = \frac{350}{120 - w} \leftarrow \{ \text{time against the wind}$$

$$(120 + w)(120 - w) \cdot \frac{490}{120 + w} = \frac{350}{120 - w} \cdot (120 + w)(120 - w)$$

$$490(120 - w) = 350(120 + w)$$

$$58{,}800 - 490w = 42{,}000 + 350w$$

$$16{,}800 = 840w$$

$$20 = w.$$

The speed of the wind is 20 mph.

Work

Problems involving "work" can be very sophisticated and require calculus and physics. The problems we will be concerned with are related to the time involved to complete the work on a particular job. These problems involve only the idea of what fraction of the job is done in one unit of time (hours, minutes, days, weeks, and so on). For example, if a man can dig a ditch in 4 hours, what part (of the ditch-digging job) can he do in one hour?

The answer is $\dfrac{1}{4}$. If the job took 5 hours, he would do $\dfrac{1}{5}$ in one hour. If the job took x hours, he would do $\dfrac{1}{x}$ in one hour.

Example 3: Calculating Work Using Proportions

a. Mike can clean his family's pool in 2 hours. His younger sister, Stacey, can do it in 3 hours. If they work together, how long will it take them to clean the pool?

Solution: Let x = number of hours working together. Then

	Hours to Clean Pool	Part Cleaned in 1 hour
Mike	2	$\dfrac{1}{2}$
Stacey	3	$\dfrac{1}{3}$
Together	x	$\dfrac{1}{x}$

$$\underbrace{\text{Part done in}}_{\text{1 hr by Mike}} + \underbrace{\text{Part done in}}_{\text{1hr by Stacey}} = \underbrace{\text{Part done in}}_{\text{1hr together}}$$

$$\frac{1}{2} \quad + \quad \frac{1}{3} \quad = \quad \frac{1}{x}$$

$$\frac{1}{2}(6x) + \frac{1}{3}(6x) = \frac{1}{x}(6x)$$

$$3x + 2x = 6$$

$$5x = 6$$

$$x = \frac{6}{5}$$

Multiply each term on both sides of the equation by $6x$, the LCM of the denominators.

Together they can clean the pool in $\dfrac{6}{5}$ hours, or 1 hour, 12 minutes.

(Note that this answer seems reasonable since it is less time than either person would take working alone.)

Continued on next page...

b. A man was told that his new Jacuzzi® pool would fill through an inlet valve in 3 hours. He knew something was wrong when the pool took 8 hours to fill. He found he had left the drain valve open. How long will it take to drain the pool once it is filled and only the drain valve is open?

Solution: Let t = time to drain pool. (**Note:** We use the information gained when the pool was filled with both valves open. In that situation, the inlet and outlet valves worked against each other.)

	Hours to Fill or Drain	Part Filled or Drained in 1 Hour
Inlet	3	$\dfrac{1}{3}$
Outlet	t	$\dfrac{1}{t}$
Together	8	$\dfrac{1}{8}$

$$\underbrace{\text{Part filled by inlet}} - \underbrace{\text{Part emptied by outlet}} = \underbrace{\text{Part filled together}}$$

$$\frac{1}{3} - \frac{1}{t} = \frac{1}{8}$$

$$\frac{1}{3}(24t) - \frac{1}{t}(24t) = \frac{1}{8}(24t)$$

$$8t - 24 = 3t$$

$$5t = 24$$

$$t = \frac{24}{5}$$

The pool will drain in $\dfrac{24}{5}$ hours, or 4 hours, 48 minutes. (Note that this is more time than the inlet valve would take to fill the pool. If the outlet valve worked faster than the inlet valve, then the pool would never have filled in the first place.)

8.6 Exercises

1. The ratio of two numbers is 7 to 9. If the sum of the numbers is 48, find the numbers.

2. The ratio of two numbers is 3 to 7. If the sum of the numbers is 60, find the numbers.

3. In a mathematics class, the ratio of women to men is 6 to 5. If there are 44 students in the class, how many men are there?

4. In a certain class, the ratio of success to failure is 7 to 2. How many can be expected to successfully complete the class if 36 people are enrolled?

5. The denominator of a fraction is 2 more than the numerator. If both numerator and denominator are increased by 3, the result is $\frac{5}{6}$. Find the original fraction.

6. In an iodine-alcohol mixture, there are 3 parts of iodine for each 8 parts of alcohol. If you want a total of 220 milliliters of mixture, how many milliliters of iodine and alcohol are needed?

7. An office has two copiers. One makes 8 copies per minute and the other makes 10 copies per minute. If both are used at the same time to make a total of 675 copies, how many copies are made by each machine? (**Hint:** Both machines are on for the same amount of time.)

8. The denominator of a fraction is one more than twice the numerator. If both numerator and denominator are increased by seven, the result is $\frac{3}{4}$. Find the original fraction.

9. Julie travels 15 miles per hour faster than Derek. She can travel 180 miles in the same amount of time that he can travel 135 miles. Find the two rates.

	Rate	Time	Distance
Derek	x	?	135
Julie	$x+15$	?	180

10. An airliner travels 1820 miles in the same amount of time it takes a private plane to travel 520 miles. If the airliner's average speed is 52 mph more than three times the average speed of the private plane, find both rates.

	Rate	Time	Distance
Private plane	x	?	520
Airliner	$3x + 52$	?	1820

11. An airplane can travel 320 mph in still air. If it travels 690 miles with the wind in the same length of time it travels 510 miles against the wind, what is the speed of the wind?

12. A boat can travel 5 miles downstream in the same length of time it travels 3 miles upstream. If the rate of the current is 3 mph, find the speed of the boat in still water.

13. After traveling 750 miles, the pilot of an airliner increased the speed by 20 mph and continued on for another 780 miles. If the same amount of time was spent at each rate, find the rates.

14. Mr. Snyder had a sales meeting scheduled 80 miles from his home. After traveling the first 30 miles at a rather slow rate due to traffic, he found that in order to make the meeting on time, he had to increase his speed by 25 mph. If he traveled the same length of time at each rate, find the rates.

15. An airliner leaves Phoenix traveling at an average rate of 420 mph. Thirty minutes later, a second airliner leaves Phoenix traveling along a parallel route. If the second plane overtakes the first at a point 1680 miles from Phoenix, find its rate. (**Hint:** The difference in the times is $\frac{1}{2}$ hr.)

	Rate	Time	Distance
First airliner	420	?	1680
Second airliner	x	?	1680

16. Mrs. Green leaves El Paso on a business trip to Los Angeles, traveling at 56 mph. Fifteen minutes later, Mr. Green discovers that she had forgotten her briefcase and, jumping into their second car, he begins to pursue her. What was his average speed if he caught her 84 miles outside of El Paso? (**Hint:** The difference in times is $\frac{1}{4}$ hr.)

	Rate	Time	Distance
Mrs. Green	56	?	84
Mr. Green	x	?	84

17. A boat traveled 9 miles downstream and returned. The total trip took $2\frac{1}{4}$ hours. If the speed of the current was 3 mph, find the speed of the boat in still water.

18. A boat travels 15 miles upstream and returns. If the boat travels 14 mph in still water, find the speed of the current if the total trip takes 2 hours, 20 minutes.

19. Miles rides the ski lift to the top of Snowy Peak, a distance of $1\frac{3}{4}$ miles. He then skis directly down the slope. If he skis five times as fast as the lift travels and the total trip takes 45 minutes, find the rate at which he skis.

20. Bonnie lives 45 miles from town. Recently she went to town and returned home. She traveled 6 mph faster returning home than she did going to town. If her total travel time was 1 hour, 35 minutes, find the two rates.

21. Tom can fix a car in 6 hours. Ellen can do it in 4 hours. How long would it take them working together?

22. One company can clear a parcel of land in 25 hours. Another company requires 30 hours to do the job. If both are hired, how long will it take them to clear the parcel working together?

23. One pipe can fill a tank in 15 minutes. The tank can be drained by another pipe in 45 minutes. If both pipes are opened, how long will it take to fill the tank?

24. One pipe can fill a tank three times as fast as another. When both pipes are used, it takes $1\frac{1}{2}$ hours to fill the tank. How long does it take each pipe alone?

25. Working together, Rick and Rod can clean the snow from the driveway in 20 minutes. It would have taken Rick, working alone, 36 minutes. How long would it have taken Rod alone?

26. A carpenter and his partner can put up a patio cover in $3\frac{3}{7}$ hours. If the partner needs 8 hours to complete the patio alone, how long will it take the carpenter working alone?

27. It takes Maria twice as long to complete a certain type of research project as it takes Judy to do the same project. They both worked on a similar type of project and they finished in 1 hour. How long would each of them take to finish this project working alone?

28. It takes Bob four hours longer to repair a car than it takes Ken. Working together, they can complete the job in $1\frac{1}{2}$ hours. How long would each of them take working alone?

29. A power plant has two coal-burning boilers. If both boilers are in operation, a load of coal is burned in 6 days. If only one boiler is used, a load of coal will last the new "fuel-efficient" boiler 5 days longer than the old boiler. How long will a load of coal last each boiler?

30. Working together, Alice and Sharon can prepare the company's sales report in $2\frac{2}{5}$ hours. Working alone, Sharon would need two hours longer to prepare the report than Alice would. How long would it take each of them working alone?

Hawkes Learning Systems: Introductory Algebra

Applications Involving Rational Expressions

8.7 Additional Applications: Variation

After completing this section, you will be able to:

1. Understand direct and inverse variation.

2. Solve applied problems.

Direct Variation

Suppose that you ride your bicycle at a steady rate of 10 miles per hour. If you ride for 1 hour, the distance you travel will be 10 miles. If you ride for 2 hours, the distance you travel will be 20 miles. This relationship can be written in the form of the formula $d = 10t$ (or $\dfrac{d}{t} = 10$). We say that distance and time **vary directly** (or are in **direct variation** or are **directly proportional**). The term directly proportional implies that the ratio is constant. In this example, 10 is the constant and is called the **constant of variation**.

When two variables vary directly:

1. an increase in the value of one variable indicates an increase in the other
2. the ratio of the two quantities is constant.

Direct Variation

*A variable quantity y **varies directly as** (or is **directly proportional to**) a variable x if there is a constant k such that*

$$\frac{y}{x} = k \quad or \quad y = kx$$

*The constant k is called the **constant of variation**.*

Example 1: Direct Variation

The length, *s*, that a hanging spring stretches varies directly as the weight placed at the end of the spring. If a weight of 5 mg stretches a certain spring 3 cm, how far will the spring stretch with a weight of 6 mg? (**Note:** We assume that the weight is not so great as to break the spring.)

Solution: Since the two variables are directly proportional, they are related as

$s = kw$ where *s* is the distance the spring stretches and *w* is the weight placed at the end of the spring.

First substitute the given information to find the value for *k*. (The value for *k* will depend on the particular spring. Springs made of different material or which are wound tighter will have different values for *k*.)

$$s = kw$$

$$3 = k \cdot 5$$

$$\frac{3}{5} = k$$

So, substitute $\frac{3}{5}$ for *k* in the equation. Thus, for this spring

$$s = \frac{3}{5} w \ (\text{ or } s = 0.6w).$$

Now, for *w* = 6mg,

$$s = 0.6(6) = 3.6.$$

The spring will stretch 3.6 cm if a weight of 6 mg is placed at its end.

Inverse Variation

When two variables vary in such a way that their product is constant, we say that the two variables **vary inversely** (or are **inversely proportional**). For example, if a gas is placed in a container (as in an automobile engine) and pressure is increased on the gas, then the volume of the gas will decrease. In fact, pressure and volume are related by the formula $V \cdot P = k \left(\text{or } V = \frac{k}{P} \right)$. Note that if a product of two variables is to remain constant, then an increase in the value of one variable must be accompanied by a decrease in the other. Or, in the case of a fraction with a constant numerator, if the denominator increases in value, then the fraction decreases in value.

Inverse Variation

*A variable quantity y **varies inversely as** (or is **inversely proportional to**) a variable x if there is a constant k such that*

$$xy = k \quad or \quad y = \frac{k}{x}$$

*The constant k is called the **constant of variation**.*

Example 2: Inverse Variation

The volume of propane in a container varies inversely to the pressure on the gas. If the propane has a volume of 200 in.^{3} under a pressure of 4 lbs per in.^{2}, what will be its volume if the pressure is increased to 5 lbs per in.^{2}?

Solution:

$$V = \frac{k}{P}$$

$$200 = \frac{k}{4} \qquad \text{Substitute 200 for } V \text{ and 4 for } P \text{ and solve for } k.$$

$$k = 800$$

Now, substitute 800 for k in the equation.

$$V = \frac{800}{P}$$

$$V = \frac{800}{5} = 160$$

The volume will be 160 in.^{3}.

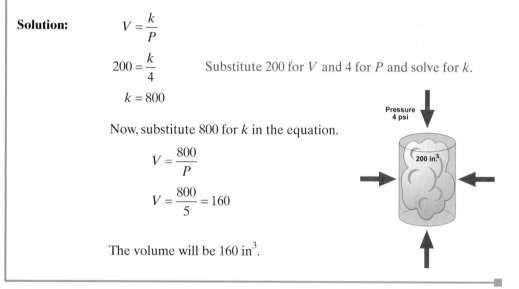

Pressure
4 psi

200 in.^{3}

8.7 Exercises

Write an equation or formula that represents the general relationship indicated in Exercises 1 – 15. Then use the given information to find the unknown value.

1. If y varies directly as x and $y = 2$ when $x = 10$, find y if $x = 7$.

2. If y is directly proportional to x and $y = 3$ when $x = 18$, find y if $x = 10$.

3. If y is directly proportional to x^2 and $y = 9$ when $x = 2$, what is y when $x = 4$?

4. If y varies inversely as x^2 and $y = -8$ when $x = 2$, find y when $x = 3$.

5. If y is inversely proportional to x and $y = 5$ when $x = 4$, find the value of y when $x = 2$.

6. If y varies inversely as x^3 and $y = 40$ when $x = \dfrac{1}{2}$, what is the value of y when $x = \dfrac{1}{3}$?

7. The total price, P, of gasoline purchased varies directly with the number of gallons purchased. If 20 gallons are purchased for $25, what will be the price of 30 gallons?

8. The total price, P, of pumpkin pies purchased is directly proportional to the number of pies purchased. If 3 pumpkin pies are bought for $19.50, what will 5 pumpkin pies cost?

9. The gravitational force, F, between an object and the earth is inversely proportional to the square of the distance from the object to the center of the earth. If an astronaut weighs 200 pounds on the surface of the earth, what will the astronaut weigh 100 miles above the earth? (Assume that the radius of the earth is 4000 miles and round to the nearest tenth.)

10. The distance a free falling object falls is directly proportional to the square of the time it falls (before it hits the ground). If an object fell 64 ft in 2 seconds, how far will it have fallen by the end of 3 seconds?

11. The volume of gas in a container varies inversely as the pressure on the gas. The volume of a particular gas in a container is 300 cm^3 when the pressure on the gas is 15 gm per cm^2. What will be the volume of the gas if the pressure is increased to 20 gm per cm^2?

12. A hanging spring will stretch 4 in if a weight of 8 lbs is placed at its end. How far will the spring stretch if the weight is increased to 10 lbs?

13. The circumference of a circle varies directly as the diameter. A circular clock with a diameter of 1 foot has a circumference of 3.14 feet. What will be the circumference of a clock with a diameter of 2 feet?

14. The area of a circle varies directly as the square of its radius. A pebble is dropped into a pool of water and the concentric circles created by the waves are observed. If the area of the circle with radius 2 ft is 12.56 ft^2, what is the area of the circle with a radius of 4 feet?

15. Several triangles are to have the same area. In this set of triangles, the height and base are inversely proportional. In one such triangle, the height is 6 in. when the base is 10 inches. Find the height of the triangle in this set with a base of 15 inches.

Lifting Force

The lifting force (or lift), P, in pounds exerted by the atmosphere on the wings of an airplane is related to the area, A, of the wings in square feet and the speed (or velocity), v, of the plane in miles per hour by the formula

$P = kAv^2$, where k is the constant of variation.

16. If the lift is 9600 lbs for a wing area of 120 ft^2 and a speed of 80 mph, find the lift of the same airplane at a speed of 100 mph.

17. If the lift is 12,000 lbs for a wing area of 110 ft^2 and a speed of 90 mph, what would be the necessary wing area to attain the same lift at 80 mph? (Round to the nearest thousandth.)

18. The lift for a wing of area 280 ft^2 is 34,300 lbs when the plane is going 210 mph. What is the lift if the speed is decreased to 180 mph?

Pressure

Boyle's Law states that if the temperature of a gas sample remains the same, the pressure of the gas is related to the volume by the formula

$P = \dfrac{k}{V}$, where k is the constant of proportionality.

19. The temperature of a gas remains the same. The pressure of the gas is 16 lbs per ft^2 when the volume is 300 ft^3. What will be the pressure if the gas is compressed into 25 ft^3?

20. A pressure of 1600 lbs per ft^2 is exerted by 2 ft^3 of air in a cylinder. If a piston is pushed into the cylinder until the pressure is 1800 lbs per ft^2, what will be the volume of the air? (Round to the nearest tenth.)

21. If 1000 ft^3 of gas exerting a pressure of 140 lbs per ft^2 must be placed in a container which has a capacity of 200 ft^3, what is the new pressure exerted by the gas?

Electricity

The resistance, R (in ohms), in a wire is given by the formula

$$R = \frac{kL}{d^2},$$

where k is the constant of variation, L is the length of the wire and d is the diameter.

22. The resistance of a wire 500 ft long with a diameter of 0.01 in. is 20 ohms. What is the resistance of a wire 1500 ft long with a diameter of 0.02 inches?

23. The resistance of a wire 100 ft long and 0.01 in. in diameter is 8 ohms. What is the resistance of a piece of the same type of wire with a length of 150 ft and a diameter of 0.015 inches?

24. The resistance is 2.6 ohms when the diameter of a wire is 0.02 in. and the wire is 10 ft long. Find the resistance of the same type of wire with a diameter of 0.01 in. and a length of 5 feet.

Gears

If a gear with T_1 teeth is meshed with a gear having T_2 teeth, the numbers of revolutions of the two gears are related by the proportion

$$\frac{T_1}{T_2} = \frac{R_2}{R_1} \quad \text{or} \quad T_1 R_1 = T_2 R_2,$$

where R_1 and R_2 are the numbers of revolutions.

25. A gear having 96 teeth is meshed with a gear having 27 teeth. Find the speed (in revolutions per minute) of the small gear if the large one is traveling at 108 revolutions per minute.

26. What size gear (how many teeth) is needed to turn a pulley on a feed grinder at 24 revolutions per minute if it is driven by a 15 tooth gear turning at 56 revolutions per minute?

27. What size gear (how many teeth) is needed to turn the reel on a harvester at 20 revolutions per minute if it is driven by a 12-tooth gear turning at 60 revolutions per minute?

Lever

If a lever is balanced with weight on opposite sides of its balance point, then the following proportion exists:

$$\frac{W_1}{W_2} = \frac{L_2}{L_1} \quad \text{or} \quad W_1L_1 = W_2L_2$$

where $L_1 + L_2 = L$, the total length of the lever

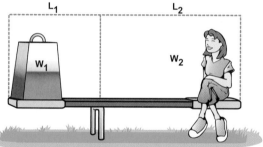

28. How much weight can be raised at one end of a bar 8 ft long by the downward force of 60 lbs when the balance point is $\frac{1}{2}$ ft from the unknown weight?

29. A force of 40 lbs at one end of a bar 5 ft long is to balance 160 lbs at the other end of the bar. Ignoring the weight of the bar, how far from the 40 lb weight should a hole be drilled in the bar for a bolt to serve as a balance point?

30. Where should the balance point of a bar 12 ft long be located if a 120 lb force is to raise a load weighing 960 lbs?

Hawkes Learning Systems: Introductory Algebra

Additional Applications: Variation

Chapter 8 Index of Key Ideas and Terms

Section 8.1   Reducing Rational Expressions

Rational Expression   page 622

A **rational expression** is an algebraic expression that can be written in the form $\dfrac{P}{Q}$ where P and Q are polynomials and $Q \neq 0$.

Fundamental Principle of Rational Expressions   page 625

If $\dfrac{P}{Q}$ is a rational expression and $Q \neq 0$ and K is a polynomial and $K \neq 0$, then $\dfrac{P}{Q} = \dfrac{P \cdot K}{Q \cdot K}$.

Opposites   page 628

In general, for $b \neq a, (a-b)$ and $(b-a)$ are **opposites**

and $\dfrac{a-b}{b-a} = \dfrac{a-b}{-1(a-b)} = -1$.

Section 8.2   Multiplication and Division with Rational Expressions

Multiplying Rational Expressions   page 633

If P, Q, R, and S are rational expressions with $Q, S \neq 0$,

then $\dfrac{P}{Q} \cdot \dfrac{R}{S} = \dfrac{P \cdot R}{Q \cdot S}$.

Dividing Rational Expressions   page 634

If P, Q, R, and S are rational expressions with $Q, R, S \neq 0$,

then $\dfrac{P}{Q} \div \dfrac{R}{S} = \dfrac{P}{Q} \cdot \dfrac{S}{R}$.

Section 8.3   Addition and Subtraction with Rational Expressions

Adding Rational Expressions   page 639

For polynomials P, Q, and R with $Q \neq 0$, $\dfrac{P}{Q} + \dfrac{R}{Q} = \dfrac{P+R}{Q}$.

Continued on next page...

Section 8.3 Addition and Subtraction with Rational Expressions (continued)

LCM for a Set of Polynomials   page 640

   To find the LCM for a set of polynomials,
   1. Completely factor each polynomial
      (including prime factors for numerical factors).
   2. Form the product of all factors that appear, using
      each factor the most number of times it appears
      in any one polynomial.

Subtracting Rational Expressions   page 642

   For polynomials P, Q, and R with $Q \neq 0$, $\dfrac{P}{Q} - \dfrac{R}{Q} = \dfrac{P - R}{Q}$.

Section 8.4 Complex Algebraic Fractions

Simplifying Complex Algebraic Fractions   pages 648 - 650
   First Method   page 648
      1. Simplify the numerator so that it is a single
         rational expression.
      2. Simplify the denominator so that it is a single
         rational expression.
      3. Divide the numerator by the denominator and
         reduce to lowest terms.
   Second Method   pages 649 - 650
      1. Find the LCM of the denominators in the fractions
         in both the numerator and the denominator.
      2. Multiply both the numerator and the denominator by this LCM.

Simplifying Complex Algebraic Expressions   page 651

**Section 8.5 Solving Proportions and Other Equations with Rational
              Expressions**

Proportions   pages 654 - 657

   A **proportion** is an equation stating that two ratios are equal.
   To solve word problems by using proportions one of the   page 655
   following conditions must be true:
      1. The numerators agree in type and the denominators agree in type.
      2. The numerators correspond and the denominators correspond.

Continued on next page...

Section 8.5 Solving Proportions and Other Equations with Rational Expressions (continued)

Solving Equations Involving Rational Expressions   pages 657 - 660
   1. Find the LCM of the denominators.
   2. Multiply both sides of the equation by this LCM and simplify.
   3. Solve the resulting equation.
   4. Check to see that no solution makes a denominator 0.
   Note: Such solutions, if there are any, are called **extraneous solutions**.

Section 8.6   Applications

Applications   pages 665 - 670
  Proportions   pages 665 - 666
  Distance-rate-time   pages 667 - 668
  Work   page 668

Section 8.7   Additional Applications: Variation

Direct Variation   pages 675 - 676
   A variable quantity y **varies directly as** (or is **directly proportional to**) a variable x if there is a constant k such that

$$\frac{y}{x} = k \ \text{ or } y = kx$$

   The constant k is called the **constant of variation**.

Inverse Variation   pages 676 - 677
   A variable quantity y **varies inversely as** (or is **inversely proportional to**) a variable x if there is a constant k such that

$$xy = k \text{ or } \ y = \frac{k}{x}$$

   The constant k is called the **constant of variation**.

Hawkes Learning Systems: Introductory Algebra

For a review of the topics and problems from Chapter 8, look at the following lessons from *Hawkes Learning Systems: Introductory Algebra*

Defining Rational Expressions
Reducing Rational Expressions
Multiplication and Division of Rational Expressions
Addition and Subtraction of Rational Expressions
Complex Algebraic Fractions
Solving Equations Involving Rational Expressions
Applications Involving Rational Expressions
Additional Applications: Variation

Chapter 8 Review

Section 8.1 Reducing Rational Expressions

In Exercises 1 – 10, state any restrictions on the variables.

1. $\dfrac{4}{9x}$ **2.** $\dfrac{23}{7y}$ **3.** $\dfrac{5x}{x-7}$ **4.** $\dfrac{3x}{x+4}$

5. $\dfrac{14}{2x+1}$ **6.** $\dfrac{-10}{3x-1}$ **7.** $\dfrac{12}{a^2+25}$ **8.** $\dfrac{3a}{a^2+u-2}$

9. $\dfrac{y^2+2y}{y^2+6y+5}$ **10.** $\dfrac{5a^2+5a}{a^2+2a+1}$

In Exercises 11 – 20, state any restriction on the variables and then reduce each rational expression to lowest terms.

11. $\dfrac{x+2}{x^2+4x+4}$ **12.** $\dfrac{y-3}{y^2-6y+9}$ **13.** $\dfrac{x^2-2x-3}{-x^2+2x+3}$ **14.** $\dfrac{3-y}{y-3}$

15. $\dfrac{4+x}{x^2-16}$ **16.** $\dfrac{-5+x}{2x^2-50}$ **17.** $\dfrac{4x-8}{x^2-2x}$ **18.** $\dfrac{2x+3}{4x^2+12x+9}$

19. $\dfrac{x^2-25}{x^2-10x+25}$ **20.** $\dfrac{6x^2+7x-5}{3x^2-4x-15}$

Section 8.2 Multiplication and Division with Rational Expressions

Perform the indicated operations in Exercises 21 – 30 and reduce to lowest terms. Assume that no denominator has a value of 0.

21. $\dfrac{x}{x-2}\cdot\dfrac{x^2-4}{x^2}$ **22.** $\dfrac{2y-6}{3y-9}\cdot\dfrac{2y}{2y-4}$ **23.** $\dfrac{5y^2}{y^2+3y}\cdot\dfrac{y^2-9}{5y-5}$

24. $\dfrac{x^2-8x+16}{x^2-16}\cdot\dfrac{2x+8}{4}$ **25.** $\dfrac{2x^2-x}{6x^2+12x}\div\dfrac{2x-1}{4x^2+8x}$ **26.** $\dfrac{x-7}{3}\div\dfrac{x^2-49}{6x+42}$

27. $\dfrac{x^2+2x-8}{4x}\div\dfrac{2x^2-5x+2}{5x^2}$ **28.** $\dfrac{y-3}{y^2-3y-4}\div\dfrac{y^2-y-2}{y+1}$

29. $\dfrac{2a^2-7a+3}{6a^2+5a+1}\div\dfrac{36a^2-1}{2a^2-5a-3}\div\dfrac{18a^2+3a-10}{2a+1}$

30. $\dfrac{16x^2}{x^2-4}\cdot\dfrac{4x^2-4x-3}{16x^2-4}\div\dfrac{6x^2+9x}{1-4x^2}$

31. The area of a rectangle (in square meters) is represented by the polynomial function $A(x) = 10x^2 + 21x + 9$. If the length of the rectangle is $(5x + 3)$ meters, find a representation for the width.

$$A(x) = 10x^2 + 21x + 9$$

$5x + 3$

32. A rectangle has dimensions x by $x + 5$ feet. Six feet are added to each side. (a) Write a function that represents that area of the original rectangle. (b) Write a function that represents the area of the new rectangle. (c) Write a function that represents the area by which the original rectangle was increased.

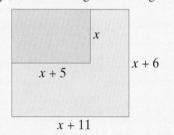

Section 8.3 Addition and Subtraction with Rational Expressions

Perform the indicated operations and reduce if possible in Exercises 33 – 48. Assume that no denominator has a value of 0.

33. $\dfrac{5x}{x+2} + \dfrac{10}{x+2}$

34. $\dfrac{3x}{x+4} - \dfrac{12}{x-4}$

35. $\dfrac{3x-2}{2x-6} + \dfrac{x-10}{2x-6}$

36. $\dfrac{2x+7}{5x+5} + \dfrac{-x-6}{5(x+1)}$

37. $\dfrac{y+1}{3y-1} + \dfrac{2y}{y+2}$

38. $\dfrac{a}{a+2} - \dfrac{1}{a-1}$

39. $\dfrac{6}{2x-3} - \dfrac{5}{3-2x}$

40. $\dfrac{4}{2-x} - \dfrac{3}{x-2}$

41. $\dfrac{x^2}{x^2+4x+4} - \dfrac{2x+8}{x^2+4x+4}$

42. $\dfrac{x+5}{x-5} - \dfrac{100}{x^2-25}$

43. $\dfrac{3x-12}{x^2+x-20} - \dfrac{x^2+5x}{x^2+9x+20}$ (**Hint:** Reduce first.)

44. $\dfrac{5x+20}{x^2+8x+16} + \dfrac{3x-12}{x^2-16}$ (**Hint:** Reduce first.)

45. $\dfrac{4x}{x^2-2x-15} - \dfrac{10}{x^2-6x+5}$

46. $\dfrac{4}{y^2-1} + \dfrac{4}{y+1} + \dfrac{2}{1-y}$

47. $\dfrac{x+3}{x^2+x-6} + \dfrac{x-2}{x^2+4x-12}$

48. $-4 + \dfrac{1-2x}{x+6} + \dfrac{x^2+1}{x^2+4x-12}$

Section 8.4 Complex Algebraic Fractions

Simplify the complex algebraic fractions in Exercises 49 – 58 so that each is a rational expression in reduced form.

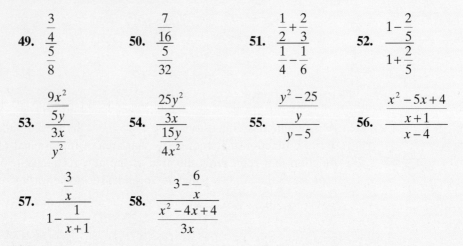

49. $\dfrac{\dfrac{3}{4}}{\dfrac{5}{8}}$ **50.** $\dfrac{\dfrac{7}{16}}{\dfrac{5}{32}}$ **51.** $\dfrac{\dfrac{1}{2}+\dfrac{2}{3}}{\dfrac{1}{4}-\dfrac{1}{6}}$ **52.** $\dfrac{1-\dfrac{2}{5}}{1+\dfrac{2}{5}}$

53. $\dfrac{\dfrac{9x^2}{5y}}{\dfrac{3x}{y^2}}$ **54.** $\dfrac{\dfrac{25y^2}{3x}}{\dfrac{15y}{4x^2}}$ **55.** $\dfrac{\dfrac{y^2-25}{y}}{y-5}$ **56.** $\dfrac{\dfrac{x^2-5x+4}{x+1}}{x-4}$

57. $\dfrac{\dfrac{3}{x}}{1-\dfrac{1}{x+1}}$ **58.** $\dfrac{3-\dfrac{6}{x}}{\dfrac{x^2-4x+4}{3x}}$

Write each of the expressions as a single fraction reduced to lowest terms in Exercises 59 – 62.

59. $\dfrac{1}{x+2}-\dfrac{5}{2x}\cdot\dfrac{4x}{x+2}$ **60.** $\dfrac{7}{x}+\dfrac{5}{x^2-3x}\cdot\dfrac{x-3}{10}$

61. $\left(\dfrac{4}{y}-\dfrac{2}{3y}\right)\div\dfrac{3y+2}{6y}$ **62.** $\left(\dfrac{4}{y}+\dfrac{2}{y+3}\right)\div\dfrac{y^2-9}{2y}$

Section 8.5 Solving Proportions and Other Equations with Rational Expressions

State any restrictions on the variables and then solve the equations in Exercises 63 – 72.

63. $\dfrac{8}{5x}=\dfrac{4}{25}$ **64.** $\dfrac{2}{15}=\dfrac{16}{3x}$

65. $\dfrac{4}{x+3}=\dfrac{1}{2}$ **66.** $\dfrac{20}{x+4}=\dfrac{4}{5}$

67. $\dfrac{3x}{x-2}+4=\dfrac{8}{x-2}$ **68.** $\dfrac{2y}{y-5}+3=\dfrac{4}{y-5}$

69. $\dfrac{3}{x+3}=1-\dfrac{5}{x+7}$ **70.** $\dfrac{5}{y-1}-\dfrac{3}{y+1}=\dfrac{28}{9(y+1)}$

71. $\dfrac{5}{y+4}+\dfrac{3}{y-1}=\dfrac{-11}{2(y+4)}$ **72.** $\dfrac{x}{x+2}=\dfrac{1}{x+1}-\dfrac{x}{x^2+3x+2}$

Section 8.6 Applications

73. The ratio of two numbers is 2 to 3. If the sum of the numbers is 45, find the numbers.

74. In a statistics class the ratio of men to women is 5 to 4. If there are 36 students in the class, how many women are in the class?

75. The numerator of a fraction is three more than twice the denominator. If both numerator and denominator are increased by five, the simplified result is $\dfrac{11}{6}$. Find the original fraction.

76. A boat can travel 10 miles downstream in the same time it travels 5 miles upstream. If the rate of the current is 5 mph, find the speed of the boat in still water.

77. An airplane leaves Burbank Airport flying at an average rate of 150 mph. Thirty minutes later the pilot's friend takes off from the same airport trying to catch him. If the second plane overtakes the first at a point 450 miles from Burbank Airport, find the average flying speed of his friend.

78. Adam rides his bike to the top of Mount High, a distance of 12 miles. He then rides back down the mountain four times as fast as he rode up the mountain. If the total ride took 2.5 hours, find the rate at which he rode up the mountain.

79. One pipe can fill a swimming pool twice as fast as another. When both pipes are used, the pool can be filled in 2 hours. How long would it take each pipe alone to fill the pool?

80. Working together, a bricklayer and his partner can lay a brick sidewalk in 4.8 hours. If he could have completed the job in 8 hours working alone, how long would his partner have needed to complete the job working alone?

81. A painter can paint the walls in a room in 2 hours. His helper can paint the same size room in 3 hours. How long would it take them working together to paint a room of this size?

82. The pilot of a corporate jet can fly the jet 900 miles in 2 hours less than she can fly her own plane a distance of 600 miles. If the jet averages a speed 3 times faster than the pilot's own plane, what is the average speed of the jet?

Section 8.7 Additional Applications: Variation

83. If y varies directly as x and $y = 4$ when $x = 20$, find y if $x = 30$.

84. If y is inversely proportional to x and $y = 8$ when $x = 10$, find y when $x = 16$.

85. A hanging spring will stretch 10 in. if a weight of 14 lbs is placed at its end. How far will the spring stretch if the weight is increased to 20 lbs?

86. The circumference of a circle varies directly as the diameter of the circle. A circular pizza with a diameter of 4 in. has a circumference of 12.56 inches. What will be the circumference of a pizza with a diameter of 6 inches?

87. The area of a circle varies directly as the square of its radius. A circular pizza with a radius of 4 in. has an area of 50.24 square inches. What is the area of a pizza with a radius of 6 inches?

88. The distance a falling object travels is directly related to the square of the time it falls. If an object falls 144 feet in 3 seconds, how far will it fall (assuming it doesn't hit the ground) in 5 seconds?

89. Time and rate of speed are inversely proportional. That is, as a rate of speed increases, the time to travel a fixed distance decreases. If it would take a car 3 hours at 40 miles per hour to travel a fixed distance, how long would it take the car to travel this same distance at 50 miles per hour?

90. A pressure of 1200 lbs per ft^2 is exerted by 3 ft^3 of gas in a cylinder. If a piston pushes into the cylinder until the volume is 2 ft^3, what will be the pressure exerted by the gas? (**Hint:** $V = \dfrac{k}{P}$)

91. The resistance, R, in a wire of fixed length, is inversely proportional to the square of the diameter, d, of the wire. A wire of diameter 0.02 cm has a resistance of 0.5 ohm. What is the resistance of a wire of the same type and the same length, but with a diameter of 0.04 cm?

92. A gear that has 48 teeth is meshed with a gear that has 16 teeth. Find the speed (in revolutions per minute) of the small gear if the large one is rotating at 54 rpm. (**Hint:** $T_1 R_1 = T_2 R_2$)

Chapter 8 Test

In Exercises 1 – 3, reduce each of the rational expressions to lowest terms, and state any restrictions on the variable.

1. $\dfrac{2x^2 + 5x - 3}{2x^2 - x}$

2. $\dfrac{8x^2 - 2x - 3}{4x^2 + x - 3}$

3. $\dfrac{15x^4 - 18x^3}{36x - 30x^2}$

4. Evaluate the rational expression for $x = 2$: $\dfrac{x - 6}{6 - x}$

5. Evaluate the rational expression for $y = -3$: $\dfrac{2y^3}{y^2 - 25}$

In Exercises 6 – 11, perform the indicated operations and reduce all answers. Also, state any restrictions on the variables.

6. $\dfrac{16x^2 + 24x + 9}{3x^2 + 14x + 8} \cdot \dfrac{21x + 14}{4x^2 + 3x}$

7. $\dfrac{3x - 3}{3x^2 - x - 2} \div \dfrac{24x^2 - 6}{6x^2 + x - 2}$

8. $\dfrac{3}{x^2 - 1} + \dfrac{7}{x^2 - x - 2}$

9. $\dfrac{3x}{x - 4} + \dfrac{7x}{x + 4} - \dfrac{x + 3}{x^2 - 16}$

10. $\dfrac{1}{x - 1} + \dfrac{2x}{x^2 - x - 12} - \dfrac{5}{x^2 - 5x + 4}$

11. $\dfrac{x^2 + 8x + 15}{x^2 + 6x + 5} \cdot \dfrac{x - 3}{2x} \div \dfrac{x + 3}{x^2}$

Simplify the complex fractions in Exercises 12 and 13.

12. $\dfrac{\dfrac{1}{x} - \dfrac{1}{x^2}}{1 - \dfrac{1}{x^2}}$

13. $\dfrac{\dfrac{6}{x - 5}}{\dfrac{x + 18}{x^2 - 2x - 15}}$

Solve the equations in Exercises 14 – 16.

14. $\dfrac{9}{3 - x} = \dfrac{8}{2x + 1}$

15. $\dfrac{3}{4} = \dfrac{2x + 5}{5 - x}$

16. $\dfrac{2x}{x - 1} - \dfrac{3}{x^2 - 1} = \dfrac{2x + 5}{x + 1}$

17. Simplify the expression in part (a) and solve the equation in part (b).

a. $\dfrac{3}{x} - \dfrac{2}{x + 1} + \dfrac{5}{2x}$

b. $\dfrac{3}{x} - \dfrac{2}{x + 1} = \dfrac{5}{2x}$

18. Two numbers are in a ratio of 5 to 7. Their sum is 156. Find the numbers.

19. Lisa can travel 228 miles in the same time that Kim travels 168 miles. If Lisa's speed is 15 mph faster than Kim's, find their rates.

20. A plumber can complete a job in 2 hours. If an apprentice helps, it takes only $1\frac{1}{3}$ hours. How long would it take the apprentice working alone?

21. The pressure on one side of a metal plate submerged horizontally in water varies directly as the depth of the water. If a plate 10 ft below the surface has a total pressure of 625 lbs on one face, how much pressure will then be on that same plate 25 ft below the surface?

22. The lifting force on the wings of an airplane is given by the formula $P = kAv^2$. If the lift on the wing of a certain plane is 16,000 lbs for a wing area of 180 ft^2 at a speed of 120 mph, find the lift of the same plane at a speed of 150 mph.

23. The resistance, R (in ohms), of a wire is given by the formula $R = \dfrac{kL}{d^2}$. The resistance of a wire with a diameter of 0.01 in. and a length of 250 ft is 10 ohms. What is the resistance of a piece of the same type of wire with a length of 300 ft and a diameter of 0.02 inches?

24. A swimming pool has two inlet pipes. One inlet pipe alone can fill the pool in 3 hours and the two pipes open together can fill the pool in 2 hours. How long would it take for the second inlet pipe alone to fill the pool?

Cumulative Review: Chapters 1 – 8

1. Perform the indicated operations and reduce.

   a. $\dfrac{5}{6} + \dfrac{3}{4}$ **b.** $\dfrac{1}{2} - \dfrac{5}{7}$ **c.** $\dfrac{1}{2} \cdot \dfrac{1}{2}$ **d.** $\dfrac{7}{32} \div \dfrac{21}{8}$

2. Find 12.5% of 40.

In Exercises 3 – 6, perform the indicated operations and simplify the expressions.

3. $(3x + 7)(x - 4)$ 4. $(2x - 5)^2$

5. $-4(x + 2) + 5(2x - 3)$ 6. $x(x + 7) - (x + 1)(x - 4)$

7. In the formula $A = P + Prt$, solve for t.

8. How long will it take for an investment of $600 at a rate of 5% to be worth $615? (**Hint:** $I = Prt$)

Factor each expression in Exercises 9 and 10.

9. $4x^2 + 16x + 15$ 10. $2x^2 + 6x - 20$

11. Given the relation $r = \{(-1, 5), (0, 2), (-1, 0), (5, 0), (6, 0)\}$.

    a. What is the domain of the relation?
    b. What is the range of the relation?
    c. Is the relation a function? Explain.

12. For the function $f(x) = x^3 - 2x^2 - 1$, find:

    a. $f(3)$ **b.** $f(0)$ **c.** $f\left(-\dfrac{2}{3}\right)$

13. Find the equation of the line that has slope $\dfrac{3}{4}$ and passes through the point $(-2, 4)$. Graph the line.

14. Find the equation of the line passing through the two points $(-5, -1)$ and $(4, 2)$. Graph the line.

15. Find the equation of the line passing through the point $(-3, 7)$ with slope undefined. Graph the line.

16. Find the equation of the line perpendicular to the line $x - 2y = 3$ and passing through the point $(-4, 0)$. Graph both lines.

In Exercises 17 and 18, use the vertical line test to determine whether each of the graphs does or does not represent a function.

17.

18.

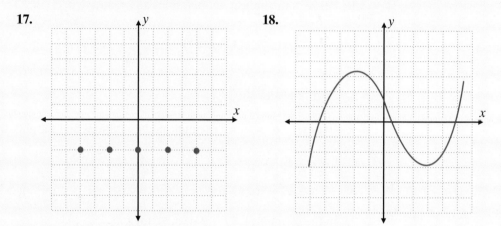

Use a TI-84 Plus graphing calculator to graph the solutions in Exercises 19 and 20.

19. $3x + y > 4$

20. $\begin{cases} x + y < 6 \\ x - 2y \le 4 \end{cases}$

In Exercises 21 and 22, use a TI-84 Plus graphing calculator to find the solutions to each system of linear equations.

21. $\begin{cases} x + y = 5.2 \\ 2x - y = -1.5 \end{cases}$

22. $\begin{cases} -x + 2y = 6 \\ 3x + 2y = -4 \end{cases}$

Solve the systems of equations in Exercises 23 – 26 algebraically. State whether the system is consistent, inconsistent, or dependent.

23. $\begin{cases} 2x - y = 10 \\ x + 2y = 5 \end{cases}$

24. $\begin{cases} y = 3x + 2 \\ -6x + 2y = 4 \end{cases}$

25. $\begin{cases} y = \dfrac{1}{2}x + 3 \\ 6x - 2y = 3 \end{cases}$

26. $\begin{cases} x + 2y = 5 \\ 2x + 4y = 9 \end{cases}$

27. Use a graphing calculator to graph the two functions $f(x) = \dfrac{2}{5}x + 1$ and $g(x) = -2x^2 + 6$. Estimate the points of intersection.

Simplify each expression in Exercises 28 – 29.

28. $\dfrac{32x^3 y^2}{4xy^2}$

29. $\dfrac{15xy^{-1}}{3x^{-2}y^{-3}}$

In Exercises 30 and 31, express each quotient as a sum of fractions and simplify if possible.

30. $\dfrac{3x^3y + 10x^2y^2 + 5xy^3}{5x^2y}$ **31.** $\dfrac{7x^2y^2 - 21x^2y^3 + 8x^2y^4}{7x^2y^2}$

Divide in Exercises 32 and 33 by using the long division algorithm.

32. $\left(x^2 - 14x - 16\right) \div \left(x + 1\right)$ **33.** $\dfrac{2x^3 + 5x^2 + 7}{x + 3}$

In Exercises 34 – 41, reduce the rational expressions to lowest terms and indicate any restrictions on the variable.

34. $\dfrac{8x^3 + 4x^2}{2x^2 - 5x - 3}$ **35.** $\dfrac{4x^2 + 5x + 9}{x - 2}$ **36.** $\dfrac{x^2 - 9}{x + 3}$

37. $\dfrac{x}{x^2 + x}$ **38.** $\dfrac{4 - x}{3x - 12}$ **39.** $\dfrac{2x + 6}{3x + 9}$

40. $\dfrac{x^2 + 3x}{x^2 + 7x + 12}$ **41.** $\dfrac{x^2 + 2x - 15}{2x^2 - 12x + 18}$

Perform the indicated operations and simplify in Exercises 42 – 51. Assume that no denominator is 0.

42. $\dfrac{x^2}{x + y} - \dfrac{y^2}{x + y}$ **43.** $\dfrac{7}{x} + \dfrac{9}{x - 4}$ **44.** $\dfrac{4}{y + 2} - \dfrac{4}{y + 3}$

45. $\dfrac{4x}{3x + 3} - \dfrac{x}{x + 1}$ **46.** $\dfrac{4x}{x - 4} \div \dfrac{12x^2}{x^2 - 16}$ **47.** $\dfrac{x^2 + 3x + 2}{x + 3} \div \dfrac{x + 1}{3x^2 + 6x}$

48. $\dfrac{2x + 1}{x^2 + 5x - 6} \cdot \dfrac{x^2 + 6x}{x}$ **49.** $\dfrac{8}{x^2 + x - 6} + \dfrac{2x}{x^2 - 3x + 2}$

50. $\dfrac{x}{x^2 + 3x - 4} + \dfrac{x + 1}{x^2 - 1}$ **51.** $\dfrac{x + 1}{x^2 + 4x + 4} \div \dfrac{x^2 - x - 2}{x^2 - 2x - 8}$

Simplify the complex algebraic fractions in Exercises 52 – 54.

52. $\dfrac{\dfrac{3}{x} + \dfrac{1}{6x}}{\dfrac{7}{3x}}$ **53.** $\dfrac{\dfrac{1}{x} - \dfrac{1}{x^2}}{\dfrac{1}{x} + \dfrac{1}{x^2}}$ **54.** $\dfrac{\dfrac{4}{3x} + \dfrac{1}{6x}}{\dfrac{1}{x^2} - \dfrac{1}{2x}}$

Solve each of the equations in Exercises 55 – 59.

55. $4(x+2)-7 = -2(3x+1)-3$ **56.** $(x+4)(x-5) = 10$

57. $0 = x^2 - 7x + 10$ **58.** $x^2 + 8x - 4 = 16$

59. $\dfrac{7}{2x-1} = \dfrac{3}{x+6}$

60. a. Simplify the following expression: $\dfrac{3}{x} - \dfrac{5}{x+3}$

     b. Solve the following equation: $\dfrac{3}{x} = \dfrac{5}{x+3}$

61. a. Simplify the following expression: $\dfrac{3x}{x+2} + \dfrac{2}{x-4} + 3$

     b. Solve the following equation: $\dfrac{3x}{x+2} + \dfrac{2}{x-4} = 3$

62. An airplane can travel 1035 miles in the same time that a train travels 270 miles. The speed of the plane is 50 mph more than three times the speed of the train. Find the speed of each.

63. A man can wax his car three times as fast as his daughter can. Together they can complete the job in 2 hours. How long would it take each of them working alone?

64. The illumination (in foot-candles, fc) of a light source varies inversely as the square of the distance from the source. If a certain light source provides an illumination of 10 fc at a distance of 20 ft, what is the illumination at a distance of 40 feet?

65. Two numbers are in a ratio of 2 to 5. If their sum is 84, find the numbers.

66. A family travels 18 miles downriver and returns. It takes 8 hours to make the round trip. Their rate in still water is twice the rate of the current. Find the rate of the current.

67. If a hanging spring stretches 5 cm when a weight of 13 gm is placed at its end, how far will the spring stretch if a weight of 20 gm is placed at its end?

Real Numbers and Radicals

Did You Know?

An important method of reasoning related to mathematical proofs is proof by contradiction. See if you can follow the reasoning in the following "proof" that $\sqrt{2}$ is an irrational number. This proof uses the fact that the square of an integer is even if and only if the integer is even.

Proof:

$\sqrt{2}$ is either an irrational number or a rational number. Suppose that $\sqrt{2}$ is a rational number and

$\dfrac{a}{b} = \sqrt{2}$ Where a and b are integers and $\dfrac{a}{b}$ is reduced. (± 1 are the only factors common to a and b.)

$\dfrac{a^2}{b^2} = 2$ Square both sides.

$a^2 = 2b^2$ This means a^2 is an even integer.

So,

$a = 2n$ Since a^2 is even, a must be even.

$a^2 = 4n^2$

$a^2 = 4n^2 = 2b^2$ Substitute

$2n^2 = b^2$ This means b^2 is an even integer.

Therefore, b is an even integer.

But if a and b are both even, 2 is a common factor. This contradicts the statement that $\dfrac{a}{b}$ is reduced. Thus our original supposition that $\sqrt{2}$ is rational is false, and $\sqrt{2}$ is an irrational number.

9.1	Real Numbers and Evaluating Radicals
9.2	Simplifying Radicals
9.3	Addition, Subtraction, and Multiplication with Radicals
9.4	Rationalizing Denominators
9.5	Solving Equations with Radicals
9.6	Rational Exponents
9.7	The Pythagorean Theorem

"The number of grains of sand on the beach at Coney Island is much less than a googol: 10,000, 000,000,000 ,000,000,000,000,000,000,000,000, 000,000,000,000,00 0,000,000,000,000,000,000, 000,000,000,000,000,000,0 00,000,000,000."

Edward Kasner

R eal numbers and real number lines have been discussed and used throughout this text. In Chapter 9, the related concepts of real numbers, rational numbers, and irrational numbers are going to be presented in some depth. These are extremely important concepts since the real number system forms the foundation for most of the mathematical ideas we use today.

Calculators have helped a great deal in making operations with both rational numbers and irrational numbers seem easier. However, you should understand that in many cases calculators give only approximate values. To fully grasp this fact, as with much of mathematics, you need to understand some of the more abstract theoretical concepts, such as those that will be introduced in Chapter 9.

9.1   Real Numbers and Evaluating Radicals

Objectives

After completing this section, you will be able to:

1. Determine if a real number is rational or irrational.

2. Understand the meaning of radical expressions.

3. Use a TI-84 Plus calculator to evaluate radical expressions.

Real Numbers

Real numbers and their properties were discussed in Chapter 1. **Real number lines** were discussed in Chapter 3 in relationship to the solutions of inequalities in one variable such as $x < 3$. In Chapter 4, ordered pairs of real numbers and the graphs of linear equations and linear inequalities in two variables were analyzed in detail. In this section, we are going to expand on our development of the real number system, and we begin by repeating the definitions of **integers** and **rational numbers**.

Integers

Integers are the whole numbers and their opposites.

$$..., -4, -3, -2, -1, 0, 1, 2, 3, 4, ...$$

> A **rational number** is any number that can be written in the form $\dfrac{a}{b}$ where a and b are integers and $b \neq 0$.

In decimal form, all rational numbers can be written as **terminating decimals** or **infinite repeating decimals**. That is, if an integer is divided by a nonzero integer, then the decimal form of the quotient will be either terminating or infinitely repeating. For example,

$$\frac{1}{4} = 0.250000... = 0.25\overline{0} \qquad \text{Repeating 0's}$$

$$\text{or} \quad \frac{1}{4} = 0.25 \qquad \text{Terminating decimal}$$

$$\frac{1}{3} = 0.3333... = 0.\overline{3}$$

$$\frac{1}{7} = 0.142857142857142857... = 0.\overline{142857}$$

Perfect Squares and Square Roots

A number is **squared** when it is multiplied by itself. For example,

$$5^2 = 5 \cdot 5 = 25,$$

$$(-30)^2 = (-30) \cdot (-30) = 900,$$

$$\text{and } (2.1)^2 = (2.1) \cdot (2.1) = 4.41.$$

If an integer is squared, the result is called a **perfect square**. The squares for the integers from 1 to 20 are shown here for easy reference.

Table 9.1 Squares of Integers from 1 to 20 (Perfect Squares)				
$1^2 = 1$	$2^2 = 4$	$3^2 = 9$	$4^2 = 16$	$5^2 = 25$
$6^2 = 36$	$7^2 = 49$	$8^2 = 64$	$9^2 = 81$	$10^2 = 100$
$11^2 = 121$	$12^2 = 144$	$13^2 = 169$	$14^2 = 196$	$15^2 = 225$
$16^2 = 256$	$17^2 = 289$	$18^2 = 324$	$19^2 = 361$	$20^2 = 400$

Now we want to reverse the process of squaring. That is, given a number, we want to find a number that when squared will result in the given number. This is called **finding the square root** of the given number. For example,

since $7^2 = 49$, then 7 is the **square root** of 49

and, since $10^2 = 100$, then 10 is the **square root** of 100.

We write

$$\sqrt{49} = 7 \quad \text{and} \quad \sqrt{100} = 10.$$

Terminology of Radicals

*The symbol $\sqrt{}$ is called a **radical sign**.*

*The number under the radical sign is called the **radicand**.*

*The complete expression, such as $\sqrt{49}$, is called a **radical** or **radical expression**.*

Every positive real number has two square roots, one positive and one negative. The positive square root is called the **principal square root**. For example, since $(-9)^2 = 81$ and $(+9)^2 = 81$, we see that 81 has the two square roots −9 and +9. We write

$$\sqrt{81} = 9 \qquad \longleftarrow \quad \text{the principal square root}$$

$$\text{and} \ -\sqrt{81} = -9. \qquad \longleftarrow \quad \text{the negative square root}$$

The number 0 has only one square root, namely 0.

Square Root

If b is a nonnegative real number and a is a real number such that

$$a^2 = b,$$

*then a is called a **square root** of b. If a is nonnegative, then we write*

$$\sqrt{b} = a \quad \text{and} \quad -\sqrt{b} = -a.$$

The radicand may or may not be a perfect square. Example 1 shows radicals that involve fractions.

Example 1: Square Roots

a. Since $6^2 = 36$, $\sqrt{36} = 6$.

b. Since $\left(\dfrac{2}{3}\right)^2 = \dfrac{4}{9}$, $\sqrt{\dfrac{4}{9}} = \dfrac{2}{3}$.

c. Since $8^2 = 64$, $\sqrt{64} = 8$.

d. 49 has two square roots, one positive and one negative. The $\sqrt{}$ sign is understood to represent the positive square root or the principal square root. Therefore,

$$\sqrt{49} = 7$$

$$\text{and} \quad -\sqrt{49} = -7 \qquad \text{The negative square root must have a "−" sign in front of the radical sign.}$$

e. $-\sqrt{\dfrac{25}{121}} = -\dfrac{5}{11}$

f. $\sqrt{100} = 10$ and $-\sqrt{100} = -10$

g. $\sqrt{4.41} = 2.1$

> **NOTES**
> The square roots of negative real numbers are **not real numbers**. (There is no real number whose square is negative.) Thus $\sqrt{-4}$ is not a real number and $\sqrt{-25}$ is not a real number. Such numbers are called **imaginary numbers** or **nonreal complex numbers**. These numbers and their properties will be studied in later courses in mathematics.

Cube Roots

A number is **cubed** when it is used as a factor 3 times. For example,

$$5^3 = 5 \cdot 5 \cdot 5 = 125 \qquad \text{and} \qquad (-30)^3 = (-30) \cdot (-30) \cdot (-30) = -27{,}000.$$

If an integer is cubed, the result is called a **perfect cube**. The cubes for the integers from 1 to 10 are shown here for easy reference.

Table 9.2 Cubes of Integers from 1 to 10 (Perfect Cubes)				
$1^3 = 1$	$2^3 = 8$	$3^3 = 27$	$4^3 = 64$	$5^3 = 125$
$6^3 = 216$	$7^3 = 343$	$8^3 = 512$	$9^3 = 729$	$10^3 = 1000$

Cube Root

If a and b are real numbers such that

$$a^3 = b$$

*then a is called the **cube root** of b. We write $\sqrt[3]{b} = a$.*

Example 2: Cube Roots

a. Since $2^3 = 8$, $\sqrt[3]{8} = 2$.

b. Since $4^3 = 64$, $\sqrt[3]{64} = 4$.

c. Since $(-6)^3 = -216$, $\sqrt[3]{-216} = -6$.   Note that the cube root of a negative number is negative.

d. Since $\left(\dfrac{1}{3}\right)^3 = \dfrac{1}{27}$, $\sqrt[3]{\dfrac{1}{27}} = \dfrac{1}{3}$.

Irrational Numbers

In Examples 1 and 2, all the roots were integers or other rational numbers. However, not all roots are integers or other rational numbers. In fact, most roots, written in decimal form, are infinite nonrepeating decimals called **irrational numbers**.

Irrational Numbers

*Irrational numbers are real numbers that are not rational numbers. Written in decimal form, irrational numbers are **infinite, nonrepeating decimals**.*

Examples of irrational numbers are:

$$\sqrt{5}, \sqrt{7}, \sqrt{39}, \sqrt[3]{2}, -\sqrt{10}, \text{ and } \pi.$$

Example 3 shows the decimal form of some irrational numbers.

Example 3: Irrational Numbers

The following numbers are all irrational numbers. Your calculator will show them rounded to nine or more decimal places; however, the digits continue without end and there is no pattern to the digits.

a. $\sqrt{2} = 1.4142136...$ See "Did You Know?" at the beginning of this chapter for a formal proof of the fact that $\sqrt{2}$ is irrational.

b. $\sqrt[3]{4} = 1.5874011...$

c. $\pi = 3.14159265358979...$ π is the ratio of the circumference to the diameter of any circle.

d. $e = 2.7182818284590...$ e is the base of natural logarithms and will be studied in detail in later courses.

Irrational numbers are just as important as rational numbers and are just as useful in solving equations. As we discussed in Chapter 1, real number lines have points corresponding to irrational numbers as well as rational numbers. For example, in decimal form,

$$-\frac{1}{3} = -0.33333... = -0.\overline{3} \qquad \text{Repeating and never-ending decimal; \textbf{a rational number}}$$

$$\sqrt{2} = 1.4142135... \qquad \text{Nonrepeating and never-ending decimal; \textbf{an irrational number}}$$

However, each number corresponds to just one point on a line (Figure 9.1).

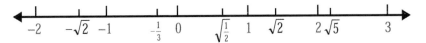

Figure 9.1

The following statement forms the basis for understanding number lines.

Every point on a number line is represented by one real number, and every real number is represented by one point on a number line.

Real numbers are numbers that are either rational or irrational. That is, every rational number and every irrational number is also a real number, just as every integer is also a rational number. (Of course, every integer is a real number, too.)

The relationships between the various types of numbers we have discussed are shown in the following diagram.

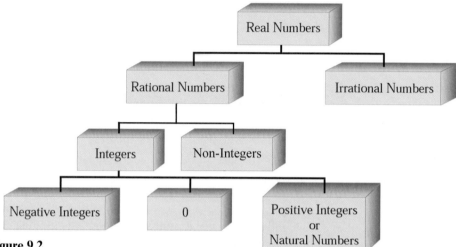

Figure 9.2

Now consider the case where $a = \sqrt{3}$. By the definition of square root, $a^2 = 3$. That is,

$$a^2 = \left(\sqrt{3}\right)^2 = 3.$$

Thus, even though $\sqrt{3}$ is an irrational number and it is an infinite nonrepeating decimal, its square is the integer 3. **The square of the square root of a positive real number is that number**.

Example 4: Find the Square

Use the definition of square root to find the square of each radical expression.

a. $\sqrt{15}$

    Solution: $\left(\sqrt{15}\right)^2 = 15$

b. $\sqrt{7}$

    Solution: $\left(\sqrt{7}\right)^2 = 7$

c. $-\sqrt{23}$

    Solution: $\left(-\sqrt{23}\right)^2 = 23$ The square of a negative number is positive.

Summary of Square Roots of Integers

In general, if a is a positive real number, then

   1. $\sqrt{a}$ *is a real number.*

   2. $\sqrt{-a}$ *is nonreal.*

Note: $\sqrt{a}$ *is rational if a is a perfect square or the quotient of perfect squares. Otherwise,* $\sqrt{a}$ *is irrational.*

Using a TI-84 Plus Graphing Calculator to Evaluate Radicals

A TI-84 Plus graphing calculator can be used to find decimal approximations for radicals and expressions containing radicals. The displays in Example 5 illustrate the advantage of being able to see the entire expression being evaluated.

Example 5: Evaluating Radical Expressions with a Calculator

Use a TI-84 Plus graphing calculator to evaluate each of the following radical expressions. In each example the steps (or keys to press) are shown. The TI-84 Plus gives answers accurate up to nine decimal places. You (and your instructor) may choose to round these answers to fewer than nine places.

a. $\sqrt{17}$

      Solution: Press the keys in the following sequence:

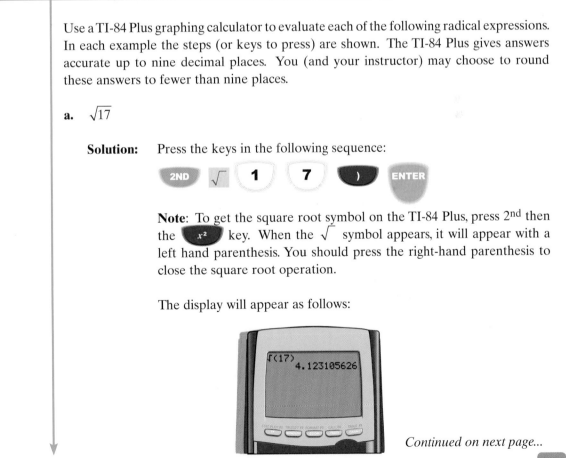

          Note: To get the square root symbol on the TI-84 Plus, press 2nd then the x^2 key. When the $\sqrt{}$ symbol appears, it will appear with a left hand parenthesis. You should press the right-hand parenthesis to close the square root operation.

          The display will appear as follows:

               √(17)
                         4.123105626

Continued on next page...

b. $\sqrt[4]{100}$ Note: This expression represents the 4th root of 100.

Solution: Press the keys in the following sequence:

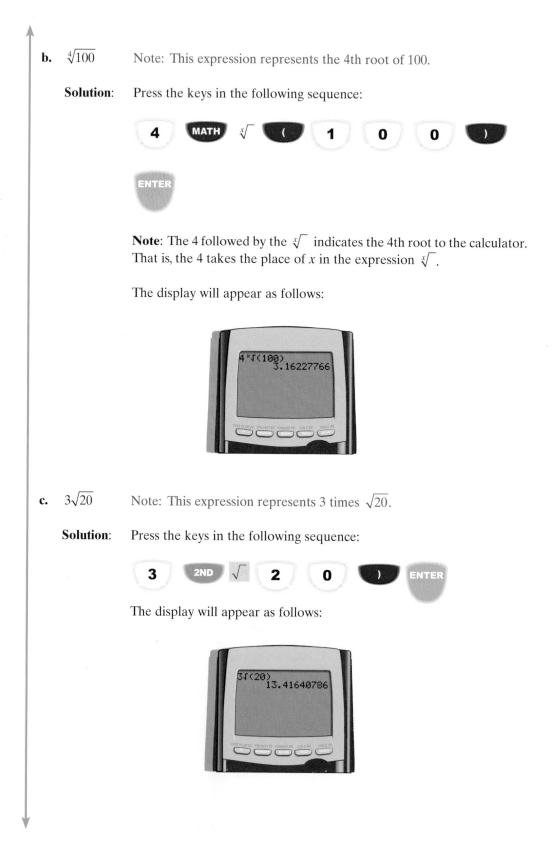

Note: The 4 followed by the $\sqrt[x]{}$ indicates the 4th root to the calculator. That is, the 4 takes the place of x in the expression $\sqrt[x]{}$.

The display will appear as follows:

4*√(100)
          3.16227766

c. $3\sqrt{20}$ Note: This expression represents 3 times $\sqrt{20}$.

Solution: Press the keys in the following sequence:

The display will appear as follows:

3√(20)
          13.41640786

d. $1 + 2\sqrt{3}$

Note: The calculator is programmed to follow the rules for order of operations.

Solution: Press the keys in the following sequence:

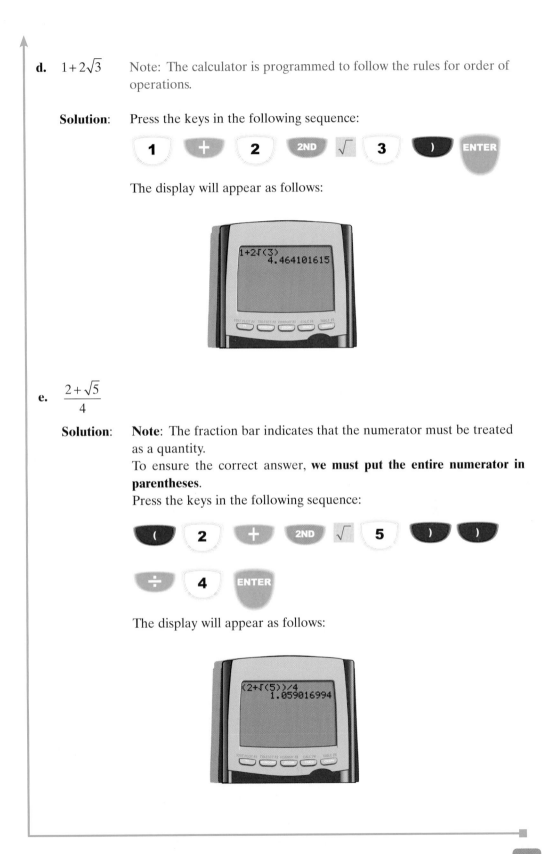

The display will appear as follows:

```
1+2√(3)
          4.464101615
```

e. $\dfrac{2 + \sqrt{5}}{4}$

Solution: **Note**: The fraction bar indicates that the numerator must be treated as a quantity.

To ensure the correct answer, **we must put the entire numerator in parentheses**.

Press the keys in the following sequence:

The display will appear as follows:

```
(2+√(5))/4
          1.059016994
```

Now you should have some idea of about how big (in decimal form) a radical number (rational or irrational) is and be able to estimate its place on a number line. Of course, you can always check your estimate with your calculator. For example, since 22 is between 16 and 25, $\sqrt{22}$ should be between 4 and 5. Symbolically,

$$16 < 22 < 25$$
$$\text{and}\quad \sqrt{16} < \sqrt{22} < \sqrt{25}$$
$$\text{so}\quad 4 < \sqrt{22} < 5.$$

You might guess $\sqrt{22}$ to be about 4.5 or 4.6. With a calculator, $\sqrt{22} = 4.69041576$ (accurate to 8 decimal places).

9.1 Exercises

Test your memory of perfect squares and cubes by simplifying the following square and cube roots in Exercises 1 – 14.

1. $\sqrt{9}$ **2.** $\sqrt{289}$ **3.** $\sqrt[3]{125}$ **4.** $\sqrt{121}$ **5.** $\sqrt{49}$

6. $\sqrt[3]{1}$ **7.** $\sqrt{169}$ **8.** $\sqrt[3]{1000}$ **9.** $\sqrt{81}$ **10.** $\sqrt{361}$

11. $\sqrt[3]{216}$ **12.** $\sqrt{36}$ **13.** $\sqrt[3]{343}$ **14.** $\sqrt[3]{512}$

15. First, use your knowledge of radicals to estimate the value of each number. Use your estimates to place all of the numbers on one number line and then use your calculator to check that each estimate makes sense.

    a. $\sqrt{32}$ **b.** $-\sqrt{18}$ **c.** $-\sqrt{\pi}$ **d.** $\sqrt[3]{20}$ **e.** $\sqrt[4]{60}$

16. First, use your knowledge of radicals to estimate the value of each number. Use your estimates to place all of the numbers on one number line and then use your calculator to check that each estimate makes sense.

    a. $-\sqrt{27}$ **b.** $\sqrt{29}$ **c.** $\sqrt[3]{-20}$ **d.** $\sqrt{40}$ **e.** $\sqrt[4]{85}$

In Exercises 17 – 33, determine whether each number is rational, irrational, or nonreal.

17. $\sqrt{4}$　　**18.** $\sqrt{17}$　　**19.** $\sqrt{169}$　　**20.** $\sqrt{\dfrac{4}{9}}$　　**21.** $\sqrt[3]{8}$　　**22.** $-\sqrt{\dfrac{1}{4}}$

23. $\sqrt{-36}$　　**24.** $\sqrt[3]{-27}$　　**25.** $\sqrt{1.69}$　　**26.** $\sqrt{5.29}$　　**27.** $-\sqrt[3]{125}$　　**28.** $\sqrt{-10}$

29. $0.1010010001\ldots$ (nonrepeating)　　　　**30.** $4.\overline{23}$ (repeating)

31. $1.\overline{93471}$ (repeating)　　　　**32.** $-6.0\overline{51}$ (repeating)

33. $3.1415926535\ldots$ (nonrepeating)

Use your calculator to find the value (accurate to four decimal places) of each of the radical expressions in Exercises 34 – 43. (Test your understanding by first estimating the value of each expression before comparing your estimate with your calculator results.)

34. $1 + 2\sqrt{2}$　　　　**35.** $3 + 2\sqrt{6}$　　　　**36.** $-1 - 5\sqrt{3}$　　　　**37.** $4 - 3\sqrt{5}$

38. $-2 + 2\sqrt{8}$　　**39.** $\dfrac{1 + \sqrt{10}}{2}$　　**40.** $\dfrac{-2 - \sqrt{12}}{3}$　　**41.** $\dfrac{-10 + \sqrt[3]{16}}{3}$

42. $\dfrac{4 - \sqrt[3]{30}}{10}$　　**43.** $\dfrac{5 + \sqrt{90}}{6}$

In Exercises 44 – 48, find the value of each indicated square.

44. $\left(\sqrt{5}\right)^2$　　**45.** $\left(\sqrt{25}\right)^2$　　**46.** $\left(\sqrt{36}\right)^2$　　**47.** $\left(-\sqrt{6}\right)^2$　　**48.** $\left(-\sqrt{2}\right)^2$

In Exercises 49 – 78, find the value of each radical expression. Use your calculator when necessary. (Test your understanding by first estimating the value of each expression before comparing your estimate with your calculator results.)

49. $\sqrt[3]{27}$　　　　**50.** $\sqrt[3]{125}$　　　　**51.** $\sqrt[3]{8}$　　　　**52.** $\sqrt[3]{64}$

53. $\sqrt[4]{16}$ **54.** $\sqrt[4]{81}$ **55.** $\sqrt[3]{1331}$ **56.** $\sqrt[4]{625}$

57. $\sqrt[3]{\dfrac{27}{8}}$ **58.** $\sqrt[5]{\dfrac{32}{243}}$ **59.** $\sqrt{65}$ **60.** $\sqrt{127}$

61. $\sqrt{37.9}$ **62.** $\sqrt{82.5}$ **63.** $\sqrt[3]{39.4}$ **64.** $\sqrt[3]{155.3}$

65. $\sqrt[4]{75}$ **66.** $\sqrt[5]{100}$ **67.** $\sqrt[5]{10,000}$ **68.** $\sqrt[3]{1000}$

69. $\sqrt[3]{216}$ **70.** $\sqrt[6]{64}$ **71.** $\sqrt[4]{256}$ **72.** $\sqrt{144}$

73. $\sqrt[3]{512}$ **74.** $\sqrt[3]{729}$ **75.** $\sqrt{175}$ **76.** $\sqrt[5]{100,000}$

77. $\sqrt[3]{130}$ **78.** $\sqrt[4]{90}$

Writing and Thinking About Mathematics

79. Discuss, in your own words, why the square root of a negative number is nonreal. Is the same reasoning true about the fourth root of a negative number?

Hawkes Learning Systems: Introductory Algebra

Real Numbers and Evaluating Radicals

9.2

Simplifying Radicals

After completing this section, you will be able to:

Simplify radicals, including square roots and cube roots.

Simplifying Square Roots

Various roots, rational and irrational, can be related to solutions of equations, and we want such numbers to be in a **simplified form** for easier calculations and algebraic manipulations. We need the two properties of radicals stated here for square roots. (Similar properties are true for other radicals.)

Properties of Square Roots

*If a and b are **positive** real numbers, then*

1. $\sqrt{ab} = \sqrt{a}\sqrt{b}$.

2. $\sqrt{\dfrac{a}{b}} = \dfrac{\sqrt{a}}{\sqrt{b}}$.

For example, we know that 144 is a perfect square and $\sqrt{144} = 12$. But using property 1 of square roots, we can also write

$$\sqrt{144} = \sqrt{36 \cdot 4} = \sqrt{36} \cdot \sqrt{4} = 6 \cdot 2 = 12.$$

Similarly,

$$\sqrt{\frac{9}{25}} = \frac{\sqrt{9}}{\sqrt{25}} = \frac{3}{5}.$$

The number 450 is not a perfect square, and to simplify $\sqrt{450}$ we can use property 1 and any of the following three approaches.

Approach I: Factor 450 as $25 \cdot 18$ because 25 is a perfect square. Then

$$\sqrt{450} = \sqrt{25 \cdot 18} = \sqrt{25} \cdot \sqrt{18} = 5\sqrt{18}$$

However, $5\sqrt{18}$ is not in simplest form because 18 has a perfect square factor, 9. Thus to complete the process, we have

$$\sqrt{450} = 5\sqrt{18} = 5\sqrt{9 \cdot 2} = 5\sqrt{9} \cdot \sqrt{2} = 5 \cdot 3\sqrt{2} = 15\sqrt{2}.$$

Approach II: Note that 225 is a perfect square factor of 450 and $450 = 225 \cdot 2$. Then

$$\sqrt{450} = \sqrt{225 \cdot 2} = \sqrt{225} \cdot \sqrt{2} = 15\sqrt{2}.$$

Approach III: Use prime factors.

$$\sqrt{450} = \sqrt{45 \cdot 10} = \sqrt{3 \cdot 3 \cdot 5 \cdot 2 \cdot 5} = \sqrt{3 \cdot 3 \cdot 5 \cdot 5} \cdot \sqrt{2} = 3 \cdot 5 \cdot \sqrt{2} = 15\sqrt{2}$$

Of these three approaches, the second appears to be the easiest because it has the fewest steps. However, "seeing" the largest square factor may be difficult. If you do not immediately see a square factor, then proceed to find other factors or prime factors as illustrated.

Simplest Form of a Square Root

*A square root is considered to be in **simplest form** when the radicand has no perfect square as a factor.*

Example 1: Simplifying Square Roots

Simplify the following square roots.

a. $\sqrt{24}$ **b.** $\sqrt{72}$ **c.** $\sqrt{\dfrac{75}{4}}$ **d.** $\sqrt{\dfrac{288}{9}}$

Solutions:

a. Factor 24 so that one factor is a square number:

$$\sqrt{24} = \sqrt{4 \cdot 6} = \sqrt{4} \cdot \sqrt{6} = 2\sqrt{6}$$

b. Find the largest square factor of 72:

$$\sqrt{72} = \sqrt{36 \cdot 2} = \sqrt{36} \cdot \sqrt{2} = 6\sqrt{2}$$

If you noticed 9 as a square factor (and not 36), you can still simplify the expression, but with a few more steps as shown here:

$$\sqrt{72} = \sqrt{9 \cdot 8} = \sqrt{9 \cdot 4 \cdot 2} = \sqrt{9} \cdot \sqrt{4} \cdot \sqrt{2} = 3 \cdot 2 \cdot \sqrt{2} = 6\sqrt{2}$$

c. $\sqrt{\dfrac{75}{4}} = \dfrac{\sqrt{75}}{\sqrt{4}} = \dfrac{\sqrt{25 \cdot 3}}{2} = \dfrac{\sqrt{25} \cdot \sqrt{3}}{2} = \dfrac{5\sqrt{3}}{2}$

d. $\sqrt{\dfrac{288}{9}} = \dfrac{\sqrt{288}}{\sqrt{9}} = \dfrac{\sqrt{144 \cdot 2}}{3} = \dfrac{\sqrt{144} \cdot \sqrt{2}}{3} = \dfrac{\overset{4}{\cancel{12}} \cdot \sqrt{2}}{\cancel{3}} = 4\sqrt{2}$

In this example, we can begin by reducing the fraction because 9 is a factor of 288:

$$\sqrt{\dfrac{288}{9}} = \sqrt{32} = \sqrt{16 \cdot 2} = \sqrt{16} \cdot \sqrt{2} = 4\sqrt{2}$$

Simplifying Square Roots with Variables

Consider simplifying the radical expression $\sqrt{x^2}$. We do not know whether x represents a positive number or a negative number. For example,

$$\text{If } x = 3, \text{then } \sqrt{x^2} = \sqrt{3^2} = \sqrt{9} = 3 = x.$$

$$\text{But, if } x = -3, \text{then } \sqrt{x^2} = \sqrt{(-3)^2} = \sqrt{9} = 3 \neq x.$$

$$\text{In fact, if } x = -3, \text{then } \sqrt{x^2} = \sqrt{(-3)^2} = \sqrt{9} = 3 = |x|.$$

Thus simplifying radical expressions with variables involves more detailed analysis than simplifying radical expressions with only constants. The following definition indicates the correct way to simplify $\sqrt{x^2}$.

Square Root of x^2

If x is a real number, then $\sqrt{x^2} = |x|$.

Note: *If $x \geq 0$ is given, then we can write $\sqrt{x^2} = x$.*

While dealing with the absolute value in simplifying square roots is correct mathematically, we can avoid some confusion by assuming that the variables under the radical sign represent only positive numbers or 0. **Therefore, for the remainder of this text, in all square root expressions such as $\sqrt{x^2}$ we will assume that $x \geq 0$ and write $\sqrt{x^2} = x$.**

Example 2: Simplifying Square Roots

Simplify each of the following radical expressions. Assume that all variables are ≥ 0.

a. $\sqrt{36x^2}$ **b.** $\sqrt{75a^2}$ **c.** $\sqrt{8x^2y^2}$

Solutions:

a. $\sqrt{36x^2} = \sqrt{36} \cdot \sqrt{x^2} = 6x$

b. $\sqrt{75a^2} = \sqrt{25 \cdot 3 \cdot a^2} = 5 \cdot \sqrt{3} \cdot a = 5a\sqrt{3}$

c. $\sqrt{8x^2y^2} = \sqrt{4 \cdot 2 \cdot x^2 \cdot y^2} = 2 \cdot \sqrt{2} \cdot x \cdot y = 2xy\sqrt{2}$

Simplifying Square Roots with Even and Odd Powers of Variables

If the exponent on a variable is larger than 2, we look for even exponents. The square root of an even power can be found by dividing the exponent by 2. For example,

because $x^2 \cdot x^2 = x^4$ and $a^3 \cdot a^3 = a^6,$

we have $\sqrt{x^4} = x^2$ and $\sqrt{a^6} = a^3.$

To find the square root of an expression with odd exponents, factor the expression into two terms, one with exponent 1 and the other with an even exponent. For example,

$$\sqrt{x^5} = \sqrt{x^4 \cdot x} = \sqrt{x^4} \cdot \sqrt{x} = x^2\sqrt{x}.$$

Square Roots of Expressions with Even and Odd Exponents

For any real number x and positive integer m,

$$\sqrt{x^{2m}} = \left| x^m \right| \quad \text{and} \quad \sqrt{x^{2m+1}} = \left| x^m \right| \sqrt{x}.$$

Note that the absolute value sign is necessary only if m is odd.

Also, note that for $\sqrt{x^{2m+1}}$ to be defined, x cannot be negative.

Comment: *If m is any integer, then 2m is even and 2m + 1 is odd.*

Example 3: Simplifying Square Roots

Simplify each of the following radical expressions. Assume that all variables are ≥ 0.

a. $\sqrt{16x^4}$ **b.** $\sqrt{49x^2y^3}$ **c.** $\sqrt{18a^6b^4}$ **d.** $\sqrt{\dfrac{9a^{15}}{64b^8}}$

Solutions:

a. $\sqrt{16x^4} = 4x^2$ Note that the even exponents are divided by 2.

b. $\sqrt{49x^2y^3} = \sqrt{49 \cdot x^2 \cdot y^2 \cdot y} = 7xy\sqrt{y}$

c. $\sqrt{18a^6b^4} = \sqrt{9 \cdot 2 \cdot a^6 \cdot b^4} = 3a^3b^2\sqrt{2}$

d. $\sqrt{\dfrac{9a^{15}}{64b^8}} = \dfrac{\sqrt{9 \cdot a^{14} \cdot a}}{\sqrt{64b^8}} = \dfrac{3a^7\sqrt{a}}{8b^4}$

Simplifying Cube Roots

Cube roots can be simplified if they have a perfect cube factor. Perfect cubes are:

$$0, \; 1, \; 8, \; 27, \; 64, \; 125, \; 216, \text{ and so on.}$$

$$0^3 \quad 1^3 \quad 2^3 \quad 3^3 \quad 4^3 \quad 5^3 \quad 6^3$$

Simplest Form of a Cube Root

*A cube root is considered to be in **simplest form** when the radicand has no perfect cube as a factor.*

When finding cube roots, we need not be concerned about positive and negative values for variables because cube roots of negative numbers are defined to be negative. For example,

we have $(-2)^3 = -8$ and therefore $\sqrt[3]{-8} = -2$.

Similarly, $(-5)^3 = -125$ and therefore, $\sqrt[3]{-125} = -5$.

Thus $\sqrt[3]{x^3} = x$ whether $x \geq 0$ or $x < 0$.

Example 4: Simplifying Cube Roots

Simplify each of the following cube roots. (Note that we are now looking for exponents that are divisible by 3.)

a. $\sqrt[3]{54x^3}$ **b.** $\sqrt[3]{40x^6y^4}$ **c.** $\sqrt[3]{125x^8y^9}$

Solutions:

a. $\sqrt[3]{54x^3} = \sqrt[3]{27 \cdot 2 \cdot x^3} = \sqrt[3]{27x^3} \cdot \sqrt[3]{2} = 3x\sqrt[3]{2}$

Note: The exponent is divided by 3 because we are taking a cube root.

b. $\sqrt[3]{40x^6y^4} = \sqrt[3]{8 \cdot 5 \cdot x^6 \cdot y^3 \cdot y} = \sqrt[3]{8x^6y^3} \cdot \sqrt[3]{5y} = 2x^2y\sqrt[3]{5y}$ Find a factor of 3 for each exponent.

c. $\sqrt[3]{125x^8y^9} = \sqrt[3]{125 \cdot x^6 \cdot x^2 \cdot y^9} = \sqrt[3]{125 \cdot x^6 \cdot y^9} \cdot \sqrt[3]{x^2} = 5x^2y^3\sqrt[3]{x^2}$

Simplifying Other Radical Expressions

If a radical expression is in the form of a fraction with a single denominator, we must still simplify the radical expression and then divide the numerator by the denominator. There are two methods as illustrated in Example 5. Both methods are correct and will give the same answer.

Example 5: Simplifying Radical Expressions

Simplify each radical expression.

a. $\dfrac{3+\sqrt{18}}{3}$ **b.** $\dfrac{8-\sqrt{20a^2}}{2}$

Solutions:

a. $\dfrac{3+\sqrt{18}}{3}$

Method 1: Simplify and factor the numerator:

$$\frac{3+\sqrt{18}}{3} = \frac{3+\sqrt{9 \cdot 2}}{3} = \frac{3+3\sqrt{2}}{3} = \frac{\cancel{3}\left(1+\sqrt{2}\right)}{\cancel{3}} = 1+\sqrt{2}$$

Method 2: Simplify the numerator and divide each term by the denominator:

$$\frac{3+\sqrt{18}}{3} = \frac{3+\sqrt{9\cdot2}}{3} = \frac{3+3\sqrt{2}}{3} = \frac{3}{3}+\frac{3\sqrt{2}}{3} = 1+\sqrt{2}$$

b. $\dfrac{8-\sqrt{20a^2}}{2}$

Method 1: Simplify and factor the numerator:

$$\frac{8-\sqrt{20a^2}}{2} = \frac{8-\sqrt{4\cdot5\cdot a^2}}{2} = \frac{8-2a\sqrt{5}}{2} = \frac{\cancel{2}\left(4-a\sqrt{5}\right)}{\cancel{2}} = 4-a\sqrt{5}$$

Method 2: Simplify the numerator and divide each term by the denominator:

$$\frac{8-\sqrt{20a^2}}{2} = \frac{8-\sqrt{4\cdot5\cdot a^2}}{2} = \frac{8-2a\sqrt{5}}{2} = \frac{8}{2}-\frac{2a\sqrt{5}}{2} = 4-a\sqrt{5}$$

Practice Problems

Simplify the following radical expressions.

1. $\sqrt{49}$ **2.** $\sqrt[3]{27}$ **3.** $-\sqrt{25}$ **4.** $\sqrt{\dfrac{3}{4}}$

5. $\sqrt{32x^2}$ **6.** $\sqrt[3]{32x^5}$ **7.** $\sqrt{\dfrac{54y^5}{25x^2}}$ **8.** $\dfrac{4-\sqrt{12}}{2}$

Answers to Practice Problems: 1. 7 **2.** 3 **3.** -5 **4.** $\dfrac{\sqrt{3}}{2}$ **5.** $4x\sqrt{2}$ **6.** $2x\sqrt[3]{4x^2}$ **7.** $\dfrac{3y^2\sqrt{6y}}{5x}$ **8.** $2-\sqrt{3}$

9.2 Exercises

Simplify each of the radical expressions in Exercises 1 – 62.

1. $\sqrt{36x^2}$

2. $\sqrt{49y^2}$

3. $\sqrt{\dfrac{1}{4}}$

4. $-\sqrt{121}$

5. $\sqrt{8x^3}$

6. $\sqrt{18a^5b^2}$

7. $-\sqrt{56}$

8. $\sqrt{162}$

9. $\sqrt{\dfrac{5x^4}{9}}$

10. $-\sqrt{\dfrac{7y^6}{16x^4}}$

11. $\sqrt{12}$

12. $-\sqrt{45}$

13. $\sqrt{288}$

14. $-\sqrt{63}$

15. $-\sqrt{72}$

16. $\sqrt{98}$

17. $-\sqrt{125}$

18. $\sqrt{\dfrac{32}{49}}$

19. $-\sqrt{\dfrac{11}{64}}$

20. $-\sqrt{\dfrac{125}{100}}$

21. $\sqrt{\dfrac{28}{25}}$

22. $\sqrt{\dfrac{147}{100}}$

23. $\sqrt{\dfrac{32a^5}{81b^{16}}}$

24. $\sqrt{\dfrac{75x^8}{121y^{12}}}$

25. $\sqrt{\dfrac{200x^8}{289}}$

26. $\sqrt{\dfrac{32x^{15}y^{10}}{169}}$

27. $\sqrt[3]{-1}$

28. $\sqrt[3]{216}$

29. $\sqrt[3]{1}$

30. $\sqrt[3]{-125}$

31. $\sqrt[3]{56}$

32. $\sqrt[3]{-128}$

33. $\sqrt[3]{-250}$

34. $\sqrt[3]{216x^6y^5}$

35. $\sqrt[3]{64x^9y^2}$

36. $\sqrt[3]{-72}$

37. $\dfrac{\sqrt[3]{81}}{6}$

38. $\dfrac{\sqrt[3]{192}}{10}$

39. $\sqrt[3]{\dfrac{375}{8}}$

40. $\sqrt[3]{\dfrac{-48}{125}}$

41. $\sqrt[3]{125x^4}$ **42.** $\sqrt[3]{72a^6b^4}$ **43.** $\sqrt[3]{\dfrac{125y^{12}}{27x^6}}$ **44.** $\dfrac{2-2\sqrt{3}}{4}$

45. $\dfrac{4+2\sqrt{6}}{2}$ **46.** $\dfrac{3+3\sqrt{6}}{6}$ **47.** $\dfrac{6-2\sqrt{3}}{8}$ **48.** $\dfrac{12+\sqrt{45}}{15}$

49. $\dfrac{7-\sqrt{98}}{14}$ **50.** $\dfrac{10-\sqrt{108}}{4}$ **51.** $\dfrac{4+\sqrt{288}}{20}$ **52.** $\dfrac{16-\sqrt{60}}{12}$

53. $\dfrac{12-\sqrt{18}}{21}$ **54.** $\dfrac{12-\sqrt{192}}{28}$ **55.** $\dfrac{14+\sqrt{147}}{35}$ **56.** $\dfrac{6+\sqrt{12x^2}}{2}$

57. $\dfrac{15-\sqrt{75a^2}}{5}$ **58.** $\dfrac{4-\sqrt{80x^3}}{8}$ **59.** $\dfrac{10+\sqrt{24a^3}}{8}$ **60.** $\dfrac{20+\sqrt{50y^4}}{5}$

61. $\dfrac{16-\sqrt{32x^4}}{20}$ **62.** $\dfrac{15+\sqrt{9y^5}}{6}$

Hawkes Learning Systems: Introductory Algebra

Simplifying Radicals

Addition, Subtraction, and Multiplication with Radicals

After completing this section, you will be able to:

1. Add and subtract radical expressions.

2. Multiply radical expressions.

Sometimes solving equations and simplifying algebraic expressions can involve operations with radicals (square roots, cube roots, fourth roots, and so on). The emphasis here will be on radicals that are square roots with some work on cube roots. **Like radicals** are terms that either have the same radicals or can be simplified so that the radicals are the same.

Like radicals are similar to like terms. For example,

$7\sqrt{2}$ and $3\sqrt{2}$ are like radicals. The radical $\sqrt{2}$ is the same in both terms.

$-4\sqrt{10}$ and $6\sqrt{10}$ are like radicals. The radical $\sqrt{10}$ is the same in both terms.

$2\sqrt{5}$ and $2\sqrt{3}$ are not like radicals. The radicals $\sqrt{5}$ and $\sqrt{3}$ are different.

Combining Like Radicals

To add (or subtract) like radicals, proceed just as in combining like terms. Use the distributive property, then add the coefficients:

$$7x + 3x = (7 + 3)x = 10x$$

$$7\sqrt{2} + 3\sqrt{2} = (7 + 3)\sqrt{2} = 10\sqrt{2}$$

Simplifying radicals will show whether or not they are like radicals. As illustrated here, it may be possible to combine radicals after they have been simplified.

$$\sqrt{75} - \sqrt{27} = \sqrt{25\cdot 3} - \sqrt{9\cdot 3} \quad \text{25 and 9 are perfect squares.}$$
$$= 5\sqrt{3} - 3\sqrt{3}$$
$$= 2\sqrt{3}$$

$$\sqrt{16x} + \sqrt{4x} = \sqrt{16 \cdot x} + \sqrt{4 \cdot x} \qquad \text{Assume } x \geq 0.$$
$$= 4\sqrt{x} + 2\sqrt{x}$$
$$= 6\sqrt{x}$$

Combining cube roots is done in the same way.

$$\sqrt[3]{16x} + \sqrt[3]{54x} = \sqrt[3]{8 \cdot 2x} + \sqrt[3]{27 \cdot 2x} \qquad \text{8 and 27 are perfect cubes.}$$
$$= \sqrt[3]{8} \cdot \sqrt[3]{2x} + \sqrt[3]{27} \cdot \sqrt[3]{2x}$$
$$= 2\sqrt[3]{2x} + 3\sqrt[3]{2x}$$
$$= 5\sqrt[3]{2x}$$

Example 1: Combining Like Radicals

Simplify the following expressions by combining like radicals. (Assume all variables are nonnegative.)

a. $5\sqrt{2} - \sqrt{2}$

Solution: $5\sqrt{2} - \sqrt{2} = (5-1)\sqrt{2} = 4\sqrt{2}$

b. $3\sqrt{5} + 8\sqrt{5} - 14\sqrt{5}$

Solution: $3\sqrt{5} + 8\sqrt{5} - 14\sqrt{5} = (3 + 8 - 14)\sqrt{5} = -3\sqrt{5}$

c. $\sqrt{18x^2} + \sqrt{8x^2}$

Solution: $\sqrt{18x^2} + \sqrt{8x^2} = \sqrt{9x^2 \cdot 2} + \sqrt{4x^2 \cdot 2} = 3x\sqrt{2} + 2x\sqrt{2} = 5x\sqrt{2}$

d. $\sqrt{5} - \sqrt{3} + \sqrt{12}$

Solution:

$$\sqrt{5} - \sqrt{3} + \sqrt{12} = \sqrt{5} - \sqrt{3} + \sqrt{4}\sqrt{3} = \sqrt{5} - \sqrt{3} + 2\sqrt{3} = \sqrt{5} + \sqrt{3}$$

e. $x\sqrt{x} + 7x\sqrt{x}$

Solution: $x\sqrt{x} + 7x\sqrt{x} = (x + 7x)\sqrt{x} = 8x\sqrt{x}$

f. $\sqrt[3]{125a} - \sqrt[3]{64a}$

Solution: $\sqrt[3]{125a} - \sqrt[3]{64a} = 5\sqrt[3]{a} - 4\sqrt[3]{a} = \sqrt[3]{a}$

Multiplying Radicals

In multiplication of expressions we could use the distributive property as follows. To find the product of radicals, we proceed just as in multiplying polynomials. To illustrate the similarities, we show multiplication of a polynomial followed by multiplication of a similar radical expression:

$$5(x+y) = 5x + 5y$$

$$\sqrt{2}\left(\sqrt{7}+\sqrt{3}\right) = \sqrt{2}\sqrt{7} + \sqrt{2}\sqrt{3} = \sqrt{14} + \sqrt{6}. \qquad \text{Remember: } \sqrt{a}\sqrt{b} = \sqrt{ab}.$$

And, with binomials, the **FOIL** method gives:

$$(x+5)(x-3) = x^2 - 3x + 5x - 15 = x^2 + 2x - 15$$

$$\left(\sqrt{2}+5\right)\left(\sqrt{2}-3\right) = \left(\sqrt{2}\right)^2 - 3\sqrt{2} + 5\sqrt{2} - 15$$

$$= 2 + 2\sqrt{2} - 15 = -13 + 2\sqrt{2}$$

$$\left[\textbf{Note: } \sqrt{2}\sqrt{2} = \sqrt{4} = 2 \text{ or } \left(\sqrt{2}\right)^2 = 2.\right]$$

In general, **if a is positive**, $\sqrt{a}\sqrt{a} = a.$ $\left[\text{Or } \left(\sqrt{a}\right)^2 = a.\right]$

Example 2: Multiplying and Simplifying

Find the following products and simplify.

a. $\sqrt{7}\left(\sqrt{7} - \sqrt{14}\right)$

　　Solution: $\sqrt{7}\left(\sqrt{7} - \sqrt{14}\right) = \sqrt{7}\sqrt{7} - \sqrt{7}\sqrt{14}$

$$= 7 - \sqrt{98}$$

$$= 7 - \sqrt{49 \cdot 2}$$

$$= 7 - 7\sqrt{2}$$

b. $\left(\sqrt{2} + 4\right)\left(\sqrt{2} - 4\right)$

　　Solution: $\left(\sqrt{2} + 4\right)\left(\sqrt{2} - 4\right) = \left(\sqrt{2}\right)^2 - 4^2 = 2 - 16 = -14$

c. $\left(\sqrt{5}+\sqrt{3}\right)\left(\sqrt{5}+\sqrt{3}\right)$

   Solution: $\left(\sqrt{5}+\sqrt{3}\right)\left(\sqrt{5}+\sqrt{3}\right) = \left(\sqrt{5}\right)^2 + 2\sqrt{5}\sqrt{3} + \left(\sqrt{3}\right)^2$

   $$= 5 + 2\sqrt{15} + 3 = 8 + 2\sqrt{15}$$

d. $\left(\sqrt{x}+1\right)\left(\sqrt{x}-1\right)$ (Assume $x \geq 0$.)

   Solution: $\left(\sqrt{x}+1\right)\left(\sqrt{x}-1\right) = \left(\sqrt{x}\right)^2 - 1^2 = x - 1$

Practice Problems

Simplify the following radical expressions.

1. $2\sqrt{3} - \sqrt{3}$

2. $-2\left(\sqrt{8}+\sqrt{2}\right)$

3. $\sqrt{75} - \sqrt{27} + \sqrt{20}$

4. $5\sqrt{x} - 4\sqrt{x} + 2\sqrt{x}$

5. $\left(\sqrt{3}+\sqrt{8}\right)\left(\sqrt{2}-\sqrt{3}\right)$

6. $\left(\sqrt{x}+3\right)\left(\sqrt{x}-3\right)$

7. $\sqrt[3]{8} + \sqrt[3]{27}$

8. $\sqrt[3]{16} + \sqrt[3]{250}$

9.3 Exercises

Simplify the radical expressions in Exercises 1 – 50. (Assume all variables are nonnegative.)

1. $3\sqrt{2} + 5\sqrt{2}$ **2.** $7\sqrt{3} - 2\sqrt{3}$ **3.** $6\sqrt{5} + \sqrt{5}$

4. $4\sqrt{11} - 3\sqrt{11}$ **5.** $8\sqrt{10} - 11\sqrt{10}$ **6.** $6\sqrt{17} - 9\sqrt{17}$

7. $4\sqrt[3]{3} + 9\sqrt[3]{3}$ **8.** $11\sqrt[3]{14} - 6\sqrt[3]{14}$ **9.** $6\sqrt{11} - 5\sqrt{11} - 2\sqrt{11}$

10. $\sqrt{7} + 6\sqrt{7} - 2\sqrt{7}$ **11.** $\sqrt{a} + 4\sqrt{a} - 2\sqrt{a}$ **12.** $2\sqrt{x} - 3\sqrt{x} + 7\sqrt{x}$

Answers to Practice Problems: **1.** $\sqrt{3}$ **2.** $-6\sqrt{2}$ **3.** $2\sqrt{3} + 2\sqrt{5}$ **4.** $3\sqrt{x}$ **5.** $1 - \sqrt{6}$ **6.** $x - 9$ **7.** 5 **8.** $7\sqrt[3]{2}$

13. $5\sqrt{x} + 3\sqrt{x} - \sqrt{x}$

14. $6\sqrt{xy} - 10\sqrt{xy} + \sqrt{xy}$

15. $3\sqrt{2} + 5\sqrt{3} - 2\sqrt{3} + \sqrt{2}$

16. $\sqrt{5} + \sqrt{4} - 2\sqrt{5} + 6$

17. $2\sqrt{a} + 7\sqrt{b} - 6\sqrt{a} + \sqrt{b}$

18. $4\sqrt{x} - 3\sqrt{x} + 2\sqrt{y} + 2\sqrt{x}$

19. $6\sqrt[3]{x} - 4\sqrt[3]{y} + 7\sqrt[3]{x} + 2\sqrt[3]{y}$

20. $5\sqrt[3]{x} + 9\sqrt[3]{y} - 10\sqrt[3]{y} + 4\sqrt[3]{x}$

21. $\sqrt{12} + \sqrt{27}$

22. $\sqrt{32} - \sqrt{18}$

23. $3\sqrt{5} - \sqrt{45}$

24. $2\sqrt{7} + 5\sqrt{28}$

25. $3\sqrt[3]{54} + 8\sqrt[3]{2}$ **26.** $2\sqrt[3]{128} + 5\sqrt[3]{-54}$ **27.** $\sqrt{50} - \sqrt{18} - 3\sqrt{12}$

28. $2\sqrt{48} - \sqrt{54} + \sqrt{27}$ **29.** $2\sqrt{20} - \sqrt{45} + \sqrt{36}$ **30.** $\sqrt{18} - 2\sqrt{12} + 5\sqrt{2}$

31. $\sqrt{8} - 2\sqrt{3} + \sqrt{27} - \sqrt{72}$

32. $\sqrt{80} + \sqrt{8} - \sqrt{45} + \sqrt{50}$

33. $5\sqrt[3]{16} - 4\sqrt[3]{24} + \sqrt[3]{-250}$

34. $\sqrt[3]{192} - 2\sqrt[3]{128} + \sqrt[3]{-81}$

35. $6\sqrt{2x} - \sqrt{8x}$

36. $5\sqrt{3x} + 2\sqrt{12x}$

37. $5y\sqrt{2y} - y\sqrt{18y}$

38. $9x\sqrt{xy} - x\sqrt{16xy}$

39. $4x\sqrt{3xy} - x\sqrt{12xy} - 2x\sqrt{27xy}$

40. $x\sqrt{32x} - x\sqrt{50x} + 2x\sqrt{18x}$

41. $\sqrt{36x^3} + \sqrt{81x^3}$

42. $\sqrt{4a^2b} + \sqrt{9a^2b}$

43. $\sqrt{16x^3y^4} - \sqrt{25x^3y^4}$

44. $\sqrt{72x^{12}y^{15}} + \sqrt{18x^{12}y^{15}} + \sqrt{2x^{12}y^{15}}$

45. $\sqrt{12x^{10}y^{20}} + \sqrt{27x^{10}y^{20}} - \sqrt{3x^{10}y^{20}}$ **46.** $\sqrt[3]{8a^{12}} + \sqrt[3]{1000a^{12}}$

47. $\sqrt[3]{-27x^{24}y^6} + \sqrt[3]{-125x^{24}y^6}$ **48.** $\sqrt[3]{27a^{15}b} + \sqrt[3]{8a^{15}b} + \sqrt[3]{64a^{15}b}$

49. $\sqrt[3]{-16x^9y^{12}} - \sqrt[3]{16x^{12}y^9} + \sqrt[3]{54x^3y^6}$ **50.** $\sqrt[3]{54x^{13}y^3} + \sqrt[3]{8x^{23}y^6} + \sqrt[3]{3x^{13}y^3}$

Multiply the expressions in Exercises 51 – 70 and simplify the result. (Assume all variables are nonnegative.)

51. $\sqrt{2}\left(3 - 4\sqrt{2}\right)$ **52.** $2\sqrt{7}\left(\sqrt{7} + 3\sqrt{2}\right)$ **53.** $3\sqrt{18} \cdot \sqrt{2}$

54. $2\sqrt{10} \cdot \sqrt{5}$ **55.** $-2\sqrt{6} \cdot \sqrt{8}$ **56.** $2\sqrt{15} \cdot 5\sqrt{6}$

57. $\sqrt{3}\left(\sqrt{2} + 2\sqrt{12}\right)$ **58.** $\sqrt{2}\left(\sqrt{3} - \sqrt{6}\right)$ **59.** $\sqrt{y}\left(\sqrt{x} + 2\sqrt{y}\right)$

60. $\sqrt{x}\left(\sqrt{x} - 3\sqrt{y}\right)$ **61.** $\left(5 + \sqrt{2}\right)\left(3 - \sqrt{2}\right)$ **62.** $\left(2\sqrt{3} + 1\right)\left(\sqrt{3} - 3\right)$

63. $\left(4\sqrt{3} + \sqrt{2}\right)\left(\sqrt{3} - 2\sqrt{2}\right)$ **64.** $\left(\sqrt{5} - \sqrt{3}\right)\left(2\sqrt{5} + 3\sqrt{3}\right)$

65. $\left(\sqrt{x} + 3\right)\left(\sqrt{x} - 3\right)$ **66.** $\left(\sqrt{a} + b\right)\left(\sqrt{a} + b\right)$

67. $\left(\sqrt{2} + \sqrt{7}\right)\left(\sqrt{2} - \sqrt{7}\right)$ **68.** $\left(\sqrt{x} + \sqrt{y}\right)\left(\sqrt{x} - \sqrt{y}\right)$

69. $\left(\sqrt{x} + 5\right)\left(\sqrt{x} - 3\right)$ **70.** $\left(\sqrt{3} + \sqrt{7}\right)\left(\sqrt{3} + 2\sqrt{7}\right)$

Hawkes Learning Systems: Introductory Algebra

Addition and Subtraction of Radicals
Multiplication of Radicals

9.4   Rationalizing Denominators

Objectives

After completing this section, you will be able to:

Rationalize the denominators of rational expressions containing radicals.

Rationalizing Denominators with One Term in the Denominator

Each of the expressions

$$\frac{5}{\sqrt{3}}, \quad \frac{\sqrt{7}}{\sqrt{8}}, \quad \text{and} \quad \frac{2}{3-\sqrt{2}}$$

contain a radical in the denominator that is an irrational number. Such expressions are not considered in simplest form because they are difficult to operate with algebraically. Calculations of sums and differences are much easier if the denominators are rational expressions. So, in simplifying, **the objective is to find an equivalent fraction that has a rational number or an expression with no radicals for a denominator**.

That is, we want to simplify the expression by **rationalizing the denominator**. The following examples illustrate the method. Each fraction is multiplied by 1 in the form $\dfrac{k}{k}$, where multiplication by k rationalizes the denominator.

Example 1: Rationalizing Denominators

a. $\dfrac{5}{\sqrt{3}}$

Multiply the numerator and denominator by $\sqrt{3}$ because $\sqrt{3} \cdot \sqrt{3} = 3$, a rational number.

$$\frac{5}{\sqrt{3}} = \frac{5 \cdot \sqrt{3}}{\sqrt{3} \cdot \sqrt{3}} = \frac{5\sqrt{3}}{3}$$

b. $\dfrac{4}{\sqrt{x}}$ (Assume $x > 0$)

Multiply the numerator and denominator by $\sqrt{x}$ because $\sqrt{x} \cdot \sqrt{x} = x$. There is no guarantee that x is rational, but the radical sign does not appear in the denominator of the result.

$$\frac{4}{\sqrt{x}} = \frac{4 \cdot \sqrt{x}}{\sqrt{x} \cdot \sqrt{x}} = \frac{4\sqrt{x}}{x}$$

c. $\dfrac{3}{7\sqrt{2}}$

Multiply the numerator and denominator by $\sqrt{2}$ because $\sqrt{2} \cdot \sqrt{2} = 2$, a rational number.

$$\dfrac{3}{7\sqrt{2}} = \dfrac{3 \cdot \sqrt{2}}{7\sqrt{2} \cdot \sqrt{2}} = \dfrac{3\sqrt{2}}{7 \cdot 2} = \dfrac{3\sqrt{2}}{14}$$

d. $\dfrac{\sqrt{7}}{\sqrt{8}}$

Multiply the numerator and denominator by $\sqrt{2}$ because $\sqrt{8} \cdot \sqrt{2} = \sqrt{16} = 4$, a rational number. (**Note:** $8 \cdot 2 = 16$ and 16 is a perfect square.)

$$\dfrac{\sqrt{7}}{\sqrt{8}} = \dfrac{\sqrt{7} \cdot \sqrt{2}}{\sqrt{8} \cdot \sqrt{2}} = \dfrac{\sqrt{14}}{\sqrt{16}} = \dfrac{\sqrt{14}}{4}$$

In Example 1d, if the numerator and denominator are multiplied by $\sqrt{8}$, the results will be the same, but one more step will be added because the fraction can be reduced.

$$\dfrac{\sqrt{7}}{\sqrt{8}} = \dfrac{\sqrt{7} \cdot \sqrt{8}}{\sqrt{8} \cdot \sqrt{8}} = \dfrac{\sqrt{56}}{8} = \dfrac{\sqrt{4} \cdot \sqrt{14}}{8} = \dfrac{2\sqrt{14}}{8} = \dfrac{\sqrt{14}}{4}$$

NOTES

Historical Note: An expression with a rational denominator such as $\dfrac{\sqrt{2}}{2}$ is much easier to calculate than its equivalent form $\dfrac{1}{\sqrt{2}}$. That is, with pencil and paper, $1.41421356 \div 2$ is a much easier calculation than $1 \div 1.41421356$. Now, with modern calculators and computers available to everyone, rational denominators are more important for higher level mathematics than for arithmetic.

Rationalizing Denominators with a Sum or Difference in the Denominator

Now, if the denominator of a fraction contains a sum or difference involving square roots, rationalizing the denominator involves the algebraic concept of the difference of squares. For example, consider the fraction

$$\dfrac{2}{3 - \sqrt{2}},$$

in which the denominator is of the form $a - b = 3 - \sqrt{2}$ where square roots are involved.

Two expressions of the form $(a - b)$ and $(a + b)$ are called **conjugates** of each other and their product is always the difference of two squares.

$$(a - b)(a + b) = a^2 - b^2$$

Consider two cases for rationalizing a denominator:

1. If the denominator is of the form $a - b$, multiply both the numerator and denominator by the conjugate $a + b$.
2. If the denominator is of the form $a + b$, multiply both the numerator and denominator by the conjugate $a - b$.

In either case, the denominator becomes, $a^2 - b^2$ (**the difference of two squares**) and the denominator is a rational number (or at least a rational expression). Thus

$$\frac{2}{3 - \sqrt{2}} = \frac{2\left(3 + \sqrt{2}\right)}{\left(3 - \sqrt{2}\right)\left(3 + \sqrt{2}\right)}$$

If $a - b = 3 - \sqrt{2}$, then $a + b = 3 + \sqrt{2}$.

$$= \frac{2\left(3 + \sqrt{2}\right)}{3^2 - \left(\sqrt{2}\right)^2}$$

The denominator is the difference of two squares.

$$= \frac{2\left(3 + \sqrt{2}\right)}{9 - 2}$$

The denominator is a rational number.

$$= \frac{2\left(3 + \sqrt{2}\right)}{7}$$

Example 2: Rationalizing Denominators

Rationalize the denominator of each of the following expressions and simplify.

a. $\sqrt{\dfrac{1}{2}}$

Solution: $\sqrt{\dfrac{1}{2}} = \dfrac{\sqrt{1}}{\sqrt{2}} = \dfrac{1 \cdot \sqrt{2}}{\sqrt{2} \cdot \sqrt{2}} = \dfrac{\sqrt{2}}{2}$

b. $\dfrac{5}{\sqrt{5}}$

Solution: $\dfrac{5}{\sqrt{5}} = \dfrac{5 \cdot \sqrt{5}}{\sqrt{5} \cdot \sqrt{5}} = \dfrac{5\sqrt{5}}{5} = \sqrt{5}$

c. $\dfrac{6}{\sqrt{27}}$

Solution: $\dfrac{6}{\sqrt{27}} = \dfrac{6 \cdot \sqrt{3}}{\sqrt{27} \cdot \sqrt{3}} = \dfrac{6 \cdot \sqrt{3}}{\sqrt{81}} = \dfrac{6 \cdot \sqrt{3}}{9} = \dfrac{2\sqrt{3}}{3}$

d. $\dfrac{-3}{\sqrt{xy}}$

Solution: $\dfrac{-3}{\sqrt{xy}} = \dfrac{-3 \cdot \sqrt{xy}}{\sqrt{xy} \cdot \sqrt{xy}} = \dfrac{-3\sqrt{xy}}{xy}$

e. $\dfrac{3}{\sqrt{5} + \sqrt{2}}$

Solution: $\dfrac{3}{\sqrt{5} + \sqrt{2}} = \dfrac{3\left(\sqrt{5} - \sqrt{2}\right)}{\left(\sqrt{5} + \sqrt{2}\right)\left(\sqrt{5} - \sqrt{2}\right)} = \dfrac{3\left(\sqrt{5} - \sqrt{2}\right)}{\left(\sqrt{5}\right)^2 - \left(\sqrt{2}\right)^2}$

$\qquad = \dfrac{3\left(\sqrt{5} - \sqrt{2}\right)}{5 - 2} = \dfrac{3\left(\sqrt{5} - \sqrt{2}\right)}{3} = \sqrt{5} - \sqrt{2}$

f. $\dfrac{x}{\sqrt{x} - 3}$

Solution: $\dfrac{x}{\sqrt{x} - 3} = \dfrac{x\left(\sqrt{x} + 3\right)}{\left(\sqrt{x} - 3\right)\left(\sqrt{x} + 3\right)}$

$\qquad = \dfrac{x\left(\sqrt{x} + 3\right)}{\left(\sqrt{x}\right)^2 - (3)^2} = \dfrac{x\left(\sqrt{x} + 3\right)}{x - 9}$

g. $\dfrac{1}{\sqrt{2} - 4}$

Solution: $\dfrac{1}{\sqrt{2} - 4} = \dfrac{1\left(\sqrt{2} + 4\right)}{\left(\sqrt{2} - 4\right)\left(\sqrt{2} + 4\right)} = \dfrac{\sqrt{2} + 4}{\left(\sqrt{2}\right)^2 - 4^2}$

$\qquad = \dfrac{\sqrt{2} + 4}{2 - 16} = \dfrac{\sqrt{2} + 4}{-14} = -\dfrac{\sqrt{2} + 4}{14}$

Note that the negative sign can be placed in front of the fraction.

Practice Problems

Rationalize each denominator and simplify.

1. $\sqrt{\dfrac{5}{2}}$ 　　**2.** $\dfrac{\sqrt{7}}{\sqrt{18}}$ 　　**3.** $\dfrac{4}{\sqrt{7} + \sqrt{3}}$ 　　**4.** $\dfrac{5}{\sqrt{x} + 2}$

Answers to Practice Problems: 1. $\dfrac{\sqrt{10}}{2}$ **2.** $\dfrac{\sqrt{14}}{6}$ **3.** $\sqrt{7} - \sqrt{3}$ **4.** $\dfrac{5\left(\sqrt{x} - 2\right)}{x - 4}$

9.4 Exercises

Rationalize each denominator in Exercises 1 – 60 and simplify.

1. $\dfrac{5}{\sqrt{2}}$

2. $\dfrac{7}{\sqrt{5}}$

3. $\dfrac{-3}{\sqrt{7}}$

4. $\dfrac{-10}{\sqrt{2}}$

5. $\dfrac{6}{\sqrt{3}}$

6. $\dfrac{8}{\sqrt{2}}$

7. $\dfrac{\sqrt{18}}{\sqrt{2}}$

8. $\dfrac{\sqrt{25}}{\sqrt{3}}$

9. $\dfrac{\sqrt{27x}}{\sqrt{3x}}$

10. $\dfrac{\sqrt{45y}}{\sqrt{5y}}$

11. $\dfrac{\sqrt{ab}}{\sqrt{9ab}}$

12. $\dfrac{\sqrt{5}}{\sqrt{12}}$

13. $\dfrac{\sqrt{4}}{\sqrt{3}}$

14. $\sqrt{\dfrac{3}{8}}$

15. $\sqrt{\dfrac{9}{2}}$

16. $\sqrt{\dfrac{3}{5}}$

17. $\sqrt{\dfrac{1}{x}}$

18. $\sqrt{\dfrac{x}{y}}$

19. $\sqrt{\dfrac{2x}{y}}$

20. $\sqrt{\dfrac{x}{4y}}$

21. $\dfrac{2}{\sqrt{2y}}$

22. $\dfrac{-10}{3\sqrt{5}}$

23. $\dfrac{21}{5\sqrt{7}}$

24. $\dfrac{x}{5\sqrt{x}}$

25. $\dfrac{-2y}{5\sqrt{2y}}$

26. $\dfrac{3}{1+\sqrt{2}}$

27. $\dfrac{2}{\sqrt{6}-2}$

28. $\dfrac{-11}{\sqrt{3}-4}$

29. $\dfrac{1}{\sqrt{5}-3}$

30. $\dfrac{7}{3-2\sqrt{2}}$

31. $\dfrac{-6}{5-3\sqrt{2}}$

32. $\dfrac{11}{2\sqrt{3}+1}$

33. $\dfrac{-\sqrt{3}}{\sqrt{2}+5}$

34. $\dfrac{\sqrt{2}}{\sqrt{7}+4}$

35. $\dfrac{7}{1-3\sqrt{5}}$

36. $\dfrac{-3\sqrt{3}}{6+\sqrt{3}}$

37. $\dfrac{1}{\sqrt{3}-\sqrt{5}}$

38. $\dfrac{-4}{\sqrt{7}-\sqrt{3}}$

39. $\dfrac{-5}{\sqrt{2}+\sqrt{3}}$

40. $\dfrac{7}{\sqrt{2}+\sqrt{5}}$

41. $\dfrac{4}{\sqrt{x}+1}$

42. $\dfrac{-7}{\sqrt{x}-3}$

43. $\dfrac{5}{6+\sqrt{y}}$

44. $\dfrac{x}{\sqrt{x}+2}$

45. $\dfrac{8}{2\sqrt{x}+3}$

46. $\dfrac{3\sqrt{x}}{\sqrt{2x}-5}$

47. $\dfrac{\sqrt{4y}}{\sqrt{5y}-\sqrt{3}}$

48. $\dfrac{\sqrt{3x}}{\sqrt{2}+\sqrt{3x}}$

49. $\dfrac{3}{\sqrt{x}-\sqrt{y}}$

50. $\dfrac{4}{2\sqrt{x}+\sqrt{y}}$

51. $\dfrac{x}{\sqrt{x}+2\sqrt{y}}$

52. $\dfrac{y}{\sqrt{x}-\sqrt{3y}}$

53. $\dfrac{\sqrt{3}+1}{\sqrt{3}-2}$

54. $\dfrac{\sqrt{2}+4}{5-\sqrt{2}}$

55. $\dfrac{\sqrt{5}-2}{\sqrt{5}+3}$

56. $\dfrac{1+\sqrt{3}}{3-\sqrt{3}}$

57. $\dfrac{\sqrt{x}+1}{\sqrt{x}-1}$

58. $\dfrac{\sqrt{x}-4}{\sqrt{x}+3}$

59. $\dfrac{\sqrt{x}+2}{\sqrt{3x}+y}$

60. $\dfrac{3-\sqrt{x}}{2\sqrt{x}+y}$

Hawkes Learning Systems: Introductory Algebra

Rationalizing Denominators
Division of Radicals

9.5   Solving Equations with Radicals

After completing this section, you will be able to:

Solve equations with radical expressions by squaring both sides of the equations.

If an equation involves radicals with radicands that include variables, then the equation is called a **radical equation**. For example,

$$\sqrt{x} = 5, \quad \sqrt{2x + 1} = 3, \quad \text{and} \quad x + 1 = \sqrt{3x - 1}$$

are all radical equations. To solve such equations, we need the following property of real numbers.

Property of Real Numbers

For real numbers a and b, if a = b, then $a^2 = b^2$.

We use this property by **squaring both sides of a radical equation that contains square roots**. However, because squaring positive numbers and squaring negative numbers can sometimes give the same result, we must be sure to check all potential solutions in the **original equation**. For example, consider the following equation:

$$x = -2 \qquad \text{Given an equation with one solution.}$$

$$(x)^2 = (-2)^2 \qquad \text{Square both sides.}$$

$$x^2 = 4 \qquad \text{Result is an equation with two}$$

$$x^2 - 4 = 0 \qquad \text{solutions.}$$

$$(x + 2)(x - 2) = 0$$

$$x = -2 \quad \text{or} \quad x = 2$$

Thus squaring both sides created an equation with more solutions than the original equation. In this example, 2 is **not** a solution of the **original equation**.

Example 1: Solving Radical Equations

Solve the following radical equations.

a. $\sqrt{2x-3} = 5$

Solution:

$$\sqrt{2x-3} = 5$$

$$\left(\sqrt{2x-3}\right)^2 = (5)^2 \qquad \text{Square both sides.}$$

$$2x-3 = 25$$

$$2x = 28$$

$$x = 14$$

Check:

$$\sqrt{2\cdot 14-3} \overset{?}{=} 5$$

$$\sqrt{28-3} \overset{?}{=} 5$$

$$\sqrt{25} \overset{?}{=} 5$$

$$5 = 5$$

Thus 14 checks and the solution is 14.

b. $\sqrt{18-x} = x-6$

Solution:

$$\sqrt{18-x} = x-6$$

$$\left(\sqrt{18-x}\right)^2 = (x-6)^2 \qquad \text{Square both sides.}$$

$$18-x = x^2 - 12x + 36$$

$$0 = x^2 - 11x + 18 \qquad \text{One side must be 0 to solve by factoring.}$$

$$0 = (x-2)(x-9) \qquad \text{Factor.}$$

There are two **potential** solutions, 2 and 9.

Check: $x = 2$: $\qquad\qquad\qquad x = 9$:

$$\sqrt{18-2} \overset{?}{=} 2-6 \qquad\qquad \sqrt{18-9} \overset{?}{=} 9-6$$

$$\sqrt{16} \overset{?}{=} -4 \qquad\qquad\qquad \sqrt{9} \overset{?}{=} 3$$

$$4 \neq -4 \qquad\qquad\qquad\qquad 3 = 3$$

So 2 is not a solution. $\qquad$ So 9 is a solution.

The only solution to the original equation is 9.

The process of squaring both sides will give a new equation. This equation will have all the solutions of the original equation and possibly, as illustrated in Example 1b, some solutions that are not solutions to the original equation. These "extra" solutions, if they exist, are called **extraneous solutions**. (See Section 8.5.)

In Example 1b, $x = 2$ is an extraneous solution. It is a solution of the quadratic equation,

$$0 = x^2 - 11x + 18$$

but it is not a solution of the original radical equation,

$$\sqrt{18 - x} = x - 6$$

Checking the solution shows that substituting $x = 2$ in the radical equation gives

$$4 \neq -4$$

We can see that squaring both sides does give a new equation that is true. Namely, $(4)^2 = (-4)^2$.

Solving radical equations by squaring does not always yield extraneous solutions, but we emphasize that all solutions must be checked in the original equation. The steps for solving an equation with square roots are summarized below.

To Solve an Equation with a Square Root Radical Expression

1. *Isolate the radical expression on one side of the equation.*
2. *Square both sides of the equation.*
3. *Solve the new equation.*
4. *Check each solution of the new equation in the original equation and eliminate any extraneous solutions.*

The following examples illustrate more possible results when solving radical equations.

Example 2: Solving and Checking Radical Equations

Solve the following radical equations.

a. $\sqrt{2x + 5} + 3 = 18$

Solution:

$$\sqrt{2x + 5} + 3 = 18$$
$$\sqrt{2x + 5} = 15 \qquad \text{Isolate the radical on one side.}$$
$$\left(\sqrt{2x + 5}\right)^2 = 15^2 \qquad \text{Square both sides.}$$
$$2x + 5 = 225$$
$$2x = 220$$
$$x = 110$$

Check: $\sqrt{2\cdot110+5}+3\overset{?}{=}18$

$$\sqrt{225}+3\overset{?}{=}18$$

$$15+3\overset{?}{=}18$$

$$18=18$$

The solution is 110.

b. $a-3=\sqrt{3a-9}$

Solution: $$a-3=\sqrt{3a-9}$$

$$\left(a-3\right)^2=\left(\sqrt{3a-9}\right)^2 \qquad \text{Square both sides.}$$

$$a^2-6a+9=3a-9$$

$$a^2-9a+18=0$$

$$\left(a-6\right)\left(a-3\right)=0 \qquad \text{Factor.}$$

$$a-6=0 \quad \text{or} \quad a-3=0$$

$$a=6 \qquad\qquad a=3$$

Both 6 and 3 check, so there are two solutions: 6 and 3.

c. $\sqrt{x+1}=-3$

Solution: We can stop right here. There is **no real solution** to this equation because the radical on the left is nonnegative and cannot possibly equal −3, a negative number. Suppose we did not notice this relationship. Then, proceeding as usual, we will find an answer that does not check.

$$\sqrt{x+1}=-3$$

$$\left(\sqrt{x+1}\right)^2=\left(-3\right)^2 \qquad \text{Square both sides.}$$

$$x+1=9$$

$$x=8$$

Continued on next page...

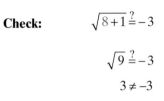

Check: $\sqrt{8+1} \overset{?}{=} -3$

$$\sqrt{9} \overset{?}{=} -3$$

$$3 \ne -3$$

So 8 does **not** check, and there are no solutions.

Practice Problems

Solve the following radical equations.

1. $\sqrt{x+5} = 6$ **2.** $\sqrt{4x+1} = 9$ **3.** $\sqrt{2x-5} = -1$ **4.** $x+1 = \sqrt{10x+21}$

9.5 Exercises

Solve the equations in Exercises 1 – 45. Check all solutions in the original equation.

1. $\sqrt{x+3} = 6$ **2.** $\sqrt{x-1} = 4$ **3.** $\sqrt{2x-1} = 3$

4. $\sqrt{3x+1} = 5$ **5.** $\sqrt{3x+4} = -5$ **6.** $\sqrt{2x-5} = -1$

7. $\sqrt{5x+4} = 7$ **8.** $\sqrt{3x-2} = 4$ **9.** $\sqrt{6-x} = 3$

10. $\sqrt{11-x} = 5$ **11.** $\sqrt{x-4} + 6 = 2$ **12.** $\sqrt{2x-7} + 5 = 3$

13. $\sqrt{4x+1} - 2 = 3$ **14.** $\sqrt{6x+4} - 3 = 5$ **15.** $\sqrt{2x+11} = 3$

16. $\sqrt{3x+7} = 1$ **17.** $\sqrt{5-2x} = 7$ **18.** $\sqrt{4-3x} = 5$

Answers to Practice Problems: 1. $x = 31$ **2.** $x = 20$ **3.** no solution **4.** $x = 10$

19. $\sqrt{5x-4}=14$ **20.** $\sqrt{7x-5}=4$ **21.** $\sqrt{2x-1}=\sqrt{x+1}$

22. $\sqrt{3x+2}=\sqrt{x+4}$ **23.** $\sqrt{x+2}=\sqrt{2x-5}$ **24.** $\sqrt{2x-5}=\sqrt{3x-9}$

25. $\sqrt{4x-3}=\sqrt{2x+5}$ **26.** $\sqrt{4x-6}=\sqrt{3x-1}$ **27.** $\sqrt{x+6}=x+4$

28. $\sqrt{x+4}=x-8$ **29.** $\sqrt{x-2}=x-2$ **30.** $\sqrt{x+8}=x-4$

31. $x+1=\sqrt{x+7}$ **32.** $x+2=\sqrt{x+4}$ **33.** $\sqrt{4x+1}=x-5$

34. $\sqrt{3x+1}=x-3$ **35.** $\sqrt{2-x}=x+4$ **36.** $x+2=\sqrt{x+14}$

37. $x-3=\sqrt{27-3x}$ **38.** $x+1=2\sqrt{x+4}$ **39.** $2x+3=\sqrt{3x+7}$

40. $3x+1=\sqrt{x+5}$ **41.** $x-2=\sqrt{3x-6}$ **42.** $\sqrt{8-4x}=x+1$

43. $\sqrt{x+3}=x+3$ **44.** $\sqrt{-x-5}=x+5$ **45.** $x+6=\sqrt{2x+12}$

Hawkes Learning Systems: Introductory Algebra

Solving Radical Equations

9.6   Rational Exponents

After completing this section, you will be able to:

1. Write radical expressions with fractional exponents in radical form.

2. Simplify expressions with fractional exponents.

n^{th} Roots

In Section 9.1, we discussed radicals involving square roots and cube roots. For example,

$$\sqrt{36} = 6 \text{ because } 6^2 = 36$$

and

$$\sqrt[3]{125} = 5 \text{ because } 5^3 = 125$$

The 3 in $\sqrt[3]{125}$ is called the **index**. In $\sqrt{36}$, the index is understood to be 2. That is, we could write $\sqrt[2]{36}$ for the square root. Examples of other roots are,

$$\sqrt[4]{81} = 3 \text{ because } 3^4 = 81$$
$$\sqrt[5]{-32} = -2 \text{ because } \left(-2\right)^5 = -32$$

and

$$\sqrt[5]{\frac{1}{32}} = \frac{1}{2} \text{ because } \left(\frac{1}{2}\right)^5 = \frac{1}{32}$$

The symbol $\sqrt[n]{b}$ is read "the nth root of b." **If $a = \sqrt[n]{b}$, then $a^n = b$.** We need to know the conditions on a and b that will guarantee that $\sqrt[n]{b}$ is a real number.

As we discussed in Section 9.1, $\sqrt{-4}$ is not classified as a real number because the square of a real number is positive or zero. Such numbers are called **complex numbers** and will be discussed in the next course in algebra. In the general case,

$$\sqrt[n]{b} \text{ is not a real number if } n \text{ is even and } b \text{ is negative.}$$

In other words, even roots of negative numbers are **not** real numbers.

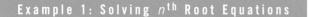

Example 1: Solving n^{th} Root Equations

    a. $\sqrt[3]{-27} = -3$ because $\left(-3\right)^3 = -27$

    b. $\sqrt[4]{0.0016} = 0.2$ because $\left(0.2\right)^4 = 0.0016$

    c. $\sqrt[5]{0.00001} = 0.1$ because $\left(0.1\right)^5 = 0.00001$

    d. $\sqrt[6]{-64}$ is not a real number

Rational (or Fractional) Exponents

The following definitions show how radicals can be expressed using fractional exponents. We will find that algebraic operations with radicals are generally easier to perform when fractional exponents are used.

Rational Exponents (Exponents of the Form $\frac{1}{n}$)

> If n is a positive integer and $\sqrt[n]{b}$ is a real number, then
>
> $$\sqrt[n]{b} = b^{\frac{1}{n}}.$$

Thus by definition,

$$\sqrt{25} = \left(25\right)^{\frac{1}{2}} = 5$$

and

$$\sqrt[3]{8} = \left(8\right)^{\frac{1}{3}} = 2.$$

Example 2: Simplifying Fractional Exponents

Simplify the following expressions

    a. $\left(49\right)^{\frac{1}{2}}$

        Solution: $\left(49\right)^{\frac{1}{2}} = \sqrt{49} = 7$

    b. $\left(-216\right)^{\frac{1}{3}}$

        Solution: $\left(-216\right)^{\frac{1}{3}} = \sqrt[3]{-216} = -6$

    c. $25^{\frac{1}{2}}$

        Solution: $25^{\frac{1}{2}} = \sqrt{25} = 5$

Continued on next page...

d. $\left(\dfrac{16}{81}\right)^{\frac{1}{4}}$

Solution: $\left(\dfrac{16}{81}\right)^{\frac{1}{4}} = \sqrt[4]{\dfrac{16}{81}} = \dfrac{2}{3}$

Now consider the problem of evaluating the expression $8^{\frac{2}{3}}$. We need the following definition for using rational exponents.

Rational Exponents (Exponents of the Form $\frac{m}{n}$)

If b is a nonnegative real number and m and n are integers with n > 0, then,

$$b^{\frac{m}{n}} = \left(b^{\frac{1}{n}}\right)^m = \left(\sqrt[n]{b}\right)^m \ or \ b^{\frac{m}{n}} = \left(b^m\right)^{\frac{1}{n}}.$$

Using this definition, we have,

$$8^{\frac{2}{3}} = \left(8^{\frac{1}{3}}\right)^2 = (2)^2 = 4 \ or \ 8^{\frac{2}{3}} = \left(8^2\right)^{\frac{1}{3}} = (64)^{\frac{1}{3}} = 4$$

and

$$32^{\frac{3}{5}} = \left(32^{\frac{1}{5}}\right)^3 = (2)^3 = 8 \ or \ 32^{\frac{3}{5}} = \left(32^3\right)^{\frac{1}{5}} = (32,768)^{\frac{1}{5}} = 8$$

Now that we have $b^{\frac{m}{n}}$ defined for rational exponents, we can consider simplifying expressions such as

$$x^{\frac{2}{3}} \cdot x^{\frac{1}{6}}, \ 36^{-\frac{1}{2}} \ and \ 2^{\frac{3}{4}} \cdot 2^{\frac{1}{4}}$$

All the previous properties of exponents apply to rational exponents (See Chapter 6).

Example 3: Simplifying Expressions with Rational Exponents

Using the properties of exponents, simplify each expression. Assume that all variables are positive.

a. $x^{\frac{2}{3}} \cdot x^{\frac{1}{6}}$

Solution: $x^{\frac{2}{3}} \cdot x^{\frac{1}{6}} = x^{\frac{2}{3} + \frac{1}{6}} = x^{\frac{4}{6} + \frac{1}{6}} = x^{\frac{5}{6}}$ Add the exponents.

b. $\dfrac{a^{\frac{3}{4}}}{a^{\frac{1}{2}}}$

Solution: $\dfrac{a^{\frac{3}{4}}}{a^{\frac{1}{2}}} = a^{\frac{3}{4} - \frac{1}{2}} = a^{\frac{3}{4} - \frac{2}{4}} = a^{\frac{1}{4}}$ Subtract the exponents.

c. $36^{\frac{-1}{2}}$

> **Solution:** $36^{-\frac{1}{2}} = \dfrac{1}{36^{\frac{1}{2}}} = \dfrac{1}{6}$ Use the property of negative exponents: $a^{-n} = \dfrac{1}{a^n}$.

d. $\left(2y^{\frac{1}{4}}\right)^3 \cdot \left(3y^{\frac{1}{8}}\right)^2$

> **Solution:** $\left(2y^{\frac{1}{4}}\right)^3 \cdot \left(3y^{\frac{1}{8}}\right)^2 = 2^3 y^{\frac{3}{4}} \cdot 3^2 y^{\frac{1}{4}}$ Multiply the exponents.
>
> $= 8y^{\frac{3}{4}} \cdot 9y^{\frac{1}{4}}$ Simplify.
>
> $= 72 y^{\frac{3}{4} + \frac{1}{4}}$ $8 \cdot 9 = 72$.
>
> $= 72y$ Add the exponents.

e. $2^{\frac{3}{4}} \cdot 2^{\frac{1}{4}}$

> **Solution:** $2^{\frac{3}{4}} \cdot 2^{\frac{1}{4}} = 2^{\frac{3}{4} + \frac{1}{4}}$ Keep the base 2.
>
> $= 2^1$ Add the exponents.
>
> $= 2$

f. $\left(3x^{\frac{1}{2}}\right)\left(2x^{\frac{5}{6}}\right)$

> **Solution:** $\left(3x^{\frac{1}{2}}\right)\left(2x^{\frac{5}{6}}\right) = 6x^{\frac{1}{2} + \frac{5}{6}} = 6x^{\frac{3}{6} + \frac{5}{6}} = 6x^{\frac{8}{6}} = 6x^{\frac{4}{3}}$

Practice Problems

Simplify the following radical expressions.

1. $25^{-\frac{1}{2}}$ **2.** $(-125)^{\frac{1}{3}}$ **3.** $a^{\frac{1}{4}} \cdot a^{\frac{1}{4}}$ **4.** $\left(5x^{\frac{2}{3}}\right)\left(3x^{\frac{5}{6}}\right)$

Answers to Practice Problems: **1.** $\dfrac{1}{5}$ **2.** -5 **3.** $a^{\frac{1}{2}}$ **4.** $15x^{\frac{3}{2}}$

9.6 Exercises

Write each of the expressions in 1 – 10 as radical expressions, then simplify.

1. $16^{\frac{1}{2}}$ **2.** $49^{\frac{1}{2}}$ **3.** $-8^{\frac{1}{3}}$ **4.** $216^{\frac{1}{3}}$

5. $81^{\frac{1}{4}}$ **6.** $16^{\frac{1}{4}}$ **7.** $\left(-32\right)^{\frac{1}{5}}$ **8.** $243^{\frac{1}{5}}$

9. $0.0004^{\frac{1}{2}}$ **10.** $0.09^{\frac{1}{2}}$

Simplify the expressions in Exercises 11 – 30.

11. $\left(\dfrac{4}{9}\right)^{\frac{1}{2}}$ **12.** $\left(\dfrac{1}{27}\right)^{\frac{1}{3}}$ **13.** $\left(\dfrac{8}{125}\right)^{\frac{1}{3}}$ **14.** $\left(\dfrac{16}{9}\right)^{\frac{1}{2}}$

15. $36^{\frac{-1}{2}}$ **16.** $64^{\frac{-1}{3}}$ **17.** $8^{\frac{-1}{3}}$ **18.** $100^{\frac{-1}{2}}$

19. $81^{\frac{-1}{4}}$ **20.** $16^{\frac{-1}{4}}$ **21.** $4^{\frac{3}{2}}$ **22.** $64^{\frac{2}{3}}$

23. $\left(-27\right)^{\frac{2}{3}}$ **24.** $36^{\frac{3}{2}}$ **25.** $9^{\frac{5}{2}}$ **26.** $8^{\frac{5}{3}}$

27. $16^{\frac{-3}{4}}$ **28.** $25^{\frac{-3}{2}}$ **29.** $\left(-8\right)^{\frac{-2}{3}}$ **30.** $\left(-125\right)^{\frac{-2}{3}}$

Using the properties of exponents, simplify each expression in Exercises 31 – 60. Assume that all variables are positive.

31. $x^{\frac{1}{4}}\cdot x^{\frac{1}{3}}$ **32.** $x^{\frac{1}{4}}\cdot x^{\frac{1}{4}}$ **33.** $5^{\frac{1}{2}}\cdot 5^{\frac{-3}{2}}$ **34.** $2^{\frac{-2}{3}}\cdot 2^{\frac{1}{3}}$

35. $x^{\frac{1}{5}} \cdot x^{\frac{2}{5}}$

36. $x^{\frac{1}{8}} \cdot x^{\frac{5}{8}}$

37. $\dfrac{x^{\frac{3}{4}}}{x^{\frac{1}{4}}}$

38. $\dfrac{x^{\frac{5}{7}}}{x^{\frac{3}{7}}}$

39. $\dfrac{x^{\frac{4}{5}}}{x^{\frac{2}{5}}}$

40. $\dfrac{x^{\frac{2}{3}}}{x^{\frac{2}{3}}}$

41. $\dfrac{2^{\frac{2}{3}}}{2^{\frac{-1}{3}}}$

42. $\dfrac{a^{\frac{5}{4}}}{a^{\frac{-1}{4}}}$

43. $a^{\frac{2}{3}} \cdot a^{\frac{1}{2}}$

44. $y^{\frac{3}{4}} \cdot y^{\frac{1}{2}}$

45. $x^{\frac{3}{4}} \cdot x^{\frac{-1}{8}}$

46. $y^{\frac{1}{2}} \cdot y^{\frac{-5}{6}}$

47. $\dfrac{x^{\frac{2}{5}}}{x^{\frac{1}{2}}}$

48. $\dfrac{x^{\frac{1}{2}}}{x^{\frac{2}{3}}}$

49. $\dfrac{6^2}{6^{\frac{1}{2}}}$

50. $\dfrac{10^3}{10^{\frac{4}{3}}}$

51. $\dfrac{x^{\frac{3}{4}}}{x^{\frac{-1}{2}}}$

52. $\dfrac{a^{\frac{2}{5}}}{a^{\frac{-1}{10}}}$

53. $\dfrac{x^{\frac{5}{3}}}{x^{\frac{-1}{3}}}$

54. $\dfrac{y^{\frac{1}{2}}}{y^{\frac{-2}{3}}}$

55. $\left(7x^{\frac{1}{3}}\right)^2 \cdot \left(4x^{\frac{1}{2}}\right)$

56. $\left(2x^{\frac{1}{2}}\right)^3 \cdot \left(3x^{\frac{1}{3}}\right)^2$

57. $\left(9x^2\right)^{\frac{1}{2}} \cdot \left(8x^3\right)^{\frac{1}{3}}$

58. $\left(x^{\frac{2}{3}}\right)^3 \cdot \left(25x^4\right)^{\frac{1}{2}}$

59. $\left(16x^8\right)^{\frac{1}{4}} \cdot \left(9x^2\right)^{\frac{-1}{2}}$

60. $\left(2x^{\frac{3}{4}}\right)^2 \cdot \left(3x^{\frac{3}{2}}\right)^{-2}$

Hawkes Learning Systems: Introductory Algebra

 Rational Exponents

The Pythagorean Theorem

Objectives

After completing this section, you will be able to:

1. Use the distance formula to find the distance between two points.

2. Understand the Pythagorean Theorem.

3. Determine if triangles are right triangles by using the Pythagorean Theorem.

4. Use the distance formula to find various properties of triangles.

Distance Between Two Points on a Vertical or Horizontal Line

A formula particularly useful in fields of study where geometry is involved, such as surveying, machining, and engineering, is that for finding the **distance between two points**. The formula is related to the Cartesian coordinate system discussed in Chapter 4.

We first discuss the distance between two points that are on a horizontal line or on a vertical line as indicated by the following two questions.

1. What is the distance between the points $P_1(2, 3)$ and $P_2(6, 3)$? (The y-coordinates are equal.)

2. What is the distance between the points $P_3(-1, -4)$ and $P_4(-1, 1)$? (The x-coordinates are equal.) See Figure 9.3.

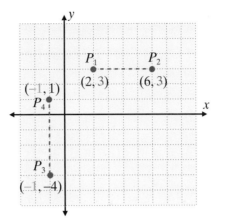

Figure 9.3

Since P_1 and P_2 lie on a horizontal line and have the same y-coordinate, the distance between the points is the difference between the x-coordinates:

$$\text{distance } (\, P_1 \text{ to } P_2\,) = 6 - 2 = 4.$$

Since P_3 and P_4 lie on a vertical line and have the same x-coordinate, the distance between the points is the difference between the y-coordinates:

$$\text{distance } (\, P_3 \text{ to } P_4\,) = 1 - (\, -4\,) = 5.$$

For the distance from P_1 to P_2, why not reverse the coordinates and take

$$\text{distance } (\, P_1 \text{ to } P_2\,) = 2 - 6 = -4?$$

The reason is that the term **distance** is taken to mean a nonnegative number (positive or 0). To guarantee a nonnegative number, take the absolute value of the difference. By taking the absolute value, the order of subtraction is no longer important.

The distance between two points, as indicated in the following formulas, is represented by d.

For $P_1(x_1, y_1)$ and $P_2(x_2, y_1)$ on a horizontal line,

$$d = |x_2 - x_1| \ \left(or \ d = |x_1 - x_2|\right)$$

For $P_1(x_1, y_1)$ and $P_2(x_1, y_2)$ on a vertical line,

$$d = |y_2 - y_1| \ \left(or \ d = |y_1 - y_2|\right)$$

Example 1: Calculating Distance on Horizontal and Vertical Lines

a. Find the distance, d, between the two points $(5, 7)$ and $(-3, 7)$.

Solution: Since the points are on a horizontal line (they have the same y-coordinate), we find the absolute value of the difference between the x-coordinates.

$$d = |{-3} - 5| = |-8| = 8$$

or

$$d = |5 - (-3)| = |8| = 8$$

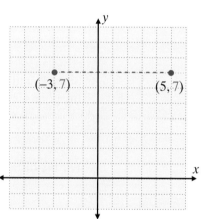

b. Find the distance, d, between the two points $(2, 8)$ and $\left(2, \dfrac{-1}{2}\right)$.

Solution: Since the points are on a vertical line (they have the same x-coordinate), we find the absolute value of the difference between the y-coordinates.

$$d = \left|8 - \left(\dfrac{-1}{2}\right)\right| = \left|8\dfrac{1}{2}\right| = 8\dfrac{1}{2}$$

or

$$d = \left|\dfrac{-1}{2} - 8\right| = \left|-8\dfrac{1}{2}\right| = 8\dfrac{1}{2}$$

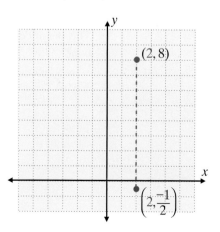

The Pythagorean Theorem: $c^2 = a^2 + b^2$

Among the most interesting and useful applications of squares and square roots are those dealing with **right triangles** and the **Pythagorean Theorem**. (Pythagoras was a famous Greek mathematician who lived in the 6th century BC.)

A **right triangle** is a triangle in which one angle is a right angle (measures 90°). Two of the sides are perpendicular and are called **legs**. The longest side is called the **hypotenuse** and is opposite the 90° angle.

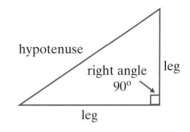

Figure 9.4

The Pythagorean Theorem

In a right triangle, the square of the hypotenuse is equal to the sum of the squares of the two legs.

$$c^2 = a^2 + b^2$$

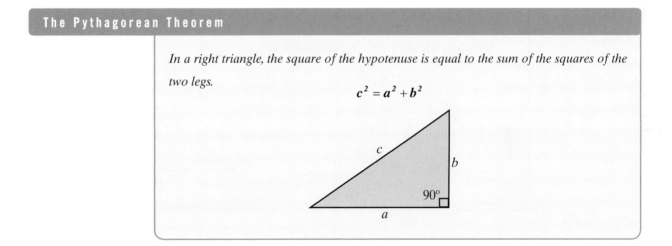

Example 2: The Pythagorean Theorem

a. What is the length of the hypotenuse of a right triangle if one leg is 8 cm long and the other leg is 6 cm long?

Solution: By the Pythagorean Theorem,

$$c^2 = 8^2 + 6^2$$

$$c^2 = 64 + 36$$

$$c^2 = 100$$

$$c = \sqrt{100} = 10 \text{ cm}$$

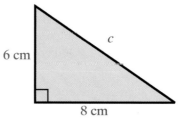

The hypotenuse is 10 cm long. (Note that only the positive square root of 100 is used because the negative square root does not make sense as a solution to the problem.)

b. In the right triangle shown here the length of one of the legs is unknown. If the hypotenuse is 13 ft and one of the legs is 12 ft, what is the length of the other leg?

Solution: By the Pythagorean Theorem,

$$a^2 + 12^2 = 13^2$$

$$a^2 + 144 = 169$$

$$a^2 = 25$$

$$a = \sqrt{25} = 5 \text{ ft}$$

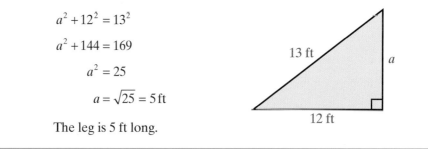

The leg is 5 ft long.

Distance Between Two Points in a plane:

$$d = \sqrt{(x_2 - x_1)^2 + (y_2 - y_1)^2}$$

If two points do not lie on a vertical line or a horizontal line, we apply the Pythagorean Theorem and calculate the length of the hypotenuse of the right triangle formed as shown in Figure 9.5. In Figure 9.5, the distance between the two points $P_1(-1, 2)$ and $P_2(5, 4)$ is the hypotenuse and the third point is at the vertex $(5, 2)$ where the right angle is located.

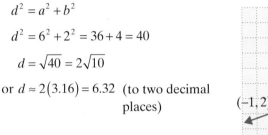

$$d^2 = a^2 + b^2$$
$$d^2 = 6^2 + 2^2 = 36 + 4 = 40$$
$$d = \sqrt{40} = 2\sqrt{10}$$

or $d \approx 2(3.16) = 6.32$ (to two decimal places)

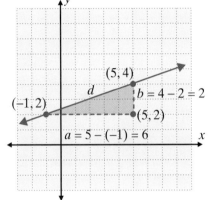

Figure 9.5

We could write directly $d = \sqrt{a^2 + b^2}$. Carrying this idea one step further, we can write a formula for d involving the coordinates of two general points $P_1(x_1, y_1)$ and $P_2(x_2, y_2)$. (See Figure 9.6.)

$$d = \sqrt{(x_2 - x_1)^2 + (y_2 - y_1)^2}$$

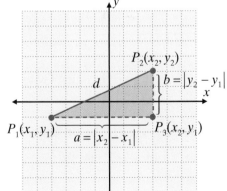

Figure 9.6

Formula for the Distance Between Two Points

The formula for the distance between two points $P_1(x_1, y_1)$ and $P_2(x_2, y_2)$ is

$$d = \sqrt{(x_2 - x_1)^2 + (y_2 - y_1)^2}$$

The absolute values, $\left| x_2 - x_1 \right|$ and $\left| y_2 - y_1 \right|$, are not needed in the formula because each of the expressions, $x_2 - x_1$ and $y_2 - y_1$, is squared. So, even though the differences might be negative, the squares, $(x_2 - x_1)^2$ and $(y_2 - y_1)^2$ will always be nonnegative. Also, the order of subtraction may be reversed because the squares will give the same value.

NOTES

In the actual calculation of *d*, be sure to add the squares before taking the square root. In general,

$$\sqrt{a^2 + b^2} \neq a + b.$$

For example, suppose that $d = \sqrt{3^2 + 4^2} = \sqrt{9 + 16}$. Then adding first gives

$$\sqrt{9 + 16} = \sqrt{25} = 5. \qquad \textbf{CORRECT}$$

Taking square roots first gives the wrong result:

$$\sqrt{9 + 16} = \sqrt{9} + \sqrt{16} = 3 + 4 = 7. \quad \textbf{INCORRECT}$$

Example 3: Distance Between Two Points

a. Find the distance between the two points $(\, 3, 4 \,)$ and $(\, -2, 7 \,)$.

Solution:
$$\begin{aligned} d &= \sqrt{(3 - (-2))^2 + (4 - 7)^2} \\ &= \sqrt{5^2 + (-3)^2} \\ &= \sqrt{25 + 9} \\ &= \sqrt{34} \end{aligned}$$

b. Find the distance between the two points $\left(\dfrac{1}{2}, \dfrac{2}{3} \right)$ and $\left(\dfrac{3}{4}, \dfrac{5}{3} \right)$.

Solution:
$$\begin{aligned} d &= \sqrt{\left(\frac{3}{4} - \frac{1}{2} \right)^2 + \left(\frac{5}{3} - \frac{2}{3} \right)^2} \\ &= \sqrt{\left(\frac{1}{4} \right)^2 + \left(\frac{3}{3} \right)^2} \\ &= \sqrt{\frac{1}{16} + 1} \\ &= \sqrt{\frac{1}{16} + \frac{16}{16}} \\ &= \sqrt{\frac{17}{16}} = \frac{\sqrt{17}}{4} \end{aligned}$$

c. Show that the triangle determined by the points $A(-2, 1)$, $B(3, 4)$, and $C(1, -4)$ is an isosceles triangle. (An isosceles triangle has two equal sides. So, the object is to determine whether or not two sides have the same length.)

Solution: The length of the line segment AB is the distance between the points A and B. We denote this distance by $|AB|$. Thus to show that the triangle ABC is isosceles, we need to show that $|AB| = |AC|$ or that $|AB| = |BC|$ or $|AC| = |BC|$. If none of these relationships are true then the triangle does not have two equal sides and is not isosceles.

$$|AB| = \sqrt{(-2-3)^2 + (1-4)^2} = \sqrt{(-5)^2 + (-3)^2}$$
$$= \sqrt{25 + 9} = \sqrt{34}$$

$$|AC| = \sqrt{(-2-1)^2 + (1-(-4))^2}$$
$$= \sqrt{(-3)^2 + (5)^2}$$
$$= \sqrt{9 + 25} = \sqrt{34}$$

Since $|AB| = |AC|$, the triangle is isosceles.

Practice Problems

Find the distance between the two given points.

1. $(-2, 8), (3, -4)$ **2.** $(-5, -1), (1, 5)$ **3.** $\left(\dfrac{3}{11}, \dfrac{4}{11}\right), (0, 0)$

4. *Determine whether or not a triangle with sides of 60 ft, 11 ft, and 61 ft is a right triangle.*

Answers to Practice Problems: 1. 13 **2.** $6\sqrt{2}$ **3.** $\dfrac{5}{11}$ **4.** Yes; $60^2 + 11^2 = 61^2$

9.7 Exercises

For Exercises 1 – 24, find the distance between the two given points.

1. $(-3, 6), (-3, 2)$ **2.** $(5, 7), (-3, 7)$ **3.** $(4, -3), (7, -3)$

4. $(-1, 2), (5, 2)$ **5.** $(3, 1), \left(\dfrac{-1}{2}, 1\right)$ **6.** $\left(\dfrac{4}{3}, 1\right), \left(\dfrac{4}{3}, \dfrac{-2}{3}\right)$

7. $(3, 1), (2, 0)$ **8.** $(4, 6), (5, -2)$ **9.** $(1, 5), (-1, 2)$

10. $(0, 0), (-3, 4)$ **11.** $(2, -7), (-3, 5)$ **12.** $(5, -3), (7, -3)$

13. $\left(\dfrac{3}{7}, \dfrac{4}{7}\right), (0, 0)$ **14.** $(-5, 2), (1, 1)$ **15.** $(4, 1), (7, 5)$

16. $(10, 7), (1, 7)$ **17.** $(-10, 3), (2, -2)$ **18.** $\left(\dfrac{7}{3}, 2\right), \left(\dfrac{-2}{3}, 1\right)$

19. $(-3, 2), (3, -6)$ **20.** $\left(\dfrac{3}{4}, 6\right), \left(\dfrac{3}{4}, -2\right)$ **21.** $(4, 0), (0, -3)$

22. $(0, -2), (4, -3)$ **23.** $\left(\dfrac{4}{5}, \dfrac{2}{7}\right), \left(\dfrac{-6}{5}, \dfrac{2}{7}\right)$ **24.** $(6, 8), (2, 5)$

25. Use the distance formula and the Pythagorean Theorem to decide if the triangle determined by the points $A(1, -2)$, $B(7, 1)$, and $C(5, 5)$ is a right triangle.

26. Use the distance formula and the Pythagorean Theorem to decide if the triangle determined by the points $A(-5, -1)$, $B(2, 1)$, and $C(-1, 6)$ is a right triangle.

In Exercises 27 and 28, show that the triangle determined by the given points is an isosceles triangle (has two equal sides).

27. $A(1,1), B(5,9), C(9,5)$ **28.** $A(1,-4), B(3,2), C(9,4)$

In Exercises 29 and 30, show that the triangle determined by the given points is an equilateral triangle (all sides equal).

29. $A(1,0), B(3, \sqrt{12}), C(5,0)$ **30.** $A(0,5), B(0,-3), C(\sqrt{48},1)$

In Exercises 31 and 32, show that the diagonals of the rectangle $ABCD$ are equal.

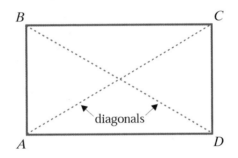

31. $A(2,-2), B(2,3), C(8,3), D(8,-2)$ **32.** $A(-1,1), B(-1,4), C(4,4), D(4,1)$

In Exercises 33 – 40, use the Pythagorean Theorem to determine whether or not each of the triangles is a right triangle.

33. **34.**

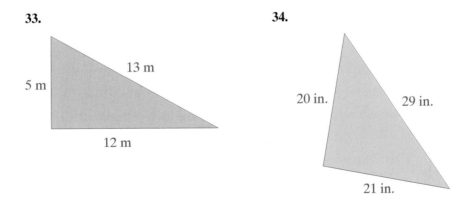

35.

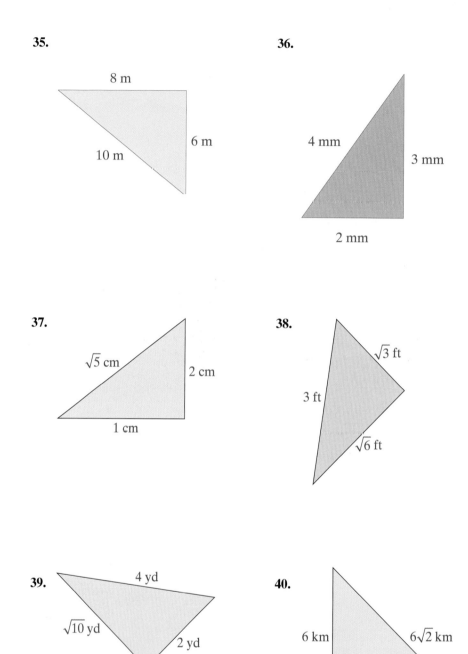

8 m

6 m

10 m

36.

4 mm

3 mm

2 mm

37.

$\sqrt{5}$ cm

2 cm

1 cm

38.

$\sqrt{3}$ ft

3 ft

$\sqrt{6}$ ft

39.

4 yd

$\sqrt{10}$ yd

2 yd

40.

6 km

$6\sqrt{2}$ km

6 km

In Exercises 41 – 46, use your calculator to find the answers.

41. A square is said to be inscribed in a circle if each corner of the square lies on the circle. (Use π = 3.14.)

  a. Find the circumference and area of a circle with diameter 30 feet.

  b. Find the perimeter and area of a square inscribed in the circle.

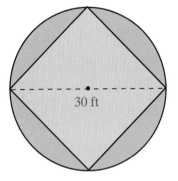

30 ft

42. The shape of a baseball infield is a square with sides 90 feet long.

  a. Find the distance (to the nearest tenth of a foot) from home plate to second base.

  b. Find the distance (to the nearest tenth of a foot) from first base to third base.

90 ft

43. The distance from home plate to the center of the pitcher's mound is 60.5 feet.

  a. Is the center of the pitcher's mound exactly half way between home plate and second base?

  b. If not, which base is it closer to, home plate or second base?

  c. Do the two diagonals of the square intersect at the center of the pitcher's mound?

44. The GE Building in New York is 850 feet tall (70 stories). At a certain time of day, the building casts a shadow 100 feet long. Find the distance from the top of the building to the tip of the shadow (to the nearest tenth of a foot).

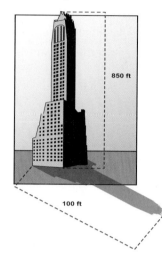

850 ft

100 ft

45. To create a square inside a square, a quilting pattern requires four triangular pieces like the one shaded in the figure shown here. If the square in the center measures 12 centimeters on a side, and the two legs of each traingle are of equal length, how long are the legs of each triangle, to the nearest tenth of a centimeter?

12 cm

46. If an airplane passes directly over your head at an altitude of 1 mile, how far (to the nearest hundredth of a mile) is the airplane from your position after it has flown 2 miles farther at the same altitude?

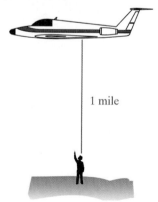

1 mile

In Exercises 47 – 49, find the perimeter of the triangle determined by the given points.

47. $A(-5,0), B(3,4), C(0,0)$

48. $A(-6,-1), B(-3,3), C(6,4)$

49. $A(-2,5), B(3,1), C(2,-2)$

Writing and Thinking About Mathematics

50. If three positive integers satisfy the Pythagorean Theorem, they are called a **Pythagorean Triple**. For example, 3, 4, and 5 are a Pythagorean Triple because $3^2 + 4^2 = 5^2$. There are an infinite number of such triples. To see how some triples can be found, fill out the following table and verify that the last three numbers are indeed Pythagorean Triples.

u	v	$2uv$	$u^2 - v^2$	$u^2 + v^2$
2	1	4	3	5
3	2	___	___	___
5	2	___	___	___
4	3	___	___	___
7	1	___	___	___
6	5	___	___	___

Hawkes Learning Systems: Introductory Algebra

The Pythagorean Theorem

Chapter 9 Index of Key Ideas and Terms

Section 9.1   Real Numbers and Evaluating Radicals

Rational Numbers   page 699

A **rational number** is any number that can be written in the form $\frac{a}{b}$ where a and b are integers and $b \neq 0$.
In decimal form, all rational numbers can be written as terminating decimals or infinite repeating decimals.

Perfect Squares   pages 699 - 700

The square of an integer is called a **perfect square**.

Square Root Terminology   page 700

The symbol $\sqrt{}$ is called a **radical sign**.
The number under the radical sign is called the **radicand**.
The complete expression, such as $\sqrt{49}$ is called a **radical** or **radical expression**.

Square Root   pages 700 - 701

If b is a nonnegative real number and a is a real number such that $a^2 = b$, then a is called a **square root** of b.
If a is nonnegative, then we write $\sqrt{b} = a$ and $-\sqrt{b} = -a$.

Cube Root   pages 701 - 702

If a and b are real numbers such that $a^3 = b$, then a is called the **cube root** of b.
We write $\sqrt[3]{b} = a$.

Irrational Numbers   pages 702 - 704

Irrational numbers are real numbers that are not rational numbers.
Written in decimal form, irrational numbers are infinite nonrepeating decimals.

Summary of Square Roots of Integers   page 705

In general, if a is a positive real number, then

1. $\sqrt{a}$ is a real number,
2. $\sqrt{-a}$ is nonreal.

Note: $\sqrt{a}$ is rational if a is a perfect square or the quotient of perfect squares. Otherwise, $\sqrt{a}$ is irrational.

Section 9.2   Simplifying Radicals

Properties of Square Roots   page 711

If a and b are positive real numbers, then

1. $\sqrt{ab} = \sqrt{a}\sqrt{b}$

2. $\sqrt{\dfrac{a}{b}} = \dfrac{\sqrt{a}}{\sqrt{b}}$

Simplest Form   pages 712 - 716

A square root is considered to be in simplest form when the radicand has no perfect square as a factor.

A cube root is considered to be in simplest form when the radicand has no perfect cube as a factor.

Square Root of x^2 pages 713 - 715

If x is a real number, then $\sqrt{x^2} = |x|$.

Note: If $x \geq 0$ is given, then we can write $\sqrt{x^2} = x$.

Square Roots of Expressions with Even and Odd Exponents   pages 714 - 715

For any real number x and positive integer m,

$$\sqrt{x^{2m}} = |x^m| \qquad \text{and} \qquad \sqrt{x^{2m+1}} = |x^m|\sqrt{x}$$

Section 9.3   Addition, Subtraction, and Multiplication with Radicals

Adding Like Radicals   pages 720 - 721

Multiplying Radicals   pages 722 - 723

Section 9.4   Rationalizing Denominators

Rationalizing Denominators with One Term   pages 726 - 727

Rationalizing Denominators with a Sum or Difference   pages 727 - 729

Multiply the numerator and denominator by the conjugate of the denominator.

CHAPTER 9   Real Numbers and Radicals

Section 9.5   Solving Equations with Radicals

To Solve an Equation with a Square Root Radical Expression   pages 734 - 736
1. Isolate the radical expression on one side of the equation.
2. Square both sides of the equation.
3. Solve the new equation.
4. Check each solution of the new equation in the original equation and eliminate any extraneous solutions.

Section 9.6   Rational Exponents

n^{th} Roots   pages 738 - 740

The symbol $\sqrt[n]{b}$ is read "the nth root of b."

$\sqrt[n]{b}$ is not a real number if n is even and b is negative.

Fractional Exponents   pages 739 - 741

If n is a positive integer and $\sqrt[n]{b}$ is a real number, then $\sqrt[n]{b} = b^{\frac{1}{n}}$.

If b is a nonnegative number and m and n are integers with $n > 0$, then

$$b^{\frac{m}{n}} = \left(b^{\frac{1}{n}} \right)^{m} = \left(\sqrt[n]{b} \right)^{m}.$$

Section 9.7   The Pythagorean Theorem

Distance Between Two Points on a Horizontal or Vertical Line   pages 744 - 746

For $P_1(x_1, y_1)$ and $P_2(x_2, y_1)$ on a horizontal line $d = |x_2 - x_1|$.

For $P_1(x_1, y_1)$ and $P_2(x_1, y_2)$ on a vertical line $d = |y_2 - y_1|$.

The Pythagorean Theorem   pages 747 - 748

In a right triangle, the square of the hypotenuse is equal to the sum of the squares of the two legs.

$$c^2 = a^2 + b^2$$

Formula for the Distance Between Two Points   pages 749 - 751

The formula for the distance between two points $P_1(x_1, y_1)$ and $P_2(x_2, y_2)$ is $d = \sqrt{(x_2 - x_1)^2 + (y_2 - y_1)^2}$.

Hawkes Learning Systems: Introductory Algebra

For a review of the topics and problems from Chapter 9, look at the following lessons from *Hawkes Learning Systems: Introductory Algebra*

Real Numbers and Evaluating Radicals
Simplifying Radicals
Addition and Subtraction of Radicals
Multiplication of Radicals
Rationalizing Denominators
Division of Radicals
Solving Equations with Radicals
Rational Exponents
The Pythagorean Theorem

Chapter 9 Review

Section 9.1 Real Numbers and Evaluating Radicals

Test your memory of perfect squares and cubes by simplifying the following square and cube roots in Exercises 1 – 4.

1. $\sqrt{196}$ 2. $\sqrt{361}$ 3. $\sqrt[3]{8}$ 4. $\sqrt[3]{729}$

In Exercises 5 – 10, determine whether each number is rational, irrational, or nonreal.

5. $16.\overline{03}$ 6. $\sqrt{-20}$ 7. $\sqrt{3}$ 8. $\sqrt{\dfrac{1}{9}}$

9. $-\sqrt{196}$ 10. $0.2020020002 \ldots$ (nonrepeating)

Use your calculator, if necessary, to find the value (accurate to four decimal places) of each of the square root expressions in Exercises 11 – 16.

11. $1 + 3\sqrt{2}$ 12. $-2 - \sqrt{10}$ 13. $\dfrac{3 + \sqrt{12}}{4}$ 14. $\dfrac{-16 + \sqrt{30}}{2}$

15. $\sqrt{4} + \sqrt{9}$ 16. $\sqrt{121} + \sqrt[3]{125}$

In Exercises 17 – 20, use your calculator to find the value (accurate to four decimal places) of each radical expression.

17. $\sqrt[3]{80.5}$ 18. $\sqrt[4]{100}$ 19. $\sqrt[5]{-32}$ 20. $\sqrt[4]{\dfrac{625}{81}}$

Section 9.2 Simplifying Radicals

Simplify each of the radical expressions in Exercises 21 – 32. Assume that all variables represent positive real numbers.

21. $\sqrt{\dfrac{1}{16}}$ 22. $-\sqrt{225}$ 23. $\sqrt{9x^3}$ 24. $\sqrt{8a^4}$

25. $\sqrt{\dfrac{27x^2}{100}}$ 26. $\sqrt{\dfrac{75a^3}{9}}$ 27. $\sqrt[3]{\dfrac{3000x^6}{343}}$ 28. $\sqrt[3]{\dfrac{8y^{12}}{27x^{15}}}$

29. $\dfrac{2 + 2\sqrt{3}}{2}$ 30. $\dfrac{5 - \sqrt{125}}{5}$ 31. $\dfrac{10 + \sqrt{200a^2}}{5}$ 32. $\dfrac{4 - \sqrt{16x^3}}{8}$

Section 9.3 Addition, Subtraction, and Multiplication with Radicals

Simplify each radical expression in Exercises 33 – 44. Assume that all variables represent nonnegative real numbers.

33. $10\sqrt{3} - 4\sqrt{3}$ **34.** $8\sqrt{5} + \sqrt{5}$ **35.** $\sqrt{a} + 4\sqrt{a} + 2\sqrt{a}$

36. $3\sqrt{x} - 4\sqrt{x} - 2\sqrt{x}$ **37.** $\sqrt{6} + \sqrt{4} - 3\sqrt{6} + 3$ **38.** $\sqrt{a} + 4\sqrt{b} + 3\sqrt{a} - \sqrt{b}$

39. $\sqrt[3]{24} + \sqrt[3]{54}$ **40.** $3\sqrt{7} + 4\sqrt{28}$ **41.** $\sqrt{25x^3} + \sqrt{49x^3}$

42. $\sqrt{16a^2 b} + \sqrt{25a^2 b}$ **43.** $\sqrt{75x^{10} y^{13}} + \sqrt{20x^{10} y^{12}}$ **44.** $\sqrt[3]{64x^{12} y^{16}} - \sqrt[3]{125x^{12} y^{16}}$

Multiply as indicated in Exercises 45 – 52 and simplify the result. Assume that all variables represent nonnegative real numbers.

45. $\sqrt{3}\left(\sqrt{3} + 4\sqrt{2}\right)$ **46.** $2\sqrt{7}\left(\sqrt{7} - 2\sqrt{3}\right)$ **47.** $\sqrt{2}\left(\sqrt{2} + 3\sqrt{6}\right)$

48. $\sqrt{x}\left(2\sqrt{y} - 3\sqrt{x}\right)$ **49.** $\left(3 + \sqrt{2}\right)\left(5 - \sqrt{2}\right)$ **50.** $\left(2\sqrt{5} + 3\right)\left(\sqrt{5} + 1\right)$

51. $\left(\sqrt{2} + \sqrt{5}\right)\left(\sqrt{2} - \sqrt{5}\right)$ **52.** $\left(\sqrt{3} + \sqrt{8}\right)\left(\sqrt{3} + 5\sqrt{2}\right)$

Section 9.4 Rationalizing Denominators

Rationalize each denominator and simplify in Exercises 53 – 64. Assume that all variables represent positive real numbers.

53. $\dfrac{7}{\sqrt{2}}$ **54.** $\dfrac{-8}{\sqrt{5}}$ **55.** $\sqrt{\dfrac{7}{12}}$ **56.** $\sqrt{\dfrac{1}{y}}$

57. $\dfrac{28}{5\sqrt{7}}$ **58.** $\dfrac{a}{3\sqrt{a}}$ **59.** $\dfrac{-10}{\sqrt{6} + 2}$ **60.** $\dfrac{1}{\sqrt{3} - 5}$

61. $\dfrac{5}{\sqrt{y} - 1}$ **62.** $\dfrac{x}{\sqrt{x} + 3}$ **63.** $\dfrac{\sqrt{x} - 1}{\sqrt{x} + 1}$ **64.** $\dfrac{2 + \sqrt{3}}{2 - \sqrt{3}}$

Section 9.5 Solving Equations with Radicals

Solve the equations in Exercises 65 – 76. Check all solutions in the original equation.

65. $\sqrt{2x - 5} = 9$ **66.** $\sqrt{3x + 4} = 7$ **67.** $\sqrt{x + 7} = -5$

68. $-11 = \sqrt{5x + 1}$ **69.** $x + 3 = \sqrt{x + 9}$ **70.** $\sqrt{x - 4} - 6 = 2$

71. $x - 3 = \sqrt{2x - 6}$ **72.** $\sqrt{x + 5} = x + 5$ **73.** $\sqrt{2x - 7} = 3 - x$

74. $\sqrt{5x - 9} = 1 - x$ **75.** $2x + 1 = \sqrt{2x + 3}$ **76.** $\sqrt{-x - 7} = x + 7$

Section 9.6 Rational Exponents

Simplify each expression in Exercises 77 – 88. Assume that all variables represent positive real numbers.

77. $16^{\frac{3}{4}}$

78. $9^{\frac{3}{2}}$

79. $8^{\frac{-2}{3}}$

80. $25^{\frac{-1}{2}}$

81. $x^{\frac{1}{2}} \cdot x^{\frac{1}{3}}$

82. $y^{\frac{3}{4}} \cdot y^{\frac{3}{16}}$

83. $\dfrac{y^{\frac{2}{3}}}{y^{\frac{-1}{2}}}$

84. $\dfrac{a^{\frac{2}{5}}}{a^{\frac{2}{3}}}$

85. $\left(2x^{\frac{1}{4}}\right)^2 \cdot \left(4x^{\frac{1}{3}}\right)^2$

86. $\left(8x^6\right)^{\frac{1}{3}} \cdot \left(16x^2\right)^{\frac{-1}{2}}$

87. $\left(5a^{\frac{1}{6}}\right)^2 \cdot \left(9a^4\right)^{\frac{1}{2}}$

88. $\left(36x^{\frac{2}{3}}\right)^{\frac{3}{2}} \cdot \left(6x^{\frac{3}{2}}\right)^{-2}$

Section 9.7 The Pythagorean Theorem

For Exercises 89 – 92, find the distance between the two given points.

89. $(-4,6)$, $(-4,7)$

90. $(-5,2)$, $(5,2)$

91. $(-4,2)$, $(-1,6)$

92. $(0,-3)$, $(7,2)$

In Exercises 93 and 94, use the **Pythagorean Theorem** to determine whether or not each triangle is a right triangle.

93.

94.

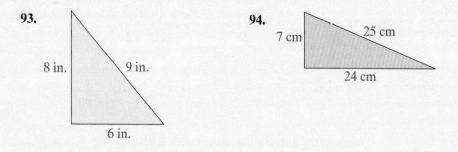

95. Show that the triangle determined by the points $A(1,-2)$, $B(4,4)$, and $C(7,1)$ is an isosceles triangle (has two equal sides).

96. A man is 6 feet tall and at a certain time of sunny day his shadow is 8 feet long. What is the distance from the top of his head to the tip of his shadow? (Assume that the ground is level.)

97. A square has a diagonal of length 40 feet. Find the length of one side of the square (to the nearest tenth of a foot).

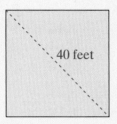

40 feet

98. Show that the triangle determined by the points $X(2,0)$, $Y(6,0)$, and $Z\left(4,\sqrt{12}\right)$ is an equilateral triangle (all sides equal).

99. The points $A(-2,2)$, $B(-2,5)$, $C(6,2)$, and $D(6,5)$ are the vertices of a rectangle. Show that the diagonals of the rectangle have the same length.

100. Find the perimeter of the triangle determined by the points $D(-6,0)$, $E(-6,3)$, and $F(0,0)$. Is this triangle a right triangle? Explain.

Chapter 9 Test

1. Write the rational number $\dfrac{3}{7}$ in the form of an infinite repeating decimal.

2. Write the irrational number $\sqrt{13}$ in decimal form accurate to 7 decimal places. (**Hint:** Use your calculator.)

Simplify the radical expressions in Exercises 3 – 14. Assume that all variables are positive.

3. $\sqrt{144}$

4. $\sqrt{25x^2}$

5. $-\sqrt{81x^4 y^3}$

6. $\sqrt[3]{27a^3}$

7. $\sqrt[3]{24x^6 y^{12}}$

8. $-\sqrt{\dfrac{16x^2}{y^8}}$

9. $\sqrt{\dfrac{4a^5}{9b^{10}}}$

10. $5\sqrt{8} + 3\sqrt{2}$

11. $2\sqrt{27x^2} + 5\sqrt{48x^2} - 4\sqrt{3x^2}$

12. $\dfrac{\sqrt{96} + 12}{24}$

13. $\dfrac{21 - \sqrt{98}}{7}$

14. $\sqrt[3]{-40} + \sqrt[3]{108}$

Multiply the expressions in Exercises 15 – 18 and simplify if possible.

15. $3\sqrt{5} \cdot 5\sqrt{5}$

16. $\sqrt{15} \cdot \left(\sqrt{3} + \sqrt{5}\right)$

17. $\left(5 + \sqrt{11}\right)\left(5 - \sqrt{11}\right)$

18. $\left(6 + \sqrt{2}\right)\left(5 - \sqrt{2}\right)$

Rationalize each denominator in Exercises 19 – 22 and simplify if possible.

19. $\dfrac{3}{\sqrt{18}}$

20. $\sqrt{\dfrac{4}{75}}$

21. $\dfrac{2}{\sqrt{3} - 1}$

22. $\dfrac{\sqrt{2}}{\sqrt{3} + \sqrt{2}}$

Solve the equations in Exercises 23 – 25. Check all solutions in the original equation.

23. $\sqrt{5x - 1} = 13$

24. $\sqrt{4x + 2} = \sqrt{12 - x}$

25. $2x - 5 = \sqrt{13 - 6x}$

Simplify the expressions in Exercises 26 – 30. Assume that all variables are positive.

26. $16^{\frac{3}{4}}$

27. $\left(\dfrac{4}{9}\right)^{\frac{-3}{2}}$

28. $a^{\frac{3}{4}} \cdot a^{\frac{-1}{2}}$

29. $\dfrac{y^{\frac{3}{4}}}{y^{\frac{-1}{2}}}$

30. $\left(16x^2\right)^{\frac{1}{2}}\left(27x^3\right)^{\frac{2}{3}}$

For Exercises 31 and 32, find the distance between the two points.

31. $(-5, 7), (3, -1)$

32. $(2, 5), (-4, -6)$

33. Determine whether or not a triangle with sides 20 yd, 21 yd, and 29 yd is a right triangle. Sketch a picture of the triangle.

34. A ladder that is 25 feet long is leaning against a building. If the bottom of the ladder is 6 feet from the base of the building, how far up the building (to the nearest tenth of a foot) is the ladder reaching?

35. Discuss, in your own words, why it is not possible to have a triangle with sides 3 cm, 4 cm, and 8 cm. (**Hint**: Try to draw this triangle to scale.)

Cumulative Review: Chapters 1 – 9

Complete the square by adding the correct term in Exercises 1 and 2. Then factor as indicated.

1. $x^2 + 8x + \underline{\hspace{1cm}} = (\quad)^2$ **2.** $x^2 - \underline{\hspace{1cm}} + 49 = (\quad)^2$

Perform the indicated operations in Exercises 3 and 4 and simplify.

3. $\dfrac{2x+10}{x^2-x-6} \cdot \dfrac{x^2+6x+8}{x+5}$ **4.** $\dfrac{x}{x+6} - \dfrac{2x-5}{x^2+4x-12}$

Graph the equations given in Exercises 5 and 6.

5. $y = \dfrac{3}{4}x - 2$ **6.** $2x + 3y = 12$

Simplify the radical expressions in Exercises 7 – 14. Assume that all variables represent positive real numbers.

7. $-\sqrt{63x^4}$ **8.** $\sqrt{\dfrac{36a^3}{25b^6}}$ **9.** $\sqrt{48x^{10}y^{20}}$ **10.** $\sqrt{54a^8b^{14}}$

11. $\sqrt[3]{40}$ **12.** $\sqrt[3]{-250}$ **13.** $\dfrac{8+\sqrt{12}}{2}$ **14.** $\dfrac{20-\sqrt{48}}{8}$

In Exercises 15 – 28, perform the indicated operations and simplify. Assume that all variables represent nonnegative real numbers.

15. $2\sqrt{50} - \sqrt{98} + 3\sqrt{8}$ **16.** $11\sqrt{2} + \sqrt{50}$ **17.** $4\sqrt{20} - \sqrt{45}$

18. $\sqrt[3]{-54} + 5\sqrt[3]{2}$ **19.** $\sqrt{50x^4} - 2\sqrt{8x^4}$ **20.** $\sqrt{12} - 5\sqrt{3} + \sqrt{48}$

21. $\sqrt{24} \cdot \sqrt{6}$ **22.** $2\sqrt{30} \cdot \sqrt{20}$ **23.** $2\sqrt{48} \cdot \sqrt{2}$

24. $\sqrt{14} \cdot \sqrt{21}$ **25.** $\left(\sqrt{3}-5\right)\left(\sqrt{3}+5\right)$ **26.** $\left(\sqrt{2}+3\right)\left(\sqrt{2}+5\right)$

27. $\left(\sqrt{x}+\sqrt{2}\right)\left(\sqrt{x}-\sqrt{2}\right)$ **28.** $\left(2\sqrt{x}+3\right)\left(\sqrt{x}-8\right)$

Rationalize each denominator in Exercises 29 – 34 and simplify if possible. Assume that all variables represent positive real numbers.

29. $\dfrac{9}{\sqrt{3}}$

30. $\sqrt{\dfrac{3}{20}}$

31. $\sqrt{\dfrac{5}{27}}$

32. $-\sqrt{\dfrac{9}{2y}}$

33. $\dfrac{2}{\sqrt{3}-5}$

34. $\dfrac{1}{\sqrt{2}+3}$

Solve the equations in Exercises 35 – 42.

35. $x(x+4)-(x+4)(x-3)=0$

36. $(2x+1)(x-4)=5$

37. $\dfrac{x-1}{2x+1}+\dfrac{1}{x+2}=\dfrac{1}{2}$

38. $x+3=\sqrt{x+15}$

39. $\sqrt{7x+2}=3$

40. $\sqrt{4x-3}=5$

41. $\sqrt{2x-9}=4-x$

42. $\sqrt{5x-3}=\sqrt{2x+3}$

43. Find the equation of the line through the two points $(-5, 3)$ and $(5, 1)$.

44. Find the equation of the line passing through the point $(6, 2)$ and perpendicular to the line $y = 3x + 5$. Graph both lines.

Solve the systems of equations in Exercises 45 and 46.

45. $\begin{cases} 2x-y=7 \\ 3x+2y=0 \end{cases}$

46. $\begin{cases} 4x+3y=15 \\ 2x-5y=1 \end{cases}$

Use a TI-84 Plus graphing calculator to graph the solution sets of the systems of linear inequalities in Exercises 47 and 48.

47. $\begin{cases} y \le 4x+1 \\ y \le -\dfrac{1}{4}x \end{cases}$

48. $\begin{cases} y > 3 \\ y > \dfrac{1}{2}x \end{cases}$

Set up equations and solve the problems in Exercises 49 – 54.

49. Two automobiles start from the same place at the same time and travel in opposite directions. One travels 12 mph faster than the other. At the end of $3\dfrac{1}{2}$ hours, they are 350 miles apart. Find the rate of each automobile.

50. The owner of a candy store sells candy for $1.50 per pound and $2.00 per pound. How many pounds of each must he use to obtain 20 pounds of a mixture of candy which sells for $1.80 per pound?

51. A man and his daughter can paint their cabin in 3 hours. Working alone, it would take the daughter 8 hours longer than it would the father. How long would it take them each working alone?

52. For lunch, Jason had one burrito and 2 tacos. Matt had 2 burritos and 3 tacos. If Jason spent $5.30 and Matt spent $9.25, find the price of one burrito and the price of one taco.

53. A boat is being pulled to the shore from a dock. When the rope to the boat is 40 meters long, the boat is 30 meters from the dock. What is the height of the dock (to the nearest tenth of a meter) above the deck of the boat?

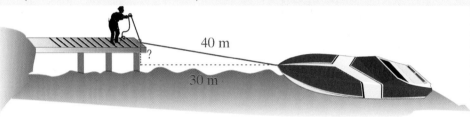

54. A ball is dropped from the top of a building that is 240 feet tall. Use the formula $h = -16t^2 + v_0 t + h_0$ to answer the following questions.
   a. When will the ball be 144 feet above the ground?
   b. How far will the ball have dropped at that time?
   c. When will the ball hit the ground?

Quadratic Equations

Did You Know?

The quadratic formula is a general method for solving second-degree equations of the form $ax^2 + bx + c = 0$, where a, b, and c can be any real numbers. The quadratic formula is a very old formula; it was known to Babylonian mathematicians around 2000 B.C. However, Babylonian and, later, Greek mathematicians always discarded negative solutions of quadratic equations because they felt that these solutions had no physical meaning. Greek mathematicians always tried to interpret their algebraic problems from a geometrical viewpoint and hence the development of the geometric method of "completing the square".

Consider the following equation: $x^2 + 6x = 7$. A geometric figure is constructed having areas x^2, $3x$, and $3x$.

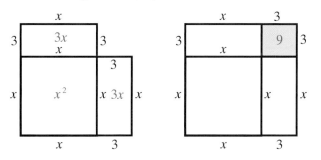

Note that to make the figure a square, one must add a 3 by 3 section (area = 9). Thus 9 must be added to both sides of the equation to restore equality. Therefore,

$$x^2 + 6x = 7 \qquad \text{Original equation}$$
$$x^2 + 6x + 9 = 7 + 9 \qquad \text{Adding 9 to complete the square}$$
$$x^2 + 6x + 9 = 16$$

10.1 Quadratic Equations: The Square Root Method

10.2 Quadratic Equations: Completing the Square

10.3 Quadratic Equations: The Quadratic Formula

10.4 Applications

10.5 Quadratic Functions: $y = ax^2 + bx + c$

So the square with side $x + 3$ now has an area of 16 square units. Therefore, the sides must be of length 4 units, which means $x + 3 = 4$. Hence $x = 1$.

"What is the difference between method and device? A method is a device you can use twice."

George Pólya in How to Solve It

Note that actually the solution set of the original equation is $\{1, -7\}$, since $(-7)^2 + 6(-7) = 49 - 42 = 7$. Thus the Greek mathematicians "lost" the negative solution because of their strictly geometric interpretation of quadratic equations. There were, therefore, many quadratic equations that the Greek mathematicians could not solve because both solutions were negative numbers or nonreal (complex) numbers. Negative solutions to equations were almost completely ignored until the early 1500's, during the Renaissance.

Quadratic equations were discussed in Chapter 7, and the solutions were found by factoring. However, in many "real-life" situations, factoring is simply not practical because the coefficients do not "cooperate" and the solutions to the equations involve square roots that yield irrational numbers or nonreal complex numbers.

Three techniques, other than factoring, for solving quadratic equations are presented in this chapter. The third technique, the **quadratic formula**, is essentially a combination of the first two in formula form. You will study and use this formula in almost every mathematics course you ever take after this one.

10.1   Quadratic Equations: *The Square Root Method*

Objectives

After completing this section, you will be able to:

1. Solve quadratic equations using the definition of square root.

2. Solve applications related to right triangles and the Pythagorean Theorem.

Review of Solving Quadratic Equations by Factoring

Factoring polynomials and solving quadratic equations by factoring were discussed in Chapter 7. While this chapter deals with three new techniques for solving quadratic equations, the method of solving by factoring is so basic and important that an example is given here for emphasis and review. (**Note:** Now would be a good time to review all of Chapter 7.)

Example 1 ▷ Factoring Method

Solve the following quadratic equations by factoring.

a. $x^2 + 7x = 18$

Solution:

$x^2 + 7x = 18$	Original equation
$x^2 + 7x - 18 = 0$	Standard form with 0 on one side
$(x+9)(x-2) = 0$	Factor.
$x + 9 = 0$ or $x - 2 = 0$	Set each factor equal to 0.
$x = -9$ $x = 2$	Solve each linear equation.

The solutions are -9 and 2.

b. $2x^2 - x = 3$

Solution:

$2x^2 - x = 3$	Original equation
$2x^2 - x - 3 = 0$	Standard form with 0 on one side
$(2x - 3)(x + 1) = 0$	Factor.
$2x - 3 = 0$ or $x + 1 = 0$	Set each factor equal to 0.
$x = \dfrac{3}{2}$ $x = -1$	Solve each linear equation.

The solutions are $\dfrac{3}{2}$ and -1.

Solving Quadratic Equations Using the Square Root Method

Solving quadratic equations by factoring is a very good technique, and we want to use it much of the time. However, there are quadratic expressions that are not easily factored and some that are not factorable at all using real numbers. For example, consider solving the equation $x^2 = 5$ by factoring. As the following steps show, the factors involve square roots that are not integers.

$x^2 = 5$	
$x^2 - 5 = 0$	Get 0 on one side.
$x^2 - \left(\sqrt{5}\right)^2 = 0$	Difference of two squares with $5 = \left(\sqrt{5}\right)^2$
$\left(x + \sqrt{5}\right)\left(x - \sqrt{5}\right) = 0$	Factor.
$x + \sqrt{5} = 0$ or $x - \sqrt{5} = 0$	Set each factor equal to 0.
$x = -\sqrt{5}$ $x = \sqrt{5}$	There are two irrational solutions.

A simpler and more direct way to solve this equation is by **taking square roots of both sides**, as shown in the following statement.

Square Root Method

For a quadratic equation in the form $x^2 = c$ where c is nonnegative,

$$x = \sqrt{c} \ \ or \ \ x = -\sqrt{c}.$$

This can be written as $x = \pm\sqrt{c}$.

NOTES In this method, we do not set a polynomial expression equal to 0. Instead we set a squared expression equal to a nonnegative real number.

Example 2: The Square Root Method

Solve the following quadratic equations by using the square root method. Write the radicals in simplest form.

a. $3x^2 = 51$

Solution: $3x^2 = 51$ Divide both sides by 3 so that the coefficient of x^2 is 1.

$$x^2 = 17$$

$$x = +\sqrt{17}$$ Keep in mind that the expression $x = \pm\sqrt{17}$ represents the two equations $x = \sqrt{17}$ and $x = -\sqrt{17}$.

b. $(x+4)^2 = 21$

Solution: $(x+4)^2 = 21$

$$x+4 = \pm\sqrt{21}$$

$$x = -4 \pm \sqrt{21}$$ There are two solutions: $-4 + \sqrt{21}$ and $-4 - \sqrt{21}$.

c. $(x-2)^2 = 50$

Solution: $(x-2)^2 = 50$

$$x-2 = \pm\sqrt{50}$$ Simplify the radical $\left(\sqrt{50} = \sqrt{25}\sqrt{2} = 5\sqrt{2}\right)$.

$$x = 2 \pm 5\sqrt{2}$$ There are two solutions: $2+5\sqrt{2}$ and $2-5\sqrt{2}$.

d. $(2x+4)^2 = 72$

Solution: $(2x+4)^2 = 72$

$$2x+4 = \pm\sqrt{72}$$

$$2x = -4 \pm 6\sqrt{2}$$ Simplify the radical $\left(\sqrt{72} = \sqrt{36}\sqrt{2} = 6\sqrt{2}\right)$.

$$x = \frac{-4 \pm 6\sqrt{2}}{2}$$

$$x = \frac{\cancel{2}\left(-2 \pm 3\sqrt{2}\right)}{\cancel{2}}$$ Factor and simplify.

$$x = -2 \pm 3\sqrt{2}$$ There are two solutions: $-2+3\sqrt{2}$ and $-2-3\sqrt{2}$.

e. $(x+7)^2 + 10 = 8$

Solution: $(x+7)^2 + 10 = 8$

$$(x+7)^2 = -2$$

There is no real solution. The square of a real number cannot be negative.

The Pythagorean Theorem

The Pythagorean Theorem was discussed in Section 9.7 and was used to develop the formula $d = \sqrt{(x_2 - x_1)^2 + (y_2 - y_1)^2}$ for finding the distance between two points. In this section, we show how the Pythagorean Theorem can be used to solve a variety of applications which involve solving quadratic equations. The theorem is stated again for easy reference and to emphasize its importance.

The Pythagorean Theorem

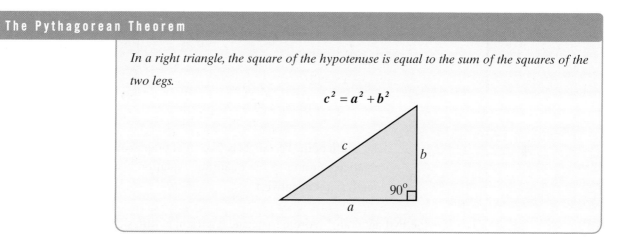

In a right triangle, the square of the hypotenuse is equal to the sum of the squares of the two legs.

$$c^2 = a^2 + b^2$$

Example 3: The Pythagorean Theorem

a. If the hypotenuse of a right triangle is 15 cm long and one leg is 10 cm long, what is the length of the other leg?

Solution: Using the Pythagorean Theorem,

$$a^2 + 10^2 = 15^2$$
$$a^2 + 100 = 225$$
$$a^2 = 125$$
$$a = \sqrt{125} = \sqrt{25} \cdot \sqrt{5} = 5\sqrt{5} \ (\approx 11.18 \, \text{cm})$$

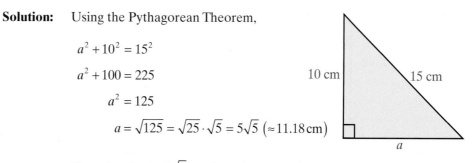

The other leg is $5\sqrt{5}$ cm long (or approximately 11.18 cm long).

(The negative solution to the quadratic equation is not considered because length is not negative.)

Continued on next page...

b. The diagonal of a rectangle is twice the width. The length of the rectangle is 6 feet. Find the width of the rectangle.

Solution: Let x = width of the rectangle
and $2x$ = length of the diagonal.

Then, by using the Pythagorean Theorem, we have

$$(2x)^2 = x^2 + 6^2$$

$$4x^2 = x^2 + 36$$

$$3x^2 = 36$$

$$x^2 = 12$$

$$x = \sqrt{12} = \sqrt{4}\cdot\sqrt{3} = 2\sqrt{3}\ (\approx 3.46).$$

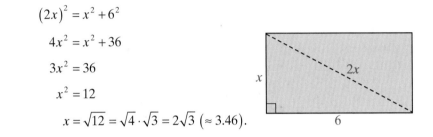

The width of the rectangle is $2\sqrt{3}$ feet or approximately 3.46 feet. (The negative solution to the quadratic equation is not considered because length is not negative.)

c. Use the Pythagorean Theorem to decide if a triangle with sides of length 8 in., 12 in., and 14 in. is a right triangle.

Solution: If the triangle is a right triangle, then the sum of the squares of the two shorter sides must equal the square of the longest side.

$$8^2 + 12^2 = 64 + 144 = 208$$
$$14^2 = 196$$

So $8^2 + 12^2 \neq 14^2$, and the triangle is **not** a right triangle.

Practice Problems

Solve the following quadratic equations by using the square root method. Write the radicals in simplest form.

1. $x^2 = 144$ *2.* $5x^2 = 320$ *3.* $(x-4)^2 = 12$ *4.* $(3x+1)^2 = 54$

Answers to Practice Problems: 1. $x = \pm 12$ **2.** $x = \pm 8$ **3.** $x = 4 \pm 2\sqrt{3}$ **4.** $x = \dfrac{-1 \pm 3\sqrt{6}}{3}$

10.1 Exercises

Solve the quadratic equations in Exercises 1 – 10 by factoring.

1. $x^2 = 11x$

2. $x^2 - 10x + 16 = 0$

3. $x^2 = -15x - 36$

4. $2x^2 + 36x + 34 = 0$

5. $9x^2 + 6x - 15 = 0$

6. $5x^2 + 17x = -6$

7. $(x+3)(x-1) = 4x$

8. $(x-7)(x-2) = 6$

9. $(2x-3)(2x+1) = 3x - 6$

10. $(x-2)(5x+4) = 3x^2 - 15x - 12$

Solve the quadratic equations in Exercises 11 – 40 by using the square root method. Write the radicals in simplest form.

11. $x^2 = 121$

12. $x^2 = 81$

13. $3x^2 = 108$

14. $5x^2 = 245$

15. $x^2 = 35$

16. $x^2 = 42$

17. $x^2 - 62 = 0$

18. $x^2 - 75 = 0$

19. $x^2 - 45 = 0$

20. $x^2 - 98 = 0$

21. $3x^2 = 54$

22. $5x^2 = 60$

23. $9x^2 = 4$

24. $4x^2 = 25$

25. $(x-1)^2 = 4$

26. $(x+3)^2 = 9$

27. $(x+2)^2 = -25$

28. $(x-5)^2 = 36$

29. $(x+1)^2 = \dfrac{1}{4}$

30. $(x-9)^2 = -\dfrac{9}{25}$

31. $(x-3)^2 = \dfrac{4}{9}$

32. $(x-2)^2 = \dfrac{1}{16}$

33. $(x-6)^2 = 18$

34. $(x+8)^2 = 75$

35. $2(x-7)^2 = 24$ **36.** $3(x+11)^2 = 60$ **37.** $(3x+4)^2 = 27$

38. $(2x+1)^2 = 48$ **39.** $(5x-2)^2 = 63$ **40.** $(4x-3)^2 = 125$

Use the Pythagorean Theorem to decide if each of the triangles determined by the three sides given in Exercises 41 – 44 is a right triangle.

41. 8 cm, 15 cm, 17 cm   **42.** 4 cm, 5 cm, 6 cm

43. $\sqrt{2}$ in., $\sqrt{2}$ in., 2 in.   **44.** 3 ft, 6 ft, $3\sqrt{5}$ ft

In each of the Exercises 45 – 50, right triangles are shown with two sides given. Determine the length of the missing side.

45. $a = 9, b = 12, c = ?$ **46.** $a = 10, b = 24, c = ?$

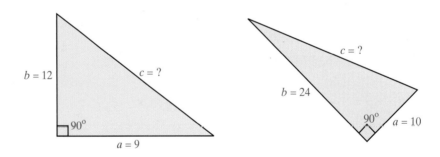

47. $a = 6, c = 12, b = ?$ **48.** $b = 10, c = 30, a = ?$

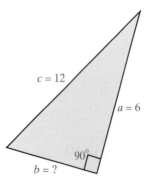

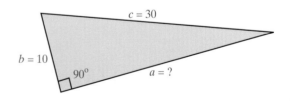

49. $a = 1, b = \sqrt{2}, c = ?$ **50.** $b = 4, c = 4\sqrt{2}, a = ?$

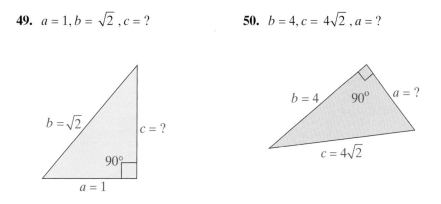

51. The hypotenuse of a right triangle is twice the length of one of the legs. The length of the other leg is $4\sqrt{3}$ feet. Find the length of the leg and the hypotenuse.

52. One leg of a right triangle is three times the length of the other. The length of the hypotenuse is 20 cm. Find the lengths of the legs.

53. The two legs of a right triangle are the same length. The hypotenuse is 6 centimeters long. Find the length of the legs.

54. The two legs of a right triangle are the same length. The hypotenuse is $4\sqrt{2}$ meters long. Find the length of the legs.

55. The top of a telephone pole is 40 feet above the ground and a guy wire is to be stretched from the top of the pole to a point on the ground 10 feet from the base of the pole. How long (to the nearest tenth of a foot) should the wire be if, for connecting purposes, 6 feet of wire is to be added after the distance is calculated?

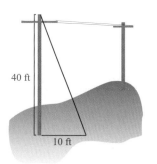

56. The library is located "around the corner" from the bank. If the library is 100 yards from the street corner and the bank is 75 yards around the corner on another street, what is the distance between the library and the bank "as the crow flies?"

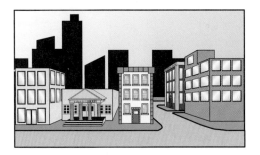

57. A ball is dropped from the top of a building that is known to be 144 feet high. The formula for finding the height of the ball at any time is $h = 144 - 16t^2$ where t is measured in seconds. How many seconds will it take for the ball to hit the ground?

58. An oil spill from a ruptured pipeline is circular in shape and covers an area of about 8 square miles. About how many miles long (to the nearest tenth of a mile) is the diameter of the oil spill? (Use $\pi = 3.14$.)

Calculator Problems

Use a calculator to solve the equations in Exercises 59 – 66. Round your answers to the nearest hundredth.

59. $x^2 = 647$ **60.** $x^2 = 378$ **61.** $19x^2 = 523$ **62.** $14x^2 = 795$

63. $6x^2 = 17.32$ **64.** $15x^2 = 229.63$ **65.** $2.1x^2 = 35.82$ **66.** $4.7x^2 = 118.34$

Use a calculator to find the two values of each expression accurate to 4 decimal places.

67. $2 \pm \sqrt{5}$ **68.** $-4 \pm \sqrt{89}$ **69.** $\dfrac{2 \pm \sqrt{7}}{2}$ **70.** $\dfrac{5 \pm \sqrt{13}}{10}$

Hawkes Learning Systems: Introductory Algebra

Quadratic Equations: The Square Root Method

10.2 Quadratic Equations: *Completing the Square*

After completing this section, you will be able to:

1. *Determine the constant terms that will make incomplete trinomials perfect square trinomials.*

2. *Solve quadratic equations by completing the square.*

Completing the Square

In Chapter 6 and Chapter 7, we discussed perfect square trinomials:

$$(x+a)^2 = x^2 + 2ax + a^2 \quad \text{and} \quad (x-a)^2 = x^2 - 2ax + a^2$$

In particular, in Section 7.4, we discussed factoring perfect square trinomials and how to add terms to binomials so that the resulting trinomial will be a perfect square. This procedure is called **completing the square**. The discussion here is how to solve quadratic equations by completing the square.

As review, we complete the square for $x^2 + 12x$ and $a^2 - 20a$. Both binomials have a leading coefficient of 1.

$$x^2 + 12x + \underline{} = ()^2$$

$$\frac{1}{2}(12) = 6 \text{ and } 6^2 = 36$$

Therefore, the number 36 will complete the square.

Thus

$$x^2 + 12x + \underline{36} = (x+6)^2$$

Similarly, for

$$a^2 - 20a + \underline{} = ()^2$$

we have

$$\frac{1}{2}(-20) = -10 \text{ and } (-10)^2 = 100$$

So

$$a^2 - 20a + \underline{100} = (a-10)^2$$

If the leading coefficient is **not 1**, we proceed to factor out the leading coefficient as follows:

$$2x^2 + 20x + \underline{\ ?\ } = 2(\qquad)^2$$

$$2\left(x^2 + 10x + \underline{\qquad}\right) = 2(\qquad)^2 \quad \text{Factor out 2.}$$

$$2\left(x^2 + 10x + \underline{25}\right) = 2\left(x + 5\right)^2 \quad \text{Complete the square of the expression inside the}$$

parentheses, $\dfrac{1}{2}(10) = 5$ and $5^2 = 25$.

$$2x^2 + 20x + 50 = 2\left(x + 5\right)^2 \quad \text{The number to be added to the original expression}$$

is $2 \cdot 25 = 50$.

Solving Quadratic Equations by Completing the Square

The following procedure is used in the examples to solve quadratic equations by completing the square.

To Solve Quadratic Equations by Completing the Square

1. Arrange terms with variables on one side and constants on the other.

2. Divide each term by the coefficient of x^2. (We want the leading coefficient to be 1.)

*3. Find the number that completes the square of the quadratic expression and add this number to **both** sides of the equation.*

4. Find the positive and negative square roots of both sides.

5. Solve for x. Remember, usually there will be two solutions.

Example 1: Completing The Square

Solve the following quadratic equations by completing the squares.

a. $x^2 - 6x + 4 = 0$

Solution: $x^2 - 6x + 4 = 0$

$\qquad\qquad x^2 - 6x = -4$ Add -4 to both sides of the equation.

$\qquad x^2 - 6x + 9 = -4 + 9$ Add 9 to both sides of the equation. The left side is now a **perfect square trinomial**. $\dfrac{1}{2}(-6) = -3$ and $\left(-3\right)^2 = 9$.

$\qquad\qquad \left(x - 3\right)^2 = 5$ Simplify.

$\qquad\qquad x - 3 = \pm\sqrt{5}$ Find square roots of both sides.

$\qquad\qquad x = 3 \pm \sqrt{5}$ Solve for x. Remember, there are two solutions: $3 + \sqrt{5}$ and $3 - \sqrt{5}$.

b. $x^2 + 5x = 7$

    Solution: $\quad x^2 + 5x = 7$

$$x^2 + 5x + \frac{25}{4} = 7 + \frac{25}{4}$$

Complete the square on the left: $\frac{1}{2} \cdot 5 = \frac{5}{2}$ and $\left(\frac{5}{2}\right)^2 = \frac{25}{4}$.

$$\left(x + \frac{5}{2}\right)^2 = \frac{53}{4}$$

Simplify: $\left(7 + \frac{25}{4} = \frac{28}{4} + \frac{25}{4} = \frac{53}{4}\right)$.

$$x + \frac{5}{2} = \pm\sqrt{\frac{53}{4}}$$

Find square roots.

$$x = -\frac{5}{2} \pm \sqrt{\frac{53}{4}}$$

Solve for x.

$$x = -\frac{5}{2} \pm \frac{\sqrt{53}}{2}$$

Special property of square roots: $\sqrt{\frac{a}{b}} = \frac{\sqrt{a}}{\sqrt{b}}$ for $a > 0$ and $b > 0$.

$$\text{or } x = \frac{-5 \pm \sqrt{53}}{2}$$

Combine the fractions.

c. $6x^2 + 12x - 9 = 0$

    Solution: $\quad 6x^2 + 12x - 9 = 0$

$$6x^2 + 12x = 9$$

Add 9 to both sides of the equation.

$$\frac{6x^2}{6} + \frac{12x}{6} = \frac{9}{6}$$

Divide each term by 6 **so that the leading coefficient will be 1**.

$$x^2 + 2x = \frac{3}{2}$$

The leading coefficient is 1.

$$x^2 + 2x + 1 = \frac{3}{2} + 1$$

Complete the square: $\frac{1}{2} \cdot 2 = 1$ and $1^2 = 1$

$$(x + 1)^2 = \frac{5}{2}$$

Simplify.

$$x + 1 = \pm\sqrt{\frac{5}{2}}$$

Find square roots.

$$x = -1 \pm \sqrt{\frac{5}{2}}$$

Solve for x.

$$x = -1 \pm \frac{\sqrt{5}}{\sqrt{2}} \cdot \frac{\sqrt{2}}{\sqrt{2}}$$

Rationalize the denominator.

Continued on next page...

$$x = -1 \pm \frac{\sqrt{10}}{2}$$ Simplify.

or $$x = \frac{-2 \pm \sqrt{10}}{2}$$ Combine the fractions with a common denominator.

d. $2x^2 + 5x - 8 = 0$

    Solution: $2x^2 + 5x - 8 - 0$

$$2x^2 + 5x = 8$$ Add 8 to both sides of the equation.

$$\frac{2x^2}{2} + \frac{5x}{2} = \frac{8}{2}$$ Divide each term by 2 **so that the leading coefficient will be 1**.

$$x^2 + \frac{5}{2}x = 4$$ Simplify.

$$x^2 + \frac{5}{2}x + \frac{25}{16} = 4 + \frac{25}{16}$$ Complete the square: $\frac{1}{2} \cdot \frac{5}{2} = \frac{5}{4}$ and $\left(\frac{5}{4}\right)^2 = \frac{25}{16}$.

$$\left(x + \frac{5}{4}\right)^2 = \frac{89}{16}$$ Simplify $\left(4 + \frac{25}{16} = \frac{64}{16} + \frac{25}{16} = \frac{89}{16}\right)$.

$$x + \frac{5}{4} = \pm\sqrt{\frac{89}{16}}$$ Find the square roots.

$$x = -\frac{5}{4} \pm \frac{\sqrt{89}}{4}$$ Solve for x. Also, $\sqrt{\frac{89}{16}} = \frac{\sqrt{89}}{\sqrt{16}} = \frac{\sqrt{89}}{4}$.

or $$x = \frac{-5 \pm \sqrt{89}}{4}$$ Combine the fractions.

e. $x^2 + 8x = -4$

    Solution: $x^2 + 8x = -4$

$$x^2 + 8x + 16 = -4 + 16$$ Complete the square.

$$(x + 4)^2 = 12$$ Simplify.

$$x + 4 = \pm\sqrt{12}$$ Find square roots.

$$x = -4 \pm \sqrt{12}$$ Solve for x.

$$x = -4 \pm 2\sqrt{3}$$ $\sqrt{12} = \sqrt{4} \cdot \sqrt{3} = 2\sqrt{3}$ by special property of radicals $\sqrt{ab} = \sqrt{a} \cdot \sqrt{b}$ for $a > 0$ and $b > 0$.

Solving quadratic equations by completing the square is a good technique and it provides practice with several algebraic operations. However, do not forget that, in general, factoring is the preferred technique whenever the factors are relatively easy to find. Some of the following exercises allow the options of solving by factoring or by completing the square.

10.2 Exercises

Add the correct constants in Exercises 1 – 10 and factor the resulting trinomial as indicated. (Constant factors of the x^2 term must be dealt with in Exercises 3, 4, 9, and 10.)

1. $x^2 + 12x + \underline{\hspace{0.5cm}} = (\hspace{1cm})^2$

2. $x^2 - 14x + \underline{\hspace{0.5cm}} = (\hspace{1cm})^2$

3. $2x^2 - 16x + \underline{\hspace{0.5cm}} = 2(\hspace{1cm})^2$

4. $5x^2 + 10x + \underline{\hspace{0.5cm}} = 5(\hspace{1cm})^2$

5. $x^2 - 3x + \underline{\hspace{0.5cm}} = (\hspace{1cm})^2$

6. $x^2 + 5x + \underline{\hspace{0.5cm}} = (\hspace{1cm})^2$

7. $x^2 + x + \underline{\hspace{0.5cm}} = (\hspace{1cm})^2$

8. $x^2 - 7x + \underline{\hspace{0.5cm}} = (\hspace{1cm})^2$

9. $2x^2 + 4x + \underline{\hspace{0.5cm}} = 2(\hspace{1cm})^2$

10. $3x^2 + 18x + \underline{\hspace{0.5cm}} = 3(\hspace{1cm})^2$

Solve each quadratic equation in Exercises 11 – 30 by completing the square.

11. $x^2 + 6x - 7 = 0$

12. $x^2 + 8x + 12 = 0$

13. $x^2 - 4x - 45 = 0$

14. $x^2 - 10x + 21 = 0$

15. $x^2 - 3x - 40 = 0$

16. $x^2 + x - 42 = 0$

17. $3x^2 + x - 4 = 0$ **18.** $2x^2 + x - 6 = 0$ **19.** $4x^2 - 4x - 3 = 0$

20. $3x^2 + 11x + 10 = 0$ **21.** $x^2 + 6x + 3 = 0$ **22.** $x^2 - 4x - 3 = 0$

23. $x^2 + 2x - 5 = 0$ **24.** $x^2 + 8x + 2 = 0$ **25.** $3x^2 - 2x - 1 = 0$

26. $2x^2 + 5x + 2 = 0$ **27.** $x^2 + x - 3 = 0$ **28.** $x^2 - 5x + 1 = 0$

29. $2x^2 + 3x - 1 = 0$ **30.** $3x^2 - 4x - 2 = 0$

Solve the quadratic equations in Exercises 31 – 55 by factoring or completing the square.

31. $x^2 - 9x + 2 = 0$ **32.** $x^2 - 8x - 20 = 0$ **33.** $x^2 + 7x - 14 = 0$

34. $x^2 + 5x + 3 = 0$ **35.** $x^2 - 11x - 26 = 0$ **36.** $x^2 + 6x - 4 = 0$

37. $2x^2 + x - 2 = 0$ **38.** $6x^2 - 2x - 2 = 0$ **39.** $3x^2 - 6x + 3 = 0$

40. $5x^2 + 15x - 5 = 0$ **41.** $4x^2 + 20x - 8 = 0$ **42.** $2x^2 - 7x + 4 = 0$

43. $6x^2 - 8x + 1 = 0$ **44.** $5x^2 - 10x + 3 = 0$ **45.** $2x^2 + 7x + 4 = 0$

46. $3x^2 - 5x - 3 = 0$ **47.** $4x^2 - 2x - 3 = 0$ **48.** $2x^2 - 9x + 7 = 0$

49. $3x^2 + 8x + 5 = 0$ **50.** $5x^2 + 11x - 1 = 0$ **51.** $2x^2 + 5x + 3 = 0$

52. $2x^2 - 12x + 10 = 0$ **53.** $3x^2 + 24x + 21 = 0$ **54.** $4x^2 + 6x - 10 = 0$

55. $6x^2 - x - 1 = 0$

Hawkes Learning Systems: Introductory Algebra

Quadratic Equations: Completing the Square

10.3   Quadratic Equations: *The Quadratic Formula*

Objectives

After completing this section, you will be able to:

1. *Write quadratic equations in standard form.*

2. *Identify the coefficients of quadratic equations in standard form.*

3. *Solve quadratic equations by using the quadratic formula.*

The Quadratic Formula: $x = \dfrac{-b \pm \sqrt{b^2 - 4ac}}{2a}$

Now we are interested in developing a formula that will be useful in solving quadratic equations of any form. **This formula will always work**, but do not forget the factoring and completing the square techniques because in some cases they are easier to apply than the formula.

General Form of a Quadratic Equation

The **general quadratic equation** is

$$ax^2 + bx + c = 0$$

where a, b, and c are real constants and $a \neq 0$.

We want to solve the general quadratic equation for x in terms of the constant coefficients $a, b,$ and c. The technique is to **complete the square** (Section 10.2).

Development of the Quadratic Formula

$ax^2 + bx + c = 0$	Begin with the general quadratic equation.
$ax^2 + bx = -c$	Add $-c$ to both sides.
$\dfrac{ax^2}{a} + \dfrac{bx}{a} = \dfrac{-c}{a}$	Divide each term by a so that the leading coefficient will be 1.
$x^2 + \dfrac{b}{a}x = \dfrac{-c}{a}$	Rewrite: $\dfrac{bx}{a} = \dfrac{b}{a}x$.
$x^2 + \dfrac{b}{a}x + \left(\dfrac{b}{2a}\right)^2 = \left(\dfrac{b}{2a}\right)^2 + \dfrac{-c}{a}$	Complete the square: $\dfrac{1}{2}\left(\dfrac{b}{a}\right) = \dfrac{b}{2a}$.

$$\left(x + \frac{b}{2a}\right)^2 = \frac{b^2}{4a^2} + \frac{-c}{a} \qquad \text{Simplify.}$$

$$\left(x + \frac{b}{2a}\right)^2 = \frac{b^2}{4a^2} + \frac{-c \cdot 4a}{a \cdot 4a} \qquad \text{Common denominator is } 4a^2.$$

$$\left(x + \frac{b}{2a}\right)^2 = \frac{b^2 - 4ac}{4a^2} \qquad \text{Simplify.}$$

$$x + \frac{b}{2a} = \pm\sqrt{\frac{b^2 - 4ac}{4a^2}} \qquad \text{Find the square roots of both sides.}$$

$$x + \frac{b}{2a} = \pm\frac{\sqrt{b^2 - 4ac}}{\sqrt{4a^2}} \qquad \text{Use the relationship } \sqrt{\frac{a}{b}} = \frac{\sqrt{a}}{\sqrt{b}} \text{ if } a, b > 0.$$

$$x + \frac{b}{2a} = \pm\frac{\sqrt{b^2 - 4ac}}{2a} \qquad \text{Simplify.}$$

$$x = \frac{-b}{2a} \pm \frac{\sqrt{b^2 - 4ac}}{2a} \qquad \text{Solve for } x.$$

$$x = \frac{-b \pm \sqrt{b^2 - 4ac}}{2a} \qquad \text{THE QUADRATIC FORMULA}$$

Or, we can write the two solutions separately as:

$$x = \frac{-b + \sqrt{b^2 - 4ac}}{2a} \quad \text{and} \quad x = \frac{-b - \sqrt{b^2 - 4ac}}{2a}$$

Quadratic Formula

The solutions of the general quadratic equation $ax^2 + bx + c = 0$, where $a \neq 0$, are

$$\boldsymbol{x = \frac{-b \pm \sqrt{b^2 - 4ac}}{2a}}.$$

NOTES

Special Note:

The expression $b^2 - 4ac$ is called the **discriminant**.

If

$b^2 - 4ac = 0 \rightarrow$ there is only one real solution.

$b^2 - 4ac > 0 \rightarrow$ there are two real solutions.

$b^2 - 4ac < 0 \rightarrow$ there are no real solutions. (The square root of a negative number is not a real number.)

Discussions of negative discriminants are given in later courses in algebra.

Solving Quadratic Equations by Using the Quadratic Formula

Now the solutions to quadratic equations can be found by going directly to the formula.

Example 1: The Quadratic Formula

Solve the following quadratic equations by using the quadratic formula:

$$x = \frac{-b \pm \sqrt{b^2 - 4ac}}{2a}.$$

a. $2x^2 + x - 2 = 0$

Solution: $a = 2, b = 1, c = -2$

$$x = \frac{-1 \pm \sqrt{1^2 - 4(2)(-2)}}{2 \cdot 2} = \frac{-1 \pm \sqrt{1 + 16}}{4} = \frac{-1 \pm \sqrt{17}}{4}$$

In many practical applications of quadratic equations, we want to know a decimal approximation of the solutions. Using a calculator, we find the following approximate values to the solutions of the equation $2x^2 + x - 2 = 0$:

$$x = \frac{-1 + \sqrt{17}}{4} \approx 0.78078 \quad \text{and} \quad x = \frac{-1 - \sqrt{17}}{4} \approx -1.2808$$

b. $-3x^2 = -5x + 1$

Solution: $-3x^2 = -5x + 1$ First, rewrite the equation so that one side is 0.
$-3x^2 + 5x - 1 = 0$

Now we see that $a = -3, b = 5,$ and $c = -1$.

Substituting in the quadratic formula gives

$$x = \frac{-5 \pm \sqrt{5^2 - 4(-3)(-1)}}{2(-3)} = \frac{-5 \pm \sqrt{25 - 12}}{-6} = \frac{-5 \pm \sqrt{13}}{-6} = \frac{5 \pm \sqrt{13}}{6}.$$

This shows that the quadratic formula works correctly even though the leading coefficient a is negative. We could also multiply all the terms on both sides of the equation by -1 and solve the new equation. **The solutions will be the same.**

$$-3x^2 + 5x - 1 = 0$$
$$3x^2 - 5x + 1 = 0 \quad\quad \text{Multiply every term by } -1.$$

Now $a = 3$, $b = -5$, and $c = 1$.

$$x = \frac{-(-5) \pm \sqrt{(-5)^2 - 4(3)(1)}}{2 \cdot 3} = \frac{5 \pm \sqrt{25 - 12}}{6} = \frac{5 \pm \sqrt{13}}{6}$$

Again, if needed, a calculator will give approximate values for the solutions:

$$x = \frac{5 + \sqrt{13}}{6} \approx 1.4343 \quad \text{and} \quad x = \frac{5 - \sqrt{13}}{6} \approx 0.2324.$$

c. $\dfrac{1}{6}x^2 - x + \dfrac{1}{2} = 0$

Solution: $6 \cdot \dfrac{1}{6}x^2 - 6 \cdot x + 6 \cdot \dfrac{1}{2} = 6 \cdot 0$ Multiply each term by 6, the least common denominator.

$x^2 - 6x + 3 = 0$ Integer coefficients are much easier to use in the formula.

So $a = 1$, $b = -6$, and $c = 3$.

$$x = \frac{-(-6) \pm \sqrt{(-6)^2 - 4(1)(3)}}{2 \cdot 1} = \frac{6 \pm \sqrt{36 - 12}}{2}$$

$$= \frac{6 \pm \sqrt{24}}{2} = \frac{6 \pm 2\sqrt{6}}{2} = \frac{2(3 \pm \sqrt{6})}{2}$$

$$= 3 \pm \sqrt{6}$$

d. $2x^2 - 25 = 0$

Solution: The square root method could be applied by adding 25 to both sides, dividing by 2, and then taking square roots. The result is the same by using the quadratic formula with $b = 0$.

$$2x^2 - 25 = 0 \ \left(\text{or } 2x^2 + 0x - 25 = 0 \right)$$

So, $a = 2$, $b = 0$, and $c = -25$.

$$x = \frac{-(0) \pm \sqrt{0^2 - 4(2)(-25)}}{2 \cdot 2} = \frac{\pm \sqrt{200}}{4} = \frac{\pm 10\sqrt{2}}{4} = \frac{\pm 5\sqrt{2}}{2}$$

e. $(3x - 1)(x + 2) = 4x$

Solution: $(3x - 1)(x + 2) = 4x$

$3x^2 + 5x - 2 = 4x$ Multiply on the left-hand side.

$3x^2 + x - 2 = 0$ Subtract $4x$ from both sides so that one side is 0.

Continued on next page...

Now we see that $a = 3$, $b = 1$, and $c = -2$.

Substituting in the quadratic formula gives

$$x = \frac{-1 \pm \sqrt{1^2 - 4(3)(-2)}}{2 \cdot 3} = \frac{-1 \pm \sqrt{25}}{6} = \frac{-1 \pm 5}{6}$$

$$x = \frac{-1 + 5}{6} = \frac{4}{6} = \frac{2}{3} \quad \text{and} \quad x = \frac{-1 - 5}{6} = \frac{-6}{6} = -1$$

Note: Whenever the solutions are rational numbers, the equation can be solved by factoring. In this example, we could have solved as follows:

$$3x^2 + x - 2 = 0$$

$$(3x - 2)(x + 1) = 0$$

$$3x - 2 = 0 \text{ and } x + 1 = 0$$

$$x = \frac{2}{3} \qquad x = -1$$

f. $4x^2 + 12x + 9 = 0$

Solution: $a = 4$, $b = 12$, $c = 9$

$$x = \frac{-12 \pm \sqrt{12^2 - 4(4)(9)}}{2 \cdot 4}$$

$$= \frac{-12 \pm \sqrt{144 - 144}}{8} = \frac{-12 \pm \sqrt{0}}{8} = -\frac{12}{8} = -\frac{3}{2}$$

Note that when the discriminant is 0, there is only one solution. This equation could also be solved by factoring in the following manner:

$4x^2 + 12x + 9 = 0$ The quadratic expression is a **perfect square trinomial**.

$$(2x + 3)^2 = 0$$

$2x + 3 = 0$ The factor $2x + 3$ is repeated.

$$2x = -3$$

$x = -\dfrac{3}{2}$ $-\dfrac{3}{2}$ is called a **double solution** or a **double root**.

Quadratic equations should be solved by factoring whenever possible. Factoring is generally the easiest method. This fact is illustrated in Examples 1e and 1f. However, the quadratic formula will always give the solutions whether they are rational, irrational, or nonreal (as shown in Example 2a). Thus the quadratic formula is a very useful tool. **In future courses in mathematics every text and instructor will assume that you know or can easily remember the quadratic formula.**

Example 2: Nonreal Solutions

Solve the following equation: $x^2 + x + 1 = 0$.

Solution: $a = 1, \ b = 1, \ c = 1$

$$x = \frac{-1 \pm \sqrt{1^2 - 4(1)(1)}}{2 \cdot 1} = \frac{-1 \pm \sqrt{1-4}}{2} = \frac{-1 \pm \sqrt{-3}}{2}$$

There is no real solution. This example illustrates the fact that not every equation has real solutions. None of the exercises in this text has this kind of answer. Such solutions are called **nonreal complex numbers** and will be discussed in detail in the next course in algebra.

Practice Problems

Solve the equations by using the quadratic formula.

1. $x^2 + 5x + 1 = 0$

2. $4x^2 - x = 1$

3. $(x+2)(x-2) = 4x$

4. $-2x^2 + 3 = 0$ (**Note:** Here $b = 0$.)

5. $5x^2 - x = 0$ (**Note:** Here $c = 0$.)

Solve the following quadratic equations and write the answers in decimal form accurate to four decimal places.

6. $x^2 + 10x + 5 = 0$

7. $5x^2 - x - 1 = 0$

Answers to Practice Problems: 1. $x = \dfrac{-5 \pm \sqrt{21}}{2}$ **2.** $x = \dfrac{1 \pm \sqrt{17}}{8}$ **3.** $x = 2 \pm 2\sqrt{2}$ **4.** $x = \dfrac{0 \pm 2\sqrt{6}}{-4}$ or $x = \dfrac{\pm\sqrt{6}}{2}$

5. $x = 0, \dfrac{1}{5}$ (This problem could be solved by factoring.)

6. $x = -0.5279, -9.4721$ **7.** $x = -0.3583, 0.5583$

10.3 Exercises

Rewrite each of the quadratic equations in Exercises 1 – 10 in the form $ax^2 + bx + c = 0$ with $a > 0$; then identify the constants $a, b,$ and c.

1. $x^2 - 3x = 2$

2. $x^2 + 2 = 5x$

3. $x = 2x^2 + 6$

4. $5x^2 = 3x - 1$

5. $4x + 3 = 7x^2$

6. $x - 4 - 3x^2$

7. $4 = 3x^2 - 9x$

8. $6x + 4 = 3x^2$

9. $x^2 + 5x = 3 - x^2$

10. $x^2 + 4x - 1 = 2x + 3x^2$

Solve the quadratic equations in Exercises 11 – 36 by using the quadratic formula.

11. $x^2 - 4x - 1 = 0$

12. $x^2 - 3x + 1 = 0$

13. $x^2 - 3x = 4$

14. $x^2 + 5x = 2$

15. $-2x^2 + x = -1$

16. $-3x^2 + x = -1$

17. $5x^2 + 3x - 2 = 0$

18. $-2x^2 + 5x - 1 = 0$

19. $9x^2 = 3x$

20. $4x^2 - 81 = 0$

21. $x^2 - 7 = 0$

22. $2x^2 + 5x - 3 = 0$

23. $x^2 + 4x = x - 2x^2$

24. $x^2 - 2x + 1 = 2 - 3x^2$

25. $3x^2 + 4x = 0$

26. $4x^2 - 10 = 0$

27. $\dfrac{2}{5}x^2 + x - 1 = 0$

28. $2x^2 + 3x - \dfrac{3}{4} = 0$

29. $-\dfrac{x^2}{2} - 2x + \dfrac{1}{3} = 0$

30. $\dfrac{x^2}{3} - x - \dfrac{1}{5} = 0$

31. $(2x + 1)(x + 3) = 2x + 6$

32. $(x + 5)(x - 1) = -3$

33. $\dfrac{5x + 2}{3x} = x - 1$

34. $-6x^2 = -3x - 1$

35. $4x^2 = 7x - 3$

36. $9x^2 - 6x - 1 = 0$

Solve the quadratic equations in Exercises 37 – 54 by using any method (factoring, completing the square, or the quadratic formula).

37. $2x^2 + 7x + 3 = 0$ **38.** $5x^2 - x - 4 = 0$ **39.** $3x^2 - 7x + 1 = 0$

40. $2x^2 - 2x - 1 = 0$ **41.** $10x^2 = x + 24$ **42.** $9x^2 + 12x = -2$

43. $3x^2 - 11x = 4$ **44.** $5x^2 = 7x + 5$ **45.** $-4x^2 + 11x - 5 = 0$

46. $-4x^2 + 12x - 9 = 0$ **47.** $10x^2 + 35x + 30 = 0$ **48.** $6x^2 + 2x = 20$

49. $25x^2 + 4 = 20$ **50.** $3x^2 - 4x + \dfrac{1}{3} = 0$ **51.** $\dfrac{3}{4}x^2 - 2x + \dfrac{1}{8} = 0$

52. $\dfrac{11}{2}x + 1 = 3x^2$ **53.** $\dfrac{3}{7}x^2 = \dfrac{1}{2}x + 1$ **54.** $\dfrac{35}{4x} = x - 1$

Solve the quadratic equations in Exercises 55 – 60 and write the answers in decimal form accurate to four decimal places.

55. $x^2 - x - 1 = 0$ **56.** $3x^2 - 8x - 13 = 0$ **57.** $(2x + 1)(2x - 1) = 3x$

58. $\dfrac{-x - 1}{3x} = x + 1$ **59.** $5x^2 - 100 = 0$ **60.** $(3x + 1)(3x - 2) = 10x$

Hawkes Learning Systems: Introductory Algebra

Quadratic Equations: The Quadratic Formula

10.4   Applications

Objectives

After completing this section, you will be able to:

Solve applied problems using quadratic equations.

Application problems are designed to teach you to read carefully, to think clearly, and to translate from English to algebraic expressions and equations. The problems do not tell you directly to add, subtract, multiply, divide, or square. You must decide on a method of attack based on the wording of the problem and your previous experience and knowledge. Study the following examples and read the explanations carefully. They are similar to some, but not all, of the problems in the exercises. Example 1 makes use of the Pythagorean Theorem that was discussed in Sections 9.7 and 10.1.

Example 1: The Pythagorean Theorem

The length of a rectangle is 7 feet longer than the width. If one diagonal measures 13 feet, what are the dimensions of the rectangle?

Solution: Draw a diagram for problems involving geometric figures whenever possible.

Let $x =$ width of the rectangle

and $x + 7 =$ length of the rectangle.

Then, by the Pythagorean Theorem:

$$\left(x+7 \right)^2 + x^2 = 13^2$$

$$x^2 + 14x + 49 + x^2 = 169$$

$$2x^2 + 14x + 49 - 169 = 0$$

$$2x^2 + 14x - 120 = 0$$

$$2\left(x^2 + 7x - 60 \right) = 0$$

$$2\left(x - 5 \right)\left(x + 12 \right) = 0$$

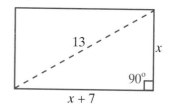

$x - 5 = 0$ or $x + 12 = 0$

$x = 5$ $\cancel{x = -12}$ A negative number does not fit

Thus $x = 5$ and $x + 7 = 12$.   the conditions of the problem.

Check: $5^2 + 12^2 \overset{?}{=} 13^2$

$25 + 144 \overset{?}{=} 169$

$169 = 169$ True Statement

The width is 5 ft and the length is 12 ft.

Example 2: Work

Working for a janitorial service, a woman and her daughter can clean a building in 5 hours. If the daughter were to do the job by herself, she would take 24 hours longer than her mother would take. How long would it take her mother to clean the building without the daughter's help?

(**Note**: For reference, this problem is similar to those discussed in Section 8.6 on work. The difference is that in this section the resulting equations are quadratic.)

Solution: Let x = time for mother alone
and $x + 24$ = time for daughter alone.

	Hours to Complete	Part Completed in 1 Hour
Mother	x	$\dfrac{1}{x}$
Daughter	$x + 24$	$\dfrac{1}{x+24}$
Together	5	$\dfrac{1}{5}$

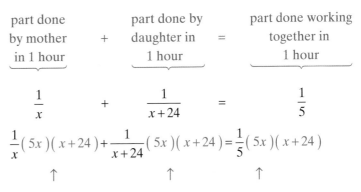

$$\frac{1}{x}(5x)(x+24) + \frac{1}{x+24}(5x)(x+24) = \frac{1}{5}(5x)(x+24)$$

↑   ↑   ↑

Multiply each term by the LCM of the denominators.

Continued on next page...

$$5(x+24)+5x = x(x+24)$$

$$5x+120+5x = x^2 +24x$$

$$0 = x^2 +24x-10x-120$$

$$0 = x^2 +14x-120$$

$$0 = (x-6)(x+20)$$

$$x-6 = 0 \text{ or } x+20 = 0$$

$$x = 6 \qquad \cancel{x = -20}$$

The mother could do the job alone in 6 hours.

Example 3: Distance-Rate-Time

A small plane travels at a speed of 200 mph in still air. Flying with a tailwind, the plane is clocked over a distance of 960 miles. Flying against a headwind, it takes 2 hours more time to complete the return trip. What was the wind velocity?

Solution: The basic formula is $d = rt$ (distance = rate $\cdot$ time).

Also, $t = \dfrac{d}{r}$ and $r = \dfrac{d}{t}$.

Let x = wind velocity.
Then $200 + x$ = speed of airplane going with the wind (tailwind),
           $200 - x$ = speed of airplane returning against the wind
                         (headwind),
and   960 = distance each way.

We know distance and can represent rate (or speed), so the formula $t = \dfrac{d}{r}$ is used for representing time.

	Rate	Time	Distance
Going	$200 + x$	$\dfrac{960}{200 + x}$	960
Returning	$200 - x$	$\dfrac{960}{200 - x}$	960

$$\underbrace{\begin{array}{c} \text{time} \\ \text{returning} \end{array}}_{} \quad - \quad \underbrace{\begin{array}{c} \text{time} \\ \text{going} \end{array}}_{} \quad = \quad \underbrace{\begin{array}{c} \text{difference} \\ \text{in time} \end{array}}_{}$$

$$\frac{960}{200 - x} \quad - \quad \frac{960}{200 + x} \quad = \quad 2$$

$$\frac{960}{200 - x}(200 - x)(200 + x) - \frac{960}{200 + x}(200 - x)(200 + x) = 2(200 - x)(200 + x)$$

$$960(200 + x) - 960(200 - x) = 2\left(40{,}000 - x^2\right)$$

$$192{,}000 + 960x - 192{,}000 + 960x = 80{,}000 - 2x^2$$

$$2x^2 + 1920x - 80{,}000 = 0$$

$$x^2 + 960x - 40{,}000 = 0$$

$$(x - 40)(x + 1000) = 0$$

$$x - 40 = 0 \quad \text{or} \quad x + 1000 = 0$$

$$x = 40 \qquad \qquad \cancel{x = -1000}$$

The wind velocity was 40 mph.

Check: $\quad \dfrac{960}{200 - 40} - \dfrac{960}{200 + 40} \overset{?}{=} 2$

$$\frac{960}{160} - \frac{960}{240} \overset{?}{=} 2$$

$$6 - 4 \overset{?}{=} 2$$

$$2 = 2 \qquad \qquad \text{True Statement}$$

Example 4: Geometry

a. A square piece of cardboard has a small square, 2 in. by 2 in., cut from each corner. The edges are then folded up to form a box with a volume of 5000 in.^{3} What are the dimensions of the box? (**Hint**: The volume is the product of the length, width, and height: $V = lwh$.)

Solution: Draw a diagram illustrating the information.

Let x – one side of the square.

$$2(x-4)(x-4) = 5000$$

$$2(x^2 - 8x + 16) = 5000$$

$$x^2 - 8x + 16 = 2500$$

$$x^2 - 8x - 2484 = 0$$

$$(x - 54)(x + 46) = 0$$

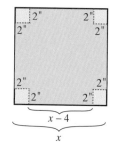

$$x - 54 = 0 \text{ or } x + 46 = 0$$

$$x = 54 \qquad x = -46$$

$$x - 4 = 50$$

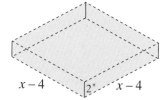

The dimensions of the box are 50 in. by 50 in. by 2 in.

b. A little league baseball field is in the shape of a square with sides of 60 feet. What is the distance (to the nearest tenth of a foot) the catcher must throw from home plate to second base?

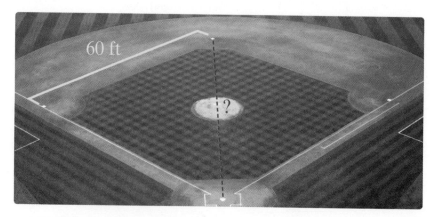

Solution: Since the distance from home plate to second base is the hypotenuse of a right triangle with sides of length 60 feet, we can use the Pythagorean Theorem as follows:

$$x^2 = 60^2 + 60^2$$

$$x^2 = 3600 + 3600$$

$$x^2 = 7200$$

$$x = \sqrt{7200} \qquad \text{We want only the positive square root.}$$

$$x = 84.9 \text{ ft} \qquad \text{Accurate to the nearest tenth of a foot.}$$

The catcher must throw about 84.9 feet.

10.4 Exercises

Find the solutions to each of the following problems by setting up and solving a quadratic equation.

1. The sum of a positive number and its square is 132. Find the number.

2. The dimensions of a rectangle can be represented by two consecutive even integers. The area of the rectangle is 528 square centimeters. Find the width and the length of the rectangle. (**Hint**: area = length · width.)

3. A rectangle has a length 5 meters less than twice its width. If the area is 63 square meters, find the dimensions of the rectangle. (**Hint**: area = length · width.)

4. The sum of a positive number and its square is 992. Find the number.

5. The area of a rectangular field is 198 square meters. If it takes 58 meters of fencing to enclose the field, what are the dimensions of the field? (**Hint**: The length plus the width is 29 meters.)

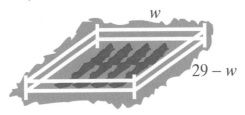

w

$29 - w$

6. The length of a rectangle exceeds the width by 5 cm. If both were increased by 3 cm, the area would be increased by 96 cm^2. Find the dimensions of the original rectangle.

[**Hint**: The original rectangle has area $w(w+5)$.]

7. The Wilsons have a rectangular swimming pool that is 10 ft longer than it is wide. The pool is completely surrounded by a concrete deck that is 6 ft wide. The total area of the pool and the deck is 1344 ft^2. Find the dimensions of the pool.

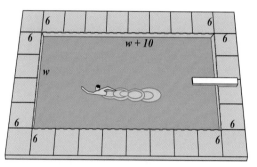

8. The product of two positive consecutive odd integers exceeds their sum by 287. Find the positive integers.

9. The sum of the squares of two positive consecutive integers is 221. Find the positive integers.

10. The difference between two positive numbers is 9. If the smaller number is added to the square of the larger number, the result is 147. Find the numbers.

11. The difference between a positive number and 3 is four times the reciprocal of the number. Find the number. (**Hint**: The reciprocal of x is $\dfrac{1}{x}$.)

12. The sum of a positive number and 5 is fourteen times the reciprocal of the number. Find the number. (**Hint**: The reciprocal of x is $\dfrac{1}{x}$.)

13. Each side of a square is increased by 10 cm. The area of the resulting square is 9 times the area of the original square. Find the length of the sides of the original square.

14. If 5 meters are added to each side of a square, the area of the resulting square is four times the area of the original square. Find the length of the sides of the original square.

15. The diagonal of a rectangle is 13 meters. The length is 2 meters more than twice the width. Find the dimensions of the rectangle.

16. The length of a rectangle is 4 meters more than its width. If the diagonal is 20 meters, what are the dimensions of the rectangle?

17. A right triangle has two equal sides. If the hypotenuse is 12 cm, what is the length of the equal sides?

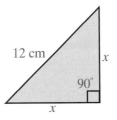

18. Mr. Prince owns a 15 unit apartment complex. If all units are rented, the rent for each apartment is $200 per month. Each time the rent is increased by $20, he will lose 1 tenant. What is the rental rate if he receives $3120 monthly in rent? (**Hint**: Let x = number of empty units.)

19. The Ski Club is planning to charter a bus to a ski resort. The cost will be $900 and each member will share the cost equally. If the club had 15 more members, the cost per person would be $10 less. How many are in the club now?

(**Hint**: If x = number in club now, $\dfrac{900}{x}$ = cost per person.)

20. A sporting goods store owner estimates that if he sells a certain model of basketball shoes for x dollars a pair, he will be able to sell $125 - x$ pairs. Find the price if his sales are $3750. Is there more than one possible answer?

21. Mr. Green traveled to a city 200 miles from his home to attend a meeting. Due to car trouble, his average speed returning was 10 mph less than his speed going. If the total driving time for the round trip was 9 hours, at what rate of speed did he travel to the city?

	Rate	Time	Distance
Going	x	?	200
Returning	$x - 10$	?	200

22. A motorboat takes a total of 2 hours to travel 8 miles downstream and 4 miles back on a river that is flowing at a rate of 2 mph. Find the rate of the boat in still water.

23. A small motorboat travels 12 mph in still water. It takes 2 hours longer to travel 45 miles going upstream than it does going downstream. Find the rate of the current. (**Hint**: $12 + c =$ rate going downstream and $12 - c =$ rate going upstream.)

24. Recently Mr. and Mrs. Roberts spent their vacation in San Francisco, which is 540 miles from their home. Being a little reluctant to return home, the Roberts took 2 hours longer on their return trip and their average speed was 9 mph slower than when they were going. What was their average rate of speed as they traveled from home to San Francisco?

25. The Blumin Garden Club planned to give their president a gift of appreciation costing $120 and to divide the cost evenly. In the meantime, 5 members dropped out of the club. If it now costs each of the remaining members $2 more than originally planned, how many members initially participated in the gift buying?

(**Hint**: If x = number in club initially, $\dfrac{120}{x}$ = cost per member.)

26. A rectangular sheet of metal is 6 in. longer than it is wide. A box is to be made by cutting out 3 in. squares at each corner and folding up the sides. If the box has a volume of 336 in.^{3}, what were the original dimensions of the sheet metal? (See Example 4.)

27. A box is to be made out of a square piece of cardboard by cutting out 2 in. squares at each corner and folding up the sides. If the box has a volume of 162 in.^{3}, how big was the piece of cardboard? (See Example 4.)

28. A woman and her daughter can paint their cabin in 3 hours. Working alone it would take the daughter 8 hours longer than it would the mother. How long would it take the mother to paint the cabin alone?

29. Two pipes can fill a tank in 8 minutes if both are turned on. If only one is used it would take 30 minutes longer for the smaller pipe to fill the tank than the larger pipe. How long will it take the smaller pipe to fill the tank?

30. A farmer and his son can plow a field with two tractors in 4 hours. If it would take the son 6 hours longer than the father to plow the field alone, how long would it take each if they worked alone?

31. A ball is thrown upward with an initial velocity of 32 feet per second from the edge of a cliff near the beach. The cliff is 50 feet above the beach and the height of the ball can be found by using the equation $h = -16t^2 + 32t + 50$ where t is measured in seconds.

a. When will the ball be 66 feet above the beach?
b. When will the ball be 30 feet above the beach?
c. In about how many seconds will the ball hit the beach?

32. The height of a projectile fired upward from the ground with a velocity of 128 feet per second is given by the formula $h = -16t^2 + 128t$.

a. When will the projectile be 256 feet above the ground?
b. Will the projectile ever be 300 feet above the ground? Explain.
c. When will the projectile be 240 feet above the ground?
d. In how many seconds will the projectile hit the ground?

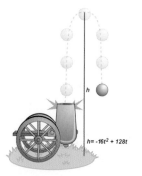

33. A ladder is 30 feet long and you want to place the base of the ladder 10 feet from the base of a building. About how far up the building (to the nearest tenth of a foot) will the ladder reach?

34. A flag pole is on top of a building and is held in place by steel cables attached to the top of the pole. If one such cable is 40 feet long and is attached at a point on the roof of the building 20 feet from the base of the flag pole, what is the length of the flag pole (to the nearest tenth of a foot)?

Writing and Thinking About Mathematics

35. Suppose that you are to solve an applied problem and the solution leads to a quadratic equation. You decide to use the quadratic formula to solve the equation. Explain what restrictions you must be aware of when you use the formula.

Hawkes Learning Systems: Introductory Algebra

Applications: Quadratic Equations

Quadratic Functions: $y = ax^2 + bx + c$

After completing this section, you will be able to:

1. Graph quadratic functions.

2. Find the vertex and the line of symmetry of a parabola.

3. Find maximum and minimum values of quadratic functions.

In Chapters 4 and 5, we discussed linear equations and linear functions in great detail. For example,

$$y = 2x + 5 \text{ is a linear function}$$

and its graph is a straight line with slope 2 and y-intercept 5. The equation can also be written in the function form

$$f(x) = 2x + 5$$

In general, we can write linear functions in the form

$$y = mx + b \quad \text{or} \quad f(x) = mx + b$$

where the slope is m and the y-intercept is b.

Quadratic Functions

Now consider the function

$$y = x^2 - 2x - 3$$

This function is **not** linear, so its graph is **not** a straight line. The nature of the graph can be seen by plotting several points, as shown in Figure 10.1.

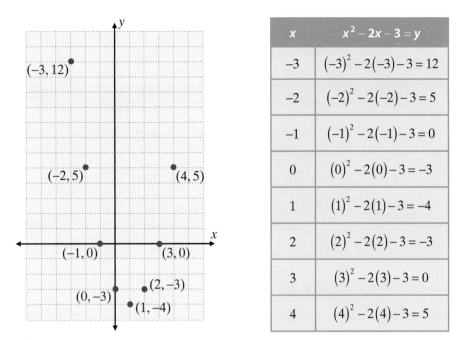

x	$x^2 - 2x - 3 = y$
-3	$(-3)^2 - 2(-3) - 3 = 12$
-2	$(-2)^2 - 2(-2) - 3 = 5$
-1	$(-1)^2 - 2(-1) - 3 = 0$
0	$(0)^2 - 2(0) - 3 = -3$
1	$(1)^2 - 2(1) - 3 = -4$
2	$(2)^2 - 2(2) - 3 = -3$
3	$(3)^2 - 2(3) - 3 = 0$
4	$(4)^2 - 2(4) - 3 = 5$

Figure 10.1

The points do not lie on a straight line. The graph can be seen by drawing a smooth curve called a **parabola** through the points, as shown in Figure 10.2. The point $(1, -4)$ is the "turning point" of the curve and is called the **vertex** of the parabola. The line $x = 1$ is the **line of symmetry** or the **axis of symmetry** for the parabola. The curve is a "mirror image" of itself on either side of the line $x = 1$.

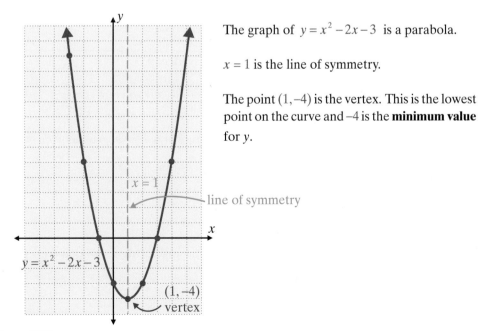

The graph of $y = x^2 - 2x - 3$ is a parabola.

$x = 1$ is the line of symmetry.

The point $(1, -4)$ is the vertex. This is the lowest point on the curve and -4 is the **minimum value** for y.

Figure 10.2

809

Parabolas (or parabolic arcs) occur frequently in describing information in real life. For example, the paths of projectiles (thrown balls, artillery shells, arrows) and a revenue function in business can be illustrated using parabolas.

Quadratic Function

A **quadratic function** is a function of the form

$$y = ax^2 + bx + c$$

where a, b, and c are real constants and $a \neq 0$.

The graph of every quadratic function is a parabola. The position of the parabola, its shape, and whether it "opens up" or "opens down" can be determined by investigating the function and its coefficients.

The following information about quadratic functions is given here without proof. A thorough development is part of the next course in algebra.

General Information on Quadratic Functions

For the **quadratic function** $y = ax^2 + bx + c$:

1. If $a > 0$, the parabola "opens upward."
2. If $a < 0$, the parabola "opens downward."
3. $x = -\dfrac{b}{2a}$ is the **line of symmetry**.
4. The **vertex** (turning point) occurs where $x = -\dfrac{b}{2a}$. Substitute this value for x in the function and find the y-value of the vertex. The vertex is the lowest point on the curve if the parabola opens upward or it is the highest point on the curve if the parabola opens downward.

Several graphs related to the basic form

$$y = ax^2 \quad \text{where } b = 0 \text{ and } c = 0$$

are shown in Figure 10.3.

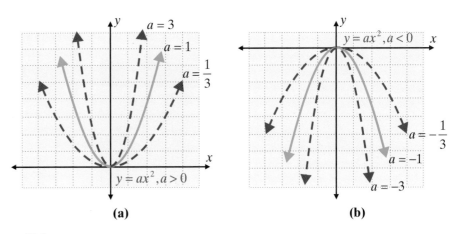

(a)   **(b)**

Figure 10.3

In the cases illustrated in Figure 10.3, where $y = ax^2$, we have $b = 0$, and

$$x = -\frac{b}{2a} = -\frac{0}{2a} = 0.$$

So the vertex occurs at the origin, $(0, 0)$. In addition, the axis of symmetry is the y-axis (the line $x = 0$).

Example 1: Graphing a Quadratic Function ($b=0$)

For the quadratic function $y = x^2 - 1$, find (a) its vertex, (b) its line of symmetry, and (c) its x-intercepts. Plot a few specific points and graph the parabola.

Solution: For $y = x^2 - 1$, we have $a = 1$, $b = 0$, and $c = -1$.

(a)   The vertex is at $x = -\dfrac{0}{2 \cdot 1} = 0$. Substituting 0 for x gives $y = 0^2 - 1 = -1$.

The vertex is the point $(0, -1)$.

(b)   The line of symmetry is $x = -\dfrac{0}{2(1)} = 0$.

(c)   The x-intercepts can be found by setting $y = 0$ and solving the resulting quadratic equation. Thus

$$x^2 - 1 = 0$$
$$x^2 = 1$$
$$x = \pm\sqrt{1}$$
$$x = \pm 1$$

Continued on next page...

Thus the graph crosses the x-axis at the points $(1, 0)$ and $(-1, 0)$. These points and others can be found as shown in the table.

x	x² − 1 = y
−2	$(-2)^2 - 1 = 3$
−1	$(-1)^2 - 1 = 0$
0	$(0)^2 - 1 = -1$
1	$(1)^2 - 1 = 0$
2	$(2)^2 - 1 = 3$

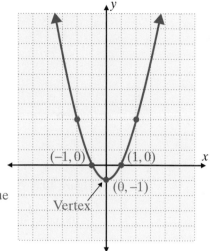

-1 is called the minimum value of the quadratic function.

Example 2: Graphing a Quadratic Function

For the quadratic function $f(x) = x^2 - 6x + 1$, find (a) its vertex, (b) its line of symmetry, and (c) its x-intercepts. Plot a few specific points and graph the parabola.

Solution: For $f(x) = x^2 - 6x + 1$, $a = 1$, $b = -6$, and $c = 1$.

(a)   The vertex is at $x = \dfrac{-6}{2 \cdot 1} = 3$. Substituting 3 for x gives

$y = 3^2 - 6 \cdot 3 + 1 = -8$. The vertex is the point $(3, -8)$.

(b)   The line of symmetry is $x = -\dfrac{-6}{2(1)} = 3$.

(c)   The x-intercepts can be found by setting $y = 0$ and solving the resulting quadratic equation. Thus, using the quadratic formula, we have

$x^2 - 6x + 1 = 0$

$$x = \frac{-(-6) \pm \sqrt{(-6)^2 - 4(1)(1)}}{2 \cdot 1} = \frac{6 \pm \sqrt{32}}{2} = \frac{6 \pm 4\sqrt{2}}{2} = 3 \pm 2\sqrt{2}.$$

The x-intercepts are at $\left(3 + 2\sqrt{2}, 0\right)$ and $\left(3 - 2\sqrt{2}, 0\right)$.

Estimating these values with a calculator gives the x-intercepts at about $(5.83, 0)$ and $(0.17, 0)$.

Thus the graph crosses the x-axis at about $(5.83, 0)$ and $(0.17, 0)$.

Other points can be found as shown in the table.

x	$x^2 - 6x + 1 = y$
1	$(1)^2 - 6(1) + 1 = -4$
3	$(3)^2 - 6(3) + 1 = -8$
5	$(5)^2 - 6(5) + 1 = -4$
$3 + 2\sqrt{2}$	$\left(3 + 2\sqrt{2}\right)^2 - 6\left(3 + 2\sqrt{2}\right) + 1 = 0$
$3 - 2\sqrt{2}$	$\left(3 - 2\sqrt{2}\right)^2 - 6\left(3 - 2\sqrt{2}\right) + 1 = 0$

-8 is the minimum value of the quadratic function.

Vertex

Example 3: Graphing a Quadratic Function ($c = 0$)

For the quadratic function $y = -2x^2 + 4x$, find (a) its vertex, (b) its line of symmetry, and (c) its x-intercepts. Plot a few specific points and graph the parabola.

Solution: For $y = -2x^2 + 4x$, $a = -2$, $b = 4$, and $c = 0$.

(a) The vertex is at $x = -\dfrac{4}{2(-2)} = 1$. Substituting 1 for x gives

$$y = -2 \cdot 1^2 + 4 \cdot 1 = 2.$$

The vertex is the point $(1, 2)$.

(b) The line of symmetry is $x = -\dfrac{4}{2(-2)} = 1$.

(c) The x-intercepts can be found by setting $y = 0$ and solving the resulting quadratic equation. Thus, by factoring, we have

$$-2x^2 + 4x = 0$$
$$-2x(x - 2) = 0$$
$$x = 0 \text{ and } x = 2$$

Continued on next page...

The x-intercepts are at $(0, 0)$ and $(2, 0)$. Thus the graph crosses the x-axis at the points $(0, 0)$ and $(2, 0)$.

These points, along with others points can be found as shown in the table.

x	$-2x^2 + 4x = y$
0	$-2(0)^2 + 4(0) = 0$
1	$-2(1)^2 + 4(1) = 2$
2	$-2(2)^2 + 4(2) = 0$

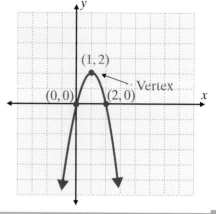

2 is called the maximum value of the quadratic function.

Example 4: Quadratic Function Application

Find the dimensions of the rectangle with maximum area if the perimeter is 40 meters.

Solution: Let A = area
l = length
w = width.

$A = lw$ and $2l + 2w = 40$
$2w = 40 - 2l$
$w = 20 - l$

So, substituting for w,

$$A = l(20 - l)$$
$$A = 20l - l^2$$
$$A = -l^2 + 20l.$$

This is a quadratic function with $a = -1$, $b = 20$, and $c = 0$. The maximum area occurs at the vertex of the corresponding parabola where

$$l = -\frac{b}{2a} = -\frac{20}{2(-1)} = 10$$

and $w = 20 - l = 20 - 10 = 10$.

Thus the maximum occurs when $l = 10$ m and $w = 10$ m. (The rectangle is, in fact, a square and the area = 100 m².)

Using a Calculator to Graph Quadratic Functions

We illustrate how to graph a quadratic function on a TI-84 Plus graphing calculator by graphing the function $y = x^2 + 2x - 3$ in the following step-by-step manner.

Step 1: Press the key marked Y=.

Step 2: Enter the function to be graphed: $\backslash Y_1 = X^2 + 2X - 3$.

(**Note:** The variable X is found by pressing X,T,θ,n.

X^2 is then found by pressing x^2.)

(Or, you can enter $\backslash Y_1 = X \wedge 2 + 2X - 3$.)

Step 3: Press GRAPH in the upper right hand corner of the keyboard.

The graph of the parabola should appear on the display as shown here.

If the graph does not appear as shown, you may need to adjust your window as follows:

a. Press WINDOW and set the numbers as needed, or

b. Press ZOOM and then press **6** to get the standard window to get the window from -10 to 10 on the x-axis and from -10 to 10 on the y-axis.

To estimate the location of various points on the graph of the parabola, you can press TRACE and then use the left and right arrows to move along the parabola. The function $Y_1 = X^2 + 2X - 3$ will appear at the top of the display and the coordinates of the point indicated by the cursor will appear at the bottom of the display. An example is shown here.

Example 5: Graphing with a Calculator

Use your graphing calculator to graph the quadratic function $y = 2x^2 - x - 15$. Then use the trace key to estimate the location of three points on the graph. (As shown in the following diagrams, set the **WINDOW** so that **xmin** = −10, **xmax** = 10, **ymin** = −20 and **ymax** = 10.)

Solution: **Step 1:** Press the key marked ▬ Y= ▬ .

Step 2: Enter the function to be graphed: $\backslash Y_1 = 2X^2 - X - 15$.

Step 3: Press ▬ GRAPH ▬ in the upper right hand corner of the keyboard.

The graph of the parabola should appear on the display as shown here.

Now, press the ▬ TRACE ▬ key and move the cursor around to estimate the location of points. Three such locations are shown here.

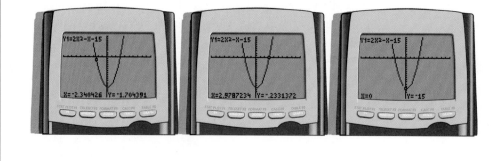

10.5 Exercises

For each quadratic function in Exercises 1 – 20, (a) find its vertex, (b) find its line of symmetry, (c) locate (or estimate) its x-intercepts, and (d) graph the function. (**Note:** If solving the quadratic equation results in nonreal solutions, then the graph does not cross the x-axis.)

1. $y = x^2 + 4$

2. $y = x^2 - 6$

3. $y = 8 - x^2$

4. $y = -2 - x^2$

5. $y = x^2 - 2x - 3$

6. $y = x^2 - 4x + 5$

7. $y = x^2 + 6x$

8. $y = x^2 - 8x$

9. $y = -x^2 - 4x + 2$

10. $y = -x^2 - 5x + 4$

11. $y = 2x^2 - 10x + 3$

12. $y = 2x^2 - 12x - 5$

13. $y = x^2 + 7x - 4$

14. $y = x^2 + 5x + 4$

15. $y = -x^2 + x - 3$

16. $y = -x^2 - 7x + 3$

17. $y = 2x^2 + 7x - 4$

18. $y = 2x^2 - x - 3$

19. $y = 3x^2 + 5x + 2$

20. $y = 3x^2 - 9x + 5$

21. The perimeter of a rectangle is 60 yards. What are the dimensions of the rectangle with maximum area?

22. The perimeter of a rectangle is 56 feet. What are the dimensions of the rectangle with maximum area?

In Exercises 23 – 26 use the formula $h = -16t^2 + v_0 t + h_0$ where h is the height (in feet) of a projectile after time, t (in seconds). The initial velocity is v_0 and the initial height is h_0.

23. A ball is thrown vertically upward from the ground with an initial velocity of 112 ft per sec.

   a. When will the ball reach its maximum height?
   b. What will be the maximum height?

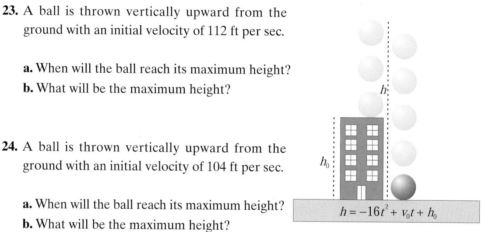

24. A ball is thrown vertically upward from the ground with an initial velocity of 104 ft per sec.

   a. When will the ball reach its maximum height?
   b. What will be the maximum height?

25. A stone is projected vertically upward from a platform that is 32 ft high at a rate of 128 ft per sec.

   a. When will the stone reach its maximum height?
   b. What will be the maximum height?

26. A stone is projected vertically upward from a platform that is 20 ft high at a rate of 160 ft per sec.

   a. When will the stone reach its maximum height?
   b. What will be the maximum height?

In business, the term **revenue** represents income. The revenue (income) is found by multiplying the number of units sold times the price per unit.
[Revenue = (price) · (units sold)]

27. A store owner estimates that by charging x dollars each for a certain lamp, he can sell $40 - x$ lamps each week. What price will yield maximum revenue?

  a. What is the revenue function $R(x)$?
  b. What price will yield a maximum revenue?

28. When fishing reels are priced at x dollars each, local consumers will buy $36 - x$ fishing reels.

  a. What is the revenue function $R(x)$?
  b. What price will yield a maximum revenue?

29. A manufacturer produces calculators. She estimates that by selling them for x dollars each, she will be able to sell $80 - 2x$ calculators each week.

  a. What is the revenue function $R(x)$?
  b. What price will yield a maximum revenue?
  c. What will be the maximum revenue?

30. A manufacturer produces radios. He estimates that by selling them for x dollars each, he will be able to sell $100 - x$ radios each month.

  a. What is the revenue function $R(x)$?
  b. What price will yield a maximum revenue?
  c. What will be the maximum revenue?

31. Use your graphing calculator to graph each of the following quadratic functions.
  a. Graph each equation one at a time.
  b. Graph all three parabolas at the same time (all three curves should appear on the display).

    i. $y = x^2 + 1$ **ii.** $y = x^2 + 3$ **iii.** $y = x^2 - 4$

32. In your own words, state the effect on the graph of $y = x^2$ by changing the constant term k in the function $y = x^2 + k$.

33. Use your graphing calculator to graph each of the following quadratic functions
  a. Graph each equation one at a time.
  b. Graph all three parabolas at the same time (all three curves should appear on the display).

    i. $y = (x - 3)^2$ **ii.** $y = (x - 5)^2$ **iii.** $y = (x + 2)^2$

34. In your own words, state the effect on the graph of $y = x^2$ by changing the constant term h in the function $y = (x - h)^2$.

35. Discuss the general relationship of the graph of a function of the form $y = (x - h)^2 + k$ to the graph of the function $y = x^2$.

Hawkes Learning Systems: Introductory Algebra

Graphing Parabolas

Chapter 10 Index of Key Ideas and Terms

Section 10.1   Quadratic Equations: The Square Root Method

Quadratic Equations   page 772

The Square Root Method   pages 773 - 774
For a quadratic equation in the form $x^2 = c$ where c is nonnegative, $x = \sqrt{c}$ or $x = -\sqrt{c}$. This can be written as $x = \pm\sqrt{c}$.

Pythagorean Theorem   pages 775 - 776
In a right triangle, the square of the hypotenuse is equal to the sum of the squares of the legs.
Symbolically, $c^2 = a^2 + b^2$, where c is the length of the hypotenuse and a and b are the lengths of the legs.

Section 10.2   Quadratic Equations: Completing the Square

Completing the Square   pages 781 - 784
To solve quadratic equations by completing the square:
1. Arrange the terms with variables on one side and constants on the other.
2. Divide each term by the coefficient of x^2. (We want the leading coefficient to be 1.)
3. Find the number that completes the square of the quadratic expression and add this number to **both** sides of the equation.
4. Find the positive and negative square roots of both sides.
5. Solve for x. Remember, usually there will be two solutions.

Section 10.3   Quadratic Equations: The Quadratic Formula

General Form of a Quadratic Equation   page 788
The general quadratic equation is $ax^2 + bx + c = 0$, where a, b, and c are real constants and $a \neq 0$.

Quadratic Formula   pages 788 - 789
The solutions of the general quadratic equation $ax^2 + bx + c = 0$, where $a \neq 0$, are $x = \dfrac{-b \pm \sqrt{b^2 - 4ac}}{2a}$.

Section 10.4 Applications

Applications

Pythagorean Theorem	pages 796 - 797
Work	pages 797 - 798
Distance-Rate-Time	pages 798 - 799
Geometry	pages 800 - 801

Section 10.5 Quadratic Functions: $y = ax^2 + bx + c$

Graphing Parabolas   pages 808 - 814

Quadratic Functions   page 810

A quadratic function is a function of the form

$$y = ax^2 + bx + c$$

where $a, b,$ and c are real constants and $a \neq 0$.

General Information on Quadratic Functions   page 810

For the quadratic function

1. If $a > 0$, the parabola "opens upward."
2. If $a < 0$, the parabola "opens downward."
3. $x = -\dfrac{b}{2a}$ is the line of symmetry.
4. The vertex (turning point) occurs where $x = -\dfrac{b}{2a}$.

Substitute this value for x in the function and find the y-value of the vertex. The vertex is the lowest point on the curve if the parabola opens upward or it is the highest point on the curve if the parabola opens downward.

Hawkes Learning Systems: Introductory Algebra

For a review of the topics and problems from Chapter 10, look at the following lessons from *Hawkes Learning Systems: Introductory Algebra*

Quadratic Equations: The Square Root Method
Quadratic Equations: Completing the Square
Quadratic Equations: The Quadratic Formula
Applications: Quadratic Equations
Graphing Parabolas

Chapter 10 Review

10.1 Quadratic Equations: *The Square Root Method*

Solve the quadratic equations in Exercises 1 – 10 by using the square root method. Write the radicals in simplest form.

1. $x^2 = 400$

2. $2x^2 = 144$

3. $9x^2 = -289$

4. $(x-6)^2 = 49$

5. $(x+3)^2 = 20$

6. $2(x+1)^2 = 80$

7. $(3x-1)^2 = 16$

8. $(4x+9)^2 = 81$

9. $(2x+5)^2 = 75$

10. $3(x-1)^2 = 240$

Use the Pythagorean Theorem to decide if each of the triangles determined by the three sides given in Exercises 11 and 12 is a right triangle.

11. 6 cm, 8 cm, 10 cm

12. 6 ft, 6 ft, $6\sqrt{2}$ ft

In Exercises 13 and 14, right triangles are shown with two sides given. Determine the length of the missing side.

13. $a = 8$ $b = 8$ $c = ?$

14. $a = ?$ $b = 20$ $c = 29$

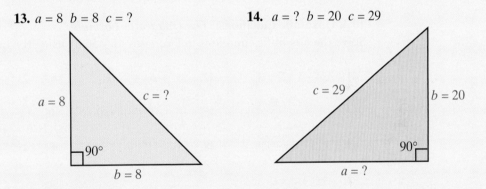

15. The hypotenuse of a right triangle is 2 in. longer than one of the legs. The length of the other leg is 10 in. Find the perimeter of the triangle.

16. A ball is dropped from the top of a building known to be 100 meters high. The formula for finding the height (in meters) of the ball at time t (in seconds) is $h = -4.9t^2 + 100$. In how many seconds (to the nearest tenth of a second) will the ball hit the ground?

Use a TI-84 Plus graphing calculator to solve the equations in Exercises 17 – 20. Round your answers to the nearest hundredth.

17. $x^2 = 250$

18. $20x^2 = 52.5$

19. $3.2x^2 = 100$

20. $5.6x^2 = -200$

10.2 Quadratic Equations: *Completing the Square*

Solve each quadratic equation in Exercises 21 – 30 by completing the square.

21. $x^2 - 8x - 9 = 0$ **22.** $x^2 + 3x + 2 = 0$ **23.** $x^2 + x - 56 = 0$

24. $4x^2 - 1 = 0$ **25.** $9x^2 - 4 = 0$ **26.** $x^2 - 4x + 1 = 0$

27. $x^2 + 8x + 4 = 0$ **28.** $x^2 + x - 4 = 0$ **29.** $3x^2 - 4x - 1 = 0$

30. $2x^2 + 3x - 2 = 0$

Solve the quadratic equations in Exercises 31 – 40 by factoring or by completing the square.

31. $16x^2 - 25 = 0$ **32.** $25x^2 - 16 = 0$ **33.** $x^2 + 11x - 26 = 0$

34. $x^2 + 8x - 20 = 0$ **35.** $6x^2 - 12x + 5 = 0$ **36.** $3x^2 - 4x - 3 = 0$

37. $4x^2 + 2x - 3 = 0$ **38.** $x^2 + 7x - 10 = 0$ **39.** $5x^2 - 10x + 1 = 0$

40. $3x^2 + 2x - 5 = 0$

10.3 Quadratic Equations: *The Quadratic Formula*

Solve the quadratic equations in Exercises 41 – 50 by using the **quadratic formula**.

41. $2x^2 - x - 2 = 0$ **42.** $3x^2 - 6x + 2 = 0$ **43.** $2x^2 - 49 = 0$

44. $\frac{1}{3}x^2 - x + \frac{1}{2} = 0$ **45.** $\frac{1}{4}x^2 + 2x - \frac{1}{3} = 0$ **46.** $(x-2)(x-3) = 6$

47. $(2x+1)(x-1) = 8$ **48.** $(3x-1)(x+2) = 6x$ **49.** $4x^2 - 12x + 9 = 0$

50. $-3x^2 + 4 = 0$

Solve the quadratic equations in Exercises 51 – 56 **by using any method** (factoring, completing the square, or the quadratic formula).

51. $x^2 + x - 2 = 0$ **52.** $7x^2 - 3x = 0$ **53.** $x^2 + 6x = 10$

54. $\frac{11}{2}x - 1 = 2x^2$ **55.** $\frac{3}{4}x^2 + 2x = 3$ **56.** $-4x^2 + 3x + 1 = 0$

Solve the quadratic equations in Exercises 57 – 60 and use a calculator to find the answers in decimal form accurate to four decimal places.

57. $x^2 - 4x - 1 = 0$ **58.** $4x^2 - 8x - 11 = 0$ **59.** $(3x+1)(3x-1) = 6x$

60. $6x^2 = 300$

10.4 Applications

61. The sum of the squares of two consecutive positive integers is 85. Find these integers.

62. The length of a rectangle is 6 feet longer than the width. The area of the rectangle is 720 square feet. Find the dimensions of the rectangle.

63. The length of a rectangle is 5 meters less than twice the width. If both dimensions are decreased by 3 meters, the area is 84 square meters. Find the length and width of the original rectangle.

64. The product of two consecutive even integers is 10 less than 5 times their sum. Find the integers.

65. A right triangle has two equal sides. If the hypotenuse is 20 meters, what is the length of each of the equal sides?

66. Christine drove her car to a mathematics teachers' conference in Sacramento, 288 miles from her home. Because she had a late start, she drove a little faster than usual. On the trip home her average speed dropped by 8 miles per hour. If her total driving time for the round trip was 8.5 hours, what was her average speed going to the conference?

67. Jeremy and Thomas rowed their boat down a river for 12 miles then rowed back to their dock. The round trip took 8 hours. If the rate of the current was 2 mph, at what rate can Jeremy and Thomas row in still water?

68. Two pipes can be used to fill a swimming pool. If both are turned on, the pool will fill in 2.4 hours. If only one of the pipes is used, the larger pipe would fill the pool 2 hours faster than the smaller pipe. How many hours would it take for the pool to be filled by each of the pipes working alone?

69. A square piece of tin has a small square, 5 cm by 5 cm, cut from each corner. The edges are then folded up to form a box with a volume of 40,500 cm^3. What are the dimensions of the box?

70. A small private plane can fly at 200 mph when the air is calm. The plane would take 24 minutes longer to fly 396 miles against the wind than flying that distance with the wind. What is the rate of the wind?

10.5 Quadratic Functions: $y = ax^2 + bx + c$

For each quadratic function in Exercises 71 – 76, **(a)** find its vertex, **(b)** find its line of symmetry, **(c)** locate (or estimate) its x-intercepts, and **(d)** graph the function.

71. $y = x^2 - 4$ **72.** $y = 16 - x^2$ **73.** $y = x^2 - 6x + 5$

74. $y = x^2 - 10x$ **75.** $y = -x^2 + 2x - 3$ **76.** $y = 3x^2 - 10x + 2$

77. The perimeter of a rectangle is 80 yards. What are the dimensions of the rectangle with maximum area?

78. Michael estimates that by charging x dollars for each of his pieces of jewelry he can sell $80 - x$ pieces each week.
  a. Write his revenue function for these items.
  b. What price should he charge to maximize his revenue from these pieces of jewelry?
  c. What will be his maximum revenue from these items?

For Exercises 79 and 80, use the formula $h = -16t^2 + v_0 t + h_0$ where h is the height (in feet) of a projectile in t seconds. Its initial velocity is v_0 and its initial height is h_0.

79. A toy rocket is projected from the ground with an initial velocity of 48 feet per second.
  a. When will the rocket reach its maximum height?
  b. What will be the maximum height?
  c. In how many seconds will the rocket hit the ground?

80. A ball is thrown upward at 64 feet per second from the edge of a cliff known to be 80 feet above the beach.
  a. When will the ball reach its maximum height?
  b. What will be the maximum height?
  c. In how many seconds will the ball hit the ground?

Chapter 10 Test

Solve the equations in Exercises 1 and 2.

1. $(x+5)^2 = 49$ **2.** $8x^2 = 96$

3. Find the missing side in the right triangle shown here.

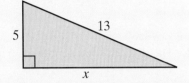

Add the correct constant in Exercises 4 – 6 in order to make the trinomial factorable as indicated.

4. $x^2 - 24x + \underline{\quad} = (\quad)^2$ **5.** $x^2 + 9x + \underline{\quad} = (\quad)^2$

6. $3x^2 + 9x + \underline{\quad} = 3(\quad)^2$

Solve the quadratic equations in Exercises 7 – 9 by factoring.

7. $x^2 - 3x + 2 = 0$ **8.** $5x^2 + 12x = 0$ **9.** $3x^2 - x - 10 = 0$

Solve the quadratic equations in Exercises 10 – 12 by completing the square.

10. $x^2 + 6x + 8 = 0$ **11.** $x^2 - 5x = 6$ **12.** $2x^2 - 8x - 4 = 0$

Solve the quadratic equations in Exercises 13 – 15 using the quadratic formula.

13. $3x^2 + 8x + 2 = 0$ **14.** $2x^2 = 4x + 3$ **15.** $\frac{1}{2}x^2 - x = 2$

Solve the quadratic equations in Exercises 16 – 18 using any method.

16. $x^2 + 6x = 2$ **17.** $3x^2 - 7x + 2 = 0$ **18.** $2x^2 - 3x = 1$

For each quadratic function in Exercises 19 – 21, (**a**) find its vertex, (**b**) find its line of symmetry, (**c**) locate (or estimate) its x-intercepts, and (**d**) graph the function. (**Note**: If solving the quadratic equation results in nonreal solutions, then the graph does not cross the x-axis.)

19. $y = x^2 - 5$ **20.** $y = -x^2 + 4x$ **21.** $y = 2x^2 + 3x + 1$

First, determine an equation for each of the Exercises 22 – 25, and then solve the problem.

22. The diagonal of a square is 36 inches long. Find the length of each side.

23. A toy rocket is projected upward from the ground with an initial velocity of 144 feet per second. The height of the rocket can be found by using the equation $h = -16t^2 + 144t$ where t is measured in seconds.
   a. In approximately how many seconds will the rocket be 96 ft above the ground?
   b. What is the maximum height of the rocket?
   c. In approximately how many seconds will the rocket hit the ground?

24. A small boat travels 8 mph in still water. It takes 10 hours longer to travel 60 miles upstream than it takes to travel 60 miles downstream. Find the rate of the current.

25. A rectangular sheet of metal is 3 inches longer than it is wide. A box is to be made by cutting out 4 inch squares at each corner and folding up the sides. If the box has a volume of 720 cubic inches, what are the dimensions of the sheet of metal? (**Hint:** Sketch a diagram of the rectangle and a diagram of the box.)

Cumulative Review: Chapters 1 – 10

Simplify each expression in Exercises 1 and 2.

1. $\left(-3x^2 y \right)^2$

2. $\dfrac{\left(2x^2 \right)\left(-6x^3 \right)}{3x^{-1}}$

In Exercises 3 and 4, simplify by combining like terms.

3. $4x + 3 - 2(x + 4) + (2 - x)$

4. $4x - [3x - 2(x + 5) + 7 - 5x]$

Solve each of the equations in Exercises 5 – 16.

5. $5 (x - 8) - 3 (x - 6) = 0$

6. $\dfrac{5}{4}x + 4 = \dfrac{3}{4}x + 7$

7. $x^2 - 8x + 15 = 0$

8. $x^2 + 5x - 36 = 0$

9. $\dfrac{1}{x} + \dfrac{2}{3x} = 1$

10. $\dfrac{3}{x-1} + \dfrac{5}{x+1} = 2$

11. $(x - 4)^2 = 9$

12. $(x + 3)^2 = 6$

13. $x^2 + 4x - 2 = 0$

14. $2x^2 - 3x - 4 = 0$

15. $x - 2 = \sqrt{x + 18}$

16. $x + 3 = 2\sqrt{2x + 6}$

In Exercises 17 – 20, solve each inequality and graph the solution set on a real number line.

17. $15 - 3x + 1 \geq x + 4$

18. $4(6 - x) \geq -2(3x + 1)$

19. $2(x - 5) - 4 < x - 3(x - 1)$

20. $x - (2x + 5) \leq 9 - (4 - x)$

Solve the system of equations in Exercises 21 – 25.

21. $\begin{cases} x + 5y = 9 \\ 2x - 3y = 1 \end{cases}$

22. $\begin{cases} 6x - y = 11 \\ 2x - 3y = -11 \end{cases}$

23. $\begin{cases} 2x - 5y = 11 \\ 5x - 3y = -1 \end{cases}$

24. $\begin{cases} x + 2y = 6 \\ y = -\dfrac{1}{2}x + 3 \end{cases}$

25. $\begin{cases} y = -3x + 4 \\ 3x + y = 7 \end{cases}$

Use a TI-84 Plus graphing calculator to graph the solution set for each system of linear inequalities in Exercises 26 and 27.

26. $\begin{cases} y < \dfrac{1}{3}x \\ y > -x + 4 \end{cases}$

27. $\begin{cases} y \leq 4x - 1 \\ y \leq -2x + 3 \end{cases}$

In Exercises 28 and 29, graph the solution set for each system of linear inequalities without the aid of a calculator.

28. $\begin{cases} x > -2 \\ y \le 5 \end{cases}$

29. $\begin{cases} 2x + y \ge 1 \\ 2x - y \ge 2 \end{cases}$

In Exercises 30 and 31, solve each formula for the indicated letter.

30. $P = a + 2b$, solve for b

31. $A = \dfrac{h}{2}(a+b)$, solve for a

Factor the expressions in Exercises 32 – 36.

32. $x^2 - 12x + 36$

33. $25x^2 - 49$

34. $3x^2 + 10x + 3$

35. $3x^2 + 6x - 72$

36. $x^3 + 4x^2 - 8x - 32$

In Exercises 37 – 42, perform the indicated operations and simplify.

37. $\left(5x^2 + 3x - 7\right) + \left(2x^2 - 3x - 7\right)$

38. $\left(2x^2 - 7x + 3\right) + \left(5x^2 - 3x + 4\right)$

39. $\dfrac{x^2}{x^2 - 4} \cdot \dfrac{x^2 - 3x + 2}{x^2 - x}$

40. $\dfrac{3x}{x^2 - 6x - 7} \div \dfrac{2x^2}{x^2 - 8x + 7}$

41. $\dfrac{2x}{x^2 - 25} - \dfrac{x+1}{x+5}$

42. $\dfrac{4x}{x^2 - 3x - 4} + \dfrac{x-2}{x^2 + 3x + 2}$

Simplify the radical expressions in Exercises 43 – 48 and then use a calculator to find the value of each expression accurate to four decimal places.

43. $\sqrt{63}$

44. $\sqrt{243}$

45. $\sqrt[3]{250}$

46. $\sqrt{\dfrac{25}{12}}$

47. $\sqrt{147} + \sqrt{48}$

48. $\left(2\sqrt{3} + 1\right)\left(\sqrt{3} - 4\right)$

Simplify the expression in Exercises 49 – 52.

49. $4x^{\frac{2}{3}} \cdot 2x^{\frac{1}{3}}$

50. $\left(4x^3\right)^2 \left(2x^2\right)^{-1}$

51. $\left(\dfrac{m^{\frac{1}{2}} n^{\frac{3}{4}}}{m^{\frac{-1}{2}} n^{\frac{1}{4}}}\right)^2$

52. $\left(\dfrac{25x^2 y^4}{36x^4 y^6}\right)^{\frac{1}{2}}$

In Exercises 53 – 55, graph the linear function and state its slope and its y-intercept.

53. $y = \dfrac{2}{3}x + 4$

54. $4x - 3y = 6$

55. $2x + 4y = -5$

For each quadratic function in Exercises 56 – 58, (**a**) find its vertex, (**b**) find its line of symmetry, (**c**) locate (or estimate) its x-intercepts, and (**d**) graph the function.
(**Note**: If solving the quadratic equation results in nonreal solutions, then the graph does not cross the x-axis.)

56. $y = x^2 - 8x + 7$ **57.** $y = 2(x+3)^2 - 6$ **58.** $y = x^2 + x + 1$

59. Write the equation $5x + 2y = 6$ in slope-intercept form. Find the slope and the y-intercept.

60. Given the two points $(-5, 2)$ and $(3, 5)$, find:
    a. the slope of the line through the two points,
    b. the distance between the two points,
    c. the midpoint of the line segment joining the points, and
    d. an equation of the line through the two points.

61. Given the function $f(x) = 5x - 3$, find (a) $f(-1)$, (b) $f(6)$, and (c) $f(0)$.

62. Given the function $F(x) = 3x^2 - 2x + 7$, find (a) $F(0)$, (b) $F(-2)$, and (c) $F(5)$.

63. Anna has money in two separate savings accounts; one pays 5% and one pays 8%. The amount invested at 5% exceeds the amount invested at 8% by $800. The total annual interest is $209. How much is invested at each rate?

64. Tim can clean the weeds from a vacant lot in 10 hours. Beth can do the same job in 8 hours. How long would it take if they worked together?

65. The speed of a boat in still water is 8 mph. It travels 6 miles upstream and returns. The total trip takes 1 hr, 36 min. Find the speed of the current.

66. How many liters of a solution that is 40% insecticide must be mixed with 40 liters of a solution that is 25% insecticide to produce a solution that is 30% insecticide?

67. The length of a rectangle is one inch less than twice the width. The length of the diagonal is 17 inches. Find the width and length of the rectangle.

68. George bought 7 pairs of socks. Some cost $2.50 per pair and some cost $3.00 per pair. The total cost was $19.50. How many pairs of each type did he buy?

69. The length of a rectangle is five more than twice the width. The area is 52 cm^2. Find the width and the length.

70. At a small college, the ratio of women to men is $9 : 7$. If there is a total enrollment of 6400 students, how many women and how many men are enrolled?

71. Louise has $2700 invested. Part of the money is invested at 5.5% and the remainder is invested at 7.2%. Her total annual interest is $168.90. How much is invested at each rate?

72. Anand traveled 6 miles downstream and returned. His speed downstream was 5 mph faster than his speed upstream. If the total trip took 1 hour, find his speed in each direction.

73. A ball is thrown upward from the edge of a building with an initial velocity of 96 feet per second. The building is 120 feet tall and the height of the ball above the ground can be found by using the equation $h = -16t^2 + 96t + 120$ where t is measured in seconds.
   a. When will the ball be 200 feet above the ground?
   b. When will the ball be 60 feet above the ground?
   c. In about how many seconds will the ball hit the ground?

74. A ladder is 40 feet long and the base of the ladder is to be set 15 feet from the base of a building under construction. About how far up the building (to the nearest tenth of a foot) will the ladder reach?

Difference of Two Cubes and Sum of Two Cubes:

$$(x-a)\left(x^2+ax+a^2\right)=x^3-a^3$$

$$(x+a)\left(x^2-ax+a^2\right)=x^3+a^3$$

Objectives

After completing this section, you will be able to:

1. Find the difference and sum of two cubes by multiplying.

2. Factor the difference and sum of two cubes.

In Section 6.6, three **special products** were developed and discussed: the difference of two squares and two types of perfect square trinomials.

Products:

I. $(x+a)(x-a)=x^2-a^2$ **Difference of Two Squares**

II. $(x+a)^2=x^2+2ax+a^2$ **Perfect Square Trinomial**

III. $(x-a)^2=x^2-2ax+a^2$ **Perfect Square Trinomial**

Then, in Chapter 7 (Section 7.4), we showed how to factor these special products by recognizing the forms of the products. That is, given a product in one of these three forms, we can immediately go to the factors simply by having done many problems and **memorizing** these forms as formulas.

Factors:

I. **Difference of Two Squares:** $x^2-a^2=(x+a)(x-a)$

II. **Perfect Square Trinomial:** $x^2+2ax+a^2=(x+a)^2$

III. **Perfect Square Trinomial:** $x^2-2ax+a^2=(x-a)^2$

Multiplying to Get the Difference and Sum of Two Cubes

Here, in Appendix I, we introduce two more forms that occur frequently in algebra: the **difference and sum of two cubes**. These two forms can be developed by applying the distributive property as follows:

$$(x-a)(x^2+ax+a^2)=x\cdot x^2+x\cdot ax+x\cdot a^2-a\cdot x^2-a\cdot ax-a\cdot a^2$$

$$=x^3+ax^2+a^2x-ax^2-a^2x-a^3$$

$$=x^3-a^3 \quad \text{difference of two cubes}$$

$$(x+a)(x^2-ax+a^2)=x\cdot x^2-x\cdot ax+x\cdot a^2+a\cdot x^2-a\cdot ax+a\cdot a^2$$

$$=x^3-ax^2+a^2x+ax^2-a^2x+a^3$$

$$=x^3+a^3 \quad \text{sum of two cubes}$$

NOTES

Important Notes:

1. In each case, the middle terms drop out and only two terms are left.

2. The expressions in parentheses are **not** perfect square trinomials. These trinomials are **not** factorable.

3. The sign in the binomial agrees with the sign in the result.

We now have the following two new formulas:

Difference of Two Cubes and Sum of Two Cubes

> **IV.** $(x-a)(x^2+ax+a^2)-x^3-a^3$ *Difference of Two Cubes*
>
> **V.** $(x+a)(x^2-ax+a^2)=x^3+a^3$ *Sum of Two Cubes*

Example 1: Difference of Two Cubes

Find each product.

a. $(x-3)(x^2+3x+9)$ **b.** $(2y-5)(4y^2+10y+25)$ **c.** $(x-4y)(x^2+4xy+16y^2)$

Solutions:

a. $(x-3)(x^2+3x+9)=x^3-3^3=x^3-27$

b. $(2y-5)(4y^2+10y+25)=(2y)^3-5^3=8y^3-125$

c. $(x-4y)(x^2+4xy+16y^2)=x^3-(4y)^3=x^3-64y^3$

Example 2: Sum of Two Cubes

Find each product.

a. $(x+5)(x^2-5x+25)$ **b.** $(2x+3)(4x^2-6x+9)$ **c.** $(3x+2y)(9x^2-6xy+4y^2)$

Solutions:

a. $(x+5)(x^2-5x+25) = x^3+5^3 = x^3+125$

b. $(2x+3)(4x^2-6x+9) = (2x)^3+3^3 = 8x^3+27$

c. $(3x+2y)(9x^2-6xy+4y^2) = (3x)^3+(2y)^3 = 27x^3+8y^3$

Factoring the Difference and Sum of Two Cubes

The forms of the products that give the difference and sum of two cubes should be memorized. These forms can be used for factoring because factoring is simply the reverse of multiplication.

IV. Difference of Two Cubes: $x^3-a^3=(x-a)(x^2+ax+a^2)$

V. Sum of Two Cubes: $x^3+a^3=(x+a)(x^2-ax+a^2)$

Example 3: Difference of Two Cubes

Factor completely. (Remember to first factor out any common monomial factor.)

a. y^3-27 **b.** $4x^3-32$ **c.** x^6-125

Solutions:

a. $y^3-27 = y^3-3^3 = (y-3)(y^2+3y+9)$

b. $4x^3-32 = 4(x^3-8) = 4(x^3-2^3) = 4(x-2)(x^2+2x+4)$

c. $x^6-125 = (x^2)^3-5^3 = (x^2-5)(x^4+5x^2+25)$

Example 4: Sum of Two Cubes

Factor completely. (Remember to first factor out any common monomial factor.)

a. $x^3 + 8y^3$ **b.** $x^6 + 343$ **c.** $16y^{12} + 250$

Solutions:

a. $x^3 + 8y^3 = x^3 + (2y)^3 = (x + 2y)(x^2 - 2xy + 4y^2)$

b. $x^6 + 343 = (x^2)^3 + 7^3 = (x^2 + 7)(x^4 - 7x^2 + 49)$

c. $16y^{12} + 250 = 2(8y^{12} + 125) = 2((2y^4)^3 + 5^3) = 2(2y^4 + 5)(4y^8 - 10y^4 + 25)$

A.1 Exercises

Find each product.

1. $(x + 2)(x^2 - 2x + 4)$ **2.** $(x + 3)(x^2 - 3x + 9)$ **3.** $(x - 5)(x^2 + 5x + 25)$

4. $(x - 1)(x^2 + x + 1)$ **5.** $(x + y)(x^2 - xy + y^2)$ **6.** $(x + 3y)(x^2 - 3xy + 9y^2)$

7. $(y + 1)(y^2 - y + 1)$ **8.** $(2x + 3)(4x^2 - 6x + 9)$ **9.** $(xy + 2)(x^2y^2 - 2xy + 4)$

10. $(x^2 - y^3)(x^4 + x^2y^3 + y^6)$ **11.** $(10x + 1)(100x^2 - 10x + 1)$

12. $(x + 10)(x^2 - 10x + 100)$ **13.** $(5x + y)(25x^2 - 5xy + y^2)$

14. $(x - 4y)(x^2 + 4xy + 16y^2)$ **15.** $(2x - 5y)(4x^2 + 10xy + 25y^2)$

16. $(x^3 - y)(x^6 + x^3y + y^2)$ **17.** $(3x + 2)(9x^2 - 6x + 4)$

18. $(1 + 4y)(1 - 4y + 16y^2)$ **19.** $(2 - 3y)(4 + 6y + 9y^2)$

20. $(x^4 - 2)(x^8 + 2x^4 + 4)$ **21.** $(x^4 + 3)(x^8 - 3x^4 + 9)$

22. $(x^3 + 7)(x^6 - 7x^3 + 49)$ **23.** $(x^2 + 4)(x^4 - 4x^2 + 16)$

24. $(x^2 + y^2)(x^4 - x^2y^2 + y^4)$ **25.** $(x^3 - 10)(x^6 + 10x^3 + 100)$

Factor completely.

26. $x^3 - 125$

27. $x^3 - 64$

28. $x^3 - 8y^3$

29. $x^3 - y^3$

30. $x^3 + 216$

31. $x^3 + 125$

32. $x^3 + y^3$

33. $x^3 + 27y^3$

34. $27x^3 - 1$

35. $8x^3 + 1$

36. $27x^3 + 8$

37. $5x^3 + 40$

38. $3x^3 + 81$

39. $54x^3 - 2y^3$

40. $3x^4 + 375xy^3$

41. $x^3y + y^4$

42. $x^4y^3 - x$

43. $x^2y^2 - x^2y^5$

44. $2x^2 - 16x^2y^3$

45. $24x^4y + 81xy^4$

46. $x^6 - 64y^3$

47. $x^6 - y^9$

48. $x^{12} - y^9$

49. $x^3 - 1000$

50. $x^3 + (x - y)^3$

51. $(x + 2)^3 + (y - 3)^3$

52. $(x + y)^3 + (x - y)^3$

53. $27x^3 + (y - 1)^3$

54. $(x - 3y)^3 - 64$

55. $(x + 1)^3 - 1$

Hawkes Learning Systems: Introductory Algebra

Special Factorizations: Cubes

Pi

As discussed in the text on page 80, π is an irrational number, and so the decimal form of π is an infinite nonrepeating decimal. Mathematicians even in ancient times realized that π is a constant value obtained from the ratio of a circle's circumference to its diameter, but they had no sense that it might be an irrational number. As early as about 1800 B.C. the Babylonians gave π a value of 3, and around 1600 B.C. the ancient Egyptians were using the approximation of $\frac{256}{81}$, what would be decimal value of about 3.1605. In the third century B.C. the Greek mathematician Archimedes used polygons approximating a circle to determine that the value of π must lie between $\frac{223}{71}(\approx 3.1408)$ and $\frac{22}{7}(\approx 3.1429)$. He was thus accurate to two decimal places. About seven hundred years later, in the fourth century A.D. Chinese mathematician Tsu Chung-Chi refined Archimedes' method and expressed the constant as $\frac{355}{113}$, which was correct to six decimal places. By 1610, Ludolph van Ceulen of Germany had also used a polygon method to find π accurate to 35 decimal places.

Knowing that the decimal expression of π would not terminate, mathematicians still sought a repeating pattern in its digits. Such a pattern would mean that π was a rational number and that there would be some ratio of two whole numbers that would produce the correct decimal representations. Finally, in 1767, Johann Heinrich Lambert provided a proof to show that π is indeed irrational and thus is nonrepeating as well as nonterminating.

Since Lambert's proof, mathematicians have still made an exercise of calculating π to more and more decimal places. The advent of the computer age in this century has made that work immeasurably easier, and on occasion you will still see newspaper articles pronouncing that mathematics researchers have reached a new high in the number of decimal places in their approximations. In 1988 that number was 201,326,000 decimal places. Within 1 year that record was more than doubled, and most recent approximations of π now reach beyond 1.24 trillion decimal places! For your understanding, appreciation and interest, the value of π is given in the table on the next page to a mere 3742 decimal places as calculated by a computer program. To show π calculated to 1.24 trillion decimal places would take every page of nearly 2500 copies of this text!

The Value of π

π ≈
3.14159265358979323846264338327950288419716939937510582097494459230781640628
20899862803482534211706798214808651328230664709384460955058223172535940812848
11174502841027019385211055596446229489549303819644288109756659334461284756482
33786783165271201909145648566923460348610454326648213393607260249141273724587
00660631558817488152092096282925409171536436789259036001133053054882046652138
41469519415116094330572703657595919530921861173819326117931051185480744623799
62749567351885752724891227938183011949129833673362440656643086021394946395224
73719070217986094370277053921717629317675238467481846766940513200056812714526
35608277857713427577896091736371787214684409012249534301465495853710507922796
89258923542019956112129021960864034418159813629774771309960518707211349999998
37297804995105973173281609631859502445945534690830264252230825334468503526193
11881710100031378387528865875332083814206171776691473035982534904287554687311
59562863882353787593751957781857780532171226806613001927876611195909216420198
93809525720106548586327886593615338182796823030195203530185296899577362259941
38912497217752834791315155748572424541506959508295331168617278558890750983817
54637464939319255060400927701671139009848824012858361603563707660104710181942
95559619894676783744944825537977472684710404753464620804668425906949129331367
70289891521047521620569660240580381501935112533824300355876402474964732639141
99272604269922796782354781636009341721641219924586315030286182974555706749838
50549458858692699569092721079750930295532116534498720275596023648066549911988
18347977535663698074265425278625518184175746728909777727938000816470600161452
49192173217214477235014144197356854816136115735255213347574184946843852332390
74941433345477624168625189835694855620992192221842725502542568876717904946016
53466804988627232791786085784383827967976681454100953883786360950680064225125
20511739298489608412848862694560424196528502221066118630674427862203919494504
71237137869609563643719172874677646575739624138908658326459958133904780275900
99465764078951269468398352595709825822620522489407726719478268482601476990902
64013639443745530506820349625245174939965143142980919065925093722169646151570
98583874105978859597729754989301617539284681382686838689427741559918559252459
53959431049972524680845987273644695848653836736222626099124608051243884390451
24413654976278079771569143599770012961608944169486855584840635342207222582848
86481584560285060168427394522674676788952521385225499546667278239864565961163
54886230577456498035593634568174324112515076069479451096596094025228879710893
14566913686722874890456010150330861792868092087476091782493858900971490967598
52613655497818931297848216829989487226588048575640142704775551323796414515237
46234364542858444795265867821051141354735739523113427166102135969536231442952
48493718711014576540359027993440374200731057853906219838744780847848968332144
57138687519435064302184531910484810053706146806749192781911979399520614196634
28754440643745123718192179998391015919561814675142691239748940907186494231961
56794520809514655022523160388193014209376213785595663893778708303906979207734
67221825625996615014215030680384477345492026054146659252014974428507325186660
02132434088190710486331734649651453905796268561005508106658796998163574736384
05257145910289706414011097120628043903975951567715770042033786993600723055876
31763594218731251471205329281918261861258673215791984148488291644706095752706
95722091756711672291098169091528017350671274858322287183520935396572512108357
91513698820914442100675103346711031412671113699086585163983150197016515116851
71437657618351556508849099898599823873455283316355076479185358932261854896321
32933089857064204675259070915481416549859461637180270981994309924488957571282
89059232332609729971208443357326548938239119325974...

Answers

Chapter R

Exercises R.1, pages 11 - 13

1. a. 2 **b.** 3 **c.** 8 **3. a.** 4 **b.** 2 **c.** 16 **5. a.** 9 **b.** 2 **c.** 81 **7. a.** 11 **b.** 2 **c.** 121 **9.** 2^4 or 4^2 **11.** 5^2 **13.** 7^2 **15.** 6^2 **17.** 10^3 **19.** 2^3
21. 6^5 **23.** 11^3 **25.** $2^3 \cdot 3^2$ **27.** $2 \cdot 3^2 \cdot 11^2$ **29.** $3^3 \cdot 7^3$ **31.** 64 **33.** 121 **35.** 400 **37.** 169 **39.** 900 **41.** 2704 **43.** 390,625
45. 1,953,125 **47.** 78,125 **49.** 21 **51.** 19 **53.** 28 **55.** 2 **57.** 18 **59.** 0 **61.** 0 **63.** 46 **65.** 9 **67.** 74 **69.** 20 **71.** 132 **73.** 176
75. 61 **77.** 1532 **79.** 132 **81.** Error. Answers will vary.

Exercises R.2, pages 22 - 23

1. A counting number greater than 1 that has exactly two different factors (or divisors)–itself and 1. **3.** A counting number
with more than two different factors (or divisors). **5.** 53, 59, 61, 67, 71 (Answers may vary.) **7.** 13, 17 **9.** 2 **11.** 8, 9
13. 4, 17 **15.** 2, 251 **17.** 10, 17 **19.** 4, 111 **21.** $2^2 \cdot 13$ **23.** $2^3 \cdot 7 \cdot 11$ **25.** $2^2 \cdot 7 \cdot 11$ **27.** 79 is prime. **29.** 17^2 **31.** 5^3
33. $2^4 \cdot 5^2$ **35.** $2^3 \cdot 3 \cdot 5$ **37.** $3 \cdot 7 \cdot 11$ **39.** $2^2 \cdot 3^2 \cdot 47$ **41. a.** 5, 10, 15, 20, 25, 30, 35, 40, 45, 50, 55, 60 **b.** 6, 12, 18, 24, 30, 36,
42, 48, 54, 60, 66, 72 **c.** 10, 20, 30, 40, 50, 60, 70, 80, 90, 100, 110, 120 **d.** 15, 30, 45, 60, 75, 90, 105, 120, 135, 150, 165, 180
43. 105 **45.** 30 **47.** 66 **49.** 140 **51.** 150 **53.** 180 **55.** 196 **57.** 210 **59.** 90 **61.** 100 **63.** 150 **65.** 364 **67.** 2520 **69.** 924
71. 7623 **73.** 2880 **75.** 7840

Exercises R.3, pages 32 - 36

1. a. A fraction in which the numerator is less than the denominator. **b.** A fraction in which the numerator is greater than the
denominator. **3.** Answers will vary. **5.** Use the fact that 1 is equal to any counting number divided by itself and divide out all
common factors found in both the numerator and the denominator. **7. a.** $\dfrac{13}{25}$ **b.** $\dfrac{12}{25}$ **9. a.** $\dfrac{2}{5}$ **b.** $\dfrac{3}{5}$ **11.** $\dfrac{3}{4} = \dfrac{3}{4} \cdot \dfrac{3}{3} = \dfrac{9}{12}$

13. $\dfrac{6}{7} = \dfrac{6}{7} \cdot \dfrac{2}{2} = \dfrac{12}{14}$ **15.** $\dfrac{3}{16} = \dfrac{3}{16} \cdot \dfrac{5}{5} = \dfrac{15}{80}$ **17.** $\dfrac{7}{26} = \dfrac{7}{26} \cdot \dfrac{2}{2} = \dfrac{14}{52}$ **19.** $\dfrac{18}{1} = \dfrac{18}{1} \cdot \dfrac{3}{3} = \dfrac{54}{3}$ **21.** $\dfrac{1}{6}$ **23.** $\dfrac{21}{8}$ **25.** $\dfrac{273}{80}$ **27.** $\dfrac{21}{2}$

29. $\dfrac{77}{4}$ **31.** $\dfrac{25}{24}$ **33.** $\dfrac{10}{3}$ **35.** 1 **37.** $\dfrac{9}{16}$ **39.** $\dfrac{1}{4}$ **41.** $\dfrac{1}{16}$ **43.** Undefined **45.** 0 **47.** $\dfrac{8}{5}$ **49.** $\dfrac{9}{20}$ **51.** $\dfrac{9}{20}$; $\dfrac{11}{20}$

53. 12 miles; 48 miles; $\dfrac{4}{5}$ **55. a.** $\dfrac{2}{5}$ **b.** $\dfrac{12}{25}$ **57. a.** more than 90 **b.** less than 90 **c.** 100 passengers **59. a.** Yes, it would pass
b. Pass by 5 votes.

Exercises R.4, pages 43 - 46

1. a. $\dfrac{1}{9}$ **b.** $\dfrac{2}{3}$ **3. a.** $\dfrac{4}{49}$ **b.** $\dfrac{4}{7}$ **5. a.** $\dfrac{25}{16}$ **b.** $\dfrac{5}{2}$ **7. a.** $\dfrac{64}{81}$ **b.** $\dfrac{16}{9}$ **9. a.** $\dfrac{9}{16}$ **b.** $\dfrac{3}{2}$ **11.** $\dfrac{3}{7}$ **13.** $\dfrac{2}{5}$ **15.** $\dfrac{8}{25}$ **17.** $\dfrac{1}{5}$

19. $\dfrac{1}{16}$ **21.** $\dfrac{17}{21}$ **23.** $\dfrac{1}{3}$ **25.** $\dfrac{5}{14}$ **27.** 0 **29.** $\dfrac{2}{25}$ **31.** $\dfrac{7}{12}$ **33.** $\dfrac{7}{90}$ **35.** $\dfrac{1}{3}$ **37.** $\dfrac{63}{50}$ **39.** $\dfrac{7}{30}$ **41.** $\dfrac{3}{16}$ **43.** $\dfrac{3}{8}$

45. $\dfrac{19}{10}$ **47.** $\dfrac{21}{16}$ **49. a.** $\dfrac{13}{30}$ **b.** \$1430 **51.** $\dfrac{19}{20}$ in. by $\dfrac{7}{10}$ in.

Exercises R.5, pages 60 - 63

1. 7.5   **3.** 14.027   **5.** $23\frac{2}{100}$ **7.** $52\frac{1}{1000}$ **9.** 0.6   **11.** 2.24   **13.** two tenths   **15.** two hundred sixty and three thousandths
17. 72.1   **19.** 0.30   **21.** 1.4   **23.** 5.76   **25.** 1.64   **27.** 25.89   **29.** 147.385   **31.** 19.464   **33.** 0.04   **35.** 0.01096   **37.** 5.00   **39.** 2.03
41. $9690   **43.** 0.237   **45.** $1296   **47.** 13.3 mph   **49. a.** 77.2 in.   **b.** 337.68 in.^{2} **51.** 3%   **53.** 300%   **55.** 108%   **57.** 0.06   **59.** 0.032
61. 1.2   **63.** 15%   **65.** 96%   **67.** 162.5%   **69.** 0.9%   **71.** 31.112   **73.** 108.040   **75.** 1.056   **77.** 26.636   **79.** 19.464

Chapter R Review, pages 69 - 71

1. $3^3 \cdot 5^2$ **2.** $13^3 \cdot 17$ **3.** $2^2 \cdot 11^2$ **4.** $7 \cdot 11^2 \cdot 19^3$ **5.** 144   **6.** 64   **7.** 279,936   **8.** 32,768   **9.** 33   **10.** 37   **11.** 39   **12.** 19   **13.** 8

14. 8   **15.** 5   **16.** 248   **17.** 130   **18.** 31   **19.** 60   **20.** 240

21. $5\overline{)375}$ quotient 75   **22.** $3\overline{)912}$ quotient 304   **23.** 2,3,5,7,11,13,17,19,23,29

24. A counting number with more than two different factors (or divisors)

25. prime   **26.** composite   **27.** composite   **28.** composite   **29.** $5^2 \cdot 7$
30. $2 \cdot 7 \cdot 11$ **31.** $3 \cdot 5 \cdot 13$ **32.** $3^2 \cdot 5^2 \cdot 11$ **33.** 1890   **34.** 270   **35.** 120
36. 468   **37.** 120   **38.** 462   **39.** 4,601,025   **40.** 207,025   **41.** 12   **42.** 24   **43.** $\frac{3}{40}$

44. $\frac{7}{8} = \frac{7}{8} \cdot \frac{3}{3} = \frac{21}{24}$ **45.** $\frac{15}{17} = \frac{15}{17} \cdot \frac{3}{3} = \frac{45}{51}$ **46.** $\frac{1}{16} = \frac{1}{16} \cdot \frac{4}{4} = \frac{4}{64}$ **47.** $\frac{5}{17}$ **48.** $\frac{6}{7}$ **49.** $\frac{10}{23}$ **50.** $\frac{2}{19}$ **51.** $\frac{5}{13}$ **52.** $\frac{13}{9}$

53. $\frac{4}{5}$ **54.** 1   **55.** $\frac{3}{10}$ **56.** $\frac{16}{135}$ **57.** $\frac{5}{11}$ **58.** $\frac{3}{32}$ **59.** $\frac{11}{52}$ **60.** $\frac{3}{52}$ **61.** $\frac{4}{3}$ **62.** $\frac{6}{5}$ **63.** $\frac{1}{4}$ **64.** $\frac{2}{5}$ **65.** $\frac{39}{100}$ **66.** $\frac{11}{50}$

67. $\frac{32}{25}$ **68.** $\frac{19}{18}$ **69.** $\frac{17}{20}$ **70.** $\frac{5}{12}$ **71.** $\frac{5}{16}$ **72.** $\frac{9}{32}$ **73.** $\frac{41}{180}$ **74.** $\frac{13}{24}$ **75.** $\frac{8}{9}$ **76.** $\frac{7}{10}$ **77.** $\frac{29}{48}$ **78.** $\frac{53}{30}$ **79.** $\frac{71}{45}$

80. $\frac{17}{10}$ **81.** 0.14   **82.** 7.6   **83.** 300.003   **84.** 0.303   **85.** Seventy-five and nine hundredths   **86.** Six and twenty-three ten-
thousandths   **87.** 173.62   **88.** 62.94   **89.** 138.165   **90.** 39.61   **91.** $33,010   **92.** 15.5 inches   **93.** 3.5%   **94.** 123%   **95.** 0.83
96. 0.005   **97.** 5%   **98.** 37.5%   **99.** 45%   **100.** 70%

Chapter R Test, pages 72 - 73

1. $3^2 \cdot 5^4$ **2.** 169; the square of 13   **3.** 18   **4.** 40   **5.** 7   **6.** 24   **7.** 2, 3, 5, 7, 11, 13, 17, 19, 23, 29, 31, 37   **8.** $2^2 \cdot 5 \cdot 47$ **9.** 360

10. 600   **11.** $\frac{2}{5}$ **12.** $\frac{1}{4}$ **13.** $\frac{5}{21}$ **14.** $\frac{1}{135}$ **15.** $\frac{12}{35}$ **16.** 4   **17.** 276.47   **18.** 49.29   **19.** 12.383   **20.** 324.65   **21.** 3.1   **22.** 6.13
23. 17.13   **24.** 6.107   **25.** Five and one hundred seven thousandths   **26.** One hundred and one thousandth   **27.** 0.0432
28. 5.7%   **29.** 3%   **30.** 102%   **31.** 25%   **32.** 2%   **33.** 64   **34.** 64   **35.** 397.32 ft^2 **36.** 20.8 miles per gallon   **37.** 50,625
38. 1714.75   **39.** 7698.3   **40.** 18.7

Chapter 1

Exercises 1.1, pages 86 - 89

1.   **3.**   **5.**   **7.**
9.   **11.**   **13.**   **15.**
17.   **19.** No Solution   **21.**   **23.**   **25.** 10   **27.** −4

29. −11.3 **31.** < **33.** > **35.** = **37.** < **39.** > **41.** < **43.** < **45.** = **47.** True **49.** True **51.** False; −9 < −8.5 **53.** True **55.** True

57. True **59.** True **61.** True **63.** False; $-\left|-3\right| > -\left|4\right|$ **65.** False; $\left|-\dfrac{5}{2}\right| > 2$ **67.** True **69.** True **71.** True

73. **75.** **77.** **79.** No solution

81. **83.** **85.** **87.**

89. **91.** **93.** 46 **95.** −6 **97.** $\dfrac{1}{3}$ **99.** If y is a negative number then

$-y$ represents a positive number. For example, if $y = -2$, then $-y = -(-2) = 2$.

Exercises 1.2, pages 93 - 95

1. 13 **3.** −4 **5.** 0 **7.** 5 **9.** −13 **11.** −8 **13.** −10 **15.** 0 **17.** 17 **19.** −29 **21.** −9 **23.** −3 **25.** 22 **27.** −54 **29.** −7 **31.** −16
33. −26 **35.** 0 **37.** −32 **39.** 5 **41.** −83 **43.** 12 **45.** −32 **47.** −11 **49.** 39 **51.** −2 is a solution **53.** −4 is a solution
55. −6 is a solution **57.** 18 is a solution **59.** −2 is not a solution **61.** −10 is a solution **63.** −72 is not a solution **65.** 84
67. 3807 **69.** −97,714 **71.** $\left|0\right| + \left|0\right| = 0$

Exercises 1.3, pages 102 - 105

1. −11 **3.** 6 **5.** −47 **7.** 0 **9.** 52 **11.** 5 **13.** −10 **15.** 7 **17.** 3 **19.** −16 **21.** 24 **23.** −15 **25.** −16 **27.** −57 **29.** −54 **31.** 1
33. −7 **35.** −26 **37.** −15 **39.** −15 **41.** −1 **43.** −3 **45.** 1 **47.** 0 **49.** −8 < −5 **51.** −3 < 3 **53.** 8 = 8 **55.** 0 > −27
57. −237 > −248 **59.** −18°F **61.** −$8 **63.** −14,777 ft **65.** 85 years old **67.** −8 is a solution **69.** −2 is a solution
71. 4 is not a solution **73.** 5 is a solution **75.** 1 is not a solution **77.** −4 is a solution **79.** −5 is a solution
81. −28 is a solution **83.** 1044 > −39 **85.** −15,254 > −35,090 **87.** Lost 7 pounds; 203 pounds **89.** Gained 51 yards

Exercises 1.4, pages 112 - 117

1. −12 **3.** 48 **5.** 56 **7.** −21 **9.** 56 **11.** 26 **13.** −70 **15.** −24 **17.** −288 **19.** 0 **21.** −9 **23.** −7.31 **25.** 4 **27.** −6 **29.** 2 **31.** 0
33. undefined **35.** −3 **37.** −17 **39.** 5 **41.** −0.6 **43.** 20 **45.** 0.375 **47.** True **49.** True **51.** True **53.** True **55.** True **57.** −12
59. 5° **61.** 80 **63.** 9.3 mg/dL **65.** 72.0 in. **67.** 31.5 hrs. **69.** Approx. 22,000 square miles. **71.** −34,459,110
73. −2671 **75.** −45.33 **77.** −10.29 **79.** 2.83 **81.** −90.365 **83.** See page 109

Exercises 1.5, pages 122 -124

1. a. −1, 0, 4 **b.** −3.56, $\dfrac{-5}{8}$, −1, 0, 0.7, 4, $\dfrac{13}{2}$ **c.** $-\sqrt{8}$, π, $\sqrt[3]{25}$ **d.** All are real numbers **3.** 3 + 7 **5.** 4 · 19 **7.** 30 + 48

9. $(2 \cdot 3) \cdot x$ **11.** $(3 + x) + 7$ **13.** $0 \cdot 6 = 0$ **15.** $x + 7$ **17.** $2x - 24$ **19.** 0 **21.** Commutative property of addition
23. Multiplicative identity **25.** Associative property of addition **27.** Commutative property of multiplication
29. Commutative property of multiplication **31.** Multiplicative inverse **33.** Additive inverse **35.** Multiplicative identity
37. Zero factor law **39.** Associative property of addition **41.** Commutative property of multiplication; $6 \cdot 4 = 4 \cdot 6 = 24$
43. Associative property of addition; $8 + (5 + (-2)) = (8 + 5) + (-2) = 11$
45. Distributive property; $5(4 + 18) = 5(4) + 90 = 110$
47. Associative property of multiplication; $(6 \cdot (-2)) \cdot 9 = 6 \cdot (-2 \cdot 9) = -108$
49. Commutative property of addition; $3 + (-34) = (-34) + 3 = -31$
51. Commutative property of addition; $2(3 + 4) = 2(4 + 3) = 14$
53. Commutative property of addition; $5 + (4 - 15) = (4 - 15) + 5 = -6$
55. Associative property of multiplication; $(3 \cdot 4) \cdot 5 = 3 \cdot (4 \cdot 5) = 60$

Chapter 1 Review, pages 129 - 132

1. **2.** **3.** **4.**

5. 6 **6.** -7.3 **7.** -4.1 **8.** 10 **9.** True **10.** True **11.** False; $|-2.3| > 0$ **12.** True **13.**

14. No Solution **15.** **16.** **17.**

18. **19.** **20.** **21.** 15 **22.** 22 **23.** 5 **24.** -3 **25.** 0

26. 0 **27.** -33 **28.** -21 **29.** -18 **30.** -25 **31.** -19 **32.** -29 **33.** 8 **34.** -30 **35.** -45 **36.** 46 **37.** -3 is a solution.
38. 2 is not a solution. **39.** -20 is a solution. **40.** 3 is a solution. **41.** -7 **42.** -2 **43.** -19 **44.** -8 **45.** -4 **46.** -31 **47.** 21
48. 50 **49.** 6 **50.** 22 **51.** -8 **52.** -17 **53.** 9 **54.** 7 **55.** -9 **56.** 37 **57.** $-33°$F **58.** -6000 feet **59.** -1 is a solution
60. -17 is not a solution **61.** -35 **62.** -25.6 **63.** 90 **64.** 91 **65.** -255 **66.** 24 **67.** 0 **68.** -546 **69.** 5 **70.** 17 **71.** undefined
72. 0 **73.** -0.4 **74.** -0.5 **75.** -20 **76.** -40 **77.** 62 **78.** 85.6 **79.** 65.8 **80.** 8 hurricanes **81.** Commutative Property of
Addition **82.** Associative Property of Addition **83.** Additive Identity **84.** Additive Inverse **85.** Associative Property of
Multiplication **86.** Commutative Property of Multiplication **87.** Multiplicative Identity **88.** Multiplicative Inverse
89. Distributive Property **90.** Distributive Property **91.** 238.43 **92.** -74.65 **93.** -59.76 **94.** $-40,473.75$ **95.** 5.13
96. -114.33 **97.** 1.00 **98.** -1.00 **99.** -16.30 **100.** 0.50

Chapter 1 Test, pages 133 - 134

1. a. $-5, -1, 0$ **b.** $-5, -1, \dfrac{-1}{3}, 0, \dfrac{1}{2}, 3\dfrac{1}{4}, 7.121212...$ **c.** $-\pi$ **d.** $-5, -\pi, -1, \dfrac{-1}{3}, 0, \dfrac{1}{2}, 3\dfrac{1}{4}, 7.121212...$ **2. a.** True, since for any

integer n, $n = \dfrac{n}{1}$ which is a rational number since the numerator and the denominator are both integers. **b.** False, because

rational numbers also include non-integers, such as $\dfrac{2}{3}$. **3. a.** $<$ **b.** $>$ **c.** $=$ **4.**

5. $|x| = x$ if x is positive or 0; $|x| = -x$ if x is negative. $|x|$ is the distance that x lies from zero on a real number line.
6. $y = -7$ or $y = 7$ **7. a.** $\{-2, -1, 0, 1, 2\}$ **b.** $\{..., -11, -10, -9, 9, 10, 11, ...\}$ **8.** -13 **9.** 23 **10.** 18 **11.** 112 **12.** -126 **13.** 0
14. -3 **15.** 7 **16.** -0.3 **17.** Additive Identity. **18.** Distributive Property **19.** Commutative Property of Multiplication
20. Associative Property of Addition **21.** Multiplicative Inverse **22.** Zero Factor Law **23.** -60 **24.** 91 **25.** 152 **26.** $15°$F
27. -22.5 **28.** 21.7

Chapter 2

Exercises 2.1, pages 141 - 142

1. $-5, \dfrac{1}{6}$ and 8 are like terms; $7x$ and $9x$ are like terms **3.** $-x^2$ and $2x^2$ are like terms; $5xy$ and $-6xy$ are like terms; $3x^2y$
and $5x^2y$ are like terms. **5.** $24, 8.3$ and -6 are like terms; $1.5xyz, -1.4xyz$ and xyz are like terms. **7.** 64 **9.** -121 **11.** $15x$
13. $3x$ **15.** $-2n$ **17.** $5y^2$ **19.** $12x^2$ **21.** $7x + 2$ **23.** $x - 3y$ **25.** $8x^2 + 3y$ **27.** $2n + 3$ **29.** $7a - 8b$ **31.** $8x + y$ **33.** $2x^2 - x$
35. $-2n^2 + 2n$ **37.** $3x^2 - xy + y^2$ **39.** $2x$ **41.** $-y$ **43. a.** $3x + 4$ **b.** 16 **45. a.** $-2x - 8$ **b.** -16 **47. a.** $5y + 1$ **b.** 16
49. a. $-3x - 7y$ **b.** -33 **51. a.** $3.6x^2$ **b.** 57.6 **53. a.** $3ab + b^2 + b^3$ **b.** 6 **55. a.** $3.7x + 1.1$ **b.** 15.9 **57. a.** $8a$ **b.** -16
59. a. $3b$ **b.** -3 **61. a.** $-x - 54$ **b.** -58

Exercises 2.2, pages 150 - 152

1. $\dfrac{5}{6} = \dfrac{5}{6} \cdot \dfrac{8}{8} = \dfrac{40}{48}$ **3.** $\dfrac{0}{9} = \dfrac{0}{9} \cdot \dfrac{7b}{7b} = \dfrac{0}{63b}$ **5.** $\dfrac{-3}{5} = \dfrac{-3}{5} \cdot \dfrac{7y}{7y} = \dfrac{-21y}{35y}$ **7.** $\dfrac{2}{5}$ **9.** $\dfrac{3x}{7}$ **11.** $\dfrac{3}{25b}$ **13.** $\dfrac{-2}{3}$ **15.** $\dfrac{-1}{2x}$ **17.** $\dfrac{-12z}{35}$

19. $\dfrac{3}{8}$ **21.** 35 **23.** $\dfrac{-1}{6}$ **25.** $\dfrac{81}{100}$ **27.** $\dfrac{-y^2}{7}$ **29.** $\dfrac{-9x^2}{5}$ **31.** $\dfrac{2b^2}{3a^2}$ **33.** $\dfrac{-50}{9}$ **35.** $\dfrac{2}{5b}$ **37.** $\dfrac{5y}{3}$ **39.** 0 **41.** Undefined

43. $\dfrac{22}{a^2}$ **45.** $\dfrac{-1}{10}$ **47.** $\dfrac{9a}{10b}$ **49.** $\dfrac{81}{320n^2}$ **51.** $\dfrac{-7x^2}{100y^2z}$ **53.** $\dfrac{247}{3}$ **55.** $\dfrac{12}{5}$ **57.** $\dfrac{-25}{24}$ **59.** 2 inches **61.** 98 lbs

63. 58,535,000 sq mi **65.** $\dfrac{35}{40}$ or $\dfrac{7}{8}$ **67. a.** more than 180 **b.** 200 passengers **69.** Division by zero is undefined. For example we could write $0 = \dfrac{0}{1}$. Then the reciprocal would be $\dfrac{1}{0}$, but this reciprocal is undefined since division by zero is undefined. Thus 0 does not have a reciprocal.

Exercises 2.3, pages 158 - 161

1. a. 1,2,3,6 **b.** 6,12,18,24,30,36 **3. a.** 1,3,5,15 **b.** 15,30,45,60,75,90 **5.** 120 **7.** $40xy$ **9.** $210x^2y^2$ **11.** $\dfrac{7}{9}$ **13.** 1 **15.** $\dfrac{-1}{6}$

17. $\dfrac{23}{15}$ **19.** $\dfrac{16}{15a}$ **21.** $\dfrac{-2}{5x}$ **23.** $\dfrac{1}{2y}$ **25.** $\dfrac{17}{36}$ **27.** $\dfrac{43}{72x}$ **29.** $\dfrac{15-y}{3y}$ **31.** $\dfrac{16+x}{4x}$ **33.** $\dfrac{4a-15}{20}$ **35.** $\dfrac{6-x}{8}$ **37.** $\dfrac{5}{6a}$

39. $\dfrac{-11}{12x}$ **41.** 0 **43.** 0 **45. a.** $6a$ **b.** $9a^2$ **c.** $\dfrac{2}{3a}$ **d.** $\dfrac{1}{9a^2}$ **e.** 1 **47. a.** $-14ab$ **b.** $49a^2b^2$ **c.** $\dfrac{2}{7ab}$ **d.** $\dfrac{1}{49a^2b^2}$ **e.** 1 **49. a.** $\dfrac{8x}{3}$

b. $\dfrac{16x^2}{9}$ **c.** 0 **d.** 1 **e.** $\dfrac{16x^2}{9}$ **51.** $\dfrac{1}{18,009,460}$ **53.** $\dfrac{92}{19}$ **55. a.** $\dfrac{13}{30}$ **b.** $1170

Exercises 2.4, pages 169 - 170

1. 108.72 **3.** −7.626 **5.** 157.184 **7.** 169.376 **9.** 15.1 **11.** 3.52 **13.** 0.38 **15.** 0.025 **17.** 0.375 **19.** 0.05 **21.** $0.\overline{6}$ **23.** $0.\overline{23}$

25. 0.4375 **27.** $-0.\overline{54}$ **29.** $1.\overline{185}$ **31.** $6.58\overline{3}$ **33.** 95.31 **35.** $1082.23 **37.** $\dfrac{13}{10}$ **39.** $\dfrac{1323}{250}$ **41.** $\dfrac{-6423}{100}$ **43.** $\dfrac{17}{20}$ **45.** $\dfrac{6}{7}$

47. $\dfrac{-103}{1000}$

Exercises 2.5, pages 176 - 178

1. a. 36 **b.** 16 **3.** −25 **5.** −10 **7.** −45 **9.** −137 **11.** 152 **13.** −189 **15.** 143 **17.** −36 **19.** −1 **21.** $\dfrac{11}{30}$ **23.** $\dfrac{1}{24}$ **25.** $\dfrac{31}{24}$

27. $\dfrac{107}{24}$ **29.** $\dfrac{41}{32}$ **31.** $\dfrac{3}{5}$ **33.** $-\dfrac{341}{30}$ **35.** $-\dfrac{15}{17}$ **37.** 24 **39.** $\dfrac{7}{10}$ **41.** $5\dfrac{1}{12}$ **43.** $\dfrac{-1081}{1600}$ **45.** 42.45 **47.** 67.77 **49.** 15.41

51. $\dfrac{7}{2}$ **53.** $\dfrac{-23}{21y}$ **55.** $\dfrac{7}{5}$ **57.** $-\dfrac{35}{4}$ **59.** $\dfrac{a}{36}$ **61.** $(3^2 - 9) = 0$ and division by 0 is undefined. **63.** It will always be larger. If you square a negative number you get a positive number. All positive numbers are larger than all negative numbers.

Exercises 2.6, pages 183 - 185

1. 4 times a number **3.** 1 more than twice a number **5.** 5.3 less than 7 times a number **7.** −2 times the difference between a number and 8 **9.** 5 times the sum of twice a number and 3 **11.** 6 times the difference between a number and 1

13. 3 times a number plus 7; 3 times the sum of a number and 7   **15.** The product of 7 and a number minus 3; 7 times the difference between a number and 3   **17.** $x + 6$ **19.** $x - 4$ **21.** $3x - 5$ **23.** $\dfrac{x-3}{7}$ **25.** $3(x - 8)$ **27.** $3x - 5$ **29.** $8 - 2x$ **31.** $8(x - 6) + 4$ **33.** $3x - 5x$ **35.** $2(x - 7) - 6$ **37.** $2(17 + x) + 9$ **39. a.** $x - 6$ **b.** $6 - x$ **41.** $24d$ **43.** $60m$ **45.** $365y$ **47.** $0.11x$ **49.** $60h + 20$ **51.** $\$20 + 0.15m$ **53.** $0.09x + 250$ **55.** $2w + 2(2w - 3) = 6w - 6$

Chapter 2 Review, pages 190 - 193

1. $13x$ **2.** $6y$ **3.** $25x^2$ **4.** $-3a^2$ **5.** $-3x + 2$ **6.** $5y - 6$ **7.** $3n + 4$ **8.** $5x - 34$ **9.** $y^2 + 7y$ **10.** $7a^2 + 4a$ **11. a.** $5x + 5$ **b.** 0
12. a. $-2y - 26$ **b.** -34 **13. a.** $-a^2 + 4a + 7$ **b.** -5 **14. a.** $x^2 + 9x - 10$ **b.** -18 **15. a.** $3a^2 - 4a + 2$ **b.** 22
16. a. $5y^2 - 8y + 26$ **b.** 74 **17. a.** $2x^2 + 3x - 13$ **b.** -14 **18. a.** $17x - 17$ **b.** -34 **19. a.** $7y^2 + 9y - 136$ **b.** 12
20. a. $80x^2 + 20x + 100$ **b.** 160 **21.** $\dfrac{-3}{16}$ **22.** $\dfrac{-1}{12}$ **23.** $\dfrac{4}{3}$ **24.** 1 **25.** $\dfrac{36}{175}$ **26.** $\dfrac{3}{10}$ **27.** $\dfrac{4}{11}$ **28.** $\dfrac{7}{12}$ **29.** $\dfrac{29}{35}$
30. $\dfrac{3}{5}$ was spent; $\dfrac{2}{5}$ remains   **31.** 180 **32.** 1050 **33.** $210xyz$ **34.** $120a^2b^2$ **35.** $\dfrac{9}{7}$ **36.** $\dfrac{3}{2}$ **37.** $-\dfrac{1}{8}$ **38.** $-\dfrac{3}{16}$ **39.** $\dfrac{53}{60}$
40. $\dfrac{7}{24}$ **41.** $\dfrac{2}{5}$ **42.** $\dfrac{5}{16}$ **43.** $\dfrac{7}{8}$ **44.** $\dfrac{11}{36}$ **45.** $\dfrac{8}{15}$ **46.** 0 **47.** $\dfrac{2x-1}{14}$ **48.** $\dfrac{3y-4}{12}$ **49.** $\dfrac{15+2a}{3a}$ **50.** $\dfrac{x-50}{5x}$ **51.** 53.2
52. 80.84 **53.** -6.222 **54.** 232.5 **55.** 3.6 **56.** 0.05 **57.** 81 **58.** 15.1 **59.** 0.4375 **60.** $0.\overline{2}$ **61.** $0.\overline{45}$ **62.** $0.1\overline{6}$ **63.** $0.\overline{307692}$
64. 4.125 **65.** $\dfrac{51}{50}$ **66.** $\dfrac{-1073}{100}$ **67.** $\dfrac{-18,901}{500}$ **68.** $\dfrac{-2829}{200}$ **69.** $\dfrac{47}{40}$ **70.** $\dfrac{-21}{40}$ **71.** 88 **72.** 16 **73.** 18 **74.** -4 **75.** 26
76. 32 **77.** $\dfrac{49}{36}$ **78.** $\dfrac{-9}{2}$ **79.** $\dfrac{-1}{360}$ **80.** $\dfrac{165}{4}$ **81.** $\dfrac{2x-1}{4}$ **82.** $\dfrac{4-6y}{7y}$ **83.** $\dfrac{-4}{7}$ **84.** $\dfrac{31}{28}$ **85.** $\dfrac{37}{100}$ **86.** $\dfrac{-11}{81}$ **87.** $\dfrac{-9}{8}$
88. $\dfrac{33}{40}$ **89.** $\dfrac{-21}{256}$ **90.** $\dfrac{7}{8}$ **91.** 3 more than 5 times a number   **92.** 5 less than 6 times a number   **93.** 8 times a number increased by 3 times the same number   **94.** 7 times a number plus 4 times the same number   **95.** -3 multiplied by the sum of a number and 2   **96.** -2 multiplied by the sum of a number and 6   **97.** 5 times the difference between a number and 3   **98.** 4 times the difference between a number and 10   **99.** 7 times a number divided by 33   **100.** 50 divided by the product of 6 and a number   **101.** $5x + 3x$ **102.** $4(x + 10)$ **103.** $2x + 17$ **104.** $x - 32$ **105.** $28 - 6x$ **106.** $72 + 8(x + 2)$
107. $2x + 3$ **108.** $\dfrac{x}{14}$ **109.** $\dfrac{22}{x+9}$ **110.** $10x - 32$

Chapter 2 Test, pages 194 - 195

1. a. $5x^2 + 7x$ **b.** 6 **2. a.** $6y - 6$ **b.** 12 **3. a.** $5x + 2$ **b.** -8 **4.** $90x^2y^2$ **5.** 17.952 **6.** 1.35 **7.** $-\dfrac{9}{2}$ **8.** $\dfrac{8}{a^2}$ **9.** $\dfrac{9}{10}$
10. $-\dfrac{2}{y}$ **11.** -44 **12.** -32 **13.** $\dfrac{29}{40}$ **14.** $\dfrac{17}{90}$ **15.** $\dfrac{35+x}{5x}$ **16.** -15.98 **17.** $0.\overline{428571}$ **18. a.** The product of 5 and a number increased by 18 **b.** 3 multiplied by the sum of a number and 6 **c.** 42 less the product of 7 and a number
19. a. $6x - 3$ **b.** $2(x + 5)$ **c.** $2(x + 15) - 4$ **20.** $-\dfrac{33}{16}$ **21.** $\$80$ **22. a.** 15 gallons **b.** No, you need $\$46.50$ to fill the tank so you are short $\$6.50$.

Chapter 2 Cumulative Review, pages 196 - 198

1. 20 **2.** -14 **3.** 9 **4.** -15 **5.** -126 **6.** 459 **7.** -39 **8.** 124 **9. a.** 18 **b.** -26 **c.** 36.7 **d.** 7.31 **10.** 108 **11.** $72xy^2$
12. Associative property of multiplication   **13.** Commutative property of multiplication   **14.** Distributive property
15. Commutative property of addition   **16.** Associative property of addition   **17.** Multiplicative identity   **18.** Multiplicative Inverse   **19.** Additive Identity   **20.** Additive Inverse   **21. a.** $\{-4, 4\}$ **b.** No Solution

22. $\{-6,-5,-4,-3,-2,-1,0,1,2,3,4,5,6\}$ 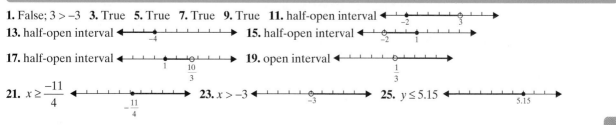 **23.** 7 **24.** 18 **25.** 90 **26.** 12 **27.** $\dfrac{5}{6}$ **28.** $\dfrac{7}{5}$ **29.** $\dfrac{2}{9}$

30. $\dfrac{43}{40}$ **31.** $\dfrac{6}{5y}$ **32.** $\dfrac{18+x}{6x}$ **33.** $\dfrac{-1}{36}$ **34.** $\dfrac{-1}{2a}$ **35.** $\dfrac{1}{6}$ **36.** $\dfrac{-7}{45}$ **37.** $\dfrac{263}{576}$ **38.** $\dfrac{3}{5}$ **39.** 0 **40.** $\dfrac{-7}{16}$ **41.** 70.2 **42.** 51.84

43. −19.703 **44.** 6.4 **45.** 5.02 **46.** 22.6 **47.** 0.5625 **48.** $0.\overline{7}$ **49.** $0.2\overline{27}$ **50.** $0.\overline{285714}$ **51.** $\dfrac{211}{100}$ **52.** $\dfrac{-97}{100}$ **53.** $\dfrac{-19,803}{1000}$

54. $\dfrac{-63}{10}$ **55. a.** $|x+y|+|x-y|$ **b.** 10 **56. a.** $3|x|+5|y|-2x+2y$ **b.** 11 **57. a.** $2x+2$ **b.** 12 **58. a.** y^3+2y^2-3y-3 **b.** 3

59. a. $2x-3$ **b.** $\dfrac{x}{6}+3x$ **c.** $2(x+10)-13$ **60. a.** The sum of 6 and 4 times a number **b.** 18 times the difference between a number and 5 **61.** $\dfrac{-5}{51}$ **62.** −5 **63.** $\dfrac{10}{3}$ cups $\left(\text{or } 3\dfrac{1}{3} \text{ cups}\right)$ **64.** 21.7 **65. a.** 1000 **b.** 2500 **66.** $13x-5$ **67.** $\dfrac{a}{6}$ **68.** $\dfrac{12+y}{4y}$

69. a. $10x$ **b.** $25x^2$ **c.** $\dfrac{2}{5x}$ **d.** $\dfrac{1}{25x^2}$ **e.** 1 **70. a.** 0 **b.** $-36y^2$ **c.** $\dfrac{1}{3y}$ **d.** $\dfrac{-1}{36y^2}$ **e.** −1

Chapter 3

Exercises 3.1, pages 209 - 212

1. $x=7$ **3.** $y=-4$ **5.** $x=-19$ **7.** $n=37$ **9.** $z=-6$ **11.** $x=5$ **13.** $y=-5.9$ **15.** $x=-1.2$ **17.** $x=9$ **19.** $y=8$ **21.** $x=20$

23. $y=10$ **25.** $x=-2$ **27.** $x=12$ **29.** $n=8$ **31.** $y=2.1$ **33.** $x=\dfrac{20}{9}$ **35.** $x=-13.3$ **37.** $y=-12$ **39.** $x=\dfrac{-2}{5}$ **41.** $x=-4$

43. $x=2.9$ **45.** $n=9.7$ **47.** $y=-13.2$ **49.** $x=-4$ **51.** 3.75 in. **53.** 16.125 m **55.** 4.5 cm **57.** No, 105° **59.** $855

61. 7 hours **63.** 3.5 hours **65.** 100.979 **67.** 27.087 **69.** −1036 **71.** −121.962

Exercises 3.2, pages 217 - 219

1. $x=-3$ **3.** $x=2$ **5.** $x=-0.12$ **7.** $y=0$ **9.** $x=-2$ **11.** $y=-6$ **13.** $n=6$ **15.** $n=8$ **17.** $x=0$ **19.** $x=\dfrac{-8}{15}$ **21.** $y=\dfrac{28}{5}$

23. $x=2$ **25.** $y=\dfrac{7}{5}$ **27.** $x=-4.5$ **29.** $-44=x$ **31.** $x=2$ **33.** $x=-4$ **35.** $0.5=y$ **37.** $x=1.5$ **39.** $x=0.2$ **41.** $l=64$ ft

43. 18 months **45.** $x=6.1$ **47.** $x=1.12$

Exercises 3.3, pages 225 - 227

1. $x=-5$ **3.** $n=3$ **5.** $y=6$ **7.** $n=0$ **9.** $x=0$ **11.** $z=-1$ **13.** $y=\dfrac{1}{5}$ **15.** $x=-4$ **17.** $x=-3$ **19.** $x=-21$ **21.** $y=0$

23. $x=-5$ **25.** $n=\dfrac{1}{6}$ **27.** $n=2$ **29.** $x=\dfrac{7}{60}$ **31.** $x=\dfrac{8}{5}$ **33.** $x=\dfrac{2}{3}$ **35.** $x=6$ **37.** $x=-11$ **39.** $x=-5$ **41.** $x=4.04$

43. $n=-1.5$ **45.** $x=0$ **47.** 36 in. **49.** 10 cm **51.** 12 ft, 38 ft **53.** −13.1 **55.** $35 **57.** $x=-50.21$ **59.** $x=1.066604651$

Exercises 3.4, pages 237 - 239

1. False; $3>-3$ **3.** True **5.** True **7.** True **9.** True **11.** half-open interval **13.** half-open interval **15.** half-open interval **17.** half-open interval **19.** open interval

21. $x \geq \dfrac{-11}{4}$ **23.** $x>-3$ **25.** $y \leq 5.15$

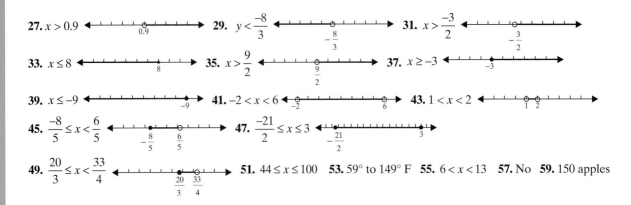

27. $x > 0.9$ **29.** $y < \dfrac{-8}{3}$ **31.** $x > \dfrac{-3}{2}$

33. $x \le 8$ **35.** $x > \dfrac{9}{2}$ **37.** $x \ge -3$

39. $x \le -9$ **41.** $-2 < x < 6$ **43.** $1 < x < 2$

45. $\dfrac{-8}{5} \le x < \dfrac{6}{5}$ **47.** $\dfrac{-21}{2} \le x \le 3$

49. $\dfrac{20}{3} \le x < \dfrac{33}{4}$ **51.** $44 \le x \le 100$ **53.** $59°$ to $149°$ F **55.** $6 < x < 13$ **57.** No **59.** 150 apples

Exercises 3.5, pages 246 - 250

1. $120 **3.** 72 days **5.** $10,000 **7.** 176 ft/sec **9.** 4 milliliters **11.** $1120 **13.** 14 **15.** 336 in. or 28 ft **17.** $337.50 **19.** $2400

21. $b = P - a - c$ **23.** $m = \dfrac{F}{a}$ **25.** $w = \dfrac{A}{l}$ **27.** $n = \dfrac{R}{p}$ **29.** $P = A - I$ **31.** $m = 2A - n$ **33.** $t = \dfrac{I}{Pr}$ **35.** $b = \dfrac{P - a}{2}$

37. $\beta = 180° - \alpha - \gamma$ **39.** $h = \dfrac{V}{lw}$ **41.** $b = \dfrac{2A}{h}$ **43.** $\pi = \dfrac{A}{r^2}$ **45.** $g = \dfrac{mv^2}{2K}$ **47.** $y = \dfrac{6 - 2x}{3}$ **49.** $x = \dfrac{11 - 2y}{5}$

51. $b = \dfrac{2A - hc}{h}$ or $b = \dfrac{2A}{h} - c$ **53.** $x = \dfrac{8R + 36}{3}$ **55.** $y = -x - 12$ **57.** $C = 0.12x + 25$ **59.** $P = 358n - 5400$

61. a. 1 **b.** -1 **c.** 1.5 **d.** -2

Exercises 3.6, pages 256 - 260

1. $x - 5 = 13 - x$; 9 **3.** $36 = 2x + 4$; 16 **5.** $7x = 2x + 35$; 7 **7.** $3x + 14 = 6 - x$; -2 **9.** $\dfrac{2x}{5} = x + 6$; -10 **11.** $4(x - 5) = x + 4$; 8

13. $\dfrac{2x + 5}{11} = 4 - x$; 3 **15.** $x - 21 = 8x$; -3 **17.** $x + x + 4 = 24$; 10 cm **19.** $x + x + 2x - 3 = 45$; 12 cm, 12 cm, 21 cm

21. $3x + 1500 = 12,000$; $3500 **23.** $n + (n + 2) = 60$; 29, 31 **25.** $n + (n + 1) + (n + 2) = 93$; 30, 31, 32

27. $171 - n = (n + 1) + (n + 2)$; 56, 57, 58 **29.** $208 - 3n = (n + 1) + (n + 2) + (n + 3) - 50$; 42, 43, 44, 45

31. $n + 2(n + 2) = 4(n + 4) - 54$; 42, 44, 46 **33.** $4n = (n + 2) + (n + 4) + 44$; 25, 27, 29 **35.** $n + (n + 1) + (n + 2) + (n + 3) = 90$;

21, 22, 23, 24 **37.** $24 + x = 2x + 3x$; 6 **39.** $x + x + 25,000 = 275,000$; lot: $125,000, house: $150,000

41. $2x = 50 - 10.50$; $19.75 **43.** $x + 3x - 1 + 2x + 5 = 64$; 10 in., 29 in., 25 in. **45.** $3n + 2(n + 4) = 6(n + 2) - 20$; 16, 18, 20

Note: Answers for 46 – 50 may vary. **47.** Find two consecutive integers such that 3 times the the second is 53 more than

the first; 25, 26 **49.** The difference between a number and $\dfrac{5}{6}$ times that number is equal to $\dfrac{5}{3}$; 10

Exercises 3.7, pages 266 - 270

1. 91% **3.** 137% **5.** 37.5% **7.** 150% **9.** 0.69 **11.** 0.113 **13.** 0.005 **15.** 0.82 **17.** $\dfrac{7}{20}$ **19.** $\dfrac{13}{10}$ **21. a.** 32% **b.** 28% **c.** 12%

23. a. 57.3% **b.** 17.1% **25.** 61.56 **27.** 40 **29.** 2180 **31.** 125% **33.** 80 **35. a.** 8% **b.** 10% **c.** b **37. a.** $990 **b.** 22%

39. $1952.90 **41.** 1.5% **43.** 40 **45. a.** $700 **b.** 33.33% **47. a.** $7134.29 **b.** $6559.23 **49. a.** 10% **b.** 11.11% **c.** The first

percentage is a percentage of his original weight, while the second percentage is a percentage of his weight after he lost

weight. **51. a.** Because the 6% comission is against the selling price, not the $141,000 the couple wanted. **b.** The selling

price is 100%. Therefore, $141,000 is 94% of the selling price (selling price(100%) – realtor fee (6%) = amount for

couple(94%)). The selling price should be $150,000.

Exercises 3.8, pages 278 - 281

1. e **3.** k **5.** c **7.** d **9.** g **11.** l **13.** m **15.** h **17.** 70 cm; 300 cm^2 **19.** 54 mm; 126 mm^2 **21.** 31.4 in.; 78.5 in.^{2}
23. 1436.03 in.^{3} **25.** 729 m^3 **27.** 56 in. **29.** 63 cm^2 **31.** 6154.4 ft^3 **33.** 125 ft^3 **35.** 272 cm^2 **37.** 529.875 in.^{3}
39. 28 cm; 48 cm^2 **41.** 212.04 m^2 **43.** 130.08 in.^{2} **45.** 6,334,233.21 cm^3

Chapter 3 Review, pages 288 - 292

1. $x = 7$ **2.** $x = 14$ **3.** $y = -11$ **4.** $y = -8$ **5.** $n = -8$ **6.** $n = -13$ **7.** $x = -5$ **8.** $x = -4$ **9.** $n = 10$ **10.** $n = 6$ **11.** $y = -18$
12. $y = -30$ **13.** $x = 18$ **14.** $y = 0.4$ **15.** $x = -3$ **16.** $x = -5$ **17.** $x = -8$ **18.** $y = 9$ **19.** $y = -7$ **20.** $x = 12$ **21.** $x = -25$
22. $y = 6$ **23.** $y = -3$ **24.** $n = \dfrac{17}{9}$ **25.** $n = \dfrac{2}{5}$ **26.** $x = 7$ **27.** $x = 28$ **28.** $x = -12.35$ **29.** $x = -8.9$ **30.** $x = 12$
31. $y = 1.02$ **32.** $y = 1.8$ **33.** $n = 0$ **34.** $n = 0$ **35.** $x = 27$ **36.** $x = 26$ **37.** $n = -12$ **38.** $n = -15$ **39.** $x = 4$ **40.** $x = \dfrac{27}{5}$
41. $y = -2$ **42.** $x = 13$ **43.** $x = \dfrac{-1}{9}$ **44.** $x = \dfrac{2}{15}$ **45.** $y = -35.6$ **46.** open interval
47. closed interval **48.** half-open interval
49. open interval **50.** half-open interval
51. $x \geq 1$ **52.** $x \leq 7$ **53.** $y < -1$
54. $y < -5$ **55.** $x > \dfrac{36}{7}$ **56.** $x < \dfrac{5}{4}$
57. $2 \leq x \leq 5$ **58.** $-1 < x < 3$ **59.** $-4 \leq x < \dfrac{4}{3}$
60. $2 \leq x < 6$ **61.** $\pi = \dfrac{L}{2rh}$ **62.** $R = P + C$ **63.** $\dfrac{2A}{h} = b$ **64.** $\alpha = 180° - \beta - \gamma$ **65.** $g = \dfrac{v - v_0}{-t}$
66. $m = \dfrac{2gK}{v^2}$ **67.** $y = 6 - 3x$ **68.** $y = 6x - 3$ **69.** $x = \dfrac{10 + 2y}{5}$ **70.** $x = \dfrac{4y + 7}{3}$ **71.** $C = 0°$ **72.** 47.5 mph **73.** 40 meters
74. $P = 0.16$ **75.** $y = -2$ **76.** 19 **77.** 6 **78.** 9 **79.** -9 **80.** 24, 25, 26 **81.** 91, 93, 95 **82.** 10, 12, 14 **83.** 46, 47, 48, 49
84. $-29, -27, -25$ **85.** $245,000 **86.** 78.2 **87.** 70% **88.** 1.2% **89.** 18 **90.** $60.30 **91.** $21.45; $281.45 **92.** 6.25%
93. 0.25 years **94.** $10,666.67 **95.** 24% **96.** $p = 2a + 2b$ **97.** $A = \pi r^2$ **98.** $V = \dfrac{4}{3}\pi r^3$ **99.** 20.2 in. **100.** 530.66 cm^2
101. 27 ft^3 **102.** 468 m^2 **103.** 78.5 in.^{2} **104.** 686.2 cm^3 **105.** $P = 48$ in.; $A = 120$ in.^{2}

Chapter 3 Test, pages 293 - 295

1. $x = -3$ **2.** $x = 5$ **3.** $x = 0$ **4.** $x = -\dfrac{9}{8}$ **5.** $x = 20$ **6.** $x = -2$ **7.** 111.6 **8.** 32% **9.** $337.50 **10.** The $6000 investment is
the better investment since it has a higher percent of profit, 6%, than the $10,000 investment, 5%. **11.** 3 m, 4 m, 5 m
12. $m = \dfrac{N - p}{rt}$ **13.** $y = \dfrac{7 - 5x}{3}$ **14.** $-20°C$ **15.** $x = \dfrac{26}{3}$ **16. a.** perimeter $= 27.42$ ft **b.** area $= 50.13$ ft^2 **17.** 4186.67 cm^3
18. $x \leq \dfrac{-10}{3}$ **19.** $x < -20$ **20.** $x > \dfrac{-13}{3}$
21. $\dfrac{-9}{4} \leq x \leq -1$ **22.** $(2y + 5) + y = -22$; $-9, -13$ **23.** $2n + 3(n + 1) = 83$; 16, 17
24. $74 + 2w = 122$; **a.** 24 in. **b.** 888 in.^{2} **25.** $(.75)x = 547.50$ **a.** $730, **b.** $605.35 **26.** $3(n + 2) = n + (n + 4) + 27$; $n = 25, 27,$
29 **27. a.** 200 ft^3 **b.** 7.4 yd^3 **28.** 100.48 in.^{3}

Chapter 3 Cumulative Review, pages 296 - 299

1. True **2.** True **3.** False; $\frac{3}{4} \le |-1|$ **4.** 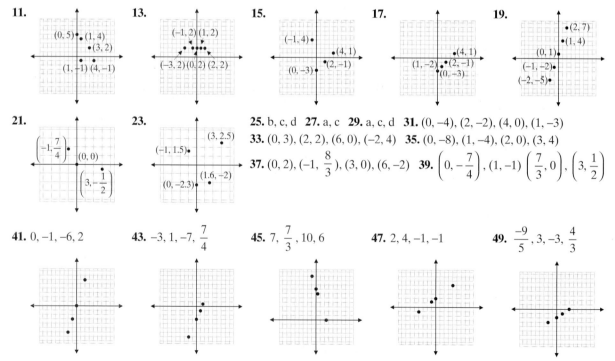 **5.** **6.** See page 30

7. a. 1400 **b.** $60x^2y^2$ **8. a.** 67 **b.** 23.3 **9.** 189 **10.** 13 **11.** 35 **12.** $\frac{49}{40}$ **13.** $\frac{-31}{5}$ **14.** 3150 **15. a.** $-2y - 12$; **b.** -18

16. a. $4x + 7$; **b.** -1 **17. a.** $x^2 + 13x$; **b.** -22 **18. a.** $x - 7$; **b.** -9 **19.** $9 - 2x$ **20.** $3(x + 10)$ **21.** $24x + 5$ **22.** $6T + E + 3F$

23. $x = -1$ **24.** $x = 3$ **25.** $x = -11$ **26.** $x = -4$ **27.** $x = \frac{31}{8}$ **28.** $y = \frac{24}{5}$ **29.** $y = \frac{15}{2}$ **30.** $y = 0$ **31.** $v = \frac{h + 16t^2}{t}$

32. $r = \frac{A - P}{Pt}$ **33.** $y = \frac{5x - 14}{3}$ **34.** 136 **35.** 22% **36.** 16.38 **37.** $x \le 2$

38. $x \ge \frac{13}{3}$ **39.** $x > \frac{9}{28}$ **40.** $-1.1 \le x \le 7.4$

41. a. 45.7 m **b.** 139.25 m^2 **42.** 1177.5 in.^{3} **43.** $80 **44.** $506.94 **45.** $x = 11$ **46.** 18, 20, 22 **47.** $-2, -1, 0, 1$

48. < 33.33 miles **49.** The value of furniture sold was $12,500, Jay's and Kay's salary was $1250. **50.** 10.71 in.

51. $86 \le x \le 100$ **52.** $10,000 **53.** These are equally good investments since the percent of profit, 8%, is the same for each investment. **54. a.** 5 in., 12 in., 13 in. **b.** 30 in^2.

Chapter 4

Exercises 4.1, pages 310 - 319

1. $\{A(-5, 1), B(-3, 3), C(-1, 1), D(1, 2), E(2, -2)\}$ **3.** $\{A(-3, -2), B(-1, -3), C(-1, 3), D(0, 0), E(2, 1)\}$

5. $\{A(-4, 4), B(-3, -4), C(0, -4), D(0, 3), E(4, 1)\}$ **7.** $\{A(-6, 2), B(-1, 6), C(0, 0), D(1, -6), E(6, 3)\}$

9. $\{A(-5, 0), B(-2, 2), C(-1, -4), D(0, 6), E(2, 0)\}$

25. b, c, d **27.** a, c **29.** a, c, d **31.** $(0, -4), (2, -2), (4, 0), (1, -3)$

33. $(0, 3), (2, 2), (6, 0), (-2, 4)$ **35.** $(0, -8), (1, -4), (2, 0), (3, 4)$

37. $(0, 2), (-1, \frac{8}{3}), (3, 0), (6, -2)$ **39.** $\left(0, -\frac{7}{4}\right), (1, -1)\left(\frac{7}{3}, 0\right), \left(3, \frac{1}{2}\right)$

41. $0, -1, -6, 2$ **43.** $-3, 1, -7, \frac{7}{4}$ **45.** $7, \frac{7}{3}, 10, 6$ **47.** $2, 4, -1, -1$ **49.** $\frac{-9}{5}, 3, -3, \frac{4}{3}$

51. $-5, 2, -10, \dfrac{8}{5}$ **53.** $-1.5, 0, 3.8, -1$ **55.** $-0.8, -2.4, -1, -7.2$

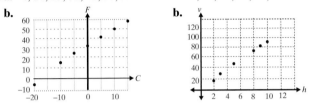

57. For example, $(0, 4), (2, 2)$, and $(4, 0)$.   **59.** For example, $(-5, 5), (-3, 2)$, and $(1, -4)$.   **61.** For example, $(-4, 1), (0, 2)$, and $(4, 3)$.   **63.** For example, $(-4, -3), (-4, 0)$, and $(-4, 2)$.   **65.** For example, $(1, 5), (3, 2)$, and $(5, -1)$.
67. a. $-4, 14, 23, 32, 41, 50, 59$ **69. a.** $18, 27, 45, 72, 81, 90$ **71.** Answers will vary.
b. 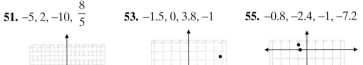   **b.**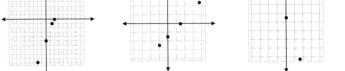

Exercises 4.2, pages 329 - 332

1. $(-14, 2), (-1, -11)$ **3.** $(2, 4), (0, 8), (4, 0)$ **5.** $(0, 5), (1, 3), (2.5, 0)$ **7.** a   **9.** d   **11.** f

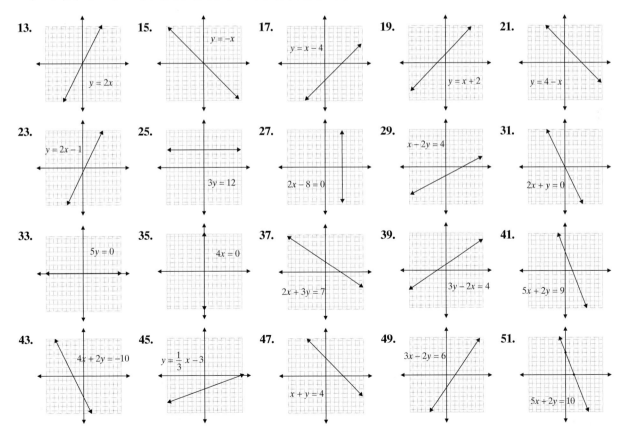

53.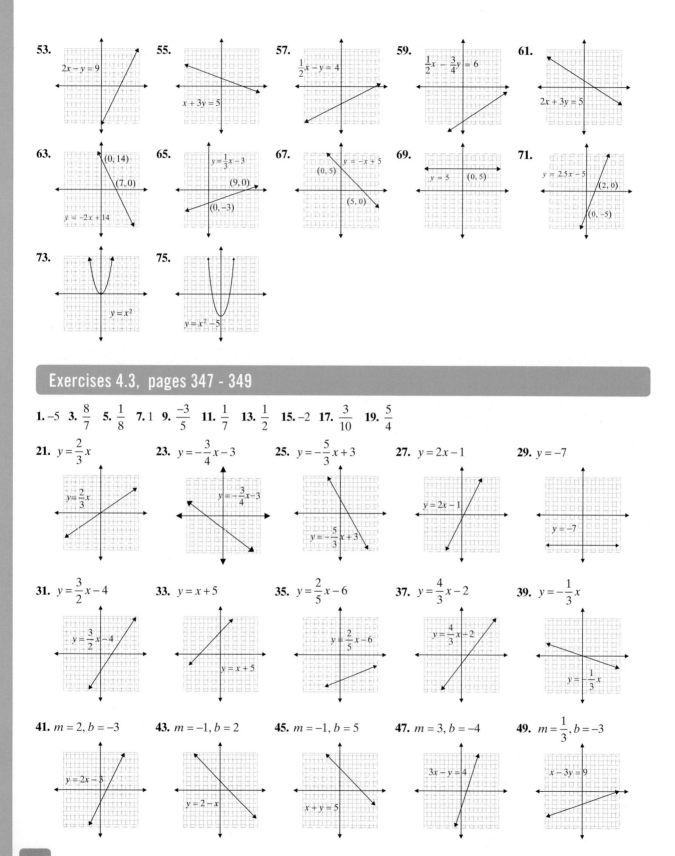

$2x - y = 9$

55.

$x + 3y = 5$

57.

$\frac{1}{2}x - y = 4$

59.

$\frac{1}{2}x - \frac{3}{4}y = 6$

61.

$2x + 3y = 5$

63.

$(0, 14)$

$(7, 0)$

$y = -2x + 14$

65.

$y = \frac{1}{3}x - 3$

$(9, 0)$

$(0, -3)$

67.

$(0, 5)$

$y = -x + 5$

$(5, 0)$

69.

$y = 5$ $(0, 5)$

71.

$y = 2.5x - 5$

$(2, 0)$

$(0, -5)$

73.

$y = x^2$

75.

$y = x^2 - 5$

Exercises 4.3, pages 347 - 349

1. -5 **3.** $\frac{8}{7}$ **5.** $\frac{1}{8}$ **7.** 1 **9.** $\frac{-3}{5}$ **11.** $\frac{1}{7}$ **13.** $\frac{1}{2}$ **15.** -2 **17.** $\frac{3}{10}$ **19.** $\frac{5}{4}$

21. $y = \frac{2}{3}x$

$y = \frac{2}{3}x$

23. $y = -\frac{3}{4}x - 3$

$y = -\frac{3}{4}x - 3$

25. $y = -\frac{5}{3}x + 3$

$y = -\frac{5}{3}x + 3$

27. $y = 2x - 1$

$y = 2x - 1$

29. $y = -7$

$y = -7$

31. $y = \frac{3}{2}x - 4$

$y = \frac{3}{2}x - 4$

33. $y = x + 5$

$y = x + 5$

35. $y = \frac{2}{5}x - 6$

$y = \frac{2}{5}x - 6$

37. $y = \frac{4}{3}x - 2$

$y = \frac{4}{3}x + 2$

39. $y = -\frac{1}{3}x$

$y = -\frac{1}{3}x$

41. $m = 2, b = -3$

$y = 2x - 3$

43. $m = -1, b = 2$

$y = 2 - x$

45. $m = -1, b = 5$

$x + y = 5$

47. $m = 3, b = -4$

$3x - y = 4$

49. $m = \frac{1}{3}, b = -3$

$x - 3y = 9$

51. $m = -\dfrac{3}{2}, b = 3$ **53.** $m = \dfrac{4}{3}, b = 1$ **55.** $m = \dfrac{-6}{5}, b = -3$ **57.** $m = \dfrac{-1}{4}, b = \dfrac{3}{2}$ **59.** $m = \dfrac{1}{2}, b = \dfrac{2}{3}$

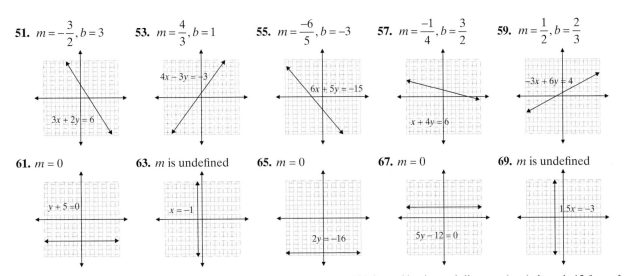

61. $m = 0$ **63.** m is undefined   **65.** $m = 0$ **67.** $m = 0$ **69.** m is undefined

71. A grade of 12% means the slope of the road is 0.12. For every 100 feet of horizontal distance (run) there is 12 feet of vertical distance (rise).

Exercises 4.4, pages 357 - 360

1. $y = 2x + 3$ **3.** $y = -\dfrac{2}{5}x + \dfrac{13}{5}$ **5.** $y = \dfrac{3}{4}x + \dfrac{17}{4}$ **7.** $y = -1$ **9.** $x = 2$ **11.** $y = \dfrac{3}{2}x - \dfrac{5}{6}$ **13.** $y = -\dfrac{1}{3}x + \dfrac{7}{3}$

15. $y = -\dfrac{2}{3}x + 6$ **17.** $y = 2$ **19.** $y = \dfrac{1}{3}x + \dfrac{11}{3}$ **21.** $x = -2$ **23.** $y = -\dfrac{3}{10}x + \dfrac{4}{5}$ **25.** $y = 5$ **27.** $x = -1$ **29.** $y = -6$

31. $x = 0$ **33.** $y = -\dfrac{4}{3}x + \dfrac{23}{3}$ **35.** $y = \dfrac{3}{2}x - \dfrac{25}{2}$ **37.** Neither; The slopes of the two equations are not equal and although they are reciprocals of each other, neither slope is negative.

39. $y = 2x - 2$ **41.** $C = 0.15x + 30$

43. a. $I = 0.09x + 200$ **b.** 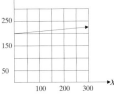   **c.** The slope is percentage of sales on which the clerk earns an income. She receives $9 for every $100 in sales.

45. a. $A = -200p + 2400$ **b.** $m = -200$ **c.** Attendance decreases by 200 people for every dollar increase in price. **d.** Since p represents admission charged to enter the park, p only makes sense for positive values of p.

47. a. $y = -\dfrac{1}{4}x + \dfrac{7}{4}$ **49. a.** $y = -\dfrac{1}{2}x$ **51. a.** $y = \dfrac{3}{2}x + \dfrac{1}{2}$

b.   **b.**   **b.**

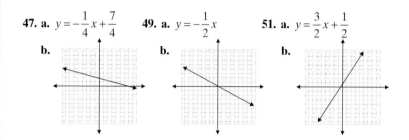

Exercises 4.5, pages 368 - 372

1. a. {(−4, 0), (−2, 4), (1, 2), (2, 5), (6, −3)} **b.** $D = \{-4, -2, 1, 2, 6\}$, $R = \{-3, 0, 2, 4, 5\}$ **c.** It is a function.

3. a. {(−5, −4), (−4, −2), (−2, −2), (1, −2), (2, 1)} **b.** $D = \{-5, -4, -2, 1, 2\}$, $R = \{-4, -2, 1\}$ **c.** It is a function.

5. a. {(−4, −2), (−4, 1), (−1, −1), (−1, 3),(3, −4)} **b.** $D = \{-4, -1, 3\}$, $R = \{-4, -2, -1, 1, 3\}$ **c.** It is not a function.

7. a. {(−5, 3), (−5, −5), (1, −2), (1, 2), (0, 5)} **b.** $D = \{-5, 0, 1\}$, $R = \{-5, -2, 2, 3, 5\}$ **c.** It is not a function.

9. a. {(−5, 5), (−3, 1), (0, −3), (3, −1), (4, 2)} **b.** $D = \{-5, -3, 0, 3, 4\}$, $R = \{-3, -1, 1, 2, 5\}$ **c.** It is a function.

11. Function   **13.** Function   **15.** Function   **17.** Function   **19.** Function

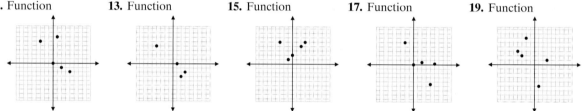

21. D: $-6 \le x \le 6$ R: $0 \le y \le 6$ **23.** D: $-6 < x \le 6$ R: {−6, −4, −2, 0, 2, 4}   **25.** Not a function.   **27.** Not a function.

29. D: All real numbers, R: $-1.5 \le y \le 1.5$ **31.** Not a function.   **33.** $f(-2) = -3, f(-1) = -1, f(0) = 1, f(1) = 3, f(5) = 11$

35. $g(-2) = 10, g(-1) = 6, g(0) = 2, g(1) = -2, g(5) = -18$ **37.** $h(-2) = 4, h(-1) = 1, h(0) = 0, h(1) = 1, h(5) = 25$

39. $F(-2) = 16, F(-1) = 9, F(0) = 4, F(1) = 1, F(5) = 9$ **41.** $H(-2) = 8, H(-1) = 7, H(0) = 0, H(1) = -7, H(5) = 85$

43. $P(-2) = 9.5, P(-1) = 6, P(0) = 3.5, P(1) = 2, P(5) = 6$

45. $f(-2) = -44, f(-1) = -15, f(0) = -2, f(1) = 1, f(5) = 33$

Exercises 4.6, pages 381 - 382

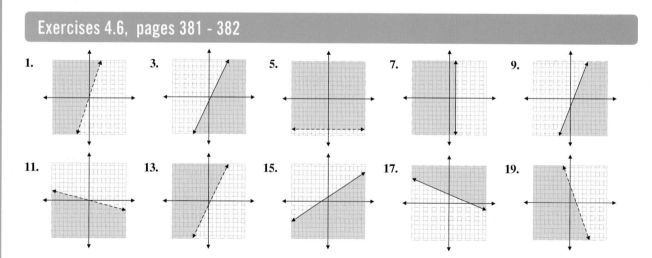

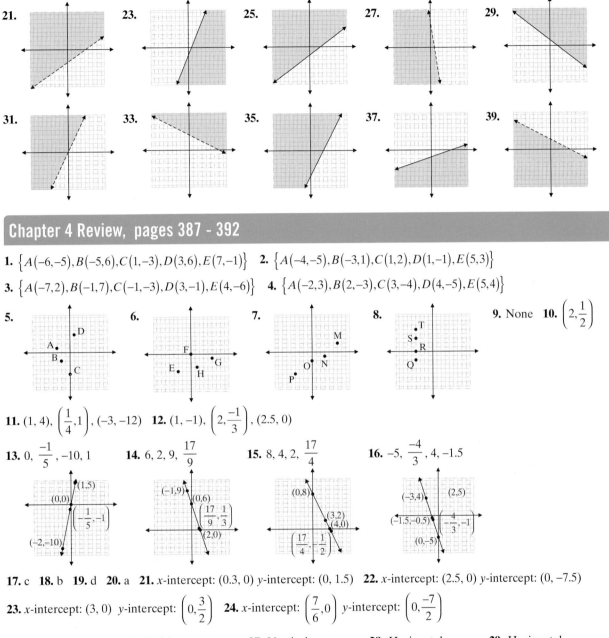

Chapter 4 Review, pages 387 - 392

1. $\{A(-6,-5), B(-5,6), C(1,-3), D(3,6), E(7,-1)\}$ **2.** $\{A(-4,-5), B(-3,1), C(1,2), D(1,-1), E(5,3)\}$

3. $\{A(-7,2), B(-1,7), C(-1,-3), D(3,-1), E(4,-6)\}$ **4.** $\{A(-2,3), B(2,-3), C(3,-4), D(4,-5), E(5,4)\}$

5. **6.** **7.** **8.** **9.** None **10.** $\left(2, \dfrac{1}{2}\right)$

11. $(1, 4)$, $\left(\dfrac{1}{4}, 1\right)$, $(-3, -12)$ **12.** $(1, -1)$, $\left(2, \dfrac{-1}{3}\right)$, $(2.5, 0)$

13. 0, $\dfrac{-1}{5}$, -10, 1 **14.** $6, 2, 9, \dfrac{17}{9}$ **15.** $8, 4, 2, \dfrac{17}{4}$ **16.** $-5, \dfrac{-4}{3}, 4, -1.5$

17. c **18.** b **19.** d **20.** a **21.** x-intercept: $(0.3, 0)$ y-intercept: $(0, 1.5)$ **22.** x-intercept: $(2.5, 0)$ y-intercept: $(0, -7.5)$

23. x-intercept: $(3, 0)$ y-intercept: $\left(0, \dfrac{3}{2}\right)$ **24.** x-intercept: $\left(\dfrac{7}{6}, 0\right)$ y-intercept: $\left(0, \dfrac{-7}{2}\right)$

25. Neither **26.** Neither **27.** Vertical **28.** Horizontal **29.** Horizontal

30. Neither **31.** -7 **32.** $\dfrac{-3}{7}$ **33.** $\dfrac{5}{6}$ **34.** 1 **35.** $\dfrac{-1}{2}$ **36.** 2 **37.** $\dfrac{4}{3}$ **38.** 0.5

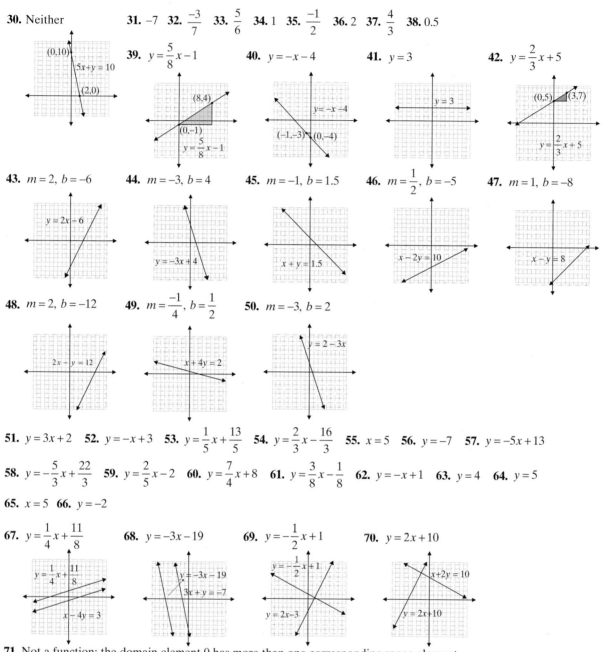

39. $y = \dfrac{5}{8}x - 1$ **40.** $y = -x - 4$ **41.** $y = 3$ **42.** $y = \dfrac{2}{3}x + 5$

43. $m = 2, b = -6$ **44.** $m = -3, b = 4$ **45.** $m = -1, b = 1.5$ **46.** $m = \dfrac{1}{2}, b = -5$ **47.** $m = 1, b = -8$

48. $m = 2, b = -12$ **49.** $m = \dfrac{-1}{4}, b = \dfrac{1}{2}$ **50.** $m = -3, b = 2$

51. $y = 3x + 2$ **52.** $y = -x + 3$ **53.** $y = \dfrac{1}{5}x + \dfrac{13}{5}$ **54.** $y = \dfrac{2}{3}x - \dfrac{16}{3}$ **55.** $x = 5$ **56.** $y = -7$ **57.** $y = -5x + 13$

58. $y = -\dfrac{5}{3}x + \dfrac{22}{3}$ **59.** $y = \dfrac{2}{5}x - 2$ **60.** $y = \dfrac{7}{4}x + 8$ **61.** $y = \dfrac{3}{8}x - \dfrac{1}{8}$ **62.** $y = -x + 1$ **63.** $y = 4$ **64.** $y = 5$

65. $x = 5$ **66.** $y = -2$

67. $y = \dfrac{1}{4}x + \dfrac{11}{8}$ **68.** $y = -3x - 19$ **69.** $y = -\dfrac{1}{2}x + 1$ **70.** $y = 2x + 10$

71. Not a function; the domain element 0 has more than one corresponding range element.
72. Not a function; the domain element –2 has more than one corresponding range element. **73.** Function **74.** Function
75. $D: -5 \le x \le 5$, $R: -6 \le y \le 6$ **76.** Not a function. **77.** $D: -8 \le x \le 8$, $R: -3 \le y \le 3$ **78.** $D: -6 \le x \le 6$, $R: -6 \le y \le 4$
79. Not a function. **80.** $D: -8 \le x \le -2$ and $D: -1 \le x \le 8$, $R: -7 \le y \le 4$ **81.** $f(-3) = -19$, $f(-1) = -13$, $f(0) = -10$,
$f(2) = -4$, $f(3) = -1$ **82.** $g(-3) = 18$, $g(-1) = 10$, $g(0) = 6$, $g(2) = -2$, $g(3) = -6$ **83.** $h(-3) = 13$, $h(-1) = 5$, $h(0) = 4$,
$h(2) = 8$, $h(3) = 13$ **84.** $f(-3) = 40$, $f(-1) = 16$, $f(0) = 7$, $f(2) = -5$, $f(3) = -8$ **85.** $F(-3) = 16$, $F(-1) = 0$, $F(0) = 1$,
$F(2) = 21$, $F(3) = 40$ **86.** $C(-3) = -3$, $C(-1) = -3$, $C(0) = 0$, $C(2) = 12$, $C(3) = 21$

87. 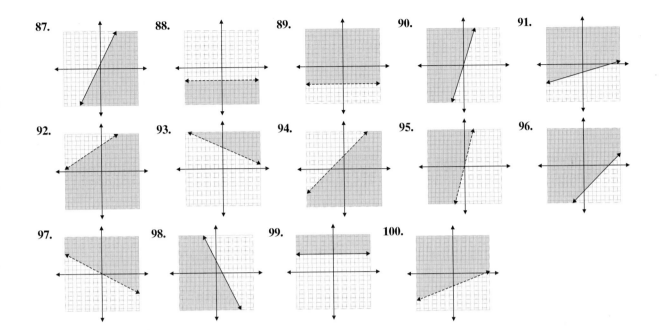 **88.** **89.** **90.** **91.**

92. **93.** **94.** **95.** **96.**

97. **98.** **99.** **100.**

Chapter 4 Test, pages 393 - 395

1. b and c   **2.** $\{A(-3, 2), B(-2, -1), C(0, -3), D(1, 2), E(3, 4), F(5, 0), G(4, -2)\}$ **3.** **4.**

5. y-intercept $= \dfrac{9}{4}$ x-intercept $= 3$ **6.**

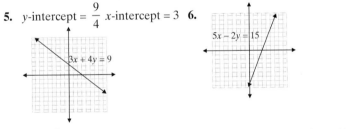

7. $m = \dfrac{-3}{5}$, $b = -3$ **8.** $m = \dfrac{-7}{4}$ **9.** $y - 2 = \dfrac{1}{4}(x + 3)$ or $y = \dfrac{1}{4}x + \dfrac{11}{4}$

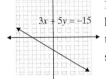

10. $y = -x + 4$; $m = -1$ **11.** $y = -2$, $m = 0$ **12. a.** $C = 4t + 9$ **b.** Because t represents the number of hours worked and one cannot work a negative number of hours.   **c.** For every hour a person rents the carpet cleaning machine he will be charged \$4.   **13.** These lines are parallel since they have the same slope, but different y-intercepts.

14. $y = 2x - 8$ **15.** $y = x - 2$ **16.** $D = \{-1, 0, 2, 5\}$ $R = \{-3, 4, 6\}$

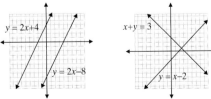

17. It is not a function because the domain element, 2, has more than one range element, -5 and 0.   **18. a.** -7 **b.** 14 **19. a.** 22 **b.** 7 **20.** It is a function. $D = \{-6, -4, -3, 3, 7\}$ $R = \{-4, -3, 0, 1, 3\}$ **21.** It is a function D : $-2 \le x \le 2$ R: $0 \le y \le 4$ **22.** Not a function   **23.** It is a function D: All real numbers $R : y \ge 0$

24. Open   **25.** Closed

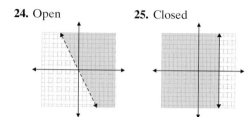

Chapter 4 Cumulative Review, pages 396 - 399

1. $x + 15$ **2.** $3x + 8$ **3.** 540 **4.** $120a^2 b^2$ **5.** 34 **6.** 13.125 **7.** 83.3 **8.** -4000 ft   **9.** False, $-15 < 5$ **10.** True

11. False, $\dfrac{7}{8} \geq \dfrac{7}{10}$ **12.** True   **13.** $x = \dfrac{8}{5}$ **14.** $x = \dfrac{-1}{2}$ **15.** $x = \dfrac{3}{8}$ **16.** $x = -3$ **17.** $d = \dfrac{C}{\pi}$ **18.** $y = \dfrac{10 - 3x}{5}$

19. $w = \dfrac{P - 2l}{2}$ **20.** $h = \dfrac{3V}{\pi r^2}$ **21.** $x \geq -6$ 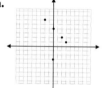   **22.** $-7 < x < 4$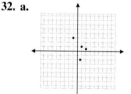

23. $x \geq 72$ 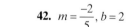   **24.** $-8 < x \leq \dfrac{112}{9}$ 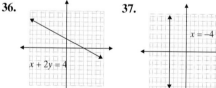   **25.**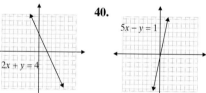

26. 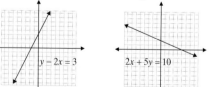   **27.** $D = \{ -3, -2, -1, 1, 3 \}$; $R = \{ 0, 1, 3 \}$ It is a function

28. $D = \{ -3, -1, 1, 4 \}$ $R = \{ -2, -1, 1, 3, 4 \}$; It is not a function   **29.** $D = \{ -2, 0, 2, 3, 4 \}$; $R = \{ -1, 0, 1, 2, 3, 4 \}$; It is not a function   **30.** $D: -5 < x \leq -1$ and $0 < x \leq 6$, $R: \{-4, -2, 2, 4, 5\}$ It is a function.

31. a.

32. a.

33. a, b, c   **34.** a, b, c, d   **35.**
$y = -4x$

b. $D: \{ -2, 0, 2, 3 \}$; $R: \{ -3, 1, 2, 4, 6 \}$

b. $D: \{ -1, 1, \dfrac{2}{3}, 2 \}$; $R: \{ -2, 1, \dfrac{1}{2}, 3 \}$

c. It is not a function

c. It is a function

36.
$x + 2y = 4$

37.
$x = -4$

38.
$y = 3$

39.
$2x + y = 4$

40.
$5x - y = 1$

41. $m = 2, b = 3$ **42.** $m = \dfrac{-2}{5}, b = 2$ **43.** $2x + 3y = 16$ **44.** $4x + 3y = 18$

$y - 2x = 3$

$2x + 5y = 10$

45. $y = -1$ **46.** $y = -\dfrac{1}{4}x + \dfrac{13}{4}$ **47.** $x = -3$ **48.** $y = \dfrac{3}{2}x - 11$ **49.** $2x + y = 0$

50. $x - 2y = -6$ **51.** $y = \dfrac{1}{3}x + 5$ **52.** $y = x + 7$ **53.** Closed   **54.** Open

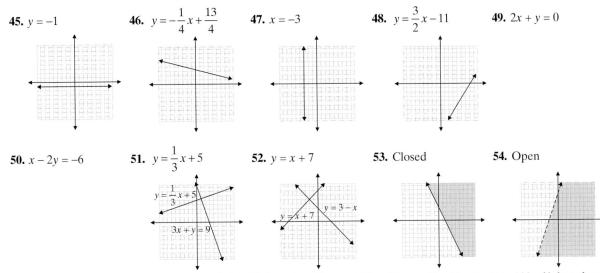

55. a. -11 **b.** 22 **56. a.** -1 **b.** 76 **57.** $20, 22, 24$ **58.** length $= 25$ cm, width $= 16$ cm **59.** -19 **60.** 54 to 100 **61.** less than 140 miles **62.** $15, 17, 19$

Chapter 5

Exercises 5.1, pages 410 - 413

1. c   **3.** a, c   **5.** $(0, 2)$ **7.** $(4, 2)$ **9.** $m_1 = -2, b_1 = 3, m_2 = -2, b_2 = \dfrac{5}{2}$ **11.** $m_1 = 3, b_1 = -8, m_2 = 3, b_2 = -6$

13. $(2, 0)$, consistent   **15.** no solution, inconsistent   **17.** $(2, 3)$, consistent   **19.** $(x, 2x - 5)$, dependent   **21.** no solution, inconsistent

23. $(-1, -1)$, consistent   **25.** $(1, 3)$, consistent   **27.** $(2, 5)$, consistent   **29.** $(8, 7)$, consistent   **31.** $(3, 0)$, consistent

33. $(4, 0)$, consistent   **35.** no solution, inconsistent   **37.** $(3, 1)$, consistent   **39.** $(-1, 2)$, consistent   **41.** $(1, -1)$, consistent

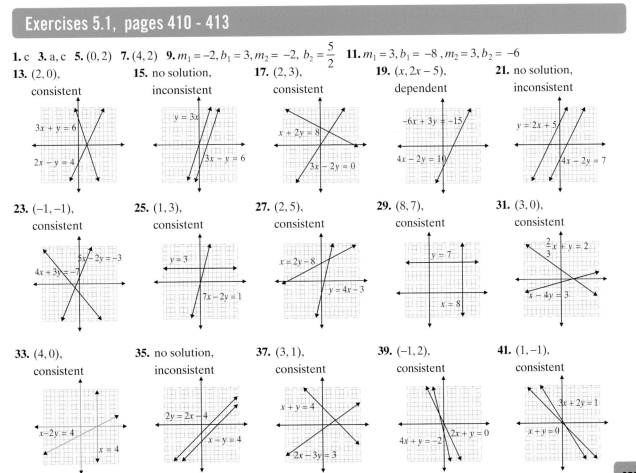

43. $x = 20, y = 5$ **45.** $x = 2, y = 8$ **47.** $(1, 4)$ **49.** $\left(\dfrac{3}{2}, 2\right)$ **51.** $\left(\dfrac{3}{2}, -3\right)$

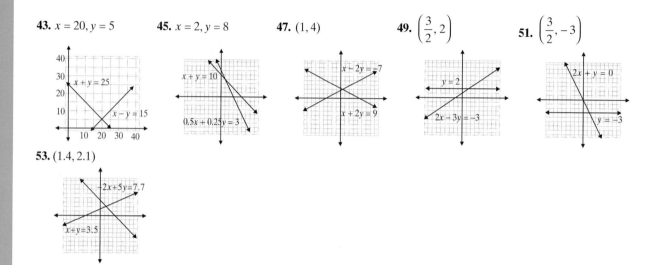

53. $(1.4, 2.1)$

Exercises 5.2, pages 417 - 419

1. Consistent, $(2, 4)$ **3.** Consistent, $(1, -2)$ **5.** Consistent, $(-6, -2)$ **7.** Consistent, $(4, 1)$ **9.** Inconsistent, no solution
11. Consistent, $(3, 2)$ **13.** Dependent, $(x, -2x + 3)$ **15.** Consistent, $(3, -2)$ **17.** Consistent, $(-8, 16)$
19. Consistent, $\left(2, \dfrac{1}{3}\right)$ **21.** Consistent, $\left(\dfrac{7}{2}, -\dfrac{1}{2}\right)$ **23.** Consistent, $\left(\dfrac{17}{16}, \dfrac{-19}{16}\right)$ **25.** Consistent, $(-2, 1)$
27. Consistent, $\left(\dfrac{-4}{5}, \dfrac{-7}{5}\right)$ **29.** Consistent, $\left(\dfrac{11}{7}, \dfrac{8}{7}\right)$ **31.** Consistent, $\left(2, \dfrac{5}{3}\right)$ **33.** Consistent, $(3, 3)$ **35.** Consistent, $(10, 20)$
37. Inconsistent, no solution   **39.** Dependent, $\left(x, \dfrac{-3}{2}x + 12\right)$ **41.** Inconsistent, no solution   **43.** $(20, 5)$ **45.** $(2, 8)$

Exercises 5.3, pages 426 - 428

1. $x = 3, y = -1$ **3.** $x = 1, y = \dfrac{-3}{2}$ **5.** Inconsistent, no solution   **7.** $x = 1, y = -5$ **9.** Dependent, $\left(x, \dfrac{1}{2}x - 2\right)$
11. $x = \dfrac{22}{7}, y = \dfrac{-2}{7}$ **13.** $x = 2, y = -2$ **15.** Dependent, $(x, 2x - 4)$ **17.** $x = 2, y = -1$ **19.** $x = -2, y = -3$ **21.** $x = 5, y = -6$
23. $x = 3, y = -1$ **25.** $x = 2, y = 4$ **27.** $x = -6, y = 2$ **29.** Dependent, $(x, 3 - x)$ **31.** Inconsistent, no solution
33. $x = \dfrac{44}{17}, y = \dfrac{-2}{17}$ **35.** $x = 6, y = 4$ **37.** $x = 4, y = 7$ **39.** $x = \dfrac{1}{2}, y = \dfrac{2}{3}$ **41.** $x = \dfrac{-45}{7}, y = \dfrac{92}{7}$ **43.** $y = 5x - 7, m = 5, b = -7$
45. $y = \dfrac{1}{9}x + \dfrac{13}{9}; m = \dfrac{1}{9}, b = \dfrac{13}{9}$ **47.** $y = \dfrac{11}{4}x - \dfrac{49}{4}; m = \dfrac{11}{4}, b = \dfrac{-49}{4}$
49. $x = 3, y = -1$ **51.** $x = 1.2, y = 1.6$ **53.** $x = 6.73, y = 0.13$

55. $x = \$4000$ at $10\%, y = \$6000$ at 6% **57.** 40 liters of 30% solution (x), 60 liters of 40% solution (y)

Exercises 5.4, pages 435 - 439

1. $x = 33$, $y = 23$ **3.** $x = 21$, $y = 15$ **5.** Rate of boat $(x) = 10$ mph, Rate of current $(y) = 2$ mph   **7.** He traveled $1\frac{1}{2}$ hrs. at the first rate (x) and 2 hrs. at the second rate (y).   **9.** His rate of speed to the city was 50 mph.   **11.** Steve (x) traveled at 28 mph and Fred (y) traveled at 7 mph   **13.** Mary's speed (x) was 58 mph and Linda's speed (y) was 50 mph.
15. He jogged about 12 miles.   **17.** speed of airliner $(x) = 406$ mph and speed of private plane $(y) = 116$ mph.
19. 18 quarters and 9 dimes.   **21.** $n = 52$ and $p = 130$.   **23.** Length is 33 m and width is 17 m.   **25.** $y = \dfrac{-3}{4}x - \dfrac{5}{2}$.
27. Anna is 12 years old and Beth is 22 years old.   **29.** 800 adults and 2700 students attended.   **31.** Length is 97.5 m and width is 32.5 m.   **33.** 5000 general admission and 7500 reserved tickets were sold.   **35.** 1 adult ticket is $3 and 1 child's ticket is $2.   **37.** $x = \$14$ for shirts $y = \$27$ for pair of slacks.   **39.** They produced 7 of Model X and 10 of Model Y.
41. The number is 49.

Exercises 5.5, pages 445 - 448

1. $5500 at 6%, $3500 at 10%   **3.** $7400 at 5.5%, $2600 at 6%   **5.** $450 at 8%, $650 at 10%   **7.** $3500 in each or $7000 total
9. $20,000 at 24%, $11,000 at 18%   **11.** $800 at 5%, $2100 at 7%   **13.** $8500 at 9%, $3500 at 11%
15. 20 pounds of 20%, 30 pounds of 70%   **17.** 20 ounces of 30%, 30 ounces of 20%   **19.** 450 pounds of 35%, 1350 pounds of 15%   **21.** 20 pounds of 40%, 30 pounds of 15%   **23.** 10 g of acid, 20 g of the 40% solution
25. 10 oz of salt, 50 oz of the 4% solution

Exercises 5.6, pages 452 - 453

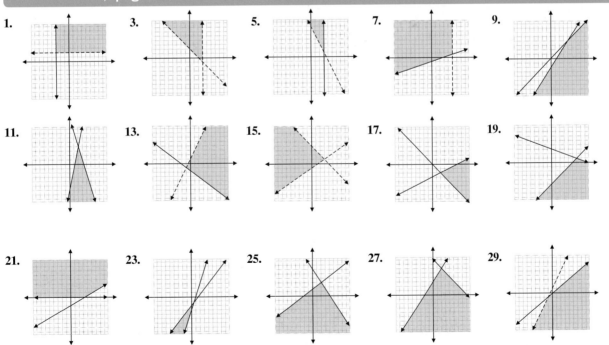

31. a. **b.** **c.**

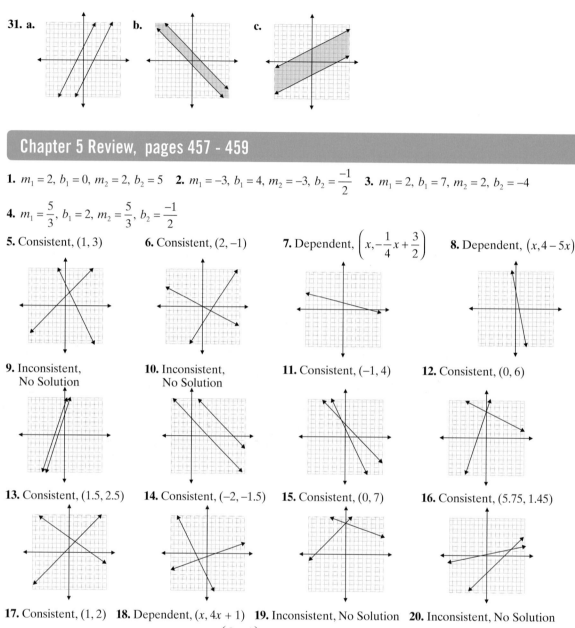

Chapter 5 Review, pages 457 - 459

1. $m_1 = 2$, $b_1 = 0$, $m_2 = 2$, $b_2 = 5$ **2.** $m_1 = -3$, $b_1 = 4$, $m_2 = -3$, $b_2 = \dfrac{-1}{2}$ **3.** $m_1 = 2$, $b_1 = 7$, $m_2 = 2$, $b_2 = -4$

4. $m_1 = \dfrac{5}{3}$, $b_1 = 2$, $m_2 = \dfrac{5}{3}$, $b_2 = \dfrac{-1}{2}$

5. Consistent, $(1, 3)$ **6.** Consistent, $(2, -1)$ **7.** Dependent, $\left(x, -\dfrac{1}{4}x + \dfrac{3}{2}\right)$ **8.** Dependent, $(x, 4 - 5x)$

9. Inconsistent, No Solution **10.** Inconsistent, No Solution **11.** Consistent, $(-1, 4)$ **12.** Consistent, $(0, 6)$

13. Consistent, $(1.5, 2.5)$ **14.** Consistent, $(-2, -1.5)$ **15.** Consistent, $(0, 7)$ **16.** Consistent, $(5.75, 1.45)$

17. Consistent, $(1, 2)$ **18.** Dependent, $(x, 4x + 1)$ **19.** Inconsistent, No Solution **20.** Inconsistent, No Solution

21. Dependent, $(x, 5x - 6)$ **22.** Consistent, $\left(\dfrac{8}{3}, \dfrac{-1}{3}\right)$ **23.** Consistent, $(0, -2)$ **24.** Consistent, $(5, 6)$ **25.** Consistent, $(2, -1)$

26. Consistent, $\left(\dfrac{49}{11}, \dfrac{46}{11}\right)$ **27.** Consistent, $\left(\dfrac{3}{7}, \dfrac{2}{7}\right)$ **28.** Consistent, $(-1, 6)$ **29.** Dependent, $(x, 3x + 6)$ **30.** Inconsistent,

No Solution **31.** Inconsistent, No Solution **32.** Dependent, $\left(x, \dfrac{1}{3}x + 10\right)$ **33.** Alice is 11 years old and John is 3 years

old. **34.** 2 quarters and 16 dimes **35.** -7 and -5 **36.** 5 and 7 **37.** He averaged 2 mph for the first 3.6 hours and 3 mph for the last 2.4 hours. **38.** length = 22.5 meters, width = 17.5 meters **39.** They sold 30 of the $110 shirts and 20 of the $65 dollar shirts. **40.** $y = 2x + 1$ **41.** $15,000 at 6% , $5000 at 8% **42.** $4000 at 8%, $8000 at 5% **43.** $7500 at each rate **44.** $22,000 in each type of investment **45.** 20 gallons at 25% salt, 40 gallons at 40% salt **46.** 40 pounds of 22% fat, 40 pounds of 10% fat **47.** 30 ounces of pure acid, 20 ounces of 10% acid **48.** 80 tons of 20% alloy, 20 tons of 60% alloy

49. 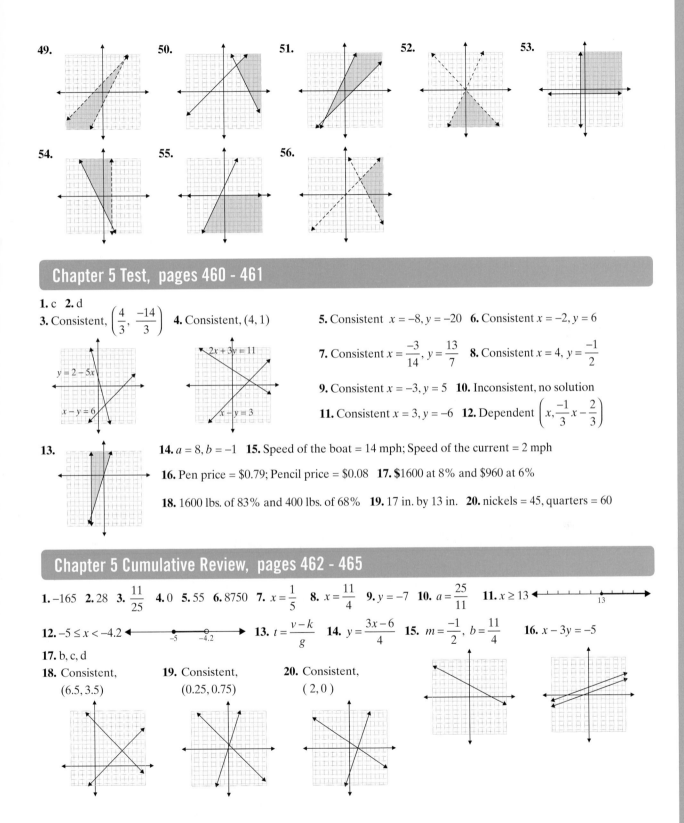 **50.** **51.** **52.** **53.**

54. **55.** **56.**

Chapter 5 Test, pages 460 - 461

1. c **2.** d

3. Consistent, $\left(\dfrac{4}{3}, \dfrac{-14}{3}\right)$ **4.** Consistent, $(4, 1)$

5. Consistent $x = -8, y = -20$ **6.** Consistent $x = -2, y = 6$

7. Consistent $x = \dfrac{-3}{14}, y = \dfrac{13}{7}$ **8.** Consistent $x = 4, y = \dfrac{-1}{2}$

9. Consistent $x = -3, y = 5$ **10.** Inconsistent, no solution

11. Consistent $x = 3, y = -6$ **12.** Dependent $\left(x, \dfrac{-1}{3}x - \dfrac{2}{3}\right)$

13.

14. $a = 8, b = -1$ **15.** Speed of the boat = 14 mph; Speed of the current = 2 mph

16. Pen price = $0.79; Pencil price = $0.08 **17.** $1600 at 8% and $960 at 6%

18. 1600 lbs. of 83% and 400 lbs. of 68% **19.** 17 in. by 13 in. **20.** nickels = 45, quarters = 60

Chapter 5 Cumulative Review, pages 462 - 465

1. -165 **2.** 28 **3.** $\dfrac{11}{25}$ **4.** 0 **5.** 55 **6.** 8750 **7.** $x = \dfrac{1}{5}$ **8.** $x = \dfrac{11}{4}$ **9.** $y = -7$ **10.** $a = \dfrac{25}{11}$ **11.** $x \geq 13$

12. $-5 \leq x < -4.2$ **13.** $t = \dfrac{v - k}{g}$ **14.** $y = \dfrac{3x - 6}{4}$ **15.** $m = \dfrac{-1}{2}, b = \dfrac{11}{4}$ **16.** $x - 3y = -5$

17. b, c, d

18. Consistent, $(6.5, 3.5)$ **19.** Consistent, $(0.25, 0.75)$ **20.** Consistent, $(2, 0)$

21. Consistent, (−1, 4)

22. Consistent, $\left(\dfrac{13}{10}, \dfrac{-9}{10} \right)$

23. Inconsistent, no solution

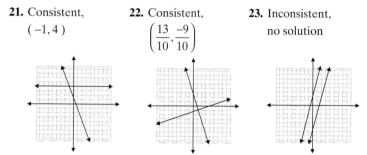

24. Consistent, (−6, 2)   **25.** Inconsistent, no solution   **26.** Dependent, $\left(x, \dfrac{-4}{3}x + \dfrac{8}{3} \right)$ **27.** Consistent, (2, −4)   **28.** Consistent,

(2, 3)   **29.** Consistent, (−1, −6)   **30.** Inconsistent, no solution   **31.** Dependent, $\left(x, 2 - \dfrac{1}{5}x \right)$ **32.** Consistent, (8, 5)

33. Consistent, $\left(\dfrac{22}{21}, \dfrac{13}{21} \right)$ **34.** Consistent, (1, 2)   **35.** Consistent, $\left(\dfrac{54}{11}, \dfrac{-5}{11} \right)$ **36.** $y = -7x + 20$ **37.** $y = \dfrac{-4}{3}x + \dfrac{7}{3}$

38. a. The measures of the angles are 20° and 70°.   **b.** The measures of the angles are 40° and 140°.   **39.** Ten 41¢ stamps and five 58¢ stamps.   **40.** speed of the boat = 10 mph and speed of the current = 2 mph.   **41.** length = 14 yards width = 11 yards.   **42.** A = 4; B = 18   **43.** Eastbound train is traveling at 40 mph, Westbound train is traveling at 45 mph.   **44.** $N = 950t - 850$ **45.** { $A(-5, -2), B(-3, -2), C(-2, 4), D(1, -1), E(2, 4)$ }; D = { −5, −3, −2, 1, 2 }; R = { −2, −1, 4 }; It is a function.   **46.** Not a function   **47.** Function   **48. a.** −22   **b.** 23   **c.** 3.5   **49. a.** −10   **b.** −1   **c.** 80

50.   **51.**   **52.**   **53.**   **54.**

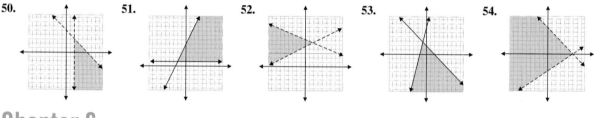

Chapter 6

Exercises 6.1, pages 478 - 480

1. 27; product rule, 1 exponent rule   **3.** 512; product rule, 0 exponent rule   **5.** $\dfrac{1}{3}$; negative exponent rule   **7.** $\dfrac{1}{25}$; negative exponent rule   **9.** 16; product rule, 0 exponent rule   **11.** −64; product rule, 0 exponent rule   **13.** −500; product rule

15. $\dfrac{3}{8}$; negative exponent rule, 1 exponent rule   **17.** $\dfrac{-3}{25}$; negative exponent rule, 1 exponent rule   **19.** x^5; product rule; 1 exponent rule   **21.** y^2; product rule, 0 exponent rule   **23.** $\dfrac{1}{x^3}$; negative exponent rule   **25.** $\dfrac{2}{x}$; negative exponent rule

27. $\dfrac{-8}{y^2}$; negative exponent rule   **29.** $\dfrac{5x^6}{y^4}$; negative exponent rule   **31.** 4; 0 exponent rule   **33.** 49; quotient rule   **35.** $\dfrac{1}{10}$; quotient rule, negative exponent rule   **37.** $\dfrac{1}{8}$; quotient rule, negative exponent rule   **39.** x^2; quotient rule   **41.** x^2; quotient rule   **43.** x^4; quotient rule   **45.** $\dfrac{1}{x^4}$; quotient rule, negative exponent rule   **47.** x^6; quotient rule   **49.** x^2; quotient rule

51. y^2; quotient rule   **53.** $3x^3$; product rule or 0 exponent rule   **55.** $10x^4$; product rule   **57.** $36x^3$; product rule, 0 exponent rule   **59.** $-14x^5$; product rule   **61.** $-12x^6$; product rule   **63.** $4y$; quotient rule   **65.** $3y^2$; quotient rule   **67.** $-2y^2$; quotient rule   **69.** $-7x^2$; quotient rule   **71.** 10^3; product rule, quotient rule   **73.** 1; 0 exponent rule   **75.** $\dfrac{15x^3}{y^2}$; product rule, negative exponent rule   **77.** $\dfrac{-2y^6}{x^4}$; quotient rule, negative exponent rule   **79.** $12a^3b^9c$; product rule   **81.** $-4a^{10}b^3c$; quotient rule

83. 1 **85.** 0.390625 **87.** 99,376.2576

Exercises 6.2, pages 491- 494

1. -81 **3.** 81 **5.** $1,000,000$ **7.** $36x^6$ **9.** $-108x^6$ **11.** -3 **13.** $\dfrac{-2y^6}{27x^{15}}$ **15.** $\dfrac{16x^2}{y^4}$ **17.** $\dfrac{9x^4}{y^6}$ **19.** $\dfrac{y^2}{x^2}$ **21.** $\dfrac{y^{10}}{4x^2}$ **23.** $\dfrac{1}{64a^6b^9}$

25. $25x^2y^4$ **27.** $16a^4b^4$ **29.** $\dfrac{1}{8a^3b^6}$ **31.** $\dfrac{y^8}{16x^8}$ **33.** $\dfrac{8a^6}{b^9}$ **35.** $\dfrac{49y^4}{x^6}$ **37.** $\dfrac{24y^5}{x^7}$ **39.** $\dfrac{4x^3}{3}$ **41.** $\dfrac{16x^{13}}{243y^6}$ **43.** $96x^2y^4z^4$

45. 8.6×10^4 **47.** 3.62×10^{-2} **49.** 1.83×10^7 **51.** 0.042 **53.** $7,560,000$ **55.** $851,500,000$ **57.** $3 \times 10^2 \cdot 1.5 \times 10^{-4}; 4.5 \times 10^{-2}$

59. $3 \times 10^{-4} \cdot 2.5 \times 10^{-6}; 7.5 \times 10^{-10}$ **61.** $\dfrac{3.9 \times 10^3}{3 \times 10^{-3}}; 1.3 \times 10^6$ **63.** $\dfrac{1.25 \times 10^2}{5 \times 10^4}; 2.5 \times 10^{-3}$ **65.** $\dfrac{2 \times 10^{-2} \times 3.9 \times 10^3}{1.3 \times 10^{-2}}; 6 \times 10^3$

67. $\dfrac{5 \times 10^{-3} \times 6.5 \times 10^2 \times 3.3 \times 10^0}{1.1 \times 10^{-3} \times 2.5 \times 10^3}; 3.9 \times 10^0$ **69.** $\dfrac{(1.4 \times 10^{-2})(9.22 \times 10^2)}{(3.5 \times 10^3)(2.0 \times 10^6)}; 1.844 \times 10^{-9}$ **71.** 4.0678×10^{16}

73. 1.6605×10^{-27} kg **75.** 1.8×10^{12} cm/min; 1.08×10^{14} cm/hr **77.** 6.5×10^{-19} grams **79.** $60,000,000,000,000$ **81.** 8.01×10^8
83. 8.5×10^7 **85.** 1×10^2 **87.** 1.864×10^2

Exercises 6.3, pages 499 - 501

1. Monomial **3.** Not a polynomial **5.** Trinomial **7.** Not a polynomial **9.** Binomial **11.** $4y$; first degree monomial;
$a_1 = 4, a_0 = 0$ **13.** $x^3 + 3x^2 - 2x$; third degree trinomial; $a_3 = 1, a_2 = 3, a_1 = -2, a_0 = 0$ **15.** $-2x^2$; second degree
monomial; $a_2 = -2, a_1 = 0, a_0 = 0$ **17.** 0; monomial of no degree; $a_0 = 0$ **19.** $6a^5 - 7a^3 - a^2$; fifth degree trinomial;
$a_5 = 6, a_4 = 0, a_3 = -7; a_2 = -1, a_1 = 0, a_0 = 0$ **21.** $2y^3 + 4y$; third degree binomial; $a_3 = 2, a_2 = 0, a_1 = 4, a_0 = 0$
23. 4; monomial of degree 0; $a_0 = 4$ **25.** $2x^3 + 3x^2 - x + 1$; third degree polynomial; $a_3 = 2, a_2 = 3, a_1 = -1, a_0 = 1$
27. $4x^4 - x^2 + 4x - 10$; fourth degree polynomial; $a_4 = 4, a_3 = 0, a_2 = -1, a_1 = 4, a_0 = -10$ **29.** $9x^3 + 4x^2 + 8x$; third degree
trinomial; $a_3 = 9, a_2 = 4, a_1 = 8, a_0 = 0$ **31.** -16 **33.** -41 **35.** 81 **37.** 13 **39.** -28 **41.** $3a^4 + 5a^3 - 8a^2 - 9a$ **43.** $3a + 11$
45. $10a + 25$ **47. a.**   **b.**   **c.**

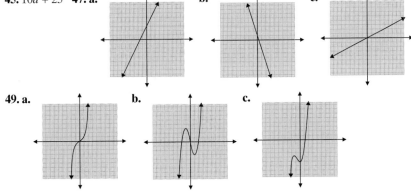

49. a.   **b.**   **c.**

Exercises 6.4, pages 505 - 507

1. $3x^2 + 7x + 2$ **3.** $2x^2 + 11x - 7$ **5.** $3x^2$ **7.** $x^2 - 5x + 17$ **9.** $-x^2 + 2x - 7$ **11.** $4x^2 + 2x - 7$ **13.** $-x^2 + 7x - 3$
15. $4x^3 + x^2 + 7$ **17.** $2x^3 + 11x^2 - 4x - 3$ **19.** $5x^2 - 5x - 1$ **21.** $x^2 + x + 6$ **23.** $-x^4 - 2x^3 - 16$ **25.** $-3x^2 - 6x - 2$
27. $9x^2 + 3x$ **29.** $3x^4 - 7x^3 + 3x^2 - 5x + 9$ **31.** $-4x^2 + 6x - 11$ **33.** $-2x^3 + 3x^2 - 2x - 8$ **35.** $2x^3 + 4x^2 + 3x - 10$

37. $8x^4 + 6x^2 + 15$ **39.** $2x^3 - 5x^2 + 11x - 11$ **41.** $4x - 13$ **43.** $-7x + 9$ **45.** $2x + 17$ **47.** $3x^3 + 13x^2 + 9$ **49.** $8x^2 - x - 2$
51. $3x - 19$ **53.** $7x - 1$ **55.** $4x^2 - x$ **57.** $x^2 + 11x - 8$ **59.** $7x^3 - x^2 - 7$

Exercises 6.5, pages 512 - 514

1. $-6x^5 - 15x^3$ **3.** $-20x^4 + 30x^2$ **5.** $-y^5 + 8y - 2$ **7.** $-4x^8 + 8x^7 - 12x^4$ **9.** $-35t^5 + 14t^4 + 7t^3$ **11.** $-x^4 - 5x^2 + 4x$
13. $6x^2 - x - 2$ **15.** $9a^2 - 25$ **17.** $-10x^2 + 39x - 14$ **19.** $x^3 + 5x^2 + 8x + 4$ **21.** $x^2 + x - 12$ **23.** $a^2 - 2a - 48$
25. $x^2 - 3x + 2$ **27.** $3t^2 - 3t - 60$ **29.** $x^3 + 11x^2 + 24x$ **31.** $2x^2 - 7x - 4$ **33.** $6x^2 + 17x - 3$ **35.** $4x^2 - 9$ **37.** $16x^2 + 8x + 1$
39. $y^3 + 2y^2 + y + 12$ **41.** $3x^2 - 8x - 35$ **43.** $-16x^3 + 50x^2 + 25x - 14$ **45.** $12x^4 + 28x^3 + 47x^2 + 34x + 48$ **47. a.** $4t^2 + 17t - 42$
b. 8 **49. a.** $5a^2 + 17a - 12$ **b.** 42 **51. a.** $3x^2 + 2x - 16$ **b.** 0 **53. a.** $6x^2 - x - 35$ **b.** -13
55. a. $25y^2 + 20y + 4$ **b.** 144 **57. a.** $y^3 - 4y^2 + 2y - 8$ **b.** -12 **59. a.** $9x^2 - 16$ **b.** 20 **61. a.** $x^3 - 8$ **b.** 0 **63. a.** $25a^2 - 60a + 36$ **b.** 16
65. a. $3x^3 - 2x^2 + 26x + 9$ **b.** 77 **67. a.** $t^3 - 6t^2 + 11t - 6$ **b.** 0 **69. a.** $y^4 + 2y^3 + y^2 - 4$ **b.** 32 **71.** $-3x - 21$ **73.** $3a^2 - 17a + 11$
75. $2y^2 - 61$

Exercises 6.6, pages 520 - 524

1. $x^2 - 9$, difference of two squares   **3.** $x^2 - 10x + 25$, perfect square trinomial   **5.** $x^2 - 36$, difference of two squares
7. $x^2 + 16x + 64$, perfect square trinomial   **9.** $2x^2 + x - 3$ **11.** $9x^2 - 24x + 16$; perfect square trinomial   **13.** $4x^2 - 1$,
difference of two squares   **15.** $9x^2 - 12x + 4$, perfect square trinomial   **17.** $x^2 + 6x + 9$, perfect square trinomial
19. $x^2 - 10x + 25$, perfect square trinomial   **21.** $25x^2 - 81$, difference of two squares   **23.** $4x^2 + 28x + 49$, perfect square
trinomial   **25.** $81x^2 - 4$, difference of two squares   **27.** $10x^4 - 11x^2 - 6$ **29.** $49x^2 + 14x + 1$, perfect square trinomial
31. $5x^2 + 11x + 2$ **33.** $4x^2 + 13x - 12$ **35.** $3x^2 - 25x + 42$ **37.** $x^2 + 10x + 25$ **39.** $x^4 - 1$ **41.** $x^4 + 6x^2 + 9$ **43.** $x^6 - 4x^3 + 4$
45. $x^4 + 3x^2 - 54$ **47.** $x^2 - \dfrac{4}{9}$ **49.** $x^2 - \dfrac{9}{16}$ **51.** $x^2 + \dfrac{6}{5}x + \dfrac{9}{25}$ **53.** $x^2 - \dfrac{5}{3}x + \dfrac{25}{36}$ **55.** $x^2 - \dfrac{1}{4}x - \dfrac{1}{8}$ **57.** $x^2 + \dfrac{5}{6}x + \dfrac{1}{6}$
59. $x^2 - 1.96$ **61.** $x^2 - 5x + 6.25$ **63.** $x^2 - 4.6225$ **65.** $x^2 + 2.48x + 1.5376$ **67.** $2.0164x^2 + 27.264x + 92.16$
69. $129.96x^2 - 12.25$ **71.** $93.24x^2 + 142.46x - 104.04$ **73. a.** $A(x) = 400 - 4x^2$ **b.** $P(x) = 80$ **75.** $A(x) = 8x + 15$
77. a. $A(x) = 150 - 4x^2$ **b.** $P(x) = 50$ **c.** $V(x) = 150x - 50x^2 + 4x^3$

Exercises 6.7, pages 532 - 533

1. $x^2 + 2x + \dfrac{3}{4}$ **3.** $2x^2 - 3x - \dfrac{3}{5}$ **5.** $2x + 5$ **7.** $x + 6 - \dfrac{3}{x}$ **9.** $2x + 3 - \dfrac{3}{2x}$ **11.** $2x - 3 - \dfrac{1}{x}$ **13.** $x + 3y - \dfrac{11y}{7x}$
15. $\dfrac{3x^2}{4} - 2y - \dfrac{y^2}{x}$ **17.** $\dfrac{5x}{8} - 1 + \dfrac{2}{y}$ **19.** $\dfrac{8x^2}{9} - xy + \dfrac{5}{9y}$ **21.** $12 + \dfrac{2}{23}$ **23.** $x + 1$ **25.** $y + 4 - \dfrac{1}{y+4}$ **27.** $x - 12 + \dfrac{42}{x+5}$
29. $4a + 3 + \dfrac{20}{a-6}$ **31.** $4x - 1 - \dfrac{1}{2x+3}$ **33.** $3x + 2 - \dfrac{2}{2x-1}$ **35.** $x - 2 - \dfrac{2}{x+2}$ **37.** $2x^2 + 5x + 2$ **39.** $t^2 + 3t + \dfrac{2}{3t+1}$
41. $2x^2 - 7x + 28 - \dfrac{118}{x+4}$ **43.** $x^2 + 2x + 4$ **45.** $2x + \dfrac{3x+3}{x^2-2}$ **47.** $x + 8 + \dfrac{8x-10}{x^2-x+1}$ **49.** $4x^2 + 6x + 9$ **51.** $4x + 3 + \dfrac{8}{x-1}$
53. $3x^2 + 2x + 5$ **55.** $2a + 3 - \dfrac{4a}{a^2+2}$ **57.** $4a^2 - 10a + 25$ **59.** $x^3 + x^2 + x + 1$

Chapter 6 Review, pages 537 - 540

1. 125 **2.** 64 **3.** $\dfrac{-3}{8}$ **4.** $\dfrac{5}{9}$ **5.** $\dfrac{1}{y^3}$ **6.** x^7 **7.** $\dfrac{1}{y^2}$ **8.** $\dfrac{4}{x}$ **9.** $-12x^4$ **10.** $2y$ **11.** $-6x^4$ **12.** 1 **13.** $-6a^4b^4c^2$

14. $\dfrac{3a^7}{b^3}$ **15.** $\dfrac{5}{xy}$ **16.** $25x^6$ **17.** $-8x^{12}$ **18.** $\dfrac{17}{x^6}$ **19.** $\dfrac{x}{y}$ **20.** $9x^2y^4$ **21.** $\dfrac{b^2}{16a^2}$ **22.** $\dfrac{9x^6}{y^4}$ **23.** m^6n^9 **24.** $-10x^4y^6$

25. $\dfrac{5}{x^3}$ **26.** $\dfrac{64}{27x^7y^6}$ **27.** $\dfrac{4}{27a^{14}b^3}$ **28.** $\dfrac{y^{10}}{x^4}$ **29.** $\dfrac{9}{a^4b^6}$ **30.** 9.7×10^4 **31.** 2.93×10^7 **32.** 5.62×10^{-2} **33.** 7.5×10^{-3}

34. 0.053 **35.** 0.000724 **36.** 823,000 **37.** 94,850,000 **38.** 1.25×10^{-3} **39.** 2×10^{-5} **40.** $6x$, first-degree monomial

41. 5.0×10^0 **42.** 1.0×10^{-6} **43.** $6x^2 - x$, second-degree binomial **44.** $x^3 - 3x^2 + 5x$, third-degree trinomial

45. $5x^3 - 4x$, third-degree binomial **46.** 5, monomial of 0 degree **47.** $8x^3 - 6x^2 - 5$, third-degree trinomial

48. $4a^3 - 4a^2 - 2a + 9$, third degree polynomial **49.** $-10a^3 + 4a^2 + 9a$, third degree trinomial **50.** 19 **51.** 14 **52.** 154

53. 78 **54.** $a^2 + 6a + 9$ **55.** $-3c^3 - 10c + 15$ **56.** -12 **57.** $4x^2 + 9x + 3$ **58.** $-x^2 + x + 3$ **59.** $2x^2 + 19x + 4$

60. $-5x^2 - 3x - 4$ **61.** $x^2 - 7x + 8$ **62.** $-2x^3 + 3x^2 + 3x + 7$ **63.** $2x^3 + 13x^2 - x - 7$ **64.** $6x^3 - 11x^2 + 9x - 27$

65. $2x^2 - x - 10$ **66.** $2x^4 + 7x^3 - 6x^2 + 10$ **67.** $6x^2 - x$ **68.** $3x^2 + 2x - 2$ **69.** $4x^2 - 7x + 2$ **70.** $10x^3 - 5x^2 + 20x + 2$

71. $5x - 18$ **72.** $-x - 13$ **73.** $-6x^3 + 10x - 22$ **74.** $5x^2 - x + 4$ **75.** $17x - 91$ **76.** $-11x^2 + 210x$ **77.** $5x^2 - 8x - 1$

78. $-x^2 - 11x + 12$ **79.** $-6x^5 - 8x^3$ **80.** $-35a^5 + 21a^4 - 7a^3$ **81.** $-6a^2 + 8a + 30$ **82.** $x^3 + 3x^2 - 6x - 8$

83. $x^3 - 11x^2 + 30x$ **84.** $-4x^2 + 8x + 60$ **85.** $2y^3 - 5y^2 + y + 2$ **86.** $x^3 + 9x^2 + 26x + 24$ **87.** $4x^2 - 17x - 42$

88. $3y^3 + 4y^2 - 29y - 8$ **89.** $x^4 + 6x^3 + 12x^2 + 9x - 10$ **90.** $2a^4 + 11a^3 + a^2 - 23a - 15$ **91.** $2x^2 + 5x + 3$

92. $6x^2 + 25x + 25$ **93.** $x^2 - 169$, difference of two squares **94.** $y^2 - 64$, difference of two squares

95. $4x^2 + 36x + 81$, perfect square trinomial **96.** $9x^2 - 6x + 1$, perfect square trinomial **97.** $y^4 + 2y^2 - 24$

98. $6x^2 + 19x - 77$ **99.** $x^6 - 100$, difference of two squares **100.** $t^2 - \dfrac{1}{16}$, difference of two squares **101.** $y^2 + \dfrac{1}{6}y - \dfrac{1}{3}$

102. $49 - 42x + 9x^2$ **103.** $3x^2 + 2x + \dfrac{1}{3}$ **104.** $x^2 + 4x + \dfrac{3}{2}$ **105.** $9x - 12 + \dfrac{3}{x}$ **106.** $2x - 4y + \dfrac{y^2}{2x}$ **107.** $a^2 - 2a + \dfrac{4}{5}$

108. $\dfrac{1}{2}x^2 + 2x - 1$ **109.** $4ab + 1$ **110.** $\dfrac{7x^2}{4} - 2xy - \dfrac{5}{4y}$ **111.** $a + 3$ **112.** $x + 5$ **113.** $y + 7$ **114.** $x - 2$

115. $4x - 13 - \dfrac{23}{x - 2}$ **116.** $4y - 1 + \dfrac{8}{2y + 3}$ **117.** $3x + 2 - \dfrac{4}{2x - 1}$ **118.** $x - 4 + \dfrac{7}{x + 4}$ **119.** $x^2 + x + 1$ **120.** $4x^2 - 6x + 9$

121. y^2 **122.** $16x^2 + 20x + 25$

Chapter 6 Test, pages 541 - 542

1. $-10a^5$ **2.** 1 **3.** $\dfrac{4x^3}{y^7}$ **4.** $\dfrac{x}{3y^2}$ **5.** $\dfrac{x^2}{4y^2}$ **6.** $4x^2y^4$ **7. a.** 135,000 **b.** 0.0000027 **8. a.** 1.25×10^8 **b.** 5.2×10^{-4}

9. $8x^2 + 3x$; second degree binomial **10.** $-x^3 + 3x^2 + 3x - 1$; third degree polynomial **11.** $5x^5 + 2x^4 - 11x + 3$; fifth degree

polynomial **12. a.** 20 **b.** -110 **13.** $-3x + 2$ **14.** $20x - 8$ **15.** $5x^3 - x^2 + 6x + 5$ **16.** $-2x^2 + x - 6$ **17.** $7x^4 + 14x^2 + 4$

18. $7x^3 - 2x^2 - 7x + 1$ **19.** $15x^7 - 20x^6 + 15x^5 - 40x^4 - 10x^2$ **20.** $49x^2 - 9$; difference of two squares **21.** $16x^2 + 8x + 1$;

perfect square trinomial   **22.** $36x^2 - 60x + 25$; perfect square trinomial   **23.** $12x^2 + 24x - 15$ **24.** $6x^3 - 69x^2 + 189x$

25. $7x^2 + 7x + 14$ **26.** $10x^4 + 4x^3 - 15x^2 - 41x - 14$ **27.** $2x + \dfrac{3}{2} - \dfrac{3}{x}$ **28.** $\dfrac{5}{3} + 2b + \dfrac{b^2}{a}$ **29.** $x - 6 - \dfrac{2}{2x + 3}$

30. $x - 9 + \dfrac{15x - 12}{x^2 + x - 3}$ **31.** $2x^2 - 3x - 13$ **32.** $6y + 27 + \dfrac{39}{y - 3}$

Chapter 6 Cumulative Review, pages 543 - 546

1. $(x + 15)$ **2.** $(3x + 8)$ **3.** 540 **4.** $120a^2b^3$ **5.** $16, 2$ **6.** $-9, -4$ **7.** $14, -4$ **8.** $4, -12$ **9.** $5x^2 + 16x$; second degree binomial $a_2 = 5, a_1 = 16, a_0 = 0$ **10.** $5x - 18$; first degree binomial $a_1 = 5, a_0 = -18$ **11.** $x - 9$; first degree binomial $a_1 = 1, a_0 = -9$

12. $x^4 - 2x^3 + 4x^2 - 10x + 40$; fourth degree polynomial $a_4 = 1, a_3 = -2, a_2 = 4 \quad a_1 = -10, a_0 = 40$ **13.** $x = \dfrac{8}{5}$

14. $x = \dfrac{-1}{2}$ **15.** $x = \dfrac{3}{8}$ **16.** $x = -3$ **17.** $d = \dfrac{C}{\pi}$ **18.** $y = \dfrac{10 - 3x}{5}$ **19.** $x \geq -6$

20. $-7 < x < 4$ **21.** $y = \dfrac{2}{3}x + \dfrac{8}{3}$ **22.** $5x + 8y + 17 = 0$ **23.** $2x + 5y = 35$

24. Parallel; the slope of both lines equals $\dfrac{-1}{3}$.

25. Perpendicular; the two slopes are -2 and $\dfrac{1}{2}$ which are negative reciprocals of one another.

26. a. 22 **b.** 57

27. Open

28. Closed

29. $x = 0, \ y = 6$

30. $x = \dfrac{-12}{7}, \ y = \dfrac{31}{7}$

31. $x = 1, y = 7$

32. $x = \dfrac{-94}{25}, \ y = \dfrac{-57}{25}$

33.

34.

35. $2x$ **36.** $64x^6y^3$ **37.** $\dfrac{49x^{10}}{y^4}$ **38.** $\dfrac{36x^4}{y^{10}}$ **39.** $\dfrac{a^6}{b^4}$

40. $\dfrac{x}{3y^4}$ **41.** 1 **42.** $\dfrac{x^8}{9y^6}$ **43.** $\dfrac{80x^{10}y^4}{3}$ **44.** $\dfrac{24a^3}{11bc^5}$

45. a. 0.00000028 **b.** $35{,}100$ **46.** $1.5 \times 10^{-3} \times 4.2 \times 10^3; 6.3 \times 10^0$ **47.** $\dfrac{8.4 \times 10^2}{2.1 \times 10^{-4}}; 4.0 \times 10^6$ **48.** $\dfrac{5.0 \times 10^{-3} \times 7.7 \times 10}{1.1 \times 10^{-2} \times 3.5 \times 10^3}; 1.0 \times 10^{-2}$

49. a. $x^2 + 4x$; **b.** second degree binomial **c.** 21 **d.** 5 **50. a.** $3x^2 + 8x$ **b.** second degree binomial. **c.** 51 **d.** 35

51. a. $-x^4 + 3x^3 + x^2 - 2x$ **b.** fourth degree polynomial. **c.** 3 **d.** -965 **52. a.** $-x^3 - 6x^2 + 2x - 4$ **b.** third degree polynomial **c.** -79 **d.** -39 **53.** $x^3 - 2x^2 - 6x - 7$ **54.** $-x^2 + x - 5$ **55.** $3x^3 + 4x^2 - 7x - 3$ **56.** $-x^3 + 7x^2 - 6$ **57.** $6x^2 - 7x + 1$

58. $-x - 17$ **59.** $-3x^3 + 12x^2 - 3x$ **60.** $5x^5 + 10x^4$ **61.** $x^2 - 36$ **62.** $x^2 + x - 12$ **63.** $9x^2 + 42x + 49$ **64.** $4x^2 - 1$

65. $x^4 - 25$ **66.** $x^4 - 4x^2 + 4$ **67.** $-x^2 + 5x - 2$ **68.** $x^2 - 8x + 16$ **69.** $2x^2 - x - 36$ **70.** $5x^2 - 27x - 18$ **71.** $6x^2 + x - 12$

72. $9x^2 - 64$ **73.** $4x^3 - 3x^2 - x$ **74.** $4x - 7 + \dfrac{3}{x}$ **75.** $\dfrac{13x}{5} + 2y + \dfrac{y^2}{x}$ **76.** $\dfrac{1}{7} - 3y + \dfrac{1}{xy} - 4y^2$ **77.** $x - 2$ **78.** $x + 2 + \dfrac{4}{4x - 3}$

79. $2x^2 - x + 3 - \dfrac{2}{x+3}$ **80.** $20, 22, 24$ **81.** 16 cm by 25 cm   **82.** -19 **83.** $38, 38, 42$ **84.** $a = 3, b = -4$ **85.** \$35,000 at 8% &

\$65,000 at 6%   **86.** 6 ounces of 10% mixture & 2 ounces of 30% mixture

Chapter 7

Exercises 7.1, pages 556 - 558

1. 5   **3.** 8   **5.** 1   **7.** $10x^3$ **9.** $13ab$ **11.** $14cd^2$ **13.** $12xy$ **15.** x^4 **17.** $-4y$ **19.** $3x^3$ **21.** $2x^2y$ **23.** $m + 9$ **25.** $x - 6$ **27.** $b + 1$
29. $3y + 4x + 1$ **31.** $11(x - 11)$ **33.** $4y(4y^2 + 3)$ **35.** $-8(a + 2b)$ **37.** $-3a(2x - 3y)$ **39.** $2x^2y(8x^2 - 7)$
41. $-14x^2y(y^2 + 1)$ **43.** $5(x^2 - 3x - 1)$ **45.** $4m^2(2x^3 - 3y + z)$ **47.** $17x^3y^4(x^2 - 3y + 2xy^2)$ **49.** $x^4y^2(15 + 24x^2y^4 - 32x^3y)$
51. $(y + 3)(7y^2 + 2)$ **53.** $(x - 4)(3x + 1)$ **55.** $(x - 2)(4x^3 - 1)$ **57.** $(2y + 3)(10y - 7)$ **59.** $(x - 2)(a - b)$
61. $(b + c)(x + 1)$ **63.** $(x^2 + 6)(x + 3)$ **65.** $(x + 6y)(x - 4)$ **67.** $(y - 4)(5x + z)$ **69.** $(2x - 3y)(z - 8)$
71. $(a + 3)(x + 5y)$ **73.** $(4y + 3)(x - 1)$ **75.** $(x - 1)(y + 1)$ **77.** Not factorable   **79.** Not factorable
81. $(3x - 4u)(y - 2v)$ **83.** $(2c - 3d)(3a + b)$

Exercises 7.2, pages 563 - 565

1. $\{1, 15\}, \{-1, -15\}, \{3, 5\}, \{-3, -5\}$ **3.** $\{1, 20\}, \{-1, -20\}, \{4, 5\}, \{-4, -5\}, \{2, 10\}, \{-2, -10\}$ **5.** $\{1, -6\}, \{6, -1\}, \{2, -3\}, \{3, -2\}$
7. $\{1, 16\}, \{-1, -16\}, \{4, 4\}, \{-4, -4\}, \{8, 2\}, \{-8, -2\}$ **9.** $\{1, -10\}, \{10, -1\}, \{5, -2\}, \{2, -5\}$ **11.** $4, 3$ **13.** $-7, 2$ **15.** $8, -1$ **17.** $-6, -6$
19. $-5, -4$ **21.** $x + 1$ **23.** $p - 10$ **25.** $a + 6$ **27.** $(x - 4)(x + 3)$ **29.** $(y + 6)(y - 5)$ **31.** not factorable   **33.** not factorable
35. $(x + 6)(x - 3)$ **37.** $(y - 12)(y - 2)$ **39.** $(x - 9)(x + 3)$ **41.** $x(x + 7)(x + 3)$ **43.** $5(x - 4)(x + 3)$
45. $10(y - 3)(y + 2)$ **47.** $4p^2(p + 1)(p + 8)$ **49.** $2x^2(x - 9)(x + 2)$ **51.** $2(x^2 - x - 36)$ **53.** $(a - 36)(a + 6)$
55. $3y^3(y - 8)(y + 1)$ **57.** $(x - 3y)(x + y)$ **59.** $20(a + b)(a + b)$ **61.** $x + 5$ inches   **63.** $x(10 - 2x)(40 - 2x)$ The height
is x, the width is $(10 - 2x)$ and the length is $(40 - 2x)$.   **65.** This is not an error, but the trinomial is not completely factored.
The completely factored form of this trinomial is $2(x + 2)(x + 3)$

Exercises 7.3, pages 575 - 577

1. $(x + 2)(x + 3)$ **3.** $(2x - 5)(x + 1)$ **5.** $(6x + 5)(x + 1)$ **7.** $-(x - 2)(x - 1)$ **9.** $(x - 5)(x + 2)$ **11.** $-(x - 14)(x + 1)$
13. Not Factorable   **15.** $-x(2x + 1)(x - 1)$ **17.** $(t - 1)(4t + 1)$ **19.** $(5a - 6)(a + 1)$ **21.** $(7x - 2)(x + 1)$
23. $(4x + 3)(x + 5)$ **25.** $(x - 2)(x + 8)$ **27.** $2(2x - 5)(3x - 2)$ **29.** $(3x - 1)(x - 2)$ **31.** $(3x - 1)(3x - 1)$
33. $(2x + 1)^2$ **35.** $(3y - 4)(4y + 3)$ **37.** not factorable   **39.** $2b(4a - 3)(a - 2)$ **41.** not factorable   **43.** $(4x - 1)(4x - 1)$
45. $(8x - 3)(8x - 3)$ **47.** $2(3x - 5)(x + 2)$ **49.** $5(2x + 3)(x + 2)$ **51.** $-2(9x^2 - 36x + 4)$ **53.** $x^2(7x^2 - 5x + 3)$
55. $-2(m - 2)(6m + 1)$ **57.** $3x(2x - 1)(x + 2)$ **59.** $9xy^3(x^2 + x + 1)$ **61.** $3x(2x - 9)(2x - 9)$

63. $4xy(3y - 4)(4y - 3)$ **65.** $7y^2(y - 4)(3y - 2)$ **67.** $(a + 3)(x + y)$ **69.** $(5 + b)(x^2 + y^2)$
71. a. 640 ft, 384 ft   **b.** 144 ft, 400 ft   **c.** 7 seconds; $0 = -16(t + 7)(t - 7)$

Exercises 7.4, pages 583 - 585

1. $(x - 1)(x + 1)$ **3.** $(x - 7)(x + 7)$ **5.** $(x + 2)^2$ **7.** $(x - 6)^2$ **9.** $(4x - 3)(4x + 3)$ **11.** $(3x - 1)(3x + 1)$
13. $(5 - 2x)(5 + 2x)$ **15.** $(x - 7)^2$ **17.** $(x + 3)^2$ **19.** $3(x - 3y)(x + 3y)$ **21.** $x(x - y)(x + y)$ **23.** $\left(x - \dfrac{1}{2}\right)\left(x + \dfrac{1}{2}\right)$

25. $\left(x-\dfrac{3}{4}\right)\left(x+\dfrac{3}{4}\right)$ **27.** $(x-1)(x+1)(x^2+1)$ **29.** $2(x-8)^2$ **31.** $a(y+1)^2$ **33.** $4y^2(x-3)^2$ **35.** not factorable

37. not factorable **39.** not factorable **41.** $(4x+8)(4x-8)$ **43.** $a(x-5)^2$ **45.** $3(7+y)(7-y)$ **47.** $\left(x+\dfrac{5}{2}\right)^2$

49. $2\left(x+\dfrac{3}{5}\right)\left(x-\dfrac{3}{5}\right)$ **51.** $x^2-6x+9=(x-3)^2$ **53.** $x^2-4x+4=(x-2)^2$ **55.** $x^2+8x+16=(x+4)^2$

57. $x^2-18x+81=(x-9)^2$ **59.** $x^2+x+\dfrac{1}{4}=\left(x+\dfrac{1}{2}\right)^2$ **61.** $x^2-9x+\dfrac{81}{4}=\left(x-\dfrac{9}{2}\right)^2$ **63.** $x^2+5x+\dfrac{25}{4}=\left(x+\dfrac{5}{2}\right)^2$

65. $x^2-3x+\dfrac{9}{4}=\left(x-\dfrac{3}{2}\right)^2$

Exercises 7.5, pages 592 - 593

1. $x=2,3$ **3.** $x=-2,\dfrac{9}{2}$ **5.** $x=-3$ **7.** $x=-5$ **9.** $x=0,2$ **11.** $x=-1,4$ **13.** $x=0,-3$ **15.** $x=4,-3$ **17.** $x=4,2$ **19.** $x=-4,3$

21. $x=\dfrac{-1}{2},3$ **23.** $x=\dfrac{-2}{3},2$ **25.** $x=4,\dfrac{-1}{2}$ **27.** $x=-2,\dfrac{4}{3}$ **29.** $x=\dfrac{3}{2}$ **31.** $x=0,\dfrac{8}{5}$ **33.** $x=\pm 2$ **35.** $x=-1$

37. $x=2$ **39.** $x=3$ **41.** $x=-5,10$ **43.** $x=-2,-6$ **45.** $x=\dfrac{1}{2}$ **47.** $x=0,2,4$ **49.** $x=0,\dfrac{-1}{2},\dfrac{-2}{3}$ **51.** $x=\pm 10$ **53.** $x=\pm 5$

55. $x=-4$ **57.** $x=3$ **59.** $x=3,-1$ **61.** $x=-2,-8$ **63.** $x=-5,2$ **65.** $x=-5,7$ **67.** $x=-6,2$ **69.** $x=-1,\dfrac{2}{3}$ **71.** $x=4,\dfrac{-3}{2}$

Exercises 7.6, pages 597 - 601

1. $x^2=7x; x=0,7$ **3.** $x^2=x+12; x=4$ **5.** $x(x+7)=78;\;\; x=-13,6$, so the numbers are -13 and -6 or 13 and 6.

7. $x^2+3x=54; x=6$, so the number is 6. **9.** $x+(x+8)^2=124; x=3$, so the numbers are 3 and 11.

11. $x^2+(x-5)^2=97;\; x=-4,9$, so the numbers are -9 and -4 or 9 and 4. **13.** $x(x+1)=72;\; x=8$, so the numbers are 8 and 9. **15.** $x^2+(x+1)^2=85;\; x=6$, so the numbers are 6 and 7. **17.** $(x+1)^2-x^2=17;\; x=8$, so the numbers are 8 and 9.

19. $x(x+2)=120;\; x=-12,10$, so the numbers are -12 and -10 or 12 and 10.

21. $w\,(2w)=72$; width = 6 in. and length = 12 in. **23.** $w(w+12)=85;\; w=5$, width = 5 m and length = 17 m.

25. $l(l-4)=117;\; l=13$, width = 9 ft and length = 13 ft. **27.** $\dfrac{1}{2}(h+5)h=42;\; h=7$, height is 7 m and base is 12 m.

29. $\dfrac{1}{2}(h-6)h=56;\; h=14$, height is 14 ft. **31.** $w(10-w)=24$; so the rectangle is 4 cm by 6 cm.

33. $x(x+13)=140;\; x=7$, so there are 7 trees in each row. **35.** $x(x+7)=144;\; x=9$, so there are 9 rows.

37. $(l+6)(l+1)=300; l=14$ m, so the rectangle is 14 m by 9 m. **39.** $w(52-2w)=320;\; w=10$ yd, 16 yd, so the corral is 10 yd by 32 yd or 16 yd by 20 yd. **41.** $w(3w)=48;\; w=4$; width = 4 ft and length = 12 ft.

43. $x(x+8)=-16; x=-4$, so the numbers are -4 and 4.

Exercises 7.7, pages 602 - 606

1. 20.4 **3.** 70 inches or 5 ft 10 in. **5.** 800 lb **7.** 6 in. **9.** 10 in. **11.** 200 ft **13.** 4 seconds and 6 seconds **15.** 12 amps or 20 amps **17.** 15 amps or 35 amps **19.** \$16 or \$20 **21.** \$1.10 per pound or \$0.90 per pound **23.** \$2.90 or \$3.50 **25.** \$4 **27.** 2260.8 in.^{2} **29.** 5 in.

Chapter 7 Review, pages 612 - 614

1. $11(x-2)$ **2.** $3(y+8)$ **3.** $-7(a+2b)$ **4.** $-4y(y-7)$ **5.** $4x^2y(4x-3)$ **6.** $a(m^2+11m+25)$ **7.** $-7xt^3(x^3-14tx^2-5t^2)$

8. $-6x^2y^3(xy+y-4)$ **9.** $(y+5)(8y+3)$ **10.** $(a+7)(3a-2)$ **11.** $(x+4)(a+b)$ **12.** $(x-10)(2a+3b)$ **13.** $(a+c)(x-1)$

14. $(x+y)(b+4)$ **15.** $(1-4y)(x+2z)$ **16.** $(z^2+5)(1+c)$ **17.** not factorable   **18.** not factorable   **19.** not factorable

20. $(1-3y)(x-z)$ **21.** $(m+6)(m+1)$ **22.** $(a-3)(a-1)$ **23.** $(x+9)(x+2)$ **24.** $(y+3)(y+5)$ **25.** $(n-6)(n-2)$

26. $(x-6)(x-4)$ **27.** $(x-2)(x+5)$ **28.** $(y+4)(y+9)$ **29.** not factorable   **30.** not factorable   **31.** $(x+5)(x+7)$

32. $(x+8)(x+9)$ **33.** $(a-10)(a+5)$ **34.** $(x-7)(x+3)$ **35.** $(x+4)(x+25)$ **36.** $(x+3)(2x+1)$ **37.** $(2x-1)(1-x)$

38. $(6x-1)(5-2x)$ **39.** $(3x-2)(4x+3)$ **40.** $(2y-1)(3y-4)$ **41.** not factorable   **42.** $3(3x+2)(7x-5)$

43. $2(2x+5)(4x-7)$ **44.** $(3x+8)(3-x)$ **45.** $-1(5x-7)(3x+2)$ **46.** not factorable   **47.** not factorable

48. $-4(x-10)(x+5)$ **49.** $7(y-4)(y+6)$ **50.** $(3x-2)(6x-1)$ **51.** $(y+7)(y-7)$ **52.** $(x+10)(x-10)$ **53.** not factorable

54. not factorable   **55.** $(5x-6)(5x+6)$ **56.** $2(x-7)(x+7)$ **57.** $3(x-7)(x+7)$ **58.** $(8a+1)(8a-1)$

59. $3(4x^2-20x-25)$ **60.** $x(21x^2-13x-2)$ **61.** $(x+6)^2$ **62.** $6(y+10)^2$ **63.** $5(x-15)^2$ **64.** $7(a-3)^2$

65. $-2a(x+11)^2$ **66.** $121, y+11$ **67.** $36, x-6$ **68.** $12.25, y-3.5$ **69.** $24y, y+12$ **70.** $14x, x-7$ **71.** $x=0,-5$ **72.** $x=6,0$

73. $x=5,\dfrac{2}{3}$ **74.** $y=\dfrac{7}{3},\dfrac{-5}{2}$ **75.** $x=\dfrac{5}{2},0,\dfrac{-5}{2}$ **76.** $x=6,0,-2$ **77.** $y=\dfrac{6}{5},0,\dfrac{-6}{5}$ **78.** $x=9,0,-9$ **79.** $a=-7$ **80.** $y=\dfrac{16}{3},4$

81. $x=4,-3$ **82.** $x=7,-5$ **83.** $x=0,-6$ **84.** $x=13,-3$ **85.** $x=5,8$ **86.** $13,10$ **87.** 9 streets   **88.** 15 x 30 yards   **89.** 11, 13

90. $-12,-13$ **91.** 20 in.   **92.** $l=20$in., $w=10$ in.   **93.** 4 in.   **94.** $6,-14$ or $14,-6$ **95.** 20 rows, 30 seats   **96.** 423.9 cm^3

97. $t=3$ or $t=0.5$ seconds   **98.** $t=5$ or $t=2.5$ seconds   **99.** \$10 or \$15   **100.** \$5

Chapter 7 Test, pages 615 - 616

1. 15   **2.** $8x^2y$ **3.** $2x^3$ **4.** $-7y^2$ **5.** $10x^3(2y+1)\ (y+1)$ **6.** $(x-5)(x-4)$ **7.** $-(x+7)(x+7)$ **8.** $6(x+1)(x-1)$

9. $2(6x-5)(x+1)$ **10.** $(x+3)(3x-8)$ **11.** $(4x-5y)(4x+5y)$ **12.** $x(x+1)(2x-3)$ **13.** $(2x-3)(3x-2)$

14. $(y+7)(2x-3)$ **15.** not factorable   **16.** $-3x(x^2-2x+2)$ **17.** $25;\ (x-5)^2$ **18.** $64;\ (x+8)^2$ **19.** $x=-2,\dfrac{5}{3}$

20. $x=8,-1$ **21.** $x=0,-6$ **22.** $x=5,-3$ **23.** $x=5,\dfrac{-3}{4}$ **24.** $x=\dfrac{-5}{4},\dfrac{3}{2}$ **25.** $x=4,\dfrac{3}{2}$ **26.** $6,20$ or $-4,-30$

27. Length = 15 cm, Width = 11 cm   **28.** $n=18,19$ **29.** $12,3$ **30. a.** $A(x)=240-(x^2+3x)=-(x^2+3x-240)$ **b.** $P(x)=64$

Chapter 7 Cumulaive Review, pages 617 - 620

1. 120   **2.** $168x^2y$ **3.** $\dfrac{55}{48}$ **4.** $\dfrac{19}{60a}$ **5.** $\dfrac{3}{10}$ **6.** $\dfrac{75x}{23}$ **7.** $-8x^2y$ **8.** $4x^2y$ **9.** $\dfrac{-xy^4}{3}$ **10.** $7x^2y$ **11. a.** $D=\{2,3,5,7.1\}$

b. $R=\{-3,-2,0,3.2\}$ **c.** It is a function because each first coordinate (domain) has only one corresponding second coordinate (range).   **12. a.** 58 **b.** 4 **c.** $\dfrac{11}{4}$

13. $x+y+2=0$ **14.** $2x-y+8=0$ **15.** $y=3x+4$ **16.** $y=-\dfrac{1}{3}x+\dfrac{2}{3}$ **17.** $-\dfrac{1}{2}x+4=y$

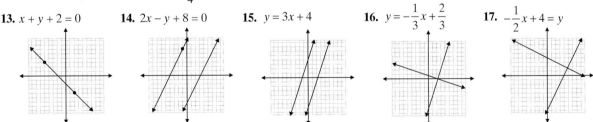

18. Function   **19.** Not a function   **20.** $x=3, y=-3$, consistent   **21.** $x=0, y=-1$, consistent   **22.** No solution; inconsistent

23. $x=\dfrac{13}{10}, y=\dfrac{3}{2}$ **24.** $x=2.4, y=-1.7$ **25.** $\left(\dfrac{-1}{5},\dfrac{-1}{5}\right),\left(\dfrac{1}{4},\dfrac{1}{4}\right)$

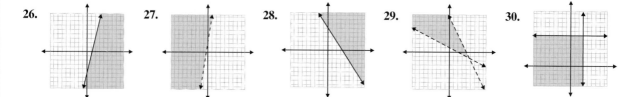

31. $13x + 1$ **32.** $x^2 + 12x + 3$ **33.** $3x^2 - 5x + 3$ **34.** $4x^2 + 5x - 8$ **35.** $x^2 - 3x - 5$ **36.** $-4x^2 - 2x + 3$

37. $2x^2 + x - 28$ **38.** $-3x^2 - 17x + 6$ **39.** $x^2 + 12x + 36$ **40.** $4x^2 - 28x + 49$ **41.** $3x - 2 + \dfrac{3}{5x}$ **42.** $2x - \dfrac{5}{4} + \dfrac{1}{y}$

43. $3x + 1 + \dfrac{22}{x-2}$ **44.** $x^2 - 8x + 38 - \dfrac{142}{x+4}$ **45.** $4(2x - 5)$ **46.** $6(x - 16)$ **47.** $(x - 6)(2x - 3)$ **48.** $(2x - 3)(3x + 4)$

49. $5(x - 2)$ **50.** $-4x(3x + 4)$ **51.** $8xy(2x - 3)$ **52.** $5x^2(2x^2 - 5x + 1)$ **53.** $(2x + 1)(2x - 1)$ **54.** $(y - 10)(y - 10)$

55. $3(x + 4y)(x - 4y)$ **56.** $(x - 9)(x + 2)$ **57.** $5(x + 4)(x + 4)$ **58.** Not Factorable   **59.** $(5x + 2)(5x + 2)$

60. $(x + 1)(3x + 2)$ **61.** $2x(x - 5)(x - 5)$ **62.** $4x(x^2 + 25)$ **63.** $(x + 2)(y + 3)$ **64.** $(x - 2)(a + b)$ **65.** $4; (x - 2)^2$

66. $81; (x + 9)^2$ **67.** $16; (x - 4)^2$ **68.** $10x; (x - 5)^2$ **69.** $5x; \left(x - \dfrac{5}{2}\right)^2$ **70.** $x = -1, 7$ **71.** $x = 0, \dfrac{-5}{3}$ **72.** $x = -3, \dfrac{3}{4}$

73. $x = 0, 7$ **74.** $x = -2, -6$ **75.** $x = -4, 7$ **76.** $x = 0, 1, -6$ **77.** $x = -10, 6$ **78.** $x = -5, 1$ **79.** $x = 0, -7$ **80.** $x = 0, -2$ **81.** $x = 0, 4$

82. $x = 5, \dfrac{-5}{2}$ **83.** $x = 1, \dfrac{-5}{4}$ **84.** $x = 0, -5, 2$ **85.** boat: 5 mph, current: 2 mph.   **86.** \$7500 at 6%, \$2500 at 8%.

87. Width = 13 in., Length = 17 in.   **88.** 12 and 3   **89.** 8 and 9 or -8 and -9 **90.** $t = 2.5, 3$ seconds

Chapter 8

Exercises 8.1, pages 630 - 632

1. $x \neq 0$ **3.** $y \neq \dfrac{4}{3}$ **5.** $x \neq 0, 4$ **7.** $y \neq -2, 5$ **9.** no restriction   **11.** $y \neq 4$ **13.** $x \neq -4$ **15.** $x \neq 0, -5$ **17.** $a \neq 7, -7$

19. no restriction   **21.** $\dfrac{2}{75}$ **23.** 9   **25.** 0   **27.** $\dfrac{-3}{32}$ **29.** 1   **31.** $\dfrac{x-3}{x}, x \neq 0$ **33.** $\dfrac{x+2}{2}, x \neq 0$ **35.** $\dfrac{1}{2}, y \neq \dfrac{-3}{2}$

37. $\dfrac{x-2}{2x+1}, x \neq \dfrac{-1}{2}$ **39.** $\dfrac{1}{x-4}, x \neq 0, 4$ **41.** $-7, x \neq 2$ **43.** $1, x \neq \dfrac{-5}{3}$ **45.** $-1, x \neq -1, 1$ **47.** $\dfrac{x}{x+2}, x \neq -2$

49. $\dfrac{x+2}{x-5}, x \neq -5, 5$ **51.** $\dfrac{x-6}{x+3}, x \neq -3$ **53.** $\dfrac{-(x-3)}{2x(x+2)}, x \neq 0, 2, -2$ **55.** $\dfrac{4x+5}{x+2}, x \neq -\dfrac{5}{4}, -2$ **57.** $\dfrac{-3}{4}, x \neq 2$

59. $\dfrac{a+4}{a+2}, a \neq -2$ **61.** $\dfrac{3(y+2)}{2(y-2)}, y \neq 2, 3$ **63.** $\dfrac{5y+2}{5y+1}, y \neq \dfrac{-1}{5}, -3$ **65.** Since $a = b, a - b = 0$, so the equation cannot be

divided by $a - b$, since division by 0 is undefined.

Exercises 8.2, pages 636 - 638

1. $\dfrac{2x}{x+2}$ **3.** $\dfrac{2x^2 - 6x}{x-1}$ **5.** $3x - 6$ **7.** $\dfrac{2x-2}{x+1}$ **9.** $\dfrac{5x}{6}$ **11.** $\dfrac{x^2 - 2x}{(x+1)(x-1)}$ **13.** $\dfrac{x+3}{x+4}$ **15.** $\dfrac{3x+1}{x+1}$ **17.** $\dfrac{5-x}{(x+7)(x-2)}$ **19.** -1

21. $\dfrac{2x^2 - x}{x+4}$ **23.** $\dfrac{4x}{x-3}$ **25.** $\dfrac{x-3}{2x-3}$ **27.** $\dfrac{3x-2}{3x+2}$ **29.** $\dfrac{x^2 - x - 12}{(3x+2)(2x+1)}$ **31.** $\dfrac{2x}{(x-4)(x-1)}$ **33.** $\dfrac{x^2 - 3x}{(x+1)^2}$ **35.** $\dfrac{x(x+5)}{2x+1}$

37. $\dfrac{(x+3)^2(x+5)^2}{(x-8)^2(x-1)(x+1)}$ **39.** $\dfrac{(2x-1)}{(6x+1)(2x+1)}$ **41.** $2x - 5$ feet

Exercises 8.3, pages 644 - 646

1. 4 **3.** 2 **5.** $\dfrac{x-3}{x+1}$ **7.** $\dfrac{1}{x-1}$ **9.** $\dfrac{x-2}{x+2}$ **11.** $\dfrac{x}{2x-1}$ **13.** $\dfrac{x-5}{x-4}$ **15.** $\dfrac{2}{x-3}$ **17.** $\dfrac{-2}{5(x-2)}$ **19.** $\dfrac{3x+8}{x(x+4)}$ **21.** $\dfrac{x^2-2x+8}{(x-4)(x+4)}$

23. $\dfrac{3}{x-2}$ **25.** $\dfrac{x+3}{x-5}$ **27.** $\dfrac{x^2+x+1}{(x+2)(x-1)}$ **29.** $\dfrac{x^2+11x+4}{(4-x)(4+x)}$ **31.** $\dfrac{4x+24}{(x-2)(x+2)}$ **33.** $\dfrac{9x+4}{12(2-x)}$ **35.** $\dfrac{-2x-17}{(x+5)(x-1)}$

37. $\dfrac{3x^2-20x}{(x-6)(x+6)}$ **39.** $\dfrac{2x^2+2x-1}{(x-1)(x-7)}$ **41.** $\dfrac{x^2+4x+2}{2(x+2)(x+3)}$ **43.** $\dfrac{x^2+3x}{-2(x-4)(x+4)}$ **45.** $\dfrac{-1}{(x+2)(x+1)(x-1)}$

47. $\dfrac{5x^2+18x}{(x+3)(x-3)(x+4)}$ **49.** $\dfrac{-2x^2-8x-3}{(x+1)(x+2)^2}$ **51.** $\dfrac{-2x^2+13x}{(x+4)(x-4)(x+1)}$ **53.** $\dfrac{3x^2-5x-7}{(2x-1)(x+3)(x-4)}$

55. $\dfrac{4x^2+17x}{(x+2)(x-2)(x+5)}$

Exercises 8.4, pages 652 - 653

1. $\dfrac{8}{7}$ **3.** $\dfrac{10}{7}$ **5.** $\dfrac{1}{2}$ **7.** 3 **9.** $\dfrac{2x}{3y}$ **11.** $16xy$ **13.** $\dfrac{8}{7x^2y^3}$ **15.** $\dfrac{3x}{x-2}$ **17.** $\dfrac{2x^2+6x}{2x-1}$ **19.** $\dfrac{x+3}{x}$ **21.** $\dfrac{x+2}{4x}$ **23.** $\dfrac{2x}{3(x+6)}$

25. $\dfrac{x}{x-1}$ **27.** $\dfrac{x}{y+x}$ **29.** $\dfrac{2x+2}{x+2}$ **31.** $\dfrac{x+2}{x+3}$ **33.** $\dfrac{-5}{x+1}$ **35.** $\dfrac{29}{4(4x+5)}$ **37.** $\dfrac{x^2-3x-6}{x(x-1)}$ **39.** $\dfrac{x^2-4x-2}{(x-4)(x+4)}$

41. a. $\dfrac{8}{5}$ **b.** 1 **c.** $\dfrac{x^4+x^3+3x^2+2x+1}{x^3+x^2+2x+1}$

Exercises 8.5, pages 661 - 664

1. $x\ne 0; x=10$ **3.** $x\ne -4; x=20$ **5.** $x\ne 1; x=-3$ **7.** $x\ne 0,2; x=4$ **9.** $x\ne 4,-7; x=-62$ **11.** $x=\dfrac{3}{2}$ **13.** $x=7$

15. $x\ne 0; x=\dfrac{38}{3}$ **17.** $x\ne 0; x=\dfrac{3}{2}$ **19.** $x\ne 6; x=5$ **21.** $x\ne 3; x=\dfrac{13}{5}$ **23.** $x\ne -4,0; x=-6,2$ **25.** $x\ne 0,-1; x=\dfrac{-3}{2}, 2$

27. $x\ne -4,3; x=-11,2$ **29.** $x\ne 3; x=3;$ therefore there is no solution. **31.** $x\ne -3,-2; x=\dfrac{-3}{2}$ **33.** $x\ne 4,\dfrac{1}{2}; x=-3$

35. $x\ne \dfrac{1}{4},-1; x=\dfrac{2}{3}$ **37.** $x\ne -5,4; x=-2$ **39.** $x\ne -1,1; x=-9,2$ **41.** $x\ne 3,-1, x=-2$ **43.** $x\ne -5,-4; x=3$ **45.** $8.50

47. 144 bulbs **49.** 504 students **51.** 600 at bats **53.** 8.8 quarts **55.** 40 ft. **57. a.** $\dfrac{2x^2+7x-5}{x^2-x}$ **b.** $x=\dfrac{5}{2}$ and $x\ne 1,0$

Exercises 8.6, pages 671 - 674

1. 21 and 27 **3.** 20 men **5.** $\dfrac{7}{9}$ **7.** 300 copies and 375 copies **9.** 45 mph, 60 mph **11.** speed of wind = 48 mph

13. His initial rate was 500 mph and his final rate was 520 mph **15.** 480 mph **17.** 9 mph **19.** 14 mph **21.** 2 hrs 24 minutes

23. 22.5 minutes **25.** 45 minutes **27.** Judy takes 1.5 hrs, Maria takes 3 hrs **29.** one takes 10 days, the other 15 days

Exercises 8.7, pages 677 - 681

1. $\dfrac{7}{5}$ **3.** 36 **5.** 10 **7.** $37.50 **9.** 190.4 lb **11.** 225 cm^3 **13.** 6.28 ft **15.** 4 in. **17.** 139.219 ft^2 **19.** 192 lbs per ft^2

21. 700 lbs per ft^2 **23.** $5\dfrac{1}{3}$ ohms **25.** 384 rpm **27.** 36 teeth **29.** 4 ft

Chapter 8 Review, pages 686 - 690

1. $x \neq 0$ **2.** $y \neq 0$ **3.** $x \neq 7$ **4.** $x \neq -4$ **5.** $x \neq \dfrac{-1}{2}$ **6.** $x \neq \dfrac{1}{3}$ **7.** no restriction **8.** $a \neq -2, 1$ **9.** $y \neq -5, -1$ **10.** $a \neq -1$

11. $\dfrac{1}{x+2}, x \neq -2$ **12.** $\dfrac{1}{y-3}, y \neq 3$ **13.** $-1, x \neq -1, 3$ **14.** $-1, y \neq 3$ **15.** $\dfrac{1}{x-4}, x \neq -4, 4$ **16.** $\dfrac{1}{2(x+5)}, x \neq -5, 5$

17. $\dfrac{4}{x}, x \neq 0, 2$ **18.** $\dfrac{1}{2x+3}, x \neq \dfrac{-3}{2}$ **19.** $\dfrac{x+5}{x-5}, x \neq 5$ **20.** $\dfrac{2x-1}{x-3}, x \neq \dfrac{-5}{3}, 3$ **21.** $\dfrac{x+2}{x}$ **22.** $\dfrac{2y}{3(y-2)}$ **23.** $\dfrac{y(y-3)}{y-1}$

24. $\dfrac{x-4}{2}$ **25.** $\dfrac{2x}{3}$ **26.** 2 **27.** $\dfrac{5x(x+4)}{4(2x-1)}$ **28.** $\dfrac{y-3}{(y-4)(y-2)(y+1)}$ **29.** $\dfrac{(a-3)^2(2a-1)(2a+1)}{(3a+1)(6a+1)(6a-1)(6a+5)(3a-2)}$

30. $\dfrac{-4x(2x+1)(2x-3)}{3(x+2)(x-2)(2x+3)}$ **31.** $2x + 3$ meters **32. a.** $A_{old}(x) = x^2 + 5x$ **b.** $A_{new}(x) = x^2 + 17x + 66$ **c.** $A(x) = 12x + 66$

33. 5 **34.** $\dfrac{3(x^2 - 8x - 16)}{(x+4)(x-4)}$ **35.** 2 **36.** $\dfrac{1}{5}$ **37.** $\dfrac{7y^2 + y + 2}{(3y-1)(y+2)}$ **38.** $\dfrac{a^2 - 2a - 2}{(a+2)(a-1)}$ **39.** $\dfrac{11}{2x-3}$ **40.** $\dfrac{7}{2-x}$ **41.** $\dfrac{x-4}{x+2}$

42. $\dfrac{x+15}{x+5}$ **43.** $\dfrac{-x^2 - 2x + 12}{(x+5)(x+4)}$ **44.** $\dfrac{8}{(x+4)}$ **45.** $\dfrac{2(2x+3)}{(x-1)(x+3)}$ **46.** $\dfrac{2}{y+1}$ **47.** $\dfrac{2x+4}{(x-2)(x+6)}$ **48.** $\dfrac{-5x^2 - 11x + 47}{x^2 + 4x - 12}$

49. $\dfrac{6}{5}$ **50.** $\dfrac{14}{5}$ **51.** 14 **52.** $\dfrac{3}{7}$ **53.** $\dfrac{3xy}{5}$ **54.** $\dfrac{20xy}{9}$ **55.** $\dfrac{y+5}{y}$ **56.** $\dfrac{x-1}{x+1}$ **57.** $\dfrac{3(x+1)}{x^2}$ **58.** $\dfrac{9}{x-2}$ **59.** $\dfrac{-9}{x+2}$

60. $\dfrac{15}{2x}$ **61.** $\dfrac{20}{3y+2}$ **62.** $\dfrac{12(y+2)}{(y+3)^2(y-3)}$ **63.** $x \neq 0; x = 10$ **64.** $x \neq 0; x = 40$ **65.** $x \neq -3; x = 5$ **66.** $x \neq -4; x = 21$

67. $x \neq 2; x = \dfrac{16}{7}$ **68.** $y \neq 5; y = \dfrac{19}{5}$ **69.** $x \neq -3, -7; x = -5, 3$ **70.** $y \neq -1, 1; y = 10$ **71.** $y \neq -4, 1; y = \dfrac{-1}{9}$

72. $x \neq -1, -2; x = 1$ **73.** 18 and 27 **74.** 16 **75.** $\dfrac{17}{7}$ **76.** 15 mph **77.** 180 mph **78.** 6 mph **79.** 3 hours, 6 hours

80. 12 hours **81.** $\dfrac{6}{5}$ hours **82.** 450 mph **83.** 6 **84.** 5 **85.** $\dfrac{100}{7}$ in. **86.** 18.84 in. **87.** 113.04 square inches **88.** 400 feet

89. $\dfrac{12}{5}$ hours **90.** 1800 lbs per ft^2 **91.** 0.125 ohm **92.** 162 rpm

Chapter 8 Test, pages 691 - 692

1. $\dfrac{x+3}{x}, x \neq 0, \dfrac{1}{2}$ **2.** $\dfrac{2x+1}{x+1}, x \neq -1, \dfrac{3}{4}$ **3.** $\dfrac{-x^2}{2}, x \neq 0, \dfrac{6}{5}$ **4.** -1 **5.** $\dfrac{27}{8}$ **6.** $\dfrac{7(4x+3)}{x(x+4)}, x \neq \dfrac{-2}{3}, \dfrac{-3}{4}, 0, -4$

7. $\dfrac{1}{2(2x+1)}, x \neq 1, \dfrac{-2}{3}, \dfrac{1}{2}, \dfrac{-1}{2}$ **8.** $\dfrac{(10x-13)}{(x-1)(x+1)(x-2)} x \neq -1, 1, 2$ **9.** $\dfrac{10x^2 - 17x - 3}{(x-4)(x+4)} x \neq -4, 4$

10. $\dfrac{(3x^2 - 8x - 27)}{(x-1)(x-4)(x+3)}, x \neq 1, 4, -3$ **11.** $\dfrac{x(x-3)}{2(x+1)}, x \neq 0, -1, -3, -5$ **12.** $\dfrac{1}{x+1}$ **13.** $\dfrac{6(x+3)}{x+18}$ **14.** $x = \dfrac{15}{26}$ **15.** $\dfrac{-5}{11}$

16. $x = 2$ **17. a.** $\dfrac{7x+11}{2x(x+1)}$ **b.** $x = \dfrac{1}{3}$ **18.** 65; 91 **19.** Lisa's speed = 57 mph; Kim's speed = 42 mph **20.** 4 hrs **21.** 1562.5 lbs

22. 25,000 lbs **23.** 3 ohms **24.** 6 hrs

Chapter 8 Cumulative Review, pages 693 - 696

1. a. $\dfrac{19}{12}$ **b.** $\dfrac{-3}{14}$ **c.** $\dfrac{1}{4}$ **d.** $\dfrac{1}{12}$ **2.** 5 **3.** $3x^2 - 5x - 28$ **4.** $4x^2 - 20x + 25$ **5.** $6x - 23$ **6.** $10x + 4$ **7.** $t = \dfrac{A-P}{Pr}$ **8.** $\dfrac{1}{2}$ year

9. $(2x+5)(2x+3)$ **10.** $2(x+5)(x-2)$ **11. a.** $D = \{-1, 0, 5, 6\}$ **b.** $R = \{0, 5, 2\}$ **c.** It is not a function because the x-coordinate

−1 has more than one corresponding *y*-coordinate.   **12. a.** 8 **b.** −1 **c.** $\dfrac{-59}{27}$

13. $y = \dfrac{3}{4}x + \dfrac{11}{2}$;   **14.** $y = \dfrac{1}{3}x + \dfrac{2}{3}$;   **15.** $x = -3$ **16.** $y = -2x - 8$ **17.** Function   **18.** Function

19.   **20.**   **21.** $x = \dfrac{37}{30}, y = \dfrac{119}{30}$ **22.** $x = \dfrac{-5}{2}, y = \dfrac{7}{4}$ **23.** Consistent; $(5, 0)$

24. Dependent; $(x, 3x+2)$ **25.** Consistent; $\left(\dfrac{9}{5}, \dfrac{39}{10}\right)$

26. Inconsistent; no solution

27. $(-1.684, 0.326)$ and $(1.484, 1.594)$

28. $8x^2$ **29.** $5x^3y^2$ **30.** $\dfrac{3}{5}x + 2y + \dfrac{y^2}{x}$ **31.** $1 - 3y + \dfrac{8}{7}y^2$ **32.** $x - 15 - \dfrac{1}{x+1}$

33. $2x^2 - x + 3 - \dfrac{2}{x+3}$ **34.** $\dfrac{4x^2}{x-3}, x \neq 3, \dfrac{-1}{2}$ **35.** Does not reduce, $x \neq 2$ **36.** $x - 3, x \neq -3$

37. $\dfrac{1}{x+1}, x \neq 0, -1$ **38.** $\dfrac{-1}{3}, x \neq 4$ **39.** $\dfrac{2}{3}, x \neq -3$ **40.** $\dfrac{x}{x+4}, x \neq -4, -3$ **41.** $\dfrac{x+5}{2(x-3)}, x \neq 3$ **42.** $x - y$ **43.** $\dfrac{4(4x-7)}{x(x-4)}$

44. $\dfrac{4}{(y+2)(y+3)}$ **45.** $\dfrac{x}{3x+3}$ **46.** $\dfrac{x+4}{3x}$ **47.** $\dfrac{3x(x+2)^2}{x+3}$ **48.** $\dfrac{2x+1}{x-1}$ **49.** $\dfrac{2x^2+14x-8}{(x+3)(x-1)(x-2)}$ **50.** $\dfrac{2(x+2)}{(x-1)(x+4)}$

51. $\dfrac{x-4}{(x+2)(x-2)}$ **52.** $\dfrac{19}{14}$ **53.** $\dfrac{x-1}{x+1}$ **54.** $\dfrac{3x}{2-x}$ **55.** $\dfrac{-3}{5}$ **56.** $x = 6, -5$ **57.** $x = 2, 5$ **58.** $x = 2, -10$ **59.** $x = -45$

60. a. $\dfrac{9-2x}{x(x+3)}$ **b.** $x = \dfrac{9}{2}$ **61. a.** $\dfrac{6x^2-16x-20}{(x-4)(x+2)}$ **b.** $x = 7$ **62.** train's speed = 60 mph; airplane's speed = 230 mph

63. The father takes $2\dfrac{2}{3}$ hrs and the daughter takes 8 hrs   **64.** 2.5 fc   **65.** 60, 24   **66.** 3 mph   **67.** $7\dfrac{9}{13}$ cm

Chapter 9

Exercises 9.1, pages 708 - 710

1. 3   **3.** 5   **5.** 7   **7.** 13   **9.** 9   **11.** 6   **13.** 7   **15. a.** 5.6568 **b.** −4.2426 **c.** −1.7725 **d.** 2.7144 **e.** 2.7831
17. rational   **19.** rational   **21.** rational   **23.** nonreal   **25.** rational   **27.** rational   **29.** irrational   **31.** rational   **33.** irrational
35. 7.8990   **37.** −2.7082   **39.** 2.0811   **41.** −2.4934   **43.** 2.4145   **45.** 25   **47.** 6   **49.** 3   **51.** 2   **53.** 2   **55.** 11   **57.** 1.5   **59.** 8.0623
61. 6.1563   **63.** 3.4028   **65.** 2.9428   **67.** 6.3096   **69.** 6   **71.** 4   **73.** 8   **75.** 13.2288   **77.** 5.0658   **79.** Because the square of a real
number is never negative. The same is also true for the fourth root of a negative number.

Exercises 9.2, pages 718 - 719

1. $6x$ **3.** $\dfrac{1}{2}$ **5.** $2x\sqrt{2x}$ **7.** $-2\sqrt{14}$ **9.** $\dfrac{\sqrt{5}x^2}{3}$ **11.** $2\sqrt{3}$ **13.** $12\sqrt{2}$ **15.** $-6\sqrt{2}$ **17.** $-5\sqrt{5}$ **19.** $\dfrac{-\sqrt{11}}{8}$ **21.** $\dfrac{2\sqrt{7}}{5}$

23. $\dfrac{4a^2\sqrt{2a}}{9b^8}$ **25.** $\dfrac{10x^4\sqrt{2}}{17}$ **27.** -1 **29.** 1 **31.** $2\sqrt[3]{7}$ **33.** $-5\sqrt[3]{2}$ **35.** $4x^3\sqrt[3]{y^2}$ **37.** $\dfrac{\sqrt[3]{3}}{2}$ **39.** $\dfrac{5\sqrt[3]{3}}{2}$ **41.** $5x\sqrt[3]{x}$ **43.** $\dfrac{5y^4}{3x^2}$

45. $2+\sqrt{6}$ **47.** $\dfrac{3-\sqrt{3}}{4}$ **49.** $\dfrac{1-\sqrt{2}}{2}$ **51.** $\dfrac{1+3\sqrt{2}}{5}$ **53.** $\dfrac{4-\sqrt{2}}{7}$ **55.** $\dfrac{2+\sqrt{3}}{5}$ **57.** $3-a\sqrt{3}$ **59.** $\dfrac{5+a\sqrt{6a}}{4}$ **61.** $\dfrac{4-x^2\sqrt{2}}{5}$

Exercises 9.3, pages 723 - 725

1. $8\sqrt{2}$ **3.** $7\sqrt{5}$ **5.** $-3\sqrt{10}$ **7.** $13\sqrt[3]{3}$ **9.** $-\sqrt{11}$ **11.** $3\sqrt{a}$ **13.** $7\sqrt{x}$ **15.** $4\sqrt{2}+3\sqrt{3}$ **17.** $8\sqrt{b}-4\sqrt{a}$ **19.** $13\sqrt[3]{x}-2\sqrt[3]{y}$

21. $5\sqrt{3}$ **23.** 0 **25.** $17\sqrt[3]{2}$ **27.** $2\sqrt{2}-6\sqrt{3}$ **29.** $6+\sqrt{5}$ **31.** $\sqrt{3}-4\sqrt{2}$ **33.** $5\sqrt[3]{2}-8\sqrt[3]{3}$ **35.** $4\sqrt{2x}$ **37.** $2y\sqrt{2y}$

39. $-4x\sqrt{3xy}$ **41.** $15x\sqrt{x}$ **43.** $-xy^2\sqrt{x}$ **45.** $4x^5y^{10}\sqrt{3}$ **47.** $-8x^8y^2$ **49.** $xy^2\sqrt[3]{2}\left(-2x^2y^2-2x^3y+3\right)$ **51.** $3\sqrt{2}-8$

53. 18 **55.** $-8\sqrt{3}$ **57.** $12+\sqrt{6}$ **59.** $2y+\sqrt{xy}$ **61.** $13-2\sqrt{2}$ **63.** $8-7\sqrt{6}$ **65.** $x-9$ **67.** -5 **69.** $x+2\sqrt{x}-15$

Exercises 9.4, pages 730 - 731

1. $\dfrac{5\sqrt{2}}{2}$ **3.** $\dfrac{-3\sqrt{7}}{7}$ **5.** $2\sqrt{3}$ **7.** 3 **9.** 3 **11.** $\dfrac{1}{3}$ **13.** $\dfrac{2\sqrt{3}}{3}$ **15.** $\dfrac{3\sqrt{2}}{2}$ **17.** $\dfrac{\sqrt{x}}{x}$ **19.** $\dfrac{\sqrt{2xy}}{y}$ **21.** $\dfrac{\sqrt{2y}}{y}$ **23.** $\dfrac{3\sqrt{7}}{5}$

25. $\dfrac{-\sqrt{2y}}{5}$ **27.** $2+\sqrt{6}$ **29.** $\dfrac{-\left(3+\sqrt{5}\right)}{4}$ **31.** $\dfrac{-6\left(5+3\sqrt{2}\right)}{7}$ **33.** $\dfrac{\sqrt{3}}{23}\left(\sqrt{2}-5\right)$ **35.** $\dfrac{-7\left(1+3\sqrt{5}\right)}{44}$ **37.** $\dfrac{-\left(\sqrt{3}+\sqrt{5}\right)}{2}$

39. $5\left(\sqrt{2}-\sqrt{3}\right)$ **41.** $\dfrac{4\left(\sqrt{x}-1\right)}{\left(x-1\right)}$ **43.** $\dfrac{5\left(6-\sqrt{y}\right)}{36-y}$ **45.** $\dfrac{8\left(2\sqrt{x}-3\right)}{4x-9}$ **47.** $\dfrac{2\sqrt{y}\left(\sqrt{5y}+\sqrt{3}\right)}{\left(5y-3\right)}$ **49.** $\dfrac{3\left(\sqrt{x}+\sqrt{y}\right)}{x-y}$

51. $\dfrac{x\left(\sqrt{x}-2\sqrt{y}\right)}{\left(x-4y\right)}$ **53.** $-\left(\sqrt{3}+1\right)\left(\sqrt{3}+2\right)$ **55.** $\dfrac{-\left(\sqrt{5}-2\right)\left(\sqrt{5}-3\right)}{4}$ **57.** $\dfrac{\left(\sqrt{x}+1\right)^2}{\left(x-1\right)}$ **59.** $\dfrac{\left(\sqrt{x}+2\right)\left(\sqrt{3x}-y\right)}{\left(3x-y^2\right)}$

Exercises 9.5, pages 736 - 737

1. $x=33$ **3.** $x=5$ **5.** no solution **7.** $x=9$ **9.** $x=-3$ **11.** no solution **13.** $x=6$ **15.** $x=-1$ **17.** $x=-22$ **19.** $x=40$

21. $x=2$ **23.** $x=7$ **25.** $x=4$ **27.** $x=-2$ **29.** $x=2, x=3$ **31.** $x=2$ **33.** $x=12$ **35.** $x=-2$ **37.** $x=6$ **39.** $x=\dfrac{-1}{4}$

41. $x=2,5$ **43.** $x=-2,-3$ **45.** $x=-4,-6$

Exercises 9.6, pages 742 - 743

1. $\sqrt{16}$, 4 **3.** $-\sqrt[3]{8}$, -2 **5.** $\sqrt[4]{81}$, 3 **7.** $\sqrt[5]{-32}$, -2 **9.** $\sqrt{0.0004}$, 0.02 **11.** $\dfrac{2}{3}$ **13.** $\dfrac{2}{5}$ **15.** $\dfrac{1}{6}$ **17.** $\dfrac{1}{2}$ **19.** $\dfrac{1}{3}$ **21.** 8 **23.** 9

25. 243 **27.** $\dfrac{1}{8}$ **29.** $\dfrac{1}{4}$ **31.** $x^{\frac{7}{12}}$ **33.** $\dfrac{1}{5}$ **35.** $x^{\frac{3}{5}}$ **37.** $x^{\frac{1}{2}}$ **39.** $x^{\frac{2}{5}}$ **41.** 2 **43.** $a^{\frac{7}{6}}$ **45.** $x^{\frac{5}{8}}$ **47.** $x^{\frac{-1}{10}}$ **49.** $6^{\frac{1}{2}}$ **51.** $x^{\frac{5}{4}}$

53. x^2 **55.** $196\cdot x^{\frac{7}{6}}$ **57.** $6x^2$ **59.** $\dfrac{2x}{3}$

Exercises 9.7, pages 752 - 757

1. 4 **3.** 3 **5.** $\dfrac{7}{2}$ **7.** $\sqrt{2}$ **9.** $\sqrt{13}$ **11.** 13 **13.** $\dfrac{5}{7}$ **15.** 5 **17.** 13 **19.** 10 **21.** 5 **23.** 2 **25.** Yes **27.** $\left|AB\right|=\left|AC\right|=4\sqrt{5}$

29. Each side is 4 units. **31.** Both diagonals are $\sqrt{61}$ units long. **33.** Yes,; $5^2+12^2=13^2$ **35.** Yes; $8^2+6^2=10^2$

37. Yes; $1^2 + 2^2 = \left(\sqrt{5}\right)^2$ **39.** No; $\left(\sqrt{10}\right)^2 + 2^2 \neq 4^2$ **41. a.** $C = 2\pi r = 94.2$ ft $A = \pi r^2 = 706.5$ ft^2 **b.** $P \approx 84.85$ ft $A = 450$ ft^2

43. a. No **b.** Home plate **c.** No **45.** 8.5 cm **47.** $10 + 4\sqrt{5}$ **49.** $\sqrt{10} + \sqrt{41} + \sqrt{65}$

Chapter 9 Review, pages 762 - 765

1. 14 **2.** 19 **3.** 2 **4.** 9 **5.** Rational **6.** Nonreal **7.** Irrational **8.** Rational **9.** Rational **10.** Irrational **11.** 5.2426

12. −5.1623 **13.** 1.6160 **14.** −5.2614 **15.** 5 **16.** 16 **17.** 4.3178 **18.** 3.1623 **19.** −2 **20.** 1.6667 **21.** $\dfrac{1}{4}$ **22.** −15

23. $3x\sqrt{x}$ **24.** $2a^2\sqrt{2}$ **25.** $\dfrac{3x\sqrt{3}}{10}$ **26.** $\dfrac{5a\sqrt{3a}}{3}$ **27.** $\dfrac{10x^2\sqrt[3]{3}}{7}$ **28.** $\dfrac{2y^4}{3x^5}$ **29.** $1 + \sqrt{3}$ **30.** $1 - \sqrt{5}$ **31.** $2 + 2a\sqrt{2}$

32. $\dfrac{1 - x\sqrt{x}}{2}$ **33.** $6\sqrt{3}$ **34.** $9\sqrt{5}$ **35.** $7\sqrt{a}$ **36.** $-3\sqrt{x}$ **37.** $5 - 2\sqrt{6}$ **38.** $4\sqrt{a} + 3\sqrt{b}$ **39.** $2\sqrt[3]{3} + 3\sqrt[3]{2}$ **40.** $11\sqrt{7}$

41. $12x\sqrt{x}$ **42.** $9a\sqrt{b}$ **43.** $x^5y^6\left(5\sqrt{3y} + 2\sqrt{5}\right)$ **44.** $-x^4y^5\sqrt[3]{y}$ **45.** $4\sqrt{6} + 3$ **46.** $14 - 4\sqrt{21}$ **47.** $6\sqrt{3} + 2$

48. $2\sqrt{xy} - 3x$ **49.** $2\sqrt{2} + 13$ **50.** $5\sqrt{5} + 13$ **51.** −3 **52.** $7\sqrt{6} + 23$ **53.** $\dfrac{7\sqrt{2}}{2}$ **54.** $\dfrac{-8\sqrt{5}}{5}$ **55.** $\dfrac{\sqrt{21}}{6}$ **56.** $\dfrac{\sqrt{y}}{y}$

57. $\dfrac{4\sqrt{7}}{5}$ **58.** $\dfrac{\sqrt{a}}{3}$ **59.** $-5\sqrt{6} + 10$ **60.** $\dfrac{-5 - \sqrt{3}}{22}$ **61.** $\dfrac{5\sqrt{y} + 5}{y - 1}$ **62.** $\dfrac{x\sqrt{x} - 3x}{x - 9}$ **63.** $\dfrac{x - 2\sqrt{x} + 1}{x - 1}$ **64.** $4\sqrt{3} + 7$

65. $x = 43$ **66.** $x = 15$ **67.** no solution **68.** no solution **69.** $x = 0$ **70.** $x = 68$ **71.** $x = 3,5$ **72.** $x = -4,-5$

73. no solution **74.** no solution **75.** $x = \dfrac{1}{2}$ **76.** $x = -7$ **77.** 8 **78.** 27 **79.** $\dfrac{1}{4}$ **80.** $\dfrac{1}{5}$ **81.** $x^{\frac{5}{6}}$ **82.** $y^{\frac{15}{16}}$ **83.** $y^{\frac{7}{6}}$

84. $\dfrac{1}{a^{\frac{4}{15}}}$ **85.** $64x^{\frac{7}{6}}$ **86.** $\dfrac{x}{2}$ **87.** $75a^{\frac{7}{3}}$ **88.** $\dfrac{6}{x^2}$ **89.** 1 **90.** 10 **91.** 5 **92.** $\sqrt{74}$ **93.** No; $6^2 + 8^2 \neq 9^2$

94. Yes; $7^2 + 24^2 = 25^2$ **95.** $AB = \sqrt{(1-4)^2 + (-2-4)^2} = 3\sqrt{5}; AC = \sqrt{(7-1)^2 + (-2-1)^2} = 3\sqrt{5}; AB = AC = 3\sqrt{5}$

96. 10 feet **97.** 28.3 feet

98. $XY = \sqrt{(6-2)^2 + (0-0)^2} = 4; XZ = \sqrt{(4-2)^2 + \left(\sqrt{12}-0\right)^2} = 4; YZ = \sqrt{(4-6)^2 + \left(\sqrt{12}-0\right)^2} = 4; XY = XZ = YZ = 4$

99. $AD = \sqrt{(-2-6)^2 + (2-5)^2} = \sqrt{73}; BC = \sqrt{(-2-6)^2 + (5-2)^2} = \sqrt{73}; AD = BC = \sqrt{73}$

100. $P = 9 + 3\sqrt{5}$; Yes, $3^2 + 6^2 = \left(3\sqrt{5}\right)^2$

Chapter 9 Test, pages 766 - 767

1. $0.\overline{428571}$ **2.** 3.6055513 **3.** 12 **4.** $5x$ **5.** $-9x^2y\sqrt{y}$ **6.** $3a$ **7.** $2x^2y^4\sqrt[3]{3}$ **8.** $\dfrac{-4x}{y^4}$ **9.** $\dfrac{2a^2\sqrt{a}}{3b^5}$ **10.** $13\sqrt{2}$

11. $22x\sqrt{3}$ **12.** $\dfrac{\sqrt{6} + 3}{6}$ **13.** $3 - \sqrt{2}$ **14.** $-2\sqrt[3]{5} + 3\sqrt[3]{4}$ **15.** 75 **16.** $3\sqrt{5} + 5\sqrt{3}$ **17.** 14 **18.** $28 - \sqrt{2}$ **19.** $\dfrac{\sqrt{2}}{2}$

20. $\dfrac{2\sqrt{3}}{15}$ **21.** $1 + \sqrt{3}$ **22.** $-2 + \sqrt{6}$ **23.** $x = 34$ **24.** $x = 2$ **25.** no solution **26.** 8 **27.** $\dfrac{27}{8}$ **28.** $a^{\frac{1}{4}}$ **29.** $y^{\frac{5}{4}}$

30. $36x^3$ **31.** $8\sqrt{2}$ **32.** $\sqrt{157}$ **33.** Yes; $20^2 + 21^2 = 29^2$

34. 24.3 ft **35.** $3 + 4 < 8$

Chapter 9 Cumulative Review, pages 768 - 770

1. $x^2 + 8x + 16 = (x+4)^2$ **2.** $x^2 - 14x + 49 = (x-7)^2$ **3.** $\dfrac{2(x+4)}{x-3}$ **4.** $\dfrac{x^2 - 4x + 5}{(x+6)(x-2)}$

5.   **6.**   **7.** $-3x^2\sqrt{7}$ **8.** $\dfrac{6a\sqrt{a}}{5b^3}$ **9.** $4x^5y^{10}\sqrt{3}$ **10.** $3a^4b^7\sqrt{6}$ **11.** $2\sqrt[3]{5}$ **12.** $-5\sqrt[3]{2}$

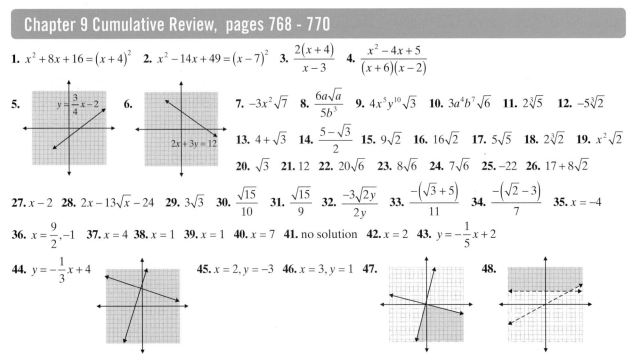

13. $4+\sqrt{3}$ **14.** $\dfrac{5-\sqrt{3}}{2}$ **15.** $9\sqrt{2}$ **16.** $16\sqrt{2}$ **17.** $5\sqrt{5}$ **18.** $2\sqrt[3]{2}$ **19.** $x^2\sqrt{2}$

20. $\sqrt{3}$ **21.** 12 **22.** $20\sqrt{6}$ **23.** $8\sqrt{6}$ **24.** $7\sqrt{6}$ **25.** -22 **26.** $17+8\sqrt{2}$

27. $x-2$ **28.** $2x - 13\sqrt{x} - 24$ **29.** $3\sqrt{3}$ **30.** $\dfrac{\sqrt{15}}{10}$ **31.** $\dfrac{\sqrt{15}}{9}$ **32.** $\dfrac{-3\sqrt{2y}}{2y}$ **33.** $\dfrac{-(\sqrt{3}+5)}{11}$ **34.** $\dfrac{-(\sqrt{2}-3)}{7}$ **35.** $x = -4$

36. $x = \dfrac{9}{2}, -1$ **37.** $x = 4$ **38.** $x = 1$ **39.** $x = 1$ **40.** $x = 7$ **41.** no solution   **42.** $x = 2$ **43.** $y = -\dfrac{1}{5}x + 2$

44. $y = -\dfrac{1}{3}x + 4$ **45.** $x = 2, y = -3$ **46.** $x = 3, y = 1$ **47.**   **48.**

49. 44 mph and 56 mph   **50.** 8 pounds at \$1.50 and 12 pounds at \$2.   **51.** 4 hrs and 12 hrs   **52.** Burrito price is \$2.60 and taco price is \$1.35   **53.** 26.5 m   **54. a.** $\sqrt{6}$ or 2.45 seconds   **b.** 96 feet   **c.** $\sqrt{15}$ or 3.87 seconds.

Chapter 10

Exercises 10.1, pages 777 - 780

1. $x = 0, 11$ **3.** $x = -12, -3$ **5.** $x = 1, -\dfrac{5}{3}$ **7.** $x = -1, 3$ **9.** $x = \dfrac{3}{4}, 1$ **11.** $x = \pm 11$ **13.** $x = \pm 6$ **15.** $x = \pm\sqrt{35}$

17. $x = \pm\sqrt{62}$ **19.** $x = \pm 3\sqrt{5}$ **21.** $x = \pm 3\sqrt{2}$ **23.** $x = \pm\dfrac{2}{3}$ **25.** $x = -1, 3$ **27.** no solution   **29.** $x = -\dfrac{3}{2}, -\dfrac{1}{2}$

31. $x = \dfrac{7}{3}, \dfrac{11}{3}$ **33.** $x = 6 \pm 3\sqrt{2}$ **35.** $x = 7 \pm 2\sqrt{3}$ **37.** $x = \dfrac{-4 \pm 3\sqrt{3}}{3}$ **39.** $x = \dfrac{2 \pm 3\sqrt{7}}{5}$ **41.** Yes   **43.** Yes   **45.** $c = 15$

47. $b = 6\sqrt{3}$ **49.** 1   **51.** The length of the leg is 4 ft and the hypotenuse is 8 ft.   **53.** $3\sqrt{2}$ cm   **55.** 47.2 ft   **57.** 3 seconds

59. $x = \pm 25.44$ **61.** $x = \pm 5.25$ **63.** $x = \pm 1.70$ **65.** $x = \pm 4.13$ **67.** 4.2361, -0.2361   **69.** 2.3229, -0.3229

Exercises 10.2, pages 785 - 787

1. $x^2 + 12x + 36 = (x+6)^2$ **3.** $2x^2 - 16x + 32 = 2(x-4)^2$ **5.** $x^2 - 3x + \dfrac{9}{4} = \left(x - \dfrac{3}{2}\right)^2$ **7.** $x^2 + x + \dfrac{1}{4} = \left(x + \dfrac{1}{2}\right)^2$

9. $2x^2 + 4x + 2 = 2(x+1)^2$ **11.** $x = -7, 1$ **13.** $x = -5, 9$ **15.** $x = -5, 8$ **17.** $x = -\dfrac{4}{3}, 1$ **19.** $x = -\dfrac{1}{2}, \dfrac{3}{2}$ **21.** $x = -3 \pm\sqrt{6}$

23. $x = -1 \pm\sqrt{6}$ **25.** $x = -\dfrac{1}{3}, 1$ **27.** $x = \dfrac{-1 \pm\sqrt{13}}{2}$ **29.** $x = \dfrac{-3 \pm\sqrt{17}}{4}$ **31.** $x = \dfrac{9 \pm\sqrt{73}}{2}$ **33.** $x = \dfrac{-7 \pm\sqrt{105}}{2}$ **35.** $x = -2, 13$

37. $x = \dfrac{-1 \pm \sqrt{17}}{4}$ **39.** $x = 1$ **41.** $x = \dfrac{-5 \pm \sqrt{33}}{2}$ **43.** $x = \dfrac{4 \pm \sqrt{10}}{6}$ **45.** $x = \dfrac{-7 \pm \sqrt{17}}{4}$ **47.** $x = \dfrac{1 \pm \sqrt{13}}{4}$ **49.** $x = \dfrac{-5}{3}, -1$

51. $x = -1, -\dfrac{3}{2}$ **53.** $x = -1, -7$ **55.** $x = -\dfrac{1}{3}, \dfrac{1}{2}$

Exercises 10.3, pages 794 - 795

1. $x^2 - 3x - 2 = 0; a = 1, b = -3, c = -2$ **3.** $2x^2 - x + 6 = 0; a = 2, b = -1, c = 6$ **5.** $7x^2 - 4x - 3 = 0; a = 7, b = -4, c = -3$

7. $3x^2 - 9x - 4 = 0; a = 3, b = -9, c = -4$ **9.** $2x^2 + 5x - 3 = 0; a = 2, b = 5, c = -3$ **11.** $x = 2 \pm \sqrt{5}$ **13.** $x = -1, 4$

15. $x = \dfrac{-1}{2}, 1$ **17.** $x = -1, \dfrac{2}{5}$ **19.** $x = 0, \dfrac{1}{3}$ **21.** $x = \pm \sqrt{7}$ **23.** $x = -1, 0$ **25.** $x = \dfrac{-4}{3}, 0$ **27.** $x = \dfrac{-5 \pm \sqrt{65}}{4}$

29. $x = \dfrac{-6 \pm \sqrt{42}}{3}$ **31.** $x = \dfrac{1}{2}, -3$ **33.** $x = \dfrac{4 \pm \sqrt{22}}{3}$ **35.** $x = 1, \dfrac{3}{4}$ **37.** $x = -3, -\dfrac{1}{2}$ **39.** $x = \dfrac{7 \pm \sqrt{37}}{6}$ **41.** $x = -\dfrac{3}{2}, \dfrac{8}{5}$

43. $x = 4, -\dfrac{1}{3}$ **45.** $x = \dfrac{11 \pm \sqrt{41}}{8}$ **47.** $x = \dfrac{-3}{2}, -2$ **49.** $x = \pm \dfrac{4}{5}$ **51.** $x = \dfrac{8 \pm \sqrt{58}}{6}$ **53.** $x = \dfrac{7 \pm \sqrt{385}}{12}$

55. $x = 1.6180, -0.6180$ **57.** $x = 1, -0.25$ **59.** $x = \pm 4.4721$

Exercises 10.4, pages 801 - 807

1. $x^2 + x = 132; x = 11$ **3.** $w(2w - 5) = 63; w = 7$, the rectangle is 7 m by 9 m **5.** $x(29 - x) = 198; x = 11$ or 18, the field

is 11 m by 18 m **7.** $(w + 12)(w + 22) = 1344; w = 20$, the pool is 20 ft by 30 ft. **9.** $x^2 + (x + 1)^2 = 221; x = 10$, the numbers

are 10 and 11. **11.** $x - 3 = \dfrac{4}{x}; x = 4$ **13.** $(x + 10)^2 = 9x^2; x = 5$, the side of the original square is 5 cm.

15. $w^2 + (2w + 2)^2 = 169; w = 5$, the rectangle is 5 m by 12 m. **17.** $2x^2 = 144$; the legs of the triangle are $6\sqrt{2}$ cm each.

19. $\dfrac{900}{x} - 10 = \dfrac{900}{x + 15}; x = 30$, there are 30 people in the club. **21.** $\dfrac{200}{x} + \dfrac{200}{x - 10} = 9; x = 50$, he traveled to the city at 50

mph. **23.** $\dfrac{45}{12 - c} - \dfrac{45}{12 + c} = 2; c = 3$, the rate of the current is 3 mph. **25.** $\dfrac{120}{x} + 2 = \dfrac{120}{x - 5}; x = 20$, there were initially 20

members. **27.** $2(x - 4)^2 = 162; x = 13$, the cardboard was 13 in. by 13 in. **29.** $\dfrac{1}{x} + \dfrac{1}{x + 30} = \dfrac{1}{8}; x = 10$; the smaller pipe

would take 40 min. **31 a.** After 1 sec. **b.** 2.5 sec. **c.** 3 sec. **33.** 28.3 ft **35.** a cannot be equal to zero and b^2 must be greater

than or equal to $4ac$ to produce a real solution. Also, all the solutions found may not apply to the problem at hand. You

must check that each answer makes sense in the context of the problem.

Exercises 10.5, pages 817 - 820

1. a. $(0, 4)$ **b.** $x = 0$ **3. a.** $(0, 8)$ **b.** $x = 0$ **5. a.** $(1, -4)$ **b.** $x = 1$ **7. a.** $(-3, -9)$ **b.** $x = -3$

 c. None **c.** $\pm 2\sqrt{2}$ **c.** −1 and 3 **c.** −6, 0

 d. **d.** **d.** **d.**

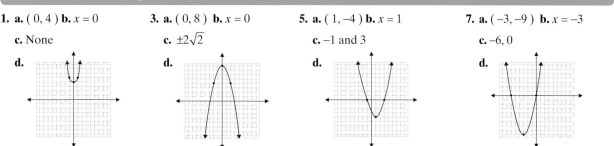

9. a. $(-2, 6)$ **b.** $x = -2$ **11. a.** $\left(\dfrac{5}{2}, -\dfrac{19}{2}\right)$ **b.** $x = \dfrac{5}{2}$ **13. a.** $\left(-\dfrac{7}{2}, -\dfrac{65}{4}\right)$ **b.** $x = -\dfrac{7}{2}$ **15. a.** $\left(\dfrac{1}{2}, -\dfrac{11}{4}\right)$ **b.** $x = \dfrac{1}{2}$

c. $-2 \pm \sqrt{6}$ **c.** $\dfrac{5 \pm \sqrt{19}}{2}$ **c.** $\dfrac{-7 \pm \sqrt{65}}{2}$ **c.** None

d.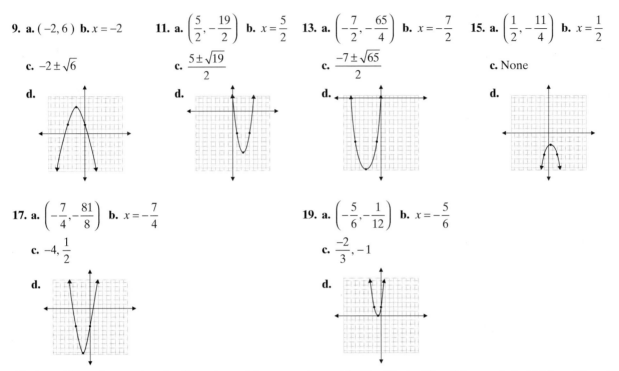

17. a. $\left(-\dfrac{7}{4}, -\dfrac{81}{8}\right)$ **b.** $x = -\dfrac{7}{4}$ **19. a.** $\left(-\dfrac{5}{6}, -\dfrac{1}{12}\right)$ **b.** $x = -\dfrac{5}{6}$

c. $-4, \dfrac{1}{2}$ **c.** $\dfrac{-2}{3}, -1$

d.   **d.**

21. $A = x(30 - x)$; $x = 15$, so the dimensions of the rectangle are 15 yd by 15 yd.   **23. a.** 3.5 seconds **b.** 196 feet   **25. a.** 4 seconds **b.** 288 feet   **27. a.** $R(x) = x(40 - x)$ **b.** \$20   **29. a.** $R(x) = x(80 - 2x)$ **b.** \$20 **c.** \$800

31. a. 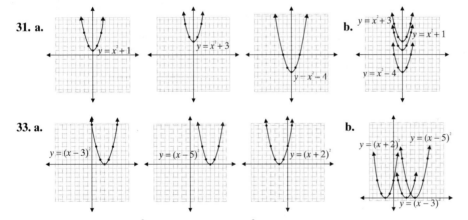   **b.**

33. a.   **b.**

35. The graph $y = (x - h)^2 + k$ is the graph $y = x^2$ shifted k units vertically and h units horizontally.

Chapter 10 Review, pages 823 - 826

1. $x = \pm 20$ **2.** $x = \pm 6\sqrt{2}$ **3.** No solution   **4.** $x = -1, 13$ **5.** $x = -3 \pm 2\sqrt{5}$ **6.** $x = -1 \pm 2\sqrt{10}$ **7.** $x = -1, \dfrac{5}{3}$ **8.** $x = -\dfrac{9}{2}, 0$

9. $x = \dfrac{-5 \pm 5\sqrt{3}}{2}$ **10.** $x = 1 \pm 4\sqrt{5}$ **11.** Yes   **12.** Yes   **13.** $c = 8\sqrt{2}$ **14.** $a = 21$ **15.** 60   **16.** 4.5 sec   **17.** ± 15.81 **18.** ± 1.62

19. ± 5.59 **20.** No solution   **21.** $x = -1, 9$ **22.** $x = -2, -1$ **23.** $x = -8, 7$ **24.** $x = \pm \dfrac{1}{2}$ **25.** $x = \pm \dfrac{2}{3}$ **26.** $x = 2 \pm \sqrt{3}$

27. $x = -4 \pm 2\sqrt{3}$ **28.** $x = \dfrac{-1 \pm \sqrt{17}}{2}$ **29.** $x = \dfrac{2 \pm \sqrt{7}}{3}$ **30.** $x = -2, \dfrac{1}{2}$ **31.** $x = \pm \dfrac{5}{4}$ **32.** $x = \pm \dfrac{4}{5}$ **33.** $x = -13, 2$

34. $x = -10, 2$ **35.** $x = \dfrac{6 \pm \sqrt{6}}{6}$ **36.** $x = \dfrac{2 \pm \sqrt{13}}{3}$ **37.** $x = \dfrac{-1 \pm \sqrt{13}}{4}$ **38.** $x = \dfrac{-7 \pm \sqrt{89}}{2}$ **39.** $x = \dfrac{5 \pm 2\sqrt{5}}{5}$ **40.** $x = \dfrac{-5}{3}, 1$

41. $x = \dfrac{1 \pm \sqrt{17}}{4}$ **42.** $x = \dfrac{3 \pm \sqrt{3}}{3}$ **43.** $x = \dfrac{\pm 7\sqrt{2}}{2}$ **44.** $x = \dfrac{3 \pm \sqrt{3}}{2}$ **45.** $x = \dfrac{-12 \pm 2\sqrt{39}}{3}$ **46.** $x = 0, 5$ **47.** $x = \dfrac{1 \pm \sqrt{73}}{4}$

48. $x = -\dfrac{2}{3}, 1$ **49.** $x = \dfrac{3}{2}$ **50.** $x = \dfrac{\pm 2\sqrt{3}}{3}$ **51.** $x = -2, 1$ **52.** $x = 0, \dfrac{3}{7}$ **53.** $x = -3 \pm \sqrt{19}$ **54.** $x = \dfrac{11 \pm \sqrt{89}}{8}$

55. $x = \dfrac{-4 \pm 2\sqrt{13}}{3}$ **56.** $x = -\dfrac{1}{4}, 1$ **57.** $x = 4.2361, -0.2361$ **58.** $x = 2.9365, -0.9365$ **59.** $x = 0.8047, -0.1381$

60. $x = 7.0711, -7.0711$ **61.** $6, 7$ **62.** Length is 30 ft, Width is 24 ft   **63.** Length is 15 m, Width is 10 m   **64.** The numbers are 2 and 0 or 10 and 8   **65.** The length of the two identical sides is $10\sqrt{2}$ **66.** 72 mph   **67.** 4 mph   **68.** 6 hours, 4 hours   **69.** 90 cm × 90 cm × 5 cm   **70.** 20 mph

71. a. $(0, -4)$ **b.** $x = 0$ **72. a.** $(0, 16)$ **b.** $x = 0$ **73. a.** $(3, -4)$ **b.** $x = 3$ **74. a.** $(5, -25)$ **b.** $x = 5$
  c. $x = \pm 2$ **c.** $x = \pm 4$ **c.** $x = 1, 5$ **c.** $x = 0, 10$

d.   **d.**   **d.**   **d.**

75. a. $(1, -2)$ **b.** $x = 1$ **76. a.** $\left(\dfrac{5}{3}, -\dfrac{19}{3}\right)$ **b.** $x = \dfrac{5}{3}$

  c. None   **c.** $x = 0.2137, 3.1196$

d.   **d.**

77. The rectangle is 20 yd × 20 yd   **78. a.** $R = x(80 - x)$ **b.** \$40
c. \$1600   **79. a.** $t = 1.5$ **b.** 36 ft   **c.** 3 sec **80. a.** $t = 2$ **b.** 144 ft
c. 5 sec

Chapter 10 Test, pages 827 - 828

1. $x = -12, 2$ **2.** $x = \pm 2\sqrt{3}$ **3.** $x = 12$ **4.** $x^2 - 24x + 144 = (x - 12)^2$ **5.** $x^2 + 9x + \dfrac{81}{4} = \left(x + \dfrac{9}{2}\right)^2$

6. $3x^2 + 9x + \dfrac{27}{4} = 3\left(x + \dfrac{3}{2}\right)^2$ **7.** $x = 1, 2$ **8.** $x = \dfrac{-12}{5}, 0$ **9.** $x = -\dfrac{5}{3}, 2$ **10.** $x = -4, -2$ **11.** $x = -1, 6$ **12.** $x = 2 \pm \sqrt{6}$

13. $x = \dfrac{-4 \pm \sqrt{10}}{3}$ **14.** $x = \dfrac{2 \pm \sqrt{10}}{2}$ **15.** $x = 1 \pm \sqrt{5}$ **16.** $x = -3 \pm \sqrt{11}$ **17.** $x = \dfrac{1}{3}, 2$ **18.** $x = \dfrac{3 \pm \sqrt{17}}{4}$

19. a. $(0, -5)$ **b.** $x = 0$ **20. a.** $(2, 4)$ **b.** $x = 2$ **21. a.** $\left(\dfrac{-3}{4}, \dfrac{-1}{8}\right)$ **b.** $x = \dfrac{-3}{4}$

  c. $\pm\sqrt{5}$ **c.** $0, 4$ **c.** $-1, \dfrac{-1}{2}$

d.   **d.**   **d.**

$y = -x^2 + 4x$

$y = 2x^2 + 3x + 1$

22. $2x^2 = 36^2$; $x = 18\sqrt{2}$ in.   **23. a.** 0.725 sec and 8.275 sec **b.** 324 ft **c.** 9 seconds   **24.** $\dfrac{60}{8-x} - \dfrac{60}{8+x} = 10$; $x = 4$ the rate of current is 4 mph

25. $4(w-5)(w-8) = 720$; $w = 20$, the metal is 20 in. by 23 in.

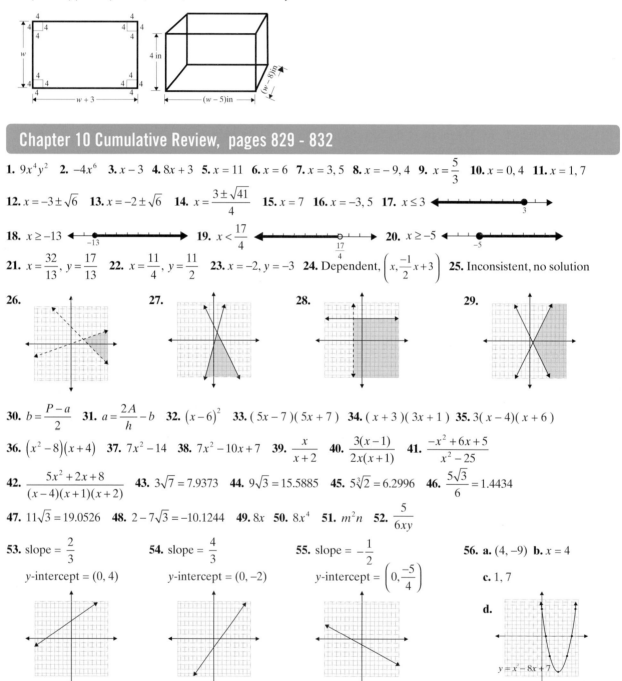

Chapter 10 Cumulative Review, pages 829 - 832

1. $9x^4y^2$ **2.** $-4x^6$ **3.** $x-3$ **4.** $8x+3$ **5.** $x=11$ **6.** $x=6$ **7.** $x=3,5$ **8.** $x=-9,4$ **9.** $x=\dfrac{5}{3}$ **10.** $x=0,4$ **11.** $x=1,7$

12. $x=-3\pm\sqrt{6}$ **13.** $x=-2\pm\sqrt{6}$ **14.** $x=\dfrac{3\pm\sqrt{41}}{4}$ **15.** $x=7$ **16.** $x=-3,5$ **17.** $x\le 3$

18. $x\ge -13$ **19.** $x<\dfrac{17}{4}$ **20.** $x\ge -5$

21. $x=\dfrac{32}{13}, y=\dfrac{17}{13}$ **22.** $x=\dfrac{11}{4}, y=\dfrac{11}{2}$ **23.** $x=-2, y=-3$ **24.** Dependent, $\left(x, \dfrac{-1}{2}x+3\right)$ **25.** Inconsistent, no solution

26.   **27.**   **28.**   **29.**

30. $b=\dfrac{P-a}{2}$ **31.** $a=\dfrac{2A}{h}-b$ **32.** $(x-6)^2$ **33.** $(5x-7)(5x+7)$ **34.** $(x+3)(3x+1)$ **35.** $3(x-4)(x+6)$

36. $(x^2-8)(x+4)$ **37.** $7x^2-14$ **38.** $7x^2-10x+7$ **39.** $\dfrac{x}{x+2}$ **40.** $\dfrac{3(x-1)}{2x(x+1)}$ **41.** $\dfrac{-x^2+6x+5}{x^2-25}$

42. $\dfrac{5x^2+2x+8}{(x-4)(x+1)(x+2)}$ **43.** $3\sqrt{7}=7.9373$ **44.** $9\sqrt{3}=15.5885$ **45.** $5\sqrt[3]{2}=6.2996$ **46.** $\dfrac{5\sqrt{3}}{6}=1.4434$

47. $11\sqrt{3}=19.0526$ **48.** $2-7\sqrt{3}=-10.1244$ **49.** $8x$ **50.** $8x^4$ **51.** m^2n **52.** $\dfrac{5}{6xy}$

53. slope $=\dfrac{2}{3}$ **54.** slope $=\dfrac{4}{3}$ **55.** slope $=-\dfrac{1}{2}$ **56. a.** $(4,-9)$ **b.** $x=4$

     y-intercept $=(0,4)$ y-intercept $=(0,-2)$ y-intercept $=\left(0, \dfrac{-5}{4}\right)$ **c.** 1, 7

     d. $y=x^2-8x+7$

57. a. $(-3, -6)$ **b.** $x = -3$ **58. a.** $\left(-\dfrac{1}{2}, \dfrac{3}{4}\right)$ **b.** $x = -\dfrac{1}{2}$

c. $-3 \pm \sqrt{3}$ **c.** None

d.   **d.**

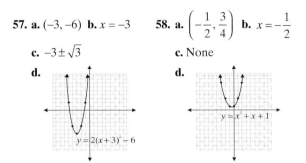

59. $y = -\dfrac{5}{2}x + 3; m = -\dfrac{5}{2}, b = 3$ **60. a.** $m = \dfrac{3}{8}$ **b.** $\sqrt{73}$ **c.** $\left(-1, \dfrac{7}{2}\right)$ **d.** $y - 2 = \dfrac{3}{8}(x + 5)$ **61. a.** -8 **b.** 27 **c.** -3 **62. a.** 7 **b.** 23

c. 72 **63.** \$2100 at 5%, \$1300 at 8% **64.** $\dfrac{40}{9}$ hours **65.** 2 mph **66.** 20 liters **67.** 8 in. by 15 in. **68.** 3 at \$2.50 and 4

at \$3.00 **69.** width is 4 cm, length is 13 cm **70.** 3600 women, 2800 men **71.** \$1500 at 5.5%, \$1200 at 7.2% **72.** 10 mph
upstream, 15 mph downstream **73. a.** 1 sec and 5 sec **b.** 6.571 seconds **c.** 7.062 seconds **74.** 37.1 ft

Appendix A.1

Appendix A.1, pages 836- 837

1. $x^3 + 8$ **3.** $x^3 - 125$ **5.** $x^3 + y^3$ **7.** $y^3 + 1$ **9.** $x^3 y^3 + 8$ **11.** $1000x^3 + 1$ **13.** $125x^3 + y^3$ **15.** $8x^3 - 125y^3$

17. $27x^3 + 8$ **19.** $8 - 27y^3$ **21.** $x^{12} + 27$ **23.** $x^6 + 64$ **25.** $x^9 - 1000$ **27.** $(x - 4)(x^2 + 4x + 16)$ **29.** $(x - y)(x^2 + xy + y^2)$

31. $(x + 5)(x^2 - 5x + 25)$ **33.** $(x + 3y)(x^2 - 3xy + 9y^2)$ **35.** $(2x + 1)(4x^2 - 2x + 1)$ **37.** $5(x + 2)(x^2 - 2x + 4)$

39. $2(3x - y)(9x^2 + 3xy + y^2)$ **41.** $y(x + y)(x^2 - xy + y^2)$ **43.** $x^2 y^2 (1 - y)(1 + y + y^2)$ **45.** $3xy(2x + 3y)(4x^2 - 6xy + 9y^2)$

47. $(x^2 - y^3)(x^4 + x^2 y^3 + y^6)$ **49.** $(x - 10)(x^2 + 10x + 100)$ **51.** $(x + y - 1)(x^2 - xy + 7x + y^2 - 8y + 19)$

53. $(3x + y - 1)(9x^2 - 3xy + 3x + y^2 - 2y + 1)$ **55.** $x(x^2 + 3x + 3)$

Index

A

Absolute value 83
ac-method of factoring 566
Addend 2
Addition
  additive identity 120
  associative property 120
  commutative property 120
  of decimals 49
  of fractions 37–38, 153, 155
  polynomials 502–503
  rational expressions 639
  with integers 92
  with radicals 720–721
Addition Principle of Equality
    201
Additive identity 120
Additive inverse 96, 120
Algebraic expressions
  evaluating 139
  simplifying 504–505
Ambiguous phrase 181
Analytic geometry 301
Applications
  amounts and costs 432–434
  Body-Mass-Index 602
  college graduates 59
  consecutive integers 255
  consumer demand 605–606
  direct variation 676
  distance-rate-time 429–431,
    667–668, 798–799
  electricity 680
  electric power 605
  gasoline prices 55
  gears 680
  geometry 596, 796–797, 800–801,
    814
  height of a projectile 604
  household budgets 42
  interest 440–441, 440–442

  levers 681
  lifting force 679
  load on a wooden beam 603
  mixture 443–445
  number problem 431
  number problems 252, 595, 597
  percent 264–266
  postage 236
  pressure 679
  proportions 666
  Pythagorean Theorem 796,
    800–801
  resistance 680
  retail 31, 51, 224
  systems of linear equations in
    two variables 429–433,
    440–444
  temperature 100
  test average 235
  time 667
  towing capacity 32
  variation 675–677
  velocity 246–247
  volume of a cylinder 606
  weight loss 101
  work 668–670, 797–798
Area 273
  of a circle 274
  of a parallelogram 274
  of a rectangle 274
  of a square 274
  of a trapezoid 274
  of a triangle 274
Arithmetic average 110
Ascending order 497
Associative Property
  of addition 120
  of multiplication 120
Attack plan for word problems
    207, 594
Average (or mean) 110
Axes, *x*- and *y*- 304
Axis of symmetry 809

B

Base 468
Base, of an exponent 3
Binomials 496
  special product of 515–519
Boundary line 373, 449
Boyle's Law 679

C

Cartesian coordinate system 302
Cartesian geometry 301
Change in value 97–98
Circle
  area of 274
  circumference of 272
  diameter of 272
  perimeter of 271
  radius of 272
Circumference of a circle 272
Closed half-plane 373
Closed interval 230
Coefficient 121, 136, 495
Coinciding lines 405
Combining Like radicals 720–721
Combining like terms 137, 502
Commutative Property
  of addition 120
  of multiplication 120
Completing the square 581–583,
    781–782
  solving quadratic equations 781
Complex algebraic fractions 647
  simplifying 647–651
Complex numbers 738
Components of an ordered pair
    302
Composite number 15
Conjugate 727
Consecutive
  even integers 254
  integers 254
  odd integers 254
Consistent system 404, 421
  two variables 404

Constant 136, 201, 495
Constant of variation 675, 677
Constant term 495
Coordinate 77
Coordinate of a point on a number line 77
Counting numbers 2, 76
Cubed 4, 701
Cube root 702
  simplest form 715
  simplifying 715

D

Decimal(s)
  adding 49
  changing a percent to a decimal 58
  changing decimals to fractions 165
  changing decimals to percent 56
  defined 47
  division 53, 163
  infinite, nonrepeating 702
  infinite repeating 699
  multiplication 52, 162
  nonrepeating 164
  repeating 163
  rounding 54
  scientific notation 488
  subtracting 50
  terminating 699
Decimal notation 47
Decimal point 47
Decimal system 1
Degree
  zero 496
Degree of a polynomial 496
Degree of the term 137, 495
Denominator 24
  not equal to zero 109
  rationalizing
    with a sum or difference in the denominator 727–729
    with one term in the denominator 726–727

Dependent system 405, 422
Dependent variable 303
Depreciation formula 248
Descending order 497
Diameter of a circle 272
Difference
  of fractions 157
  of integers 99
  of polynomials 503–504
  of rational expressions 642
  of two squares 517
Directly proportional 675
Direct variation 675
Discriminant 789
Distance
  between points in a plane 749
  between points on a horizontal line 746
  between points on a vertical line 746
Distributive property 120
  for use in multiplying polynomials 508–511
Dividend 527
Divisibility, tests for 21
Division
  by zero 24
  by zero is undefined 109
  fractions 148
  of decimals 53
  polynomials 525–531
  Principle of Equality 204
  rational expressions 634
  real numbers 108
  with decimals 163
  with fractions 30
Division algorithm 527
Divisor 14
Domain 361
Domain axis 362

E

Egyptian numerals 1
Elimination method for solving linear systems 420

Equation(s) 200
  consistent 404
  dependent 405
  equivalent 201
  first-degree 201
  inconsistent 404
  linear 201
  of a line 320
  of the form
    $ax + b = c$ 213
    $ax + b = dx + c$ 220
    $ax = c$ 204
  point-slope form 352
  quadratic 586
  radical 732
  rational expressions 658
  slope-intercept form 343
  solution of 200, 302
  solution set of 200
  solving equations with a square root radical expression 734
  solving for specified variables 213
  solving for specified variables of 204, 220
    $x + b = c$ 201
  standard form 320
Equivalent equations 201
Evaluating expressions 139
Evaluating formulas 241
Evaluating polynomials 498
Evaluating rational expressions 624
Even consecutive integers 254
Even whole numbers 16
Exactly divisible 16
Exponential expression 3
Exponents 3, 468
  0 as an 472
  base of 468
  cubed 4
  fractional 739–740
  general properties 477, 481
  integer 476
  negative 475
  Power of a Product Rule 483

Power of a Quotient 485
Power Rule 482
Product Rule 469
Quotient Rule 473
rational 739–740
rules for 481, 487
squared 4
summary of rules for 477, 481
zero 472
Expressions
algebraic 136
evaluating 139
rational 622
reduced rational 647, 651
simplifying algebraic 504–505
Extraneous solutions 658

F

Factor(s) 3, 14
common monomial 550–553
greatest common 549
Factoring
by common monomial term
550–553
by completing the square
581–583
by grouping 553–555
by the *ac*-method (grouping)
566–570
difference of two squares 578
perfect square trinomials 579
solving quadratic equations
586–590
trinomials by factoring out a
monomial first 561–562
trinomials by trial and error
570–575
trinomials with leading coefficient 1 559–561
Field properties 120
First-degree equations 201
solving 201
First-degree inequality
solving 231

First component of ordered pairs
302
FOIL method 515–516
Formulas
angles 241
Celsius and Fahrenheit 240
distance 240
evaluating 241
force 241
interest 440
in geometry 240, 241
lateral surface area 241
lifting force 679
simple interest 240
temperature 240
Fraction(s) 24
adding
different denominators 38
like denominators 37
with different denominators
155
with like denominators 153
and decimals 163
as a rational number 143
changing decimals to fractions
165
changing fractions to percent 59
changing percent to a fraction
58
complex algebraic 647
dividing 30, 148
Fundamental Principle of 146
improper 25
in lowest terms 28
multiplying 26, 145
proper 25
raising to higher terms 27, 146
reducing 28, 146
subtracting 40, 157
Fractional exponents 739–740
Function(s) 362
defined 362
domain 362
linear 366
notation 366
quadratic 810

range 362
vertical line test 363
Fundamental Principle of Frac-
tions 146
Fundamental Principle of Rational
Expressions 625

G

GCF 549
Geometry
formulas in 271–277
Graphs 77
of a number 77
of a quadratic function 811–814
of linear equations 320–326
of linear inequalities 374
points in a plane 305
systems of linear equations 403
Greater than, symbol for 81
Greatest common factor 549

H

Half-open interval 230
Half-plane 449
open and closed 373
Higher terms of a fraction 27
Horizontal axis 304
Horizontal line 324
Horizontal lines
slope of 340–341
Hypotenuse 747

I

Identity
additive 120
multiplicative 27, 120, 145
Imaginary numbers 701
Improper fraction 25
Inconsistent system 404, 422
Independent variable 303
Index 738
Inequality
first-degree 231
graphing linear 374
linear 231

solving 231
symbols of 81
Infinite, number of elements 135
Infinite repeating decimals 699
Integers 78, 118
  addition 92
  consecutive 254
  negative 78
  perfect square 517, 715
  positive 78
  subtraction 99
Intercepts, x- and y- 322–323
Interest
  applications 440–441
  simple 440
Intersection
  consistent system 404
Intersection of two half-planes
  449
Intervals 228
  closed 230
  half-open 230
  open 230
Inverse
  additive 96, 120
  multiplicative 120
Inversely proportional 677
Inverse variation 677
Irrational numbers 80, 118, 164,
  702
Irreducible 562

K

Key words, list of 179

L

LCD 38
LCM 19, 154
  for a set of polynomials 640
Leading coefficient 496
Least common denominator 38
Least common multiple 19, 154
Leg 747
Less than, symbol for 81
Like radicals 720

Like terms 137
Line(s)
  coinciding 405
  horizontal 324–326
  of symmetry 809
  parallel 337, 405
  vertical 324–326
Linear equation 201, 321
  defined 201
  graphing 320–326
  point-slope form 352
  slope-intercept form 343
  solving for specified variables
    204, 213, 220
  standard form 320
  systems of 402–403
Linear equations in two variables
  applications 429–433
  consistent 404
  ordered pairs 402
  solving by graphing 403–408
Linear function 366
Linear inequality
  graphing 374
  solving 231
Lower terms of a fraction 28

M

Mathematicians
  Al-Khowarizmi, Mohammed
    ibn-Musa 199
  Alembert, Jean le Rond 199
  Bhaskara the Learned 75
  Cantor, Georg 135
  Cardano, Girolamo 547
  Descartes, René 301, 302
  Einstein, Albert 75
  Euclid 199
  Fontana, Niccolo 547
  Gauss, Karl F. 467
  Kasner, Edward 697
  Kowa, Seki 401
  Kronecker, Leopold 135
  Pólya, George 206, 771
  Pythagoras 621

  Robert of Chester 199
  Whitehead, Alfred North 547
Maximum/minimum values 809
Mean 110
Mixed number 48
Mixture
  applications 443–445
Monomials 495, 496
  degree of 495
  division by 525
  multiplication of a polynomial
    by 508
Multiples of a number 19, 154
Multiplication
  associative property 120
  commutative property 120
  distributive property 120
  identity element of 27
  of binomials 515
  of decimals 52
  of polynomials 508–511
  Principle of Equality 204
  radicals 722–723
  rational expressions 633
  real numbers 108
  with decimals 162
  with fractions 26, 145
Multiplicative identity 27, 145
Multiplicative inverse 120

N

Natural numbers 2, 76
Negative exponents 475
Negative integers 78
Net change 100
Nonrepeating decimals 80
Notation
  function 366
  radical 700
  scientific 488–490
Numbers
  absolute value 83
  complex 738
  composite 15
  counting 2, 76

Numbers *(continued)*
  decimal 47
  even 16
  imaginary 701
  integer 78
  integers 118, 698
  irrational 80, 118, 164, 702
  mixed 48
  natural 2, 76
  ncgative 77
  nonreal complex 701
  odd 16
  positive 77
  prime 15
  rational 79, 118, 143, 699
  real 80, 118, 698
  whole 2, 76
Number line 76
Number system
  decimal 1
  Egyptian 1
  Roman numerals 1
Numerator 24
Numerical coefficient 136, 495

O

Odd consecutive integers 254
Odd whole numbers 16
One-to-one correspondence
  between points in a plane 304
  between points on a number line
      229
Open half-plane 373
Open interval 230
Operations
  key words 179
  order of 171
Opposite of an integer 96
Opposite of a number 77
Ordered pairs 302, 402
  first and second components 302
  graphing 304
  one-to-one correspondence be-
      tween points in a plane 304
Order of operations 8, 171

Origin 304

P

Parabola 809
  axis (line) of symmetry 809
  maximum/minimum values 809
  vertex of 809
Parallelogram
  area of 274
  perimeter of 271
Parallel lines 337, 355, 405
Percent 56, 261
  basic formula 262
  changing decimals to percent 56
  changing fractions to percent 59
  changing percent to a decimal
      58
  changing percent to a fraction
      58
Percent sign 56
Percent symbol 56
Perfect cube 701
  table of perfect cubes from 1 to
      10 701
Perfect square 5, 699
  table of perfect squares from 1 to
      20 699
Perfect square trinomial 518–519
Perimeter 271
  of a circle 271
  of a parallelogram 271
  of a rectangle 271
  of a square 271
  of a trapezoid 271
  of a triangle 271
Perpendicular lines 355
Pi(π) 272
Place value system 47
Point-slope form 352
Polynomials 496
  addition 502–503
  ascending order 497
  binomial 496
  degree of 496
  descending order 497

dividing 525–531
evaluating 498
factoring by grouping 553–555
factoring by the *ac*-method
    566–570
factoring by trial and error
    570–574
FOIL method 515–516
irreducible 567, 575
leading coefficient 496
monomial 495, 496
multiplying 508–511
prime 562
simplifying 497
subtracting 503–504
trinomial 496
Zero-Factor Property 587
Positive integers 78
Power of a Product Rule 483
Power of a Quotient 485
Power Rule for Exponents 482
Prime factorization 17
Prime number 15
Prime polynomial 562
Principal square root 700
Problem solving, basic steps 207,
    251
Product 3
Product rule for exponents 469
Properties
  additive identity 120
  additive inverse 120
  associative 120
  commutative 120
  distributive 120
  multiplicative identity 120
  of square roots 711
  zero-factor 120
Proper fraction 25
Proportion(s) 654
  applications with 666, 675–677
  solving 655–656
Pythagorean Theorem 747, 775

Q

Quadrant 304
Quadratic equations 586
  applications 594–597, 602
  discriminant 789
  double solution (root) 588
  general form 788
  solving by completing the square 782–784
  solving by factoring 586–591
  solving by using the quadratic formula 790–793
  solving using the square root method 773–774
  standard form 586
Quadratic formula 789
Quadratic function 810
  general information 810
Quotient 181, 527
Quotient Rule for Exponents 473

R

Radicals
  adding 720–721
  combining 720–721
  like 720
  multiplying 722–723
  notation 700
  simplifying 711–716
Radical equation 732
  solving 733
Radical sign 700
Radicand 700
Radius of a circle 272
Range 361
Range axis 362
Ratio 654
Rationalizing the denominator
  with a sum or difference in the denominator 727–729
  with one term in the denominator 726–727
Rational exponents 739–740
Rational expressions 622

adding 639
arithmetic rules for 625
determining values where undefined 623
dividing 634
evaluating 624
fundamental principle 625
multiplying 633
opposites 628
reducing 625–629
restrictions 623
solving 658–659
subtracting 642
Rational numbers 79, 118, 143, 699
  infinite repeating decimals 699
  terminating decimals 699
Real number(s) 80, 118, 698
  division 108
  graphing 77
  intervals of 228
  multiplication 108
Real number line 76, 80, 118, 698
Real number system 107
Reciprocal 30, 147
Rectangle
  area of 274
  perimeter of 271
Rectangular pyramid
  volume of 276
Rectangular solid
  volume of 276
Reducing a fraction 146
Relation 361
  domain of 361
  range of 361
Remainder 527
Repeating decimals 163, 699
Restrictions, on a rational expression 623
Right circular cone
  volume of 276
Right circular cylinder
  volume of 276
Right triangle 747
Roman numeral system 1

Rounding 54
Rules for order of operations 171

S

Scientific notation 488
Second component of ordered pairs 302
Similar terms 137
Simplest form of a square root 712
Simplifying
  algebraic expressions 504–505
  complex algebraic fractions 647–651
  other radical expressions 716–717
  radicals 711
  square roots
    with even and odd powers of variables 714–715
    with variables 713–714
Simultaneous linear equations 402
Slope 333–339
  for horizontal lines 341
  for vertical lines 341
  negative 339
  parallel 337
  positive 339
Slope-intercept form 343
Solution 200
  extraneous 658
  linear inequalities 231
  of radical equations 733
  of systems of equations 402
    by addition 420–425
    by graphing 403–406
    by substitution 414–417
  of the form
    $ax + b = c$ 213
    $ax = c$ 204
    $x + b = c$ 201
  to first-degree equations 213
  to first-degree inequalities 231

Solution of a system
  definition 402
  ordered pairs 402
Solution set 200, 449
Solving quadratic equations
  by completing the square
    782–784
  by using the quadratic formula
    790–793
  using the square root method
    773–774
Special products 515–519
  procedure for factoring 578–580
Sphere
  volume of 276
Square(s)
  area of 274
  difference of two 517
  perfect 517
  perimeter of 271
Squared 4
Square root 700
  even and odd exponents 714
  of x^2 713
  properties 711
  simplest form 712
Square root method 773
Standard form 320
  of a linear equation 321
Subscript 336
Substitution
  as a method of solving systems of
    equations 414–417
  in expressions 139
Subtraction
  of decimals 50
  of fractions 40
  with fractions 157
  with integers 99
  with polynomials 503–504
  with rational expressions 642
Sum 2, 502
System
  decimal 1
  place value 47

Systems of equations 402
  consistent 404, 421
  dependent 405, 405–406, 422
  inconsistent 404, 422
  solutions by addition 420–425
  solutions by graphing 402–408
  solutions by substitution 414–
    417
  solve by graphing 403
Systems of linear inequalities 449

T

Terminating decimals 699
Terms 495
  coefficient of 136, 495
  constant 136
  degree of 495
  like (similar) 136
Tests for divisibility 21
Translating
  algebraic expressions into Eng-
    lish 182–183
  English into algebraic expres-
    sions 179–181
Trapezoid
  area of 274
  perimeter of 271
Trial and error method for factor-
    ing 570–575
Triangle
  area of 274
  hypotenuse 747
  legs 747
  perimeter of 271
  right 747
Trinomials 496
  factoring 559–561, 561–562,
    566–571, 570–575, 579
  perfect square 579

U

Unlike terms 137

V

Variable 4, 79
  dependent 303
  independent 303
Variation 675–676
Varies directly 675
Varies inversely 677
Vertex 809
Vertical axis 304
Vertical line 325
  slope of 341
Vertical line test 363
Volume 275
  of a rectangular pyramid 276
  of a rectangular solid 276
  of a right circular cone 276
  of a right circular cylinder 276
  of a sphere 276

W

Whole numbers 2, 76
Word problems
  basic plan 594
  basic steps 207, 251
Writing algebraic expressions 179
Writing decimal numbers 48

X

x-axis 304
x-intercept 322

Y

y-axis 304
y-intercept 322

Z

Zero
  as an exponent 472
  a monomial of no degree 496
  division by 24, 109
  neither positive nor negative 78
Zero-Factor Property 587
Zero Factor Law 120

CHAPTER 6   Exponents and Polynomials

Properties of Exponents:

For nonzero real numbers a and b and integers m and n;

The Exponent 1: $a = a^1$ (a is any real number.)

The Exponent 0: $a^0 = 1$ $(a \neq 0)$

Product Rule: $a^m \cdot a^n = a^{m+n}$

Quotient Rule: $\dfrac{a^m}{a^n} = a^{m-n}$

Negative Exponents Rule: $a^{-n} = \dfrac{1}{a^n}$ and $\dfrac{1}{a^{-n}} = a^n$

Power Rule: $\left(a^m\right)^n = a^{mn}$

Power Rule for Products: $(ab)^n = a^n b^n$

Power Rule for Quotients: $\left(\dfrac{a}{b}\right)^n = \dfrac{a^n}{b^n}$

Scientific Notation:

$N = a \times 10^n$ where N is a decimal number, $1 \leq a < 10$, and n is an integer.

Classification of Polynomials:

Monomial: polynomial with one term

Binomial: polynomial with two terms

Trinomial: polynomial with three terms

Degree:
The **degree of a polynomial** is the largest of the degrees of its terms.

Leading Coefficient: The coefficient of the term with the largest degree.

Division Algorithm:

$$\frac{P(x)}{D(x)} = Q(x) + \frac{R(x)}{D(x)}, \ (D(x) \neq 0)$$

FOIL Method:

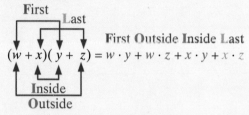

$$(w + x)(y + z) = w \cdot y + w \cdot z + x \cdot y + x \cdot z$$

Special Products of Binomials:

1. $(X + A)(X - A) = X^2 - A^2$:   Difference of two squares

2. $(X + A)^2 = X^2 + 2AX + A^2$:   Perfect square trinomial

3. $(X - A)^2 = X^2 - 2AX + A^2$:   Perfect square trinomial

CHAPTER 7   Factoring Polynomials and Solving Quadratic Equations

Finding the GCF:

1. Find the prime factorization of all integers and integer coefficients.
2. List all factors common to all terms, including variables.
3. Choose the greatest power of each factor common to all terms.
4. Multiply these powers to find the GCF. (If there are no common factors, the GCF is 1.)

Factoring Trinomials with Leading Coefficient 1 ($x^2 + bx + c$):

Find factors of c whose sum is b.

The *ac*-Method:

1. Multiply ac.
2. Find two integers whose product is ac and whose sum is b. (If this is not possible, the trinomial is not factorable.)
3. Rewrite the middle term, bx, using the two numbers found in Step 2 as coefficients.
4. Factor by grouping the first two terms and the last two terms.
5. Factor out the common binomial factor.

Trial-and-Error Method Guidelines:

1. If the sign of the constant term is positive ($+$), the signs in both factors will be the same, either both positive or both negative.
2. If the sign of the constant term is negative ($-$), the signs of the factors will be different, one positive and one negative.

Quadratic Equation:

An equation that can be written in the form $ax^2 + bx + c = 0$ where a, b, and c are real numbers and $a \neq 0$ is called a **quadratic equation**.

Zero Factor Property:

If a and b are real numbers, and $a \cdot b = 0$, then $a = 0$ or $b = 0$ or both.